xyztextbooks

ONLINE RESOURCES

A one-year All-Access Pass to XYZ Textbooks' eContent is included with the purchase of a new textbook. This All-Access Pass includes

- MathTV Videos
- eBooks
- Worksheets
- MathTV Mobile
- XYZ Homework (if required by your instructor)

TO ACTIVATE YOUR ACCOUNT

Go to: www.xyztextbooks.com, click on the "Students" link and then follow the on-screen instructions. You will also need your Student Access Code.

Your Student Access Code

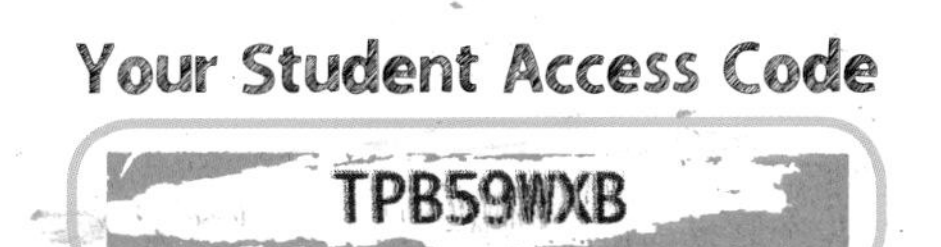

The Student Access Code above can be used by only one person.

Your All-Access Pass Information

My Email:

My Password:

Course ID#: (Provided by your instructor)

Your student access code is not your password. You will use the STUDENT ACCESS CODE only the first time you create your account. Once your code has been verified, you will not be asked for it again.

College Algebra

Building Skills and Modeling Situations

Charles P. McKeague
Katherine Yoshiwara
Denny Burzynski

SECOND PRINTING — November 2012
- Miscellaneous corrections made

xyztextbooks

College Algebra: Building Skills and Modeling Situations

Charles P. McKeague, Kathy Yoshiwara, Denny Burzynski

Publisher: XYZ Textbooks

Editor: Anne Scanlan-Rohrer
Two Ravens Editorial Services

Project Manager: Matthew Hoy

Manuscript Editor: Kate Pawlik

Editorial Assistants: Kendra Paulding, Joshua Wilbur

Composition: Rachel Hintz, Aaron Kroeger

Sales: Amy Jacobs, Richard Jones, Rachael Hillman, Bruce Spears, Katherine Hofstra

Cover Design: Kyle Schoenberger, Rachel Hintz

Cover Image: © VisualCommunications

Printed in the United States of America

ISBN-13: 978-1-936368-16-7 / ISBN-10: 1-936368-16-1

For product information and technology assistance, contact us at
XYZ Textbooks, 1-877-745-3499

For permission to use material from this text or product,
e-mail: **info@mathtv.com**

XYZ Textbooks
1339 Marsh Street
San Luis Obispo, CA 93401
USA

For your course and learning solutions, visit **www.xyztextbooks.com**

Preface to the Instructor

We have designed this book to help solve problems that you may encounter in the classroom.

Solutions to Your Problems

Problem: **Some students may ask, "What are we going to use this for?"**
Solution: Chapter and section openings feature real-world examples, which show students how the material they are learning appears in the world around them.

Problem: **Many students do not read the book.**
Solution: At the end of each section, under the heading *Getting Ready for Class*, are four questions for students to answer from the reading. Even a minimal attempt to answer these questions enhances the students' in-class experience.

Problem: **Some students may not see how the topics are connected.**
Solution: At the conclusion of the problem set for each section are a series of problems under the heading *Getting Ready for the Next Section*. These problems are designed to bridge the gap between topics covered previously, and topics introduced in the next section. Students intuitively see how topics lead into, and out of, each other.

Problem: **Some students lack good study skills, but may not know how to improve them.**
Solution: Study skills and success skills appear throughout the book, as well as online at MathTV.com. Students learn the skills they need to become successful in this class, and in their other courses as well.

Problem: **Students do well on homework, then forget everything a few days later.**
Solution: We have designed this textbook so that no topic is covered and then discarded. Throughout the book, features such as *Getting Ready for the Next Section*, *Maintaining Your Skills*, the *Chapter Summary*, and the *Chapter Test* continually reinforce the skills students need to master. If students need still more practice, there are a variety of worksheets online at MathTV.com.

Problem: **Some students just watch the videos at MathTV.com, but are not actively involved in learning.**
Solution: The Matched Problems worksheets (available online at MathTV.com) contain problems similar to the video examples. Assigning the Matched Problems worksheets ensures that students will be actively involved with the videos.

Other Helpful Solutions

Blueprint for Problem Solving: Students can use these step-by-step methods for solving common application problems.

Getting Ready for Calculus: Throughout the text students are given a glimpse of algebraic concepts that underpin what they'll be seeing when they progress on to calculus.

Using Technology: Scattered throughout the book are optional exercises that demonstrate how students can use graphing calculators and online resources such as Wolfram|Alpha to enhance their understanding of the topics being covered.

Preface to the Student

I often find my students asking themselves the question "Why can't I understand this stuff the first time?" The answer is "You're not expected to." Learning a topic in mathematics isn't always accomplished the first time around. There are many instances when you will find yourself reading over new material a number of times before you can begin to work problems. That's just the way things are in mathematics. If you don't understand a topic the first time you see it, that doesn't mean there is something wrong with you. Understanding mathematics takes time. The process of understanding requires reading the book, studying the examples, working problems, and getting your questions answered.

How to Be Successful in Mathematics

1. **If you are in a lecture class, be sure to attend all class sessions on time.** You cannot know exactly what goes on in class unless you are there. Missing class and then expecting to find out what went on from someone else is not the same as being there yourself.
2. **Read the book.** It is best to read the section that will be covered in class beforehand. Reading in advance, even if you do not understand everything you read, is still better than going to class with no idea of what will be discussed.
3. **Work problems every day and check your answers.** The key to success in mathematics is working problems. The more problems you work, the better you will become at working them. The answers to the odd-numbered problems are given in the back of the book. When you have finished an assignment, be sure to compare your answers with those in the book. If you have made a mistake, find out what it is, and correct it.
4. **Do it on your own.** Don't be misled into thinking someone else's work is your own. Having someone else show you how to work a problem is not the same as working the same problem yourself. It is okay to get help when you are stuck. As a matter of fact, it is a good idea. Just be sure you do the work yourself.
5. **Review every day.** After you have finished the problems your instructor has assigned, take another 15 minutes and review a section you have already completed. The more you review, the longer you will retain the material you have learned.
6. **Don't expect to understand every new topic the first time you see it.** Sometimes you will understand everything you are doing, and sometimes you won't. That's just the way things are in mathematics. Expecting to understand each new topic the first time you see it can lead to disappointment and frustration. The process of understanding takes time. It requires that you read the book, work problems, and get your questions answered.
7. **Spend as much time as it takes for you to master the material.** No set formula exists for the exact amount of time you need to spend on mathematics to master it. You will find out as you go along what is or isn't enough time for you. If you end up spending 2 or more hours on each section in order to master the material there, then that's how much time it takes; trying to get by with less will not work.
8. **Relax.** It's probably not as difficult as you think.

Acknowledgments

Coauthors

I am extremely pleased to have Katherine Yoshiwara and Denny Burzynski as coauthors on this project. I have known each of them for many years and have always been impressed with their writing, their presentations at mathematics conferences, and their commitment to mathematics education. Each of them has spent countless hours in service to our mathematics community.

XYZ Textbooks Crew

Production Team: Matthew Hoy, Rachel Hintz, Aaron Kroeger Our fantastic production team shepherded this book from handwritten manuscript to the final form you see today. Their attention to detail and ideas for making this book user-friendly for students is greatly appreciated.

Photo Researcher: Kathleen Olson Whenever we needed just the right photo to illustrate a concept or idea, she knew just where to look. When we were stumped, she was always ready with a new idea.

Editing and Proofreading: Kate Pawlik, Judy Barclay Their eye for detail and ferreting out even the most seemingly trivial error never ceased to amaze us. This book is far better off both mathematically and grammatically due to their invaluable assistance.

Office Staff and Customer Support: Katherine Hofstra, Rachael Hillman Our office staff are reliable, pleasant, and efficient. Plus they are lots of fun to work with.

Sales Department: Amy Jacobs, Rich Jones, Bruce Spears, Rachael Hillman Our award-winning, responsive sales staff is always conscientious and hard-working.

Technology Department: Stephen Aiena The brains behind the XYZ Textbooks' family of websites, including MathTV.com and XYZHomework.com.

XYZ Homework: Patrick McKeague, Matthew Hoy, Stephen Aiena, Mike Landrum From the big concepts to the little details, our XYZ Homework team has provided a solid, dependable online homework system that just works.

MathTV Student Peer Instructors: Gordon Kirby, Cynthia Ruiz, Edwin Martinez, Lauren Reeves, Joshua Wilbur These students and their genuine love for math have brightened my days. The videos they've made have helped countless students improve their math skills.

Focus Group Participants

Many thanks to the following instructors who participated in one of our online focus groups:

Peter Arvanites, *SUNY Rockland Community College*
Don Barnes, *Hazard Community & Technical College*
Lill Birdsall, *American River College*
Elsie Campbell, *Angelo State University*
Gregory Daubenmire, *Las Positas College*
Andrew Dohm, *Southwestern Michigan College*
Franklin Edward, *Faulkner State Community College*
Elaine Fitt, *Bucks County Community College*
Alicia Frost, *Santiago Canyon College*
Susan Hamby, *Lee College*
Alan Hayashi, *Oxnard College*
Kevin Hopkins, *Southwest Baptist University*
Bernice Keels, *Central Carolina Technical College*
Mary Legner, *Riverside Community College*
Teresita Lemus, *Nova Southeastern University*
Janine Lloyd, *Mass Bay Community College*
Valsala Mohanakumar, *Hillsborough Community College- Dale Mabry*
Kelly Nipp, *Leech Lake Tribal College*
Lyn Noble, *Florida State College at Jacksonville*
Leesa Pohl, *Donnelly College*
Mayada Shahrokhi, *Lone Star College-Cy Fair*
Mike Shirazi, *Germanna Community College*
Angela Simons, *Century College*
Jennifer Smith, *Oklahoma State University Institute of Technology*
Karen Tabor, *College of the Desert*
Philip Veer, *Johnson County Community College*
Victor Vega-Vazquez, *College of Coastal Georgia*
Tyler Wallace, *Big Bend Community College*
William Weber Jr., *University of Wyoming*

Wolfram|Alpha

We are pleased to offer Wolfram|Alpha problems in this textbook. Wolfram|Alpha is a trademark of Wolfram Alpha LLC. All logos, illustrations and search results are used with permission and are the copyrighted material of Wolfram Alpha LLC. The problems presented in this text using Wolfram|Alpha were not suggested by Wolfram|Alpha, nor does Wolfram|Alpha endorse these problems.

Contents

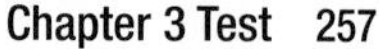

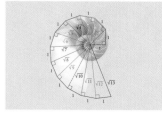

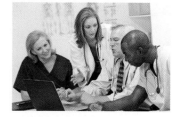

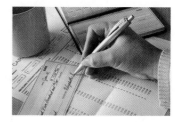

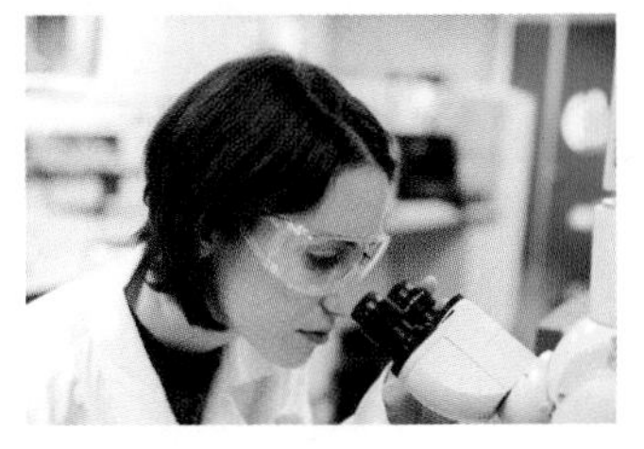

7 Systems of Equations 481

8 Sequences and Series (Online Only)

9 Conic Sections (Online Only)

10 Statistics and Probability (Online Only)

1 Algebra Review

Chapter Outline

© Steve Debenport/iStockphoto

The following table and graph show how the concentration of a popular antidepressant changes over time, once the patient stops taking it. In this particular case, the concentration in the patient's system is 80 ng/mL (nanograms per milliliter) when the patient stops taking the antidepressant, and the half-life of the antidepressant is 5 days.

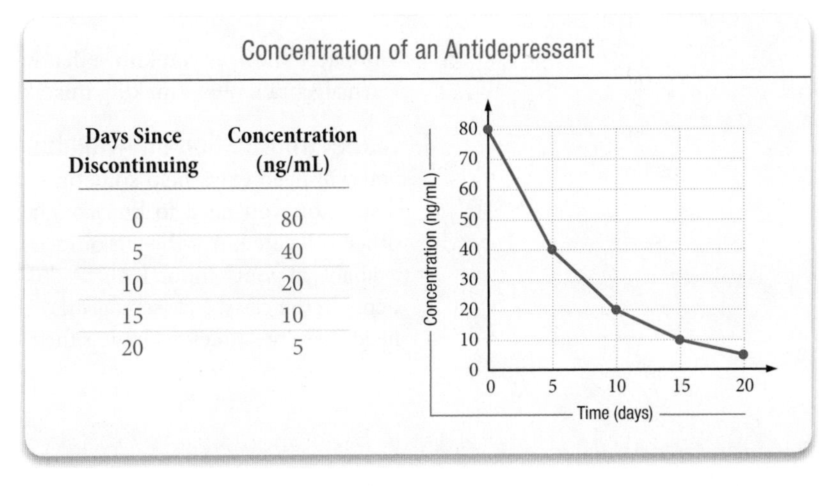

Concentration of an Antidepressant

Days Since Discontinuing	Concentration (ng/mL)
0	80
5	40
10	20
15	10
20	5

The half-life of a medication tells how quickly the medication is eliminated from a person's system: Medications with a long half-life are eliminated slowly, whereas those with a short half-life are more quickly eliminated. Half-life is the key to constructing the preceding table and graph.

In this chapter we will begin working with tables and charts. Many of the charts you will see here are taken directly from the media. As we progress through the book we will learn more about tables and charts and how they are connected.

Study Skills

Some of the students enrolled in my college algebra classes develop difficulties early in the course. Their difficulties are not associated with their ability to learn mathematics; they all have the potential to pass the course. Students who get off to a poor start do so because they have not developed the study skills necessary to be successful in algebra. Here is a list of things you can do to begin to develop effective study skills.

1. **Put Yourself on a Schedule** The general rule is that you spend 2 hours on homework for every hour you are in class. Make a schedule for yourself in which you set aside 2 hours each day to work on algebra. Once you make the schedule, stick to it. Don't just complete your assignments and stop. Use all the time you have set aside. If you complete an assignment and have time left over, read the next section in the book, and then work more problems.

2. **Find Your Mistakes and Correct Them** There is more to studying algebra than just working problems. You must always check your answers with the answers in the back of the book. When you have made a mistake, find out what it is and correct it. Making mistakes is part of the process of learning mathematics. In the prologue to *The Book of Squares*, Leonardo Fibonacci (ca. 1170–ca. 1250) had this to say about the content of his book:

   > I have come to request indulgence if in any place it contains something more or less than right or necessary; for to remember everything and be mistaken in nothing is divine rather than human . . .

   Fibonacci knew, as you know, that human beings make mistakes. You cannot learn algebra without making mistakes.

3. **Gather Information on Available Resources** You need to anticipate that you will need extra help sometime during the course. One resource is your instructor; you need to know your instructor's office hours and where the office is located. Another resource is the math lab or study center, if they are available at your school. It also helps to have the phone numbers of other students in the class, in case you miss class. You want to anticipate that you will need these resources, so now is the time to gather them together.

Definitions, Properties, and Simplifying Expressions

INTRODUCTION The diagram below is called a *bar chart*. This one shows the net price of a popular intermediate algebra textbook for each of the first five editions. (The net price is the price the bookstore pays for the book.)

From the chart, we can find many relationships between numbers. We may notice that the price of the third edition was less than the price of the fourth edition. In mathematics we use symbols to represent relationships between quantities. If we let $P(n)$ represent the price of the *nth* edition, then the relationship just mentioned, between the price of the third edition and the price of the fourth edition, can be written this way:

$$P(3) < P(4)$$

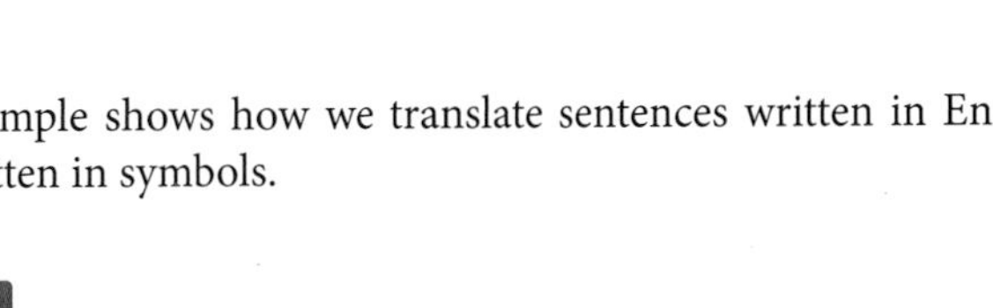

The following example shows how we translate sentences written in English into expressions written in symbols.

VIDEO EXAMPLES

SECTION 1.1

EXAMPLE 1

In English	*In Symbols*
The sum of x and 5 is less than 2.	$x + 5 < 2$
The product of 3 and x is 21.	$3x = 21$
The quotient of y and 6 is 4.	$\frac{y}{6} = 4$
Twice the difference of b and 7 is greater than 5.	$2(b - 7) > 5$
The difference of twice b and 7 is greater than 5.	$2b - 7 > 5$

Order of Operations

It is important when evaluating arithmetic expressions in mathematics that each expression have only one answer in reduced form. Consider the expression

$$3 \cdot 7 + 2$$

If we find the product of 3 and 7 first, then add 2, the answer is 23. On the other hand, if we first combine the 7 and 2, then multiply by 3, we have 27. The problem seems to have two distinct answers depending on whether we multiply first or add first. To avoid this situation, we will decide that multiplication in a situation like this will always be done before addition. In this case, only the first answer, 23, is correct.

Here is the complete set of rules for evaluating expressions.

RULE *Order of Operations*

When evaluating a mathematical expression, we will perform the operations in the following order, beginning with the expression in the innermost parentheses or brackets and working our way out.

1. Simplify all numbers with exponents, working from left to right if more than one of these expressions is present.
2. Then, do all multiplications and divisions left to right.
3. Perform all additions and subtractions left to right.

EXAMPLES Simplify each expression using the rule for order of operations.

2. $5 + 3(2 + 4) = 5 + 3(6)$ — Simplify inside parentheses
$= 5 + 18$ — Then, multiply
$= 23$ — Add

3. $40 - 20 \div 5 + 8 = 40 - 4 + 8$ — Divide first
$= 36 + 8$ — Then, add and subtract left to right
$= 44$

4. $(-2 - 3)(5 - 9) = (-5)(-4)$ — Simplify inside parentheses
$= 20$ — Multiply

5. $2 - 5(7 - 4) - 6 = 2 - 5(3) - 6$ — Simplify inside parentheses
$= 2 - 15 - 6$ — Then, multiply
$= -19$ — Finally, subtract, left to right

6. $2(4 - 7)^3 + 3(-2 - 3)^2 = 2(-3)^3 + 3(-5)^2$ — Simplify inside parentheses
$= 2(-27) + 3(25)$ — Evaluate numbers with exponents
$= -54 + 75$ — Multiply
$= 21$ — Add

Our next examples involve fractions. The fraction bar works like parentheses to separate the numerator from the denominator.

7. $\dfrac{-8 - 8}{-5 - 3} = \dfrac{-16}{-8}$ — Simplify numerator and denominator separately
$= 2$ — Divide

8. $\dfrac{-5(-4) + 2(-3)}{2(-1) - 5} = \dfrac{20 - 6}{-2 - 5}$
$= \dfrac{14}{-7}$ — Simplify numerator and denominator separately
$= -2$ — Divide

Finding the Value of an Algebraic Expression

An algebraic expression is a combination of numbers, variables, and operation symbols. For example, each of the following is an algebraic expression:

$$5x \qquad x^2 + 5 \qquad 4a^2b^3 \qquad 5t^2 - 6t + 3$$

EXAMPLE 9 Find the value of the expression $3x + 4y + 5$ when x is 6 and y is 7.

SOLUTION We substitute the given values of x and y into the expression and simplify the result:

When	$x = 6$ and $y = 7$
the expression	$3x + 4y + 5$
becomes	$3 \cdot 6 + 4 \cdot 7 + 5 = 18 + 28 + 5$
	$= 51$

■

The Real Numbers

The real number line is constructed by drawing a straight line and labeling a convenient point with the number 0. Positive numbers are in increasing order to the right of 0; negative numbers are in decreasing order to the left of 0. The point on the line corresponding to 0 is called the *origin*.

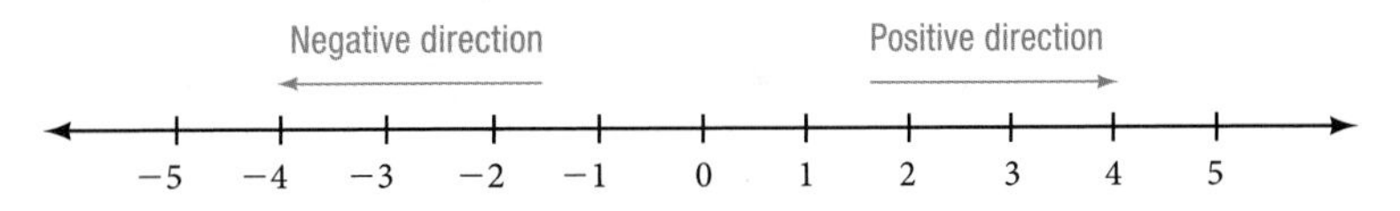

The numbers on the number line increase in size as we move to the right. When we compare the size of two numbers on the number line, the one on the left is always the smaller number.

The numbers associated with the points on the line are called *coordinates* of those points. Every point on the line has a number associated with it. The set of all these numbers makes up the set of real numbers.

(dēf DEFINITION *Real number*

A *real number* is any number that is the coordinate of a point on the real number line.

EXAMPLE 10 Locate the numbers $-4.5, -0.75, \frac{1}{2}, \sqrt{2}, \pi$, and 4.1 on the real number line.

SOLUTION

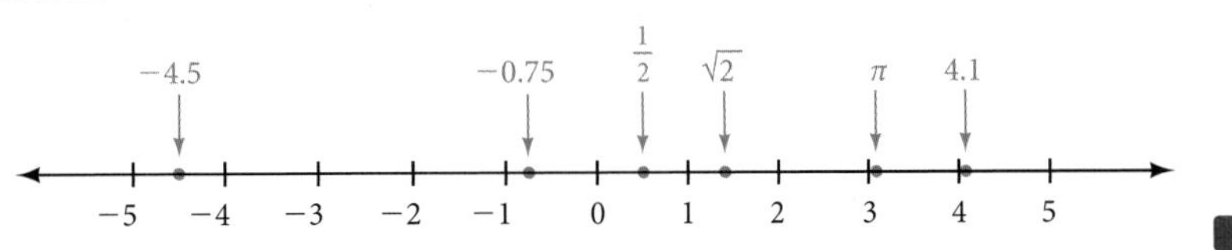

■

The Absolute Value of a Real Number

DEFINITION *Absolute value*

The *absolute value* of a number (also called its ***magnitude***) is the distance the number is from 0 on the number line. If x represents a real number, then the absolute value of x is written $|x|$.

This definition of absolute value is geometric in form because it defines absolute value in terms of the number line. Here is an alternate definition of absolute value that is algebraic in form because it only involves symbols.

Note It is important to recognize that if x is a real number, $-x$ is not necessarily negative. For example, if x is 5, then $-x$ is -5. On the other hand, if x were -5, then $-x$ would be $-(-5)$, which is 5.

ALTERNATE DEFINITION *Absolute value*

If x represents a real number, then the *absolute value* of x is written $|x|$, and is given by

$$|x| = \begin{cases} x & \text{if } x \geq 0 \\ -x & \text{if } x < 0 \end{cases}$$

If the original number is positive or 0, then its absolute value is the number itself. If the number is negative, its absolute value is its opposite (which must be positive).

EXAMPLES Write each expression without absolute value symbols.

11. $|5| = 5$ **12.** $|-2| = 2$ **13.** $-|-3| = -3$ ■

Properties of Real Numbers

For all the properties listed in this section, a, b, and c represent real numbers.

PROPERTY *Commutative Property of Addition*

In symbols: $a + b = b + a$

In words: The *order* of the numbers in a sum does not affect the result.

PROPERTY *Commutative Property of Multiplication*

In symbols: $a \cdot b = b \cdot a$

In words: The *order* of the numbers in a product does not affect the result.

PROPERTY *Associative Property of Addition*

In symbols: $a + (b + c) = (a + b) + c$

In words: The *grouping* of the numbers in a sum does not affect the result.

PROPERTY *Associative Property of Multiplication*

In symbols: $a(bc) = (ab)c$

In words: The *grouping* of the numbers in a product does not affect the result.

The following examples illustrate how the associative properties can be used to simplify expressions that involve both numbers and variables.

EXAMPLES Simplify by using the associative property.

14. $5(4x) = (5 \cdot 4)x$ Associative property

$= 20x$ Multiplication

15. $\frac{1}{4}(4a) = \left(\frac{1}{4} \cdot 4\right)a$ Associative property

$= 1a$ Multiplication

$= a$

Our next property involves both addition and multiplication. It is called the *distributive property* and is stated as follows.

Note Although the properties we are listing are stated for only two or three real numbers, they hold for as many numbers as needed. For example, the distributive property holds for expressions like $3(x + y + z + 2)$. That is,

$3(x + y + z + 2) = 3x + 3y + 3z + 6$

PROPERTY *Distributive Property*

In symbols: $a(b + c) = ab + ac$

In words: Multiplication *distributes* over addition.

EXAMPLES Apply the distributive property to each expression and then simplify the result.

16. $\frac{1}{2}(3x + 6) = \frac{1}{2}(3x) + \frac{1}{2}(6)$ Distributive property

$= \frac{3}{2}x + 3$ Multiplication

17. $2(3y + 4) + 2 = 2(3y) + 2(4) + 2$ Distributive property

$= 6y + 8 + 2$ Multiplication

$= 6y + 10$ Addition

18. $a\left(1 + \frac{1}{a}\right) = a \cdot 1 + a \cdot \frac{1}{a} = a + 1$

19. $3\left(\frac{1}{3}x + 5\right) = 3 \cdot \frac{1}{3}x + 3 \cdot 5 = x + 15$

20. $6\left(\frac{1}{3}x + \frac{1}{2}y\right) = 6 \cdot \frac{1}{3}x + 6 \cdot \frac{1}{2}y = 2x + 3y$

Combining Similar Terms

The distributive property can also be used to combine similar terms. (For now, a term is the product of a number with one or more variables.) Similar terms are terms with the same variable part. The terms $3x$ and $5x$ are similar, as are $2y$, $7y$, and $-3y$, because the variable parts are the same.

EXAMPLES Use the distributive property to combine similar terms.

21. $3x + 5x = (3 + 5)x$ Distributive property

$= 8x$ Addition

22. $3y + y = (3 + 1)y$ Distributive property

$= 4y$ Addition

Simplifying Expressions

We can use the commutative, associative, and distributive properties together to simplify expressions.

EXAMPLE 23 Simplify $7x + 4 + 6x + 3$.

SOLUTION We begin by applying the commutative and associative properties to group similar terms:

$7x + 4 + 6x + 3 = (7x + 6x) + (4 + 3)$	Commutative and associative properties
$= (7 + 6)x + (4 + 3)$	Distributive property
$= 13x + 7$	Addition

EXAMPLE 24 Simplify $4 + 3(2y + 5) + 8y$.

SOLUTION Because our rule for order of operations indicates that we are to multiply before adding, we must distribute the 3 across $2y + 5$ first:

$4 + 3(2y + 5) + 8y = 4 + 6y + 15 + 8y$	Distributive property
$= (6y + 8y) + (4 + 15)$	Commutative and associative properties
$= (6 + 8)y + (4 + 15)$	Distributive property
$= 14y + 19$	Addition

EXAMPLE 25 Simplify $3(2y - 1) + y$.

SOLUTION We begin by multiplying the 3 and $2y - 1$. Then, we combine similar terms:

$3(2y - 1) + y = 6y - 3 + y$	Distributive property
$= 7y - 3$	Combine similar terms

EXAMPLE 26 Simplify $8 - 3(4x - 2) + 5x$.

SOLUTION First, we distribute the -3 across the $4x - 2$.

$$8 - 3(4x - 2) + 5x = 8 - 12x + 6 + 5x \quad \text{Distributive property}$$
$$= -7x + 14 \quad \text{Combine similar terms}$$

EXAMPLE 27 Simplify $5(2a + 3) - (6a - 4)$.

SOLUTION We begin by applying the distributive property to remove the parentheses. The expression $-(6a - 4)$ can be thought of as $-1(6a - 4)$. Thinking of it in this way allows us to apply the distributive property.

$$-1(6a - 4) = -1(6a) - (-1)(4) = -6a + 4$$

Here is the complete problem:

$$5(2a + 3) - (6a - 4) = 10a + 15 - 6a + 4 \quad \text{Distributive property}$$
$$= 4a + 19 \quad \text{Combine similar terms}$$

The remaining properties of real numbers have to do with the numbers 0 and 1.

Note 0 and 1 are called the *additive identity* and *multiplicative identity*, respectively. Combining 0 with a number, under addition, does not change the identity of the number. Likewise, combining 1 with a number, under multiplication, does not alter the identity of the number. We see that 0 is to addition what 1 is to multiplication.

PROPERTY *Additive Identity Property*

There exists a unique number 0 such that

In symbols: $a + 0 = a$ and $0 + a = a$

In words: Adding zero to any number is the number.

PROPERTY *Multiplicative Identity Property*

There exists a unique number 1 such that

In symbols: $a \cdot 1 = a$ and $1 \cdot a = a$

In words: Any number multiplied by 1 is the number.

PROPERTY *Additive Inverse Property*

For each real number a, there exists a unique real number $-a$ such that

In symbols: $a + (-a) = 0$

In words: Opposites add to 0.

PROPERTY *Multiplicative Inverse Property*

For every real number a, except 0, there exists a unique real number $\frac{1}{a}$ such that

In symbols: $a\left(\frac{1}{a}\right) = 1$

In words: Reciprocals multiply to 1.

Sets

DEFINITION *Set*

A *set* is a collection of objects or things. The objects in the set are called *elements*, or *members*, of the set.

Sets are usually denoted by capital letters and elements of sets by lowercase letters. We use braces, $\{\ \}$, to enclose the elements of a set.

To show that an element is contained in a set we use the symbol $\in$. That is,

$x \in A$ is read "x is an element (member) of set A"

For example, if A is the set $\{1, 2, 3\}$, then $2 \in A$. On the other hand, $5 \notin A$ means 5 is not an element of set A.

DEFINITION *Subset*

Set A is a *subset* of set B, written $A \subset B$, if every element in A is also an element of B. That is,

$$A \subset B \quad \text{if and only if} \quad A \text{ is contained in } B$$

Note The diagrams shown here are called *Venn diagrams* after John Venn (1834–1923). They can be used to visualize operations with sets. The region inside the circle labeled A is set A; the region inside the circle labeled B is set B.

DEFINITION *Empty/null set*

The set with no members is called the *empty*, or *null, set*. It is denoted by the symbol $\varnothing$.

The empty set is considered a subset of every set.

Operations with Sets

Two basic operations are used to combine sets—union and intersection.

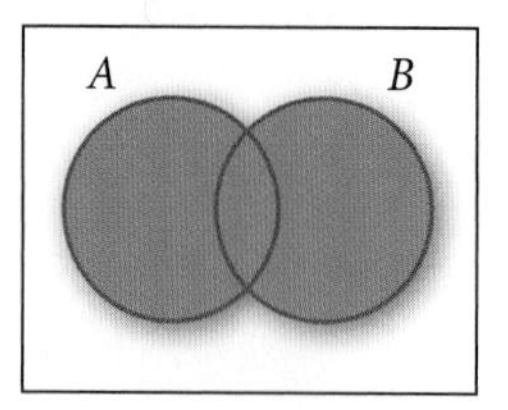

$A \cup B$

FIGURE 1 *The union of two sets*

DEFINITION *Union*

The *union* of two sets A and B, written $A \cup B$, is the set of all elements that are either in A or in B, or in both A and B. The key word here is *or*. For an element to be in $A \cup B$, it must be in A or B. In symbols, the definition looks like this:

$$x \in A \cup B \quad \text{if and only if} \quad x \in A \text{ or } x \in B$$

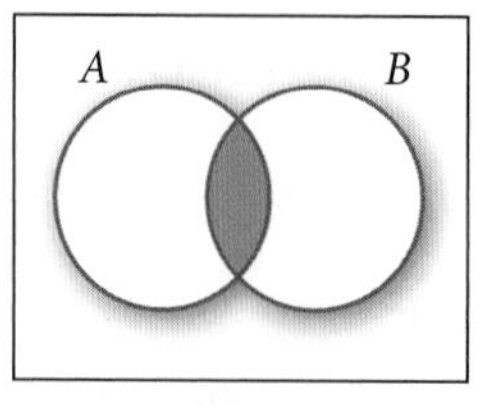

$A \cap B$

FIGURE 2 *The intersection of two sets*

DEFINITION *Intersection*

The *intersection* of two sets A and B, written $A \cap B$, is the set of elements in both A and B. The key word in this definition is the word *and*. For an element to be in $A \cap B$, it must be in both A and B. In symbols,

$$x \in A \cap B \quad \text{if and only if} \quad x \in A \text{ and } x \in B$$

EXAMPLES Let $A = \{1, 3, 5\}$, $B = \{0, 2, 4\}$, and $C = \{1, 2, 3, \ldots\}$. Then

28. $A \cup B = \{0, 1, 2, 3, 4, 5\}$

29. $A \cap B = \varnothing$ *(A and B have no elements in common)*

30. $A \cap C = \{1, 3, 5\} = A$

31. $B \cup C = \{0, 1, 2, 3, \ldots\}$

Another notation we can use to describe sets is called *set builder notation*. Here is how we write our definition for the union of two sets A and B using set builder notation:

$$A \cup B = \{x \mid x \in A \quad \text{or} \quad x \in B\}$$

The right side of this statement is read "the set of all x such that x is a member of A or x is a member of B." As you can see, the vertical line after the first x is read "such that."

Subsets of the Real Numbers

Next, we consider some of the more important subsets of the real numbers. Each set listed here is a subset of the real numbers:

Counting (or natural) numbers $= 1, 2, 3, \ldots$

Whole numbers $= 0, 1, 2, 3, \ldots$

Integers $= \ldots, -3, -2, -1, 0, 1, 2, 3, \ldots$

Rational numbers $= \left\{ \frac{a}{b} \mid a \text{ and } b \text{ are integers, } b \neq 0 \right\}$

Remember, the notation used to write the rational numbers is read "the set of numbers a/b, such that a and b are integers and b is not equal to 0." Any number that can be written in the form

$$\frac{\text{Integer}}{\text{Integer (except 0)}}$$

is a rational number. Rational numbers are numbers that can be written as the ratio of two integers. Each of the following is a rational number:

$\frac{3}{4}$ *Because it is the ratio of the integers 3 and 4*

-8 *Because it can be written as the ratio of −8 to 1*

0.75 *Because it is the ratio of 75 to 100 (or 3 to 4 if you reduce to lowest terms)*

$0.333\ldots$ *Because it can be written as the ratio of 1 to 3*

Irrational numbers $= \{x \mid x \text{ is real, but not rational}\}$

The following are irrational numbers:

$$\sqrt{2}, \quad -\sqrt{3}, \quad 4 + 2\sqrt{3}, \quad \pi, \quad \frac{\sqrt{3}}{2}$$

Note We can find decimal approximations to some irrational numbers by using a calculator or a table. For example, on an eight-digit calculator

$$\sqrt{2} = 1.4142136$$

This is not exactly $\sqrt{2}$ but simply an approximation to it. There is no decimal that gives $\sqrt{2}$ exactly.

EXAMPLE 32 For the set $\{-5, -3.5, 0, \frac{3}{4}, \sqrt{3}, \sqrt{5}, 9\}$, list the numbers that are:

a. whole numbers

b. integers

c. rational numbers

d. irrational numbers

e. real numbers

SOLUTION

a. Whole numbers: 0, 9

b. Integers: $-5, 0, 9$

c. Rational numbers: $-5, -3.5, 0, \frac{3}{4}, 9$

d. Irrational numbers: $\sqrt{3}, \sqrt{5}$

e. They are all real numbers.

The following diagram gives a visual representation of the relationships among subsets of the real numbers. Note that we cannot list all of the members of most sets, so we give some examples.

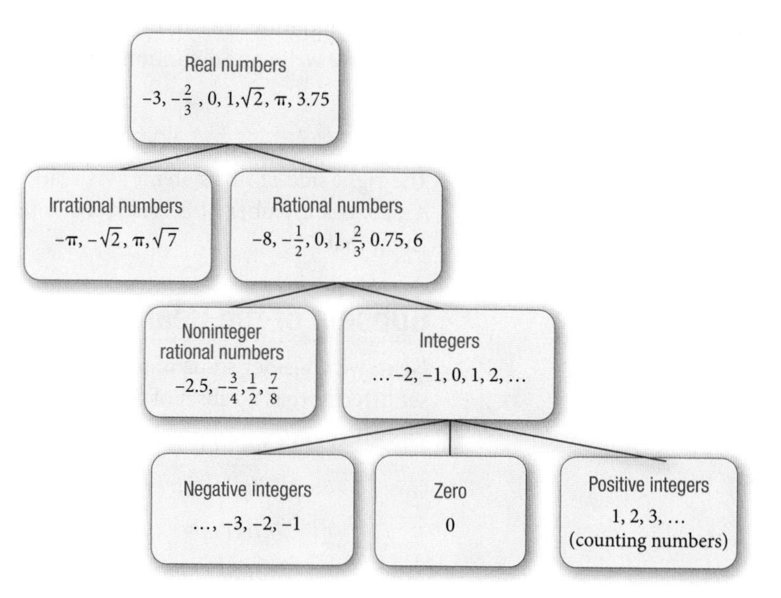

GETTING READY FOR CLASS

Each section of the book will end with some problems and questions like the ones that follow. They are for you to answer after you have read through the section, but before you go to class. All of them require that you give written responses, in complete sentences. Writing about mathematics is a valuable exercise. If you write with the intention of explaining and communicating what you know to someone else, you will find that you understand the topic you are writing about even better than you did before you started writing.

After reading through the preceding section, respond in your own words and in complete sentences.

A. What is a real number?
B. Why do we have a rule for the order of operations?
C. What is an irrational number?
D. What is the commutative property?

Problem Set 1.1

Translate each of the following statements into symbols.

1. The sum of x and 5 is 2.

2. The sum of y and -3 is 9.

3. The difference of 6 and x is y.

4. The difference of x and 6 is $-y$.

5. The product of 2 and t is less than y.

6. The product of $5x$ and y is equal to z.

7. The sum of x and y is less than the difference of x and y.

8. Twice the sum of a and b is 15.

9. Half the difference of t and s is less than twice the sum of t and s.

10. Half the product of r and s is less than twice the difference of r and s.

The problems that follow are intended to give you practice using the rule for order of operations. Simplify each expression.

11. **a.** $3 \cdot 5 + 4$ **b.** $3 \cdot (5 + 4)$ **c.** $3 \cdot 5 + 3 \cdot 4$

12. **a.** $6 + 3 \cdot 4 - 2$ **b.** $6 + 3(4 - 2)$ **c.** $(6 + 3)(4 - 2)$

13. **a.** $(7 - 4)(7 + 4)$ **b.** $7^2 - 4^2$ **c.** $(7 - 4)^2$

14. **a.** $(5 + 7)^2$ **b.** $5^2 + 7^2$ **c.** $5^2 + 2 \cdot 5 \cdot 7 + 7^2$

15. **a.** $2 + 3 \cdot 2^2 + 3^2$ **b.** $2 + 3(2^2 + 3^2)$ **c.** $(2 + 3)(2^2 + 3^2)$

16. **a.** $40 - 10 \div 5 + 1$ **b.** $(40 - 10) \div 5 + 1$ **c.** $(40 - 10) \div (5 + 1)$

Simplify each expression as much as possible. You will see each of these problems later in the book.

17. $1(-2) - 2(-16) + 1(9)$

18. $1(1) - 3(-2) + (-2)(-2)$

19. $-4(0)(-2) - (-1)(1)(1) - 1(2)(3)$

20. $3(-2)^2 + 2(-2) - 1$

21. $\dfrac{0 - 4}{0 - 2}$

22. $\dfrac{-4 - 4}{-4 - 2}$

23. $\dfrac{-6 + 6}{-6 - 3}$

24. $\dfrac{-2 - 4}{2 - 2}$

25. Find the value of each expression when x is 5.

a. $x + 2$ **b.** $2x$ **c.** x^2 **d.** 2^x

26. Find the value of each expression when x is 3.

a. $x + 5$ **b.** $5x$ **c.** x^5 **d.** 5^x

27. Find the value of each expression when x is 10.

a. $x^2 + 2x + 1$ **b.** $(x + 1)^2$ **c.** $x^2 + 1$ **d.** $(x - 1)^2$

28. Find the value of each expression when x is 8.

a. $x^2 - 6x + 9$ **b.** $(x - 3)^2$ **c.** $x^2 - 3$ **d.** $x^2 - 9$

29. Find the value of $b^2 - 4ac$ if

a. $a = 2$, $b = 5$, and $c = 3$ **b.** $a = 10$, $b = 60$, and $c = 30$

30. Find the value of $6x + 5y + 4$ if

a. $x = 3$ and $y = 2$ **b.** $x = 2$ and $y = 3$ **c.** $x = 0$ and $y = 0$

31. Fill in the table

		Sum	Difference	Product	Quotient
a	b	$a + b$	$a - b$	ab	$\frac{a}{b}$
3	12				
−3	12				
3	−12				
−3	−12				

32. Fill in the table

		Sum	Difference	Product	Quotient
a	b	$a + b$	$a - b$	ab	$\frac{a}{b}$
8	2				
−8	2				
8	−2				
−8	−2				

Write each of the following without absolute value symbols.

33. $|-2|$ **34.** $|-\sqrt{2}|$ **35.** $-|4|$ **36.** $-|-10|$

Apply the distributive property to each expression. Simplify when possible.

37. $3(x + 6)$ **38.** $2(6x + 4)$ **39.** $5(3a + 2b)$

40. $\frac{1}{3}(4x + 6)$ **41.** $\frac{1}{5}(10 + 5y)$ **42.** $(5t + 1)8$

The problems below are problems you will see later in the book. Apply the distributive property, then simplify if possible.

43. $3(3x + y - 2z)$ **44.** $10(0.3x + 0.7y)$ **45.** $100(0.06x + 0.07y)$

46. $3\left(x + \frac{1}{3}\right)$ **47.** $2\left(x - \frac{1}{2}\right)$ **48.** $x\left(1 + \frac{2}{x}\right)$

49. $a\left(1 - \frac{3}{a}\right)$ **50.** $8\left(\frac{1}{8}x + 3\right)$ **51.** $6\left(\frac{1}{2}x - \frac{1}{3}y\right)$

52. $12\left(\frac{1}{4}x + \frac{2}{3}y\right)$ **53.** $12\left(\frac{2}{3}x - \frac{1}{4}y\right)$ **54.** $20\left(\frac{2}{5}x + \frac{1}{4}y\right)$

Apply the distributive property to each expression. Simplify when possible.

55. $3(5x + 2) + 4$ **56.** $4(3x + 2) + 5$

57. $5(1 + 3t) + 4$ **58.** $2(1 + 5t) + 6$

Use the commutative, associative, and distributive properties to simplify the following.

59. $5a + 7 + 8a + a$ **60.** $2(5x + 1) + 2x$

61. $3 + 4(5a + 3) + 4a$

62. $5x + 3(x + 2) + 7$

63. $5(x + 2y) + 4(3x + y)$

64. $5b + 3(4b + a) + 6a$

65. $3(5x + 4) - x$

66. $6 - 7(3 - m)$

67. $7 - 2(3x - 1) + 4x$

68. $5(3y + 1) - (8y - 5)$

69. $4(2 - 6x) - (3 - 4x)$

70. $10 - 4(2x + 1) - (3x - 4)$

Let $A = \{0, 2, 4, 6\}$, $B = \{1, 2, 3, 4, 5\}$, and $C = \{1, 3, 5, 7\}$. List the elements of each of the following sets.

71. $A \cup B$

72. $A \cup C$

73. $A \cap B$

74. $A \cap C$

75. $B \cap C$

76. $B \cup C$

77. $A \cup (B \cap C)$

78. $C \cup (A \cap B)$

For the set $\{-6, -5.2, -\sqrt{7}, -\pi, 0, 1, 2, 2.3, \frac{9}{2}, \sqrt{17}\}$ list all the elements named in each of the following problems.

79. Counting numbers

80. Whole numbers

81. Rational numbers

82. Integers

83. Irrational numbers

84. Real numbers

85. **Running the Mile** The chart shows the five fastest winning times for the Fifth Avenue Mile run in New York City. The times are given in minutes and seconds, to the nearest hundredth of a second.

Fastest on Fifth

Continental Airlines Fifth Avenue Mile

Craig Mottram, AUS	3:49.90
Alan Webb, USA	3:51.40
Elkanah Angwenyi, KEN	3:54.30
Anthony Famiglietti, USA	3:57.10
Rui Silva, POR	3:57.40

Source: www.coolrunning.com, 2005

a. How much faster was Alan Webb's time than Elkanah Angwenyi's time?

b. How much faster was Craig Mottram's time than Rui Silva's time?

86. **Solar Power** The chart shows the annual sums of solar thermal collector shipments in the U.S.

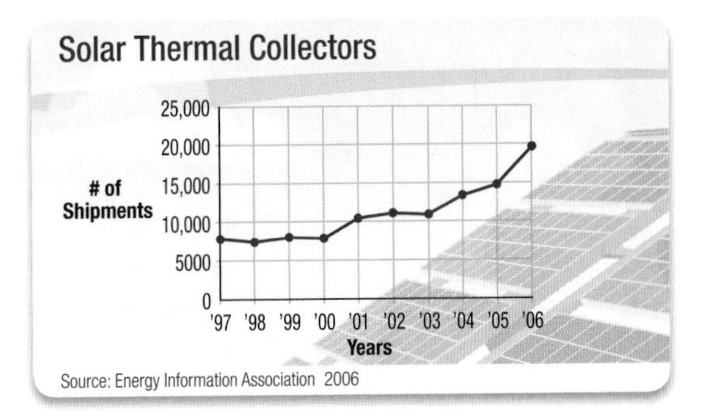

a. For which year was the sum of solar thermal collector shipments closest to 12,500 units?

b. Estimate the increase in solar thermal collector shipments from 2005 to 2006.

87. **Fermat's Last Theorem** The postage stamp shown here was issued by France in 2001 to commemorate the 400th anniversary of the birth of the French mathematician Pierre de Fermat. The stamp shows Fermat's last theorem, which states that if n is an integer greater than 2, then there are no positive integers x, y, and z that will make the formula $x^n + y^n = z^n$ true.

However, there are many ways to make the formula $x^n + y^n = z^n$ true when n is 1 or 2. Show that this formula is true for each case below.

a. $n = 1, x = 5, y = 8$, and $z = 13$

b. $n = 1, x = 2, y = 3$, and $z = 5$

c. $n = 2, x = 3, y = 4$, and $z = 5$

d. $n = 2, x = 7, y = 24$, and $z = 25$

© tillsonburg/iStockPhoto

88. **Drinking and Driving** The chart shows that if the risk of getting in an accident with no alcohol in your system is 1, then the risk of an accident with a blood-alcohol level of 0.24 is 147 times as high.

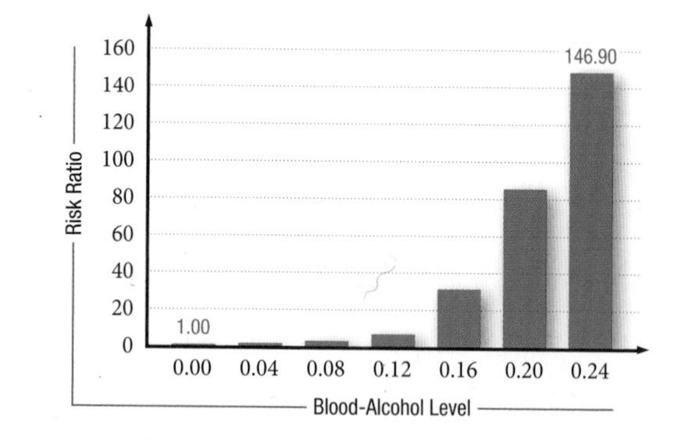

a. A person driving with a blood-alcohol level of 0.20 is how many times more likely to get in an accident than if she was driving with a blood-alcohol level of 0?

b. If the probability of getting in an accident while driving for an hour on surface streets in a certain city is 0.02%, what is the probability of getting in an accident in the same circumstances with a blood-alcohol level of 0.20?

Exponents and Scientific Notation

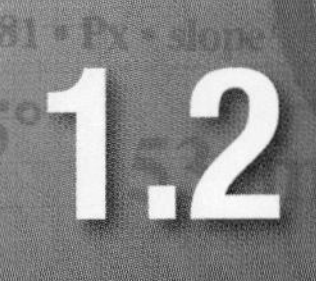

INTRODUCTION The following figure shows a square and a cube, each with a side of length 1.5 centimeters. To find the area of the square, we raise 1.5 to the second power: 1.5^2. To find the volume of the cube, we raise 1.5 to the third power: 1.5^3.

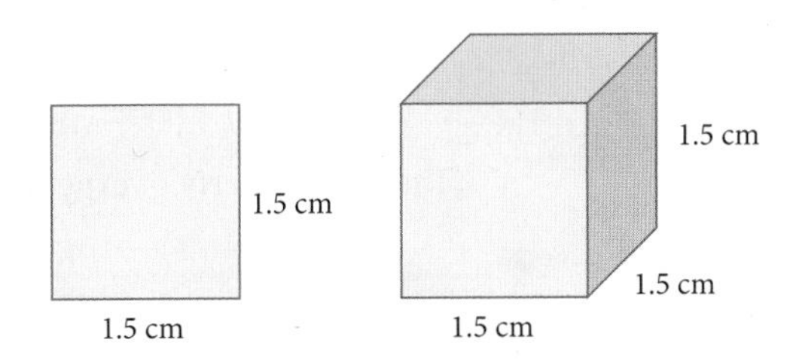

Because the area of the square is 1.5^2, we say second powers are *squares*; that is, x^2 is read "x squared." Likewise, because the volume of the cube is 1.5^3, we say third powers are *cubes*; that is, x^3 is read "x cubed." Exponents and the vocabulary associated with them are topics we will study in this section.

Properties of Exponents

[Δ≠Σ PROPERTY *Property 1 for Exponents*

If a is a real number and r and s are integers, then

$$a^r \cdot a^s = a^{r+s}$$

[Δ≠Σ PROPERTY *Property 2 for Exponents*

If a is a real number and r and s are integers, then

$$(a^r)^s = a^{r \cdot s}$$

[Δ≠Σ PROPERTY *Property 3 for Exponents*

If a and b are any two real numbers and r is an integer, then

$$(ab)^r = a^r \cdot b^r$$

Here are some examples that use combinations of the first three properties of exponents to simplify expressions involving exponents.

VIDEO EXAMPLES

SECTION 1.2

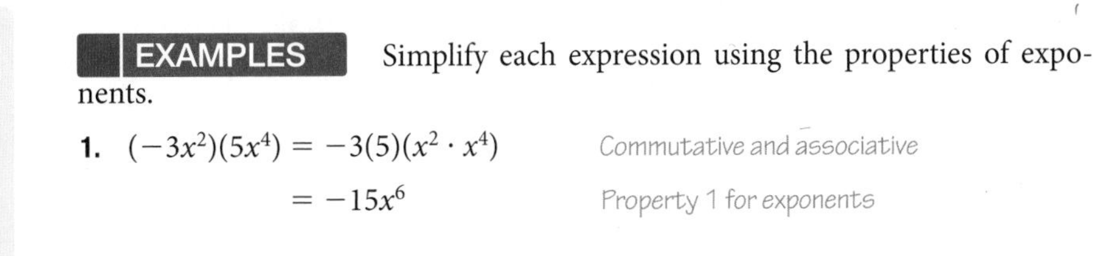

EXAMPLES Simplify each expression using the properties of exponents.

1. $(-3x^2)(5x^4) = -3(5)(x^2 \cdot x^4)$ Commutative and associative

$= -15x^6$ Property 1 for exponents

2. $(-2x^2)^3(4x^5) = (-2)^3(x^2)^3(4x^5)$ — Property 3

$= -8x^6 \cdot (4x^5)$ — Property 2

$= (-8 \cdot 4)(x^6 \cdot x^5)$ — Commutative and associative

$= -32x^{11}$ — Property 1

3. $(x^2)^4(x^2y^3)^2(y^4)^3 = x^8 \cdot x^4 \cdot y^6 \cdot y^{12}$ — Properties 2 and 3

$= x^{12}y^{18}$ — Property 1 ■

Note This property is actually a definition. That is, we are defining negative integer exponents as indicating reciprocals. Doing so gives us a way to write an expression with a negative exponent as an equivalent expression with a positive exponent.

The next property of exponents deals with negative integer exponents.

PROPERTY *Property 4 for Exponents*

If a is any nonzero real number and r is a positive integer, then

$$a^{-r} = \frac{1}{a^r}$$

EXAMPLES Write with positive exponents, and then simplify.

4. $5^{-2} = \frac{1}{5^2} = \frac{1}{25}$

5. $(-2)^{-3} = \frac{1}{(-2)^3} = \frac{1}{-8} = -\frac{1}{8}$

6. $\left(\frac{3}{4}\right)^{-2} = \frac{1}{\left(\frac{3}{4}\right)^2} = \frac{1}{\frac{9}{16}} = \frac{16}{9}$ ■

PROPERTY *Property 5 for Exponents*

If a and b are any two real numbers with $b \neq 0$, and r is an integer, then

$$\left(\frac{a}{b}\right)^r = \frac{a^r}{b^r}$$

PROPERTY *Property 6 for Exponents*

If a is any nonzero real number, and r and s are any two integers, then

$$\frac{a^r}{a^s} = a^{r-s}$$

Notice again that we have specified r and s to be any integers. Our definition of negative exponents is such that the properties of exponents hold for all integer exponents, whether positive or negative integers. Here is a proof of Property 6.

Proof of Property 6

Our proof is centered on the fact that division by a number is equivalent to multiplication by the reciprocal of the number.

$$\frac{a^r}{a^s} = a^r \cdot \frac{1}{a^s}$$ Dividing by a^s is equivalent to multiplying by $\frac{1}{a^s}$

$$= a^r a^{-s}$$ Property 4

$$= a^{r+(-s)}$$ Property 1

$$= a^{r-s}$$ Definition of subtraction

EXAMPLE 7 Apply Property 6 to each expression, and then simplify the result. All answers that contain exponents should contain positive exponents only.

a. $\frac{2^8}{2^3} = 2^{8-3} = 2^5 = 32$

b. $\frac{x^2}{x^{18}} = x^{2-18} = x^{-16} = \frac{1}{x^{16}}$

c. $\frac{a^6}{a^{-8}} = a^{6-(-8)} = a^{14}$

d. $\frac{m^{-5}}{m^{-7}} = m^{-5-(-7)} = m^2$

PROPERTY ***Property 7 for Exponents***

If a is any real number, then

$$a^1 = a$$

and

$$a^0 = 1 \quad \text{(as long as } a \neq 0\text{)}$$

EXAMPLE 8 Simplify.

a. $(2x^2y^4)^0 = 1$

b. $(2x^2y^4)^1 = 2x^2y^4$

Here are some examples that use many of the properties of exponents. There are a number of ways to proceed on problems like these. You should use the method that works best for you.

EXAMPLES Simplify.

9. $\frac{(x^3)^{-2}(x^4)^5}{(x^{-2})^7} = \frac{x^{-6}x^{20}}{x^{-14}}$ Property 2

$$= \frac{x^{14}}{x^{-14}}$$ Property 1

$$= x^{28}$$ Property 6: $x^{14-(-14)} = x^{28}$

10. $\frac{6a^5b^{-6}}{12a^3b^{-9}} = \frac{6}{12} \cdot \frac{a^5}{a^3} \cdot \frac{b^{-6}}{b^{-9}}$ Write as separate fractions

$$= \frac{1}{2}a^2b^3$$ Property 6

Note Example 10 can also be written as $\frac{a^2b^3}{2}$. Either answer is correct.

11. $\dfrac{(4x^{-5}y^3)^2}{(x^4y^{-6})^{-3}} = \dfrac{16x^{-10}y^6}{x^{-12}y^{18}}$ Properties 2 and 3

$= 16x^2y^{-12}$ Property 6

$= 16x^2 \cdot \dfrac{1}{y^{12}}$ Property 4

$= \dfrac{16x^2}{y^{12}}$ Multiplication ■

Scientific Notation

Scientific notation is a method for writing very large or very small numbers in a more manageable form. Here is the definition.

DEFINITION *Scientific notation*

A number is written in *scientific notation* if it is written as the product of a number between 1 and 10 and an integer power of 10. A number written in scientific notation has the form

$$n \times 10^r$$

where $1 \leq n < 10$ and $r =$ an integer.

EXAMPLE 12 Write 376,000 in scientific notation.

SOLUTION We must rewrite 376,000 as the product of a number between 1 and 10 and a power of 10. To do so, we move the decimal point five places to the left so that it appears between the 3 and the 7. Then we multiply this number by 10^5. The number that results has the same value as our original number and is written in scientific notation.

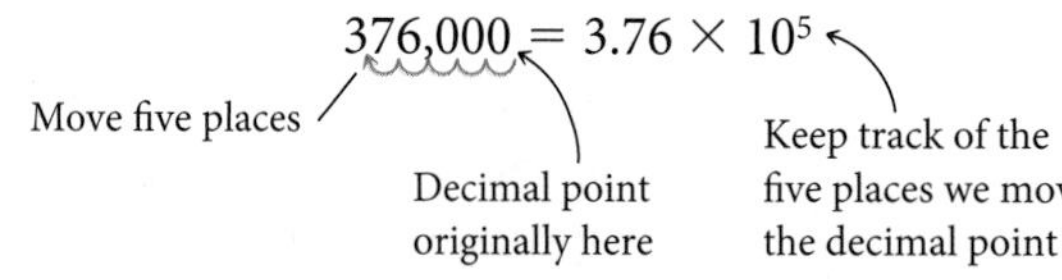

If a number written in expanded form is greater than or equal to 10, then when the number is written in scientific notation the exponent on 10 will be positive. A number that is less than one and greater than zero will have a negative exponent when written in scientific notation. ■

EXAMPLE 13 Write 4.52×10^3 in expanded form.

SOLUTION Because 10^3 is 1000, we can think of this as simply a multiplication problem. That is,

$$4.52 \times 10^3 = 4.52 \times 1{,}000 = 4{,}520$$ ■

The following table lists some additional examples of numbers written in expanded form and in scientific notation. In each case, note the relationship between the number of places the decimal point is moved and the exponent on 10.

Number Written in Expanded Form		Number Written in Scientific Notation
376,000	=	3.76×10^5
49,500	=	4.95×10^4
3,200	=	3.2×10^3
591	=	5.91×10^2
46	=	4.6×10^1
8	=	8×10^0
0.47	=	4.7×10^{-1}
0.093	=	9.3×10^{-2}
0.00688	=	6.88×10^{-3}
0.0002	=	2×10^{-4}
0.000098	=	9.8×10^{-5}

EXAMPLE 14 Simplify each expression and write all answers in scientific notation.

a. $(2 \times 10^8)(3 \times 10^{-3}) = (2)(3) \times (10^8)(10^{-3})$

$= 6 \times 10^5$

b. $\frac{4.8 \times 10^9}{2.4 \times 10^{-3}} = \frac{4.8}{2.4} \times \frac{10^9}{10^{-3}}$

$= 2 \times 10^{9-(-3)}$

$= 2 \times 10^{12}$

c. $\frac{(6.8 \times 10^5)(3.9 \times 10^{-7})}{7.8 \times 10^{-4}} = \frac{(6.8)(3.9)}{7.8} \times \frac{(10^5)(10^{-7})}{10^{-4}}$

$= 3.4 \times 10^2$

GETTING READY FOR CLASS

After reading through the preceding section, respond in your own words and in complete sentences.

A. Explain the difference between -2^4 and $(-2)^4$.
B. Explain the difference between 2^5 and 2^{-5}.
C. If a positive base is raised to a negative exponent, can the result be a negative number?
D. State Property 1 for exponents in your own words.

Problem Set 1.2

Evaluate each of the following.

1. 4^2 **2.** -4^2 **3.** -0.3^3

4. 2^5 **5.** $\left(\frac{1}{2}\right)^3$ **6.** $\left(-\frac{5}{6}\right)^2$

Use the properties of exponents to simplify each of the following as much as possible.

7. $x^5 \cdot x^4$ **8.** $(2^3)^2$ **9.** $-3a^2(2a^4)$ **10.** $5a^7(-4a^6)$

Write each of the following with positive exponents. Then simplify as much as possible.

11. 3^{-2} **12.** $(-2)^{-5}$

13. $\left(\frac{3}{4}\right)^{-2}$ **14.** $\left(\frac{1}{3}\right)^{-2} + \left(\frac{1}{2}\right)^{-3}$

Problems 15–22 are problems you will see later in the book. Multiply.

15. $8x^3 \cdot 10y^6$ **16.** $5y^2 \cdot 4x^2$ **17.** $8x^3 \cdot 9y^3$ **18.** $4y^3 \cdot 3x^2$

19. $3x \cdot 5y$ **20.** $3xy \cdot 5z$ **21.** $4x^6y^6 \cdot 3x$ **22.** $16x^4y^4 \cdot 3y$

Divide. (Assume all variables are nonzero.)

23. $\frac{10x^5}{5x^2}$ **24.** $\frac{-15x^4}{5x^2}$ **25.** $\frac{20x^3}{5x^2}$ **26.** $\frac{25x^7}{-5x^2}$

27. $\frac{8x^3y^5}{-2x^2y}$ **28.** $\frac{-16x^2y^2}{-2x^2y}$ **29.** $\frac{4x^4y^3}{-2x^2y}$ **30.** $\frac{10a^4b^2}{4a^2b^2}$

Use the properties of exponents to simplify each expression. Write all answers with positive exponents only. (Assume all variables are nonzero.)

31. $\frac{x^{-1}}{x^9}$ **32.** $\frac{x^{-3}}{x^5}$ **33.** $\frac{a^4}{a^{-6}}$ **34.** $\frac{a^5}{a^{-2}}$

35. $\frac{t^{-10}}{t^{-4}}$ **36.** $\frac{t^{-8}}{t^{-5}}$ **37.** $\left(\frac{x^5}{x^3}\right)^6$ **38.** $\left(\frac{x^7}{x^4}\right)^5$

39. $\frac{(x^5)^6}{(x^3)^4}$ **40.** $\frac{(x^7)^3}{(x^4)^5}$ **41.** $\frac{(x^{-2})^3(x^3)^{-2}}{x^{10}}$ **42.** $\frac{(x^{-4})^3(x^3)^{-4}}{x^{10}}$

43. $\frac{5a^8b^3}{20a^5b^{-4}}$ **44.** $\frac{7a^6b^{-2}}{21a^2b^{-5}}$ **45.** $\frac{(3x^{-2}y^8)^4}{(9x^4y^{-3})^2}$ **46.** $\frac{(6x^{-3}y^{-5})^2}{(3x^{-4}y^{-3})^4}$

Write each number in scientific notation.

47. 378,000 **48.** 3,780,000 **49.** 4,900 **50.** 490

51. 0.00037 **52.** 0.000037 **53.** 0.00495 **54.** 0.0495

Write each number in expanded form.

55. 5.34×10^3 **56.** 5.34×10^2 **57.** 7.8×10^6 **58.** 7.8×10^4

59. 3.44×10^{-3} **60.** 3.44×10^{-5} **61.** 4.9×10^{-1} **62.** 4.9×10^{-2}

Simplify each of the following expressions. Write all answers in scientific notation.

63. $(4 \times 10^{10})(2 \times 10^{-6})$

64. $(3 \times 10^{-12})(3 \times 10^{4})$

65. $\dfrac{8 \times 10^{14}}{4 \times 10^{5}}$

66. $\dfrac{6 \times 10^{8}}{2 \times 10^{3}}$

67. $\dfrac{(5 \times 10^{6})(4 \times 10^{-8})}{8 \times 10^{4}}$

68. $\dfrac{(6 \times 10^{-7})(3 \times 10^{9})}{5 \times 10^{6}}$

69. $\dfrac{(2.4 \times 10^{-3})(3.6 \times 10^{-7})}{(4.8 \times 10^{6})(1 \times 10^{-9})}$

70. $\dfrac{(7.5 \times 10^{-6})(1.5 \times 10^{9})}{(1.8 \times 10^{4})(2.5 \times 10^{-2})}$

71. Complete the following table.

Expanded Form	Scientific Notation $n \times 10^r$
0.000357	3.57×10^{-4}
0.00357	
0.0357	
0.357	
3.57	
35.7	
357	
3,570	
35,700	

72. Complete the following table.

Expanded Form	Scientific Notation $n \times 10^r$
0.000123	1.23×10^{-4}
	1.23×10^{-3}
	1.23×10^{-2}
	1.23×10^{-1}
	1.23×10^{0}
	1.23×10^{1}
	1.23×10^{2}
	1.23×10^{3}
	1.23×10^{4}

NASA

Galilean Moons The planet Jupiter has about 60 known moons. In the year 1610 Galileo first discovered the four largest moons of Jupiter: Io, Europa, Ganymede, and Callisto. These moons are known as the Galilean moons. Each moon has a unique period, or the time it takes to make a trip around Jupiter. Fill in the tables below.

73.

Jupiter's Moon	Period (seconds)	
Io	153,000	
Europa		3.07×10^{5}
Ganymede	618,000	
Callisto		1.44×10^{6}

74.

Jupiter's Moon	Distance from Jupiter (Kilometers)	
Io	422,000	
Europa		6.71×10^{5}
Ganymede	1,070,000	
Callisto		1.88×10^{6}

75. Large Numbers If you are 20 years old, you have been alive for more than 630,000,000 seconds. Write this last number in scientific notation.

NASA/JPL-CalTech

76. Our Galaxy The galaxy in which the Earth resides is called the Milky Way galaxy. It is a spiral galaxy that contains approximately 200,000,000,000 stars (our Sun is one of them). Write this number in words and in scientific notation.

77. Light Year A light year, the distance light travels in 1 year, is approximately 5.9×10^{12} miles. The Andromeda galaxy is approximately 2.5×10^{6} light years from our galaxy. Find the distance in miles between our galaxy and the Andromeda galaxy.

78. Distance to the Sun The distance from the Earth to the Sun is approximately 9.3×10^{7} miles. If light travels 1.1×10^{7} miles in 1 minute, how many minutes does it take the light from the Sun to reach the Earth?

79. Credit Card Debt Outstanding credit-card debt in the United States is over $772 billion.

a. Write the number 422 billion in scientific notation.

b. If there are approximately 90 million households with at least one credit card, find the average credit-card debt per household, to the nearest dollar.

© Christophe Testi/iStockphoto

80. Computer Science We all use the language of computers to indicate how much memory our computers hold or how much information we can put on a storage device such as a USB drive. Scientific notation gives us a way to compare the actual numbers associated with the words we use to describe data storage in computers. The smallest amount of data that a computer can hold is measured in bits. A byte is the next largest unit and is equal to 8, or 2^3, bits. Fill in the table below.

Number of Bytes

Unit	Exponential Form	Scientific Notation
Kilobyte	$2^{10} = 1{,}024$	
Megabyte	$2^{20} \approx 1{,}049{,}000$	
Gigabyte	$2^{30} \approx 1{,}074{,}000{,}000$	
Terabyte	$2^{40} \approx 1{,}099{,}500{,}000{,}000$	

Polynomials: Sums, Differences, and Products

1.3

Polynomials in General

DEFINITION *Term/monomial*

A *term,* or *monomial,* is a constant or the product of a constant and one or more variables raised to whole-number exponents.

The following are monomials, or terms:

$$-16 \qquad 3x^2y \qquad -\frac{2}{5}a^3b^2c \qquad xy^2z$$

The numerical part of each monomial is called the *numerical coefficient,* or just *coefficient* for short. For the preceding terms, the coefficients are -16, 3, $-\frac{2}{5}$, and 1. Notice that the coefficient for xy^2z is understood to be 1.

DEFINITION *Polynomial*

A *polynomial* is any finite sum of terms. Because subtraction can be written in terms of addition, finite differences are also included in this definition.

The following are polynomials:

$$2x^2 - 6x + 3 \qquad -5x^2y + 2xy^2 \qquad 4a - 5b + 6c + 7d$$

Polynomials can be classified further according to the number of terms present. If a polynomial consists of two terms, it is said to be a *binomial.* If it has three terms, it is called a *trinomial.* And, as stated, a polynomial with only one term is said to be a *monomial.*

DEFINITION *Degree*

The *degree* of a polynomial with one variable is the highest power to which the variable is raised in any one term.

VIDEO EXAMPLES

SECTION 1.3

EXAMPLES

1. $6x^2 + 2x - 1$ — A trinomial of degree 2
2. $5x - 3$ — A binomial of degree 1
3. $7x^6 - 5x^3 + 2x - 4$ — A polynomial of degree 6
4. $-7x^4$ — A monomial of degree 4
5. 15 — A monomial of degree 0

Polynomials with one variable are usually written in decreasing powers of the variable. When this is the case, the coefficient of the first term is called the *leading coefficient.* In Example 1, the leading coefficient is 6. In Example 2, it is 5. The leading coefficient in Example 3 is 7.

DEFINITION *Similar/like terms*

Two or more terms that differ only in their numerical coefficients are called *similar*, or *like*, terms. Because similar terms differ only in their coefficients, they have identical variable parts.

Adding Polynomials

To add two polynomials, we simply apply the commutative and associative properties to group similar terms together and then use the distributive property as we have in the following example.

EXAMPLE 6 Add $5x^2 - 4x + 2$ and $3x^2 + 9x - 6$.

SOLUTION

$$
\begin{aligned}
&(5x^2 - 4x + 2) + (3x^2 + 9x - 6) \\
&\quad = (5x^2 + 3x^2) + (-4x + 9x) + (2 - 6) && \text{Commutative and associative properties} \\
&\quad = (5 + 3)x^2 + (-4 + 9)x + (2 - 6) && \text{Distributive property} \\
&\quad = 8x^2 + 5x + (-4) \\
&\quad = 8x^2 + 5x - 4
\end{aligned}
$$

EXAMPLE 7 Find the sum of $-8x^3 + 7x^2 - 6x + 5$ and $10x^3 + 3x^2 - 2x - 6$.

SOLUTION We can add the two polynomials using the method of Example 6, or we can arrange similar terms in columns and add vertically. Using the column method, we have

$$
\begin{array}{r}
-8x^3 + 7x^2 - 6x + 5 \\
10x^3 + 3x^2 - 2x - 6 \\
\hline
2x^3 + 10x^2 - 8x - 1
\end{array}
$$

Subtracting Polynomials

To find the difference of two polynomials, we need to use the fact that the opposite of a sum is the sum of the opposites. That is,

$$-(a + b) = -a + (-b)$$

EXAMPLE 8 Subtract $4x^2 + 2x - 3$ from $9x^2 - 3x + 5$.

SOLUTION First, we write the problem in terms of subtraction. Then, we subtract by adding the opposite of each term in the polynomial that follows the subtraction sign:

$$
\begin{aligned}
&(9x^2 - 3x + 5) - (4x^2 + 2x - 3) \\
&\quad = 9x^2 - 3x + 5 + (-4x^2) + (-2x) + 3 && \text{The opposite of a sum is the sum of the opposites}
\end{aligned}
$$

$= (9x^2 - 4x^2) + (-3x - 2x) + (5 + 3)$ Commutative and associative properties

$= 5x^2 - 5x + 8$ Combine similar terms

When one set of grouping symbols is contained within another, it is best to begin the process of simplification within the innermost grouping symbols and work outward from there.

EXAMPLE 9 Simplify: $4x - 3[2 - (3x + 4)]$.

SOLUTION Removing the innermost parentheses first, we have

$$
\begin{aligned}
4x - 3[2 - (3x + 4)] &= 4x - 3(2 - 3x - 4) \\
&= 4x - 3(-3x - 2) \\
&= 4x + 9x + 6 \\
&= 13x + 6
\end{aligned}
$$

The Value of a Polynomial

EXAMPLE 10 Find the value of $5x^3 - 3x^2 + 4x - 5$ when x is 2.

SOLUTION We begin by substituting 2 for x in the original polynomial:

When $x = 2$

the polynomial $5x^3 - 3x^2 + 4x - 5$

becomes $5 \cdot 2^3 - 3 \cdot 2^2 + 4 \cdot 2 - 5$

$$
\begin{aligned}
&= 5 \cdot 8 - 3 \cdot 4 + 4 \cdot 2 - 5 \\
&= 40 - 12 + 8 - 5 \\
&= 31
\end{aligned}
$$

Multiplying Polynomials

The distributive property is the key to multiplying polynomials. The simplest type of multiplication occurs when we multiply a polynomial by a monomial.

EXAMPLE 11 Find the product of $4x^3$ and $5x^2 - 3x + 1$.

SOLUTION To multiply, we apply the distributive property:

$4x^3(5x^2 - 3x + 1)$

$= 4x^3(5x^2) + 4x^3(-3x) + 4x^3(1)$ Distributive property

$= 20x^5 - 12x^4 + 4x^3$

Note We multiply coefficients and add exponents.

The distributive property can also be applied to multiply a polynomial by a polynomial. Let's consider the case where both polynomials have two terms.

EXAMPLE 12 Multiply $2x - 3$ and $x + 5$.

SOLUTION Distributing the $2x - 3$ across the sum $x + 5$ gives us

$$\begin{aligned}
&(2x-3)(x+5) \\
&\quad = (2x-3)x + (2x-3)5 && \text{Distributive property} \\
&\quad = 2x(x) + (-3)x + 2x(5) + (-3)5 && \text{Distributive property} \\
&\quad = 2x^2 - 3x + 10x - 15 \\
&\quad = 2x^2 + 7x - 15 && \text{Combine like terms}
\end{aligned}$$

Notice the third line in Example 12. It consists of all possible products of terms in the first binomial and those of the second binomial. We can generalize this into a rule for multiplying two polynomials.

RULE ***Multiplying Two Polynomials***

Multiply each term in the first polynomial by each term in the second polynomial.

Multiplying polynomials can be accomplished by a method that looks very similar to long multiplication with whole numbers. Reworking Example 12, we have

$$\begin{array}{rll}
2x - 3 & & \\
x + 5 & & \\
\hline
2x^2 - 3x & & \text{Multiply } 2x-3 \text{ by } x \\
10x - 15 & & \text{Multiply } 2x-3 \text{ by } 5 \\
\hline
2x^2 + 7x - 15 & & \text{Add in columns}
\end{array}$$

EXAMPLE 13 Multiply $2x - 3y$ and $3x^2 - xy + 4y^2$ vertically.

SOLUTION

$$\begin{array}{rll}
3x^2 - \quad xy + 4y^2 & & \\
2x - 3y & & \\
\hline
6x^3 - 2x^2y + 8xy^2 & & \text{Multiply } 3x^2 - xy + 4y^2 \text{ by } 2x \\
- 9x^2y + 3xy^2 - 12y^3 & & \text{Multiply } 3x^2 - xy + 4y^2 \text{ by } -3y \\
\hline
6x^3 - 11x^2y + 11xy^2 - 12y^3 & & \text{Add similar terms}
\end{array}$$

Multiplying Binomials—The FOIL Method

The product of two binomials occurs very frequently in algebra. Because this type of product is so common, we have a special method of multiplication that applies only to products of binomials.

Consider the product of $2x - 5$ and $3x - 2$. Distributing $3x - 2$ over $2x$ and -5, we have

$$\begin{aligned}
(2x-5)(3x-2) &= (2x)(3x-2) + (-5)(3x-2) \\
&= (2x)(3x) + (2x)(-2) + (-5)(3x) + (-5)(-2) \\
&= 6x^2 - 4x - 15x + 10 \\
&= 6x^2 - 19x + 10
\end{aligned}$$

Looking closely at the second and third lines, we notice the following relationships:

1. $6x^2$ comes from multiplying the *first* terms in each binomial:

$$(2x - 5)(3x - 2) \qquad 2x(3x) = 6x^2 \qquad \textit{First} \text{ terms}$$

2. $-4x$ comes from multiplying the *outside* terms in the product:

$$(2x - 5)(3x - 2) \qquad 2x(-2) = -4x \qquad \textit{Outside} \text{ terms}$$

3. $-15x$ comes from multiplying the *inside* terms in the product:

$$(2x - 5)(3x - 2) \qquad -5(3x) = -15x \qquad \textit{Inside} \text{ terms}$$

4. 10 comes from multiplying the *last* two terms in the product:

$$(2x - 5)(3x - 2) \qquad -5(-2) = 10 \qquad \textit{Last} \text{ terms}$$

Once we know where the terms in the answer come from, we can reduce the number of steps used in finding the product:

$$\begin{aligned}(2x - 5)(3x - 2) &= \underset{\text{First}}{6x^2} - \underset{\text{Outside}}{4x} - \underset{\text{Inside}}{15x} + \underset{\text{Last}}{10}\\ &= 6x^2 - 19x + 10\end{aligned}$$

This is called the *FOIL method*.

EXAMPLES Multiply using the FOIL method.

14. $(4a - 5b)(3a + 2b) = \underset{\text{F}}{12a^2} + \underset{\text{O}}{8ab} - \underset{\text{I}}{15ab} - \underset{\text{L}}{10b^2}$

$= 12a^2 - 7ab - 10b^2$

15. $(3 - 2t)(4 + 7t) = \underset{\text{F}}{12} + \underset{\text{O}}{21t} - \underset{\text{I}}{8t} - \underset{\text{L}}{14t^2}$

$= 12 + 13t - 14t^2$

16. $\left(2x + \frac{1}{2}\right)\left(4x - \frac{1}{2}\right) = \underset{\text{F}}{8x^2} - \underset{\text{O}}{x} + \underset{\text{I}}{2x} - \underset{\text{L}}{\frac{1}{4}}$

$= 8x^2 + x - \frac{1}{4}$

17. $(a^5 + 3)(a^5 - 7) = \underset{\text{F}}{a^{10}} - \underset{\text{O}}{7a^5} + \underset{\text{I}}{3a^5} - \underset{\text{L}}{21}$

$= a^{10} - 4a^5 - 21$

18. $(2x + 3)(5y - 4) = \underset{\text{F}}{10xy} - \underset{\text{O}}{8x} + \underset{\text{I}}{15y} - \underset{\text{L}}{12}$ ■

The Square of a Binomial

EXAMPLE 19 Find $(4x - 6)^2$.

SOLUTION Applying the definition of exponents and then the FOIL method, we have

$$\begin{aligned}(4x-6)^2 &= (4x-6)(4x-6)\\ &= \underset{F}{16x^2} - \underset{O}{24x} - \underset{I}{24x} + \underset{L}{36}\\ &= 16x^2 - 48x + 36\end{aligned}$$

The preceding example is the square of a binomial. This type of product occurs frequently enough in algebra that we have special formulas for squares of binomials:

$$(a+b)^2 = (a+b)(a+b) = a^2 + ab + ab + b^2 = a^2 + 2ab + b^2$$
$$(a-b)^2 = (a-b)(a-b) = a^2 - ab - ab + b^2 = a^2 - 2ab + b^2$$

Observing the results in both cases, we have the following rule.

[Δ≠Σ ***The Square of a Binomial***

The square of a binomial is the sum of the square of the first term, twice the product of the two terms, and the square of the last term. That is:

$$(a+b)^2 = \underset{\text{Square of first term}}{a^2} + \underset{\text{Twice the product of the two terms}}{2ab} + \underset{\text{Square of last term}}{b^2}$$

$$(a-b)^2 = a^2 - 2ab + b^2$$

EXAMPLES Use the preceding formulas to expand each binomial square.

20. $(x+7)^2 = x^2 + 2(x)(7) + 7^2 = x^2 + 14x + 49$

21. $(3t-5)^2 = (3t)^2 - 2(3t)(5) + 5^2 = 9t^2 - 30t + 25$

22. $(4x+2y)^2 = (4x)^2 + 2(4x)(2y) + (2y)^2 = 16x^2 + 16xy + 4y^2$

23. $(5-a^3)^2 = 5^2 - 2(5)(a^3) + (a^3)^2 = 25 - 10a^3 + a^6$

24. $\left(x+\frac{1}{3}\right)^2 = x^2 + 2(x)\left(\frac{1}{3}\right) + \left(\frac{1}{3}\right)^2 = x^2 + \frac{2}{3}x + \frac{1}{9}$

Products That Are the Difference of Two Squares

Another frequently occurring kind of product is found when multiplying two binomials that differ only in the sign between their terms.

EXAMPLE 25 Multiply $(3x - 5)$ and $(3x + 5)$.

SOLUTION Applying the FOIL method, we have

$$(3x - 5)(3x + 5) = 9x^2 + 15x - 15x - 25 \quad \text{The middle terms add to 0}$$

F O I L

$$= 9x^2 - 25$$

The outside and inside products in Example 25 are opposites and therefore add to 0. Here is a generalization of Example 25:

$$(a - b)(a + b) = a^2 + ab - ab - b^2 \quad \text{The middle terms add to 0}$$

$$= a^2 - b^2$$

[Δ≠Σ] *The Product of the Sum and Difference of Two Terms*

To multiply two binomials that differ only in the sign between their two terms, simply subtract the square of the second term from the square of the first term:

$$(a - b)(a + b) = a^2 - b^2$$

The expression $a^2 - b^2$ is called the *difference of two squares*. Once we memorize and understand this rule, we can multiply binomials of this form with a minimum of work.

EXAMPLES Find the following products.

26. $(x - 5)(x + 5) = x^2 - 25$

27. $(2a - 3)(2a + 3) = 4a^2 - 9$

28. $(x^2 + 4)(x^2 - 4) = x^4 - 16$

29. $(x^3 - 2a)(x^3 + 2a) = x^6 - 4a^2$

GETTING READY FOR CLASS

After reading through the preceding section, respond in your own words and in complete sentences.

A. What are similar terms?

B. Explain in words how you subtract one polynomial from another.

C. Explain why $(x + 3)^2 \neq x^2 + 9$.

D. When will the product of two binomials result in a binomial?

Problem Set 1.3

Identify which of the following are monomials, binomials, or trinomials. Give the degree of each, and name the leading coefficient.

1. $5x^2 - 3x + 2$
2. $2x^2 + 4x - 1$
3. $3x - 5$
4. $5y + 3$
5. $8a^2 + 3a - 5$
6. $9a^2 - 8a - 4$
7. $4x^3 - 6x^2 + 5x - 3$
8. $9x^4 + 4x^3 - 2x^2 + x$
9. $-\frac{3}{4}$
10. -16
11. $4x - 5 + 6x^3$
12. $9x + 2 + 3x^3$

Simplify each of the following by combining similar terms.

13. $2x^2 - 3x + 10x - 15$
14. $12a^2 + 8ab - 15ab - 10b^2$
15. $(3x^2 + 4x - 5) - (2x^2 - 5x - 9)$
16. $(-4x^2 + 5x - 3) - (2x^2 + x - 7)$
17. $(-2a^2 + 4a - 5) - (-3a^2 - a - 9)$
18. $(-6a^2 + a - 4) - (-2a^2 - 3a - 4)$
19. $(5x^2 - 6x + 1) - (4x^2 + 7x - 2)$
20. $(11x^2 - 8x) - (4x^2 - 2x - 7)$
21. Subtract $2x^2 - 4x$ from $2x^2 - 7x$.
22. Subtract $-3x + 6$ from $-3x + 9$.
23. Subtract $x^2 - 2x + 1$ from $2x^2 - 8x - 4$.
24. Subtract $-5x^2 + x - 4$ from $-2x^2 - 3x + 1$.

Multiply.

25. $2x(6x^2 - 5x + 4)$
26. $-3x(5x^2 - 6x - 4)$
27. $(a - b)(a^2 + ab + b^2)$
28. $(a + b)(a^2 - ab + b^2)$
29. $(2x + y)(4x^2 - 2xy + y^2)$
30. $(x - 3y)(x^2 + 3xy + 9y^2)$
31. $(2a + 3)(3a + 2)$
32. $(5a - 4)(2a + 1)$
33. $(5 - 3t)(4 + 2t)$
34. $(7 - t)(6 - 3t)$
35. $(x^3 + 3)(x^3 - 5)$
36. $(x^3 + 4)(x^3 - 7)$
37. $\left(3t + \frac{1}{3}\right)\left(6t - \frac{2}{3}\right)$
38. $\left(5t - \frac{1}{5}\right)\left(10t + \frac{3}{5}\right)$

Find the following special products.

39. $(2a - 3)^2$
40. $(3a + 2)^2$
41. $(5x + 2y)^2$
42. $(3x - 4y)^2$
43. $(5 - 3t^3)^2$
44. $(7 - 2t^4)^2$
45. $(2a + 3b)(2a - 3b)$
46. $(6a - 1)(6a + 1)$
47. $(3r^2 + 7s)(3r^2 - 7s)$
48. $(5r^2 - 2s)(5r^2 + 2s)$

Here are a variety of problems that you will see again as we progress through the book. Simplify.

49. $-4x + 9x$

50. $-6x - 2x$

51. $5x^2 + 3x^2$

52. $7x^2 + 3x^2$

53. $-8x^3 + 10x^3$

54. $4x^3 - 7x^3$

55. $2x + 3 - 2x - 8$

56. $9x - 4 - 9x - 10$

57. $-1(2x - 3)$

58. $-1(-3x + 1)$

59. $-3(-3x - 2)$

60. $-4(-5x + 3)$

61. **a.** $(3x - 5) - (3a - 5)$
b. $(2x + 3) - (2a + 3)$

62. **a.** $(x^2 - 4) - (a^2 - 4)$
b. $(x^2 - 1) - (a^2 - 1)$

63. $2x^3 - 3x + 10x - 15$

64. $12a^2 + 8ab - 15ab - 10b^2$

65. $(6x^3 - 2x^2y + 8xy^2) + (-9x^2y + 3xy^2 - 12y^3)$

66. $(3x^3 - 15x^2 + 18x) + (2x^2 - 10x + 12)$

67. $4x^3(-3x)$

68. $5x^2(-4x)$

69. $4x^3(5x^2)$

70. $5x^2(3x^2)$

71. $(a^3)^2$

72. $(a^4)^2$

73. $-1(5 - x)$

74. $-1(a - b)$

75. $-1(7 - x)$

76. $-1(6 - y)$

77. $5\left(x - \frac{1}{5}\right)$

78. $7\left(x + \frac{1}{7}\right)$

79. $x\left(1 - \frac{1}{x}\right)$

80. $a\left(1 + \frac{1}{a}\right)$

81. $12\left(\frac{1}{4}x + \frac{2}{3}y\right)$

82. $20\left(\frac{2}{5}x + \frac{1}{4}y\right)$

83. $(x + 3)^2 - 2(x + 3) - 8$

84. $(x - 2)^2 - 3(x - 2) - 10$

85. $(2a - 3)^2 - 9(2a - 3) + 20$

86. $(3a - 2)^2 + 2(3a - 2) - 3$

87. $2(4a + 2)^2 - 3(4a + 2) - 20$

88. $6(2a + 4)^2 - (2a + 4) - 2$

© Rob Friedman/iStockPhoto

89. Height of an Object The formula for the height of an object that has been thrown straight up with a velocity of 64 feet/second is

$$h = -16t^2 + 64t$$

Find the height after 1 second and after 3 seconds.

90. Interest If you deposit P dollars in an account with an annual interest rate r that is compounded twice a year, then at the end of a year the amount of money in that account is given by the formula

$$A = P\left(1 + \frac{r}{2}\right)^2$$

Expand the right side of this formula, then simplify.

Review of Factoring

In this section we review the different methods of factoring. This section is important because factoring is one of the most important skills you need to be successful in this course.

HOW TO *Factor a Polynomial*

Step 1: If the polynomial has a greatest common factor (GCF) other than 1, then factor out the greatest common factor.

Step 2: If the polynomial has two terms (it is a binomial), then see if it is the difference of two squares or the sum or difference of two cubes, and then factor accordingly. Remember, if it is the sum of two squares it will not factor.

Step 3: If the polynomial has three terms (a trinomial), then it is either a perfect square trinomial, which will factor into the square of a binomial, or it is not a perfect square trinomial, in which case we try to write it as the product of two binomials. Remember, not all trinomials are factorable.

Step 4: If the polynomial has more than three terms, then try to factor it by grouping.

Step 5: As a final check, see if any of the factors you have written can be factored further. If you have overlooked a common factor, you can catch it here.

Here are some examples illustrating how we use the steps on our list.

VIDEO EXAMPLES

SECTION 1.4

EXAMPLE 1 Factor $2x^5 - 8x^3$.

SOLUTION First we check to see if the greatest common factor is other than 1. Since the greatest common factor is $2x^3$, we begin by factoring it out. Once we have done so, we notice that the binomial that remains is the difference of two squares, which we factor according to the formula $a^2 - b^2 = (a + b)(a - b)$.

$2x^5 - 8x^3 = 2x^3(x^2 - 4)$ Factor out the greatest common factor, $2x^3$

$= 2x^3(x + 2)(x - 2)$ Factor the difference of two squares ■

EXAMPLE 2 Factor $2x^5 + 8x^3$.

SOLUTION This problem is very similar to the problem in Example 1. We factor out the greatest common factor, $2x^3$, first.

$2x^5 + 8x^3 = 2x^3(x^2 + 4)$ Factor out $2x^3$

In this case we cannot factor further because the sum of two squares, $x^2 + 4$, does not factor. ■

EXAMPLE 3 Factor $3x^4 - 18x^3 + 27x^2$.

SOLUTION Step 1 is to factor out the greatest common factor $3x^2$. After we have done so, we notice the trinomial that remains is a perfect square trinomial, which will factor as the square of a binomial.

$$3x^4 - 18x^3 + 27x^2 = 3x^2(x^2 - 6x + 9) \quad \text{Factor out } 3x^2$$

$$= 3x^2(x - 3)^2 \quad x^2 - 6x + 9 \text{ is the square of } x - 3$$

■

EXAMPLE 4 Factor $6a^2 - 11a + 4$.

SOLUTION Here we have a trinomial that does not have a greatest common factor other than 1. Since it is not a perfect square trinomial, we factor it by trial and error. Without showing all the different possibilities, here is the answer.

$$6a^2 - 11a + 4 = (3a - 4)(2a - 1)$$

■

EXAMPLE 5 Factor $6a^2 - 11a - 4$.

SOLUTION This trinomial is very similar to the trinomial in Example 4. However, this trinomial cannot be factored further. It is a *prime* polynomial. Convince yourself that this trinomial is prime. ■

EXAMPLE 6 Factor $2x^4 + 16x$.

SOLUTION This binomial has a greatest common factor of $2x$. The binomial that remains after the $2x$ has been factored from each term is the sum of two cubes, which we factor according to the formula $a^3 + b^3 = (a + b)(a^2 - ab + b^2)$.

$$2x^4 + 16x = 2x(x^3 + 8) \quad \text{Factor } 2x \text{ from each term}$$

$$= 2x(x + 2)(x^2 - 2x + 4) \quad \text{The sum of two cubes}$$

■

EXAMPLE 7 Factor $2ab^5 + 8ab^4 + 2ab^3$.

SOLUTION The greatest common factor is $2ab^3$. We begin by factoring it from each term. After that we find that the trinomial that remains cannot be factored further.

$$2ab^5 + 8ab^4 + 2ab^3 = 2ab^3(b^2 + 4b + 1)$$

■

EXAMPLE 8 Factor $4x^2 - 6x + 2ax - 3a$.

SOLUTION Our polynomial has four terms, so we factor by grouping.

$$4x^2 - 6x + 2ax - 3a = 2x(2x - 3) + a(2x - 3)$$

$$= (2x - 3)(2x + a)$$

■

EXAMPLE 9 Factor $4x^4 - 35x^2 - 9$.

SOLUTION This is a trinomial in x^2. We factor it into the product of two binomials, and then we continue to factor.

$$4x^4 - 35x^2 - 9 = (x^2 - 9)(4x^2 + 1)$$

$$= (x + 3)(x - 3)(4x^2 + 1)$$

Notice that $4x^2 + 1$ does not factor further because it is the sum of two squares. You must convince yourself that this is true.

EXAMPLE 10 Factor $2x^2(x - 3) - 5x(x - 3) - 3(x - 3)$.

SOLUTION We begin by factoring out the greatest common factor $(x - 3)$. Then we factor the trinomial that remains.

$$2x^2(x - 3) - 5x(x - 3) - 3(x - 3) = (x - 3)(2x^2 - 5x - 3)$$

$$= (x - 3)(2x + 1)(x - 3)$$

$$= (2x + 1)(x - 3)^2$$

GETTING READY FOR CLASS

After reading through the preceding section, respond in your own words and in complete sentences.

A. How do you know when you've factored completely?
B. If a polynomial has four terms, what method of factoring should you try?
C. What is the first step in factoring a polynomial?
D. What do we call a polynomial that does not factor?

Problem Set 1.4

Factor the following trinomials.

1. $x^2 - 2x - 24$ **2.** $x^2 + 2x - 24$ **3.** $x^2 - 5x - 6$

4. $x^2 + 5x - 6$ **5.** $x^2 - 5x + 6$ **6.** $x^2 - x - 6$

7. $x^2 - 10x + 25$ **8.** $4x^2 + 4x + 1$ **9.** $2x^2 - 5x - 3$

10. $20x^2 - 93x + 34$ **11.** $21x^2 - 23x + 6$ **12.** $42x^2 + 23x - 10$

Factor the following binomials as the difference of two squares.

13. $x^2 - 16$ **14.** $x^2 - 25$ **15.** $a^2 - 1$ **16.** $a^2 - 9$

17. $a^2 - 16b^2$ **18.** $a^2 - 25b^2$ **19.** $9x^2 - 49$ **20.** $49x^2 - 144$

21. $16x^4 - 49$ **22.** $4x^4 - 25$ **23.** $t^4 - 81$ **24.** $t^4 - 16$

Factor the following binomials as the sum or difference of two cubes.

25. $a^3 + b^3$ **26.** $a^3 - b^3$ **27.** $x^3 - 8$ **28.** $x^3 + 125$

29. $x^3 + 1$ **30.** $x^3 - 27$ **31.** $8x^3 + 1$ **32.** $27x^3 + 64$

Factor each of the following by first factoring out the greatest common factor and then factoring the polynomial that remains, if possible.

33. $60x^2 - 130x + 60$ **34.** $90x^2 + 60x - 80$ **35.** $x^3 + 5x^2 + 6x$

36. $x^3 - 5x^2 - 6x$ **37.** $2x^3 - 5x^2 - 3x$ **38.** $6x^3 - 5x^2 - x$

39. $x^3 - 2x^2 - 24x$ **40.** $x^3 + 2x^2 - 24x$ **41.** $6x + 24$

42. $3x^2 - 3xy$ **43.** $100x^2 - 300x$ **44.** $10x^2 + 100x$

45. $20a^2 - 45$ **46.** $50a - 2ax^2$ **47.** $9a^3 - 16a$

48. $16a^3 - 25a$ **49.** $12y - 2xy - 2x^2y$ **50.** $6y - 4xy - 2x^2y$

Use factoring by grouping to factor each of the following.

51. $ax + 2x + 3a + 6$ **52.** $ay + 2y - 4a - 8$ **53.** $x^2 - 3ax - 2x + 6a$

54. $x^2 - 3ax + 2x - 6a$ **55.** $x^3 + 2x^2 - 9x - 18$ **56.** $x^3 + 5x^2 - 4x - 20$

57. $x^3 + 3x^2 - 4x - 12$ **58.** $x^3 - 3x^2 - 4x + 12$ **59.** $4x^3 + 12x^2 - 9x - 27$

60. $9x^3 + 18x^2 - 4x - 8$ **61.** $2x^3 + x^2 - 18x - 9$ **62.** $3x^3 - x^2 - 12x + 4$

Factor completely, if possible.

63. $4x^2 - 31x - 8$ **64.** $6x^2 - 55xy + 9y^2$ **65.** $x^2 + 49$

66. $25 + a^2$ **67.** $150x^3 + 65x^2 - 280x$ **68.** $360x^3 - 490x$

69. $24x^2 + 2x - 5$ **70.** $12x^2 - 49x + 4$ **71.** $x^6 - 1$

72. $x^6 - 64$ **73.** $r^2 - \frac{1}{9}$ **74.** $4a^2 + 2a + \frac{1}{4}$

75. $125t^3 + \frac{1}{27}$ **76.** $\frac{1}{25} + \frac{1}{10}t^2 + \frac{1}{16}t^4$ **77.** $15t^2 + t - 16$

78. $48x^2 - 74x + 3$ **79.** $100x^2 - 100x - 600$ **80.** $100x^2 - 100x - 1200$

81. $4x^3 + 16xy^2$ **82.** $50 - 2a^2$

83. $30x^2 + 97x + 77$ **84.** $96a^2 + 44a - 35$

85. $x^2(a + 5) + 6x(a + 5) + 9(a + 5)$ **86.** $a^2(x - 2) + 4a(x - 2) + 4(x - 2)$

87. $12a^2(x - 7) - 75(x - 7)$ **88.** $18a^2(2x + 3) - 50(2x + 3)$

Rational Expressions

We will begin this section with the definition of a rational expression. We will then state the two basic properties associated with rational expressions and go on to apply one of the properties to reduce rational expressions to lowest terms.

$$\text{Rational expressions} = \left\{ \frac{P}{Q} \mid P \text{ and } Q \text{ are polynomials, } Q \neq 0 \right\}$$

Some examples of rational expressions are

$$\frac{2x - 3}{x + 5} \qquad \frac{x^2 - 5x - 6}{x^2 - 1} \qquad \frac{a - b}{b - a}$$

Basic Properties

For rational expressions, multiplying the numerator and denominator by the same nonzero expression may change the form of the rational expression, but it will always produce an expression equivalent to the original one. The same is true when dividing the numerator and denominator by the same nonzero quantity.

PROPERTY *Properties of Rational Expressions*

If P, Q, and K are polynomials with $Q \neq 0$ and $K \neq 0$, then

$$\frac{P}{Q} = \frac{PK}{QK} \quad \text{and} \quad \frac{P}{Q} = \frac{P/K}{Q/K}$$

We reduce rational expressions to lowest terms by first factoring the numerator and denominator and then dividing both numerator and denominator by any factors they have in common.

VIDEO EXAMPLES

SECTION 1.5

EXAMPLE 1 Reduce $\frac{x^2 - 9}{x - 3}$ to lowest terms.

SOLUTION Factoring, we have

$$\frac{x^2 - 9}{x - 3} = \frac{(x + 3)(x - 3)}{x - 3}$$

The numerator and denominator have the factor $x - 3$ in common. Dividing the numerator and denominator by $x - 3$, we have

$$\frac{(x + 3)\cancel{(x - 3)}}{\cancel{x - 3}} = \frac{x + 3}{1} = x + 3$$

Note The lines drawn through the $(x - 3)$ in the numerator and denominator indicate that we have divided through by $(x - 3)$. As the problems become more involved, these lines will help keep track of which factors have been divided out and which have not.

For the problem in Example 1, there is an implied restriction on the variable x: It cannot be 3. If x were 3, the expression $\frac{(x^2 - 9)}{(x - 3)}$ would become $\frac{0}{0}$, an expression that we cannot associate with a real number. For all problems involving rational expressions, we restrict the variable to only those values that result in a nonzero denominator. When we state the relationship

$$\frac{x^2 - 9}{x - 3} = x + 3$$

we are assuming that it is true for all values of x except $x = 3$.

EXAMPLE 2 Reduce $\frac{y^2 - 5y - 6}{y^2 - 1}$ to lowest terms.

SOLUTION We factor the numerator and the denominator and then divide out the common factor.

$$\frac{y^2 - 5y - 6}{y^2 - 1} = \frac{(y - 6)(y + 1)}{(y - 1)(y + 1)}$$
$$= \frac{y - 6}{y - 1}$$

■

EXAMPLE 3 Reduce $\frac{2a^3 - 16}{4a^2 - 12a + 8}$ to lowest terms.

SOLUTION Factor first, then divide numerator and denominator by factors they have in common.

$$\frac{2a^3 - 16}{4a^2 - 12a + 8} = \frac{2(a^3 - 8)}{4(a^2 - 3a + 2)}$$
$$= \frac{2(a - 2)(a^2 + 2a + 4)}{4(a - 2)(a - 1)}$$
$$= \frac{a^2 + 2a + 4}{2(a - 1)}$$

■

The next example involves what we call a trick. The trick is to reverse the order of the terms in a difference by factoring -1 from each term in either the numerator or the denominator. The next example illustrates how this is done.

EXAMPLE 4 Reduce to lowest terms: $\frac{a - b}{b - a}$

SOLUTION The relationship between $a - b$ and $b - a$ is that they are opposites. We can show this fact by factoring -1 from each term in the numerator:

$$\frac{a - b}{b - a} = \frac{-1(-a + b)}{b - a}$$ Factor -1 from each term in the numerator

$$= \frac{-1(b - a)}{b - a}$$ Reverse the order of the terms in the numerator

$$= -1$$ Divide out common factor $b - a$

■

Multiplication and Division of Rational Expressions

Multiplication with fractions is the simplest of the four basic operations. To multiply two fractions, we simply multiply numerators and multiply denominators. That is, if a, b, c, and d are real numbers, with $b \neq 0$ and $d \neq 0$, then

$$\frac{a}{b} \cdot \frac{c}{d} = \frac{ac}{bd}$$

EXAMPLE 5 Multiply $\frac{6}{7} \cdot \frac{14}{18}$.

SOLUTION

$$\frac{6}{7} \cdot \frac{14}{18} = \frac{6(14)}{7(18)}$$ Multiply numerators and denominators

$$= \frac{2 \cdot 3(2 \cdot 7)}{7(2 \cdot 3 \cdot 3)} \qquad \text{Factor}$$

$$= \frac{2}{3} \qquad \text{Divide out common factors}$$

We multiply fractions whose numerators and denominators are monomials by multiplying numerators and multiplying denominators and then reducing to lowest terms. Here is how it looks.

EXAMPLE 6 Multiply $\frac{8x^3}{27y^8} \cdot \frac{9y^3}{12x^2}$.

SOLUTION We multiply numerators and denominators without actually carrying out the multiplication:

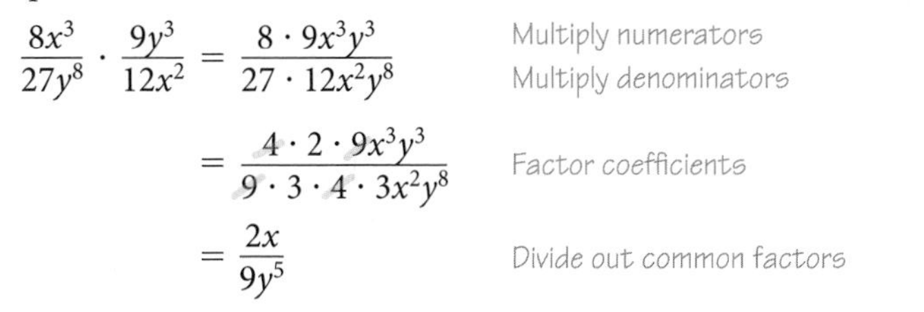

$$\frac{8x^3}{27y^8} \cdot \frac{9y^3}{12x^2} = \frac{8 \cdot 9x^3y^3}{27 \cdot 12x^2y^8} \qquad \text{Multiply numerators, Multiply denominators}$$

$$= \frac{4 \cdot 2 \cdot 9x^3y^3}{9 \cdot 3 \cdot 4 \cdot 3x^2y^8} \qquad \text{Factor coefficients}$$

$$= \frac{2x}{9y^5} \qquad \text{Divide out common factors}$$

The product of two rational expressions is the product of their numerators over the product of their denominators.

EXAMPLE 7 Multiply $\frac{2y^2 - 4y}{2y^2 - 2} \cdot \frac{y^2 - 2y - 3}{y^2 - 5y + 6}$.

SOLUTION

$$\frac{2y^2 - 4y}{2y^2 - 2} \cdot \frac{y^2 - 2y - 3}{y^2 - 5y + 6} = \frac{2y(y-2)(y-3)(y+1)}{2(y+1)(y-1)(y-3)(y-2)}$$

$$= \frac{y}{y-1}$$

Notice in both of the preceding examples that we did not actually multiply the polynomials as we did previously. It would be senseless to do that because we would then have to factor each of the resulting products to reduce them to lowest terms.

The quotient of two rational expressions is the product of the first and the reciprocal of the second.

EXAMPLE 8 Divide $\frac{x^2 - y^2}{x^2 - 2xy + y^2} \div \frac{x^3 + y^3}{x^3 - x^2y}$.

SOLUTION We begin by writing the problem as the product of the first and the reciprocal of the second and then proceed as in the previous two examples:

$$\frac{x^2 - y^2}{x^2 - 2xy + y^2} \div \frac{x^3 + y^3}{x^3 - x^2y}$$

$$= \frac{x^2 - y^2}{x^2 - 2xy + y^2} \cdot \frac{x^3 - x^2y}{x^3 + y^3} \qquad \text{Multiply by the reciprocal of the divisor}$$

$$= \frac{(x-y)(x+y)(x^2)(x-y)}{(x-y)(x-y)(x+y)(x^2 - xy + y^2)} \qquad \text{Factor}$$

$$= \frac{x^2}{x^2 - xy + y^2} \qquad \text{Divide out common factors}$$

EXAMPLE 9 Multiply $(4x^2 - 36) \cdot \frac{12}{4x + 12}$.

SOLUTION We can think of $4x^2 - 36$ as having a denominator of 1. Thinking of it in this way allows us to proceed as we did in the previous examples.

$$(4x^2 - 36) \cdot \frac{12}{4x + 12}$$

$$= \frac{4x^2 - 36}{1} \cdot \frac{12}{4x + 12} \quad \text{Write } 4x^2 - 36 \text{ with denominator 1}$$

$$= \frac{4(x - 3)(x + 3)12}{4(x + 3)} \quad \text{Factor}$$

$$= 12(x - 3) \quad \text{Divide out common factors}$$

EXAMPLE 10 Multiply $3(x - 2)(x - 1) \cdot \frac{5}{x^2 - 3x + 2}$.

SOLUTION This problem is very similar to the problem in Example 9. Writing the first rational expression with a denominator of 1, we have

$$\frac{3(x - 2)(x - 1)}{1} \cdot \frac{5}{x^2 - 3x + 2} = \frac{3(x - 2)(x - 1)5}{(x - 2)(x - 1)}$$

$$= 3 \cdot 5$$

$$= 15$$

Addition and Subtraction with the Same Denominator

To add two expressions that have the same denominator, we simply add numerators and put the sum over the common denominator.

EXAMPLE 11 Add $\frac{x}{x^2 - 1} + \frac{1}{x^2 - 1}$.

SOLUTION Because the denominators are the same, we simply add numerators:

$$\frac{x}{x^2 - 1} + \frac{1}{x^2 - 1} = \frac{x + 1}{x^2 - 1} \quad \text{Add numerators}$$

$$= \frac{x + 1}{(x - 1)(x + 1)} \quad \text{Factor denominator}$$

$$= \frac{1}{x - 1} \quad \text{Divide out common factor } x + 1$$

Addition and Subtraction with Different Denominators

Before we look at an example of addition of fractions with different denominators, we need to review the definition for the least common denominator (LCD).

DEFINITION *Least common denominator*

The *least common denominator* for a set of denominators is the smallest expression that is divisible by each of the denominators.

The first step in combining two fractions is to find the LCD. Once we have the common denominator, we rewrite each fraction as an equivalent fraction with the common denominator. After that, we simply add or subtract as we did in Example 11.

Example 12 is a review of the step-by-step procedure used to add two fractions with different denominators.

EXAMPLE 12 Add $\frac{3}{14} + \frac{7}{30}$.

SOLUTION

Step 1: ***Find the LCD.***

To do this, we first factor both denominators into prime factors.

Factor 14: $14 = 2 \cdot 7$

Factor 30: $30 = 2 \cdot 3 \cdot 5$

Because the LCD must be divisible by 14, it must have factors of 2 and 7. It must also be divisible by 30 and, therefore, have factors of 2, 3, and 5. We do not need to repeat the 2 that appears in both the factors of 14 and those of 30. Therefore,

$$\text{LCD} = 2 \cdot 3 \cdot 5 \cdot 7 = 210$$

Step 2: ***Change to equivalent fractions.***

Because we want each fraction to have a denominator of 210 and at the same time keep its original value, we multiply each by 1 in the appropriate form.

Change 3/14 to a fraction with denominator 210:

$$\frac{3}{14} \cdot \frac{15}{15} = \frac{45}{210}$$

Change 7/30 to a fraction with denominator 210:

$$\frac{7}{30} \cdot \frac{7}{7} = \frac{49}{210}$$

Step 3: ***Add numerators of equivalent fractions found in step 2:***

$$\frac{45}{210} + \frac{49}{210} = \frac{94}{210}$$

Step 4: ***Reduce to lowest terms, if necessary:***

$$\frac{94}{210} = \frac{47}{105}$$

■

The main idea in adding fractions is to write each fraction again with the LCD for a denominator. In doing so, we must be sure not to change the value of either of the original fractions.

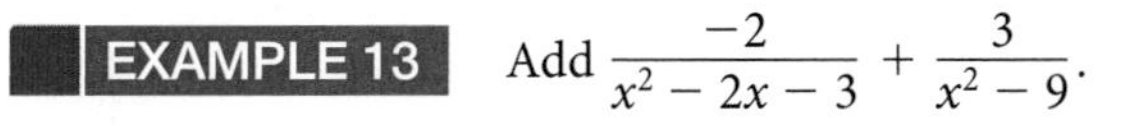

EXAMPLE 13 Add $\frac{-2}{x^2 - 2x - 3} + \frac{3}{x^2 - 9}$.

SOLUTION

Step 1: ***Factor each denominator and build the LCD from the factors:***

$$x^2 - 2x - 3 = (x - 3)(x + 1)$$
$$x^2 - 9 = (x - 3)(x + 3)$$
$$LCD = (x - 3)(x + 3)(x + 1)$$

Step 2: ***Change each rational expression to an equivalent expression that has the LCD for a denominator:***

$$\frac{-2}{x^2-2x-3} = \frac{-2}{(x-3)(x+1)} \cdot \frac{(x+3)}{(x+3)} = \frac{-2x-6}{(x-3)(x+3)(x+1)}$$

$$\frac{3}{x^2-9} = \frac{3}{(x-3)(x+3)} \cdot \frac{(x+1)}{(x+1)} = \frac{3x+3}{(x-3)(x+3)(x+1)}$$

Step 3: ***Add numerators of the rational expressions found in step 2:***

$$\frac{-2x-6}{(x-3)(x+3)(x+1)} + \frac{3x+3}{(x-3)(x+3)(x+1)} = \frac{x-3}{(x-3)(x+3)(x+1)}$$

Step 4: ***Reduce to lowest terms by dividing out the common factor x − 3:***

$$\frac{x-3}{(x-3)(x+3)(x+1)} = \frac{1}{(x+3)(x+1)}$$

■

EXAMPLE 14 Subtract $\frac{x+4}{2x+10} - \frac{5}{x^2-25}$.

SOLUTION We begin by factoring each denominator:

$$\frac{x+4}{2x+10} - \frac{5}{x^2-25} = \frac{x+4}{2(x+5)} - \frac{5}{(x+5)(x-5)}$$

The LCD is $2(x+5)(x-5)$. Completing the problem, we have

$$= \frac{x+4}{2(x+5)} \cdot \frac{(x-5)}{(x-5)} - \frac{5}{(x+5)(x-5)} \cdot \frac{2}{2}$$

$$= \frac{x^2-x-20}{2(x+5)(x-5)} - \frac{10}{2(x+5)(x-5)}$$

$$= \frac{x^2-x-30}{2(x+5)(x-5)}$$

To see if this expression will reduce, we factor the numerator into $(x-6)(x+5)$.

$$= \frac{(x-6)(x+5)}{2(x+5)(x-5)}$$

$$= \frac{x-6}{2(x-5)}$$

■

Complex Fractions

A *complex fraction* is a fraction that has one or more fractions or rational expressions in the numerator and/or denominator.

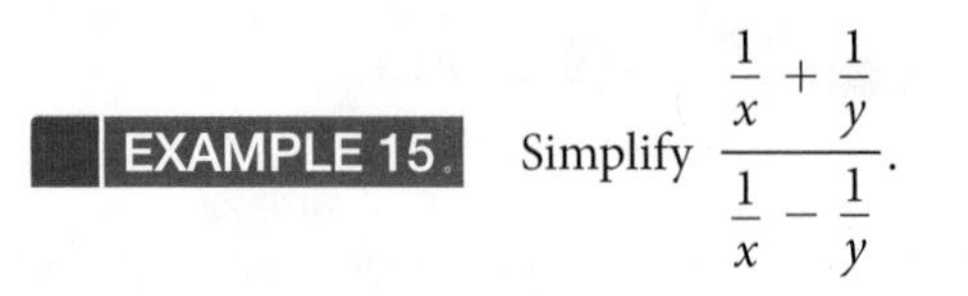

EXAMPLE 15 Simplify $\dfrac{\frac{1}{x}+\frac{1}{y}}{\frac{1}{x}-\frac{1}{y}}$.

SOLUTION We begin by multiplying both the numerator and denominator by the quantity xy, which is the LCD for all the fractions:

$$\frac{\frac{1}{x}+\frac{1}{y}}{\frac{1}{x}-\frac{1}{y}} = \frac{\left(\frac{1}{x}+\frac{1}{y}\right)\cdot xy}{\left(\frac{1}{x}-\frac{1}{y}\right)\cdot xy}$$

$$= \frac{\frac{1}{x}(xy)+\frac{1}{y}(xy)}{\frac{1}{x}(xy)-\frac{1}{y}(xy)}$$ Apply the distributive property to distribute xy over both terms in the numerator and denominator

$$= \frac{y+x}{y-x}$$

EXAMPLE 16 Simplify $\dfrac{1-\frac{4}{x^2}}{1-\frac{1}{x}-\frac{6}{x^2}}$.

SOLUTION Again, we multiply the numerator and denominator by the LCD, x^2:

$$\frac{1-\frac{4}{x^2}}{1-\frac{1}{x}-\frac{6}{x^2}} = \frac{x^2\left(1-\frac{4}{x^2}\right)}{x^2\left(1-\frac{1}{x}-\frac{6}{x^2}\right)}$$ Multiply numerator and denominator by x^2

$$= \frac{x^2\cdot 1 - x^2\cdot\frac{4}{x^2}}{x^2\cdot 1 - x^2\cdot\frac{1}{x} - x^2\cdot\frac{6}{x^2}}$$ Distributive property

$$= \frac{x^2-4}{x^2-x-6}$$ Simplify

$$= \frac{(x-2)(x+2)}{(x-3)(x+2)}$$ Factor

$$= \frac{x-2}{x-3}$$ Reduce

GETTING READY FOR CLASS

After reading through the preceding section, respond in your own words and in complete sentences.

A. What is a rational expression?
B. When is a rational expression undefined?
C. Why is factoring important when multiplying and dividing rational expressions?
D. Briefly describe how you would add two rational expressions that have the same denominator.

Problem Set 1.5

Reduce each rational expression to lowest terms.

1. $\dfrac{12x - 9y}{3x^2 + 3xy}$

2. $\dfrac{a^2 - 4a - 12}{a^2 + 8a + 12}$

3. $\dfrac{20x^2 - 93x + 34}{4x^2 - 9x - 34}$

4. $\dfrac{250a + 100ax + 10ax^2}{50a - 2ax^2}$

5. $\dfrac{(x - 4)^3(x + 3)}{(x + 3)^2(x - 4)}$

6. $\dfrac{x^3 - 1}{x^2 - 1}$

7. $\dfrac{ad - ad^2}{d - 1}$

8. $\dfrac{x^2 - 3ax - 2x + 6a}{x^2 - 3ax + 2x - 6a}$

9. $\dfrac{6 - x}{x - 6}$

10. $\dfrac{1 - y}{y^2 - 1}$

Perform the indicated operations. Be sure to write all answers in lowest terms.

11. $\dfrac{x^2 - 16}{x^2 - 25} \cdot \dfrac{x - 5}{x - 4}$

12. $\dfrac{y - 1}{y^2 - y - 6} \cdot \dfrac{y^2 + 5y + 6}{y^2 - 1}$

13. $\dfrac{x^2 + 5x + 1}{4x - 4} \cdot \dfrac{x - 1}{x^2 + 5x + 1}$

14. $\dfrac{y}{x} \div \dfrac{xy}{xy - 1}$

15. $\dfrac{1}{x^2 - 9} \div \dfrac{1}{(x - 3)^2}$

16. $\dfrac{y - 3}{y^2 - 6y + 9} \div \dfrac{y - 3}{4}$

17. $\dfrac{a^2 + 7a + 12}{a - 5} \div \dfrac{a^2 + 9a + 18}{a^2 - 7a + 10}$

18. $\dfrac{9t^2 - 1}{6t^2 + 7t - 3} \div \dfrac{27t^3 + 1}{8t^3 + 27}$

19. $\dfrac{3a^2 + 7ab - 20b^2}{a^2 + 5ab + 4b^2} \div \dfrac{3a^2 - 17ab + 20b^2}{3a - 12b}$

20. $\dfrac{ax + bx + 2a + 2b}{ax - 3a + bx - 3b} \cdot \dfrac{ax - bx - 3a + 3b}{ax - bx - 2a + 2b}$

21. $(4x + 8) \cdot \dfrac{x}{x + 2}$

22. $(x^2 - 49) \cdot \dfrac{5}{x + 7}$

23. $(x^2 - 3x + 2) \cdot \dfrac{-1}{x - 2}$

24. $(y + 1)(y + 4)(y - 1) \cdot \dfrac{3}{y^2 - 1}$

Paying Attention to Instructions The next two problems are intended to give you practice reading, and paying attention to, the instructions that accompany the problems you are working. Working these problems is an excellent way to get ready for a test or a quiz.

25. Work each problem according to the instructions given.

a. Simplify: $\dfrac{16 - 1}{64 - 1}$

b. Reduce: $\dfrac{25x^2 - 9}{125x^3 - 27}$

c. Multiply: $\dfrac{25x^2 - 9}{125x^3 - 27} \cdot \dfrac{5x - 3}{5x + 3}$

d. Divide: $\dfrac{25x^2 - 9}{125x^3 - 27} \div \dfrac{5x - 3}{25x^2 + 15x + 9}$

26. Work each problem according to the instructions given.

a. Simplify: $\dfrac{64 - 49}{64 + 112 + 49}$

b. Reduce: $\dfrac{9x^2 - 49}{9x^2 + 42x + 49}$

c. Multiply: $\dfrac{9x^2 - 49}{9x^2 + 42x + 49} \cdot \dfrac{3x + 7}{3x - 7}$

d. Divide: $\dfrac{9x^2 - 49}{9x^2 + 42x + 49} \div \dfrac{3x + 7}{3x - 7}$

Combine the following fractions.

27. $\dfrac{5}{6} + \dfrac{1}{3}$

28. $\dfrac{3}{4} + \dfrac{2}{3}$

29. $\dfrac{6}{28} - \dfrac{5}{42}$

Combine the following rational expressions. Reduce all answers to lowest terms.

30. $\dfrac{5x}{5x + 2} + \dfrac{2}{5x + 2}$

31. $\dfrac{8}{y + 8} + \dfrac{y}{y + 8}$

32. $\dfrac{2x - 4}{x + 2} - \dfrac{x - 6}{x + 2}$

Combine the following rational expressions. Reduce all answers to lowest terms.

33. $\dfrac{x + 1}{x - 2} - \dfrac{4x + 7}{5x - 10}$

34. $\dfrac{4x + 2}{3x + 12} - \dfrac{x - 2}{x + 4}$

35. $\dfrac{x + 7}{2x + 12} + \dfrac{6}{x^2 - 36}$

36. $\dfrac{1}{a + b} + \dfrac{3ab}{a^3 + b^3}$

37. $\dfrac{-3}{a^2 + a - 2} + \dfrac{5}{a^2 - a - 6}$

38. $\dfrac{1}{27x^3 - 1} - \dfrac{1}{9x^2 - 1}$

39. $\dfrac{9}{9x^2 + 6x - 8} - \dfrac{6}{9x^2 - 4}$

40. $\dfrac{3a}{a^2 + 7a + 10} - \dfrac{2a}{a^2 + 6a + 8}$

41. $3 - \dfrac{2}{2x + 3}$

42. $7 + \dfrac{3}{5 - t}$

43. $x - \dfrac{5}{3x + 4} + 1$

44. $\dfrac{1}{x} + \dfrac{x}{3x + 9} - \dfrac{3}{x^2 + 3x}$

Paying Attention to Instructions The next two problems are intended to give you practice reading, and paying attention to, the instructions that accompany the problems you are working.

45. Work each problem according to the instructions given.

a. Multiply: $\dfrac{3}{8} \cdot \dfrac{1}{6}$

b. Divide: $\dfrac{3}{8} \div \dfrac{1}{6}$

c. Add: $\dfrac{3}{8} + \dfrac{1}{6}$

d. Multiply: $\dfrac{x + 3}{x - 3} \cdot \dfrac{5x + 15}{x^2 - 9}$

e. Divide: $\dfrac{x + 3}{x - 3} \div \dfrac{5x + 15}{x^2 - 9}$

f. Subtract: $\dfrac{x + 3}{x - 3} - \dfrac{5x + 15}{x^2 - 9}$

46. Work each problem according to the instructions given.

a. Multiply: $\frac{16}{49} \cdot \frac{1}{28}$ **b.** Divide: $\frac{16}{49} \div \frac{1}{28}$

c. Subtract: $\frac{16}{49} - \frac{1}{28}$ **d.** Multiply: $\frac{3x-2}{3x+2} \cdot \frac{15x+6}{9x^2-4}$

e. Divide: $\frac{3x-2}{3x+2} \div \frac{15x+6}{9x^2-4}$ **f.** Subtract: $\frac{3x+2}{3x-2} - \frac{15x+6}{9x^2-4}$

Simplify each of the following as much as possible.

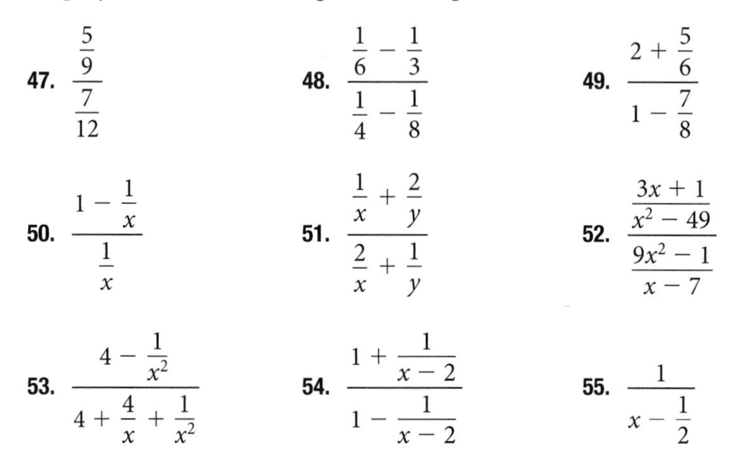

47. $\dfrac{\frac{5}{9}}{\frac{7}{12}}$ **48.** $\dfrac{\frac{1}{6}-\frac{1}{3}}{\frac{1}{4}-\frac{1}{8}}$ **49.** $\dfrac{2+\frac{5}{6}}{1-\frac{7}{8}}$

50. $\dfrac{1-\frac{1}{x}}{\frac{1}{x}}$ **51.** $\dfrac{\frac{1}{x}+\frac{2}{y}}{\frac{2}{x}+\frac{1}{y}}$ **52.** $\dfrac{\frac{3x+1}{x^2-49}}{\frac{9x^2-1}{x-7}}$

53. $\dfrac{4-\frac{1}{x^2}}{4+\frac{4}{x}+\frac{1}{x^2}}$ **54.** $\dfrac{1+\frac{1}{x-2}}{1-\frac{1}{x-2}}$ **55.** $\dfrac{1}{x-\frac{1}{2}}$

SPOTLIGHT ON SUCCESS *Student Instructor Cynthia*

Each time we face our fear, we gain strength, courage, and confidence in the doing.
—Unknown

I must admit, when it comes to math, it takes me longer to learn the material compared to other students. Because of that, I was afraid to ask questions, especially when it seemed like everyone else understood what was going on. Because I wasn't getting my questions answered, my quiz and exam scores were only getting worse. I realized that I was already paying a lot to go to college and that I couldn't afford to keep doing poorly on my exams. I learned how to overcome my fear of asking questions by studying the material before class, and working on extra problem sets until I was confident enough that at least I understood the main concepts. By preparing myself beforehand, I would often end up answering the question myself. Even when that wasn't the case, the professor knew that I tried to answer the question on my own. If you want to be successful, but you are afraid to ask a question, try putting in a little extra time working on problems before you ask your instructor for help. I think you will find, like I did, that it's not as bad as you imagined it, and you will have overcome an obstacle that was in the way of your success.

Roots and Radicals

1.6

INTRODUCTION The Pythagorean theorem can be used to construct the attractive spiral shown here.

This spiral is called the Spiral of Roots because each of the diagonals is the positive square root of one of the positive integers. ■

The Spiral of Roots

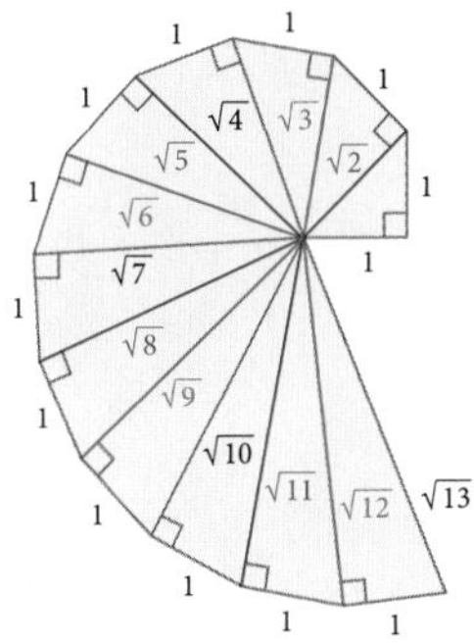

Previously, we reviewed notation (exponents) to give us the square, cube, or any other power of a number. For instance, if we wanted the square of 3, we wrote $3^2 = 9$. If we wanted the cube of 3, we wrote $3^3 = 27$. We begin this section by reviewing the notation that will take us in the reverse direction, that is, from the square of a number, say 25, back to the original number, 5.

DEFINITION *Positive square root*

If x is a nonnegative real number, then the expression $\sqrt{x}$ is called the *positive square root* of x and is such that

$$(\sqrt{x})^2 = x$$

In words: $\sqrt{x}$ is the positive number we square to get x.

VIDEO EXAMPLES

SECTION 1.6

The negative square root of x, $-\sqrt{x}$, is defined in a similar manner.

EXAMPLE 1 The positive square root of 64 is 8 because 8 is the positive number with the property $8^2 = 64$. The negative square root of 64 is -8 because -8 is the negative number whose square is 64. We can summarize both these facts by saying

$$\sqrt{64} = 8 \qquad \text{and} \qquad -\sqrt{64} = -8$$ ■

Note It is a common mistake to assume that an expression like $\sqrt{25}$ indicates both square roots, 5 and -5. The expression $\sqrt{25}$ indicates only the positive square root of 25, which is 5. If we want the negative square root, we must use a negative sign: $-\sqrt{25} = -5$.

The following is a table of the most common roots used in this book. Any of the roots that are unfamiliar should be memorized.

Square Roots		Cube Roots	Fourth Roots
$\sqrt{0} = 0$	$\sqrt{49} = 7$	$\sqrt[3]{0} = 0$	$\sqrt[4]{0} = 0$
$\sqrt{1} = 1$	$\sqrt{64} = 8$	$\sqrt[3]{1} = 1$	$\sqrt[4]{1} = 1$
$\sqrt{4} = 2$	$\sqrt{81} = 9$	$\sqrt[3]{8} = 2$	$\sqrt[4]{16} = 2$
$\sqrt{9} = 3$	$\sqrt{100} = 10$	$\sqrt[3]{27} = 3$	$\sqrt[4]{81} = 3$
$\sqrt{16} = 4$	$\sqrt{121} = 11$	$\sqrt[3]{64} = 4$	
$\sqrt{25} = 5$	$\sqrt{144} = 12$	$\sqrt[3]{125} = 5$	
$\sqrt{36} = 6$	$\sqrt{169} = 13$		

Notation An expression like $\sqrt[3]{8}$ that involves a root is called a *radical expression*. In the expression $\sqrt[3]{8}$, the ^{3} is called the *index*, the $\sqrt{}$ is the *radical sign*, and 8 is called the *radicand*. The index of a radical must be a positive integer greater than 1. If no index is written, it is assumed to be 2.

Roots and Negative Numbers

When dealing with negative numbers and radicals, the only restriction concerns negative numbers under even roots. We can have negative signs in front of radicals and negative numbers under odd roots and still obtain real numbers. Here are some examples to help clarify this.

EXAMPLES Simplify each expression, if possible.

2. $\sqrt[3]{-8} = -2$ because $(-2)^3 = -8$.

3. $\sqrt{-4}$ is not a real number because there is no real number whose square is -4.

4. $-\sqrt{25} = -5$, because -5 is the negative square root of 25.

5. $\sqrt[5]{-32} = -2$ because $(-2)^5 = -32$.

6. $\sqrt[4]{-81}$ is not a real number because there is no real number we can raise to the fourth power and obtain -81. ■

Variables Under a Radical

From the preceding examples, it is clear that we must be careful that we do not try to take an even root of a negative number. For this reason, we will assume that all variables appearing under a radical sign represent nonnegative numbers, unless we are told otherwise.

EXAMPLES Assume all variables represent nonnegative numbers, and simplify each expression as much as possible.

7. $\sqrt{25a^4b^6} = 5a^2b^3$ because $(5a^2b^3)^2 = 25a^4b^6$.

8. $\sqrt[3]{x^6y^{12}} = x^2y^4$ because $(x^2y^4)^3 = x^6y^{12}$.

9. $\sqrt[4]{81r^8s^{20}} = 3r^2s^5$ because $(3r^2s^5)^4 = 81r^8s^{20}$. ■

Rational Numbers as Exponents

We will now develop a second kind of notation involving exponents that will allow us to designate square roots, cube roots, and so on in another way.

(def DEFINITION

If x is a real number and n is a positive integer greater than 1, then

$$x^{1/n} = \sqrt[n]{x} \qquad (x \geq 0 \text{ when } n \text{ is even})$$

In words: The quantity $x^{1/n}$ is the nth root of x.

With this definition, we have a way of representing roots with exponents. Here are some examples.

EXAMPLES Write each expression as a root and then simplify, if possible.

10. $8^{1/3} = \sqrt[3]{8} = 2$

11. $36^{1/2} = \sqrt{36} = 6$

12. $-25^{1/2} = -\sqrt{25} = -5$

13. $(-25)^{1/2} = \sqrt{-25}$, which is not a real number

14. $\left(\frac{4}{9}\right)^{1/2} = \sqrt{\frac{4}{9}} = \frac{2}{3}$ ■

The properties of exponents developed earlier were applied to integer exponents only. We will now extend these properties to include rational exponents also. We do so without proof.

PROPERTY ***Properties of Exponents***

If a and b are real numbers and r and s are rational numbers, and a and b are nonnegative whenever r and s indicate even roots, then

1. $a^r \cdot a^s = a^{r+s}$	**4.** $a^{-r} = \frac{1}{a^r}$	$(a \neq 0)$
2. $(a^r)^s = a^{rs}$	**5.** $\left(\frac{a}{b}\right)^r = \frac{a^r}{b^r}$	$(b \neq 0)$
3. $(ab)^r = a^r b^r$	**6.** $\frac{a^r}{a^s} = a^{r-s}$	$(a \neq 0)$

Here are Examples 8 and 9 again, but this time we will work them using rational exponents.

EXAMPLES Write each radical with a rational exponent, then simplify.

15.
$$\begin{aligned}\sqrt[3]{x^6y^{12}} &= (x^6y^{12})^{1/3}\\ &= (x^6)^{1/3}(y^{12})^{1/3}\\ &= x^2y^4\end{aligned}$$

16.
$$\begin{aligned}\sqrt[4]{81r^8s^{20}} &= (81r^8s^{20})^{1/4}\\ &= 81^{1/4}(r^8)^{1/4}(s^{20})^{1/4}\\ &= 3r^2s^5\end{aligned}$$
■

So far, the numerators of all the rational exponents we have encountered have been 1. The next theorem extends the work we can do with rational exponents to rational exponents with numerators other than 1.

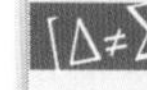

Theorem: Rational Exponents

If a is a real number, m is an integer, and n is a positive integer, and a is nonnegative whenever n is even, then

$$a^{m/n} = (a^{1/n})^m = (a^m)^{1/n}$$

Proof We can prove our theorem using the properties of exponents. Because $m/n = m(1/n)$, we have

$$a^{m/n} = a^{m(1/n)} = (a^m)^{1/n} \qquad \Big| \qquad a^{m/n} = a^{(1/n)(m)} = (a^{1/n})^m$$

Here are some examples that illustrate how we use this theorem.

EXAMPLES Simplify as much as possible.

17. $8^{2/3} = (8^{1/3})^2$ Rational Exponents Theorem
$= 2^2$ Definition of rational exponents
$= 4$ The square of 2 is 4

18. $9^{-3/2} = (9^{1/2})^{-3}$ Rational Exponents Theorem
$= 3^{-3}$ Definition of fractional exponents
$= \frac{1}{3^3}$ Property 4 for exponents
$= \frac{1}{27}$ The cube of 3 is 27

19. $\left(\frac{27}{8}\right)^{-4/3} = \left[\left(\frac{27}{8}\right)^{1/3}\right]^{-4}$ Rational Exponents Theorem
$= \left(\frac{3}{2}\right)^{-4}$ Definition of fractional exponents
$= \left(\frac{2}{3}\right)^{4}$ Property 4 for exponents
$= \frac{16}{81}$ The fourth power of $\frac{2}{3}$ is $\frac{16}{81}$ ■

The following examples show the application of the properties of exponents to rational exponents.

EXAMPLES Assume all variables represent positive quantities, and simplify as much as possible.

20. $x^{1/3} \cdot x^{5/6} = x^{1/3 + 5/6}$ Property 1
$= x^{2/6 + 5/6}$ LCD is 6
$= x^{7/6}$ Add fractions

21. $(y^{2/3})^{3/4} = y^{(2/3)(3/4)}$ Property 2

$= y^{1/2}$ Multiply fractions: $\frac{2}{3} \cdot \frac{3}{4} = \frac{6}{12} = \frac{1}{2}$

22. $\dfrac{z^{1/3}}{z^{1/4}} = z^{1/3 - 1/4}$ Property 6

$= z^{4/12 - 3/12}$ LCD is 12

$= z^{1/12}$ Subtract fractions

Operations with Radicals

Here are the first two properties of radicals. For these two properties, we will assume a and b are nonnegative real numbers whenever n is an even number.

Note There is not a property for radicals that says the nth root of a sum is the sum of the nth roots. That is,

$$\sqrt[n]{a+b} \neq \sqrt[n]{a} + \sqrt[n]{b}$$

PROPERTY *Property 1 for Radicals*

$$\sqrt[n]{ab} = \sqrt[n]{a}\sqrt[n]{b}$$

In words: The nth root of a product is the product of the nth roots.

Proof of Property 1

$\sqrt[n]{ab} = (ab)^{1/n}$ Definition of rational exponents

$= a^{1/n}b^{1/n}$ Exponents distribute over products

$= \sqrt[n]{a}\sqrt[n]{b}$ Definition of rational exponents

PROPERTY *Property 2 for Radicals*

$$\sqrt[n]{\frac{a}{b}} = \frac{\sqrt[n]{a}}{\sqrt[n]{b}} = \quad (b \neq 0)$$

In words: The nth root of a quotient is the quotient of the nth roots.

The proof of Property 2 is similar to the proof of Property 1.

These two properties of radicals allow us to change the form of, and simplify, radical expressions without changing their value.

RULE *Simplified Form for Radical Expressions*

A radical expression is in ***simplified form*** if

1. None of the factors of the radicand (the quantity under the radical sign) can be written as powers greater than or equal to the index—that is, no perfect squares can be factors of the quantity under a square root sign, no perfect cubes can be factors of what is under a cube root sign, and so forth.
2. There are no fractions under the radical sign.
3. There are no radicals in the denominator.

Satisfying the first condition for simplified form actually amounts to taking as much out from under the radical sign as possible. The following examples illustrate the first condition for simplified form.

EXAMPLE 23 Write $\sqrt{50}$ in simplified form.

SOLUTION The largest perfect square that divides 50 is 25. We write 50 as $25 \cdot 2$ and apply Property 1 for radicals:

$$
\begin{aligned}
\sqrt{50} &= \sqrt{25 \cdot 2} && 50 = 25 \cdot 2 \\
&= \sqrt{25}\sqrt{2} && \text{Property 1} \\
&= 5\sqrt{2} && \sqrt{25} = 5
\end{aligned}
$$

We have taken as much as possible out from under the radical sign—in this case, factoring 25 from 50 and then writing $\sqrt{25}$ as 5. ■

As we progress through the book you will see more and more expressions that involve the product of a number and a radical. Here are some examples:

$$3\sqrt{2} \qquad \frac{1}{2}\sqrt{5} \qquad 5\sqrt{7} \qquad 3x\sqrt{2x} \qquad 2ab\sqrt{5a}$$

All of these are products. The first expression $3\sqrt{2}$ is the product of 3 and $\sqrt{2}$. That is,

$$3\sqrt{2} = 3 \cdot \sqrt{2}$$

The 3 and the $\sqrt{2}$ are not stuck together in some mysterious way. The expression $3\sqrt{2}$ is simply the product of two numbers, one of which is rational, and the other is irrational.

EXAMPLE 24 Simplify each expression.

a. $\dfrac{\sqrt{12}}{6}$ **b.** $\dfrac{5\sqrt{18}}{15}$ **c.** $\dfrac{6+\sqrt{8}}{2}$ **d.** $\dfrac{-1+\sqrt{45}}{2}$

SOLUTION In each case, we simplify the radical first, then we factor and reduce to lowest terms.

a.
$$
\begin{aligned}
\frac{\sqrt{12}}{6} &= \frac{2\sqrt{3}}{6} && \text{Simplify the radical: } \sqrt{12} = \sqrt{4 \cdot 3} = \sqrt{4}\sqrt{3} = 2\sqrt{3} \\
&= \frac{2\sqrt{3}}{2 \cdot 3} && \text{Factor denominator} \\
&= \frac{\sqrt{3}}{3} && \text{Divide out common factors}
\end{aligned}
$$

b.
$$
\begin{aligned}
\frac{5\sqrt{18}}{15} &= \frac{5 \cdot 3\sqrt{2}}{15} && \sqrt{18} = \sqrt{9 \cdot 2} = \sqrt{9}\sqrt{2} = 3\sqrt{2} \\
&= \frac{5 \cdot 3\sqrt{2}}{3 \cdot 5} && \text{Factor denominator} \\
&= \sqrt{2} && \text{Divide out common factors}
\end{aligned}
$$

c. $\dfrac{6+\sqrt{8}}{2} = \dfrac{6+2\sqrt{2}}{2}$ $\quad\sqrt{8} = \sqrt{4\cdot 2} = \sqrt{4}\sqrt{2} = 2\sqrt{2}$

$= \dfrac{2(3+\sqrt{2})}{2}$ $\quad$ Factor numerator

$= 3 + \sqrt{2}$ $\quad$ Divide out common factors

d. $\dfrac{-1+\sqrt{45}}{2} = \dfrac{-1+3\sqrt{5}}{2}$ $\quad\sqrt{45} = \sqrt{9\cdot 5} = \sqrt{9}\sqrt{5} = 3\sqrt{5}$ ■

EXAMPLE 25 Write in simplified form: $\sqrt{48x^4y^3}$, where $x, y \geq 0$.

SOLUTION The largest perfect square that is a factor of the radicand is $16x^4y^2$. Applying Property 1 again, we have

$$\begin{aligned}\sqrt{48x^4y^3} &= \sqrt{16x^4y^2 \cdot 3y}\\ &= \sqrt{16x^4y^2}\sqrt{3y}\\ &= 4x^2y\sqrt{3y}\end{aligned}$$ ■

EXAMPLE 26 Write $\sqrt[3]{40a^5b^4}$ in simplified form.

SOLUTION We now want to factor the largest perfect cube from the radicand. We write $40a^5b^4$ as $8a^3b^3 \cdot 5a^2b$ and proceed as we did previously.

$$\begin{aligned}\sqrt[3]{40a^5b^4} &= \sqrt[3]{8a^3b^3 \cdot 5a^2b}\\ &= \sqrt[3]{8a^3b^3}\sqrt[3]{5a^2b}\\ &= 2ab\sqrt[3]{5a^2b}\end{aligned}$$ ■

Rationalizing the Denominator 1

EXAMPLE 27 Write $\sqrt{\dfrac{5}{6}}$ in simplified form.

SOLUTION First we separate the numerator and denominator into separate radicals using Property 2.

$$\sqrt{\frac{5}{6}} = \frac{\sqrt{5}}{\sqrt{6}}$$

The resulting expression satisfies the second condition for simplified form because neither radical contains a fraction. It does, however, violate Condition 3 because it has a radical in the denominator. Getting rid of the radical in the denominator is called *rationalizing the denominator* and is accomplished, in this case, by multiplying the numerator and denominator by $\sqrt{6}$:

$$\begin{aligned}\frac{\sqrt{5}}{\sqrt{6}} &= \frac{\sqrt{5}}{\sqrt{6}} \cdot \frac{\sqrt{6}}{\sqrt{6}}\\ &= \frac{\sqrt{30}}{\sqrt{6^2}}\\ &= \frac{\sqrt{30}}{6}\end{aligned}$$ ■

EXAMPLES Rationalize the denominator. ($x, y > 0$.)

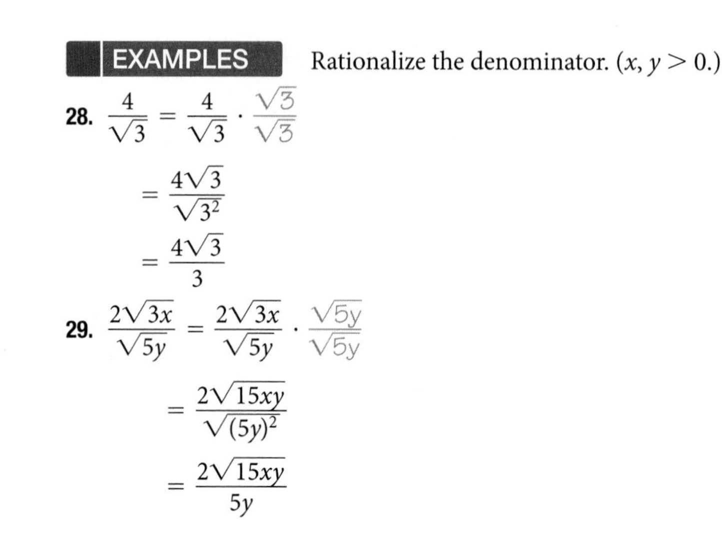

28. $\dfrac{4}{\sqrt{3}} = \dfrac{4}{\sqrt{3}} \cdot \dfrac{\sqrt{3}}{\sqrt{3}}$

$= \dfrac{4\sqrt{3}}{\sqrt{3^2}}$

$= \dfrac{4\sqrt{3}}{3}$

29. $\dfrac{2\sqrt{3x}}{\sqrt{5y}} = \dfrac{2\sqrt{3x}}{\sqrt{5y}} \cdot \dfrac{\sqrt{5y}}{\sqrt{5y}}$

$= \dfrac{2\sqrt{15xy}}{\sqrt{(5y)^2}}$

$= \dfrac{2\sqrt{15xy}}{5y}$ ■

When the denominator involves a cube root, we must multiply by a radical that will produce a perfect cube under the cube root sign in the denominator, as our next example illustrates.

EXAMPLE 30 Rationalize the denominator in $\dfrac{7}{\sqrt[3]{4}}$.

SOLUTION Because $4 = 2^2$, we can multiply both numerator and denominator by $\sqrt[3]{2}$ and obtain $\sqrt[3]{2^3}$ in the denominator.

$$\frac{7}{\sqrt[3]{4}} = \frac{7}{\sqrt[3]{2^2}}$$

$$= \frac{7}{\sqrt[3]{2^2}} \cdot \frac{\sqrt[3]{2}}{\sqrt[3]{2}}$$

$$= \frac{7\sqrt[3]{2}}{\sqrt[3]{2^3}}$$

$$= \frac{7\sqrt[3]{2}}{2}$$ ■

Square Root of a Perfect Square

So far in this chapter, we have assumed that all our variables are nonnegative when they appear under a square root symbol. There are times, however, when this is not the case.

Consider the following two statements:

$$\sqrt{3^2} = \sqrt{9} = 3 \quad \text{and} \quad \sqrt{(-3)^2} = \sqrt{9} = 3$$

Whether we operate on 3 or -3, the result is the same: Both expressions simplify to 3. The other operation we have worked with in the past that produces the same result is absolute value. That is,

$$|3| = 3 \quad \text{and} \quad |-3| = 3$$

This leads us to the next property of radicals.

[Δ≠Σ PROPERTY ***Property 3 for Radicals***

If a is a real number, then $\sqrt{a^2} = |a|$.

The result of this discussion and Property 3 is simply this:

If we know a is positive, then $\sqrt{a^2} = a$.

If we know a is negative, then $\sqrt{a^2} = |a|$.

If we don't know if a is positive or negative, then $\sqrt{a^2} = |a|$.

EXAMPLES Simplify each expression. Do *not* assume the variables represent positive numbers.

31. $\sqrt{9x^2} = 3|x|$

32. $\sqrt{5a^2b^2} = |ab|\sqrt{5}$

33. $\sqrt{x^2 - 6x + 9} = \sqrt{(x-3)^2} = |x - 3|$

34. $\sqrt{x^3 - 5x^2} = \sqrt{x^2(x-5)} = |x|\sqrt{x-5}$ ■

As you can see, we must use absolute value symbols when we take a square root of a perfect square, unless we know the base of the perfect square is a positive number. The same idea holds for higher even roots, but not for odd roots. With odd roots, no absolute value symbols are necessary.

EXAMPLES Simplify each expression.

35 $\sqrt[3]{(-2)^3} = \sqrt[3]{-8} = -2$

36. $\sqrt[3]{(-5)^3} = \sqrt[3]{-125} = -5$ ■

We can extend this discussion to all roots as follows:

PROPERTY *Extending Property 3 for Radicals*

If a is a real number, then

$\sqrt[n]{a^n} = |a|$ if n is even

$\sqrt[n]{a^n} = a$ if n is odd

Addition and Subtraction of Radical Expressions

Previously, we found we could add similar terms when combining polynomials. The same idea applies to addition and subtraction of radical expressions.

DEFINITION *Similar radicals*

Two radicals are said to be *similar radicals* if they have the same index and the same radicand.

The expressions $5\sqrt[3]{7}$ and $-8\sqrt[3]{7}$ are similar since the index is ^{3} in both cases and the radicands are 7. The expressions $3\sqrt[4]{5}$ and $7\sqrt[3]{5}$ are not similar because they have different indices, and the expressions $2\sqrt[5]{8}$ and $3\sqrt[5]{9}$ are not similar because the radicands are not the same.

We add and subtract radical expressions in the same way we add and subtract polynomials — by combining similar terms under the distributive property.

EXAMPLE 37 Combine $5\sqrt{3} - 4\sqrt{3} + 6\sqrt{3}$.

SOLUTION All three radicals are similar. We apply the distributive property to get

$$\begin{aligned} 5\sqrt{3} - 4\sqrt{3} + 6\sqrt{3} &= (5 - 4 + 6)\sqrt{3} \\ &= 7\sqrt{3} \end{aligned}$$

EXAMPLE 38 Combine $3\sqrt{8} + 5\sqrt{18}$.

SOLUTION The two radicals do not seem to be similar. We must write each in simplified form before applying the distributive property.

$$\begin{aligned} 3\sqrt{8} + 5\sqrt{18} &= 3\sqrt{4 \cdot 2} + 5\sqrt{9 \cdot 2} \\ &= 3\sqrt{4}\,\sqrt{2} + 5\sqrt{9}\,\sqrt{2} \\ &= 3 \cdot 2\sqrt{2} + 5 \cdot 3\sqrt{2} \\ &= 6\sqrt{2} + 15\sqrt{2} \\ &= (6 + 15)\sqrt{2} \\ &= 21\sqrt{2} \end{aligned}$$

The result of Example 38 can be generalized to the following rule for sums and differences of radical expressions.

RULE

To add or subtract radical expressions, put each in simplified form and apply the distributive property, if possible. We can add only similar radicals. We must write each expression in simplified form for radicals before we can tell if the radicals are similar.

Multiplication and Division of Radical Expressions

Next, we will look at multiplication and division of expressions that contain radicals. As you will see, multiplication of expressions that contain radicals is very similar to multiplication of polynomials. The division problems in this section are just an extension of the work we did previously when we rationalized denominators.

EXAMPLE 39 Multiply $(3\sqrt{5})(2\sqrt{7})$.

SOLUTION We can rearrange the order and grouping of the numbers in this product by applying the commutative and associative properties. Following this, we apply Property 1 for radicals and multiply:

$$\begin{aligned} (3\sqrt{5})(2\sqrt{7}) &= (3 \cdot 2)(\sqrt{5}\sqrt{7}) && \text{Commutative and associative properties} \\ &= (3 \cdot 2)(\sqrt{5 \cdot 7}) && \text{Property 1 for radicals} \\ &= 6\sqrt{35} && \text{Multiplication} \end{aligned}$$

In practice, it is not necessary to show the first two steps.

EXAMPLE 40 Multiply $\sqrt{3}(2\sqrt{6} - 5\sqrt{12})$.

SOLUTION Applying the distributive property, we have

$$\begin{aligned}\sqrt{3}(2\sqrt{6} - 5\sqrt{12}) &= \sqrt{3} \cdot 2\sqrt{6} - \sqrt{3} \cdot 5\sqrt{12} \\ &= 2\sqrt{18} - 5\sqrt{36}\end{aligned}$$

Writing each radical in simplified form gives

$$\begin{aligned}2\sqrt{18} - 5\sqrt{36} &= 2\sqrt{9}\sqrt{2} - 5\sqrt{36} \\ &= 6\sqrt{2} - 30\end{aligned}$$

EXAMPLE 41 Multiply $(\sqrt{3} + \sqrt{5})(4\sqrt{3} - \sqrt{5})$.

SOLUTION The same principle that applies when multiplying two binomials applies to this product. We must multiply each term in the first expression by each term in the second one. Any convenient method can be used. Let's use the FOIL method.

$$\begin{aligned}(\sqrt{3} + \sqrt{5})(4\sqrt{3} - \sqrt{5}) &= \overset{F}{\sqrt{3} \cdot 4\sqrt{3}} - \overset{O}{\sqrt{3}\cdot\sqrt{5}} + \overset{I}{\sqrt{5} \cdot 4\sqrt{3}} - \overset{L}{\sqrt{5}\cdot\sqrt{5}} \\ &= 4 \cdot 3 - \sqrt{15} + 4\sqrt{15} - 5 \\ &= 12 + 3\sqrt{15} - 5 \\ &= 7 + 3\sqrt{15}\end{aligned}$$

EXAMPLE 42 Expand and simplify $(\sqrt{x} + 3)^2$.

SOLUTION 1 We can write this problem as a multiplication problem and proceed as we did in Example 41:

$$\begin{aligned}(\sqrt{x} + 3)^2 &= (\sqrt{x} + 3)(\sqrt{x} + 3) \\ &= \overset{F}{\sqrt{x} \cdot \sqrt{x}} + \overset{O}{3\sqrt{x}} + \overset{I}{3\sqrt{x}} + \overset{L}{3 \cdot 3} \\ &= x + 3\sqrt{x} + 3\sqrt{x} + 9 \\ &= x + 6\sqrt{x} + 9\end{aligned}$$

SOLUTION 2 We can obtain the same result by applying the formula for the square of a sum: $(a + b)^2 = a^2 + 2ab + b^2$.

$$\begin{aligned}(\sqrt{x} + 3)^2 &= (\sqrt{x})^2 + 2(\sqrt{x})(3) + 3^2 \\ &= x + 6\sqrt{x} + 9\end{aligned}$$

EXAMPLE 43 Multiply $(\sqrt{6} + \sqrt{2})(\sqrt{6} - \sqrt{2})$.

SOLUTION We notice the product is of the form $(a + b)(a - b)$, which always gives the difference of two squares, $a^2 - b^2$:

$$\begin{aligned}(\sqrt{6} + \sqrt{2})(\sqrt{6} - \sqrt{2}) &= (\sqrt{6})^2 - (\sqrt{2})^2 \\ &= 6 - 2 \\ &= 4\end{aligned}$$

In Example 43, the two expressions $(\sqrt{6} + \sqrt{2})$ and $(\sqrt{6} - \sqrt{2})$ are called *conjugates*. In general, the conjugate of $\sqrt{a} + \sqrt{b}$ is $\sqrt{a} - \sqrt{b}$. If a and b are integers, multiplying conjugates of this form always produces a rational number. That is, if a and b are positive integers, then

$$\begin{aligned}(\sqrt{a} + \sqrt{b})(\sqrt{a} - \sqrt{b}) &= \sqrt{a}\sqrt{a} - \sqrt{a}\sqrt{b} + \sqrt{a}\sqrt{b} - \sqrt{b}\sqrt{b} \\ &= a - \sqrt{ab} + \sqrt{ab} - b \\ &= a - b\end{aligned}$$

which is rational if a and b are integers.

Rationalizing the Denominator 2 (Division with Radical Expressions)

EXAMPLE 44 Divide $\frac{6}{\sqrt{5} - \sqrt{3}}$. (Rationalize the denominator.)

SOLUTION Because the product of two conjugates is a rational number, we multiply the numerator and denominator by the conjugate of the denominator.

$$\begin{aligned}\frac{6}{\sqrt{5} - \sqrt{3}} &= \frac{6}{\sqrt{5} - \sqrt{3}} \cdot \frac{(\sqrt{5} + \sqrt{3})}{(\sqrt{5} + \sqrt{3})} \\ &= \frac{6\sqrt{5} + 6\sqrt{3}}{(\sqrt{5})^2 - (\sqrt{3})^2} \\ &= \frac{6\sqrt{5} + 6\sqrt{3}}{5 - 3} \\ &= \frac{6\sqrt{5} + 6\sqrt{3}}{2}\end{aligned}$$

The numerator and denominator of this last expression have a factor of 2 in common. We can reduce to lowest terms by factoring 2 from the numerator and then dividing both the numerator and denominator by 2:

$$\begin{aligned}&= \frac{2(3\sqrt{5} + 3\sqrt{3})}{2} \\ &= 3\sqrt{5} + 3\sqrt{3}\end{aligned}$$

GETTING READY FOR CLASS

After reading through the preceding section, respond in your own words and in complete sentences.

A. Which of the following is a real number, $\sqrt{-9}$ or $\sqrt{9}$?

B. Explain in your own words the meaning of $25^{1/2}$.

C. Explain how you would put $\sqrt{50}$ into simplified form for radicals.

D. How would you rationalize the denominator in the expression $\frac{1}{\sqrt{2}}$?

Problem Set 1.6

Throughout this problem set assume all variables under an even root represent nonnegative numbers, except where noted otherwise.

Find each of the following roots, if possible.

1. $-\sqrt{49}$ **2.** $-\sqrt[3]{27}$ **3.** $\sqrt[4]{16}$ **4.** $-\sqrt[4]{-16}$

5. $\sqrt{0.04}$ **6.** $\sqrt{0.81}$ **7.** $\sqrt[3]{0.008}$ **8.** $\sqrt[3]{0.125}$

Simplify each expression.

9. $\sqrt[3]{27a^{12}}$ **10.** $\sqrt[3]{8a^{15}}$ **11.** $\sqrt[3]{x^3y^6}$ **12.** $\sqrt[3]{x^6y^3}$

13. $\sqrt[5]{32x^{10}y^5}$ **14.** $\sqrt[5]{32x^5y^{10}}$ **15.** $\sqrt[4]{16a^{12}b^{20}}$ **16.** $\sqrt[4]{81a^{24}b^8}$

Simplify each of the following as much as possible.

17. $25^{3/2}$ **18.** $9^{3/2}$ **19.** $16^{3/4}$ **20.** $81^{3/4}$ **21.** $81^{-3/4}$

22. $4^{-3/2}$ **23.** $\left(\frac{25}{36}\right)^{-1/2}$ **24.** $\left(\frac{16}{49}\right)^{-1/2}$ **25.** $\left(\frac{81}{16}\right)^{-3/4}$ **26.** $\left(\frac{27}{8}\right)^{-2/3}$

Use the properties of exponents to simplify each of the following as much as possible. Assume all bases are positive. Write your answers with positive exponents.

27. $(a^{3/4})^{4/3}$ **28.** $(a^{2/3})^{3/4}$ **29.** $\frac{x^{1/5}}{x^{3/5}}$ **30.** $\frac{x^{2/7}}{x^{5/7}}$

31. $\frac{x^{5/6}}{x^{2/3}}$ **32.** $\frac{x^{7/8}}{x^{8/7}}$ **33.** $\frac{(y^{2/3})^{3/4}}{(y^{1/3})^{3/5}}$ **34.** $\left(\frac{a^{-1/5}}{b^{1/3}}\right)^{15}$

35. **a.** $\sqrt[3]{8}$ **b.** $\sqrt[3]{0.008}$ **c.** $\sqrt[3]{8,000}$ **d.** $\sqrt[3]{8 \times 10^{-6}}$

36. **a.** $\sqrt{16a^4b^8}$ **b.** $\sqrt[3]{16a^4b^8}$ **c.** $\sqrt[4]{16a^4b^8}$

Use Property 1 for radicals to write each of the following expressions in simplified form.

37. $\sqrt{8}$ **38.** $\sqrt{32}$ **39.** $\sqrt{288}$ **40.** $\sqrt{128}$

41. $\sqrt[3]{54}$ **42.** $\sqrt[3]{24}$ **43.** $\sqrt[3]{128}$ **44.** $\sqrt[3]{162}$

45. $\sqrt{18x^3}$ **46.** $\sqrt{27x^5}$ **47.** $\sqrt[4]{32y^7}$ **48.** $\sqrt[5]{32y^7}$

49. $\sqrt[3]{40x^4y^7}$ **50.** $\sqrt[3]{128x^6y^2}$ **51.** $\sqrt{48a^2b^3c^4}$ **52.** $\sqrt{72a^4b^3c^2}$

53. Simplify each expression.

a. $\frac{\sqrt{20}}{4}$ **b.** $\frac{3\sqrt{20}}{15}$ **c.** $\frac{4+\sqrt{12}}{2}$ **d.** $\frac{2+\sqrt{9}}{5}$

54. Simplify each expression.

a. $\frac{\sqrt{12}}{4}$ **b.** $\frac{2\sqrt{32}}{8}$ **c.** $\frac{9+\sqrt{27}}{3}$ **d.** $\frac{-6-\sqrt{64}}{2}$

Rationalize the denominator in each of the following expressions.

55. $\sqrt{\frac{1}{2}}$ **56.** $\sqrt{\frac{1}{3}}$ **57.** $\frac{4}{\sqrt[3]{2}}$ **58.** $\frac{5}{\sqrt[3]{3}}$

59. **a.** $\frac{1}{\sqrt{2}}$ **b.** $\frac{1}{\sqrt[3]{2}}$ **c.** $\frac{1}{\sqrt[4]{2}}$

60. **a.** $\frac{1}{\sqrt{3}}$ **b.** $\frac{1}{\sqrt[3]{9}}$ **c.** $\frac{1}{\sqrt[4]{27}}$

Simplify each expression. Do *not* assume the variables represent positive numbers.

61. $\sqrt{25x^2}$ **62.** $\sqrt{49x^2}$ **63.** $\sqrt{27x^3y^2}$ **64.** $\sqrt{40x^3y^2}$

65. $\sqrt{x^2 - 10x + 25}$ **66.** $\sqrt{x^2 - 16x + 64}$

Combine the following expressions.

67. $3\sqrt{5} + 4\sqrt{5}$ **68.** $6\sqrt{3} - 5\sqrt{3}$

69. $5\sqrt[3]{10} - 4\sqrt[3]{10}$ **70.** $6\sqrt[4]{2} + 9\sqrt[4]{2}$

71. $\sqrt{20} - \sqrt{80} + \sqrt{45}$ **72.** $\sqrt{8} - \sqrt{32} - \sqrt{18}$

73. $5x\sqrt{8} + 3\sqrt{32x^2} - 5\sqrt{50x^2}$ **74.** $2\sqrt{50x^2} - 8x\sqrt{18} - 3\sqrt{72x^2}$

Multiply.

75. $\sqrt{6}\sqrt{3}$ **76.** $\sqrt{6}\sqrt{2}$

77. $(2\sqrt{3})(5\sqrt{7})$ **78.** $(3\sqrt{5})(2\sqrt{7})$

79. $\sqrt{3}(\sqrt{2} - 3\sqrt{3})$ **80.** $\sqrt{2}(5\sqrt{3} + 4\sqrt{2})$

81. $(\sqrt{x} + 5)(\sqrt{x} - 3)$ **82.** $(\sqrt{x} + 4)(\sqrt{x} + 2)$

83. $(\sqrt{3} + 4)^2$ **84.** $(\sqrt{5} - 2)^2$

85. $(\sqrt{3} - \sqrt{2})(\sqrt{3} + \sqrt{2})$ **86.** $(\sqrt{5} - \sqrt{2})(\sqrt{5} + \sqrt{2})$

87. $(5 - \sqrt{x})(5 + \sqrt{x})$ **88.** $(3 - \sqrt{x})(3 + \sqrt{x})$

Rationalize the denominator in each of the following.

89. $\frac{\sqrt{5}}{\sqrt{5} + 1}$ **90.** $\frac{\sqrt{7}}{\sqrt{7} - 1}$ **91.** $\frac{\sqrt{x}}{\sqrt{x} - 3}$ **92.** $\frac{\sqrt{x}}{\sqrt{x} + 2}$

93. $\frac{\sqrt{5}}{2\sqrt{5} - 3}$ **94.** $\frac{\sqrt{7}}{3\sqrt{7} - 2}$ **95.** $\frac{3}{\sqrt{x} - \sqrt{y}}$ **96.** $\frac{2}{\sqrt{x} + \sqrt{y}}$

Paying Attention to Instructions The next two problems are intended to give you practice reading, and paying attention to, the instructions that accompany the problems you are working.

97. Work each problem according to the instructions given.

a. Add: $(\sqrt{x} + 2) + (\sqrt{x} - 2)$ **b.** Multiply: $(\sqrt{x} + 2)(\sqrt{x} - 2)$

c. Square: $(\sqrt{x} + 2)^2$ **d.** Divide: $\frac{\sqrt{x} + 2}{\sqrt{x} - 2}$

98. Work each problem according to the instructions given.

a. Add: $(\sqrt{x} - 3) + (\sqrt{x} + 3)$ **b.** Multiply: $(\sqrt{x} - 3)(\sqrt{x} + 3)$

c. Square: $(\sqrt{x} + 3)^2$ **d.** Divide: $\frac{\sqrt{x} + 3}{\sqrt{x} - 3}$

Complex Numbers

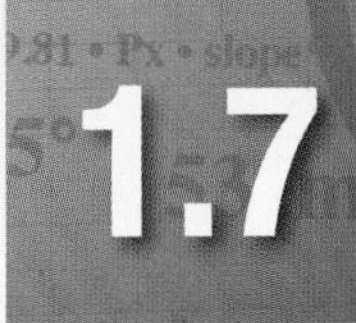

The equation $x^2 = -9$ has no real number solutions because the square of a real number is always positive. We have been unable to work with square roots of negative numbers like $\sqrt{-25}$ and $\sqrt{-16}$ for the same reason. Complex numbers allow us to expand our work with radicals to include square roots of negative numbers and to solve equations like $x^2 = -9$ and $x^2 = -64$. Our work with complex numbers is based on the following definition.

DEFINITION *The number i*

The *number i* is such that $i = \sqrt{-1}$ (which is the same as saying $i^2 = -1$).

The number i, as we have defined it here, is not a real number. Because of the way we have defined i, we can use it to simplify square roots of negative numbers.

 *Square Roots of Negative Numbers*

If a is a positive number, then $\sqrt{-a}$ can always be written as $i\sqrt{a}$. That is,

$$\sqrt{-a} = i\sqrt{a} \qquad \text{if } a \text{ is a positive number}$$

To justify our rule, we simply square the quantity $i\sqrt{a}$ to obtain $-a$. Here is what it looks like when we do so:

$$\begin{aligned}(i\sqrt{a})^2 &= i^2 \cdot (\sqrt{a})^2 \\ &= -1 \cdot a \\ &= -a\end{aligned}$$

This means that $i\sqrt{a}$ is the number we square to get $-a$.

VIDEO EXAMPLES

SECTION 1.7

EXAMPLES Write each square root in terms of the number i.

1. $\sqrt{-25} = i\sqrt{25} = i \cdot 5 = 5i$

2. $\sqrt{-49} = i\sqrt{49} = i \cdot 7 = 7i$

3. $\sqrt{-12} = i\sqrt{12} = i \cdot 2\sqrt{3} = 2i\sqrt{3}$

4. $\sqrt{-17} = i\sqrt{17}$

Note In Examples 3 and 4, we wrote i before the radical simply to avoid confusion. If we were to write the answer to 3 as $2\sqrt{3}i$, some people would think the i was under the radical sign, but it is not.

If we assume all the properties of exponents hold when the base is i, we can write any power of i as i, -1, $-i$, or 1. Using the fact that $i^2 = -1$, we have

$$\begin{aligned}i^1 &= i \\ i^2 &= -1 \\ i^3 &= i^2 \cdot i = -1(i) = -i \\ i^4 &= i^2 \cdot i^2 = -1(-1) = 1\end{aligned}$$

Because $i^4 = 1$, i^5 will simplify to i, and we will begin repeating the sequence i, -1, $-i$, 1 as we simplify higher powers of i: Any power of i simplifies to i, -1, $-i$, or 1. The easiest way to simplify higher powers of i is to write them in terms of i^2. For instance, to simplify i^{21}, we would write it as

$$(i^2)^{10} \cdot i \qquad \text{because} \qquad 2 \cdot 10 + 1 = 21$$

Then, because $i^2 = -1$, we have

$$(-1)^{10} \cdot i = 1 \cdot i = i$$

EXAMPLES Simplify as much as possible.

5. $i^{30} = (i^2)^{15} = (-1)^{15} = -1$

6. $i^{11} = (i^2)^5 \cdot i = (-1)^5 \cdot i = (-1)i = -i$

7. $i^{40} = (i^2)^{20} = (-1)^{20} = 1$

(dēf DEFINITION *Complex number*

A *complex number* is any number that can be put in the form

$$a + bi$$

where a and b are real numbers and $i = \sqrt{-1}$. The form $a + bi$ is called ***standard form*** for complex numbers. The number a is called the ***real part*** of the complex number. The number b is called the ***imaginary part*** of the complex number.

Every real number is a complex number. For example, 8 can be written as $8 + 0i$. Likewise, $-\frac{1}{2}$, π, $\sqrt{3}$, and -9 are complex numbers because they can all be written in the form $a + bi$:

$$-\frac{1}{2} = -\frac{1}{2} + 0i \qquad \pi = \pi + 0i \qquad \sqrt{3} = \sqrt{3} + 0i \qquad -9 = -9 + 0i$$

The rest of the complex numbers that are not real numbers, are divided into two additional categories: *compound numbers* and *pure imaginary numbers*. The diagram below shows all three subsets of the complex numbers, along with examples of the type of numbers that fall into those subsets.

Subsets of the Complex Numbers

All numbers of the form $a + bi$ fall into one of the following categories. Each category is a subset of the complex numbers.

Real Numbers	Compound Numbers	Pure Imaginary Numbers
When $a \neq 0$ and $b = 0$ Examples include: $-10, 0, 1, \sqrt{3}, \frac{5}{8}, \pi$	When neither a nor b is 0 Examples include: $5 + 4i, \frac{1}{3} + 4i, \sqrt{5} - i,$ $-6 + i\sqrt{5}$	When $a = 0$ and $b \neq 0$ Examples include: $-4i, i\sqrt{3}, -5i\sqrt{7}, \frac{3}{4}i$

Note: The definition for compound numbers is from Jim Metz of Kapiolani Community College in Hawaii. Some textbooks use the phrase *imaginary numbers* to represent both the compound numbers and the pure imaginary numbers. In those books, the pure imaginary numbers are a subset of the imaginary numbers. We like the definition from Mr. Metz because it keeps the three subsets from overlapping.

Equality for Complex Numbers

Two complex numbers are equal if and only if their real parts are equal and their imaginary parts are equal. That is, for real numbers a, b, c, and d,

$$a + bi = c + di \quad \text{if and only if} \quad a = c \quad \text{and} \quad b = d$$

EXAMPLE 8 Find x and y if $3x + 4i = 12 - 8yi$.

SOLUTION Because the two complex numbers are equal, their real parts are equal and their imaginary parts are equal:

$$\begin{aligned} 3x &= 12 \quad \text{and} \quad & 4 &= -8y \\ x &= 4 & y &= -\frac{1}{2} \end{aligned}$$

EXAMPLE 9 Find x and y if $(4x - 3) + 7i = 5 + (2y - 1)i$.

SOLUTION The real parts are $4x - 3$ and 5. The imaginary parts are 7 and $2y - 1$:

$$\begin{aligned} 4x - 3 &= 5 \quad \text{and} \quad & 7 &= 2y - 1 \\ 4x &= 8 & 8 &= 2y \\ x &= 2 & y &= 4 \end{aligned}$$

Addition and Subtraction of Complex Numbers

To add two complex numbers, add their real parts and their imaginary parts. That is, if a, b, c, and d are real numbers, then

$$(a + bi) + (c + di) = (a + c) + (b + d)i$$

EXAMPLES Add or subtract as indicated.

10. $(3 + 4i) + (7 - 6i) = (3 + 7) + (4 - 6)i = 10 - 2i$

11. $(7 + 3i) - (5 + 6i) = (7 - 5) + (3 - 6)i = 2 - 3i$

12. $(5 - 2i) - (9 - 4i) = (5 - 9) + (-2 + 4)i = -4 + 2i$

Multiplication of Complex Numbers

Because complex numbers have the same form as binomials, we find the product of two complex numbers the same way we find the product of two binomials.

EXAMPLE 13 Multiply $(3 - 4i)(2 + 5i)$.

SOLUTION Multiplying each term in the second complex number by each term in the first, we have

$$\begin{aligned}(3 - 4i)(2 + 5i) &= \overset{F}{3 \cdot 2} + \overset{O}{3 \cdot 5i} - \overset{I}{2 \cdot 4i} - \overset{L}{4i(5i)} \\ &= 6 + 15i - 8i - 20i^2\end{aligned}$$

Combining similar terms and using the fact that $i^2 = -1$, we can simplify as follows:

$$\begin{aligned}6 + 15i - 8i - 20i^2 &= 6 + 7i - 20(-1) \\ &= 6 + 7i + 20 \\ &= 26 + 7i\end{aligned}$$

The product of the complex numbers $3 - 4i$ and $2 + 5i$ is the complex number $26 + 7i$. ■

EXAMPLE 14 Multiply $2i(4 - 6i)$.

SOLUTION Applying the distributive property gives us

$$\begin{aligned}2i(4 - 6i) &= 2i \cdot 4 - 2i \cdot 6i \\ &= 8i - 12i^2 \\ &= 12 + 8i\end{aligned}$$

■

EXAMPLE 15 Expand $(3 + 5i)^2$.

SOLUTION We treat this like the square of a binomial. Remember, $(a + b)^2 = a^2 + 2ab + b^2$:

$$\begin{aligned}(3 + 5i)^2 &= 3^2 + 2(3)(5i) + (5i)^2 \\ &= 9 + 30i + 25i^2 \\ &= 9 + 30i - 25 \\ &= -16 + 30i\end{aligned}$$

■

EXAMPLE 16 Multiply $(2 - 3i)(2 + 3i)$.

SOLUTION This product has the form $(a - b)(a + b)$, which we know results in the difference of two squares, $a^2 - b^2$:

$$\begin{aligned}(2 - 3i)(2 + 3i) &= 2^2 - (3i)^2 \\ &= 4 - 9i^2 \\ &= 4 + 9 \\ &= 13\end{aligned}$$

■

The product of the two complex numbers $2 - 3i$ and $2 + 3i$ is the real number 13. The two complex numbers $2 - 3i$ and $2 + 3i$ are called *complex conjugates*. The fact that their product is a real number is very useful.

DEFINITION *Complex conjugates*

The complex numbers $a + bi$ and $a - bi$ are called *complex conjugates*. One important property they have is that their product is the real number $a^2 + b^2$. Here's why:

$$\begin{aligned}(a + bi)(a - bi) &= a^2 - (bi)^2 \\ &= a^2 - b^2i^2 \\ &= a^2 - b^2(-1) \\ &= a^2 + b^2\end{aligned}$$

Division with Complex Numbers

The fact that the product of two complex conjugates is a real number is the key to division with complex numbers.

EXAMPLE 17 Divide $\dfrac{2 + i}{3 - 2i}$.

SOLUTION We want a complex number in standard form that is equivalent to the quotient $\frac{2+i}{3-2i}$. We need to eliminate i from the denominator. Multiplying the numerator and denominator by $3 + 2i$ will give us what we want:

$$\begin{aligned}\frac{2 + i}{3 - 2i} &= \frac{2 + i}{3 - 2i} \cdot \frac{(3 + 2i)}{(3 + 2i)} \\ &= \frac{6 + 4i + 3i + 2i^2}{9 - 4i^2} \\ &= \frac{6 + 7i - 2}{9 + 4} \\ &= \frac{4 + 7i}{13} \\ &= \frac{4}{13} + \frac{7}{13}i\end{aligned}$$

Dividing the complex number $2 + i$ by $3 - 2i$ gives the complex number $\frac{4}{13} + \frac{7}{13}i$.

GETTING READY FOR CLASS

After reading through the preceding section, respond in your own words and in complete sentences.

A. What is the number i?
B. What is a complex number?
C. What kind of number will always result when we multiply complex conjugates?
D. Explain how to divide complex numbers.

Problem Set 1.7

Write the following in terms of i, and simplify as much as possible.

1. $\sqrt{-36}$
2. $\sqrt{-49}$
3. $-\sqrt{-25}$
4. $-\sqrt{-81}$
5. $\sqrt{-72}$
6. $\sqrt{-48}$
7. $-\sqrt{-12}$
8. $-\sqrt{-75}$

Write each of the following as i, -1, $-i$, or 1.

9. i^{28}
10. i^{31}
11. i^{26}
12. i^{37}
13. i^{75}
14. i^{42}

Find x and y so each of the following equations is true.

15. $2x + 3yi = 6 - 3i$
16. $4x - 2yi = 4 + 8i$
17. $2 - 5i = -x + 10yi$
18. $4 + 7i = 6x - 14yi$
19. $2x + 10i = -16 - 2yi$
20. $4x - 5i = -2 + 3yi$
21. $(2x - 4) - 3i = 10 - 6yi$
22. $(4x - 3) - 2i = 8 + yi$
23. $(7x - 1) + 4i = 2 + (5y + 2)i$
24. $(5x + 2) - 7i = 4 + (2y + 1)i$

Combine the following complex numbers.

25. $(2 + 3i) + (3 + 6i)$
26. $(4 + i) + (3 + 2i)$
27. $(3 - 5i) + (2 + 4i)$
28. $(7 + 2i) + (3 - 4i)$
29. $(5 + 2i) - (3 + 6i)$
30. $(6 + 7i) - (4 + i)$
31. $(3 - 5i) - (2 + i)$
32. $(7 - 3i) - (4 + 10i)$
33. $[(3 + 2i) - (6 + i)] + (5 + i)$
34. $[(4 - 5i) - (2 + i)] + (2 + 5i)$
35. $[(7 - i) - (2 + 4i)] - (6 + 2i)$
36. $[(3 - i) - (4 + 7i)] - (3 - 4i)$
37. $(3 + 2i) - [(3 - 4i) - (6 + 2i)]$
38. $(7 - 4i) - [(-2 + i) - (3 + 7i)]$
39. $(4 - 9i) + [(2 - 7i) - (4 + 8i)]$
40. $(10 - 2i) - [(2 + i) - (3 - i)]$

Find the following products.

41. $3i(4 + 5i)$
42. $2i(3 + 4i)$
43. $6i(4 - 3i)$
44. $11i(2 - i)$
45. $(3 + 2i)(4 + i)$
46. $(2 - 4i)(3 + i)$
47. $(4 + 9i)(3 - i)$
48. $(5 - 2i)(1 + i)$
49. $(1 + i)^3$
50. $(1 - i)^3$
51. $(2 - i)^3$
52. $(2 + i)^3$
53. $(2 + 5i)^2$
54. $(3 + 2i)^2$
55. $(1 - i)^2$
56. $(1 + i)^2$
57. $(3 - 4i)^2$
58. $(6 - 5i)^2$
59. $(2 + i)(2 - i)$
60. $(3 + i)(3 - i)$
61. $(6 - 2i)(6 + 2i)$
62. $(5 + 4i)(5 - 4i)$
63. $(2 + 3i)(2 - 3i)$
64. $(2 - 7i)(2 + 7i)$
65. $(10 + 8i)(10 - 8i)$
66. $(11 - 7i)(11 + 7i)$

Find the following quotients. Write all answers in standard form for complex numbers.

67. $\dfrac{2 - 3i}{i}$
68. $\dfrac{3 + 4i}{i}$
69. $\dfrac{5 + 2i}{-i}$
70. $\dfrac{4 - 3i}{-i}$
71. $\dfrac{4}{2 - 3i}$
72. $\dfrac{3}{4 - 5i}$
73. $\dfrac{6}{-3 + 2i}$
74. $\dfrac{-1}{-2 - 5i}$
75. $\dfrac{2 + 3i}{2 - 3i}$
76. $\dfrac{4 - 7i}{4 + 7i}$
77. $\dfrac{5 + 4i}{3 + 6i}$
78. $\dfrac{2 + i}{5 - 6i}$

Chapter 1 Summary

The numbers in brackets refer to the section(s) in which the topic can be found.

Order of Operations [1.1]

When evaluating a mathematical expression, we will perform the operations in the following order, beginning with the expression in the innermost parentheses or brackets and working our way out.

1. Simplify all numbers with exponents, working from left to right if more than one of these numbers is present.
2. Then, do all multiplications and divisions left to right.
3. Finally, perform all additions and subtractions left to right.

EXAMPLES

1. $10 + (2 \cdot 3^2 - 4 \cdot 2)$
$= 10 + (2 \cdot 9 - 4 \cdot 2)$
$= 10 + (18 - 8)$
$= 10 + 10$
$= 20$

Sets [1.1]

A *set* is a collection of objects or things. The objects in the set are called *elements*, or *members*, of the set.

The *union* of two sets A and B, written $A \cup B$, is all the elements that are in A *or* are in B.

The *intersection* of two sets A and B, written $A \cap B$, is the set consisting of all elements common to both A *and* B.

Set A is a *subset* of set B, written $A \subset B$, if all elements in set A are also in set B.

2. If $A = \{0, 1, 2\}$ and $B = \{2, 3\}$, then $A \cup B = \{0, 1, 2, 3\}$ and $A \cap B = \{2\}$.

Special Sets [1.1]

Counting numbers $= \{1, 2, 3, \ldots\}$
Whole numbers $= \{0, 1, 2, 3, \ldots\}$
Integers $= \{\ldots, -3, -2, -1, 0, 1, 2, 3, \ldots\}$
Rational numbers $= \left\{ \frac{a}{b} \mid a \text{ and } b \text{ are integers, } b \neq 0 \right\}$
Irrational numbers $= \{x \mid x \text{ is real, but not rational}\}$
Real numbers $= \{x \mid x \text{ is rational or } x \text{ is irrational}\}$

3. 5 is a counting number, a whole number, an integer, a rational number, and a real number.
$\frac{3}{4}$ is a rational number and a real number.
$\sqrt{2}$ is an irrational number and a real number.

Absolute Value [1.1]

The *absolute value* of a real number is its distance from 0 on the number line. If $|x|$ represents the absolute value of x, then

$$|x| = \begin{cases} x & \text{if} \quad x \geq 0 \\ -x & \text{if} \quad x < 0 \end{cases}$$

The absolute value of a real number is never negative.

4. $|5| = 5$
$|-5| = 5$

Properties of Real Numbers [1.1]

	For Addition	*For Multiplication*
Commutative	$a + b = b + a$	$ab = ba$
Associative	$a + (b + c) = (a + b) + c$	$a(bc) = (ab)c$
Identity	$a + 0 = a$	$a \cdot 1 = a$
Inverse	$a + (-a) = 0$	$a\left(\frac{1}{a}\right) = 1$
Distributive	$a(b + c) = ab + ac$	

Properties of Exponents [1.2]

5. These expressions illustrate the properties of exponents.

a. $x^2 \cdot x^3 = x^{2+3} = x^5$

b. $(x^2)^3 = x^{2 \cdot 3} = x^6$

c. $(3x)^2 = 3^2 \cdot x^2 = 9x^2$

d. $2^{-3} = \frac{1}{2^3} = \frac{1}{8}$

e. $\left(\frac{x}{5}\right)^2 = \frac{x^2}{5^2} = \frac{x^2}{25}$

f. $\frac{x^7}{x^5} = x^{7-5} = x^2$

g. $3^1 = 3$
$3^0 = 1$

If a and b represent real numbers and r and s represent rational numbers, then

1. $a^r \cdot a^s = a^{r+s}$
2. $(a^r)^s = a^{r \cdot s}$
3. $(ab)^r = a^r \cdot b^r$
4. $a^{-r} = \frac{1}{a^r} \quad (a \neq 0)$
5. $\left(\frac{a}{b}\right)^r = \frac{a^r}{b^r} \quad (b \neq 0)$
6. $\frac{a^r}{a^s} = a^{r-s} \quad (a \neq 0)$
7. $a^1 = a$
   $a^0 = 1 \quad (a \neq 0)$

Scientific Notation [1.2]

6. $49{,}800{,}000 = 4.98 \times 10^7$
$0.00462 = 4.62 \times 10^{-3}$

A number is written in scientific notation when it is written as the product of a number between 1 and 10 and an integer power of 10; that is, when it has the form

$$n \times 10^r$$

where $1 \leq n < 10$ and $r =$ an integer.

Operations with Polynomials [1.3]

7. $(3x^2 + 2x - 5) + (4x^2 - 7x + 2)$
$= 7x^2 - 5x - 3$

8. $(3x - 5)(x + 2)$
$3x^2 + 6x - 5x - 10$
$3x^2 + x - 10$

To add two polynomials, combine the coefficients of similar terms.

To multiply two polynomials, multiply each term in the first polynomial by each term in the second polynomial.

Strategy for Factoring a Polynomial [1.4]

9. a. $2x^5 - 8x^3 = 2x^3(x^2 - 4)$
$= 2x^3(x + 2)(x - 2)$

b. $3x^4 - 18x^3 + 27x^2$
$= 3x^2(x^2 - 6x + 9)$
$= 3x^2(x - 3)^2$

c. $6x^3 - 12x^2 - 48x$
$= 6x(x^2 - 2x - 8)$
$= 6x(x - 4)(x + 2)$

d. $x^2 + ax + bx + ab$
$= x(x + a) + b(x + a)$
$= (x + a)(x + b)$

Step 1: If the polynomial has a greatest common factor other than 1, then factor out the greatest common factor.

Step 2: If the polynomial has two terms (it is a binomial), then see if it is the difference of two squares or the sum or difference of two cubes, and then factor accordingly. Remember, if it is the sum of two squares, it will not factor.

Step 3: If the polynomial has three terms (a trinomial), then it is either a perfect square trinomial that will factor into the square of a binomial, or it is not a perfect square trinomial, in which case you use the trial and error method developed in Section 1.4.

Step 4: If the polynomial has more than three terms, then try to factor it by grouping.

Step 5: As a final check, see if any of the factors you have written can be factored further. If you have overlooked a common factor, you can catch it here.

Rational Expressions [1.5]

10. a. $\frac{3}{4}$ is a rational number. $\frac{x-3}{x^2-9}$ is a rational expression.

A *rational expression* is any quantity that can be expressed as the ratio of two polynomials:

$$\text{Rational expressions} = \left\{ \frac{P}{Q} \,\middle|\, P \text{ and } Q \text{ are polynomials, } Q \neq 0 \right\}$$

b. $\frac{x-3}{x^2-9} = \frac{\cancel{x-3}}{(\cancel{x-3})(x+3)}$
$= \frac{1}{x+3}$

Reducing to Lowest Terms

To reduce a rational expression to lowest terms, we first factor the numerator and denominator and then divide the numerator and denominator by any factors they have in common.

c. $\frac{x+1}{x^2-4} \cdot \frac{x+2}{3x+3}$
$= \frac{(\cancel{x+1})(\cancel{x+2})}{(x-2)(\cancel{x+2})(3)(\cancel{x+1})}$
$= \frac{1}{3(x-2)}$

Multiplication

To multiply two rational numbers or rational expressions, multiply numerators and multiply denominators. In symbols,

$$\frac{P}{Q} \cdot \frac{R}{S} = \frac{PR}{QS} \qquad (Q \neq 0 \text{ and } S \neq 0)$$

In practice, we don't really multiply, but rather, we factor and then divide out common factors.

d. $\frac{x^2-y^2}{x^3+y^3} \div \frac{x-y}{x^2-xy+y^2}$
$= \frac{x^2-y^2}{x^3+y^3} \cdot \frac{x^2-xy+y^2}{x-y}$
$= \frac{(\cancel{x+y})(\cancel{x-y})(\cancel{x^2-xy+y^2})}{(\cancel{x+y})(\cancel{x^2-xy+y^2})(\cancel{x-y})}$
$= 1$

Division

To divide one rational expression by another, we use the definition of division to rewrite our division problem as an equivalent multiplication problem. To divide by a rational expression we multiply by its reciprocal. In symbols,

$$\frac{P}{Q} \div \frac{R}{S} = \frac{P}{Q} \cdot \frac{S}{R} = \frac{PS}{QR} \qquad (Q \neq 0, S \neq 0, R \neq 0)$$

Addition and Subtraction

If P, Q, and R represent polynomials, $R \neq 0$, then

$$\frac{P}{R} + \frac{Q}{R} = \frac{P+Q}{R} \quad \text{and} \quad \frac{P}{R} - \frac{Q}{R} = \frac{P-Q}{R}$$

When adding or subtracting rational expressions with different denominators, we must find the LCD for all denominators and change each rational expression to an equivalent expression that has the LCD.

e. $\frac{2}{x-3} + \frac{3}{5}$

$= \frac{2}{x-3} \cdot \frac{5}{5} + \frac{3}{5} \cdot \frac{x-3}{x-3}$

$= \frac{3x+1}{5(x-3)}$

Complex Fractions

A rational expression that contains, in its numerator or denominator, other rational expressions is called a *complex fraction.* One method of simplifying a complex fraction is to multiply the numerator and denominator by the LCD for all denominators.

f. $\frac{\frac{1}{x} + \frac{1}{y}}{\frac{1}{x} - \frac{1}{y}} = \frac{xy\left(\frac{1}{x} + \frac{1}{y}\right)}{xy\left(\frac{1}{x} - \frac{1}{y}\right)}$

$= \frac{y+x}{y-x}$

Roots and Radicals [1.6]

Square Roots

Every positive real number x has two square roots. The *positive square root* of x is written $\sqrt{x}$, and the *negative square root* of x is written $-\sqrt{x}$. Both the positive and the negative square roots of x are numbers we square to get x; that is,

$$\left.\begin{array}{r} (\sqrt{x})^2 = x \\ \text{and} \quad (-\sqrt{x})^2 = x \end{array}\right\} \text{for } x \geq 0$$

11. a. The number 49 has two square roots, 7 and -7. They are written like this:

$\sqrt{49} = 7 \qquad -\sqrt{49} = -7$

Higher Roots

In the expression $\sqrt[n]{a}$, n is the *index*, a is the *radicand*, and $\sqrt{}$ is the *radical sign*. The expression $\sqrt[n]{a}$ is such that

$(\sqrt[n]{a})^n = a \qquad a \geq 0$ when n is even

b. $\sqrt[3]{8} = 2$

$\sqrt[3]{-27} = -3$

Rational Exponents

Rational exponents are used to indicate roots. The relationship between rational exponents and roots is as follows:

$$a^{1/n} = \sqrt[n]{a} \qquad \text{and} \qquad a^{m/n} = (a^{1/n})^m = (a^m)^{1/n}$$

$$a \geq 0 \text{ when } n \text{ is even}$$

c. $25^{1/2} = \sqrt{25} = 5$

$8^{2/3} = (\sqrt[3]{8})^2 = 2^2 = 4$

$9^{3/2} = (\sqrt{9})^3 = 3^3 = 27$

Properties of Radicals

If a and b are nonnegative real numbers whenever n is even, then

1. $\sqrt[n]{ab} = \sqrt[n]{a}\sqrt[n]{b}$
2. $\sqrt[n]{\frac{a}{b}} = \frac{\sqrt[n]{a}}{\sqrt[n]{b}} \qquad (b \neq 0)$

d. $\sqrt{4 \cdot 5} = \sqrt{4}\sqrt{5} = 2\sqrt{5}$

$\sqrt{\frac{7}{9}} = \frac{\sqrt{7}}{\sqrt{9}} = \frac{\sqrt{7}}{3}$

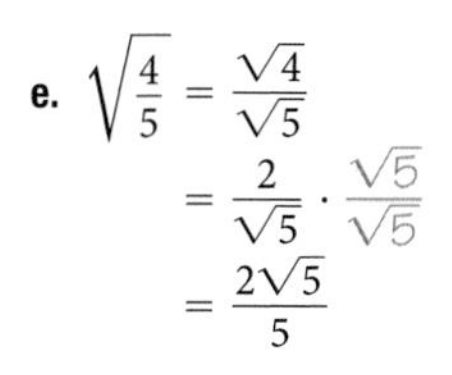

Simplified Form for Radicals

A radical expression is said to be in *simplified form*

1. If there is no factor of the radicand that can be written as a power greater than or equal to the index;
2. If there are no fractions under the radical sign; and
3. If there are no radicals in the denominator.

f. $5\sqrt{3} - 7\sqrt{3} = (5 - 7)\sqrt{3}$
$= -2\sqrt{3}$

$\sqrt{20} + \sqrt{45} = 2\sqrt{5} + 3\sqrt{5}$
$= (2 + 3)\sqrt{5}$
$= 5\sqrt{5}$

Addition and Subtraction of Radical Expressions

We add and subtract radical expressions by using the distributive property to combine similar radicals. Similar radicals are radicals with the same index and the same radicand.

g. $(\sqrt{x} + 2)(\sqrt{x} + 3)$
$= \sqrt{x}\sqrt{x} + 3\sqrt{x} + 2\sqrt{x} + 2 \cdot 3$
$= x + 5\sqrt{x} + 6$

Multiplication of Radical Expressions

We multiply radical expressions in the same way that we multiply polynomials. We can use the distributive property and the FOIL method.

h. $\frac{3}{\sqrt{2}} = \frac{3}{\sqrt{2}} \cdot \frac{\sqrt{2}}{\sqrt{2}} = \frac{3\sqrt{2}}{2}$

$\frac{3}{\sqrt{5} - \sqrt{3}} = \frac{3}{\sqrt{5} - \sqrt{3}} \cdot \frac{\sqrt{5} + \sqrt{3}}{\sqrt{5} + \sqrt{3}}$

$= \frac{3\sqrt{5} + 3\sqrt{3}}{5 - 3}$

$= \frac{3\sqrt{5} + 3\sqrt{3}}{2}$

Rationalizing the Denominator

When a fraction contains a square root in the denominator, we rationalize the denominator by multiplying numerator and denominator by

1. The square root itself if there is only one term in the denominator, or
2. The conjugate of the denominator if there are two terms in the denominator.

Rationalizing the denominator is also called division of radical expressions.

Complex Numbers [1.7]

12. $3 + 4i$ is a complex number.

Addition
$(3 + 4i) + (2 - 5i) = 5 - i$

Multiplication
$(3 + 4i)(2 - 5i)$
$= 6 - 15i + 8i - 20i^2$
$= 6 - 7i + 20$
$= 26 - 7i$

Division
$\frac{2}{3 + 4i} = \frac{2}{3 + 4i} \cdot \frac{3 - 4i}{3 - 4i}$
$= \frac{6 - 8i}{9 + 16}$
$= \frac{6}{25} - \frac{8}{25}i$

A *complex number* is any number that can be put in the form

$$a + bi$$

where a and b are real numbers and $i = \sqrt{-1}$. The *real part* of the complex number is a, and b is the *imaginary part*.

Chapter 1 Test

1. Write the following in symbols. [1.1]

The difference of $2a$ and $3b$ is less than their sum.

2. If $A = \{1, 2, 3, 4\}$ and $B = \{2, 4, 6\}$, find $A \cup B$. [1.1]

Simplify each of the following as much as possible. [1.1]

3. $3(2 - 4)^3 - 5(2 - 7)^2$

4. $\dfrac{-4(-1) - (-10)}{5 - (-2)}$

Simplify. (Assume all variables are nonzero.) [1.2]

5. $x^4 \cdot x^7 \cdot x^{-3}$

6. 2^{-5}

7. $\dfrac{a^{-5}}{a^{-7}}$

8. Write 6,530,000 in scientific notation. [1.2]

9. Perform the indicated operations and write your answer in scientific notation. [1.2]

$$\frac{(6 \times 10^{-4})(4 \times 10^9)}{8 \times 10^{-3}}$$

Simplify. [1.3]

10. $\left(\frac{3}{4}x^3 - x^2 - \frac{3}{2}\right) - \left(\frac{1}{4}x^3 + 2x - \frac{1}{2}\right)$

11. $3 - 4[2x - 3(x + 6)]$

12. $(3y - 7)(2y + 5)$

13. $(2x - 5)(x^2 + 4x - 3)$

14. $(8 - 3t^3)^2$

15. $(1 - 6y)(1 + 6y)$

Factor completely. [1.4]

16. $12x^4 + 26x^2 - 10$

17. $7ax^2 - 14ay - b^2x^2 + 2b^2y$

18. $t^3 + \frac{1}{8}$

19. $81 - x^4$

20. Reduce to lowest terms. [1.5]

$$\frac{2x^2 - 5x + 3}{2x^2 - x - 3}$$

Multiply and divide as indicated. [1.5]

21. $\dfrac{a^2 - 16}{5a - 15} \cdot \dfrac{10(a - 3)^2}{a^2 - 7a + 12}$

22. $\dfrac{x^3 - 8}{2x^2 - 9x + 10} \div \dfrac{x^2 + 2x + 4}{2x^2 + x - 15}$

Add and subtract as indicated. [1.5]

23. $\frac{1}{x} + \frac{2}{x-3}$

24. $\frac{4x}{x^2+6x+5} - \frac{3x}{x^2+5x+4}$

Simplify each complex fraction. [1.5]

25. $\dfrac{3 - \dfrac{1}{a+3}}{3 + \dfrac{1}{a+3}}$

26. $\dfrac{1 - \dfrac{9}{x^2}}{1 + \dfrac{1}{x} - \dfrac{6}{x^2}}$

Simplify each of the following. (Assume all variable bases are positive integers and all variable exponents are positive real numbers throughout this test.) [1.6]

27. $27^{-2/3}$

28. $\left(\frac{25}{49}\right)^{-1/2}$

29. $a^{3/4} \cdot a^{-1/3}$

Write in simplified form. Assume all variables represent positive integers. [1.6]

30. $\sqrt{125x^3y^5}$

31. $\sqrt[3]{40x^7y^8}$

32. $\sqrt{\frac{2}{3}}$

33. Combine $3\sqrt{12} - 4\sqrt{27}$ [1.6]

Multiply. [1.6]

34. $(\sqrt{x}+7)(\sqrt{x}-4)$

35. $(3\sqrt{2}-\sqrt{3})^2$

36. Rationalize the denominator. [1.6]

$$\frac{5}{\sqrt{3}-1}$$

Perform the indicated operations. [1.7]

37. $(3+2i) - [(7-i) - (4+3i)]$

38. $(2-3i)(4+3i)$

39. $(5-4i)^2$

40. $\frac{2-3i}{2+3i}$

Introducing

MATHEMATICAL MODELING

In the next chapter we start our work with modeling. The diagram below gives a general idea of what mathematical modeling involves. We take situations in the world around us, like trends, ideas, relationships, and we model them with mathematics. The diagram shows some of these situations.

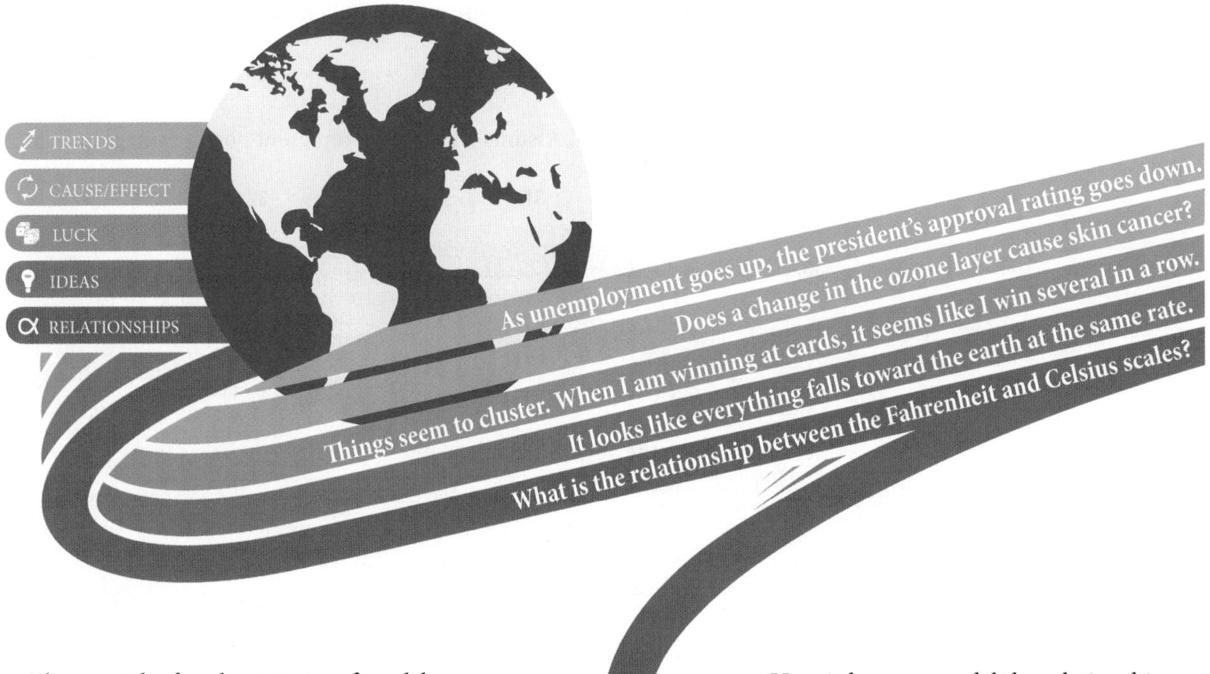

TRENDS
CAUSE/EFFECT
LUCK
IDEAS
RELATIONSHIPS
As unemployment goes up, the president's approval rating goes down.
Does a change in the ozone layer cause skin cancer?
Things seem to cluster. When I am winning at cards, it seems like I win several in a row.
It looks like everything falls toward the earth at the same rate.
What is the relationship between the Fahrenheit and Celsius scales?

These are the four basic types of models we will use throughout the book.

MODEL TYPES

IN WORDS | NUMERIC | SYMBOLIC | VISUAL

Here is how we model the relationships between the two temperature scales. Each of the four model types say the same thing about the relationship between the two temperature scales. And, any of the model types can be used to create any of the other model types.

IN WORDS

The two temperature scales are calibrated so that water freezes at 0 °C and 32 °F. Likewise, water boils at 100 °C and 212 °F. The relationship between them is linear.

NUMERIC

Degrees Celsius	Degrees Fahrenheit
0	32
25	77
50	122
75	167
100	212

SYMBOLIC

$$F = \frac{9}{5}C + 32$$

VISUAL

Fahrenheit temperatures (°F)

F

250 200 150 100 50 0

0 25 50 75 100 125 C

Celsius temperatures (°C)

2 Solving Equations and Inequalities

© YinYang/iStockPhoto

Chapter Outline

2.1 Linear and Quadratic Equations
2.2 More Quadratic Equations
2.3 Additional Equations and Formulas
2.4 More Applications and Modeling
2.5 Linear, Quadratic, and Rational Inequalities
2.6 Equations and Inequalities with Absolute Value

A recent newspaper article gave the following guideline for college students taking out loans to finance their education: The maximum monthly payment on the amount borrowed should not exceed 8% of their monthly starting salary. In this situation, the maximum monthly payment can be described mathematically with the formula

$$M = 0.08s$$

where s is the monthly starting salary.

Using this formula, we can construct a table and a line graph that show the maximum student loan payments that can be made for a variety of starting salaries.

Note When you see this icon next to an example or problem in this chapter, you will know that we are using linear and quadratic equations, inequalities, and formulas to model situations in the world around us.

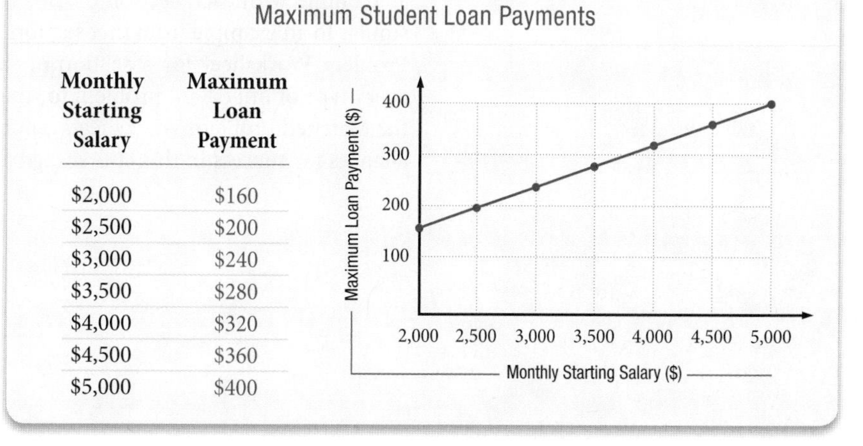

Maximum Student Loan Payments

Monthly Starting Salary	Maximum Loan Payment
\$2,000	\$160
\$2,500	\$200
\$3,000	\$240
\$3,500	\$280
\$4,000	\$320
\$4,500	\$360
\$5,000	\$400

The equation, the line graph, and the table above are each a mathematical model for the recommended maximum monthly student loan payment. Each model gives the same information in a little different form. In this chapter we will work with formulas like $M = 0.08s$ and start our work on table building. Then, in the next chapter, we will start our work with graphing.

Study Skills

If you have successfully completed Chapter 1, then you have made a good start at developing the study skills necessary to succeed in all math classes. Some of the study skills for this chapter are a continuation of the skills from Chapter 1, while others are new to this chapter.

1. **Continue to Set and Keep a Schedule** Sometimes I find students do well in Chapter 1 and then become overconfident. They will begin to put in less time with their homework. Don't do it. Keep to the same schedule.

2. **Increase Effectiveness** You want to become more and more effective with the time you spend on your homework. Increase those activities that are the most beneficial and decrease those that have not given you the results you want.

3. **List Difficult Problems** Begin to make lists of problems that give you the most difficulty. These are the problems that you continue to get wrong. They may also be the key to your success in this course.

4. **Begin to Develop Confidence with Modeling Problems** It seems that the main difference between people who are good at working application problems and those who are not is confidence. People with confidence know that no matter how long it takes them, they will eventually be able to solve the problem. Those without confidence begin by saying to themselves, "I'll never be able to work this problem." If you are in this second category, then instead of telling yourself that you can't do application problems, decide to do whatever it takes to master them. The more application problems you work, the better you will become at them.

5. **Use the Matched Problems Worksheets** These worksheets are available online with each section of the eBook. Each problem on a worksheet is similar to an example from that section. Every time you complete the Matched Problem Worksheet for a section of the book, you know you have worked every type of important problem for that section. If you consistently complete the Matched Problem Worksheets before going to class, you will increase your chances for success in this course significantly.

Linear and Quadratic Equations

INTRODUCTION The study of design, architecture, or construction is not complete without an introduction to the ideas presented in the book *A Pattern Language*, published by the Center for Environmental Structure. The book presents a language for design and construction of houses, neighborhoods, and cities, based on what is best for the people who will live there. Here is a passage from that book concerning the size of public places.

> For public squares, courts, pedestrian streets, any place where crowds are drawn together, estimate the mean number of people in the place at any given moment (P), and make the area of the place between $150P$ and $300P$ square feet.

This passage gives us a formula for the area A of a public square that averages P people at any given time. In the language of algebra, the minimum size of such a public square is given by the formula

$$A = 150P$$

and the maximum size is given by the formula

$$A = 300P$$

Formulas like these are one of the topics we will study in this section. ■

Linear Equations

A *linear equation in one variable* is any equation that can be put in the form

$$ax + b = c$$

where a, b, and c are constants and $a \neq 0$. For example, all of the equations

$$5x + 3 = 2 \qquad 2x = 7 \qquad 2x + 5 = 0$$

are linear because they can be put in the form $ax + b = c$. In the first equation, $5x$, 3, and 2 are called *terms* of the equation: $5x$ is a variable term; 3 and 2 are constant terms.

DEFINITION *Solution set*

The *solution set* for an equation is the set of all numbers that, when used in place of the variable, make the equation a true statement.

DEFINITION *Equivalent equations*

Two or more equations with the same solution set are called *equivalent equations.*

The equations $2x - 5 = 9$, $x - 1 = 6$, and $x = 7$ are all equivalent equations because the solution set for each is $\{7\}$.

Properties of Equality

The first property of equality states that adding the same quantity to both sides of an equation preserves equality. Or, more importantly, adding the same amount to both sides of an equation *never changes* the solution set. This property is called the *addition property of equality* and is stated in symbols as follows:

> **Note** Because subtraction is defined in terms of addition and division is defined in terms of multiplication, we do not need to introduce separate properties for subtraction and division. The solution set for an equation will never be changed by subtracting the same amount from both sides or by dividing both sides by the same nonzero quantity.

PROPERTY *Addition Property of Equality*

For any three algebraic expressions A, B, and C,

$$\text{if} \quad A = B$$
$$\text{then} \quad A + C = B + C$$

In words: Adding the same quantity to both sides of an equation will not change the solution set.

Our second property is called the *multiplication property of equality* and is stated as follows:

PROPERTY *Multiplication Property of Equality*

For any three algebraic expressions A, B, and C, where $C \neq 0$,

$$\text{if} \quad A = B$$
$$\text{then} \quad AC = BC$$

In words: Multiplying both sides of an equation by the same nonzero quantity will not change the solution set.

VIDEO EXAMPLES

SECTION 2.1

EXAMPLE 1 Find the solution set for $3a - 5 = -6a + 1$.

SOLUTION To solve for a, we must isolate it on one side of the equation. Let's decide to isolate a on the left side. We start by adding $6a$ to both sides of the equation.

$$3a - 5 = -6a + 1$$
$$3a + 6a - 5 = -6a + 6a + 1 \qquad \text{Add } 6a \text{ to both sides}$$
$$9a - 5 = 1$$
$$9a - 5 + 5 = 1 + 5 \qquad \text{Add 5 to both sides}$$
$$9a = 6$$
$$\frac{1}{9}(9a) = \frac{1}{9}(6) \qquad \text{Multiply both sides by } \frac{1}{9}$$
$$a = \frac{2}{3} \qquad \frac{1}{9}(6) = \frac{6}{9} = \frac{2}{3}$$

The solution set is $\left\{\frac{2}{3}\right\}$.

> **Note** We know that multiplication by a number and division by its reciprocal always produce the same result. Because of this fact, instead of multiplying each side of our equation by $\frac{1}{9}$, we could just as easily divide each side by 9. If we did so, the last two lines in our solution would look like this:

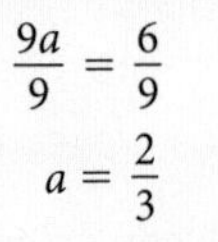

$$\frac{9a}{9} = \frac{6}{9}$$
$$a = \frac{2}{3}$$

EXAMPLE 2 Solve $\frac{2}{3}x + \frac{1}{2} = -\frac{3}{8}$.

SOLUTION We can solve this equation by applying our properties and working with fractions, or we can begin by eliminating the fractions. Let's work the problem using both methods.

Method 1: ***Working with the fractions***

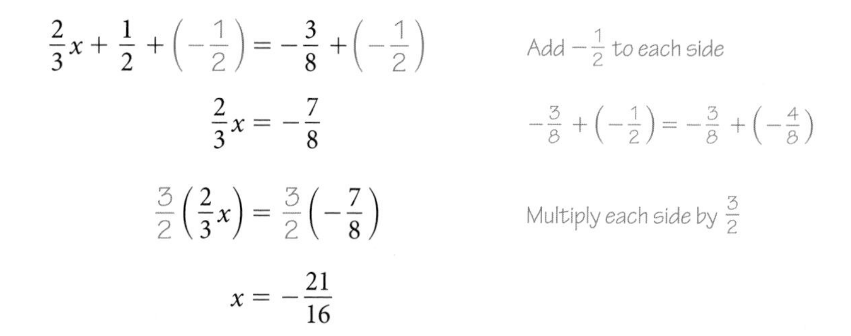

$$\frac{2}{3}x + \frac{1}{2} + \left(-\frac{1}{2}\right) = -\frac{3}{8} + \left(-\frac{1}{2}\right) \qquad \text{Add } -\frac{1}{2} \text{ to each side}$$

$$\frac{2}{3}x = -\frac{7}{8} \qquad -\frac{3}{8} + \left(-\frac{1}{2}\right) = -\frac{3}{8} + \left(-\frac{4}{8}\right)$$

$$\frac{3}{2}\left(\frac{2}{3}x\right) = \frac{3}{2}\left(-\frac{7}{8}\right) \qquad \text{Multiply each side by } \frac{3}{2}$$

$$x = -\frac{21}{16}$$

Method 2: ***Eliminating the fractions in the beginning***

Our original equation has denominators of 3, 2, and 8. The least common denominator, abbreviated LCD, for these three denominators is 24, and it has the property that all three denominators will divide it evenly. Therefore, if we multiply both sides of our equation by 24, each denominator will divide into 24, and we will be left with an equation that does not contain any denominators other than 1.

$$24\left(\frac{2}{3}x + \frac{1}{2}\right) = 24\left(-\frac{3}{8}\right) \qquad \text{Multiply each side by the LCD 24}$$

$$24\left(\frac{2}{3}x\right) + 24\left(\frac{1}{2}\right) = 24\left(-\frac{3}{8}\right) \qquad \text{Distributive property on the left side}$$

$$16x + 12 = -9 \qquad \text{Multiply}$$

$$16x = -21 \qquad \text{Add } -12 \text{ to each side}$$

$$x = -\frac{21}{16} \qquad \text{Multiply each side by } \frac{1}{16}$$

As the third line above indicates, multiplying each side of the equation by the LCD eliminates all the fractions from the equation. Both methods yield the same solution.

EXAMPLE 3 Solve the equation $0.06x + 0.05(10{,}000 - x) = 560$.

SOLUTION We can solve the equation in its original form by working with the decimals, or we can eliminate the decimals first by using the multiplication property of equality and solve the resulting equation. Let's work with the decimals.

$$0.06x + 0.05(10{,}000 - x) = 560 \qquad \text{Original equation}$$

$$0.06x + 0.05(10{,}000) - 0.05x = 560 \qquad \text{Distributive property}$$

$$0.01x + 500 = 560 \qquad \text{Simplify the left side}$$

$$0.01x + 500 + (-500) = 560 + (-500) \qquad \text{Add } -500 \text{ to each side}$$

$$0.01x = 60$$

$$\frac{0.01x}{0.01} = \frac{60}{0.01} \quad \text{Divide each side by 0.01}$$
$$x = 6{,}000$$

Check: Substituting 6,000 for x in the original equation, we have

$$0.06(6{,}000) + 0.05(10{,}000 - 6{,}000) \stackrel{?}{=} 560$$
$$0.06(6{,}000) + 0.05(4{,}000) \stackrel{?}{=} 560$$
$$360 + 200 \stackrel{?}{=} 560$$
$$560 = 560 \quad \text{A true statement}$$

Note We are placing question marks over the equal signs because we don't know yet if the expressions on the left will be equal to the expressions on the right.

Here is a list of steps to use as a guideline for solving linear equations in one variable.

HOW TO *Solve Linear Equations in One Variable*

Step 1 Simplify each side of the equation separately. When you are finished, you will have at most two terms on each side of the equation.

Step 2: Use the addition property of equality to get all variable terms on one side of the equation and all constant terms on the other side.

Step 3: Use the multiplication property of equality to get the variable by itself on one side of the equation.

Step 4: Check your solution in the original equation to be sure that you have not made a mistake in the solution process.

EXAMPLE 4 Solve the equation $8 - 3(4x - 2) + 5x = 35$.

SOLUTION We must begin by distributing the -3 across the quantity $4x - 2$. (It would be a mistake to subtract 3 from 8 first, because the rule for order of operations indicates we are to do multiplication before subtraction.) After we have simplified the left side of our equation, we apply the addition property and the multiplication property. In this example, we will show only the result:

Step 1:
$$8 - 3(4x - 2) + 5x = 35 \quad \text{Original equation}$$
$$8 - 12x + 6 + 5x = 35 \quad \text{Distributive property}$$
$$-7x + 14 = 35 \quad \text{Simplify}$$

Step 2:
$$-7x = 21 \quad \text{Add } -14 \text{ to each side}$$

Step 3:
$$x = -3 \quad \text{Multiply by } -\tfrac{1}{7}$$

Step 4: When x is replaced by -3 in the original equation, a true statement results. Therefore, -3 is the solution to our equation.

Solving Equations by Factoring

Next we will use our knowledge of factoring to solve equations. Most of the equations we will see are *quadratic equations.*

DEFINITION *Quadratic equations*

Any equation that can be written in the form

$$ax^2 + bx + c = 0$$

where a, b, and c are constants and a is not 0 ($a \neq 0$), is called a *quadratic equation.* The form $ax^2 + bx + c = 0$ is called *standard form* for quadratic equations.

Note For a quadratic equation written in standard form, the first term ax^2 is called the *quadratic term,* the second term bx is the *linear term,* and the last term c is called the *constant term.*

Each of the following is a quadratic equation:

$$2x^2 = 5x + 3 \qquad 5x^2 = 75 \qquad 4x^2 - 3x + 2 = 0$$

The number 0 is a special number, and is the key to solving quadratic equations. If we multiply two expressions and get 0, then one, or both, of the expressions must have been 0. In other words, the only way to multiply and get 0 for an answer is to multiply by 0. This fact allows us to solve certain quadratic equations. We state this fact as follows:

PROPERTY *Zero-Factor Property*

For all real numbers r and s,

$$r \cdot s = 0 \quad \text{if and only if} \quad r = 0 \quad \text{or} \quad s = 0 \quad \text{(or both)}$$

EXAMPLE 5 Solve $x^2 - 2x - 24 = 0$.

SOLUTION We begin by factoring the left side as $(x - 6)(x + 4)$ and get

$$(x - 6)(x + 4) = 0$$

Now both $(x - 6)$ and $(x + 4)$ represent real numbers. We notice that their product is 0. By the zero-factor property, one or both of them must be 0:

$$x - 6 = 0 \quad \text{or} \quad x + 4 = 0$$

Note The degree of an equation in one variable is the highest power to which the variable is raised in any one term.

We have used factoring and the zero-factor property to rewrite our original second-degree equation as two first-degree equations connected by the word *or.* Completing the solution, we solve the two first-degree equations:

$$x - 6 = 0 \quad \text{or} \quad x + 4 = 0$$
$$x = 6 \quad \text{or} \quad x = -4$$

We check our solutions in the original equation as follows:

Check $x = 6$

$$6^2 - 2(6) - 24 \stackrel{?}{=} 0$$
$$36 - 12 - 24 \stackrel{?}{=} 0$$
$$0 = 0$$

Check $x = -4$

$$(-4)^2 - 2(-4) - 24 \stackrel{?}{=} 0$$
$$16 + 8 - 24 \stackrel{?}{=} 0$$
$$0 = 0$$

In both cases the result is a true statement, which means that both 6 and -4 are solutions to the original equation.

To generalize, here are the steps used in solving a quadratic equation by factoring.

HOW TO *Solve an Equation by Factoring*

Step 1: Write the equation in standard form.
Step 2: Factor the left side.
Step 3: Use the zero-factor property to set each factor equal to 0.
Step 4: Solve the resulting linear equations.
Step 5: Check the solutions in the original equation.

EXAMPLE 6 Solve $100x^2 = 300x$.

SOLUTION We begin by writing the equation in standard form and factoring:

$$100x^2 = 300x$$

$$100x^2 - 300x = 0 \qquad \text{Standard form}$$

$$100x(x - 3) = 0 \qquad \text{Factor}$$

Using the zero-factor property to set each factor to 0, we have:

$$100x = 0 \quad \text{or} \quad x - 3 = 0$$

$$x = 0 \quad \text{or} \quad x = 3$$

The two solutions are 0 and 3.

EXAMPLE 7 Solve $(x - 2)(x + 1) = 4$.

SOLUTION We begin by multiplying the two factors on the left side. (Notice that it would be incorrect to set each of the factors on the left side equal to 4. The fact that the product is 4 does not imply that either of the factors must be 4.)

$$(x - 2)(x + 1) = 4$$

$$x^2 - x - 2 = 4 \qquad \text{Multiply the left side}$$

$$x^2 - x - 6 = 0 \qquad \text{Standard form}$$

$$(x - 3)(x + 2) = 0 \qquad \text{Factor}$$

$$x - 3 = 0 \quad \text{or} \quad x + 2 = 0 \qquad \text{Zero-factor property}$$

$$x = 3 \quad \text{or} \quad x = -2$$

Although the next equation is not quadratic, the method we use to solve it is the same as we have been using to solve quadratics.

EXAMPLE 8 Solve for x: $x^3 + 2x^2 - 9x - 18 = 0$

SOLUTION We start with factoring by grouping.

$$x^3 + 2x^2 - 9x - 18 = 0$$

$$x^2(x + 2) - 9(x + 2) = 0$$

$$(x + 2)(x^2 - 9) = 0$$

$$(x + 2)(x - 3)(x + 3) = 0 \qquad \text{The difference of two squares}$$

$$x + 2 = 0 \quad \text{or} \quad x - 3 = 0 \quad \text{or} \quad x + 3 = 0$$

$$x = -2 \quad \text{or} \quad x = 3 \quad \text{or} \quad x = -3$$

We have three solutions: -2, 3, and -3.

Formulas

A *formula* in mathematics is an equation that contains more than one variable. Some formulas are probably already familiar to you. For example, the formula for the area A of a rectangle with length l and width w is $A = lw$.

To begin our work with formulas, we will consider some examples in which we are given numerical replacements for all but one of the variables.

© fotoVoyager/iStockPhoto

EXAMPLE 9 A boat is traveling upstream against a current. If the speed of the boat in still water is r and the speed of the current is c, then the formula for the distance traveled by the boat is $d = (r - c) \cdot t$, where t is the length of time. Find c if $d = 52$ miles, $r = 16$ miles per hour, and $t = 4$ hours.

SOLUTION Substituting 52 for d, 16 for r, and 4 for t into the formula, we have

$$52 = (16 - c) \cdot 4$$

$$13 = 16 - c \qquad \text{Divide each side by 4}$$

$$-3 = -c \qquad \text{Add } -16 \text{ to each side}$$

$$3 = c \qquad \text{Divide each side by } -1$$

The speed of the current is 3 miles per hour.

EXAMPLE 10 If an object is projected into the air with an initial vertical velocity v (in feet/second), its height h (in feet) after t seconds will be given by

$$h = vt - 16t^2$$

Find t if $v = 64$ feet/second and $h = 48$ feet.

SOLUTION Substituting $v = 64$ and $h = 48$ into the formula, we have

$$48 = 64t - 16t^2$$

which is a quadratic equation. We write it in standard form and solve by factoring:

$$16t^2 - 64t + 48 = 0$$

$$t^2 - 4t + 3 = 0 \qquad \text{Divide each side by 16}$$

$$(t - 1)(t - 3) = 0 \qquad \text{Factor}$$

$$t - 1 = 0 \quad \text{or} \quad t - 3 = 0$$

$$t = 1 \quad \text{or} \quad t = 3$$

Here is how we interpret our results: If an object is projected upward with an initial vertical velocity of 64 feet/second, it will be 48 feet above the ground after 1 second and after 3 seconds. That is, it passes 48 feet going up and also coming down.

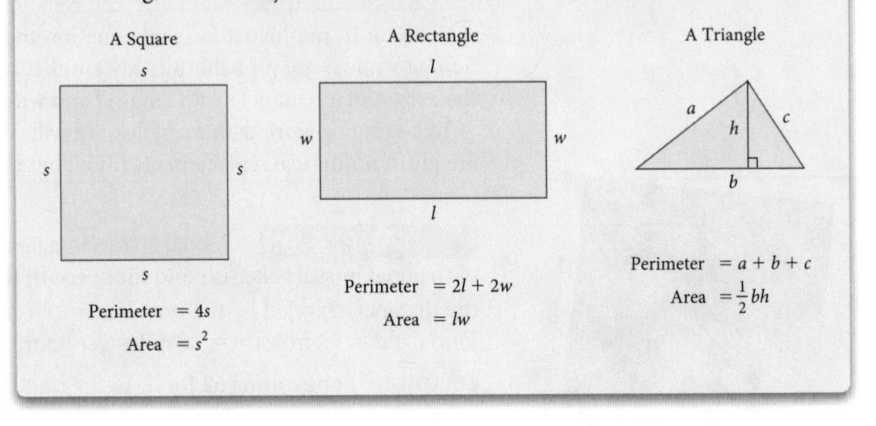

EXAMPLE 11 Given the formula $P = 2w + 2l$, solve for w.

SOLUTION To solve for w, we must isolate it on one side of the equation. We can accomplish this if we delete the $2l$ term and the coefficient 2 from the right side of the equation.

To begin, we add $-2l$ to both sides:

$$P + (-2l) = 2w + 2l + (-2l)$$

$$P - 2l = 2w$$

To delete the 2 from the right side, we can multiply both sides by $\frac{1}{2}$:

$$\frac{1}{2}(P - 2l) = \frac{1}{2}(2w)$$

$$\frac{P - 2l}{2} = w$$

The two formulas

$$P = 2w + 2l \quad \text{and} \quad w = \frac{P - 2l}{2}$$

give the relationship between P, l, and w. They look different, but they both say the same thing about P, l, and w. The first formula gives P in terms of l and w, and the second formula gives w in terms of P and l.

EXAMPLE 12 Solve for x: $ax - 3 = bx + 5$.

SOLUTION In this example, we must begin by collecting all the variable terms on the left side of the equation and all the constant terms on the other side.

$$ax - 3 = bx + 5$$

$$ax - bx - 3 = 5 \qquad \text{Add } -bx \text{ to each side}$$

$$ax - bx = 8 \qquad \text{Add 3 to each side}$$

At this point, we need to apply the distributive property to write the left side as $(a - b)x$. After that, we divide each side by $a - b$:

$$(a - b)x = 8 \quad \text{Distributive property}$$

$$x = \frac{8}{a - b} \quad \text{Divide each side by } a - b$$

Identities and Equations with No Solution

There are two special cases associated with solving linear equations in one variable, which are illustrated in the following examples.

EXAMPLE 13 Solve for x: $2(3x - 4) = 3 + 6x$

SOLUTION Applying the distributive property to the left side gives us

$$6x - 8 = 3 + 6x \quad \text{Distributive property}$$

Now, if we add $-6x$ to each side, we are left with

$$-8 = 3$$

which is a false statement. This means that there is no solution to our equation. Any number we substitute for x in the original equation will lead to a similar false statement.

EXAMPLE 14 Solve for x: $-15 + 3x = 3(x - 5)$

SOLUTION We start by applying the distributive property to the right side.

$$-15 + 3x = 3x - 15 \quad \text{Distributive property}$$

If we add $-3x$ to each side, we are left with the true statement

$$-15 = -15$$

In this case, our result tells us that any number we use in place of x in the original equation will lead to a true statement. Therefore, all real numbers are solutions to our equation. We say the original equation is an *identity* because the left side is always identically equal to the right side.

Modeling: Estimating Temperature Conversions

A good mathematical model is one that can be easily used. In some cases, making the model easy to use requires estimating. The next example is one of those situations.

© Nikada/iStockPhoto

EXAMPLE 15 Mr. McKeague traveled to Buenos Aires with a group of friends. It was a hot day when they arrived. One of the bank kiosks indicated the temperature was 25 °C. Someone asked what that would be on the Fahrenheit scale (the scale they were familiar with), and Budd, one of his friends said, "just multiply by 2 and add 30."

a. What was the temperature in °F according to Budd's approximation?

b. What is the actual temperature in °F?

c. Why does Budd's estimate work?

d. Write a formula for Budd's estimate.

SOLUTION

a. According to Budd, we multiply by 2 and add 30, so

$$2 \cdot 25 + 30 = 50 + 30 = 80\ ^\circ\mathrm{F}$$

b. Using the formula $F = \frac{9}{5}C + 32$, with $C = 25$, we have

$$F = \frac{9}{5}(25) + 32 = 45 + 32 = 77\ ^\circ\mathrm{F}$$

c. Budd's estimate works because $\frac{9}{5}$ is approximately 2 and 30 is close to 32.

d. In symbols, Budd's estimate is $F = 2 \cdot C + 30$.

Modeling: Profit, Revenue, and Cost

The most common formula in business is the formula that gives profit as the difference between revenue and cost. It is a simple formula that gives one number to tell how well a company is doing. Here is the formula:

$$\text{Profit} = \text{Revenue} - \text{Cost}$$

where revenue is the total amount of money a company brings in by selling their product and cost is the total cost to produce the product they sell. Revenue itself can be broken down further by another formula common in the business world. The revenue obtained from selling x items is the product of the number of items sold and the price per item. That is,

$$\text{Revenue} = (\text{Number of items sold})(\text{Price of each item})$$

For example, if 100 items are sold for \$9 each, the revenue is $100(9) = \$900$. Likewise, if 500 items are sold for \$11 each, then the revenue is $500(11) = \$5{,}500$. In general, if x is the number of items sold and p is the selling price of each item, then we can write

$$R = xp$$

EXAMPLE 16 A manufacturer of hard shell cases for the iPhone knows that the number of cases she can sell each week is related to the price of the cases by the equation $x = 1{,}300 - 100p$, where x is the number of cases and p is the price per case. What price should she charge for each case if she wants the weekly revenue to be \$4,000?

SOLUTION The formula for total revenue is $R = xp$. Because we want R in terms of p, we substitute $1{,}300 - 100p$ for x in the equation $R = xp$:

If	$R = xp$
and	$x = 1{,}300 - 100p$
then	$R = (1{,}300 - 100p)p$

We want to find p when R is 4,000. Substituting for R in the formula gives us

$$4{,}000 = (1{,}300 - 100p)p$$
$$4{,}000 = 1{,}300p - 100p^2$$

This is a quadratic equation. To write it in standard form, we add $100p^2$ and $-1{,}300p$ to each side, giving us

$$100p^2 - 1{,}300p + 4{,}000 = 0$$
$$p^2 - 13p + 40 = 0 \qquad \text{Divide each side by 100}$$
$$(p - 5)(p - 8) = 0$$
$$p - 5 = 0 \qquad \text{or} \qquad p - 8 = 0$$
$$p = 5 \qquad \text{or} \qquad p = 8$$

If she sells the cases for \$5 each or for \$8 each, she will have a weekly revenue of \$4,000.

Using Technology 2.1

There are a number of websites that we can use to help us to understand the concepts we are studying and to extend our knowledge of those concepts. One very popular and powerful site is Wolfram|Alpha. Let's use this website to solve one of the equations we have already solved, and then another equation that would be impossible for us to solve. In Example 8 we solved the cubic equation

$$x^3 + 2x^2 - 9x - 18 = 0$$

The solutions were -2, 3, and -3.

EXAMPLE 17 Use Wolfram|Alpha to solve the equation

$$x^3 + 2x^2 - 9x - 18 = 0$$

SOLUTION Online, go to www.wolframalpha.com. Here is what you will see:

Then enter the information below, into the box on the website, exactly as it is shown.

solve x^3+2x^2-9x-18=0

You will see the three solutions we obtained in Example 8, along with a graph of the equation. Here is that graph:

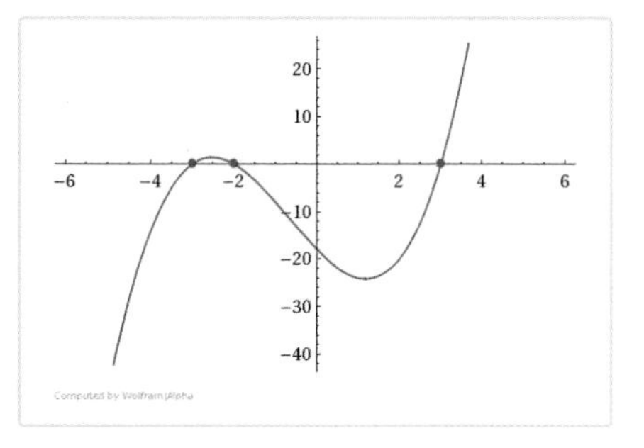

Wolfram Alpha LLC. 2012. Wolfram|Alpha
http://www.wolframalpha.com/
(access June 18, 2012)

We will cover graphing and functions in the next chapter. As you will see when we get there, the solutions to the equation $x^3 + 2x^2 - 9x - 18 = 0$ give us the points where the graph of $y = x^3 + 2x^2 - 9x - 18$ crosses the x-axis.

If we simply change the constant term in our equation to 19, instead of 18, the equation becomes unsolvable for us, but not for Wolfram|Alpha.

EXAMPLE 18 Use Wolfram|Alpha to solve the equation

$$x^3 + 2x^2 - 9x - 19 = 0$$

SOLUTION Online, go to www.wolframalpha.com, and enter the equation as you did for Example 17. If you are still on the website, you can simply edit your previous input, changing 18 to 19. Here are the three solutions we obtain: $x \approx -2.77733$, $x \approx -2.25561$, and $x \approx 3.03293$. We would not be able to find them by factoring, because the left side of our equation is not factorable with the techniques we have.

GETTING READY FOR CLASS

After reading through the preceding section, respond in your own words and in complete sentences.

A. What is the first step in solving the equation $100x^2 = 300x$?
B. How do you use the zero-factor property to help solve a quadratic equation by factoring?
C. What is a formula in mathematics?
D. Give two equivalent forms of the rate equation $d = rt$.

Problem Set 2.1

Solve each equation.

1. $7y - 4 = 2y + 11$
2. $-\frac{2}{5}x + \frac{2}{15} = \frac{2}{3}$
3. $0.14x + 0.08(10{,}000 - x) = 1{,}220$
4. $5(y + 2) - 4(y + 1) = 3$
5. $x^2 - 5x - 6 = 0$
6. $9a^3 = 16a$
7. $(x + 6)(x - 2) = -7$
8. $4x^3 + 12x^2 - 9x - 27 = 0$

9. Solve each equation.
   a. $9x - 25 = 0$
   b. $9x^2 - 25 = 0$
   c. $9x^2 - 25 = 56$
   d. $9x^2 - 25 = 30x - 50$

10. Solve each equation.
    a. $5x - 6 = 0$
    b. $(5x - 6)^2 = 0$
    c. $25x^2 - 36 = 0$
    d. $25x^2 - 36 = 28$

Solve each equation, if possible.

11. $3x - 6 = 3(x + 4)$
12. $4y + 2 - 3y + 5 = 3 + y + 4$
13. $2(4t - 1) + 3 = 5t + 4 + 3t$
14. $7x - 3(x - 2) = -4(5 - x)$
15. $7(x + 2) - 4(2x - 1) = 18 - x$
16. $2x^2 + x - 1 = (2x + 3)(x - 1)$

Now that you have practiced solving a variety of equations, we can turn our attention to the type of equation you will see as you progress through the book. Each equation appears later in the book exactly as you see it below.

Solve each equation.

17. $-\frac{3}{5}a + 2 = 8$
18. $x^3 + 3x^2 - 4x - 12 = 0$
19. $0.07x = 1.4$
20. $50 = \frac{K}{48}$
21. $2x - 3(3x - 5) = -6$
22. $2\left(-\frac{29}{22}\right) - 3y = 4$
23. $12(x + 3) + 12(x - 3) = 3(x^2 - 9)$
24. $3x + (x - 2) \cdot 2 = 6$
25. $2(2x - 3) + 2x = 45$
26. $2x + 1.5(75 - x) = 127.5$
27. $0.12x + 0.10(15{,}000 - x) = 1{,}600$
28. $(x + 2)(x) = 2^3$

Solve each of the following formulas for the indicated variable.

29. $d = rt$ for t
30. $d = (r + c)t$ for r
31. $A = \frac{1}{2}bh$ for b
32. $I = prt$ for r
33. $PV = nRT$ for R
34. $A = P + Prt$ for t
35. $F = \frac{9}{5}C + 32$ for C
36. $h = vt - 16t^2$ for v
37. $A = a + (n - 1)d$ for n
38. $2x - 3y = 6$ for y
39. $-2x - 7y = 14$ for y
40. $7x - 2y - 6 = 0$ for y
41. $ax - 5 = cx - 2$ for x
42. $A = P + Prt$ for P
43. $-3x + 4y = 12$ for y
44. $by - 9 = dy + 3$ for y

Problems 45 through 48 are problems that you will see later in the text. Solve each formula for y.

45. $x = 4y + 1$

46. $y - 1 = \frac{1}{4}(x - 3)$

47. $y + 1 = -\frac{2}{3}(x - 3)$

48. $y - 2 = \frac{1}{3}(x - 1)$

Paying Attention to Instructions The next four problems are intended to give you practice reading, and paying attention to, the instructions that accompany the problems you are working. Working these problems is an excellent way to get ready for a test or a quiz.

49. Work each problem according to the instructions given.

a. Solve: $8x - 5 = 0$

b. Add: $(8x - 5) + (2x - 3)$

c. Multiply: $(8x - 5)(2x - 3)$

d. Solve: $16x^2 - 34x + 15 = 0$

50. Work each problem according to the instructions given.

a. Subtract: $(3x + 5) - (7x - 4)$

b. Solve: $3x + 5 = 7x - 4$

c. Multiply: $(3x + 5)(7x - 4)$

d. Solve: $21x^2 + 23x - 20 = 0$

51. Work each problem according to the instructions given.

a. Solve: $-4x + 5 = 20$

b. Find the value of $-4x + 5$ when x is 3

c. Solve for y: $-4x + 5y = 20$

d. Solve for x: $-4x + 5y = 20$

52. Work each problem according to the instructions given.

a. Solve: $2x + 1 = -4$

b. Find the value of $2x + 1$ when x is 8

c. Solve for y: $2x + y = 20$

d. Solve for x: $2x + y = 20$

Getting Ready for Calculus The problems below are representative of the type of problems you will need to be familiar with in order to be successful in calculus. They are not calculus problems, but are algebra problems that occur in the process of solving calculus problems. These particular problems are taken from *Applied Calculus* by Denny Burzynski and Guy Sanders, published by XYZ Textbooks.

Solve.

53. $(t^2 - 4)(t^2 + 1) = 0$

54. $4t^3 - 6t = 0$

55. $x^3 - 9x = 0$

56. $6x - 6 = 0$

57. $3x^2 - 9 = 0$

58. $3x^2 - 6x = 0$

59. $3x^2 + 9x - 12 = 0$

60. $3x^2 + 30x + 48 = 0$

61. $40 - 26x + 3x^2 = 0$

62. $6 - x - x^2 = x - 2$

63. Use the formula $3x - 4y = 12$ to find y if

a. x is 0

b. x is -2

c. x is 4

d. x is -4

Problems 64 through 70 are problems that you will see later in the text.

64. If $x - 2y = 4$ and $x = \frac{8}{5}$, find y.

65. Let $x = 0$ and $y = 0$ in $y = a(x - 80)^2 + 70$ and solve for a.

66. Find R if $p = 1.5$ and $R = (900 - 300p)p$.

67. Find P if $P = -0.1x^2 + 27x + 1{,}700$ and

 a. $x = 100$ b. $x = 170$

68. Find h if $h = 16 + 32t - 16t^2$ and

 a. $t = \frac{1}{4}$ b. $t = \frac{7}{4}$

69. Use the formula $d = (r - c)t$ to find r if $d = 49$, $c = 4$, and $t = 3.5$.

70. If $d = Kt^2$, find K *if* $t = 2$ and $d = 64$.

Modeling Practice

71. **Revenue** A company manufactures and sells DVDs. The revenue obtained by selling x DVDs is given by the formula

$$R = 11.5x - 0.05x^2$$

Solve the equation below to find the number of DVDs they must sell to receive $650 in revenue.

$$650 = 11.5x - 0.05x^2$$

72. **Price and Revenue** The relationship between the number of pencil sharpeners x a company can sell each week and the price of each sharpener p is given by the equation $x = 1{,}800 - 100p$. At what price should the sharpeners be sold if the weekly revenue is to be $7,200?

73. **Height of a Bullet** A bullet is fired into the air with an initial upward velocity of 80 feet per second from the top of a building 96 feet high. The equation that gives the height of the bullet at any time t is $h = 96 + 80t - 16t^2$. At what times will the bullet be 192 feet in the air?

74. **Height of an Arrow** An arrow is shot into the air with an upward velocity of 48 feet per second from a hill 32 feet high. The equation that gives the height of the arrow at any time t is $h = 32 + 48t - 16t^2$. Find the times at which the arrow will be 64 feet above the ground.

75. **Current** It takes a boat 2 hours to travel 18 miles upstream against the current. If the speed of the boat in still water is 15 miles per hour, what is the speed of the current?

76. **Current** It takes a boat 6.5 hours to travel 117 miles upstream against the current. If the speed of the current is 5 miles per hour, what is the speed of the boat in still water?

© Nicole Waring/iStockPhoto

77. **Wind** An airplane takes 4 hours to travel 864 miles while flying against the wind. If the speed of the airplane on a windless day is 258 miles per hour, what is the speed of the wind?

78. **Wind** A cyclist takes 3 hours to travel 39 miles while pedaling against the wind. If the speed of the wind is 4 miles per hour, how fast would the cyclist be able to travel on a windless day?

79. Budd's Estimate On the flight home from the trip to Argentina mentioned in Example 15, the temperature outside the plane was -45 °C.

a. Use Budd's estimate to find the approximate temperature in degrees Fahrenheit.

b. What is the actual temperature in degrees Fahrenheit?

80. Budd's Estimate On a cold winter day in Argentina the temperature was 0 °C.

a. Use Budd's estimate to find the approximate temperature in degrees Fahrenheit.

b. What is the actual temperature in degrees Fahrenheit?

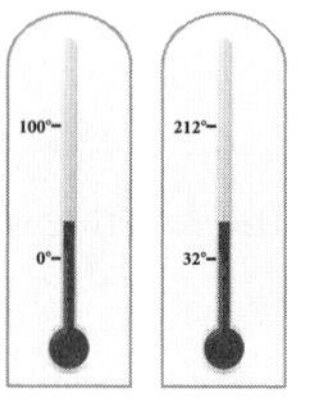

81. Budd's Estimate in Reverse Budd's method of estimating the conversion from degrees Celsius to degrees Fahrenheit is to multiply by two and then add 30. How would Budd estimate the conversion in the other direction, from degrees Fahrenheit to degrees Celsius?

82. Budd's Estimate in Reverse Use your results from Problem 81 to estimate the temperature in degrees Celsius, when the temperature is 68 °F.

Getting Ready for the Next Section

Problems under this heading, "Getting Ready for the Next Section", are problems that you must be able to work in order to understand the material in the next section. In this case, the problems below are variations on the type of problems you have already worked in Chapter 1. They are exactly the type of problems you will see in the explanations and examples in the next section.

Simplify.

83. $49 - 4(6)(-5)$

84. $49 - 4(6)(2)$

85. $(-27)^2 - 4(0.1)(1{,}700)$

86. $25 - 4(4)(-10)$

87. $-7 + \dfrac{169}{12}$

88. $-7 - \dfrac{169}{12}$

Factor.

89. $27t^3 - 8$

90. $125t^3 + 1$

Write as a complex number with i.

91. $2 + \sqrt{-16}$

92. $8+\sqrt{-48}$

Reduce to lowest terms.

93. $\dfrac{6 + 2\sqrt{2}}{2}$

94. $\dfrac{8 + 4i\sqrt{3}}{8}$

More Quadratic Equations

INTRODUCTION Table 1 is taken from the trail map given to skiers at the Northstar California Resort in Lake Tahoe, California. The table gives the length of each chair lift at Northstar, along with the change in elevation from the beginning of the lift to the end of the lift.

Right triangles are good mathematical models for chair lifts. In this section, we will use our knowledge of right triangles, along with the new material developed in the section, to solve problems involving chair lifts and a variety of other examples.

TABLE 1 From the Trail Map for the Northstar California Resort

Lift	Vertical Rise (feet)	Length (feet)
Big Springs Gondola	480	4,100
Bear Paw Double	120	790
Echo Triple	710	4,890
Aspen Express Quad	900	5,100
Forest Double	1,170	5,750
Lookout Double	960	4,330
Comstock Express Quad	1,250	5,900
Rendezvous Triple	650	2,900
Schaffer Camp Triple	1,860	6,150
Chipmunk Tow Lift	28	280
Bear Cub Tow Lift	120	750

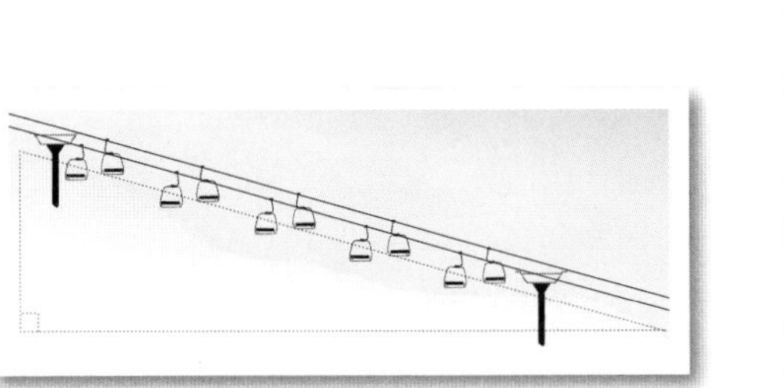

Completing the Square

Consider the equation

$$x^2 = 16$$

We could solve it by writing it in standard form, factoring the left side, and proceeding as we did previously. We can shorten our work considerably, however, if we simply notice that x must be either the positive square root of 16 or the negative square root of 16. That is,

$$\begin{aligned} \text{If} \quad & x^2 = 16 \\ \text{Then} \quad & x = \sqrt{16} \quad \text{or} \quad x = -\sqrt{16} \\ & x = 4 \quad \text{or} \quad x = -4 \end{aligned}$$

We can generalize this result as follows.

PROPERTY *Square Root Property for Equations*

If $a^2 = b$, where b is a real number, then $a = \sqrt{b}$ or $a = -\sqrt{b}$.

Notation The expression $a = \sqrt{b}$ or $a = -\sqrt{b}$ can be written in shorthand form as $a = \pm\sqrt{b}$. The symbol $\pm$ is read "plus or minus."

EXAMPLE 1 Solve $(2x - 3)^2 = 25$.

SOLUTION

$$(2x - 3)^2 = 25$$

$$2x - 3 = \pm\sqrt{25} \qquad \text{Square root property for equations}$$

$$2x - 3 = \pm 5 \qquad \sqrt{25} = 5$$

$$2x = 3 \pm 5 \qquad \text{Add 3 to both sides}$$

$$x = \frac{3 \pm 5}{2} \qquad \text{Divide both sides by 2}$$

The last equation can be written as two separate statements:

$$x = \frac{3 + 5}{2} \quad \text{or} \quad x = \frac{3 - 5}{2}$$

$$= \frac{8}{2} \qquad\qquad = \frac{-2}{2}$$

$$= 4 \quad \text{or} \quad = -1$$

The solution set is $\{4, -1\}$.

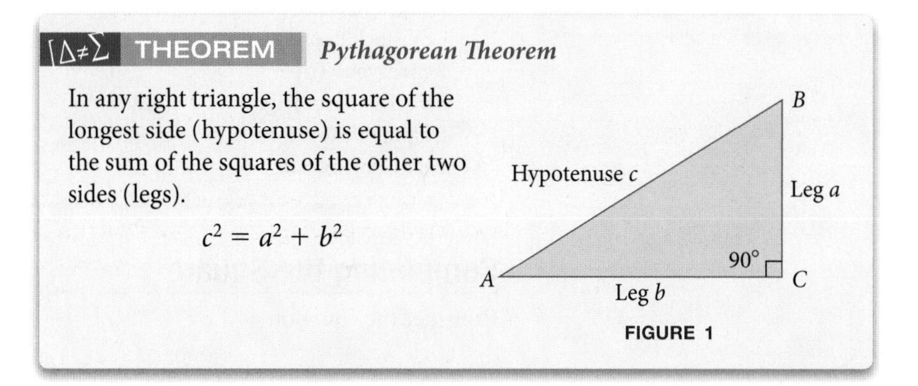

THEOREM *Pythagorean Theorem*

In any right triangle, the square of the longest side (hypotenuse) is equal to the sum of the squares of the other two sides (legs).

$$c^2 = a^2 + b^2$$

FIGURE 1

Modeling: Length of a Chair Lift

EXAMPLE 2 Table 1 in the introduction to this section gives the vertical rise of the Forest Double chair lift as 1,170 feet and the length of the chair lift as 5,750 feet. To the nearest foot, find the horizontal distance covered by a person riding this lift.

SOLUTION Figure 2 is a model of the Forest Double chair lift. A rider gets on the lift at point A and exits at point B. The length of the lift is AB.

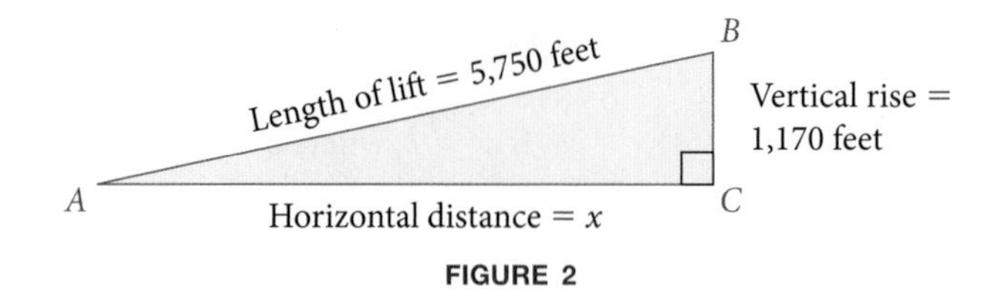

FIGURE 2

To find the horizontal distance covered by a person riding the chair lift, we use the Pythagorean theorem.

$5{,}750^2 = x^2 + 1{,}170^2$	Pythagorean theorem
$33{,}062{,}500 = x^2 + 1{,}368{,}900$	Simplify squares
$x^2 = 33{,}062{,}500 - 1{,}368{,}900$	Solve for x^2
$x^2 = 31{,}693{,}600$	Simplify the right side
$x = \sqrt{31{,}693{,}600}$	Square root property for equations
$= 5{,}630$ feet	to the nearest foot

A rider getting on the lift at point A and riding to point B will cover a horizontal distance of approximately 5,630 feet.

Completing the Square

The method of completing the square is simply a way of transforming any quadratic equation into an equation of the form found in Example 1.

The key to understanding the method of completing the square lies in recognizing the relationship between the last two terms of any perfect square trinomial whose leading coefficient is 1.

Consider the following list of perfect square trinomials and their corresponding binomial squares:

$$x^2 - 6x + 9 = (x - 3)^2$$
$$x^2 + 8x + 16 = (x + 4)^2$$
$$x^2 - 10x + 25 = (x - 5)^2$$
$$x^2 + 12x + 36 = (x + 6)^2$$

In each case, the leading coefficient is 1. A more important observation comes from noticing the relationship between the linear and constant terms (middle and last terms) in each trinomial. Observe that the constant term in each case is the square of half the coefficient of x in the middle term. For example, in the last expression, the constant term 36 is the square of half of 12, where 12 is the coefficient of x in the middle term. (Notice also that the second terms in all the binomials on the right side are half the coefficients of the middle terms of the trinomials on the left side.) We can use these observations to build our own perfect square trinomials and, in doing so, solve more quadratic equations.

EXAMPLE 3 Solve $x^2 - 6x + 5 = 0$ by completing the square.

SOLUTION We begin by adding -5 to both sides of the equation. We want just $x^2 - 6x$ on the left side so that we can add on our own final term to get a perfect square trinomial:

$$x^2 - 6x + 5 = 0$$

$$x^2 - 6x \quad = -5 \qquad \text{Add } -5 \text{ to both sides}$$

Now we can add 9 to both sides and the left side will be a perfect square:

$$x^2 - 6x + 9 = -5 + 9$$

$$(x - 3)^2 = 4$$

Note The equation in Example 3 can be solved quickly by factoring:

$$x^2 - 6x + 5 = 0$$
$$(x - 5)(x - 1) = 0$$
$$x - 5 = 0 \quad \text{or} \quad x - 1 = 0$$
$$x = 5 \quad \text{or} \quad x = 1$$

The reason we didn't solve it by factoring is we want to practice completing the square on some simple equations.

The final line is in the form of the equations we solved previously:

$$x - 3 = \pm 2$$

$$x = 3 \pm 2 \qquad \text{Add 3 to both sides}$$

$$x = 3 + 2 \quad \text{or} \quad x = 3 - 2$$

$$x = 5 \quad \text{or} \quad x = 1$$

The two solutions are 5 and 1.

EXAMPLE 4 Solve by completing the square: $x^2 + 5x - 2 = 0$

SOLUTION We must begin by adding 2 to both sides. (The left side of the equation, as it is, is not a perfect square, because it does not have the correct constant term. We will simply "move" that term to the other side and use our own constant term.)

$$x^2 + 5x = 2 \qquad \text{Add 2 to each side}$$

We complete the square by adding the square of half the coefficient of the linear term to both sides:

$$x^2 + 5x + \frac{25}{4} = 2 + \frac{25}{4} \qquad \text{Half of 5 is } \frac{5}{2}\text{, the square of which is } \frac{25}{4}$$

$$\left(x + \frac{5}{2}\right)^2 = \frac{33}{4} \qquad 2 + \frac{25}{4} = \frac{8}{4} + \frac{25}{4} = \frac{33}{4}$$

$$x + \frac{5}{2} = \pm\sqrt{\frac{33}{4}} \qquad \text{Square root property for equations}$$

$$x + \frac{5}{2} = \pm\frac{\sqrt{33}}{2} \qquad \text{Simplify the radical}$$

$$x = -\frac{5}{2} \pm \frac{\sqrt{33}}{2} \qquad \text{Add } -\frac{5}{2} \text{ to both sides}$$

$$x = \frac{-5 \pm \sqrt{33}}{2}$$

The solution set is $\left\{\frac{-5 + \sqrt{33}}{2}, \frac{-5 - \sqrt{33}}{2}\right\}$.

We can use a calculator to get decimal approximations to these solutions. If $\sqrt{33} \approx 5.74$, then

$$\frac{-5 + 5.74}{2} = 0.37 \quad \text{and} \quad \frac{-5 - 5.74}{2} = -5.37$$

HOW TO ***Solve a Quadratic Equation by Completing the Square***

To summarize the method used in the preceding two examples, we list the following steps:

Step 1: Write the equation in the form $ax^2 + bx = c$.

Step 2: If the leading coefficient is not 1, divide both sides by the coefficient so that the resulting equation has a leading coefficient of 1. That is, if $a \neq 1$, then divide both sides by a.

Step 3: Add the square of half the coefficient of the linear term to both sides of the equation.

Step 4: Write the left side of the equation as the square of a binomial, and simplify the right side if possible.

Step 5: Apply the Square Root Property for Equations, and solve as usual.

The Quadratic Formula

Next, we will use the method of completing the square to derive the quadratic formula. The *quadratic formula* is a very useful tool in mathematics. It allows us to solve all types of quadratic equations.

THEOREM *The Quadratic Theorem*

For any quadratic equation in the form $ax^2 + bx + c = 0$, $a \neq 0$, the two solutions are

$$x = \frac{-b + \sqrt{b^2 - 4ac}}{2a} \quad \text{and} \quad x = \frac{-b - \sqrt{b^2 - 4ac}}{2a}$$

Proof We will prove the quadratic theorem by completing the square on $ax^2 + bx + c = 0$:

$$ax^2 + bx + c = 0$$

$$ax^2 + bx = -c \qquad \text{Add } -c \text{ to both sides}$$

$$x^2 + \frac{b}{a}x = -\frac{c}{a} \qquad \text{Divide both sides by } a$$

To complete the square on the left side, we add the square of $\frac{1}{2}$ of $\frac{b}{a}$ to both sides $\left(\frac{1}{2} \text{ of } \frac{b}{a} \text{ is } \frac{b}{2a}\right)$.

$$x^2 + \frac{b}{a}x + \left(\frac{b}{2a}\right)^2 = -\frac{c}{a} + \left(\frac{b}{2a}\right)^2$$

We now simplify the right side as a separate step. We combine the two terms by writing each with the least common denominator $4a^2$:

$$-\frac{c}{a} + \left(\frac{b}{2a}\right)^2 = -\frac{c}{a} + \frac{b^2}{4a^2} = \frac{4a}{4a}\left(\frac{-c}{a}\right) + \frac{b^2}{4a^2} = \frac{-4ac + b^2}{4a^2}$$

It is convenient to write this last expression as

$$\frac{b^2 - 4ac}{4a^2}$$

Continuing with the proof, we have

$$x^2 + \frac{b}{a}x + \left(\frac{b}{2a}\right)^2 = \frac{b^2 - 4ac}{4a^2}$$

$$\left(x + \frac{b}{2a}\right)^2 = \frac{b^2 - 4ac}{4a^2} \qquad \text{Write left side as a binomial square}$$

$$x + \frac{b}{2a} = \pm\frac{\sqrt{b^2 - 4ac}}{2a} \qquad \text{Square root property for equations}$$

$$x = -\frac{b}{2a} \pm \frac{\sqrt{b^2 - 4ac}}{2a} \qquad \text{Add } -\frac{b}{2a} \text{ to both sides}$$

$$= \frac{-b \pm \sqrt{b^2 - 4ac}}{2a}$$

Our proof is now complete. What we have is this: If our equation is in the form $ax^2 + bx + c = 0$ (standard form), where $a \neq 0$, the two solutions are always given by the formula

$$x = \frac{-b \pm \sqrt{b^2 - 4ac}}{2a}$$

This formula is known as the *quadratic formula*. If we substitute the coefficients a, b, and c of any quadratic equation in standard form into the formula, we need only perform some basic arithmetic to arrive at the solution set.

EXAMPLE 5 Solve $x^2 - 5x - 6 = 0$ by using the quadratic formula.

SOLUTION To use the quadratic formula, we must make sure the equation is in standard form; identify a, b, and c; substitute them into the formula; and work out the arithmetic.

For the equation $x^2 - 5x - 6 = 0$, $a = 1$, $b = -5$, and $c = -6$:

$$\begin{aligned} x &= \frac{-b \pm \sqrt{b^2 - 4ac}}{2a} \\ &= \frac{-(-5) \pm \sqrt{(-5)^2 - 4(1)(-6)}}{2(1)} \\ &= \frac{5 \pm \sqrt{49}}{2} \\ &= \frac{5 \pm 7}{2} \end{aligned}$$

$$x = \frac{5+7}{2} \quad \text{or} \quad x = \frac{5-7}{2}$$

$$x = \frac{12}{2} \qquad x = -\frac{2}{2}$$

$$x = 6 \qquad x = -1$$

The two solutions are 6 and -1.

Note: Whenever the solutions to our quadratic equations turn out to be rational numbers, as in Example 5, it means the original equation could have been solved by factoring. (We didn't solve the equation in Example 5 by factoring because we were trying to get some practice with the quadratic formula.)

EXAMPLE 6 Solve $x^2 - 6x = -7$.

SOLUTION We begin by writing the equation in standard form:

$$x^2 - 6x = -7$$

$$x^2 - 6x + 7 = 0 \qquad \text{Add 7 to each side}$$

Using $a = 1$, $b = -6$, and $c = 7$ in the quadratic formula

$$x = \frac{-b \pm \sqrt{b^2 - 4ac}}{2a}$$

we have:

$$\begin{aligned} x &= \frac{-(-6) \pm \sqrt{(-6)^2 - 4(1)(7)}}{2(1)} \\ &= \frac{6 \pm \sqrt{36 - 28}}{2} \\ &= \frac{6 \pm \sqrt{8}}{2} \\ &= \frac{6 \pm 2\sqrt{2}}{2} \end{aligned}$$

The two terms in the numerator have a 2 in common. We reduce to lowest terms by factoring the 2 from the numerator and then dividing numerator and denominator by 2:

$$\begin{aligned} &= \frac{2(3 \pm \sqrt{2})}{2} \\ &= 3 \pm \sqrt{2} \end{aligned}$$

The two solutions are $3 + \sqrt{2}$ and $3 - \sqrt{2}$. This time, let's check our first solution in the original equation $x^2 - 6x = -7$.

Checking $x = 3 + \sqrt{2}$, we have:

$$(3 + \sqrt{2})^2 - 6(3 + \sqrt{2}) \stackrel{?}{=} -7$$

$$9 + 6\sqrt{2} + 2 - 18 - 6\sqrt{2} \stackrel{?}{=} -7 \qquad \text{Multiply}$$

$$11 - 18 + 6\sqrt{2} - 6\sqrt{2} \stackrel{?}{=} -7 \qquad \text{Add 9 and 2}$$

$$-7 + 0 \stackrel{?}{=} -7 \qquad \text{Subtraction}$$

$$-7 = -7 \qquad \text{A true statement}$$

Our second solution, $x = 3 - \sqrt{2}$, checks in a similar manner.

EXAMPLE 7 Solve for x: $\frac{1}{10}x^2 - \frac{1}{5}x = -\frac{1}{2}$.

SOLUTION It will be easier to apply the quadratic formula if we clear the equation of fractions. Multiplying both sides of the equation by the LCD 10 gives us:

$$x^2 - 2x = -5$$

Next, we add 5 to both sides to put the equation into standard form:

$$x^2 - 2x + 5 = 0 \qquad \text{Add 5 to both sides}$$

Applying the quadratic formula with $a = 1$, $b = -2$, and $c = 5$, we have:

$$x = \frac{-(-2) \pm \sqrt{(-2)^2 - 4(1)(5)}}{2(1)} = \frac{2 \pm \sqrt{-16}}{2} = \frac{2 \pm 4i}{2}$$

Dividing the numerator and denominator by 2, we have the two solutions:

$$x = 1 \pm 2i$$

The two solutions are $1 + 2i$ and $1 - 2i$.

EXAMPLE 8 Solve $(2x - 3)(2x - 1) = -4$.

SOLUTION We multiply the binomials on the left side and then add 4 to each side to write the equation in standard form. From there we identify a, b, and c and apply the quadratic formula:

$$(2x - 3)(2x - 1) = -4$$

$$4x^2 - 8x + 3 = -4 \qquad \text{Multiply binomials on left side}$$

$$4x^2 - 8x + 7 = 0 \qquad \text{Add 4 to each side}$$

Placing $a = 4$, $b = -8$, and $c = 7$ in the quadratic formula we have:

$$x = \frac{-(-8) \pm \sqrt{(-8)^2 - 4(4)(7)}}{2(4)}$$

$$= \frac{8 \pm \sqrt{64 - 112}}{8}$$

$$= \frac{8 \pm \sqrt{-48}}{8}$$

$$= \frac{8 \pm 4i\sqrt{3}}{8} \qquad \sqrt{-48} = i\sqrt{48} = i\sqrt{16}\sqrt{3} = 4i\sqrt{3}$$

> *Note:* It would be a mistake to try to reduce this final expression further. Sometimes first-year algebra students will try to divide the 2 in the denominator into the 2 in the numerator, which is a mistake. Remember, when we reduce to lowest terms, we do so by dividing the numerator and denominator by any factors they have in common. In this case 2 is not a factor of the numerator. This expression is in lowest terms.

To reduce this final expression to lowest terms, we factor a 4 from the numerator and then divide the numerator and denominator by 4:

$$= \frac{4(2 \pm i\sqrt{3})}{4 \cdot 2}$$

$$= \frac{2 \pm i\sqrt{3}}{2}$$

FACTS FROM GEOMETRY ***Two Special Triangles***

The triangles shown in Figures 3 and 4 occur frequently in mathematics.

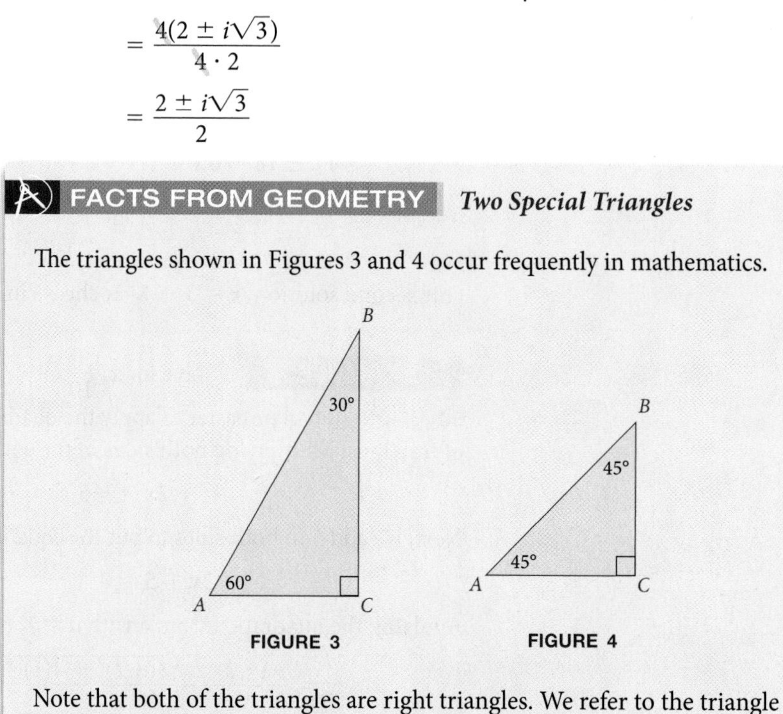

FIGURE 3 FIGURE 4

Note that both of the triangles are right triangles. We refer to the triangle in Figure 3 as a 30°–60°–90° triangle, and the triangle in Figure 4 as a 45°–45°–90° triangle.

EXAMPLE 9 If the shortest side in a 30° – 60° – 90° triangle is 1 inch, find the lengths of the other two sides.

SOLUTION In Figure 5, triangle ABC is a 30° – 60° – 90° triangle in which the shortest side AC is 1 inch long. Triangle DBC is also a 30° – 60° – 90° triangle in which the shortest side DC is 1 inch long.

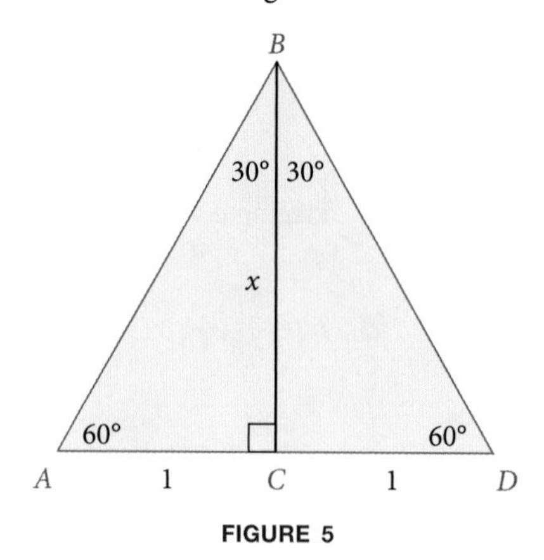

FIGURE 5

Notice that the large triangle ABD is an equilateral triangle because each of its interior angles is 60°. Each side of triangle ABD is 2 inches long. Side AB in triangle ABC is therefore 2 inches. To find the length of side BC, we use the Pythagorean theorem.

$$BC^2 + AC^2 = AB^2$$
$$x^2 + 1^2 = 2^2$$
$$x^2 + 1 = 4$$
$$x^2 = 3$$
$$x = \sqrt{3} \text{ inches}$$

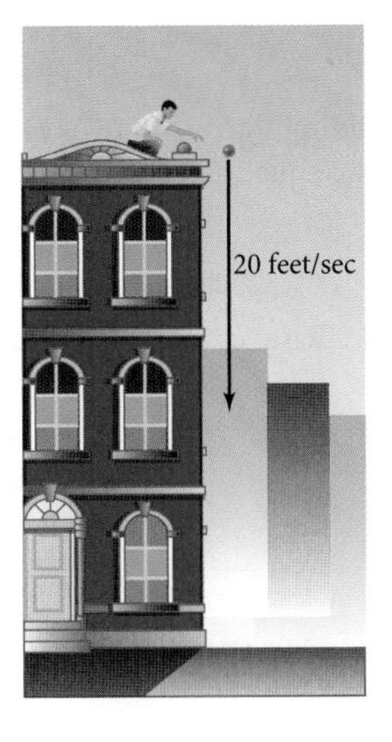

Note that we write only the positive square root because x is the length of a side in a triangle and is therefore a positive number.

Formulas

EXAMPLE 10 If an object is thrown downward with an initial velocity of 20 feet per second, the distance s, in feet, it travels in t seconds is given by the function $s = 20t + 16t^2$. How long does it take the object to fall 40 feet?

SOLUTION We let $s = 40$, and solve for t:

When $\quad s = 40$

the function $\quad s = 20t + 16t^2$

becomes $\quad 40 = 20t + 16t^2$

or $\quad 16t^2 + 20t - 40 = 0$

$\quad 4t^2 + 5t - 10 = 0 \quad$ Divide by 4

Using the quadratic formula, we have

$$t = \frac{-5 \pm \sqrt{25 - 4(4)(-10)}}{2(4)}$$
$$= \frac{-5 \pm \sqrt{185}}{8}$$
$$= \frac{-5 + \sqrt{185}}{8} \quad \text{or} \quad t = \frac{-5 - \sqrt{185}}{8}$$

The second solution is impossible because it is a negative number and time t must be positive. It takes

$$t = \frac{-5 + \sqrt{185}}{8} \quad \text{or approximately} \quad \frac{-5 + 13.60}{8} \approx 1.08 \text{ seconds}$$

for the object to fall 40 feet.

Modeling: Profit, Revenue, and Cost

EXAMPLE 11 A company produces and sells copies of an accounting program for home computers. The total weekly cost (in dollars) to produce x copies of the program is $C = 8x + 500$, and the weekly revenue for selling all x copies of the program is $R = 35x - 0.1x^2$. How many programs must be sold each week for the weekly profit to be \$1,200?

SOLUTION Substituting the given expressions for R and C in the equation $P = R - C$, we have a polynomial in x that represents the weekly profit $P(x)$:

$$\begin{aligned} P &= R - C \\ &= 35x - 0.1x^2 - (8x + 500) \\ &= 35x - 0.1x^2 - 8x - 500 \\ &= -500 + 27x - 0.1x^2 \end{aligned}$$

Setting this expression equal to 1,200, we have a quadratic equation to solve that gives us the number of programs x that need to be sold each week to bring in a profit of $1,200:

$$1{,}200 = -500 + 27x - 0.1x^2$$

We can write this equation in standard form by adding the opposite of each term on the right side of the equation to both sides of the equation. Doing so produces the following equation:

$$0.1x^2 - 27x + 1{,}700 = 0$$

Applying the quadratic formula to this equation with $a = 0.1$, $b = -27$, and $c = 1{,}700$, we have

$$\begin{aligned} x &= \frac{27 \pm \sqrt{(-27)^2 - 4(0.1)(1{,}700)}}{2(0.1)} \\ &= \frac{27 \pm \sqrt{729 - 680}}{0.2} \\ &= \frac{27 \pm \sqrt{49}}{0.2} \\ &= \frac{27 \pm 7}{0.2} \end{aligned}$$

Writing this last expression as two separate expressions, we have our two solutions:

$$x = \frac{27 + 7}{0.2} \quad \text{or} \quad x = \frac{27 - 7}{0.2}$$

$$= \frac{34}{0.2} \qquad\qquad = \frac{20}{0.2}$$

$$= 170 \qquad\qquad = 100$$

The weekly profit will be $1,200 if the company produces and sells 100 programs or 170 programs.

What is interesting about this last example is that it has rational solutions, meaning it could have been solved by factoring. But looking back at the equation, factoring does not seem like a reasonable method of solution because the coefficients are either very large or very small. So, there are times when using the quadratic formula is a faster method of solution, even though the equation you are solving is factorable.

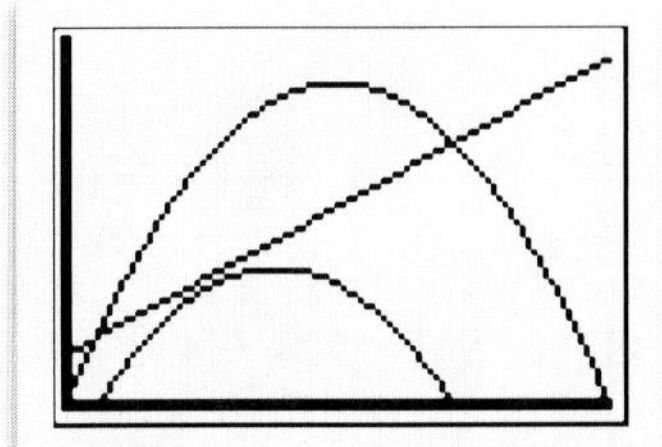

Note: When you use your graphing calculator, you're going to see images like the one above. Because it is sometimes difficult to see details with these pixelated images, we've created images like the one below that are easier to read.

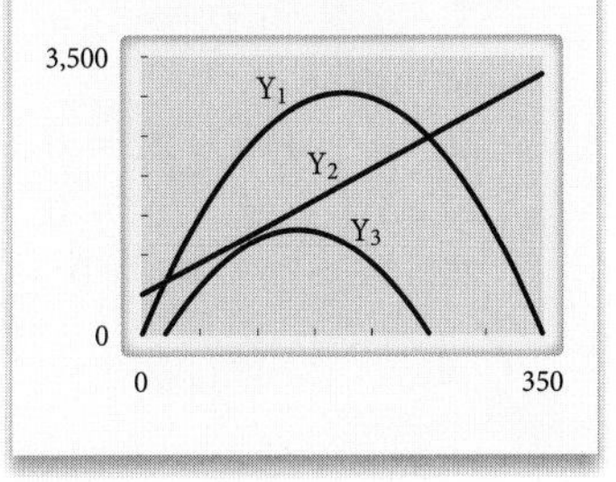

TECHNOLOGY NOTE *Graphing Calculators*

More About Example 10

We can solve the problem discussed in Example 10 by graphing the equation $Y1 = 20X + 16X^2$ in a window with X from 0 to 2 (because X is taking the place of t and we know t is a positive quantity) and Y from 0 to 50 (because we are looking for X when Y1 is 40). Graphing Y1 gives a graph similar to the graph in Figure 6. Using the Zoom and Trace features at $Y1 = 40$ gives us $X = 1.08$ to the nearest hundredth, matching the results we obtained by solving the original equation algebraically.

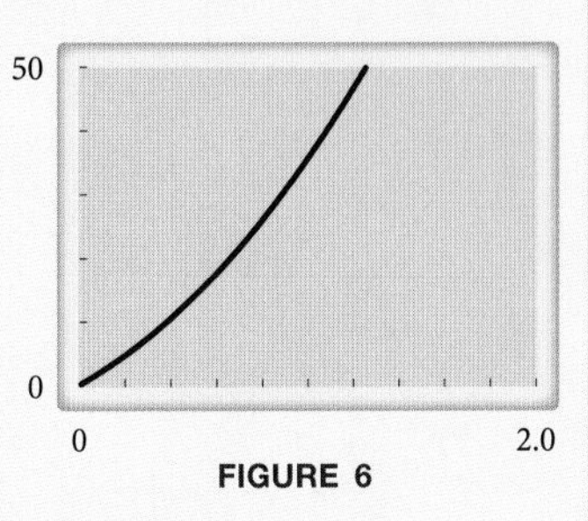

FIGURE 6

More About Example 11

To visualize the equations in Example 11, we set up our calculator this way:

$Y_1 = 35X - .1X^2$ Revenue function

$Y_2 = 8X + 500$ Cost function

$Y_3 = Y_1 - Y_2$ Profit function

Window: X from 0 to 350, Y from 0 to 3,500

Graphing these equations produces graphs similar to the ones shown in Figure 7. The lowest graph is the graph of the profit equation. Using the [ZOOM] and [TRACE] features on the lowest graph at $Y3 = 1{,}200$ produces two corresponding values of X, 170 and 100, which match the results in Example 8.

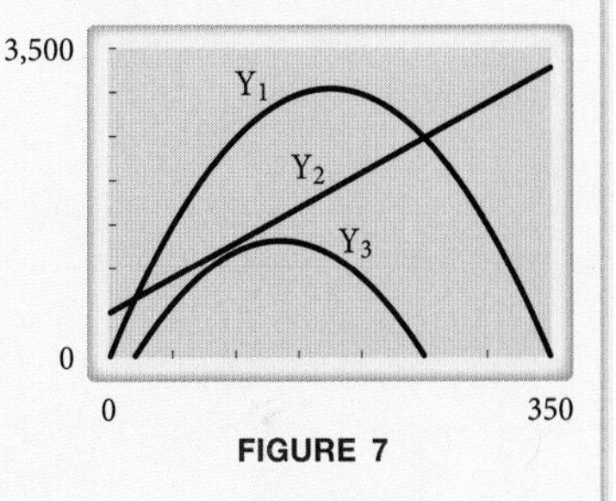

FIGURE 7

We will continue this discussion of the relationship between graphs of functions and solutions to equations in the Using Technology material in the next section.

GETTING READY FOR CLASS

After reading through the preceding section, respond in your own words and in complete sentences.

A. What kind of equation do we solve using the method of completing the square?

B. Explain in words how you would complete the square on $x^2 - 16x = 4$.

C. What is the relationship between the shortest side and the longest side in a 30°–60°–90° triangle?

D. What is the quadratic formula?

Problem Set 2.2

Solve the following equations.

1. $x^2 = 16$ **2.** $a^2 = -49$ **3.** $y^2 = \frac{5}{9}$ **4.** $x^2 + 8 = 0$

5. $9a^2 - 20 = 0$ **6.** $(3y + 7)^2 = 1$ **7.** $(3a - 5)^2 = -49$

8. $(6x - 7)^2 = -75$ **9.** $x^2 - 12x + 36 = -8$ **10.** $9a^2 - 12a + 4 = -9$

Copy each of the following, and fill in the blanks so the left side of each is a perfect square trinomial. That is, complete the square.

11. $x^2 + 6x + \_\_\_ = (x + \_\_\_)^2$ **12.** $x^2 - 2x + \_\_\_ = (x - \_\_\_)^2$

13. $a^2 - 8a + \_\_\_ = (a - \_\_\_)^2$ **14.** $x^2 + 3x + \_\_\_ = (x + \_\_\_)^2$

15. $y^2 - y + \_\_\_ = (y - \_\_\_)^2$ **16.** $x^2 - \frac{3}{4}x + \_\_\_ = (x - \_\_\_)^2$

17. $x^2 - \frac{4}{5}x + \_\_\_ = (x - \_\_\_)^2$

Solve each of the following quadratic equations by completing the square.

18. $x^2 - 2x = 8$ **19.** $x^2 - 6x = 16$ **20.** $a^2 + 10a + 22 = 0$

21. $y^2 + 6y - 1 = 0$ **22.** $x^2 - 5x - 2 = 0$ **23.** $3x^2 - 9x - 12 = 0$

24. $5t^2 + 12t - 1 = 0$ **25.** $7x^2 - 5x + 2 = 0$ **26.** $2x^2 + 6x - 1 = 0$

27. $25x^2 - 20x = 1$ **28.** $4x^2 - 6x + 1 = 0$

29. Is $x = -3 + \sqrt{2}$ a solution to $x^2 - 6x = 7$?

30. Is $x = 2 - \sqrt{5}$ a solution to $x^2 - 4x = 1$?

31. Solve each equation.

a. $5x - 7 = 0$ **b.** $5x - 7 = 8$ **c.** $(5x - 7)^2 = 8$

32. Solve each equation.

a. $5x + 11 = 0$ **b.** $5x + 11 = 9$ **c.** $(5x + 11)^2 = 9$

Solve each equation. Use factoring or the quadratic formula, whichever is appropriate. (Try factoring first. If you have any difficulty factoring, then go right to the quadratic formula.)

33. $x^2 + 5x - 6 = 0$ **34.** $a^2 + 4a + 1 = 0$ **35.** $\frac{1}{6}x^2 + \frac{1}{2}x + \frac{1}{3} = 0$

36. $\frac{x^2}{2} + \frac{2}{3} = -\frac{2x}{3}$ **37.** $2y^2 + 10y = 0$ **38.** $50x^2 - 20x = 0$

39. $\frac{t^2}{3} - \frac{t}{2} = -\frac{3}{2}$ **40.** $0.02x^2 - 0.03x + 0.05 = 0$

41. $2x - 3 = 3x^2$ **42.** $100x^2 - 600x + 900 = 0$

43. $\frac{1}{4}r^2 = \frac{2}{5}r + \frac{1}{10}$ **44.** $(x - 3)(x + 1) = -6$

45. $(x - 4)^2 + (x + 2)(x + 1) = 9$ **46.** $\frac{x^2}{6} + \frac{5}{6} = -\frac{x}{3}$

47. Solve the formula $16t^2 - vt - h = 0$ for t.

48. Solve the formula $16t^2 + vt + h = 0$ for t.

49. Solve the formula $kx^2 + 8x + 4 = 0$ for x.

50. Solve the formula $k^2x^2 + kx + 4 = 0$ for x.

51. Which two of the expressions below are equivalent?

a. $\dfrac{6 + 2\sqrt{3}}{4}$ **b.** $\dfrac{3 + \sqrt{3}}{2}$ **c.** $6 + \dfrac{\sqrt{3}}{2}$

52. Which two of the expressions below are equivalent?

a. $\dfrac{8 - 4\sqrt{2}}{4}$ **b.** $2 - 4\sqrt{2}$ **c.** $2 - \sqrt{2}$

53. Solve $3x^2 - 5x = 0$

a. by factoring **b.** by the quadratic formula

54. Solve $3x^2 + 23x - 70 = 0$

a. by factoring **b.** by the quadratic formula

Modeling Practice

55. Geometry If the shortest side in a 30°–60°–90° triangle is $\frac{1}{2}$ inch long, find the lengths of the other two sides.

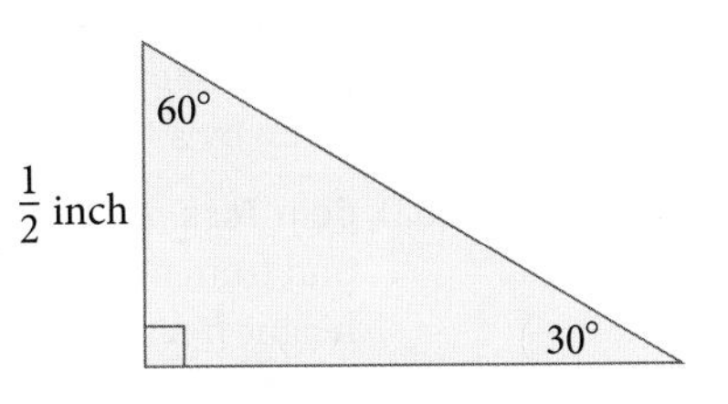

56. Geometry If the length of the shorter sides of a 45°–45°–90° triangle is x, find the length of the hypotenuse, in terms of x.

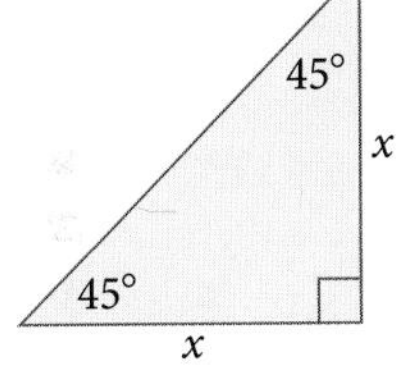

57. Chair Lift Use Table 1 from the introduction to this section to find the horizontal distance covered by a person riding the Bear Paw Double chair lift. Round your answer to the nearest foot.

58. Fermat's Last Theorem As mentioned in a previous chapter, the postage stamp shows Fermat's last theorem, which states that if n is an integer greater than 2, then there are no positive integers x, y, and z that will make the formula $x^n + y^n = z^n$ true. Use the formula $x^n + y^n = z^n$ to

a. find z if $n = 2$, $x = 6$, and $y = 8$. **b.** find y if $n = 2$, $x = 5$, and $z = 13$.

59. **Length of an Escalator** An escalator in a department store is made to carry people a vertical distance of 20 feet between floors. How long is the escalator if it makes an angle of 45° with the ground? (See Figure 8.)

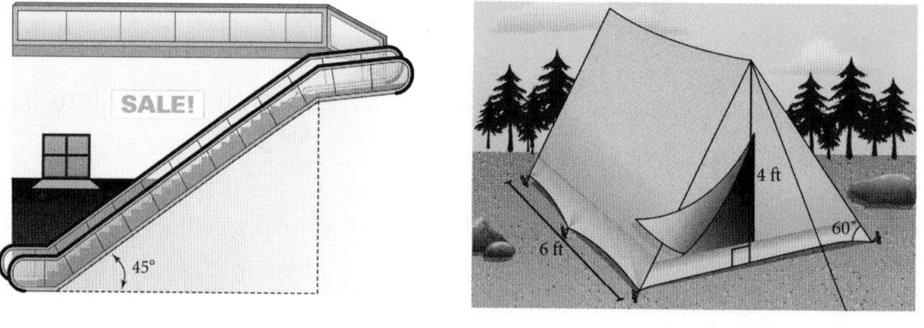

FIGURE 8

FIGURE 9

60. **Dimensions of a Tent** A two-person tent is to be made so the height at the center is 4 feet. If the sides of the tent are to meet the ground at an angle of 60° and the tent is to be 6 feet in length, how many square feet of material will be needed to make the tent? (Figure 9; assume that the tent has a floor and is closed at both ends.) Give your answer to the nearest tenth of a square foot.

61. **Falling Object** An object is thrown downward with an initial velocity of 5 feet per second. The relationship between the distance s it travels and time t is given by $s = 5t + 16t^2$. How long does it take the object to fall 74 feet?

62. **Coin Toss** A coin is tossed upward with an initial velocity of 32 feet per second from a height of 16 feet above the ground. The equation giving the object's height h at any time t is $h = 16 + 32t - 16t^2$. Does the object ever reach a height of 32 feet?

63. **Profit** The total cost (in dollars) for a company to manufacture and sell x items per week is $C = 60x + 300$, whereas the revenue brought in by selling all x items is $R = 100x - 0.5x^2$. How many items must be sold to obtain a weekly profit of $300?

64. **Profit** Suppose a company manufactures and sells x picture frames each month with a total cost of $C = 1{,}200 + 3.5x$ dollars. If the revenue obtained by selling x frames is $R = 9x - 0.002x^2$, find the number of frames it must sell each month if its monthly profit is to be $2,300.

Getting Ready for the Next Section

Factor

65. $x^2 + 4x - 5$ **66.** $27t^3 - 8$ **67.** $4y^2 + 7y - 2$ **68.** $t^2 + 9t + 18$

Solve

69. $3t - 2 = 0$ **70.** $x^3 = \frac{1}{4}$ **71.** $x^2 = -2$ **72.** $5x - 4 = 6$

73. Expand and multiply $(t + 5)^2$ **74.** Expand and multiply $(\sqrt{x} - 3)^2$

Multiply and simplify

75. $8(a - 4) \cdot \frac{6}{a - 4}$ **76.** $3(x - 2) \cdot \frac{x}{x - 2}$ **77.** $(x - 5) \cdot \frac{y - 4}{x - 5}$

Use the quadratic formula to solve

78. $x^2 + 2x + 6 = 0$ **79.** $9t^2 + 6t + 4 = 0$

Additional Equations and Formulas

We are now in a position to put our knowledge of solving equations to work to solve a variety of equations.

VIDEO EXAMPLES

SECTION 2.3

EXAMPLE 1 Solve $(x + 3)^2 - 2(x + 3) - 8 = 0$.

SOLUTION We can see that this equation is quadratic in form by replacing $x + 3$ with another variable, say, y. Replacing $x + 3$ with y we have

$$y^2 - 2y - 8 = 0$$

We can solve this equation by factoring the left side and then setting each factor equal to 0.

$$\begin{aligned}
y^2 - 2y - 8 &= 0 \\
(y - 4)(y + 2) &= 0 && \text{Factor} \\
y - 4 = 0 \quad \text{or} \quad y + 2 &= 0 && \text{Set factors to 0} \\
y = 4 \quad \text{or} \quad y &= -2
\end{aligned}$$

Because our original equation was written in terms of the variable x, we want our solutions in terms of x also. Replacing y with $x + 3$ and then solving for x, we have

$$\begin{aligned}
x + 3 = 4 \quad \text{or} \quad x + 3 &= -2 \\
x = 1 \quad \text{or} \quad x &= -5
\end{aligned}$$

The solutions to our original equation are 1 and -5.

The method we have just shown lends itself well to other types of equations that are quadratic in form, as we will see. In this example, however, there is another method that works just as well. Let's solve our original equation again, but this time, let's begin by expanding $(x + 3)^2$ and $2(x + 3)$.

$$\begin{aligned}
(x + 3)^2 - 2(x + 3) - 8 &= 0 \\
x^2 + 6x + 9 - 2x - 6 - 8 &= 0 && \text{Multiply} \\
x^2 + 4x - 5 &= 0 && \text{Combine similar terms} \\
(x - 1)(x + 5) &= 0 && \text{Factor} \\
x - 1 = 0 \quad \text{or} \quad x + 5 &= 0 && \text{Set factors to 0} \\
x = 1 \quad \text{or} \quad x &= -5
\end{aligned}$$

As you can see, either method produces the same result.

EXAMPLE 2 Solve $27t^3 - 8 = 0$.

SOLUTION It would be a mistake to add 8 to each side of this equation and then take the cube root of each side because we would lose two of our solutions. Instead, we factor the left side, and then set the factors equal to zero:

$$\begin{aligned}
27t^3 - 8 &= 0 && \text{Equation in standard form} \\
(3t - 2)(9t^2 + 6t + 4) &= 0 && \text{Factor as the difference of two cubes} \\
3t - 2 = 0 \quad \text{or} \quad 9t^2 + 6t + 4 &= 0 && \text{Set each factor equal to 0}
\end{aligned}$$

The first equation leads to a solution of $t = \frac{2}{3}$. The second equation does not factor, so we use the quadratic formula with $a = 9$, $b = 6$, and $c = 4$:

$$t = \frac{-6 \pm \sqrt{36 - 4(9)(4)}}{2(9)}$$

$$= \frac{-6 \pm \sqrt{36 - 144}}{18}$$

$$= \frac{-6 \pm \sqrt{-108}}{18}$$

$$= \frac{-6 \pm 6i\sqrt{3}}{18}$$ $\sqrt{-108} = i\sqrt{36 \cdot 3} = 6i\sqrt{3}$

$$= \frac{6(-1 \pm i\sqrt{3})}{6 \cdot 3}$$ *Factor 6 from the numerator and denominator*

$$= \frac{-1 \pm i\sqrt{3}}{3}$$ *Divide out common factor 6*

The three solutions to our original equation are

$$\frac{2}{3}, \quad \frac{-1 + i\sqrt{3}}{3}, \quad \text{and} \quad \frac{-1 - i\sqrt{3}}{3}$$

EXAMPLE 3 Solve $4x^4 + 7x^2 = 2$.

SOLUTION This equation is quadratic in x^2. We can make it easier to look at by using the substitution $y = x^2$. (The choice of the letter y is arbitrary. We could just as easily use the substitution $m = x^2$.) Making the substitution $y = x^2$ and then solving the resulting equation we have

$$4y^2 + 7y = 2$$

$$4y^2 + 7y - 2 = 0$$ *Standard form*

$$(4y - 1)(y + 2) = 0$$ *Factor*

$$4y - 1 = 0 \quad \text{or} \quad y + 2 = 0$$ *Set factors to 0*

$$y = \frac{1}{4} \quad \text{or} \quad y = -2$$

Now we replace y with x^2 to solve for x:

$$x^2 = \frac{1}{4} \quad \text{or} \quad x^2 = -2$$

$$x = \pm\sqrt{\frac{1}{4}} \quad \text{or} \quad x = \pm\sqrt{-2}$$ *Square root property for Equations*

$$x = \pm\frac{1}{2} \quad \text{or} \quad = \pm i\sqrt{2}$$

The solution set is $\left\{\frac{1}{2}, -\frac{1}{2}, i\sqrt{2}, -i\sqrt{2}\right\}$.

Equations with Rational Expressions

The first step in solving an equation that contains one or more rational expressions is to find the LCD for all denominators in the equation. We then multiply both sides of the equation by the LCD to clear the equation of all fractions. That is, after we have multiplied through by the LCD, each term in the resulting equation will have a denominator of 1.

EXAMPLE 4 Solve $\frac{6}{a-4} = \frac{3}{8}$.

SOLUTION The LCD for $a - 4$ and 8 is $8(a - 4)$. Multiplying both sides by this quantity yields

$$8(a-4) \cdot \frac{6}{a-4} = 8(a-4) \cdot \frac{3}{8}$$
$$48 = (a-4) \cdot 3$$
$$48 = 3a - 12$$
$$60 = 3a$$
$$20 = a$$

The solution set is 20, which checks in the original equation.

When we multiply both sides of an equation by an expression containing the variable, we must be sure to check our solutions. The multiplication property of equality does not allow multiplication by 0. If the expression we multiply by contains the variable, then it has the possibility of being 0. In the last example, we multiplied both sides by $8(a - 4)$. This gives a restriction $a \neq 4$ for any solution we come up with.

EXAMPLE 5 Solve $\frac{x}{x-2} + \frac{2}{3} = \frac{2}{x-2}$.

SOLUTION The LCD is $3(x - 2)$. We are assuming $x \neq 2$ when we multiply both sides of the equation by $3(x - 2)$:

$$3(x-2) \cdot \left(\frac{x}{x-2} + \frac{2}{3}\right) = 3(x-2) \cdot \frac{2}{x-2}$$
$$3x + (x-2) \cdot 2 = 3 \cdot 2$$
$$3x + 2x - 4 = 6$$
$$5x - 4 = 6$$
$$5x = 10$$
$$x = 2$$

Note In the process of solving the equation, we multiplied both sides by $3(x - 2)$, solved for x, and got $x = 2$ for our solution. But when x is 2, the quantity $3(x - 2) = 3(2 - 2) = 3(0) = 0$, which means we multiplied both sides of our equation by 0, which is not allowed under the multiplication property of equality.

The only possible solution is $x = 2$. Checking this value back in the original equation gives

$$\frac{2}{2-2} + \frac{2}{3} \stackrel{?}{=} \frac{2}{2-2}$$
$$\frac{2}{0} + \frac{2}{3} \stackrel{?}{=} \frac{2}{0}$$

The first and last terms are undefined. The proposed solution, $x = 2$, does not check in the original equation. The solution set is the empty set. There is no solution to the original equation.

When the proposed solution to an equation is not actually a solution, it is called an *extraneous* solution. In the last example, $x = 2$ is an extraneous solution.

EXAMPLE 6 Solve $\frac{1}{x+2} - \frac{1}{x} = \frac{1}{3}$.

SOLUTION To solve this equation, we must first put it in standard form. To do so, we must clear the equation of fractions by multiplying each side by the LCD for all denominators, which is $3x(x + 2)$. Multiplying both sides by the LCD, we have

$$3x(x+2)\left(\frac{1}{x+2} - \frac{1}{x}\right) = \left(\frac{1}{3}\right)3x(x+2) \quad \text{Multiply each side by the LCD}$$

$$3x(x+2) \cdot \frac{1}{x+2} - 3x(x+2) \cdot \frac{1}{x} = \frac{1}{3} \cdot 3x(x+2)$$

$$3x - 3(x+2) = x(x+2)$$

$$3x - 3x - 6 = x^2 + 2x \quad \text{Multiplication}$$

$$-6 = x^2 + 2x \quad \text{Simplify left side}$$

$$0 = x^2 + 2x + 6 \quad \text{Add 6 to each side}$$

Because the right side of our last equation is not factorable, we use the quadratic formula. From our last equation, we have $a = 1$, $b = 2$, and $c = 6$. Using these numbers for a, b, and c in the quadratic formula gives us

$$x = \frac{-2 \pm \sqrt{4 - 4(1)(6)}}{2(1)}$$

$$= \frac{-2 \pm \sqrt{4 - 24}}{2} \quad \text{Simplify inside the radical}$$

$$= \frac{-2 \pm \sqrt{-20}}{2} \quad 4 - 24 = -20$$

$$= \frac{-2 \pm 2i\sqrt{5}}{2} \quad \sqrt{-20} = i\sqrt{20} = i\sqrt{4}\sqrt{5} = 2i\sqrt{5}$$

$$= \frac{2(-1 \pm i\sqrt{5})}{2} \quad \text{Factor 2 from the numerator}$$

$$= -1 \pm i\sqrt{5} \quad \text{Divide numerator and denominator by 2}$$

Because neither of the two solutions, $-1 + i\sqrt{5}$ nor $-1 - i\sqrt{5}$, will make any of the denominators in our original equation 0, they are both solutions.

EXAMPLE 7 Solve $3 + \frac{1}{x} = \frac{10}{x^2}$.

SOLUTION To clear the equation of denominators, we multiply both sides by x^2:

$$x^2\left(3 + \frac{1}{x}\right) = x^2\left(\frac{10}{x^2}\right)$$

$$3(x^2) + \left(\frac{1}{x}\right)(x^2) = \left(\frac{10}{x^2}\right)(x^2)$$

$$3x^2 + x = 10$$

Rewrite in standard form, and solve:

$$3x^2 + x - 10 = 0$$

$$(3x - 5)(x + 2) = 0$$

$$3x - 5 = 0 \quad \text{or} \quad x + 2 = 0$$

$$x = \frac{5}{3} \quad \text{or} \quad x = -2$$

The solution set is $\left[-2, \frac{5}{3}\right]$. Both solutions check in the original equation. Remember: We have to check all solutions any time we multiply both sides of the equation by an expression that contains the variable, just to be sure we haven't multiplied by 0.

Formulas

EXAMPLE 8 Solve for y: $\dfrac{y-4}{x-5} = 3$

SOLUTION We clear the formula of fractions by multiplying each side of the formula by $x - 5$.

$$\frac{y-4}{x-5} = 3 \qquad \text{Original formula}$$

$$(x-5)\cdot\frac{y-4}{x-5} = 3\cdot(x-5) \qquad \text{Multiply each side by } (x-5)$$

$$y - 4 = 3x - 15 \qquad \text{Simplify each side}$$

$$y = 3x - 11 \qquad \text{Add 4 to each side}$$

EXAMPLE 9 Solve for y: $x = \dfrac{y-4}{y-2}$

SOLUTION To solve for y, we first multiply each side by $y - 2$ to obtain

$$x(y-2) = y - 4$$

$$xy - 2x = y - 4 \qquad \text{Distributive property}$$

$$xy - y = 2x - 4 \qquad \text{Collect all terms containing } y \text{ on the left side}$$

$$y(x-1) = 2x - 4 \qquad \text{Factor } y \text{ from each term on the left side}$$

$$y = \frac{2x-4}{x-1} \qquad \text{Divide each side by } x - 1$$

Modeling: Cycle-Plane and Headwind

EXAMPLE 10 Francine plans a 60-mile training run on her cycle-plane. The time required for the training run, in terms of the wind speed, x, is given by

$$t = \frac{60}{15 - x}$$

If it takes Francine 9 hours to cover 60 miles, what is the speed of the wind?

SOLUTION We can answer this question by solving the equation

$$\frac{60}{15 - x} = 9$$

To start, we multiply each side of the equation by the denominator of the fraction. This will clear the fraction and give us an equivalent equation without fractions. We multiply both sides of the equation by $(15 - x)$ to obtain

© Bob Rhine/NASA Dryden Flight Research Center

$$(15 - x)\frac{60}{15 - x} = 9(15 - x)$$

$$60 = 9(15 - x)$$

From here we proceed as usual.

$$60 = 135 - 9x \quad \text{Apply the distributive law}$$

$$-75 = -9x \quad \text{Subtract 135 from both sides}$$

$$8.\overline{3} = x \quad \text{Divide by } -9$$

The wind speed was $8.\overline{3}$ or $8\frac{1}{3}$ miles per hour.

Equations Involving Radicals

The first step in solving an equation that contains a radical is to eliminate the radical from the equation. To do so, we need an additional property.

PROPERTY ***Squaring Property of Equality***

If both sides of an equation are squared, the solutions to the original equation are solutions to the resulting equation.

We will never lose solutions to our equations by squaring both sides. We may, however, introduce ***extraneous solutions***. Extraneous solutions satisfy the equation obtained by squaring both sides of the original equation, but do not satisfy the original equation.

We know that if two real numbers a and b are equal, then so are their squares:

$$\text{If} \quad a = b$$

$$\text{then} \quad a^2 = b^2$$

On the other hand, extraneous solutions are introduced when we square opposites. That is, even though opposites are not equal, their squares are. For example,

$$5 = -5 \quad \text{A false statement}$$

$$(5)^2 = (-5)^2 \quad \text{Square both sides}$$

$$25 = 25 \quad \text{A true statement}$$

We are free to square both sides of an equation any time it is convenient. We must be aware, however, that doing so may introduce extraneous solutions. We must, therefore, check all our solutions in the original equation if at any time we square both sides of the original equation.

EXAMPLE 11 Solve for x: $\sqrt{3x + 4} = 5$

SOLUTION We square both sides and proceed as usual:

$$\sqrt{3x + 4} = 5$$

$$(\sqrt{3x + 4})^2 = 5^2$$

$$3x + 4 = 25$$

$$3x = 21$$

$$x = 7$$

Checking $x = 7$ in the original equation, we have

$$\sqrt{3(7) + 4} \stackrel{?}{=} 5$$
$$\sqrt{21 + 4} \stackrel{?}{=} 5$$
$$\sqrt{25} \stackrel{?}{=} 5$$
$$5 = 5$$

The solution $x = 7$ satisfies the original equation.

EXAMPLE 12 Solve $\sqrt{5x - 1} + 3 = 7$.

SOLUTION We must isolate the radical on the left side of the equation. If we attempt to square both sides without doing so, the resulting equation will also contain a radical. Adding -3 to both sides, we have

$$\sqrt{5x - 1} + 3 = 7$$
$$\sqrt{5x - 1} = 4$$

We can now square both sides and proceed as usual:

$$(\sqrt{5x - 1})^2 = 4^2$$
$$5x - 1 = 16$$
$$5x = 17$$
$$x = \frac{17}{5}$$

Checking $x = \frac{17}{5}$, we have

$$\sqrt{5\left(\frac{17}{5}\right) - 1} + 3 \stackrel{?}{=} 7$$
$$\sqrt{17 - 1} + 3 \stackrel{?}{=} 7$$
$$\sqrt{16} + 3 \stackrel{?}{=} 7$$
$$4 + 3 \stackrel{?}{=} 7$$
$$7 = 7$$

EXAMPLE 13 Solve $t + 5 = \sqrt{t + 7}$.

SOLUTION This time, squaring both sides of the equation results in a quadratic equation:

$$(t + 5)^2 = (\sqrt{t + 7})^2 \quad \text{Square both sides}$$
$$t^2 + 10t + 25 = t + 7$$
$$t^2 + 9t + 18 = 0 \quad \text{Standard form}$$
$$(t + 3)(t + 6) = 0 \quad \text{Factor the left side}$$
$$t + 3 = 0 \quad \text{or} \quad t + 6 = 0 \quad \text{Set factors equal to 0}$$
$$t = -3 \quad \text{or} \quad t = -6$$

We must check each solution in the original equation:

Check $t = -3$

$$-3 + 5 \stackrel{?}{=} \sqrt{-3 + 7}$$
$$2 \stackrel{?}{=} \sqrt{4}$$
$$2 = 2 \quad \text{A true statement}$$

Check $t = -6$

$$-6 + 5 \stackrel{?}{=} \sqrt{-6 + 7}$$
$$-1 \stackrel{?}{=} \sqrt{1}$$
$$-1 = 1 \quad \text{A false statement}$$

Because $t = -6$ does not check, our only solution is $t = -3$.

EXAMPLE 14 Solve $\sqrt{x - 3} = \sqrt{x} - 3$.

SOLUTION We begin by squaring both sides. Note carefully what happens when we square the right side of the equation, and compare the square of the right side with the square of the left side. You must convince yourself that these results are correct. (The note in the margin will help if you are having trouble convincing yourself that what is written below is true.)

$$(\sqrt{x - 3})^2 = (\sqrt{x} - 3)^2$$
$$x - 3 = x - 6\sqrt{x} + 9$$

Note It is very important that you realize that the square of $(\sqrt{x} - 3)$ is not $x + 9$. Remember, when we square a difference with two terms, we use the formula

$$(a - b)^2 = a^2 - 2ab + b^2$$

Applying this formula to $(\sqrt{x} - 3)^2$ we have

$$(\sqrt{x} - 3)^2 = (\sqrt{x})^2 - 2(\sqrt{x})(3) + 3^2$$
$$= x - 6\sqrt{x} + 9$$

Now we still have a radical in our equation, so we will have to square both sides again. Before we do, though, let's isolate the remaining radical.

$$x - 3 = x - 6\sqrt{x} + 9$$
$$-3 = -6\sqrt{x} + 9 \quad \text{Add } -x \text{ to each side}$$
$$-12 = -6\sqrt{x} \quad \text{Add } -9 \text{ to each side}$$
$$2 = \sqrt{x} \quad \text{Divide each side by } -6$$
$$4 = x \quad \text{Square each side}$$

Our only possible solution is $x = 4$, which we check in our original equation as follows:

$$\sqrt{4 - 3} \stackrel{?}{=} \sqrt{4} - 3$$
$$\sqrt{1} \stackrel{?}{=} 2 - 3$$
$$1 = -1 \quad \text{A false statement}$$

Substituting 4 for x in the original equation yields a false statement. Because 4 was our only possible solution, there is no solution to our equation.

It is also possible to raise both sides of an equation to powers greater than 2. We only need to check for extraneous solutions when we raise both sides of an equation to an even power. Raising both sides of an equation to an odd power will not produce extraneous solutions.

GETTING READY FOR CLASS

After reading through the preceding section, respond in your own words and in complete sentences.

A. Explain how a least common denominator can be used to simplify an equation.
B. What is an extraneous solution?
C. What is the squaring property of equality?
D. Under what conditions do we obtain extraneous solutions to equations that contain radical expressions?

Problem Set 2.3

Solve each of the following equations.

1. $(x + 4)^2 - (x + 4) - 6 = 0$

2. $3(x - 5)^2 + 14(x - 5) - 5 = 0$

3. $x^4 + 2x^2 - 8 = 0$

4. $x^4 - 11x^2 = -30$

5. $(3a - 2)^2 + 2(3a - 2) = 3$

6. $6(2a + 4)^2 = (2a + 4) + 2$

7. $3t^4 = -2t^2 + 8$

8. $25x^4 - 9 = 0$

9. $x^3 - 125 = 0$

10. $x^3 + 1 = 0$

11. $\frac{1}{2a} = \frac{2}{a} - \frac{3}{8}$

12. $\frac{5}{2x} = \frac{2}{x} - \frac{1}{12}$

13. $\frac{2}{x + 5} = \frac{2}{5} - \frac{x}{x + 5}$

14. $2 + \frac{5}{x} = \frac{3}{x^2}$

15. $\frac{y}{2} - \frac{4}{y} = -\frac{7}{2}$

16. $\frac{x + 6}{x + 3} = \frac{3}{x + 3} + 2$

17. $\frac{5}{a + 1} = \frac{4}{a + 2}$

18. $10 - \frac{3}{x^2} = -\frac{1}{x}$

19. $\frac{5}{x - 1} + \frac{2}{x - 1} = \frac{4}{x + 1}$

20. $\frac{2}{x + 5} + \frac{3}{x + 4} = \frac{2x}{x^2 + 9x + 20}$

21. $\frac{2}{x} - \frac{1}{x + 1} = \frac{-2}{5x + 5}$

22. $\frac{t + 3}{t^2 - 2t} = \frac{10}{t^2 - 4}$

23. $\frac{1}{y + 2} - \frac{2}{y - 3} = \frac{-2y}{y^2 - y - 6}$

24. $\frac{1}{a + 3} - \frac{a}{a^2 - 9} = \frac{2}{3 - a}$

25. $\frac{2x - 3}{5x + 10} + \frac{3x - 2}{4x + 8} = 1$

26. $\frac{y + 3}{y^2 - y} - \frac{8}{y^2 - 1} = 0$

27. Solve each equation.

a. $6x - 2 = 0$

b. $\frac{6}{x} - 2 = 0$

c. $\frac{x}{6} - 2 = -\frac{1}{2}$

d. $\frac{6}{x} - 2 = -\frac{1}{2}$

e. $\frac{6}{x^2} + 6 = \frac{20}{x}$

28. Solve each equation.

a. $5x - 2 = 0$

b. $5 - \frac{2}{x} = 0$

c. $\frac{x}{2} - 5 = -\frac{3}{4}$

d. $\frac{2}{x} - 5 = -\frac{3}{4}$

e. $-\frac{3}{x} + \frac{2}{x^2} = 5$

Paying Attention to Instructions The next four problems are intended to give you practice reading, and paying attention to, the instructions that accompany the problems you are working. Working these problems is an excellent way to get ready for a test or a quiz.

29. Work each problem according to the instructions given.

a. Divide: $\frac{6}{x^2 - 2x - 8} \div \frac{x + 3}{x + 2}$

b. Add: $\frac{6}{x^2 - 2x - 8} + \frac{x + 3}{x + 2}$

c. Solve: $\frac{6}{x^2 - 2x - 8} + \frac{x + 3}{x + 2} = 2$

30. Work each problem according to the instructions given.

a. Divide: $\dfrac{-10}{x^2 - 25} \div \dfrac{x - 4}{x - 5}$

b. Add: $\dfrac{-10}{x^2 - 25} + \dfrac{x - 4}{x - 5}$

c. Solve: $\dfrac{-10}{x^2 - 25} + \dfrac{x - 4}{x - 5} = \dfrac{4}{5}$

Solve for y.

31. $\dfrac{y + 1}{x - 0} = 4$ **32.** $\dfrac{y + 2}{x - 4} = -\dfrac{1}{2}$ **33.** $\dfrac{y + 3}{x - 7} = 0$

34. $\dfrac{y - 1}{x - 0} = -3$ **35.** $\dfrac{y - 2}{x - 6} = \dfrac{2}{3}$ **36.** $\dfrac{y - 3}{x - 1} = 0$

37. $\dfrac{x}{8} + \dfrac{y}{2} = 1$ **38.** $\dfrac{x}{7} + \dfrac{y}{9} = 1$ **39.** $\dfrac{x}{5} + \dfrac{y}{-3} = 1$ **40.** $\dfrac{x}{16} + \dfrac{y}{-2} = 1$

41. $x = \dfrac{y - 3}{y - 1}$ **42.** $x = \dfrac{y - 2}{y - 3}$ **43.** $x = \dfrac{2y + 1}{3y + 1}$ **44.** $x = \dfrac{3y + 2}{5y + 1}$

Solve each of the following equations.

45. $\sqrt{3x + 1} = 4$ **46.** $\sqrt{6x + 1} = -5$ **47.** $\sqrt{3y - 1} = 2$

48. $\sqrt{8x + 3} = -6$ **49.** $\sqrt{3x + 1} - 4 = 1$ **50.** $\sqrt{5a - 3} + 6 = 2$

51. $\sqrt[4]{4x + 1} = 3$ **52.** $\sqrt[3]{5x + 7} = 2$ **53.** $\sqrt[3]{2a + 7} = -2$

54. $\sqrt{y + 3} = y - 3$

55. $\sqrt{a + 10} = a - 2$ **56.** $\sqrt{3x + 4} = -\sqrt{2x + 3}$

57. $\sqrt{7a - 1} = \sqrt{2a + 4}$ **58.** $\sqrt[4]{6x + 7} = \sqrt[4]{x + 2}$

59. $x - 1 = \sqrt{6x + 1}$ **60.** $t + 7 = \sqrt{2t + 13}$

The following equations will require that you square both sides twice before all the radicals are eliminated.

61. $\sqrt{x + 3} = \sqrt{x} - 3$ **62.** $\sqrt{x - 1} = \sqrt{x} - 1$

63. $\sqrt{x + 5} = \sqrt{x - 3} + 2$ **64.** $\sqrt{x - 3} - 4 = \sqrt{x} - 3$

65. Solve each equation.

a. $\sqrt{y} - 4 = 6$ **b.** $\sqrt{y - 4} = 6$

c. $\sqrt{y - 4} = -6$ **d.** $\sqrt{y - 4} = y - 6$

66. Solve each equation.

a. $\sqrt{2y} + 15 = 7$ **b.** $\sqrt{2y + 15} = 7$

c. $\sqrt{2y + 15} = y$ **d.** $\sqrt{2y + 15} = y + 6$

Modeling Problems

67. Falling Object An object is tossed into the air with an upward velocity of 8 feet per second from the top of a building h feet high. The time it takes for the object to hit the ground below is given by the formula $16t^2 - 8t - h = 0$. Solve this formula for t.

© Jeremy Edwards/iStockPhoto

68. **Falling Object** An object is tossed into the air with an upward velocity of 6 feet per second from the top of a building h feet high. The time it takes for the object to hit the ground below is given by the formula $16t^2 - 6t - h = 0$. Solve this formula for t.

69. **Saint Louis Arch** The shape of the famous "Gateway to the West" arch in Saint Louis can be modeled by a parabola. The equation for one such parabola is:

$$y = -\frac{1}{150}x^2 + \frac{21}{5}x$$

where x is given in feet.

a. Sketch the graph of the arch's equation on a coordinate axis.

b. Approximately how far do you have to walk to get from one side of the arch to the other?

70. **Pollution Remediation** A small lake in a state park has become polluted by runoff from a factory upstream. The cost for removing p percent of the pollution from the lake is given, in thousands of dollars, by

$$C = \frac{25p}{100 - p}$$

How much of the pollution can be removed for $25,000?

71. **Geometry** From plane geometry and the principle of similar triangles, the relationship between y_1, y_2, and h shown in Figure 1 can be expressed as

$$\frac{1}{h} = \frac{1}{y_1} + \frac{1}{y_2}$$

Two poles are 12 feet high and 8 feet high. If a cable is attached to the top of each one and stretched to the bottom of the other, what is the height above the ground at which the two wires will meet?

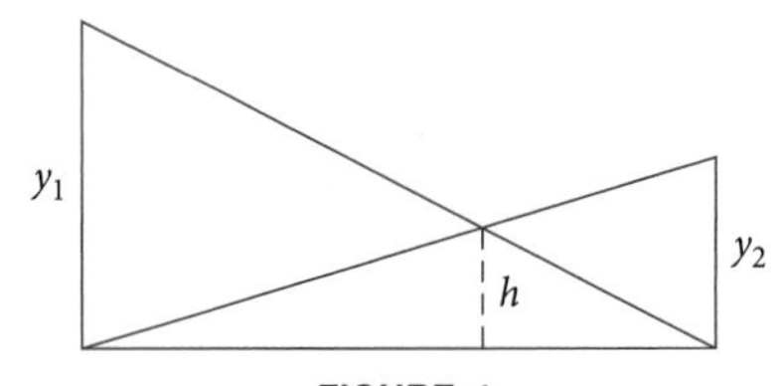

FIGURE 1

72. Kayak Race In a kayak race, the participants must paddle a kayak 450 meters down a river and then return 450 meters up the river to the starting point (Figure 2). Susan has correctly deduced that the total time t (in seconds) depends on the speed c (in meters per second) of the water according to the following expression:

$$t = \frac{450}{v + c} + \frac{450}{v - c}$$

where v is the speed of the kayak relative to the water (the speed of the kayak in still water).

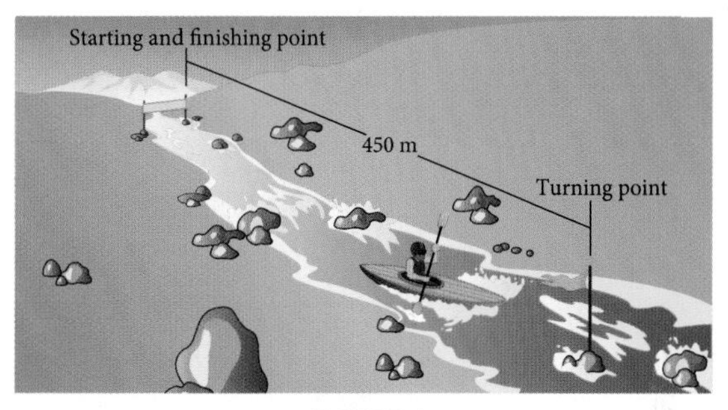

FIGURE 2

a. Fill in the following table.

Time	Speed of Kayak Relative to the Water	Current of the River
t (sec)	v (m/sec)	c (m/sec)
240		1
300		2
	4	3
	3	1
540	3	
	3	3

b. If the kayak race were conducted in the still waters of a lake, do you think that the total time of a given participant would be greater than, equal to, or smaller than the time in the river? Justify your answer.

c. Suppose Peter can paddle his kayak at 4.1 meters per second and that the speed of the current is 4.1 meters per second. What will happen when Peter makes the turn and tries to come back up the river? How does this situation show up in the equation for total time?

Getting Ready for the Next Section

Multiply.

73. $39.3 \cdot 60$

74. $1{,}100 \cdot 60 \cdot 60$

Divide. Round to the nearest tenth, if necessary.

75. $65{,}000 \div 5{,}280$

76. $3{,}960{,}000 \div 5{,}280$

Multiply.

77. $2x\left(\frac{1}{x} + \frac{1}{2x}\right)$

78. $3x\left(\frac{1}{x} + \frac{1}{3x}\right)$

Solve.

79. $12(x + 3) + 12(x - 3) = 3(x^2 - 9)$

80. $40 + 2x = 60 - 3x$

81. $\frac{1}{10} - \frac{1}{12} = \frac{1}{x}$

82. $\frac{1}{x} + \frac{1}{2x} = 2$

More Applications and Modeling 2.4

Stomp Rocket ®

INTRODUCTION The air-powered Stomp Rocket can be propelled over 200 feet using a blast of air. The harder you stomp on the Launch Pad, the farther the rocket flies.

If the rocket is launched straight up into the air with a velocity of 112 feet per second, then the formula

$$h = -16t^2 + 112t$$

gives the height h of the rocket t seconds after it is launched. We can use this formula to find the height of the rocket 3.5 seconds after launch:

When $t = 3.5$,

$$h = -16(3.5)^2 + 112(3.5) = 196$$

This means that after 3.5 seconds, the rocket reaches a height of 196 feet. ■

Models, Applications, or Word Problems?

You may have solved problems similar to the ones you will find in this section in other math classes. In those classes they may have been called *applications*, or *word problems*. Regardless of what we call them, they are all concerned with taking problems written in words and translating them into equations and formulas. Translating events or situations into mathematical structures is the heart of *mathematical modeling* and all the problems here could fall under that category too. These problems are important problems because they connect the world around us with the mathematics we are studying.

You may find that some of the examples and problems are more realistic than others. Because we are just beginning our work with these types of problems, even the ones that seem unrealistic are good practice. The method, or strategy, that we use to solve application problems in this section is called the *Blueprint for Problem Solving*. It is an outline that will overlay the solution process we use on all application problems.

[Δ≠Σ] *Blueprint for Problem Solving*

Step 1: ***Read*** the problem, and then mentally ***list*** the items that are known and the items that are unknown.

Step 2: ***Assign a variable*** to one of the unknown items. (In most cases, this will amount to letting $x =$ the item that is asked for in the problem.) Then ***translate*** the other ***information*** in the problem to expressions involving the variable.

Step 3: ***Reread*** the problem, and then ***write an equation,*** using the items and variable listed in Steps 1 and 2, that describes the situation.

Step 4: ***Solve the equation*** found in Step 3.

Step 5: ***Write your answer*** using a complete sentence.

Step 6: ***Reread*** the problem, and ***check*** your solution with the original words in the problem.

A number of substeps occur within each of the steps in our blueprint. For instance, with Steps 1 and 2 it is always a good idea to draw a diagram or picture if it helps you visualize the relationship among the items in the problem.

EXAMPLE 1 The length of a rectangle is 3 inches less than twice the width. The perimeter is 45 inches. Find the length and width.

SOLUTION When working problems that involve geometric figures, a sketch of the figure helps organize and visualize the problem.

Step 1: ***Read and list.***
Known items: The figure is a rectangle. The length is 3 inches less than twice the width. The perimeter is 45 inches.
Unknown items: The length and the width

Step 2: ***Assign a variable and translate information.***
Because the length is given in terms of the width (the length is 3 less than twice the width), we let x = the width of the rectangle. The length is 3 less than twice the width, so it must be $2x - 3$. The diagram in Figure 1 is a visual description of the relationships we have listed so far.

x

$2x - 3$

FIGURE 1

Step 3: ***Reread and write an equation.***
The equation that describes the situation is

Twice the length	+	twice the width	=	perimeter
$2(2x - 3)$	$+$	$2x$	$=$	45

Step 4: ***Solve the equation.***

$$2(2x - 3) + 2x = 45$$
$$4x - 6 + 2x = 45$$
$$6x - 6 = 45$$
$$6x = 51$$
$$x = 8.5$$

Step 5: ***Write the answer.***
The width is 8.5 inches. The length is $2x - 3 = 2(8.5) - 3 = 14$ inches.

Step 6: ***Reread and check.***
If the length is 14 inches and the width is 8.5 inches, then the perimeter must be $2(14) + 2(8.5) = 28 + 17 = 45$ inches. Also, the length, 14, is 3 less than twice the width.

Remember as you read through the steps in the solutions to the examples in this section that Step 1 is done mentally. Read the problem and then *mentally* list the items that you know and the items that you don't know. The purpose of Step 1 is to give you direction as you begin to work application problems. Finding the solution to an application problem is a process; it doesn't happen all at once. The first step is to read the problem with a purpose in mind. That purpose is to mentally note the items that are known and the items that are unknown.

Suppose we know that the sum of two numbers is 50. If we let x represent one of the two numbers, how can we represent the other? Let's suppose for a moment that x turns out to be 30. Then the other number will be 20, because their sum is 50. That is, if two numbers add up to 50, and one of them is 30, then the other must be $50 - 30 = 20$. Generalizing this to any number x, we see that if two numbers have a sum of 50, and one of the numbers is x, then the other must be $50 - x$. The following table shows some additional examples:

If Two Numbers Have a Sum of	And One of Them Is	Then the Other Must Be
50	x	$50 - x$
10	y	$10 - y$
12	n	$12 - n$

EXAMPLE 2 Suppose a person invests a total of \$10,000 in two accounts. One account earns 5% annually, and the other earns 6% annually. If the total interest earned from both accounts in a year is \$560, how much is invested in each account?

SOLUTION

Step 1: ***Read and list.***

Known items: Two accounts. One pays interest of 5%, and the other pays 6%. The total invested is \$10,000. The total interest earned is \$560.

Unknown items: The number of dollars invested in each individual account

Step 2: ***Assign a variable and translate information.***

If we let $x =$ the amount invested at 6%, then $10{,}000 - x$ is the amount invested at 5%. The total interest earned from both accounts is \$560. The amount of interest earned on x dollars at 6% is $0.06x$, whereas the amount of interest earned on $10{,}000 - x$ dollars at 5% is $0.05(10{,}000 - x)$.

	Dollars at 6%	Dollars at 5%	Total
Number of	x	$10{,}000 - x$	10,000
Interest on	$0.06x$	$0.05(10{,}000) - x$	560

Step 3: ***Reread and write an equation.***

The last line gives us the equation we are after:

$$0.06x + 0.05(10{,}000 - x) = 560$$

Step 4: ***Solve the equation.***

To make the equation a little easier to solve, we begin by multiplying both sides by 100 to move the decimal point two places to the right.

$$6x + 5(10{,}000 - x) = 56{,}000$$

$$6x + 50{,}000 - 5x = 56{,}000$$

$$x + 50{,}000 = 56{,}000$$

$$x = 6{,}000$$

Step 5: ***Write the answer.***
The amount of money invested at 6% is \$6,000. The amount of money invested at 5% is \$10,000 − \$6,000 = \$4,000.

Step 6: ***Reread and check.***
To check our results, we find the total interest from the two accounts:

The interest earned on \$6,000 at 6% is 0.06(6,000) = 360
The interest earned on \$4,000 at 5% is 0.05(4,000) = 200

The total interest = \$560

EXAMPLE 3 The lengths of the three sides of a right triangle are given by three consecutive integers. Find the lengths of the three sides.

SOLUTION

Step 1: ***Read and list.***
Known items: A right triangle. The three sides are three consecutive integers.
Unknown items: The three sides.

Step 2: ***Assign a variable and translate information.***
Let x = First integer (shortest side)
Then $x + 1$ = Next consecutive integer
$x + 2$ = Last consecutive integer (longest side)

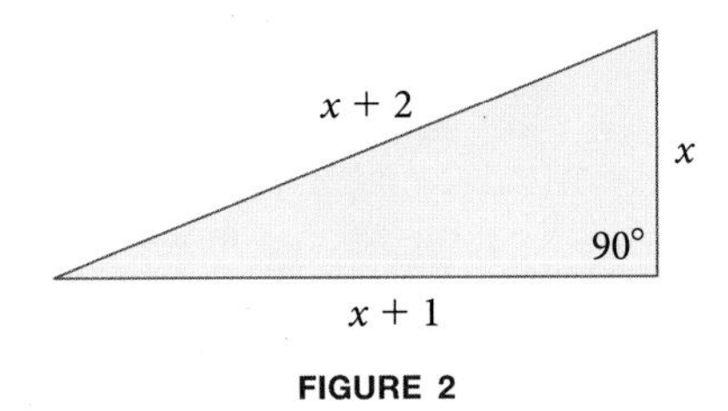

FIGURE 2

Step 3: ***Reread and write an equation.*** By the Pythagorean theorem, we have

$$(x + 2)^2 = (x + 1)^2 + x^2$$

Step 4: ***Solve the equation.***

$$x^2 + 4x + 4 = x^2 + 2x + 1 + x^2$$
$$x^2 - 2x - 3 = 0$$
$$(x - 3)(x + 1) = 0$$
$$x = 3 \quad \text{or} \quad x = -1$$

Step 5: ***Write the answer.*** Because x is the length of a side in a triangle, it must be a positive number. Therefore, $x = -1$ cannot be used. The shortest side is 3. The other two sides are 4 and 5.

Step 6: ***Reread and check.*** The three sides are given by consecutive integers. The square of the longest side is equal to the sum of the squares of the two shorter sides ($5^2 = 3^2 + 4^2$).

EXAMPLE 4 Two families from the same neighborhood plan a ski trip together. The first family is driving a newer vehicle and makes the 455-mile trip at a speed 5 miles per hour faster than the second family who is traveling in an older vehicle. The second family takes a half-hour longer to make the trip. What are the speeds of the two families?

SOLUTION The following table will be helpful in finding the equation necessary to solve this problem.

	d(distance)	r(rate)	t(time)
First Family			
Second Family			

If we let x be the speed of the second family, then the speed of the first family will be $x + 5$. Both families travel the same distance of 455 miles. Putting this information into the table we have

	d	r	t
First Family	455	$x + 5$	
Second Family	455	x	

To fill in the last two spaces in the table, we use the relationship $d = r \cdot t$. Since the last column of the table is the time, we solve the equation $d = r \cdot t$ for t and get

$$t = \frac{d}{r}$$

Taking the distance and dividing by the rate (speed) for each family, we complete the table.

	d	r	t
First Family	455	$x + 5$	$\frac{455}{x+5}$
Second Family	455	x	$\frac{455}{x}$

Reading the problem again, we find that the time for the second family is longer than the time for the first family by one-half hour. In other words, the time for the second family can be found by adding one-half hour to the time for the first family, or

$$\frac{455}{x+5} + \frac{1}{2} = \frac{455}{x}$$

Multiplying both sides by the LCD of $2x(x + 5)$ gives

$$2x \cdot (455) + x(x + 5) \cdot 1 = 455 \cdot 2(x + 5)$$

$$910x + x^2 + 5x = 910x + 4{,}550$$

$$x^2 + 5x - 4{,}550 = 0$$

$$(x + 70)(x - 65) = 0$$

$$x = -70 \quad \text{or} \quad x = 65$$

Since we cannot have a negative speed, the only solution is $x = 65$. Then

$$x + 5 = 65 + 5 = 70$$

The speed of the first family is 70 miles per hour, and the speed of the second family is 65 miles per hour.

EXAMPLE 5 An inlet pipe can fill a pool in 10 hours, while the drain can empty it in 12 hours. If the pool is empty and both the inlet pipe and drain are open, how long will it take to fill the pool?

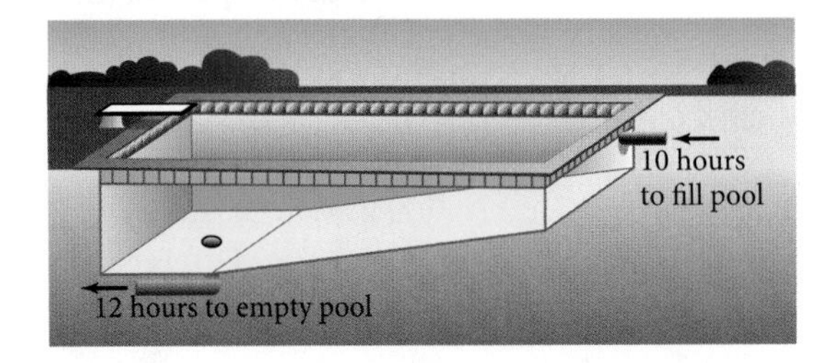

SOLUTION It is helpful to think in terms of how much work is done by each pipe in 1 hour.

Let x = the time it takes to fill the pool with both pipes open.

If the inlet pipe can fill the pool in 10 hours, then in 1 hour it is $\frac{1}{10}$ full. If the outlet pipe empties the pool in 12 hours, then in 1 hour it is $\frac{1}{12}$ empty. If the pool can be filled in x hours with both the inlet pipe and the drain open, then in 1 hour it is $\frac{1}{x}$ full when both pipes are open.

Here is the equation:

In 1 hour

$$\begin{bmatrix}\text{Amount filled}\\ \text{by inlet pipe}\end{bmatrix} - \begin{bmatrix}\text{Amount emptied}\\ \text{by the drain}\end{bmatrix} = \begin{bmatrix}\text{Fraction of pool}\\ \text{filled with both pipes}\end{bmatrix}$$

$$\frac{1}{10} - \frac{1}{12} = \frac{1}{x}$$

Multiplying through by $60x$, we have

$$60x \cdot \frac{1}{10} - 60x \cdot \frac{1}{12} = 60x \cdot \frac{1}{x}$$

$$6x - 5x = 60$$

$$x = 60$$

It takes 60 hours to fill the pool if both the inlet pipe and the drain are open.

Table Building

EXAMPLE 6 A piece of string 12 inches long is to be formed into a rectangle. Build a table that gives the length of the rectangle if the width is 1, 2, 3, 4, or 5 inches. Then find the area of each of the rectangles formed.

SOLUTION Because the formula for the perimeter of a rectangle is $P = 2l + 2w$, and our piece of string is 12 inches long, the formula we will use to find the lengths for the given widths is $12 = 2l + 2w$. To solve this formula for l, we divide each side by 2 and then subtract w. The result is $l = 6 - w$. Table 1 organizes our work so that the formula we use to find l for a given value of w is shown, and we have added a last column to give us the areas of the rectangles formed. The units for the first three columns are inches, and the units for the numbers in the last column are square inches.

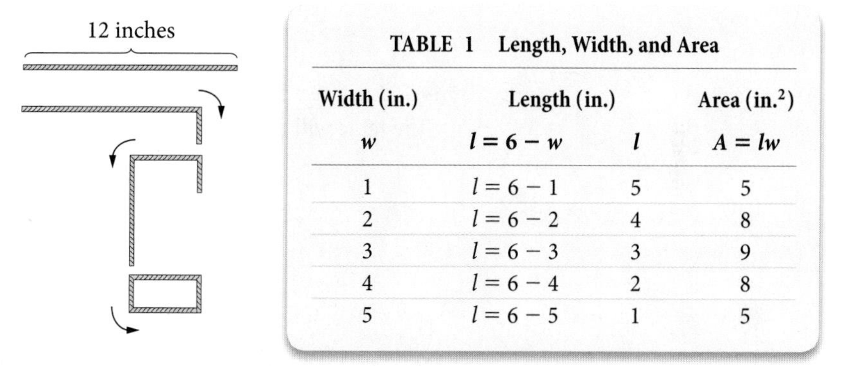

TABLE 1 Length, Width, and Area

Width (in.)	Length (in.)		Area (in.^{2})
w	$l = 6 - w$	l	$A = lw$
1	$l = 6 - 1$	5	5
2	$l = 6 - 2$	4	8
3	$l = 6 - 3$	3	9
4	$l = 6 - 4$	2	8
5	$l = 6 - 5$	1	5

Figures 3 and 4 show two *bar charts* constructed from the information in Table 1.

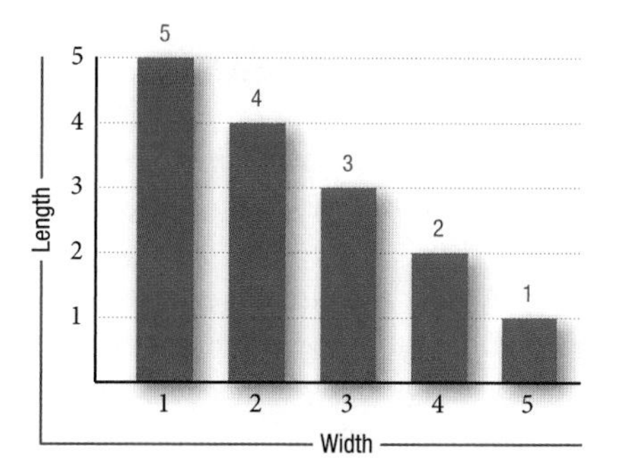

FIGURE 3 *Length and width of rectangles with perimeters fixed at 12 inches*

FIGURE 4 *Area and width of rectangles with perimeters fixed at 12 inches*

TECHNOLOGY NOTE *Graphing Calculators*

A number of graphing calculators have table-building capabilities. We can let the calculator variable X represent the widths of the rectangles in Example 6. To find the lengths, we set variable Y1 equal to 6 − X. The area of each rectangle can be found by setting variable Y2 equal to X * Y1. To have the calculator produce the table automatically, we use a table minimum of 0 and a table increment of 1. Here is a summary of how the graphing calculator is set up:

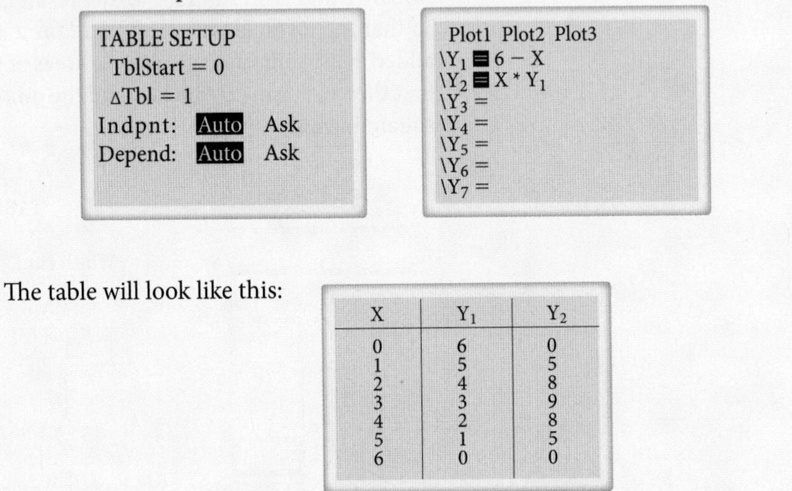

The table will look like this:

X	Y_1	Y_2
0	6	0
1	5	5
2	4	8
3	3	9
4	2	8
5	1	5
6	0	0

GETTING READY FOR CLASS

After reading through the preceding section, respond in your own words and in complete sentences.

A. What is the first step in solving an application problem?

B. What is the biggest obstacle between you and success in solving application problems?

C. Write an application problem for which the solution depends on solving the equation $2x + 2 \cdot 3 = 18$.

D. What is the last step in solving an application problem? Why is this step important?

Problem Set 2.4

Solve each application problem. Be sure to follow the steps in the Blueprint for Problem Solving.

Geometry Problems

1. **Triangle** The longest side of a triangle is two times the shortest side, while the medium side is 3 meters more than the shortest side. The perimeter is 27 meters. Find the dimensions.

2. **Rectangle** The length of a rectangle is 1 foot more than twice the width. The perimeter is 20 feet. Find the dimensions.

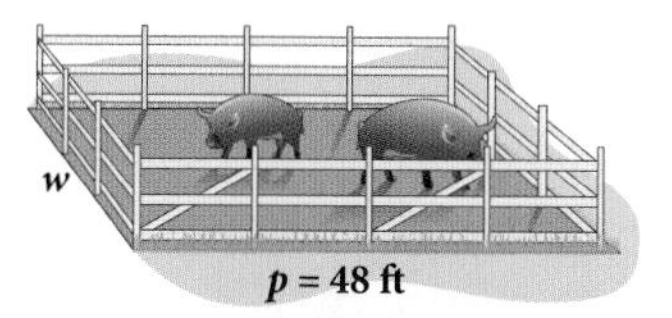

3. **Livestock Pen** A livestock pen is built in the shape of a rectangle that is twice as long as it is wide. The perimeter is 48 feet. If the material used to build the pen is \$1.75 per foot for the longer sides and \$2.25 per foot for the shorter sides (the shorter sides have gates, which increase the cost per foot), find the cost to build the pen.

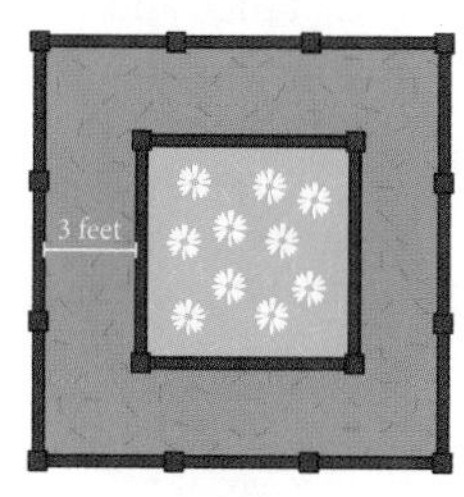

FIGURE 5

4. **Garden** A garden is in the shape of a square with a perimeter of 42 feet. The garden is surrounded by two fences. One fence is around the perimeter of the garden, whereas the second fence is 3 feet from the first fence, as Figure 5 indicates. If the material used to build the two fences is \$1.28 per foot, what was the total cost of the fences?

5. **Right Triangle** A 25-foot ladder is leaning against a building. The base of the ladder is 7 feet from the side of the building. How high does the ladder reach along the side of the building?

6. **Right Triangle** The lengths of the three sides of a right triangle are given by three consecutive even integers. Find the lengths of the three sides.

7. **Rectangle** The length of a rectangle is 4 yards more than twice the width. If the area is 70 square yards, find the width and the length.

8. **Triangle** The height of a triangle is 4 feet less than twice the base. If the area is 48 square feet, find the base and the height.

Interest Problems

9. **Investing** A man invests \$12,000 in two accounts. If one account pays 10% per year and the other pays 7% per year, how much was invested in each account if the total interest earned in the first year was \$960?

	Dollars at 10%	Dollars at 7%	Total
Number of			
Interest on			

10. **Investing** A total of \$11,000 is invested in two accounts. One of the two accounts pays 9% per year, and the other account pays 11% per year. If the total interest paid in the first year is \$1,150, how much was invested in each account?

11. Investing Travis has a total of \$6,000 invested in two accounts. The total amount of interest he earns from the accounts in the first year is \$410. If one account pays 6% per year and the other pays 8% per year, how much did he invest in each account?

Distance, Rate and Time Problems

12. Distance Between Planes Two airplanes leave from an airport at the same time. One travels due south at a speed of 480 miles per hour, and the other travels due west at a speed of 360 miles per hour. How long until the distance between the two airplanes is 2,400 miles?

13. Speed of a Current A boat, which moves at 18 miles per hour in still water, travels 14 miles downstream in the same amount of time it takes to travel 10 miles upstream. Find the speed of the current.

a. Let x be the speed of the current. Complete the distance and rate columns in the table.

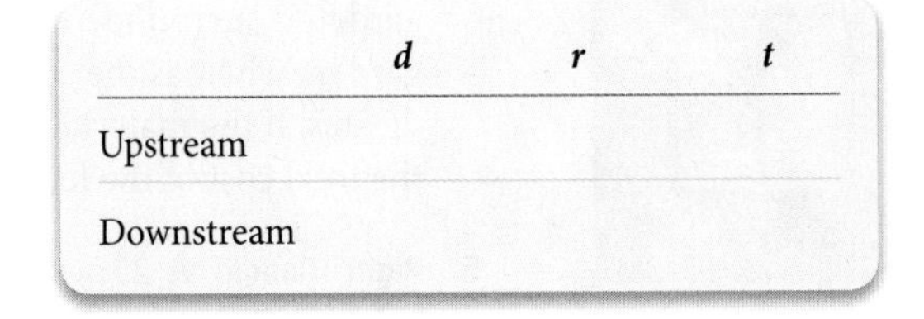

	d	r	t
Upstream			
Downstream			

b. Now use the distance and rate information to complete the time column.

c. What does the problem tell us about the two times? Use this fact to write an equation involving the two expressions for time.

d. Solve the equation. Write your answer as a complete sentence.

14. Speed of a Current A motorboat travels at 4 miles per hour in still water. It goes 12 miles upstream and 12 miles back again in a total of 8 hours. Find the speed of the current of the river.

15. Speeds of a Train and a Car A train travels 30 miles per hour faster than a car. If the train covers 120 miles in the same time the car covers 80 miles, what are the speeds of each of them?

a. Let x be the speed of the car. Complete the distance and rate columns in the table.

	d	r	t
Car			
Train			

b. Now use the distance and rate information to complete the time column.

c. What does the problem tell us about the two times? Use this fact to write an equation involving the two expressions for time.

d. Solve the equation. Write your answer as a complete sentence.

© Dmitry Naumov/iStockPhoto

16. **Car vs. Bicycle** Lou leaves for a cross-country excursion on a bicycle traveling at 20 miles per hour. His friends are driving the trip and will meet him at several rest stops along the way. The first stop is scheduled 30 miles from the original starting point. If the people driving leave 15 minutes after Lou from the same place, how fast will they have to drive to reach the first rest stop at the same time as Lou?

17. **Speed of a Truck** A bakery delivery truck leaves the bakery at 5:00 A.M. each morning on its 140-mile route. One day the driver gets a late start and does not leave the bakery until 5:30 A.M. To finish her route on time the driver drives 5 miles per hour faster than usual. At what speed does she usually drive?

Work Problems

18. **Filling a Water Tank** A water tank can be filled by an inlet pipe in 8 hours. It takes twice that long for the outlet pipe to empty the tank. How long will it take to fill the tank if both pipes are open?

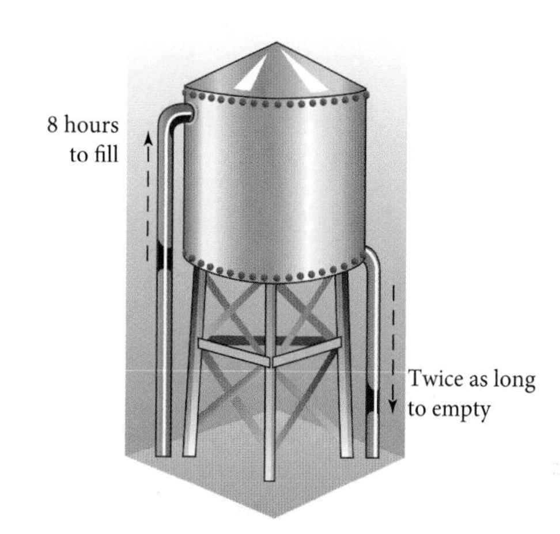

19. **Filling a Pool** It takes 10 hours to fill a pool with the inlet pipe. It can be emptied in 15 hours with the outlet pipe. If the pool is half full to begin with, how long will it take to fill it from there if both pipes are open?

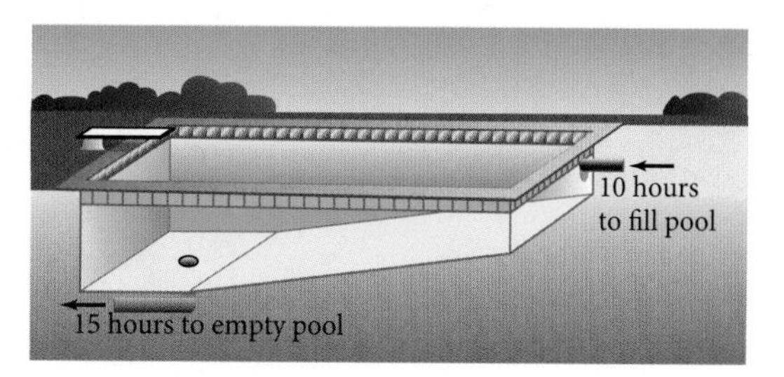

20. **Filling a Water Tank** A water tank is being filled by two inlet pipes. Pipe A can fill the tank in $4\frac{1}{2}$ hours, but both pipes together can fill the tank in 2 hours. How long does it take to fill the tank using only pipe B?

Table Building

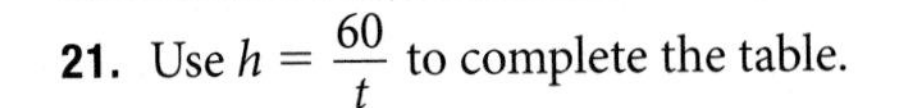

21. Use $h = \dfrac{60}{t}$ to complete the table.

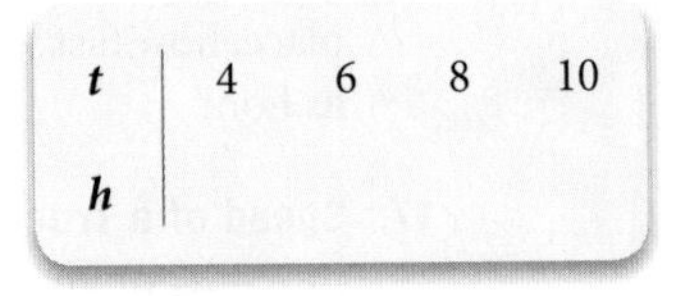

t	4	6	8	10
h				

22. Model Rocket A small rocket is projected straight up into the air with a velocity of 128 feet per second. The formula that gives the height h of the rocket t seconds after it is launched is

$$h = -16t^2 + 128t$$

Use this formula to find the height of the rocket after 1, 2, 3, 4, 5, and 6 seconds.

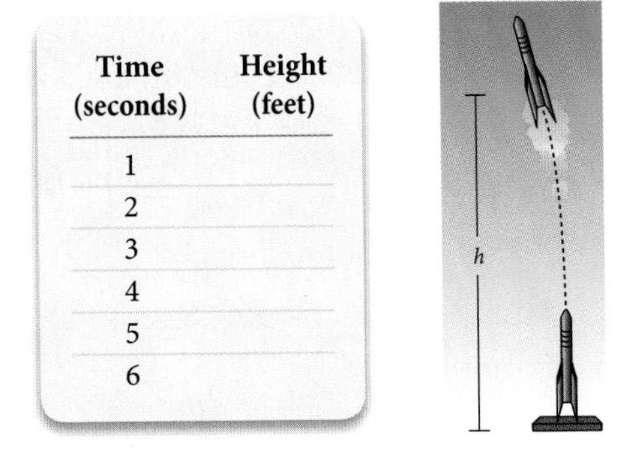

Time (seconds)	Height (feet)
1	
2	
3	
4	
5	
6	

23. Speed of a Bullet To determine the average speed of a bullet when fired from a rifle, the time is measured from when the gun is fired until the bullet hits a target that is 1,000 feet away. Use the formula $r = \frac{d}{t}$ with $d = 1{,}000$ feet to complete the following table.

Time (seconds)	Rate (feet per second)
1.00	
0.80	
0.64	
0.50	
0.40	
0.32	

24. Effects of Wind A plane that can travel 300 miles per hour in still air is traveling in a wind stream with a speed of 20 miles per hour. The distance the plane will travel against the wind is given by the formula $d = (r - w) \cdot t$, and the distance it will travel with the wind is given by the formula $d = (r + w) \cdot t$. Use these formulas with $r = 300$ and $w = 20$ to complete the following table.

Time (hours)	Distance against the Wind (miles)	Distance with the Wind (miles)
0.50		
1.00		
1.50		
2.00		
2.50		
3.00		

Maximum Heart Rate In exercise physiology, a person's maximum heart rate, in beats per minute, is found by subtracting his age in years from 220. So, if A represents your age in years, then your maximum heart rate is

$$M = 220 - A$$

Use this formula to complete the following tables.

25.

Age (years)	Maximum Heart Rate (beats per minute)
18	
19	
20	
21	
22	
23	

26.

Age (years)	Maximum Heart Rate (beats per minute)
15	
20	
25	
30	
35	
40	

Training Heart Rate A person's training heart rate, in beats per minute, is his resting heart rate plus 60% of the difference between his maximum heart rate and his resting heart rate. If resting heart rate is R and maximum heart rate is M, then the formula that gives training heart rate is

$$T = R + 0.6(M - R)$$

Use this formula along with the results of Problems 25 and 26 to fill in the following two tables.

27. For a 20-year-old person

Resting Heart Rate (beats per minute)	Training Heart Rate (beats per minute)
60	
62	
64	
68	
70	
72	

28. For a 40-year-old person

Resting Heart Rate (beats per minute)	Training Heart Rate (beats per minute)
60	
62	
64	
68	
70	
72	

Getting Ready for the Next Section

Solve.

29. $-2x - 3 = 7$

30. $3x + 3 = 2x - 1$

31. $3(2x - 4) - 7x = -3x$

32. $3(2x + 5) = -3x$

33. $(x - 3)(x + 2) = 0$

34. $x^2 - 2x - 8 = 0$

35. $6x^2 - x = 2$

36. $x^2 - 6x + 9 = 0$

Linear, Quadratic, and Rational Inequalities

INTRODUCTION The ʻApapane, a native Hawaiian bird, feeds mainly on the nectar of the ʻohiʻa lehua blossom, although the adult diet also includes insects and spiders. The ʻApapane live in high altitude regions where the ʻohiʻa blossoms are found and where the birds are protected from mosquitoes which transmit avian malaria and avian pox. Predators of the ʻApapane include the rat, feral cat, mongoose, and owl. According to the U.S. Geological Survey:

> Annual survival rates based on 1,584 recaptures of 429 banded individuals:
> 0.72 ± 0.11 for adults and 0.13 ± 0.07 for juveniles.

We can write the survival rate for the adults as an inequality:

$$0.61 \le r \le 0.83$$

Inequalities are what we will study in this section. ■

Solving Inequalities

A linear inequality in one variable is any inequality that can be put in the form

$$ax + b < c \qquad (a, b, \text{and } c \text{ constants}, a \neq 0)$$

where the inequality symbol ($<$) can be replaced with any of the other three inequality symbols ($\le$, $>$, or $\ge$).

Some examples of *linear inequalities* are

$$3x - 2 \ge 7 \qquad -5y < 25 \qquad 3(x - 4) > 2x$$

Our first property for inequalities is similar to the addition property we used when solving equations.

Note Because subtraction is defined as addition of the opposite, our new property holds for subtraction as well as addition. That is, we can subtract the same quantity from each side of an inequality and always be sure that we have not changed the solution.

[Δ≠Σ PROPERTY ***Addition Property for Inequalities***

For any algebraic expressions, A, B, and C,

$$\text{if} \qquad A < B$$
$$\text{then} \qquad A + C < B + C$$

In words: Adding the same quantity to both sides of an inequality will not change the solution set.

VIDEO EXAMPLES

SECTION 2.5

EXAMPLE 1 Solve $3x + 3 < 2x - 1$, and graph the solution.

SOLUTION We use the addition property for inequalities to write all the variable terms on one side and all constant terms on the other side:

$$
\begin{aligned}
3x + 3 &< 2x - 1 \\
3x + (-2x) + 3 &< 2x + (-2x) - 1 && \text{Add } -2x \text{ to each side} \\
x + 3 &< -1 \\
x + 3 + (-3) &< -1 + (-3) && \text{Add } -3 \text{ to each side} \\
x &< -4
\end{aligned}
$$

The solution set is all real numbers that are less than -4. To show this we can use *set notation* and write

$$\{x \mid x < -4\}$$

We can graph the solution set on the number line using an open circle at -4 to show that -4 is not part of the solution set. This is the format you may have used when graphing inequalities in your previous math class.

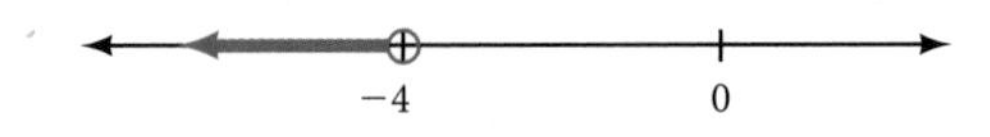

Here is an equivalent graph that uses a parenthesis opening left, instead of an open circle, to represent the end point of the graph.

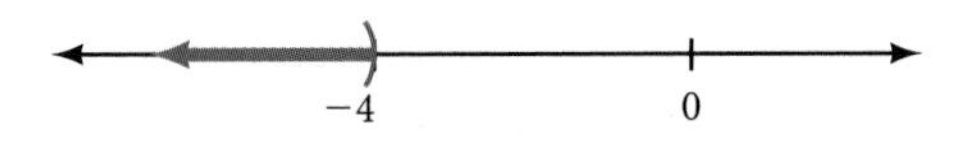

This graph gives rise to the following notation, called *interval notation*, that is an alternative way to write the solution set.

$$(-\infty, -4)$$

The preceding expression indicates that the solution set is all real numbers from negative infinity up to, but not including, -4.

We have three equivalent representations for the solution set to our original inequality. Here are all three together.

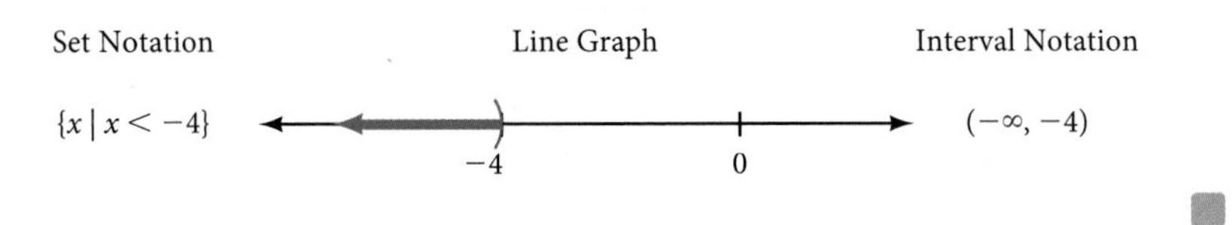

Set Notation	Line Graph	Interval Notation
$\{x \mid x < -4\}$	(graph: −4, 0)	$(-\infty, -4)$

Note The English mathematician John Wallis (1616–1703) was the first person to use the ∞ symbol to represent infinity. When we encounter the interval $(3, \infty)$, we read it as "the interval from 3 to infinity," and we mean the set of real numbers that are greater than three. Likewise, the interval $(-\infty, -4)$ is read "the interval from negative infinity to -4," which is all real numbers less than -4.

Interval Notation and Graphing

The table below shows the connection between set notation, interval notation, and number line graphs. We have included the graphs with open and closed circles for those of you who have used this type of graph previously. In this book, we will continue to show our graphs using the parentheses/brackets method.

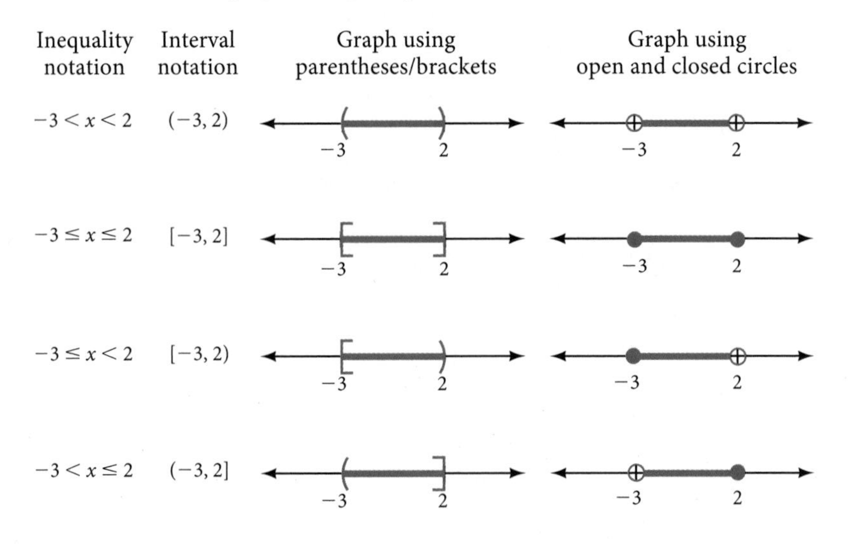

Inequality notation	Interval notation	Graph using parentheses/brackets	Graph using open and closed circles
$-3 < x < 2$	$(-3, 2)$	−3, 2	−3, 2
$-3 \le x \le 2$	$[-3, 2]$	−3, 2	−3, 2
$-3 \le x < 2$	$[-3, 2)$	−3, 2	−3, 2
$-3 < x \le 2$	$(-3, 2]$	−3, 2	−3, 2

Before we state the multiplication property for inequalities, we will take a look at what happens to an inequality statement when we multiply both sides by a positive number and what happens when we multiply by a negative number.

We begin by writing three true inequality statements:

$$3 < 5 \qquad -3 < 5 \qquad -5 < -3$$

We multiply both sides of each inequality by a positive number, say, 4:

$$4(3) < 4(5) \qquad 4(-3) < 4(5) \qquad 4(-5) < 4(-3)$$

$$12 < 20 \qquad -12 < 20 \qquad -20 < -12$$

Notice in each case that the resulting inequality symbol points in the same direction as the original inequality symbol. Multiplying both sides of an inequality by a positive number preserves the *sense* of the inequality.

Let's take the same three original inequalities and multiply both sides by -4:

$$3 < 5 \qquad -3 < 5 \qquad -5 < -3$$

$$-4(3) > -4(5) \qquad -4(-3) > -4(5) \qquad -4(-5) > -4(-3)$$

$$-12 > -20 \qquad 12 > -20 \qquad 20 > 12$$

Notice in this case that the resulting inequality symbol always points in the opposite direction from the original one. Multiplying both sides of an inequality by a negative number *reverses* the sense of the inequality. Keeping this in mind, we will now state the multiplication property for inequalities.

Note Because division is defined as multiplication by the reciprocal, we can apply our new property to division as well as to multiplication. We can divide both sides of an inequality by any nonzero number as long as we reverse the direction of the inequality when the number we are dividing by is negative.

PROPERTY ***Multiplication Property for Inequalities***

Let A, B, and C represent algebraic expressions.

If	$A < B$		
then	$AC < BC$	if	C is positive ($C > 0$)
or	$AC > BC$	if	C is negative ($C < 0$)

In words: Multiplying both sides of an inequality by a positive number always produces an equivalent inequality. Multiplying both sides of an inequality by a negative number reverses the sense of the inequality.

The multiplication property for inequalities does not limit what we can do with inequalities. We are still free to multiply both sides of an inequality by any nonzero number we choose. If the number we multiply by happens to be *negative*, then we *must also reverse* the direction of the inequality.

EXAMPLE 2 Find the solution set for $-2y - 3 \le 7$.

SOLUTION We begin by adding 3 to each side of the inequality:

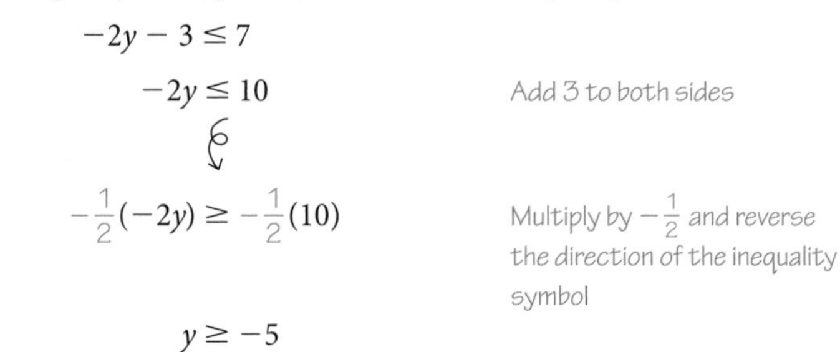

$$-2y - 3 \le 7$$

$$-2y \le 10 \qquad \text{Add 3 to both sides}$$

$$-\frac{1}{2}(-2y) \ge -\frac{1}{2}(10) \qquad \text{Multiply by } -\tfrac{1}{2} \text{ and reverse the direction of the inequality symbol}$$

$$y \ge -5$$

The solution set is all real numbers that are greater than or equal to -5. The following are three equivalent ways to represent this solution set.

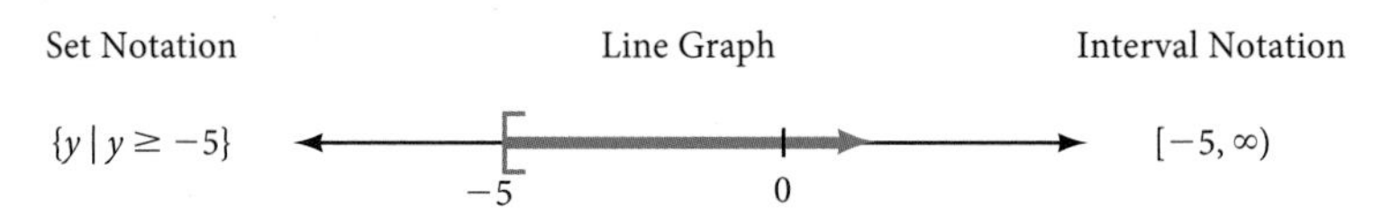

Notice how a bracket is used with interval notation to show that -5 is part of the solution set.

When our inequalities become more complicated, we use the same basic steps we used when we were solving equations. That is, we simplify each side of the inequality before we apply the addition property or multiplication property. When we have solved the inequality, we graph the solution on a number line.

Note In Examples 2 and 3, notice that each time we multiplied both sides of the inequality by a negative number we also reversed the direction of the inequality symbol. Failure to do so would cause our graph to lie on the wrong side of the endpoint.

EXAMPLE 3 Solve $3(2x - 4) - 7x \le -3x$.

SOLUTION We begin by using the distributive property to separate terms. Next, simplify both sides.

$$3(2x - 4) - 7x \le -3x \qquad \text{Original inequality}$$

$$6x - 12 - 7x \le -3x \qquad \text{Distributive property}$$

$$-x - 12 \le -3x \qquad 6x - 7x = (6 - 7)x = -x$$

$$-12 \le -2x \qquad \text{Add } x \text{ to both sides}$$

$$-\frac{1}{2}(-12) \ge -\frac{1}{2}(-2x) \qquad \text{Multiply both sides by } -\tfrac{1}{2} \text{ and reverse the direction of the inequality symbol}$$

$$6 \ge x$$

This last line is equivalent to $x \le 6$. The solution set can be represented with any of the three following items.

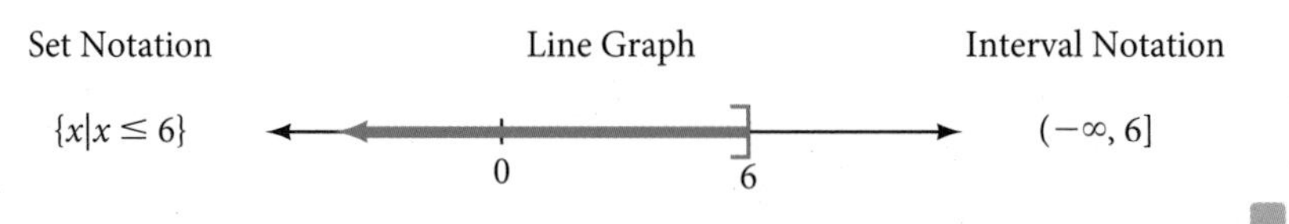

Compound Inequalities

The *union* of two sets A and B is said to be the set of all elements that are in either A or B. The word *or* is the key word in the definition. The *intersection* of two sets A and B is the set of all elements contained in both A and B, the key word here being *and*. We can use the words *and* and *or*, together with our methods of graphing inequalities, to graph some *compound inequalities*.

EXAMPLE 4 Graph: $\{x \mid x \leq -2 \text{ or } x > 3\}$

SOLUTION The two inequalities connected by the word *or* are referred to as a *compound inequality*. We begin by graphing each inequality separately:

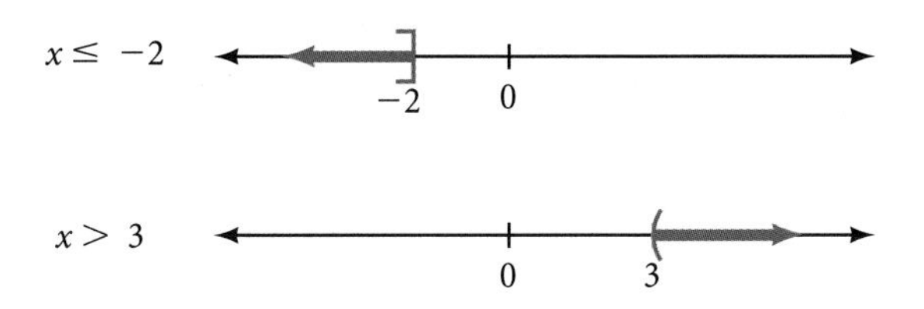

Because the two inequalities are connected by the word *or*, we graph their union. That is, we graph all points on either graph:

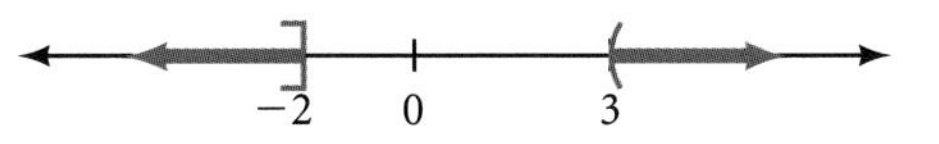

To represent this set of numbers with interval notation we use two intervals connected with the symbol for the union of two sets. Here is the equivalent set of numbers described with interval notation:

$$(-\infty, -2] \cup (3, \infty)$$

Note The square bracket indicates -2 is included in the solution set, and the parenthesis indicates 3 is not included.

EXAMPLE 5 Graph: $\{x \mid x > -1 \text{ and } x < 2\}$

SOLUTION We first graph each inequality separately:

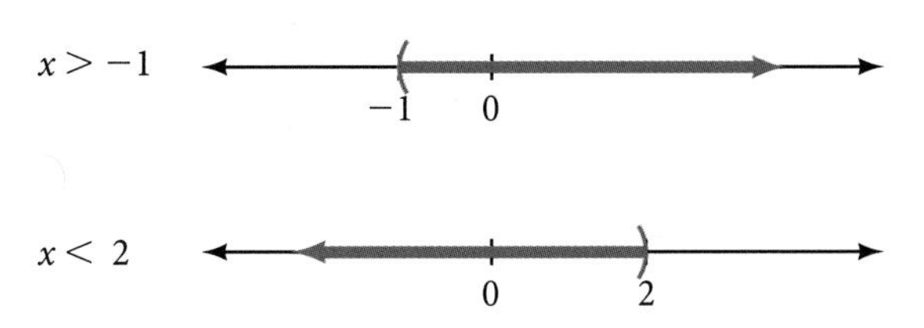

Because the two inequalities are connected by the word *and*, we graph their intersection—the part they have in common:

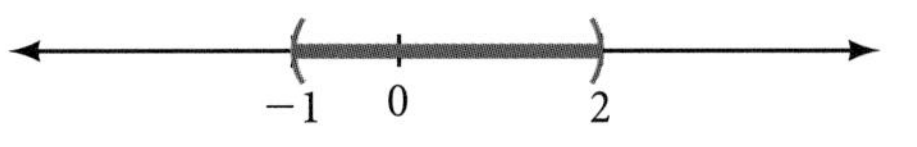

This graph corresponds to the interval $(-1, 2)$, which is called an *open interval* because neither endpoint is included in the interval.

Notation Sometimes compound inequalities that use the word *and* as the connecting word can be written in a shorter form. For example, the compound inequality $-3 \le x$ and $x \le 4$ can be written $-3 \le x \le 4$. The word *and* does not appear when an inequality is written in this form. It is implied. Inequalities of the form $-3 \le x \le 4$ are called ***continued inequalities***. This new notation is useful because writing it takes fewer symbols. The graph of $-3 \le x \le 4$ is

The corresponding interval is $[-3, 4]$, which is called a ***closed interval*** because both endpoints are included in the interval.

EXAMPLE 6 Solve and graph $-3 \le 2x - 5 \le 3$.

SOLUTION We can extend our properties for addition and multiplication to cover this situation. If we add a number to the middle expression, we must add the same number to the outside expressions. If we multiply the center expression by a number, we must do the same to the outside expressions, remembering to reverse the direction of the inequality symbols if we multiply by a negative number. We begin by adding 5 to all three parts of the inequality:

$$-3 \le 2x - 5 \le 3$$

$$2 \le 2x \le 8 \qquad \text{Add 5 to all three members}$$

$$1 \le x \le 4 \qquad \text{Multiply through by } \tfrac{1}{2}$$

Here are three ways to write this solution set:

Set Notation	Line Graph	Interval Notation
$\{x \mid 1 \le x \le 4\}$	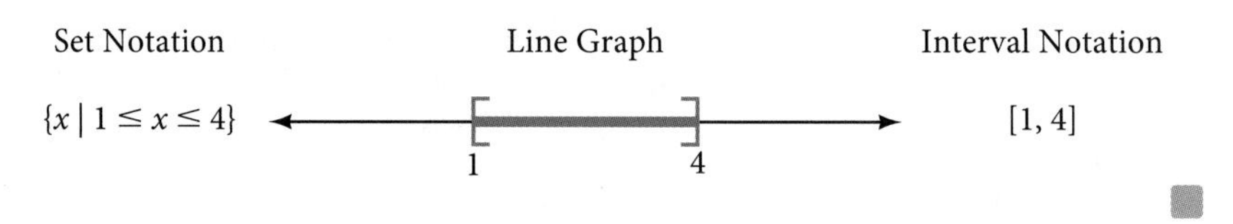	$[1, 4]$

EXAMPLE 7 Solve the compound inequality.

$$3t + 7 \le -4 \quad \text{or} \quad 3t + 7 \ge 4$$

SOLUTION We solve each half of the compound inequality separately, then we graph the solution set:

$$3t + 7 \le -4 \quad \text{or} \quad 3t + 7 \ge 4$$

$$3t \le -11 \quad \text{or} \quad 3t \ge -3 \qquad \text{Add } -7$$

$$t \le -\frac{11}{3} \quad \text{or} \quad t \ge -1 \qquad \text{Multiply by } \tfrac{1}{3}$$

The solution set can be written in any of the following ways:

Set Notation	Line Graph	Interval Notation
$\{t \mid t \le -\frac{11}{3} \text{ or } t \ge -1\}$	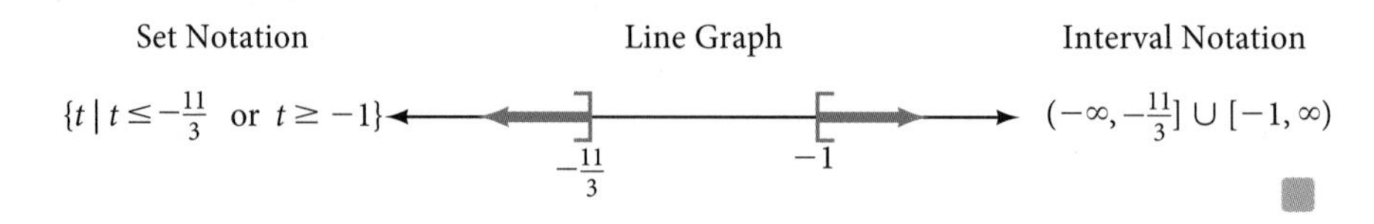	$(-\infty, -\frac{11}{3}] \cup [-1, \infty)$

Quadratic Inequalities

Quadratic inequalities in one variable are inequalities of the form

$$ax^2 + bx + c < 0 \qquad ax^2 + bx + c > 0$$

$$ax^2 + bx + c \le 0 \qquad ax^2 + bx + c \ge 0$$

where a, b, and c are constants, with $a \neq 0$. The technique we will use to solve inequalities of this type involves graphing. Suppose, for example, we want to find the solution set for the inequality $x^2 - x - 6 > 0$. We begin by factoring the left side to obtain

$$(x - 3)(x + 2) > 0$$

We have two real numbers $x - 3$ and $x + 2$ whose product $(x - 3)(x + 2)$ is greater than zero. That is, their product is positive. The only way the product can be positive is either if both factors, $(x - 3)$ and $(x + 2)$, are positive or if they are both negative. To help visualize where $x - 3$ is positive and where it is negative, we draw a real number line and label it accordingly:

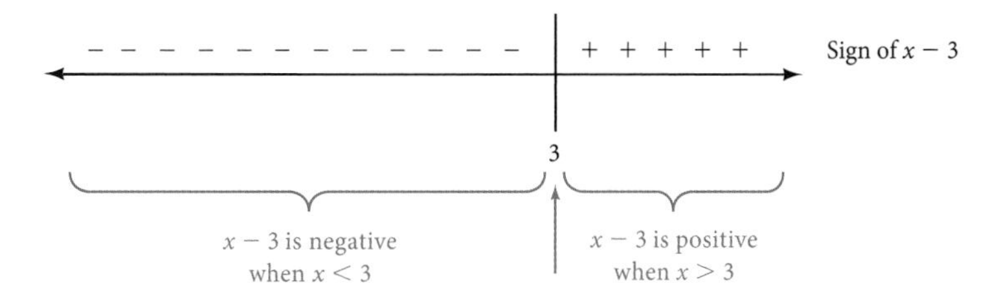

Here is a similar diagram showing where the factor $x + 2$ is positive and where it is negative:

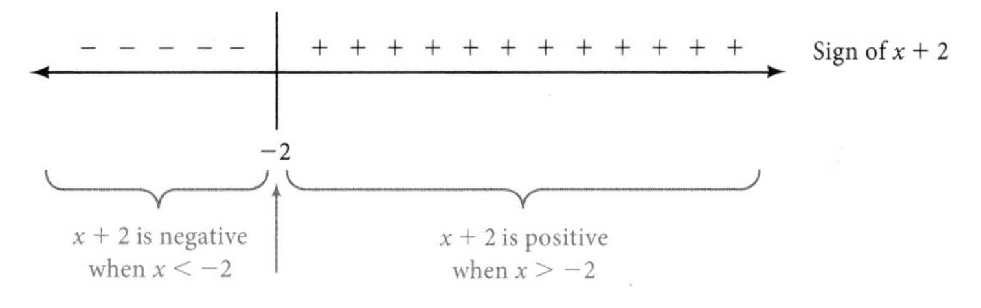

Drawing the two number lines together and eliminating the unnecessary numbers, we have

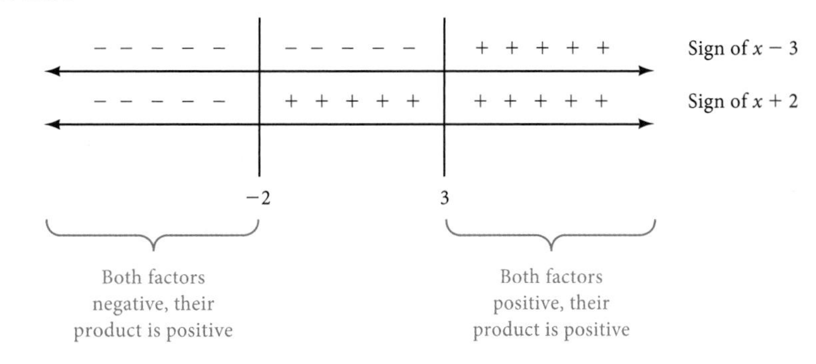

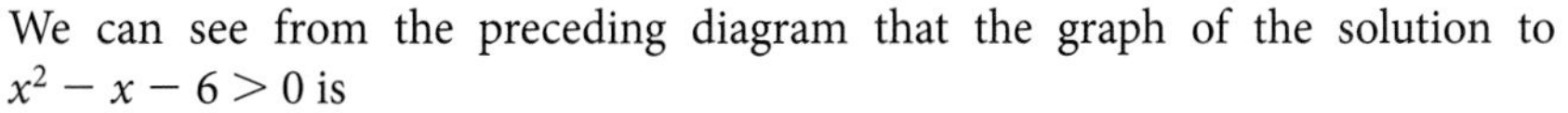

We can see from the preceding diagram that the graph of the solution to $x^2 - x - 6 > 0$ is

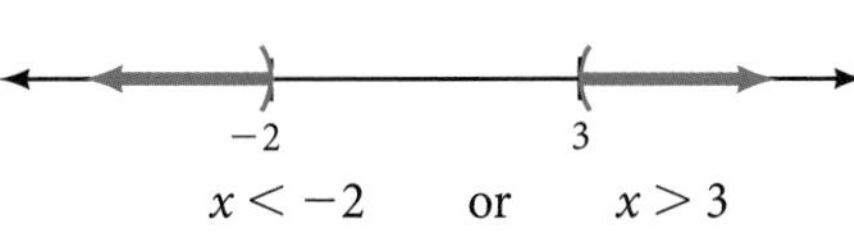

EXAMPLE 8 Solve for x: $x^2 - 2x - 8 \le 0$.

SOLUTION We begin by factoring:

$$x^2 - 2x - 8 \le 0$$

$$(x - 4)(x + 2) \le 0$$

The product $(x - 4)(x + 2)$ is negative or zero. The factors must have opposite signs. We draw a diagram showing where each factor is positive and where each factor is negative:

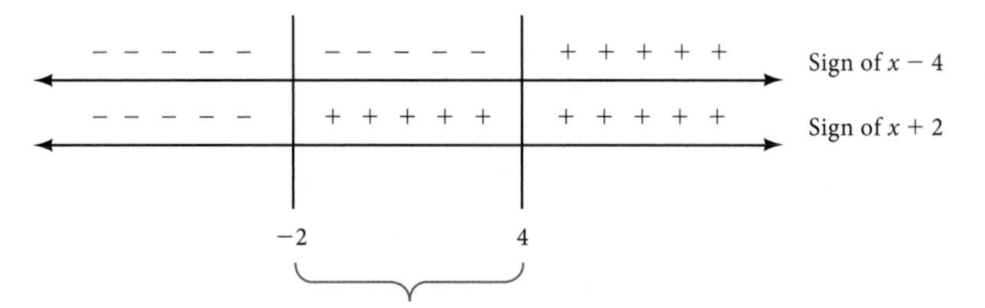

From the diagram, we have the graph of the solution set:

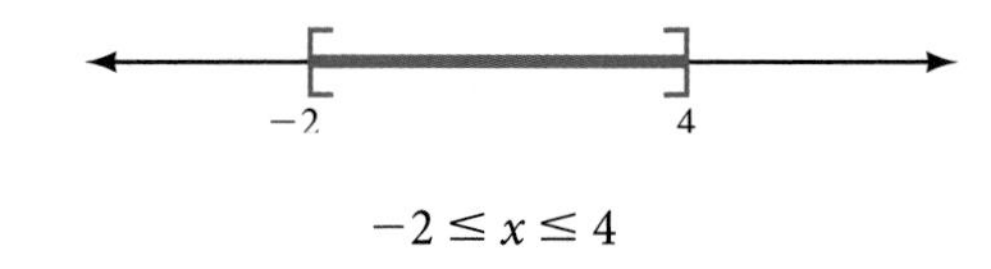

$$-2 \le x \le 4$$

EXAMPLE 9 Solve $x^2 - 6x + 9 \ge 0$.

SOLUTION

$$x^2 - 6x + 9 \ge 0$$

$$(x - 3)^2 \ge 0$$

This is a special case in which both factors are the same. Because $(x - 3)^2$ is always positive or zero, the solution set is all real numbers. That is, any real number that is used in place of x in the original inequality will produce a true statement.

Our next two examples involve inequalities that contain rational expressions.

EXAMPLE 10 Solve: $\dfrac{x - 4}{x + 1} \le 0$.

SOLUTION The inequality indicates that the quotient of $(x - 4)$ and $(x + 1)$ is negative or 0 (less than or equal to 0). We can use the same reasoning we used to solve the examples above, because quotients are positive or negative under the same conditions that products are positive or negative. Here is the diagram that shows where each factor is positive and where each factor is negative:

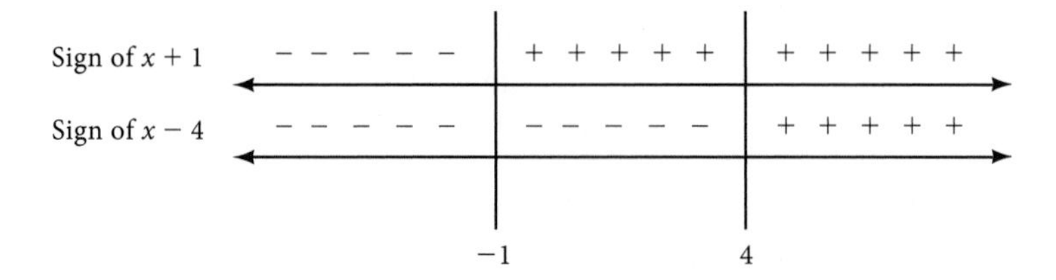

Between -1 and 4 the factors have opposite signs, making the quotient negative. Thus, the region between -1 and 4 is where the solutions lie, because the original inequality indicates the quotient $\frac{x-4}{x+1}$ is negative. The solution set and its graph are shown here:

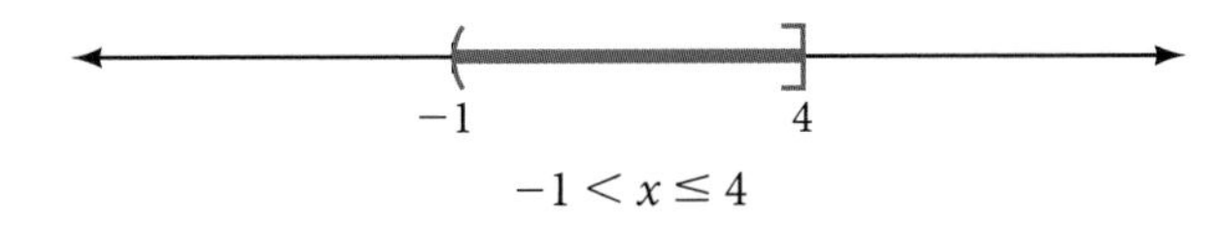

$$-1 < x \le 4$$

Notice that the left endpoint is open—that is, it is not included in the solution set—because $x = -1$ would make the denominator in the original inequality 0. It is important to check all endpoints of solution sets to inequalities that involve rational expressions.

EXAMPLE 11 Solve: $\dfrac{3}{x-2} - \dfrac{2}{x-3} > 0.$

SOLUTION We begin by adding the two rational expressions on the left side. The common denominator is $(x-2)(x-3)$:

$$\frac{3}{x-2} \cdot \frac{(x-3)}{(x-3)} - \frac{2}{x-3} \cdot \frac{(x-2)}{(x-2)} > 0$$

$$\frac{3x - 9 - 2x + 4}{(x-2)(x-3)} > 0$$

$$\frac{x-5}{(x-2)(x-3)} > 0$$

This time the quotient involves three factors. Here is the diagram that shows the signs of the three factors:

	$x < 2$	2 to 3	3 to 5	$x > 5$
Sign of $(x-5)$	− − − −	− − − −	− − − − − −	+ + + +
Sign of $(x-3)$	− − − −	− − − −	+ + + + + +	+ + + +
Sign of $(x-2)$	− − − −	+ + + +	+ + + + + +	+ + + +

Boundary points: 2, 3, 5

The original inequality indicates that the quotient is positive. For this to happen, either all three factors must be positive, or exactly two factors must be negative. Looking back at the diagram, we see the regions that satisfy these conditions are between 2 and 3 or above 5. Here is our solution set:

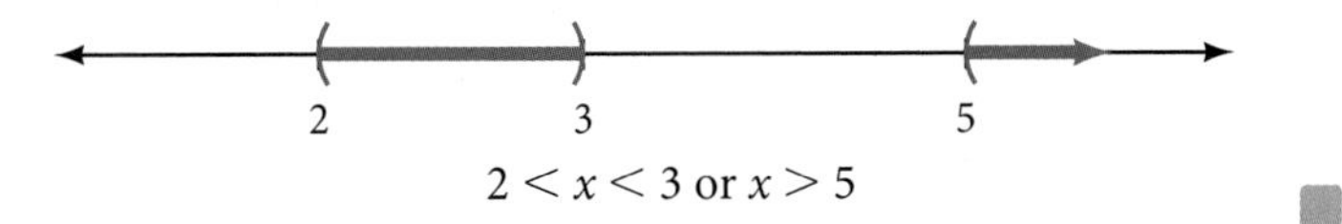

$$2 < x < 3 \text{ or } x > 5$$

Modeling with Inequalities

EXAMPLE 12 In the introduction to this section, we stated that the USGS give the survival rates for the 'Apapane as

0.72 ± 0.11 for adults $\qquad$ 0.13 ± 0.07 for juveniles

Write each survival rate as an inequality involving percent.

SOLUTION If we let a represent the survival rate for adults, then the survival rate for a is 0.72 ± 0.11, which means that a is between $0.72 - 0.11$ and $0.72 + 0.11$. Writing this fact as an inequality, we have

$$0.72 - 0.11 \le a \le 0.72 + 0.11$$

$$0.61 \le a \le 0.83$$

$$61\% \le a \le 83\%$$

If we let j represent the survival rates for juveniles, and proceed in the same manner, we have

$$0.13 - 0.07 \le j \le 0.13 + 0.07$$

$$0.06 \le j \le 0.20$$

$$6\% \le j \le 20\%$$

We have modeled the survival rates of the 'Apapane with inequalities and percent.

EXAMPLE 13 A company that manufactures ink cartridges for printers finds that they can sell x cartridges each week at a price of p dollars each, according to the formula $x = 1{,}300 - 100p$. What price should they charge for each cartridge if they want to sell at least 300 cartridges a week?

SOLUTION Because x is the number of cartridges they sell each week, an inequality that corresponds to selling at least 300 cartridges a week is

$$x \ge 300$$

Substituting $1{,}300 - 100p$ for x gives us an inequality in the variable p.

$$1{,}300 - 100p \ge 300$$

$$-100p \ge -1{,}000 \qquad \text{Add } -1{,}300 \text{ to each side}$$

$$p \le 10 \qquad \text{Divide each side by } -100\text{, and reverse the direction of the inequality symbol}$$

To sell at least 300 cartridges each week, the price per cartridge should be no more than \$10. That is, selling the cartridges for \$10 or less will produce weekly sales of 300 or more cartridges.

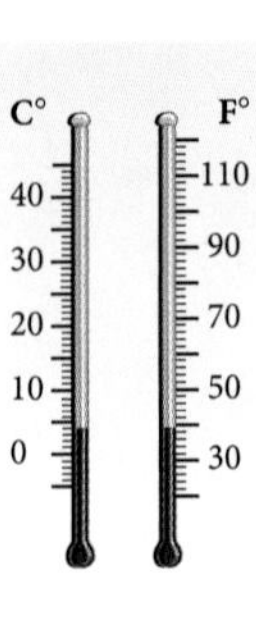

EXAMPLE 14 The formula $F = \frac{9}{5}C + 32$ gives the relationship between the Celsius and Fahrenheit temperature scales. If the temperature range on a certain day is 86° to 104° Fahrenheit, what is the temperature range in degrees Celsius?

SOLUTION From the given information, we can write $86 \le F \le 104$. However, because F is equal to $\frac{9}{5}C + 32$, we can also write:

$$86 \le \frac{9}{5}C + 32 \le 104$$

$$54 \le \frac{9}{5}C \le 72 \qquad \text{Add } -32 \text{ to each number}$$

$$\frac{5}{9}(54) \le \frac{5}{9}\left(\frac{9}{5}C\right) \le \frac{5}{9}(72) \qquad \text{Multiply each number by } \frac{5}{9}$$

$$30 \le C \le 40$$

A temperature range of 86° to 104° Fahrenheit corresponds to a temperature range of 30° to 40° Celsius.

GETTING READY FOR CLASS

After reading through the preceding section, respond in your own words and in complete sentences.

A. What is the addition property for inequalities?

B. When we use interval notation to denote a section of the real number line, when do we use parentheses ( ) and when do we use brackets [ ]?

C. Explain the difference between the multiplication property of equality and the multiplication property for inequalities.

D. When solving an inequality, when do we change the direction of the inequality symbol?

Problem Set 2.5

Solve each of the following inequalities, and graph each solution.

1. $5x \geq -115$
2. $\frac{1}{3}x > 4$
3. $-7x \geq 35$
4. $-\frac{2}{3}x < -8$
5. $-20 \geq 4x$
6. $-1 \leq -\frac{1}{5}x$

Solve each of the following inequalities, and graph each solution.

7. $-2x - 5 \leq 15$
8. $\frac{1}{2} - \frac{m}{10} > -\frac{1}{5}$
9. $\frac{9}{5} > -\frac{1}{5} - \frac{1}{2}x$
10. $-20 > 50 - 30y$
11. $\frac{3}{4}x - 2 > 7$
12. $8 - \frac{1}{3}y \geq 20$
13. $5 - \frac{1}{3}x > \frac{1}{4}x + 2$
14. $2.0 - 0.7x < 1.3 - 0.3x$

Simplify each side first, then solve the following inequalities. Write your answers with interval notation.

15. $3(2y - 4) > 0$
16. $-(a - 2) - 5a \leq 3a + 7$
17. $\frac{1}{4}t - \frac{1}{3}(2t - 5) < 0$
18. $1 < 3 - 4(3a - 1)$
19. $-\frac{1}{2}(2x + 1) \leq -\frac{3}{8}(x + 2)$
20. $-3(1 - 2x) - 3(x - 4) < -3 - 4x$
21. $\frac{1}{4}x - \frac{1}{2}(3x + 1) \geq 2$
22. $20x + 4{,}800 > 18{,}000$

Solve the following continued inequalities. Use both a line graph and interval notation to write each solution set.

23. $-3 \leq m + 1 \leq 3$
24. $-60 < 50a - 40 < 60$
25. $0.1 \leq 0.4a + 0.1 \leq 0.3$
26. $5 < \frac{1}{4}x + 1 < 9$

Graph the solution sets for the following compound inequalities. Then write each solution set using interval notation.

27. $3x + 2 < -3$ or $3x + 2 > 3$
28. $7y - 5 \leq -2$ or $7y - 5 \geq 2$
29. $3x - 1 > 2x + 4$ or $5x - 2 < 3x + 4$
30. $2x - 5 \leq -1$ or $-3x - 6 < -15$

Solve each of the following inequalities, and graph the solution set.

31. $x^2 + x - 6 < 0$
32. $x^2 - x - 12 \geq 0$
33. $x^2 - 5x > 6$
34. $4x^2 \geq -5x + 6$
35. $x^2 - 16 \geq 0$
36. $9x^2 - 4 < 0$
37. $3x^2 + x - 10 \geq 0$
38. $x^2 - 4x + 4 < 0$
39. $(x - 2)(x - 3)(x - 4) < 0$
40. $(x + 1)(x + 2)(x + 3) \geq 0$
41. $\frac{x + 4}{x - 1} \leq 0$
42. $\frac{5x}{x + 1} - \frac{3}{x + 1} < 0$
43. $\frac{2}{x - 3} + 1 \geq 0$
44. $\frac{x - 1}{(x + 2)(x - 5)} < 0$
45. $\frac{4}{x + 3} - \frac{3}{x + 2} > 0$
46. $\frac{x + 1}{2x - 2} - \frac{2}{x^2 - 1} \leq 0$

The next two problems are intended to give you practice reading, and paying attention to, the instructions that accompany the problems you are working.

47. Work each problem according to the instructions given.

a. Evaluate when $x = 0$: $-\frac{1}{2}x + 1$ **b.** Solve: $-\frac{1}{2}x + 1 = -7$

c. Is 0 a solution to $-\frac{1}{2}x + 1 < -7$? **d.** Solve: $-\frac{1}{2}x + 1 < -7$

48. Work each problem according to the instructions given.

a. Evaluate when $x = 0$: $-\frac{2}{3}x - 5$ **b.** Solve: $-\frac{2}{3}x - 5 = 1$

c. Is 0 a solution to $-\frac{2}{3}x - 5 > 1$? **d.** Solve: $-\frac{2}{3}x - 5 > 1$

Translate each of the following phrases into an equivalent inequality statement.

49. x is greater than -2 and at most 4 **50.** x is less than 9 and at least -3

51. x is less than -4 or at least 1 **52.** x is at most 1 or more than 6

Modeling Practice

The problems that follow use inequalities to model real-world situations.

53. Art Supplies A store selling art supplies finds that they can sell x sketch pads each week at a price of p dollars each, according to the formula $x = 900 - 300p$. What price should they charge if they want to sell

a. At least 300 pads each week? **b.** More than 600 pads each week?

c. Less than 525 pads each week? **d.** At most 375 pads each week?

54. Temperature Range Each of the following temperature ranges is in degrees Fahrenheit. Use the formula $F = \frac{9}{5}C + 32$ to find the corresponding temperature range in degrees Celsius. Write your answer as a continued inequality as in Example 14.

a. 95° to 113° **b.** 68° to 86° **c.** $-13°$ to 14° **d.** $-4°$ to 23°

55. Student Loan When considering how much debt to incur in student loans, you learn that it is wise to keep your student loan payment to 8% or less of your starting monthly income. Suppose you anticipate a starting annual salary of \$24,000. Set up and solve an inequality that represents the amount of monthly debt for student loans that would be considered manageable.

56. Survival Rates for Sea Gulls Here is part of a report concerning the survival rates of Western Gulls that appeared on the website of Cornell University.

Survival of eggs to hatching is 70%–80%; of hatched chicks to fledglings 50%–70%; of fledglings to age of first breeding <50%.

Write the survival rates using inequalities without percent.

57. **Dimensions of a Rectangle** The length of a rectangle is 3 inches more than twice the width. If the area is to be at least 44 square inches, what are the possibilities for the width?

58. **Dimensions of a Rectangle** The length of a rectangle is 5 inches less than three times the width. If the area is to be less than 12 square inches, what are the possibilities for the width?

59. **Revenue** A manufacturer of MP3 players knows that the weekly revenue produced by selling x MP3 players is given by the equation $R = 1{,}300p - 100p^2$, where p is the price of each MP3 player (in dollars). What price should be charged for each MP3 player if the weekly revenue is to be at least $4,000?

60. **Revenue** A manufacturer of small calculators knows that the weekly revenue produced by selling x calculators is given by the equation $R = 1{,}700p - 100p^2$, where p is the price of each calculator (in dollars). What price should be charged for each calculator if the revenue is to be at least $7,000 each week?

61. **Union Dues** A labor union has 10,000 members. For every $10 increase in union dues, membership is decreased by 200 people. If the current dues are $100, what should be the new dues (to the nearest multiple of $10) so income from dues is greatest, and what is that income? *Hint:* Because Income = (membership)(dues), we can let x = the number of $10 increases in dues, and then this will give us income of $y = (10{,}000 - 200x)(100 + 10x)$.

62. **Bookstore Receipts** The owner of a used book store charges $2 for quality paperbacks and usually sells 40 per day. For every 10-cent increase in the price of these paperbacks, he thinks that he will sell two fewer per day. What is the price he should charge (to the nearest 10 cents) for these books to maximize his income, and what would be that income? *Hint:* Let x = the number of 10-cent increases in price.

63. **Fast-Lube** The owner of a quick oil-change business charges $20 per oil change and has 40 customers per day. If each increase of $2 results in 2 fewer daily customers, what price should the owner charge (to the nearest $2) for an oil change if the income from this business is to be as great as possible?

64. **Computer Sales** A gaming computer manufacturer charges $2,200 for its base model and sells 1,500 computers per month at this price. For every $200 increase in price, it is believed that 75 fewer computers will be sold. What price should the company place on its base model of computer (to the nearest $200) to have the greatest income?

Getting Ready for the Next Section

To understand all of the explanations and examples in the next section you must be able to work the problems below.

Solve each equation.

65. $2a - 1 = -7$

66. $3x - 6 = 9$

67. $\frac{2}{3}x - 3 = 7$

68. $\frac{2}{3}x - 3 = -7$

69. $x - 5 = x - 7$

70. $x + 3 = x + 8$

71. $x - 5 = -x - 7$

72. $x + 3 = -x + 8$

Solve each inequality. Do not graph the solution set.

73. $2x - 5 < 3$

74. $-3 < 2x - 5$

75. $-4 \leq 3a + 7$

76. $3a + 2 \leq 4$

77. $4t - 3 \leq -9$

78. $4t - 3 \geq 9$

SPOTLIGHT ON SUCCESS *University of North Alabama*

Pride is a personal commitment.
It is an attitude which separates excellence from mediocrity.
—William Blake

The University of Northern Alabama places its Pride Rock, a 60-pound granite stone engraved with a lion's paw print, behind the north end zone at all home football games. The rock reminds current Lion players of the proud athletic traditions that has been established at the school, and to take pride in their efforts on the field.

Photo courtesy UNA

The same idea holds true for your work in your math class. Take pride in it. When you turn in an assignment, it should be accurate and easy for the instructor to read. It shows that you care about your progress in the course and that you take pride in your work. The work that you turn in to your instructor is a reflection of you. As the quote from William Blake indicates, pride is a personal commitment; a decision that you make, yourself. And once you make that commitment to take pride in the work you do in your math class, you have directed yourself toward excellence, and away from mediocrity.

Equations and Inequalities with Absolute Value

2.6

© bluestocking/iStockPhoto

INTRODUCTION In a student survey conducted by the University of Minnesota, it was found that 30% of students were solely responsible for their finances. The survey was reported to have a margin of error plus or minus 3.74%. This means that the difference between the sample estimate of 30% and the actual percent of students who are responsible for their own finances is most likely less than 3.74%. We can write this as an inequality:

$$|x - 0.30| \le 0.0374$$

where x represents the true percent of students who are responsible for their own finances.

You may recall that the ***absolute value*** of x, $|x|$, is the distance between x and 0 on the number line. The absolute value of a number measures its distance from 0.

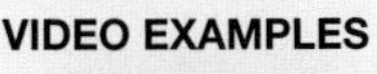
VIDEO EXAMPLES

SECTION 2.6

EXAMPLE 1 Solve for x: $|x| = 5$.

SOLUTION Using the definition of absolute value, we can read the equation as, "The distance between x and 0 on the number line is 5." If x is 5 units from 0, then x can be 5 or -5:

$$\text{If } |x| = 5 \quad \text{then } x = 5 \quad \text{or} \quad x = -5$$

In general, then, we can see that any equation of the form $|a| = b$ is equivalent to the equations $a = b$ or $a = -b$, as long as $b > 0$.

EXAMPLE 2 Solve $|2a - 1| = 7$.

SOLUTION We can read this question as "$2a - 1$ is 7 units from 0 on the number line." The quantity $2a - 1$ must be equal to 7 or -7:

$$|2a - 1| = 7$$

$$2a - 1 = 7 \quad \text{or} \quad 2a - 1 = -7$$

We have transformed our absolute value equation into two equations that do not involve absolute value.

$$2a - 1 = 7 \quad \text{or} \quad 2a - 1 = -7$$

$$2a = 8 \quad \text{or} \quad 2a = -6 \qquad \text{Add 1 to both sides}$$

$$a = 4 \quad \text{or} \quad a = -3 \qquad \text{Multiply by } \tfrac{1}{2}$$

Our solution set is $\{4, -3\}$.

To check our solutions, we put them into the original absolute value equation:

	When $a = 4$		When $a = -3$
the equation	$\lvert 2a - 1 \rvert = 7$	the equation	$\lvert 2a - 1 \rvert = 7$
becomes	$\lvert 2(4) - 1 \rvert \stackrel{?}{=} 7$	becomes	$\lvert 2(-3) - 1 \rvert \stackrel{?}{=} 7$
	$\lvert 7 \rvert \stackrel{?}{=} 7$		$\lvert -7 \rvert \stackrel{?}{=} 7$
	$7 = 7$		$7 = 7$

EXAMPLE 3 Solve $\left|\frac{2}{3}x - 3\right| + 5 = 12$.

SOLUTION To use the definition of absolute value to solve this equation, we must isolate the absolute value on the left side of the equal sign. To do so, we add -5 to both sides of the equation to obtain

$$\left|\frac{2}{3}x - 3\right| = 7$$

Now that the equation is in the correct form, we can write

$$\frac{2}{3}x - 3 = 7 \quad \text{or} \quad \frac{2}{3}x - 3 = -7$$

$$\frac{2}{3}x = 10 \quad \text{or} \quad \frac{2}{3}x = -4 \qquad \text{Add 3 to both sides}$$

$$x = 15 \quad \text{or} \quad x = -6 \qquad \text{Multiply by } \frac{3}{2}$$

The solution set is $\{15, -6\}$.

EXAMPLE 4 Solve $|3a - 6| = -4$.

SOLUTION The solution set is $\varnothing$ because the right side is negative but the left side cannot be negative. No matter what we try to substitute for the variable a, the quantity $|3a - 6|$ will always be positive or zero. It can never be -4.

Note Recall that $\varnothing$ is the symbol we use to denote the empty set. When we use it to indicate the solutions to an equation, then we are saying the equation has no solution.

Consider the statement $|a| = |b|$. What can we say about a and b? We know they are equal in absolute value. By the definition of absolute value, they are the same distance from 0 on the number line. They must be equal to each other or opposites of each other. In symbols, we write:

$$|a| = |b| \Leftrightarrow a = b \quad \text{or} \quad a = -b$$

$\uparrow$	$\uparrow$		$\uparrow$
Equal in absolute value	Equals	or	Opposites

Note $\Leftrightarrow$ means "if and only if" and "is equivalent to"

EXAMPLE 5 Solve $|3a + 2| = |2a + 3|$.

SOLUTION The quantities $3a + 2$ and $2a + 3$ have equal absolute values. They are, therefore, the same distance from 0 on the number line. They must be equals or opposites:

$$|3a + 2| = |2a + 3|$$

Equals		Opposites
$3a + 2 = 2a + 3$	or	$3a + 2 = -(2a + 3)$
$a + 2 = 3$		$3a + 2 = -2a - 3$
$a = 1$		$5a + 2 = -3$
		$5a = -5$
		$a = -1$

The solution set is $\{1, -1\}$.

It makes no difference in the outcome of the problem if we take the opposite of the first or second expression. It is very important, once we have decided which one to take the opposite of, that we take the opposite of both its terms and not just the first term. That is, the opposite of $2a + 3$ is $-(2a + 3)$, which we can think of as $-1(2a + 3)$. Distributing the -1 across *both* terms, we have

$$-1(2a + 3) = -2a - 3$$

EXAMPLE 6 Solve $|x - 5| = |x - 7|$.

SOLUTION As was the case in Example 5, the quantities $x - 5$ and $x - 7$ must be equal or they must be opposites, because their absolute values are equal:

Equals		Opposites
$x - 5 = x - 7$	or	$x - 5 = -(x - 7)$
$-5 = -7$		$x - 5 = -x + 7$
No solution here		$2x - 5 = 7$
		$2x = 12$
		$x = 6$

Because the first equation leads to a false statement, it will not give us a solution. (If either of the two equations were to reduce to a true statement, it would mean all real numbers would satisfy the original equation.) In this case, our only solution is $x = 6$.

Inequalities Involving Absolute Value

Again, the absolute value of x, which is denoted $|x|$, represents the distance that x is from 0 on the number line. We will begin by considering three absolute value expressions and their verbal translations:

Expression	In Words
$\|x\| = 7$	x is exactly 7 units from 0 on the number line.
$\|a\| < 5$	a is less than 5 units from 0 on the number line.
$\|y\| \geq 4$	y is greater than or equal to 4 units from 0 on the number line.

Once we have translated the expression into words, we can use the translation to graph the original equation or inequality. The graph is then used to write a final equation or inequality that does not involve absolute value.

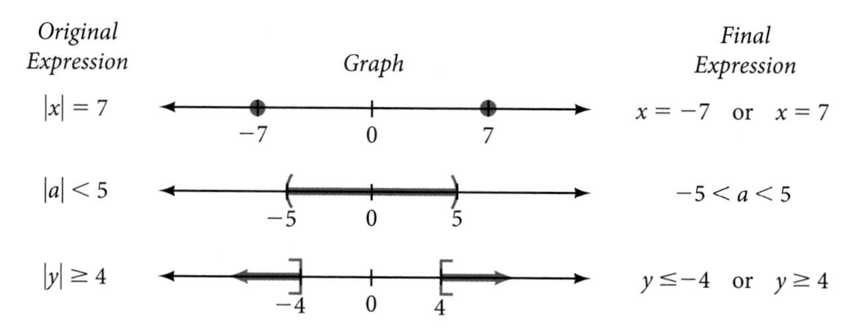

Although we will not always write out the verbal translation of an absolute value inequality, it is important that we understand the translation. Our second expression, $|a| < 5$, means a is within 5 units of 0 on the number line. The graph of this relationship is

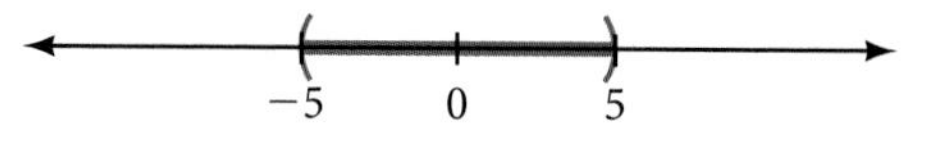

which can be written with the following continued inequality:

$$-5 < a < 5$$

We can follow this same kind of reasoning to solve more complicated absolute value inequalities.

EXAMPLE 7 Graph the solution set: $|2x - 5| < 3$.

SOLUTION The absolute value of $2x - 5$ is the distance that $2x - 5$ is from 0 on the number line. We can translate the inequality as, "$2x - 5$ is less than 3 units from 0 on the number line." That is, $2x - 5$ must appear between -3 and 3 on the number line.

A picture of this relationship is

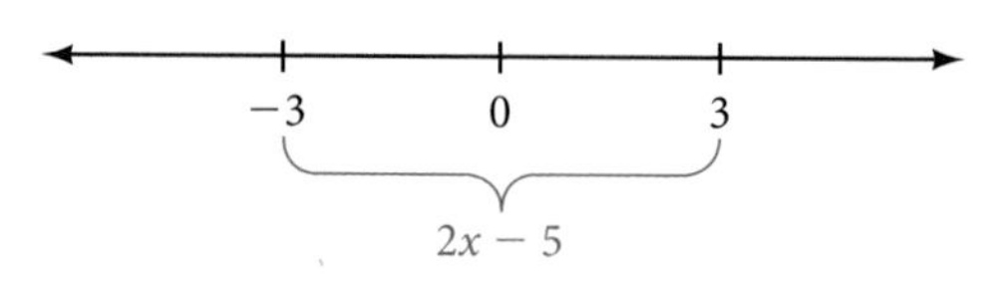

Using the picture, we can write an inequality without absolute value that describes the situation:

$$-3 < 2x - 5 < 3$$

Next, we solve the continued inequality by first adding 5 to all three members and then multiplying all three by $\frac{1}{2}$.

$$-3 < 2x - 5 < 3$$

$$2 < 2x < 8 \qquad \text{Add 5 to all three expressions}$$

$$1 < x < 4 \qquad \text{Multiply each expression by } \frac{1}{2}$$

The graph of the solution set is

We can see from the solution that for the absolute value of $2x - 5$ to be within 3 units of 0 on the number line, x must be between 1 and 4.

EXAMPLE 8 Graph the solution set: $|4t - 3| \geq 9$.

SOLUTION The quantity $4t - 3$ is greater than or equal to 9 units from 0. It must be either above +9 or below −9.

$$4t - 3 \leq -9 \quad \text{or} \quad 4t - 3 \geq 9$$

$$4t \leq -6 \quad \text{or} \quad 4t \geq 12 \qquad \text{Add 3}$$

$$t \leq -\frac{6}{4} \quad \text{or} \quad t \geq \frac{12}{4} \qquad \text{Multiply by } \frac{1}{4}$$

$$t \leq -\frac{3}{2} \quad \text{or} \quad t \geq 3$$

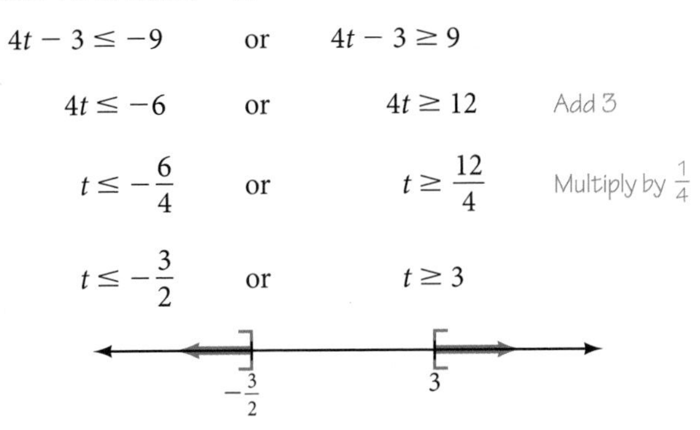

We can use the results of our examples to summarize the information we have related to absolute value equations and inequalities.

Rewriting Absolute Value Equations and Inequalities

If c is a positive real number, then each of the following statements on the left is equivalent to the corresponding statement on the right.

With Absolute Value	Without Absolute Value
$\lvert x \rvert = c$	$x = -c$ or $x = c$
$\lvert x \rvert < c$	$-c < x < c$
$\lvert x \rvert > c$	$x < -c$ or $x > c$
$\lvert ax + b \rvert = c$	$ax + b = -c$ or $ax + b = c$
$\lvert ax + b \rvert < c$	$-c < ax + b < c$
$\lvert ax + b \rvert > c$	$ax + b < -c$ or $ax + b > c$

EXAMPLE 9 Solve and graph $|2x + 3| + 4 < 9$.

SOLUTION Before we can apply the method of solution we used in the previous examples, we must isolate the absolute value on one side of the inequality. To do so, we add -4 to each side.

$$|2x + 3| + 4 < 9$$

$$|2x + 3| + 4 + (-4) < 9 + (-4)$$

$$|2x + 3| < 5$$

From this last line, we know that $2x + 3$ must be between -5 and $+5$.

$$-5 < 2x + 3 < 5$$

$$-8 < 2x < 2 \quad \text{Add } -3 \text{ to each expression}$$

$$-4 < x < 1 \quad \text{Multiply each expression by } \frac{1}{2}$$

Here are three equivalent ways to write our solution:

EXAMPLE 10 Solve and graph $|4 - 2t| > 2$.

SOLUTION The inequality indicates that $4 - 2t$ is less than -2 or greater than $+2$. Writing this without absolute value symbols, we have

$$4 - 2t < -2 \quad \text{or} \quad 4 - 2t > 2$$

To solve these inequalities we begin by adding -4 to each side.

$$4 + (-4) - 2t < -2 + (-4) \quad \text{or} \quad 4 + (-4) - 2t > 2 + (-4)$$

$$-2t < -6 \quad \text{or} \quad -2t > -2$$

Next we must multiply both sides of each inequality by $-\frac{1}{2}$. When we do so, we must also reverse the direction of each inequality symbol.

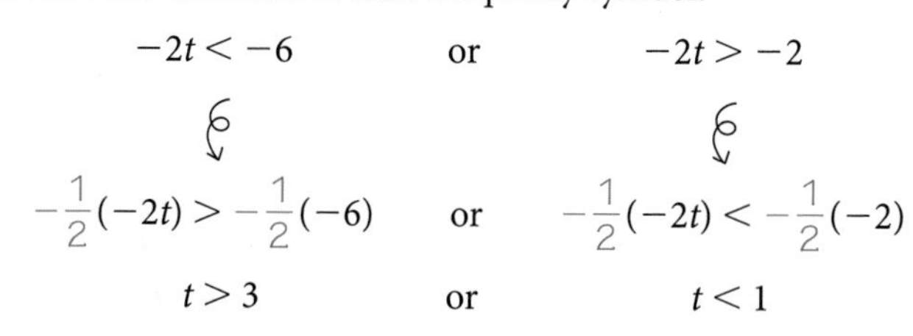

$$-2t < -6 \quad \text{or} \quad -2t > -2$$

$$-\frac{1}{2}(-2t) > -\frac{1}{2}(-6) \quad \text{or} \quad -\frac{1}{2}(-2t) < -\frac{1}{2}(-2)$$

$$t > 3 \quad \text{or} \quad t < 1$$

Although in situations like this we are used to seeing the "less than" symbol written first, the meaning of the solution is clear. We want to graph all real numbers that are either greater than 3 or less than 1. Here is the graph.

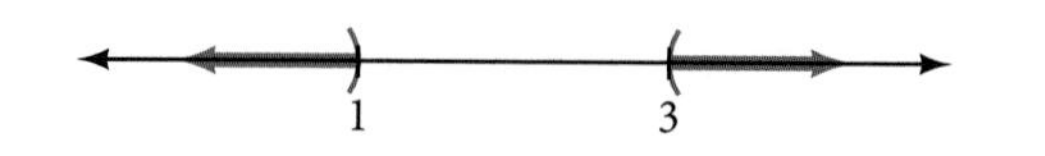

Because absolute value always results in a nonnegative quantity, we sometimes come across special solution sets when a negative number appears on the right side of an absolute value inequality.

EXAMPLE 11 Solve $|7y - 1| < -2$.

SOLUTION The *left* side is never negative because it is an absolute value. The *right* side is negative. We have a positive quantity less than a negative quantity, which is impossible. The solution set is the empty set, $\varnothing$. There is no real number to substitute for y to make this inequality a true statement.

EXAMPLE 12 Solve $|6x + 2| > -5$.

SOLUTION This is the opposite case from that in Example 11. No matter what real number we use for x on the *left* side, the result will always be positive, or zero. The *right* side is negative. We have a positive quantity (or zero) greater than a negative quantity. Every real number we choose for x gives us a true statement. The solution set is the set of all real numbers.

GETTING READY FOR CLASS

After reading through the preceding section, respond in your own words and in complete sentences.

A. Why do some of the equations in this section have two solutions instead of one?

B. Translate $|x| = 6$ into words using the definition of absolute value.

C. Explain in words what the inequality $|x - 5| < 2$ means with respect to distance on the number line.

D. Why is there no solution to the inequality $|2x - 3| < 0$?

Problem Set 2.6

Solve each equation.

1. $|a| - 5 = 2$
2. $|y| + 3 = 1$
3. $|a + 2| = \frac{7}{5}$
4. $\left|\frac{2}{7}a + \frac{3}{4}\right| = 1$
5. $800 = |400x - 200|$
6. $|2x - 5| = -7$
7. $\left|\frac{4}{5}x - 5\right| = 15$
8. $\left|2 - \frac{1}{3}a\right| = 10$
9. $|3x + 1| = 4$
10. $|9 - 4x| = 1$
11. $|5x - 3| - 4 = 3$
12. $|8 - 7y| + 9 = 1$
13. $2 + |2t - 6| = 10$
14. $\left|4 - \frac{2}{7}x\right| + 2 = 14$
15. $7 = \left|\frac{3}{5}x + \frac{1}{5}\right| + 2$
16. $1 = -3 + \left|2 - \frac{1}{4}y\right|$
17. $|2(2x + 3)| - 5 = -1$
18. $|3 + 4(3x + 1)| = 7$
19. $4 = -1 + \left|6 - \frac{4}{5}a\right|$
20. $5 = |6(k - 2) + 1|$
21. $|5a + 2| = |4a + 7|$
22. $\left|\frac{1}{10}x - \frac{1}{2}\right| = \left|\frac{1}{5}x + \frac{1}{10}\right|$
23. $|y - 5| = |y - 4|$
24. $|5x - 8| = |5x + 8|$
25. $|0.07 - 0.01x| = |0.08 - 0.02x|$
26. $|x - 4| = |4 - x|$
27. $\left|\frac{x}{3} - 1\right| = \left|1 - \frac{x}{3}\right|$
28. $|-0.4a + 0.6| = |1.3 - 0.2a|$

Paying Attention to Instructions The next two problems are intended to give you practice reading, and paying attention to, the instructions that accompany the problems you are working. Working these problems is an excellent way to get ready for a test or a quiz.

29. Work each problem according to the instructions given.
    - **a.** Solve: $4x - 5 = 0$
    - **b.** Solve: $|4x - 5| = 0$
    - **c.** Solve: $4x - 5 = 3$
    - **d.** Solve: $|4x - 5| = 3$
    - **e.** Solve: $|4x - 5| = |2x + 3|$
30. Work each problem according to the instructions given.
    - **a.** Solve: $3x + 6 = 0$
    - **b.** Solve: $|3x + 6| = 0$
    - **c.** Solve: $3x + 6 = 4$
    - **d.** Solve: $|3x + 6| = 4$
    - **e.** Solve: $|3x + 6| = |7x + 4|$

Solve each inequality and graph the solution set.

31. $|x| \leq 7$
32. $|x| > 4$
33. $|x| - 3 < -1$
34. $|t| + 5 > 8$
35. $|y| > -3$
36. $|x| \leq -4$
37. $|x + 4| < 2$
38. $|a - 6| \geq 3$
39. $|a + 2| \geq -5$
40. $|2x + 6| < 2$
41. $|5y - 1| \geq 4$

42. $|2k - 5| \geq 3$ **43.** $|x + 4| - 3 < -1$ **44.** $|2a - 6| - 1 \geq 2$

Solve each inequality and write your answer using interval notation.

45. $|a + 4| < 6$ **46.** $|2x - 5| \leq 3$ **47.** $|y - 3| > 6$

48. $|-3x + 1| \geq 7$ **49.** $|-3x + 4| \leq 7$ **50.** $|-4x + 2| < 6$

Solve each inequality and graph the solution set.

51. $|5 - x| > 3$ **52.** $|7 - x| > 2$

53. $\left|3 - \frac{3}{4}x\right| \geq 9$ **54.** $\left|3 - \frac{1}{3}x\right| > 1$

Solve each inequality.

55. $|x + 1| < 0.01$ **56.** $|2x - 1| \geq \frac{1}{8}$ **57.** $|2x + 5| < \frac{1}{2}$

58. $\left|\frac{2x - 5}{3}\right| \geq \frac{1}{6}$ **59.** $\left|\frac{2x - 3}{4}\right| < 0.35$ **60.** $\left|\frac{4x - 3}{2}\right| \leq \frac{1}{3}$

61. $\left|2x - \frac{1}{5}\right| < 0.3$ **62.** $\left|3x - \frac{3}{5}\right| < 0.2$

63. Write the continued inequality $-8 \leq x \leq 8$ as a single inequality involving absolute value.

64. Write $-3 \leq x + 2 \leq 3$ as a single inequality involving absolute value.

Paying Attention to Instructions The next two problems are intended to give you practice reading, and paying attention to, the instructions that accompany the problems you are working.

65. Work each problem according to the instructions given.

a. Evaluate when $x = 0$: $|5x + 3|$ **b.** Solve: $|5x + 3| = 7$

c. Is 0 a solution to $|5x + 3| > 7$? **d.** Solve: $|5x + 3| > 7$

66. Work each problem according to the instructions given.

a. Evaluate when $x = 0$: $|-2x - 5|$ **b.** Solve: $|-2x - 5| = 1$

c. Is 0 a solution to $|-2x - 5| > 1$? **d.** Solve: $|-2x - 5| > 1$

Modeling Practice

67. Speed Limits The interstate speed limit for cars is 75 miles per hour in several states. To discourage passing, minimum speeds are also posted, so that the difference between the fastest and slowest moving traffic is no more than 20 miles per hour. Therefore, the speed, x, of a car must satisfy the inequality $55 \leq x \leq 75$. Write this inequality as an absolute value inequality.

68. **Wavelengths of Light** When white light from the sun passes through a prism, it is broken down into bands of light that form colors. The wavelength, v, (in nanometers) of some common colors are approximately:

Blue:	$424 < v < 491$
Green:	$491 < v < 575$
Yellow:	$575 < v < 585$
Orange:	$585 < v < 647$
Red:	$647 < v < 700$

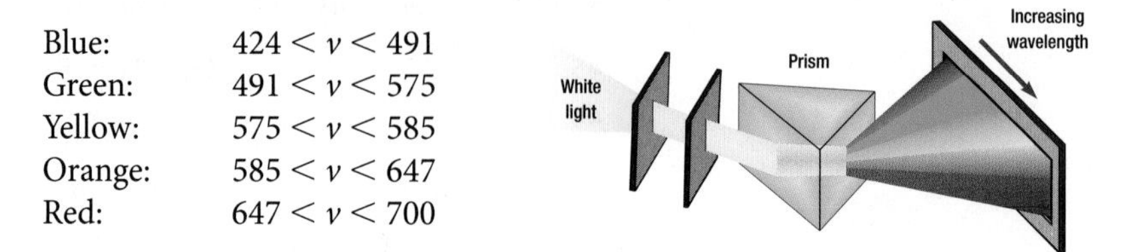

When a fireworks display made of copper is burned, it lets out light with wavelengths, v, that satisfy the relationship $|v - 455| < 23$. Write this inequality without absolute values, find the range of possible values for v, and then using the preceding list of wavelengths, determine the color of that copper fireworks display.

Maintaining Your Skills

Simplify each expression. Assume all variables represent nonzero real numbers, and write your answer with positive exponents only.

69. 3^{-2}

70. $\dfrac{x^6}{x^{-4}}$

71. $\dfrac{15x^3y^8}{5xy^{10}}$

72. $(2a^{-3}b^4)^2$

73. $\dfrac{(3x^{-3}y^5)^{-2}}{(9xy^{-2})^{-1}}$

74. $(3x^4y)^2(5x^3y^4)^3$

Write each number in scientific notation.

75. 54,000

76. 0.0359

Write each number in expanded form.

77. 6.44×10^3

78. 2.5×10^{-2}

Simplify each expression as much as possible. Write all answers in scientific notation.

79. $(3 \times 10^8)(4 \times 10^{-5})$

80. $\dfrac{8 \times 10^5}{2 \times 10^{-8}}$

Chapter 2 Summary

Strategy for Solving Linear Equations in One Variable [2.1]

1. Solve: $3(2x - 1) = 9$.

$$\begin{aligned} 3(2x - 1) &= 9 \\ 6x - 3 &= 9 \\ 6x - 3 + 3 &= 9 + 3 \\ 6x &= 12 \\ \frac{1}{6}(6x) &= \frac{1}{6}(12) \\ x &= 2 \end{aligned}$$

Step 1: **a.** Use the distributive property to separate terms, if necessary.

b. If fractions are present, consider multiplying both sides by the LCD to eliminate the fractions. If decimals are present, consider multiplying both sides by a power of 10 to clear the equation of decimals.

c. Combine similar terms on each side of the equation.

Step 2: Use the addition property of equality to get all variable terms on one side of the equation and all constant terms on the other side. A variable term is a term that contains the variable (for example, $5x$). A constant term is a term that does not contain the variable (the number 3, for example).

Step 3: Use the multiplication property of equality to get the variable by itself on one side of the equation.

Step 4: Check your solution in the original equation to be sure that you have not made a mistake in the solution process.

Solve an Equation by Factoring [2.1]

2. Solve: $100x^2 = 300x$.

$$\begin{aligned} 100x^2 - 300x &= 0 \\ 100x(x - 3) &= 0 \\ x = 0 \quad \text{or} \quad x &= 3 \end{aligned}$$

Step 1: Write the equation in standard form.
Step 2: Factor the left side.
Step 3: Use the zero-factor property to set each factor equal to 0.
Step 4: Solve the resulting linear equations.
Step 5: Check the solutions in the original equation.

Formulas [2.1]

3. Solve for w:

$$\begin{aligned} P &= 2l + 2w \\ P - 2l &= 2w \\ \frac{P - 2l}{2} &= w \end{aligned}$$

A *formula* in algebra is an equation involving more than one variable. To solve a formula for one of its variables, simply isolate that variable on one side of the equation.

The Square Root Property for Equations [2.2]

4. If $(x - 3)^2 = 25$
then $x - 3 = \pm 5$

$$\begin{aligned} x &= 3 \pm 5 \\ x &= 8 \text{ or } x = -2 \end{aligned}$$

If $a^2 = b$, where b is a real number, then

$$a = \sqrt{b} \qquad \text{or} \qquad a = -\sqrt{b}$$

which can be written as $a = \pm\sqrt{b}$.

To Solve a Quadratic Equation by Completing the Square [2.2]

5. Solve $x^2 - 6x - 6 = 0$

$$x^2 - 6x = 6$$
$$x^2 - 6x + 9 = 6 + 9$$
$$(x - 3)^2 = 15$$
$$x - 3 = \pm\sqrt{15}$$
$$x = 3 \pm \sqrt{15}$$

Step 1: Write the equation in the form $ax^2 + bx = c$.
Step 2: If $a \neq 1$, divide through by the constant a so the coefficient of x^2 is 1.
Step 3: Complete the square on the left side by adding the square of $\frac{1}{2}$ the coefficient of x to both sides.
Step 4: Write the left side of the equation as the square of a binomial. Simplify the right side if possible.
Step 5: Apply the square root property for equations, and solve as usual.

The Quadratic Theorem [2.2]

6. If $2x^2 + 3x - 4 = 0$, then

$$x = \frac{-3 \pm \sqrt{9 - 4(2)(-4)}}{2(2)}$$
$$= \frac{-3 \pm \sqrt{41}}{4}$$

For any quadratic equation in the form $ax^2 + bx + c = 0, a \neq 0$, the two solutions are

$$x = \frac{-b \pm \sqrt{b^2 - 4ac}}{2a}$$

This last equation is known as the *quadratic formula.*

Equations Quadratic in Form [2.3]

7. The equation $x^4 - x^2 - 12 = 0$ is quadratic in x^2. Letting $y = x^2$ we have

$$y^2 - y - 12 = 0$$
$$(y - 4)(y + 3) = 0$$
$$y = 4 \quad \text{or} \quad y = -3$$

Resubstituting x^2 for y, we have

$$x^2 = 4 \quad \text{or} \quad x^2 = -3$$
$$x = \pm 2 \quad \text{or} \quad x = \pm i\sqrt{3}$$

There are a variety of equations whose form is quadratic. We solve most of them by making a substitution so the equation becomes quadratic, and then solving the equation by factoring or the quadratic formula. For example,

The equation	*is quadratic in*
$(2x - 3)^2 + 5(2x - 3) - 6 = 0$	$2x - 3$
$4x^4 - 7x^2 - 2 = 0$	x^2
$2x - 7\sqrt{x} + 3 = 0$	$\sqrt{x}$

Equations Involving Rational Expressions [2.3]

8. Solve $\frac{x}{2} + 3 = \frac{1}{3}$.

$$6\left(\tfrac{x}{2}\right) + 6 \cdot 3 = 6 \cdot \tfrac{1}{3}$$
$$3x + 18 = 2$$
$$x = -\frac{16}{3}$$

To solve an equation involving rational expressions, we first find the LCD for all denominators appearing on either side of the equation. We then multiply both sides by the LCD to clear the equation of all fractions and solve as usual.

Squaring Property of Equality [2.3]

9.

$$\sqrt{2x + 1} = 3$$
$$(\sqrt{2x + 1})^2 = 3^2$$
$$2x + 1 = 9$$
$$x = 4$$

We may square both sides of an equation any time it is convenient to do so, as long as we check all resulting solutions in the original equation.

Blueprint for Problem Solving [2.4]

Step 1: ***Read*** the problem, and then mentally ***list*** the items that are known and the items that are unknown.

Step 2: ***Assign a variable*** to one of the unknown items. (In most cases this will amount to letting $x =$ the item that is asked for in the problem.) Then ***translate*** the other ***information*** in the problem to expressions involving the variable.

Step 3: ***Reread*** the problem, and then ***write an equation,*** using the items and variables listed in steps 1 and 2, that describes the situation.

Step 4: ***Solve the equation*** found in step 3.

Step 5: ***Write your answer*** using a complete sentence.

Step 6: ***Reread*** the problem, and ***check*** your solution with the original words in the problem.

10. The perimeter of a rectangle is 32 inches. If the length is 3 times the width, find the dimensions.

Step 1: This step is done mentally.

Step 2: Let $x =$ the width. Then the length is $3x$.

Step 3: The perimeter is 32; therefore

$$2x + 2(3x) = 32$$

Step 4:

$$8x = 32$$
$$x = 4$$

Step 5: The width is 4 inches. The length is $3(4) = 12$ inches.

Step 6: The perimeter is $2(4) + 2(12)$, which is 32. The length is 3 times the width.

Addition Property for Inequalities [2.5]

For expressions A, B, and C,

if $\quad A < B$

then $\quad A + C < B + C$

Adding the same quantity to both sides of an inequality never changes the solution set.

11. Adding 5 to both sides of the inequality $x - 5 < -2$ gives

$$x - 5 + 5 < -2 + 5$$
$$x < 3$$

0 3

Multiplication Property for Inequalities [2.5]

For expressions A, B, and C,

if $\quad A < B$

then $\quad AC < BC \quad$ if $\quad C > 0$ (C is positive)

or $\quad AC > BC \quad$ if $\quad C < 0$ (C is negative)

We can multiply both sides of an inequality by the same nonzero number without changing the solution set as long as each time we multiply by a negative number we also reverse the direction of the inequality symbol.

12. Multiplying both sides of $-2x \geq 6$ by $-\frac{1}{2}$ gives

$$-2x \geq 6$$
$$-\tfrac{1}{2}(-2x) \leq -\tfrac{1}{2}(6)$$
$$x \leq -3$$

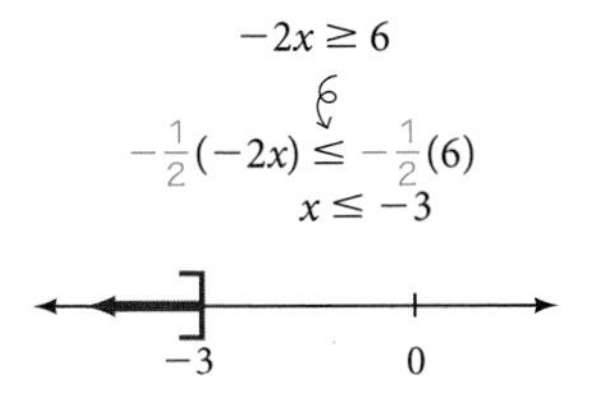

Quadratic Inequalities [2.5]

We solve quadratic inequalities by manipulating the inequality to get 0 on the right side and then factoring the left side. We then make a diagram that indicates where the factors are positive and where they are negative. From this sign diagram and the original inequality we graph the appropriate solution set.

13. Solve $x^2 - 2x - 8 > 0$. We factor and draw the sign diagram:

$$(x - 4)(x + 2) > 0$$

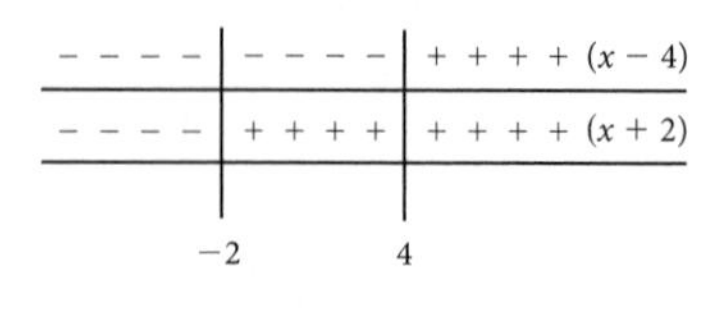

The solution is $x < -2$ or $x > 4$.

Absolute Value Equations [2.6]

To solve an equation that involves absolute value, we isolate the absolute value on one side of the equation and then rewrite the absolute value equation as two separate equations that do not involve absolute value. In general, if b is a positive real number, then

$$|a| = b \quad \text{is equivalent to} \quad a = b \quad \text{or} \quad a = -b$$

14. To solve

$$|2x - 1| + 2 = 7$$

we first isolate the absolute value on the left side by adding -2 to each side to obtain

$$|2x - 1| = 5$$

$$2x - 1 = 5 \quad \text{or} \quad 2x - 1 = -5$$

$$2x = 6 \quad \text{or} \quad 2x = -4$$

$$x = 3 \quad \text{or} \quad x = -2$$

Absolute Value Inequalities [2.6]

To solve an inequality that involves absolute value, we first isolate the absolute value on the left side of the inequality symbol. Then we rewrite the absolute value inequality as an equivalent continued or compound inequality that does not contain absolute value symbols. In general, if b is a positive real number, then

$$|a| < b \quad \text{is equivalent to} \quad -b < a < b$$

and

$$|a| > b \quad \text{is equivalent to} \quad a < -b \quad \text{or} \quad a > b$$

15. To solve

$$|x - 3| + 2 < 6$$

we first add -2 to both sides to obtain

$$|x - 3| < 4$$

which is equivalent to

$$-4 < x - 3 < 4$$

$$-1 < x < 7$$

Chapter 2 Test

Solve the following equations. [2.1, 2.2, 2.3]

1. $5 - \frac{4}{7}a = -11$

2. $\frac{1}{5}x - \frac{1}{2} - \frac{1}{10}x + \frac{2}{5} = \frac{3}{10}x + \frac{1}{2}$

3. $3x^2 = 5x + 2$

4. $100x^3 = 500x^2$

5. $5(x - 1) - 2(2x + 3) = 5x - 4$

6. $0.07 - 0.02(3x + 1) = -0.04x + 0.01$

7. $(x + 1)(x + 2) = 12$

8. $x^3 + 2x^2 - 16x - 32 = 0$

9. $(2x - 6)^2 = -8$

10. $(y + 1)(y - 3) = -6$

11. $8t^3 - 125 = 0$

12. $4x^4 - 7x^2 - 2 = 0$

13. $(2t + 1)^2 - 5(2t + 1) + 6 = 0$

14. $\frac{1}{x} + 3 = \frac{4}{3}$

15. $\frac{x}{x - 3} + 3 = \frac{3}{x - 3}$

16. $1 - \frac{1}{x} = \frac{6}{x^2}$

17. $\sqrt{3x + 1} = x - 3$

18. $\sqrt{x + 3} = \sqrt{x + 4} - 1$

Solve for the indicated variable. [2.1, 2.2, 2.3]

19. $P = 2l + 2w$; for w

20. $A = \frac{1}{2}h(b + B)$; for B

Solve for y. [2.1, 2.2, 2.3]

21. $5x - 2y = 10$

22. $\frac{y - 5}{x - 4} = 3$

Solve each of the following. [2.4]

23. **Geometry** A rectangle is twice as long as it is wide. The perimeter is 36 inches. Find the dimensions.

24. **Geometry** Two angles are supplementary. If the larger angle is 15° more than twice the smaller angle, find the measure of each angle.

25. **Right Triangle** The longest side of a right triangle is 4 inches more than the shortest side. The third side is 2 inches more than the shortest side. Find the length of each side.

26. **Velocity and Height** If an object is thrown straight up into the air with an initial velocity of 32 feet per second, then its height h (in feet) above the ground at any time t (in seconds) is given by the formula $h = 32t - 16t^2$. Find the times at which the object is on the ground by letting $h = 0$ in the equation and solving for t.

Solve the following inequalities. Write the solution set using interval notation, then graph the solution set. [2.5]

27. $-5t \le 30$

28. $5 - \frac{3}{2}x > -1$

29. $1.6x - 2 < 0.8x + 2.8$

30. $3(2y + 4) \ge 5(y - 8)$

31. $x^2 - x - 6 \le 0$

32. $2x^2 + 5x > 3$

Solve the following equations. [2.6]

33. $\left|\frac{1}{4}x - 1\right| = \frac{1}{2}$

34. $|3 - 2x| + 5 = 2$

Solve the following inequalities and graph the solutions. [2.6]

35. $|6x - 1| > 7$

36. $|3x - 5| - 4 \le 3$

37. $|5 - 4x| \ge -7$

38. $|4t - 1| < -3$

Using Art to Depict Mathematics

MODELING EXTRA

It is always interesting to see how people in other disciplines represent mathematics. This is especially true with artists. The art shown here was created by artist Deb Spatafore for an art and mathematics show celebrating Pi Day (3/14/2010). Deb used art to demonstrate Leonardo Da Vinci's ingenious solution to the problem of finding a rectangle that has the same area as a circle. Here is how Leonardo Da Vinci solved the problem.

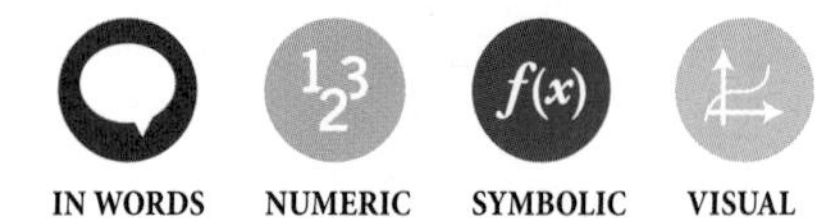

> *For a given circle, form a wheel with a width equal to one-half the radius of the circle. Roll the wheel through one complete revolution. The rectangle formed from the path has the same area as the original circle.*

Figure 1 illustrates Da Vinci's solution. It shows a circle with radius 6 cm, along with the wheel formed according to Da Vinci's instructions. As the wheel rolls through one complete revolution, the rectangular footprint left by the rolling wheel has the same area of the circle used to form the wheel. Why is this true?

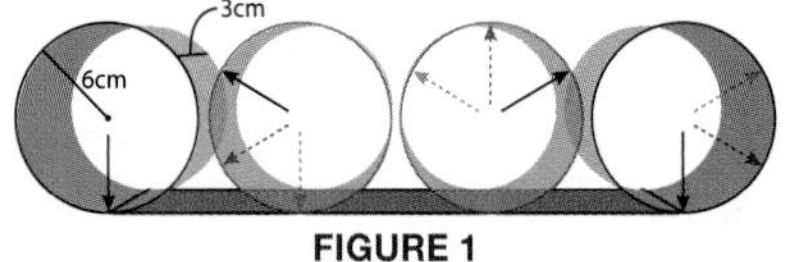

FIGURE 1

Figure 2 shows the same information, generalized so that it applies to any circle, not just one with radius 6 cm. Again, as the wheel rolls through one complete revolution, the rectangular footprint left by the rolling wheel has the same area of the circle used to form the wheel. If you can show why this is true, you will have proved Leonardo Da Vinci's proposition for all circles. Show why this is true.

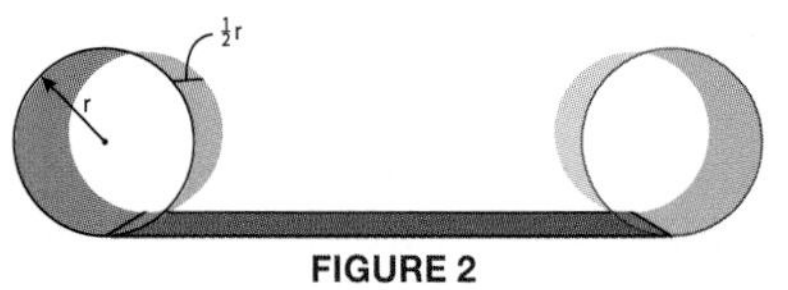

FIGURE 2

In Deb Spatafore's art, she cuts three sides of a rectangle from each photo of the Mona Lisa, and then rolls the strip up to form the circle. In each case, the circle and the rectangle demonstrate Da Vinci's property. This procedure is illustrated in Figure 3.

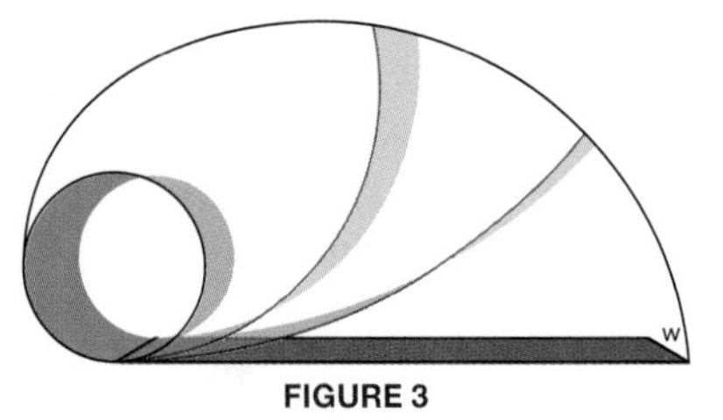

FIGURE 3

Deb starts with the rectangle and then forms the circle from it. She works in the reverse order from the description given by Da Vinci, but the result is the same. She can make any size cut she wants to form the width of the rectangle. How does she know how long to cut the length of the rectangle, so that the resulting circle and rectangle demonstrate Da Vinci's property?

3 Functions and Graphs

Chapter Outline

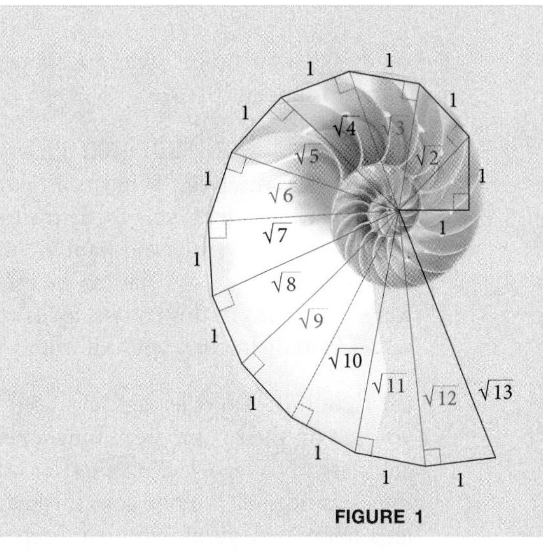

FIGURE 1

The diagram above is called the *spiral of roots*. The spiral of roots mimics the shell of the chambered nautilus, an animal that has survived largely unchanged for millions of years. The mathematical diagram is constructed using the Pythagorean theorem. Table 1 gives the lengths of the diagonals in the spiral of roots, accurate to the nearest hundredth. If we take each of the diagonals in the spiral of roots and place it above the corresponding whole number on the x-axis and then connect the tops of all these segments with a smooth curve, we have the graph shown in Figure 2. This curve is also the graph of the equation $y = \sqrt{x}$.

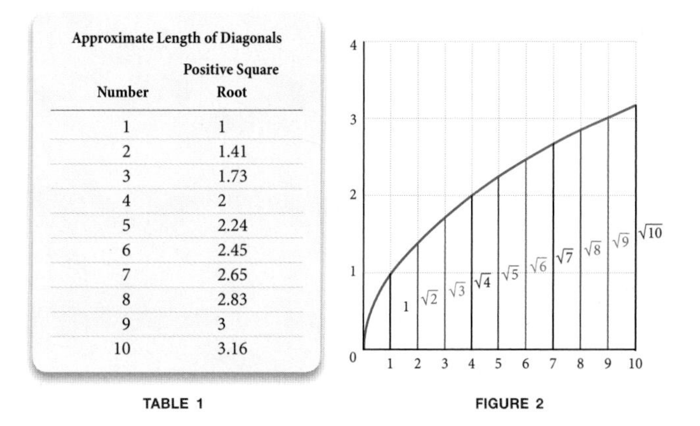

Approximate Length of Diagonals

Number	Positive Square Root
1	1
2	1.41
3	1.73
4	2
5	2.24
6	2.45
7	2.65
8	2.83
9	3
10	3.16

TABLE 1

FIGURE 2

Note When you see this icon next to an example or problem in this chapter, you will know that we are using the topics in this chapter to model situations in the world around us.

In this chapter we begin our work with functions. The diagrams, table, and text above all refer to the *square root function*. Figures 1 and 2 give us a **visual** representation of the square root function; Table 1 is a **numeric** representation, and the equation $y = \sqrt{x}$ is a **symbolic** representation. If you were to describe this relationship to a friend by saying "If we input a whole number, or any positive number, this function gives us (outputs) the square root of that number," you would be giving a **verbal** description as well. Giving different representations for the same function is something we will do throughout this chapter.

Study Skills

The study skills for this chapter are about attitude. They are points of view that point toward success.

1. **Be Focused, Not Distracted** I have students who begin their assignments by asking themselves, "Why am I taking this class?" If you are asking yourself similar questions, you are distracting yourself from doing the things that will produce the results you want in this course. Don't dwell on questions and evaluations of the class that can be used as excuses for not doing well. If you want to succeed in this course, focus your energy and efforts toward success, rather than distracting yourself from your goals.

2. **Be Resilient** Don't let setbacks keep you from your goals. You want to put yourself on the road to becoming a person who can succeed in this class, or any class in college. Failing a test or quiz, or having a difficult time on some topics, is normal. No one goes through college without some setbacks. Don't let a temporary disappointment keep you from succeeding in this course. A low grade on a test or quiz is simply a signal that you need to reevaluate your study habits.

3. **Intend to Succeed** I have a few students who simply go through the motions of studying without intending to master the material. It is more important to them to look like they are studying than to actually study. You need to study with the intention of being successful in the course. Intend to master the material, no matter what it takes.

Paired Data and Graphing

INTRODUCTION Table 1 gives the net price of a popular intermediate algebra text at the beginning of each year in which a new edition was published. (The net price is the price the bookstore pays for the book, not the price you pay for it.)

© ssstep/iStockPhoto

TABLE 1 Price of a Textbook

Edition	Year Published	Net Price ($)
First	1994	30.50
Second	1998	39.25
Third	2002	47.50
Fourth	2006	55.00
Fifth	2010	45.00

The information in Table 1 is represented visually in Figures 1 and 2. The diagram in Figure 1 is called a *bar chart*. The diagram in Figure 2 is called a *line graph*. The data in Table 1 is called *paired data* because each number in the year column is paired with a specific number in the price column.

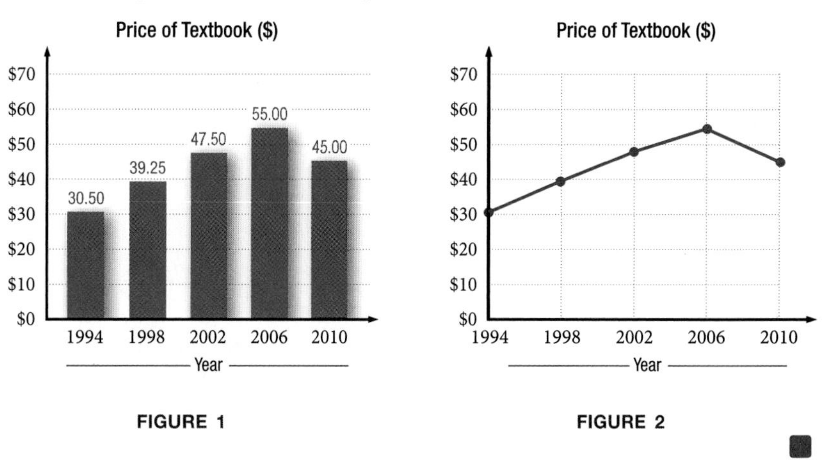

FIGURE 1

FIGURE 2

In this section we place our work with charts and graphs in a more formal setting. Our foundation will be the *rectangular coordinate system*, because it gives us a link between algebra and geometry. With it we notice relationships between certain equations and different lines and curves.

Ordered Pairs

Paired data play an important role in equations that contain two variables. Working with these equations is easier if we standardize the terminology and notation associated with paired data. So here is a definition that will do just that.

> **DEFINITION**
>
> A pair of numbers enclosed in parentheses and separated by a comma, such as $(-2, 1)$, is called an *ordered pair* of numbers. The first number in the pair is called the *x-coordinate* of the ordered pair; the second number is called the *y-coordinate.* For the ordered pair $(-2, 1)$, the x-coordinate is -2 and the y-coordinate is 1.

Rectangular Coordinate System

A *rectangular coordinate system* is made by drawing two real number lines at right angles to each other. The two number lines, called *axes*, cross each other at 0. This point is called the *origin*. Positive directions are to the right and up. Negative directions are to the left and down. The rectangular coordinate system is shown in Figure 3.

> *Note* A rectangular coordinate system allows us to connect algebra and geometry by associating geometric shapes (the curves shown in the diagrams) with algebraic equations. The French philosopher and mathematician René Descartes (1596 – 1650) is usually credited with the invention of the rectangular coordinate system, which is often referred to as the Cartesian coordinate system in his honor. However, the connection between algebra and geometry was also shown by the Persian poet and mathematician, Omar Khayyam, as early as 1070. As a philosopher, Descartes is responsible for the statement, "I think, therefore, I am."

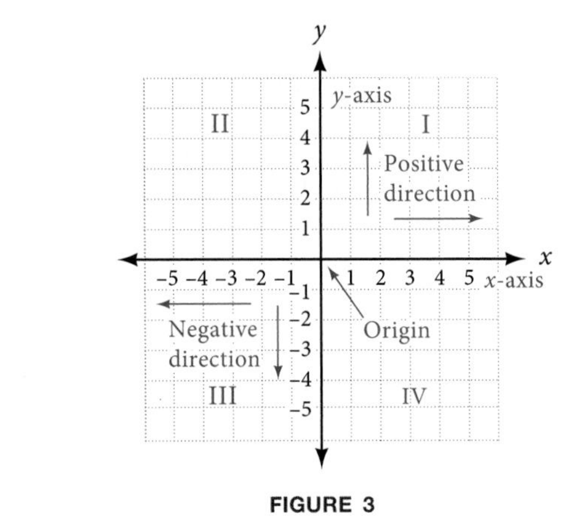

FIGURE 3

The horizontal number line is called the *x-axis*, and the vertical number line is called the *y-axis*. The two number lines divide the coordinate system into four *quadrants*, which we number I through IV in a counterclockwise direction. Points on the axes are not considered as being in any quadrant.

Graphing Ordered Pairs

To graph the ordered pair (a, b) on a rectangular coordinate system, we start at the origin and move a units right or left (right if a is positive, left if a is negative). Then we move b units up or down (up if b is positive, down if b is negative). The point where we end up is the graph of the ordered pair (a, b).

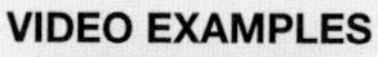

SECTION 3.1

Note From Example 1, we see that any point in quadrant I has both its x- and y-coordinates positive $(+, +)$. Points in quadrant II have negative x-coordinates and positive y-coordinates $(-, +)$. In quadrant III, both coordinates are negative $(-, -)$. In quadrant IV, the form is $(+, -)$.

EXAMPLE 1 Plot (graph) the ordered pairs $(2, 5)$, $(-2, 5)$, $(-2, -5)$, and $(2, -5)$.

SOLUTION To graph the ordered pair $(2, 5)$, we start at the origin and move 2 units to the right, then 5 units up. We are now at the point whose coordinates are $(2, 5)$. We graph the other three ordered pairs in a similar manner (see Figure 4).

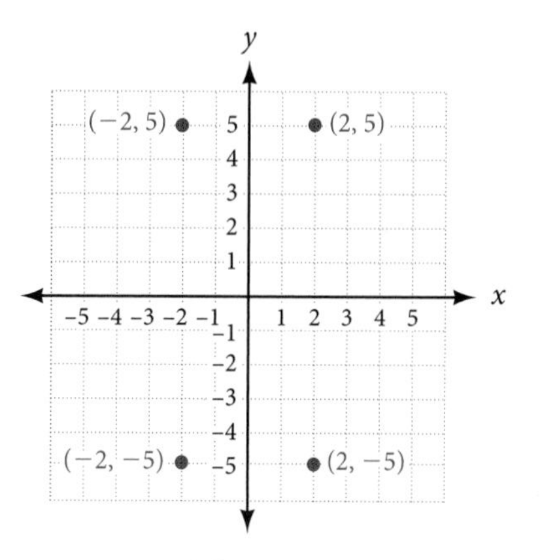

FIGURE 4

EXAMPLE 2 Graph the ordered pairs $(1, -3)$, $\left(\frac{1}{2}, 2\right)$, $(3, 0)$, $(0, -2)$, $(-1, 0)$, and $(0, 5)$.

SOLUTION

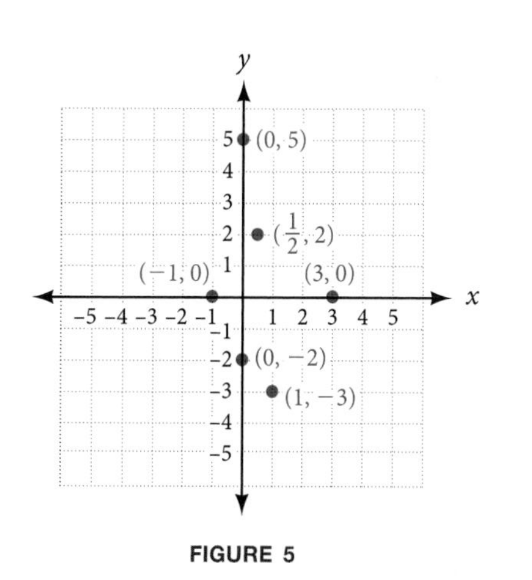

FIGURE 5

From Figure 5, we see that any point on the x-axis has a y-coordinate of 0 (it has no vertical displacement), and any point on the y-axis has an x-coordinate of 0 (no horizontal displacement).

Graphing Equations

We can plot a single point from an ordered pair, but to draw a line or a curve, we need more points.

To graph an equation in two variables, we simply graph its solution set. That is, we draw a line or smooth curve through all the points whose coordinates satisfy the equation.

EXAMPLE 3 Graph the equation $y = -\frac{1}{3}x$.

SOLUTION The graph of this equation will be a straight line. We need to find three ordered pairs that satisfy the equation. To do so, we can let x equal any numbers we choose and find corresponding values of y. However, because every value of x we substitute into the equation is going to be multiplied by $-\frac{1}{3}$, let's use numbers for x that are divisible by 3, like -3, 0, and 3. That way, when we multiply them by $-\frac{1}{3}$, the result will be an integer.

$$\text{Let } x = -3; \qquad y = -\frac{1}{3}(-3) = 1$$

The ordered pair $(-3, 1)$ is one solution.

$$\text{Let } x = 0; \qquad y = -\frac{1}{3}(0) = 0$$

The ordered pair $(0, 0)$ is a second solution.

$$\text{Let } x = 3; \qquad y = -\frac{1}{3}(3) = -1$$

The ordered pair $(3, -1)$ is a third solution.

In table form

x	y
-3	1
0	0
3	-1

Plotting the ordered pairs $(-3, 1)$, $(0, 0)$, and $(3, -1)$ and drawing a straight line through their graphs, we have the graph of the equation $y = -\frac{1}{3}x$ as shown in Figure 6.

Note It takes only two points to determine a straight line. We have included a third point for "insurance." If all three points do not line up in a straight line, we have made a mistake.

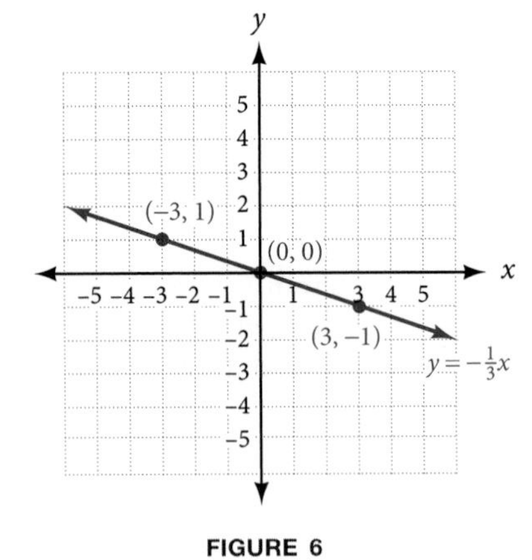

FIGURE 6

Example 3 illustrates again the connection between algebra and geometry that we mentioned earlier in this section. Descartes's rectangular coordinate system allows us to associate the equation $y = -\frac{1}{3}x$ (an algebraic concept) with a specific straight line (a geometric concept). The study of the relationship between equations in algebra and their associated geometric figures is called *analytic geometry*.

Lines Through the Origin

As you can see from Figure 6, the graph of the equation $y = -\frac{1}{3}x$ is a straight line that passes through the origin. The same will be true of the graph of any equation that has the same form as $y = -\frac{1}{3}x$. Here are three more equations that have that form, along with their graphs.

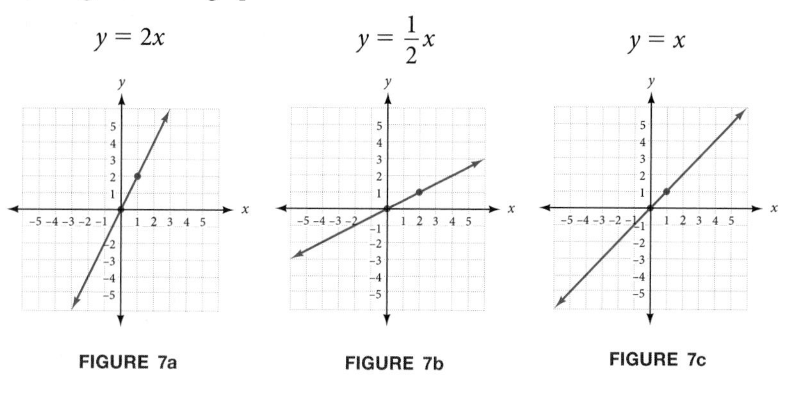

FIGURE 7a FIGURE 7b FIGURE 7c

Here is a summary of this discussion.

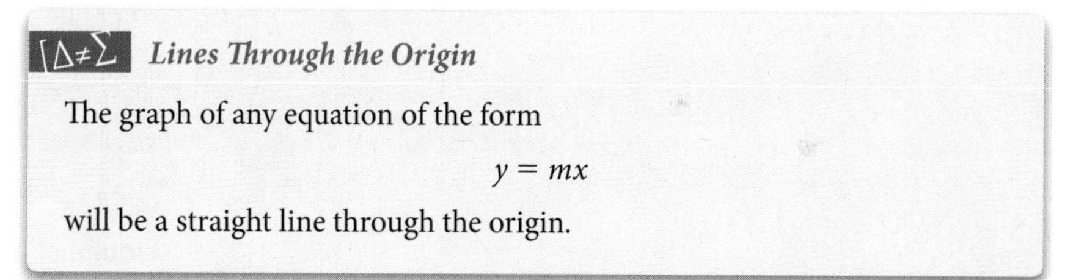

Lines Through the Origin

The graph of any equation of the form

$$y = mx$$

will be a straight line through the origin.

Intercepts

Two important points on the graph of a straight line, if they exist, are the points where the graph crosses the axes.

DEFINITION *Intercepts*

An *x-intercept* of the graph of an equation is the x-coordinate of a point where the graph crosses the x-axis. The *y-intercept* is defined similarly.

Because any point on the x-axis has a y-coordinate of 0, we can find the x-intercept by letting $y = 0$ and solving the equation for x. We find the y-intercept by letting $x = 0$ and solving for y.

EXAMPLE 4 Find the x- and y-intercepts for $2x + 3y = 6$; then graph the equation.

SOLUTION To find the y-intercept, we let $x = 0$.

When $x = 0$

we have $2(0) + 3y = 6$

$3y = 6$

$y = 2$

The y-intercept is 2 so the graph crosses the y-axis at the point (0, 2). To find the x-intercept, we let $y = 0$.

When $y = 0$

we have $2x + 3(0) = 6$

$2x = 6$

$x = 3$

The x-intercept is 3, so the graph crosses the x-axis at the point (3, 0). We use these results to graph the solution set for $2x + 3y = 6$. The graph is shown in Figure 8.

Note Graphing straight lines by finding the intercepts works best when the coefficients of x and y are factors of the constant term.

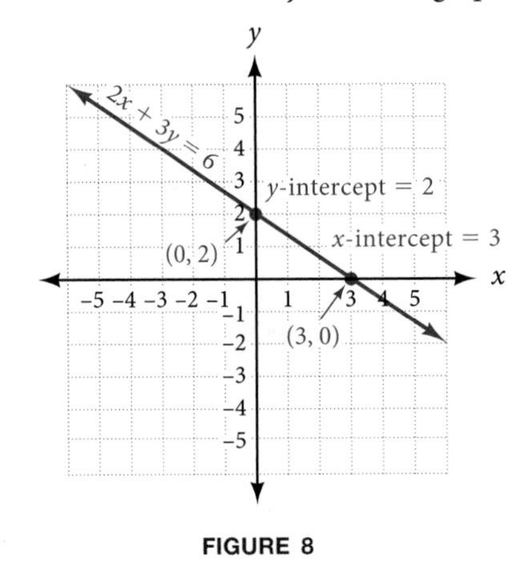

FIGURE 8

Our next two examples involve curves rather than lines. The intercepts are useful in these situations, as well.

EXAMPLE 5 Find the intercepts for $y = x^2 - 4$; then graph the equation.

SOLUTION We find the intercepts using the same method used in Example 4.

x-intercept

When $y = 0$, we have

$0 = x^2 - 4$

$0 = (x + 2)(x - 2)$

$x + 2 = 0$ or $x - 2 = 0$

$x = -2$ $\quad$ $x = 2$

The x-intercepts are -2 and 2

y-intercept

When $x = 0$, we have

$y = 0^2 - 4 = -4$

The y-intercept is -4

We use the intercepts and a table of additional values to sketch the graph.

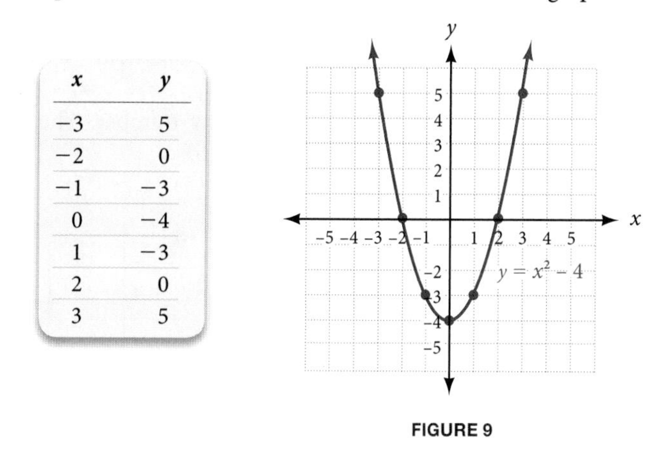

x	y
−3	5
−2	0
−1	−3
0	−4
1	−3
2	0
3	5

FIGURE 9

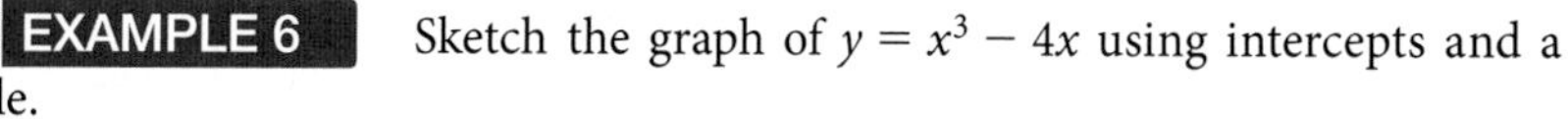

EXAMPLE 6 Sketch the graph of $y = x^3 - 4x$ using intercepts and a table.

SOLUTION The starting table below is used to get a rough idea of what the graph looks like (see Figure 10). The detail table is used to confirm that the graph has the correct shape.

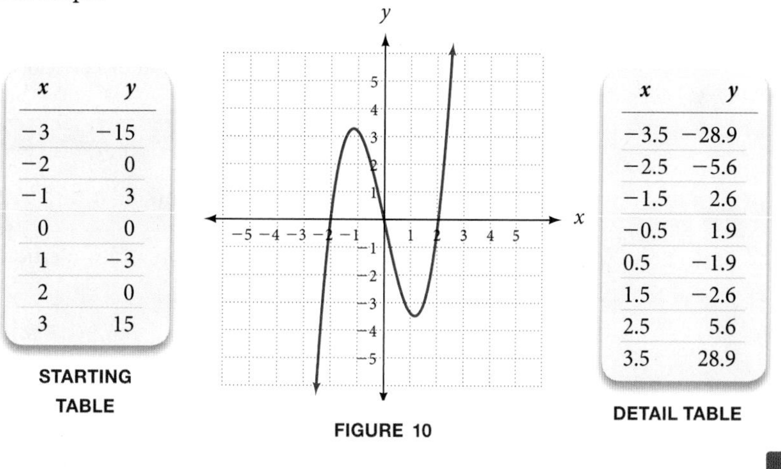

STARTING TABLE

x	y
−3	−15
−2	0
−1	3
0	0
1	−3
2	0
3	15

DETAIL TABLE

x	y
−3.5	−28.9
−2.5	−5.6
−1.5	2.6
−0.5	1.9
0.5	−1.9
1.5	−2.6
2.5	5.6
3.5	28.9

FIGURE 10

Horizontal and Vertical Lines

EXAMPLE 7 Graph each of the following lines.

a. $x = 3$ **b.** $y = -2$

SOLUTION

a. The line $x = 3$ is the set of all points whose x-coordinate is 3. The variable y does not appear in the equation, so the y-coordinate can be any number. Note that we can write our equation as a linear equation in two variables by writing it as $x + 0y = 3$. Because the product of 0 and y will always be 0, y can be any number. The graph of $x = 3$ is the vertical line shown in Figure 11a.

b. The line $y = -2$ is the set of all points whose y-coordinate is -2. The variable x does not appear in the equation, so the x-coordinate can be any number. Again, we can write our equation as a linear equation in two variables by writing it as $0x + y = -2$. Because the product of 0 and x will always be 0, x can be any number. The graph of $y = -2$ is the horizontal line shown in Figure 11b.

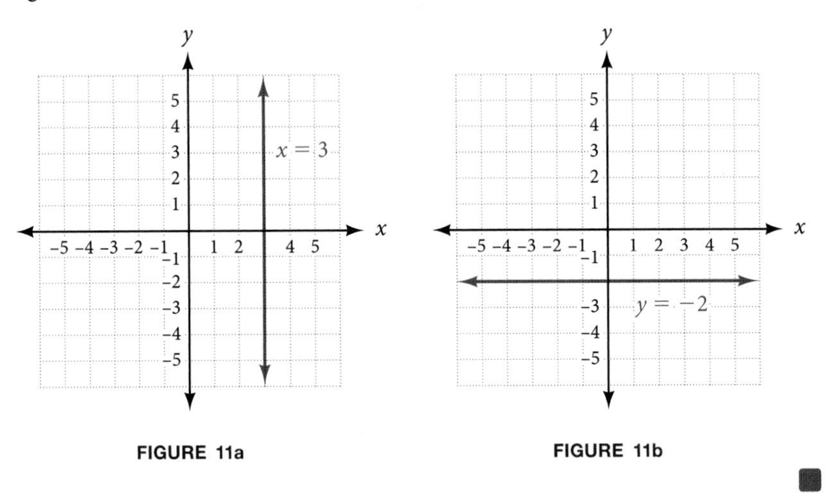

FIGURE 11a **FIGURE 11b**

Here is a summary of what we know about lines that pass through the origin, and lines that are horizontal or vertical.

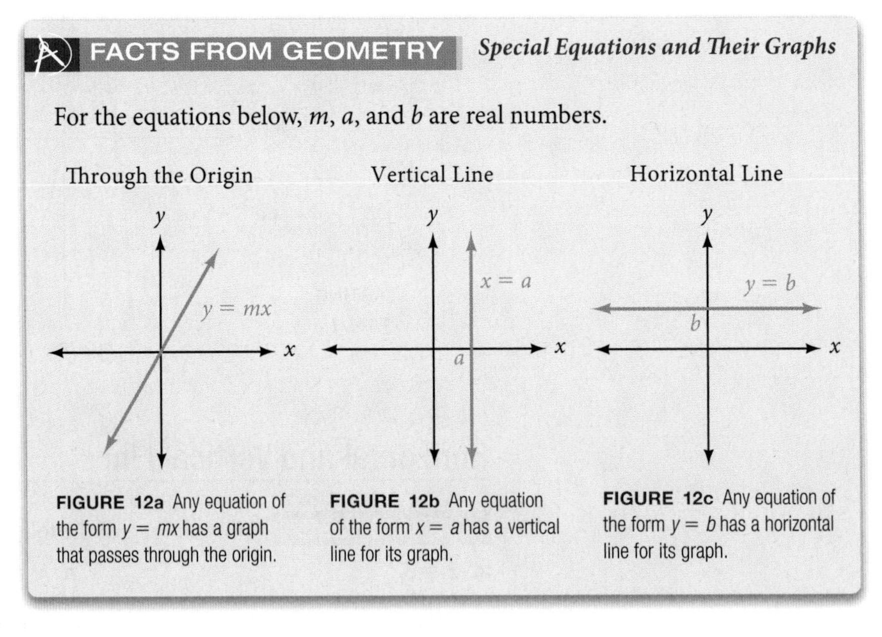

FACTS FROM GEOMETRY *Special Equations and Their Graphs*

For the equations below, m, a, and b are real numbers.

FIGURE 12a Any equation of the form $y = mx$ has a graph that passes through the origin.

FIGURE 12b Any equation of the form $x = a$ has a vertical line for its graph.

FIGURE 12c Any equation of the form $y = b$ has a horizontal line for its graph.

Modeling: Bicycle Rental Program

© mbbirdy/iStockPhoto

EXAMPLE 8 In May 2005, the city of Lyon, France, started a bicycle rental program. Over 3,000 bicycles are available at 350 computerized stations around the city. Subscribers pay an annual 15 euro fee (about \$19.61) and get a PIN to access the bicycles. The bicycles rent for 3 euro per hour and can be returned to any station. (www.velov.grandlyon.com, March 15, 2012)

Your community decides to set up a similar program, charging a \$15 subscription fee and \$5 an hour for rental. (A fraction of an hour is charged as the corresponding fraction of \$5.)

a. Make a table of values showing the cost, C, of renting a bike for various lengths of time, t.

b. Plot the points on a graph. Draw a line through the data points.

c. Write an equation for C in terms of t.

SOLUTION

a. There is an initial fee of \$15, and a rental fee of \$5 per hour. To find the cost of a rental, we multiply the number of hours by \$5 per hour, and add the result to the \$15 subscription fee. For example the cost for a one-hour rental is

$$\text{Cost} = (\$15 \text{ subscription}) + \$5 \times (\textit{one hour})$$
$$C = 15 + 5(1) = 20$$

A one-hour bike rental costs \$20. Here are more bike rental costs.

Length of Rental (Hours)	Cost of Rental (Dollars)		(t, C)
0	15	$C = 15 + 5(\mathbf{0})$	(0, 15)
1	20	$C = 15 + 5(\mathbf{1})$	(1, 20)
2	25	$C = 15 + 5(\mathbf{2})$	(2, 25)
3	30	$C = 15 + 5(\mathbf{3})$	(3, 30)

b. Each pair of values represents a point on the graph. The first value gives the horizontal coordinate of the point, and the second value gives the vertical coordinate. The points lie on a straight line, as shown in the figure. The line extends infinitely in only one direction, because negative values of t do not make sense here.

c. To write an equation, let C represent the cost of the rental and use t for the number of hours:

$$\text{Cost} = (\$15 \text{ subscription}) + \$5 \cdot (\text{number of hours})$$
$$C = 15 + 5t$$

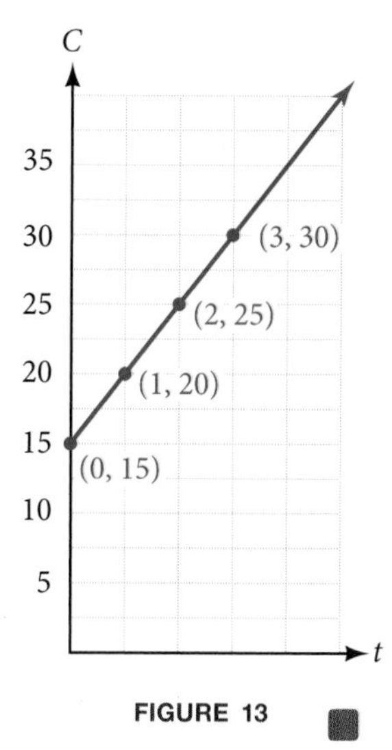

FIGURE 13

Using Technology 3.1

Graphing with Trace and Zoom

All graphing calculators have the ability to graph a function and then trace over the points on the graph, giving their coordinates. Furthermore, all graphing calculators can zoom in and out on a graph that has been drawn.

EXAMPLE 9 Use your graphing calculator to

a. Graph $y = -x + 8$.

b. Name three points on the graph.

c. Use the zoom feature to zoom out to a window twice as large as your original window.

SOLUTION

a. To graph an equation on a graphing calculator, we first set the graph window. Use the notation

$$\text{Window: } \quad -5 \le x \le 4 \text{ and } -3 \le y \le 2$$

to stand for a window in which Xmin $= -5$, Xmax $= 4$, Ymin $= -3$, and Ymax $= 2$.

Set your calculator to the following window:

$$\text{Window: } \quad -10 \le x \le 10 \text{ and } -10 \le y \le 10$$

Graph the equation $Y = -X + 8$ and compare your results with this graph:

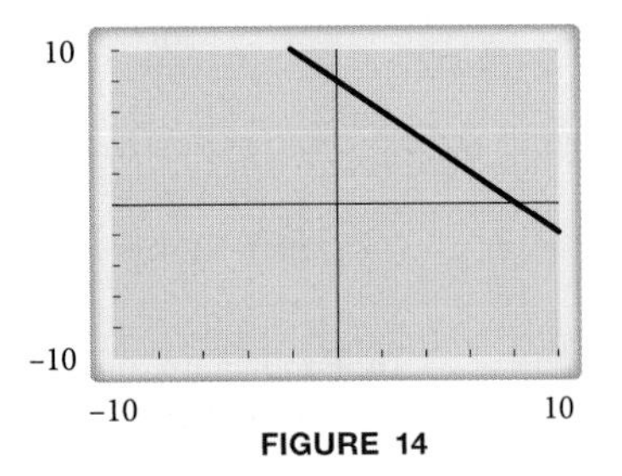

FIGURE 14

b. If we trace to the x-intercept we get

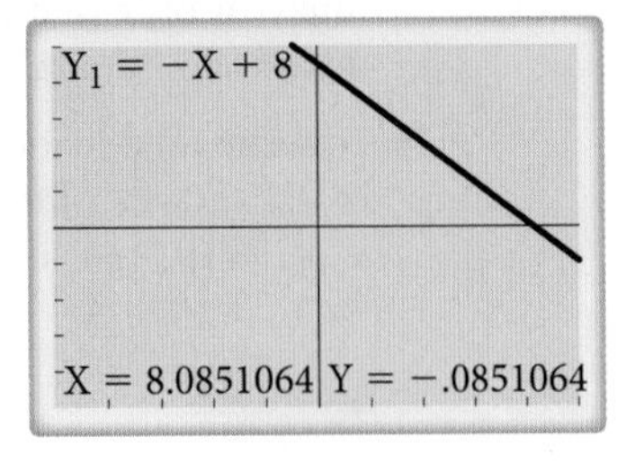

FIGURE 15

c. After zooming out your graphing window should look like this:

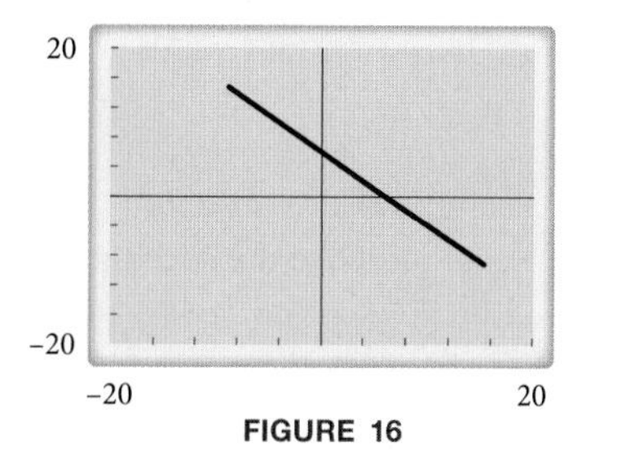

FIGURE 16

Solving for y First

To graph the equation from Example 4, $2x + 3y = 6$, on a graphing calculator, you must first solve it for y. When you do so, you will get $y = -\frac{2}{3}x + 2$, which results in the graph in Figure 8. Use this window:

$$\text{Window:} \quad -6 \leq x \leq 6 \text{ and } -6 \leq y \leq 6$$

Hint on Tracing

If you are going to use the Trace feature and you want the x-coordinates to be exact numbers, set your window so the range of X inputs is a multiple of the number of horizontal pixels on your calculator screen. On the TI-82/83, the screen is 94 pixels wide. Here are a few convenient trace windows:

To trace with x to the nearest tenth, use $-4.7 \leq x \leq 4.7$ or $0 \leq x \leq 9.4$.

To trace with x to the nearest integer, use $-47 \leq x \leq 47$ or $0 \leq x \leq 94$.

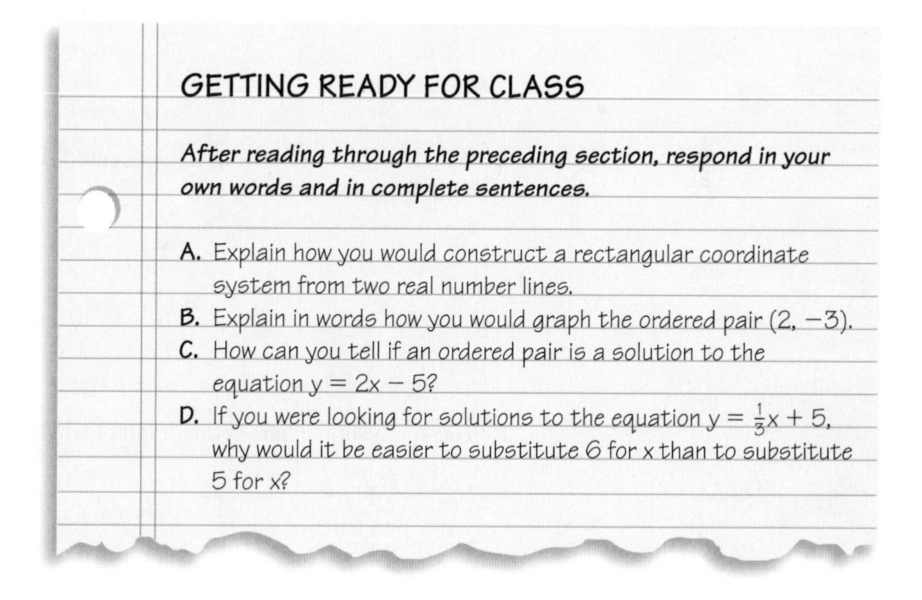

GETTING READY FOR CLASS

After reading through the preceding section, respond in your own words and in complete sentences.

A. Explain how you would construct a rectangular coordinate system from two real number lines.

B. Explain in words how you would graph the ordered pair $(2, -3)$.

C. How can you tell if an ordered pair is a solution to the equation $y = 2x - 5$?

D. If you were looking for solutions to the equation $y = \frac{1}{3}x + 5$, why would it be easier to substitute 6 for x than to substitute 5 for x?

Problem Set 3.1

Graph each ordered pair.

1. **a.** $(-1, 2)$ **b.** $(-1, -2)$ **c.** $(5, 0)$ **d.** $(0, 2)$ **e.** $(-5, -5)$ **f.** $\left(\frac{1}{2}, 2\right)$

2. **a.** $(1, 2)$ **b.** $(1, -2)$ **c.** $(0, -3)$ **d.** $(4, 0)$ **e.** $(-4, -1)$ **f.** $\left(3, \frac{1}{4}\right)$

Give the coordinates of each point.

3. **4.**

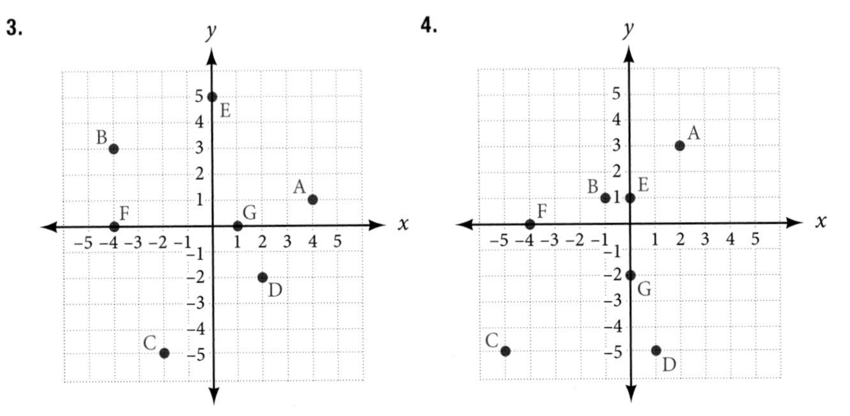

5. Which of the following tables could be produced from the equation $y = 2x - 6$?

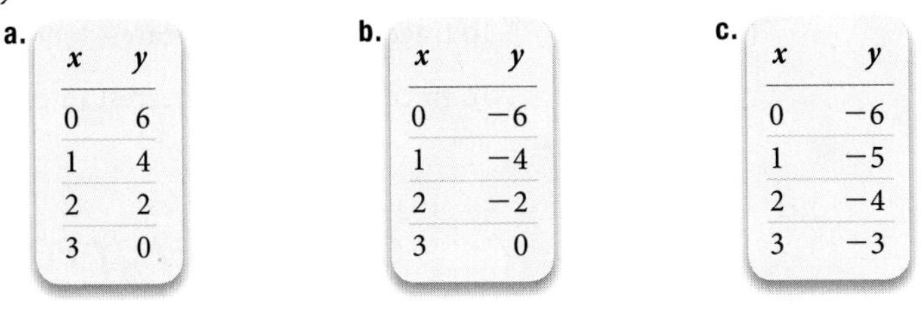

a.

x	y
0	6
1	4
2	2
3	0

b.

x	y
0	−6
1	−4
2	−2
3	0

c.

x	y
0	−6
1	−5
2	−4
3	−3

6. Which of the following tables could be produced from the equation $3x - 5y = 15$?

a.

x	y
0	5
−3	0
10	3

b.

x	y
0	−3
5	0
10	3

c.

x	y
0	−3
−5	0
10	−3

7. The graph shown here is the graph of which of the following equations?

a. $y = \frac{3}{2}x - 3$

b. $y = \frac{2}{3}x - 2$

c. $y = -\frac{2}{3}x + 2$

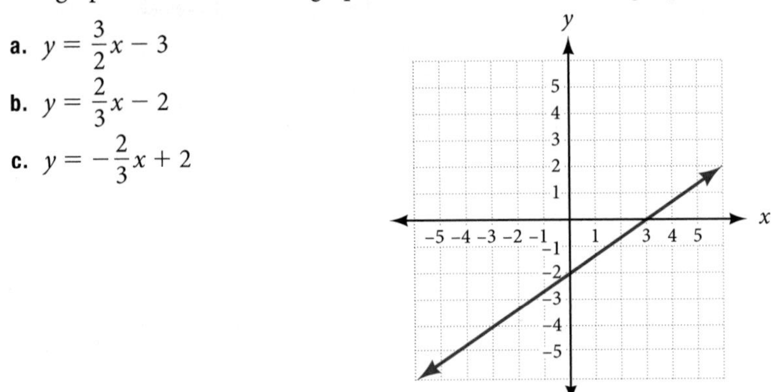

8. The graph shown here is the graph of which of the following equations?

a. $3x - 2y = 8$

b. $2x - 3y = 8$

c. $2x + 3y = 8$

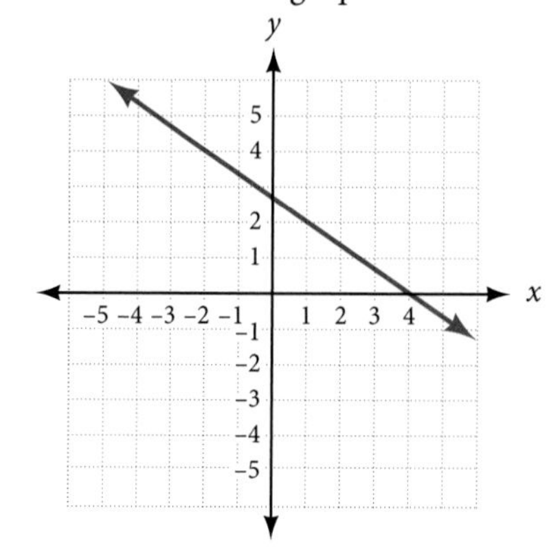

9. Graph the straight line $0.02x + 0.03y = 0.06$.

10. Graph the straight line $0.05x - 0.03y = 0.15$.

11. Graph each of the following lines.

a. $y = 2x$ **b.** $x = -3$ **c.** $y = 2$

12. Graph each of the following lines.

a. $y = 3x$ **b.** $x = -2$ **c.** $y = 4$

13. Graph each of the following lines.

a. $y = -\frac{1}{2}x$ **b.** $x = 4$ **c.** $y = -3$

14. Graph each of the following lines.

a. $y = -\frac{1}{3}x$ **b.** $x = 1$ **c.** $y = -5$

Find the intercepts for each graph. Then use them to help sketch the graph.

15. $y = x^2 - 9$ **16.** $y = x^2$ **17.** $y = 2x - 4$ **18.** $y = 4x - 2$

19. $y = \frac{1}{2}x + 1$ **20.** $y = -\frac{1}{2}x + 1$ **21.** $y = 3x$ **22.** $y = -\frac{1}{3}x$

23. $y = x^2 - x$ **24.** $y = x^2 + 3$ **25.** $y = x - 3$ **26.** $y = x + 2$

Paying Attention to Instructions The next two problems are intended to give you practice reading, and paying attention to, the instructions that accompany the problems you are working.

27. Work each problem according to the instructions given:

a. Solve: $4x + 12 = -16$

b. Find x when y is 0: $4x + 12y = -16$

c. Find y when x is 0: $4x + 12y = -16$

d. Graph: $4x + 12y = -16$

e. Solve for y: $4x + 12y = -16$

28. Work each problem according to the instructions given:

a. Solve: $3x - 8 = -12$

b. Find x when y is 0: $3x - 8y = -12$

c. Find y when x is 0: $3x - 8y = -12$

d. Graph: $3x - 8y = -12$

e. Solve for y: $3x - 8y = -12$

Modeling Practice

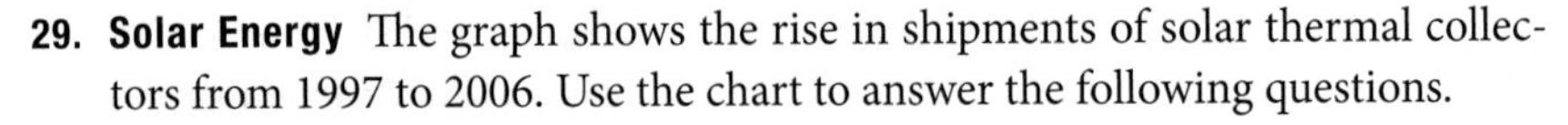

29. Solar Energy The graph shows the rise in shipments of solar thermal collectors from 1997 to 2006. Use the chart to answer the following questions.

a. Does the graph contain the point (2000, 7,500)?

b. Does the graph contain the point (2004, 15,000)?

c. Does the graph contain the point (2005, 15,000)?

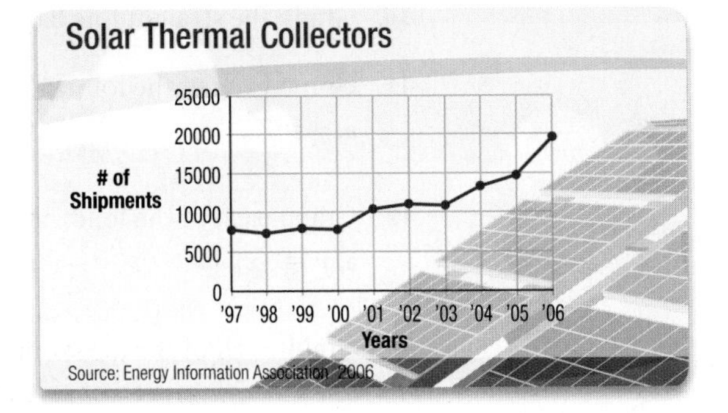

30. Health Care Costs The graph shows the projected rise in the cost of health care from 2002 to 2014. Using years as x and billions of dollars as y, write five ordered pairs that describe the information in the graph.

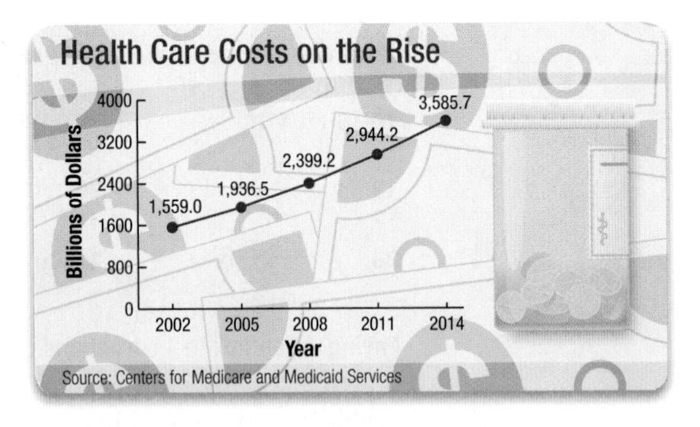

31. Hourly Wages Suppose you have a job that pays $7.50 per hour, and you work anywhere from 0 to 40 hours per week. Table 2 gives the amount of money you will earn in 1 week for working various hours. Construct a graph from the information in Table 2.

TABLE 2 Weekly Wages

Hours Worked	Pay ($)
0	0
10	75
20	150
30	225
40	300

32. Softball Toss Chaudra is tossing a softball into the air with an underhand motion. It takes exactly 2 seconds for the ball to come back to her. Table 3 shows the distance the ball is above her hand at quarter-second intervals. Construct a graph from the information in the table.

TABLE 3 Tossing a softball into the air

Time (sec)	Distance (ft)
0	0
0.25	7
0.5	12
0.75	15
1	16
1.25	15
1.5	12
1.75	7
2	0

33. Intensity of Light Table 4 gives the intensity of light that falls on a surface at various distances from a 100-watt light bulb. Construct a graph from the information in Table 4.

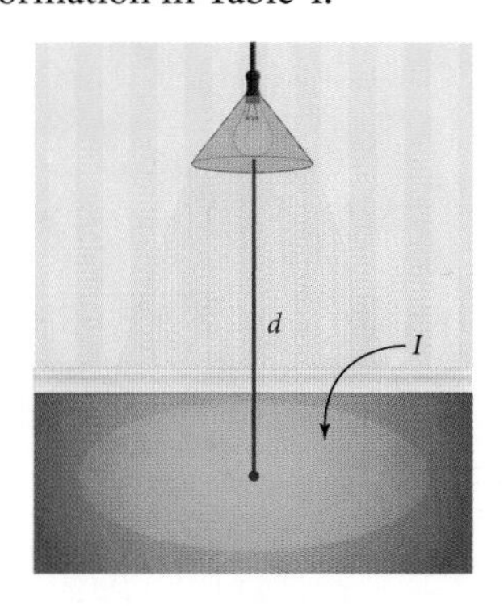

TABLE 4 Light intensity from a 100-watt light bulb

Distance Above Surface (ft)	Intensity (lumens/sq ft)
1	120.0
2	30.0
3	13.3
4	7.5
5	4.8
6	3.3

34. Value of a Painting A piece of abstract art was purchased in 1990 for \$125. Table 5 shows the value of the painting at various times, assuming that it doubles in value every 5 years. Construct a graph from the information in the table.

TABLE 5 Value of a Painting

Year	Value (\$)
1990	125
1995	250
2000	500
2005	1,000
2010	2,000

35. Reading Graphs The graph shows the number of people in line at a theater box office to buy tickets for a movie that starts at 7:30. The box office opens at 6:45.

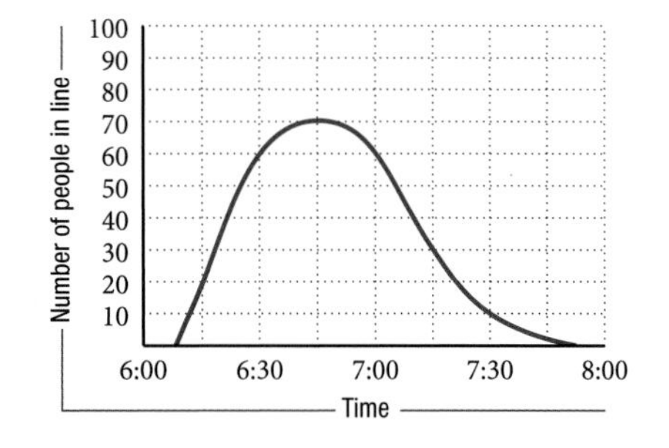

a. How many people are in line at 6:30?
b. How many people are in line when the box office opens?
c. How many people are in line when the show starts?
d. At what times are there 60 people in line?
e. How long after the show starts is there no one left in line?

36. Kentucky Derby The graph gives the monetary bets placed at the Kentucky Derby for specific years. If x represents the year in question and y represents the total wagering for that year, write five ordered pairs that describe the information in the graph.

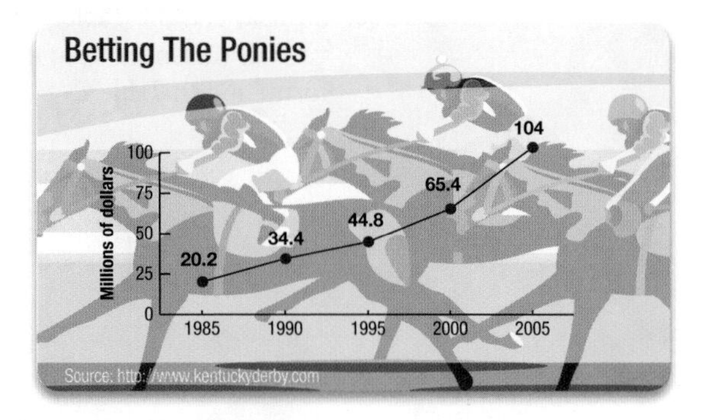

Using Technology Exercises

Graph each equation using the indicated window.

37. $y = \frac{1}{2}x - 3$ $\quad -10 \le x \le 10$ and $-10 \le y \le 10$

38. $y = \frac{1}{2}x^2 - 3$ $\quad -10 \le x \le 10$ and $-10 \le y \le 10$

39. $y = \frac{1}{2}x^2 - 3$ $\quad -4.7 \le x \le 4.7$ and $-10 \le y \le 10$

40. $y = x^3$ $\quad -10 \le x \le 10$ and $-10 \le y \le 10$

41. $y = x^3 - 5$ $\quad -4.7 \le x \le 4.7$ and $-10 \le y \le 10$

Getting Ready for the Next Section

Complete each table using the given equation.

42. $y = 8.5x$

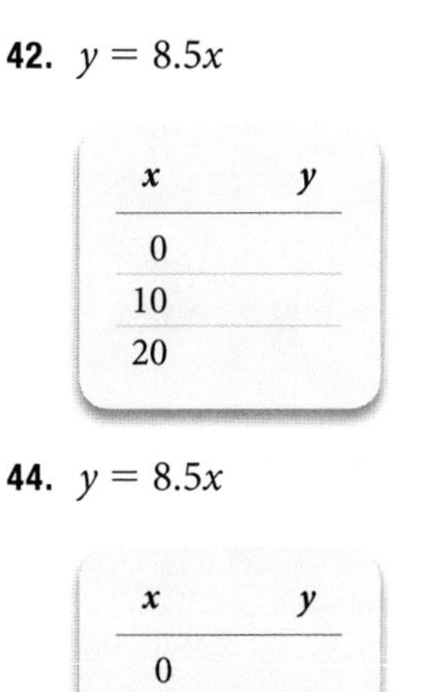

x	y
0	
10	
20	

43. $h = 32t - 16t^2$

t	h
0	
1	
−1	

44. $y = 8.5x$

x	y
0	
$\frac{1}{2}$	
1	

45. $h = 32t - 16t^2$

t	h
−3	
0	
3	

46. $y = \frac{1}{x}$

x	y
−3	
−1	
0	
1	
3	

47. $y = \sqrt{x}$

x	y
−1	
0	
1	
4	

Introduction to Functions and Relations

INTRODUCTION The ad shown here appeared in the Help Wanted section of the local newspaper the day I was writing this section of the book. If you held the job described in the ad, you would earn \$8.50 for every hour you worked. The amount of money you make in one week depends on the number of hours you work that week. In mathematics, we say that your weekly earnings are a *function* of the number of hours you work.

YARD PERSON
Full time 40 hrs. with weekend work required. Cleaning & loading trucks. \$8.50/hr.
Valid CDL with clean record & drug screen required.
Submit current MVR to KCI, 225 Suburban Rd., SLO 93405 (805) 555-3304.
BABY SITTER

An Informal Look at Functions

Suppose you have a job that pays \$8.50 per hour and that you work anywhere from 0 to 40 hours per week. If we let the variable x represent hours and the variable y represent the money you make, then the relationship between x and y can be written as

$$y = 8.5x \qquad \text{for} \qquad 0 \le x \le 40$$

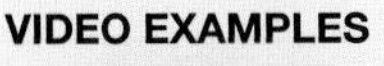

SECTION 3.2

EXAMPLE 1 Construct a table and graph for the function

$$y = 8.5x \qquad \text{for} \qquad 0 \le x \le 40$$

SOLUTION Table 1 gives some of the paired data that satisfy the equation $y = 8.5x$. Figure 1 is the graph of the equation with the restriction $0 \le x \le 40$.

TABLE 1 Weekly Wages

Hours Worked	Rule	Pay
x	$y = 8.5x$	y
0	$y = 8.5(0)$	0
10	$y = 8.5(10)$	85
20	$y = 8.5(20)$	170
30	$y = 8.5(30)$	255
40	$y = 8.5(40)$	340

y
350
300
250
200
150
100
50
0
Pay (\$)
0 10 20 30 40 x
Hours worked

Ordered Pairs

(0, 0)
(10, 85)
(20, 170)
(30, 255)
(40, 340)

FIGURE 1
Weekly wages at \$8.50 per hour

The equation $y = 8.5x$ with the restriction $0 \le x \le 40$, Table 1, and Figure 1 are three ways to describe the same relationship between the number of hours you work in one week and your gross pay for that week. In all three, we *input* values of x, and then use the function rule to *output* values of y.

Domain and Range of a Function

We began this discussion by saying that the number of hours worked during the week was from 0 to 40, so these are the values that x can assume. From the line graph in Figure 1, we see that the values of y range from 0 to 340. We call the complete set of values that x can assume the *domain* of the function. The values that are assigned to y are called the *range* of the function.

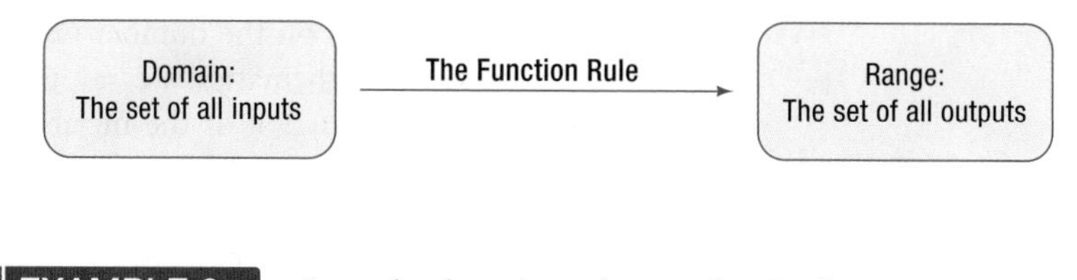

EXAMPLE 2 State the domain and range for the function

$$y = 8.5x, \quad 0 \le x \le 40$$

SOLUTION From the previous discussion, we have

$$\text{Domain} = \{x \mid 0 \le x \le 40\}$$
$$\text{Range} = \{y \mid 0 \le y \le 340\}$$

Function Maps

Another way to visualize the relationship between x and y is with the diagram in Figure 2, which we call a *function map*.

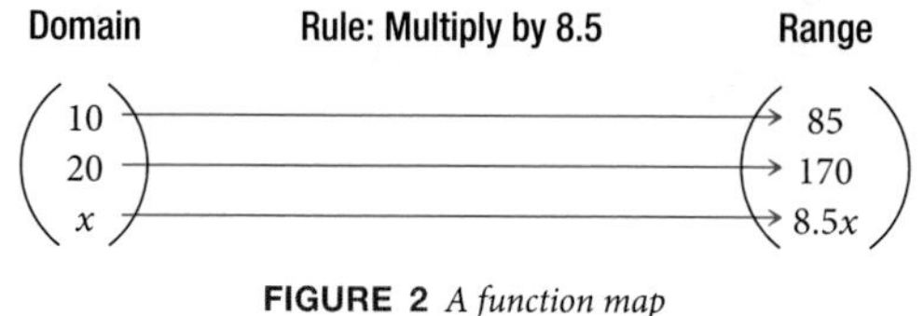

FIGURE 2 *A function map*

Although the diagram in Figure 2 does not show all the values that x and y can assume, it does give us a visual description of how x and y are related. It shows that values of y in the range come from values of x in the domain according to a specific rule (multiply by 8.5 each time).

A Formal Look at Functions

We are now ready for the formal definition of a function.

DEFINITION *Function*

A *function* is a rule that pairs each element in one set, called the *domain*, with exactly one element from a second set, called the *range*.

In other words, a function is a rule for which each input is paired with exactly one output.

Modeling: Softball Toss

EXAMPLE 3 Kendra tosses a softball into the air with an underhand motion. The distance of the ball above her hand is given by the function in symbolic form as

$$h = 32t - 16t^2 \qquad \text{for} \qquad 0 \le t \le 2$$

where h is the height of the ball in feet and t is the time in seconds. Construct a table that gives the height of the ball at quarter-second intervals, starting with $t = 0$ and ending with $t = 2$, then graph the function.

SOLUTION We construct Table 2 using the following values of t: 0, $\frac{1}{4}$, $\frac{1}{2}$, $\frac{3}{4}$, 1, $\frac{5}{4}$, $\frac{3}{2}$, $\frac{7}{4}$, 2. Then we construct the graph in Figure 3 from the table. The graph appears only in the first quadrant because neither t nor h can be negative.

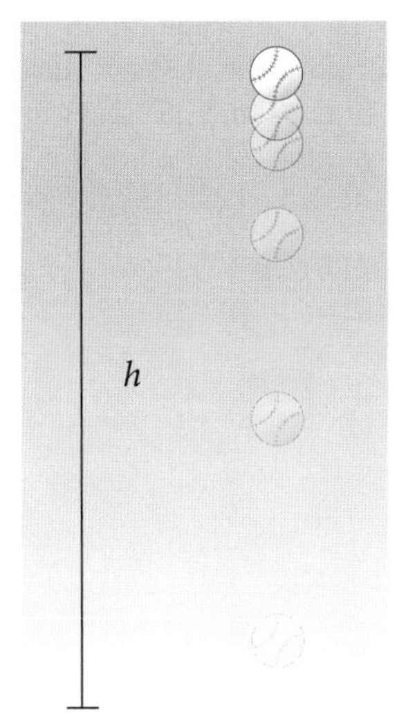

TABLE 2 Tossing a Softball into the Air

Input		Output
Time (sec) t	**Function Rule** $h = 32t - 16t^2$	**Distance (ft)** h
0	$h = 32(0) - 16(0)^2 = 0 - 0 = 0$	0
$\frac{1}{4}$	$h = 32\left(\frac{1}{4}\right) - 16\left(\frac{1}{4}\right)^2 = 8 - 1 = 7$	7
$\frac{1}{2}$	$h = 32\left(\frac{1}{2}\right) - 16\left(\frac{1}{2}\right)^2 = 16 - 4 = 12$	12
$\frac{3}{4}$	$h = 32\left(\frac{3}{4}\right) - 16\left(\frac{3}{4}\right)^2 = 24 - 9 = 15$	15
1	$h = 32(1) - 16(1)^2 = 32 - 16 = 16$	16
$\frac{5}{4}$	$h = 32\left(\frac{5}{4}\right) - 16\left(\frac{5}{4}\right)^2 = 40 - 25 = 15$	15
$\frac{3}{2}$	$h = 32\left(\frac{3}{2}\right) - 16\left(\frac{3}{2}\right)^2 = 48 - 36 = 12$	12
$\frac{7}{4}$	$h = 32\left(\frac{7}{4}\right) - 16\left(\frac{7}{4}\right)^2 = 56 - 49 = 7$	7
2	$h = 32(2) - 16(2)^2 = 64 - 64 = 0$	0

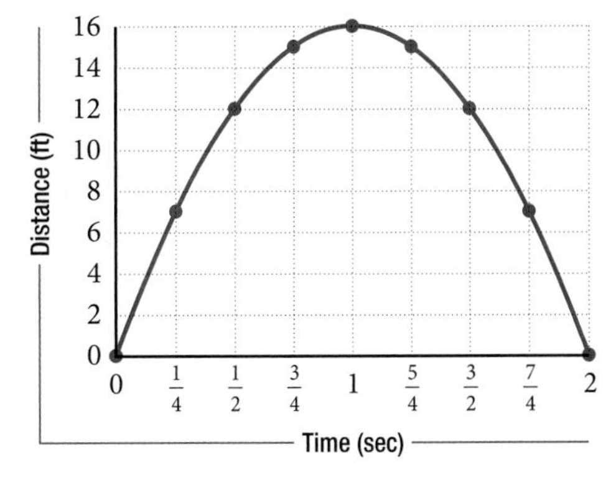

FIGURE 3

Here is a summary of what we know about functions as it applies to this example: We input values of t and output values of h according to the function rule

$$h = 32t - 16t^2 \qquad \text{for} \qquad 0 \le t \le 2$$

The domain is given by the inequality that follows the equation; it is

$$\text{Domain} = \{t \mid 0 \le t \le 2\}$$

The range is the set of all outputs that are possible by substituting the values of t from the domain into the equation. From our table and graph, it seems that the range is

$$\text{Range} = \{h \mid 0 \le h \le 16\}$$

TECHNOLOGY NOTE *More About Example 3*

Most graphing calculators can easily produce the information in Table 2. Simply set Y1 equal to $32X - 16X^2$. Then set up the table so it starts at 0 and increases by an increment of 0.25 each time. (On a TI-82/83, press [2nd] [WINDOW] to set up the table.)

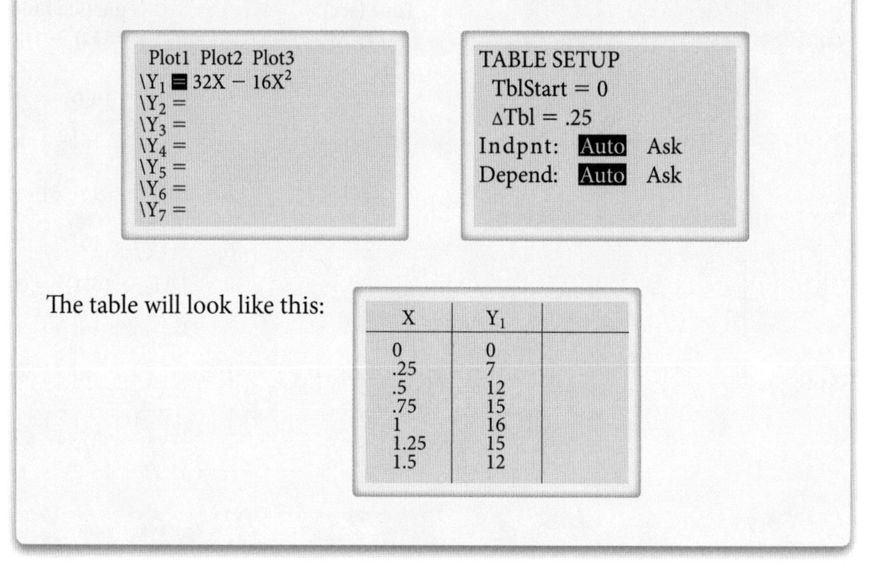

Functions as Ordered Pairs

As you can see from the examples we have done to this point, the function rule produces ordered pairs of numbers. We use this result to write an alternate definition for a function.

ALTERNATE DEFINITION *Function*

A *function* is a set of ordered pairs in which no two different ordered pairs have the same first coordinate. The set of all first coordinates is called the *domain* of the function. The set of all second coordinates is called the *range* of the function.

The restriction on first coordinates in the alternate definition keeps us from assigning a number in the domain to more than one number in the range.

A Relationship That Is Not a Function

You may be wondering if any sets of paired data fail to qualify as functions. The answer is yes, as the next example reveals.

Modeling: Used Car Prices

EXAMPLE 4 Table 3 shows the prices of used Ford Mustangs that were listed in the local newspaper. The diagram in Figure 4 is called a *scatter diagram*. It gives a visual representation of the data in Table 3. Why is this data not a function?

TABLE 3 Used Mustang Prices

Year x	Price ($) y
2010	18,999
2010	18,420
2010	16,980
2009	17,600
2009	16,840
2008	15,888
2007	12,900
2007	11,995
2006	10,985

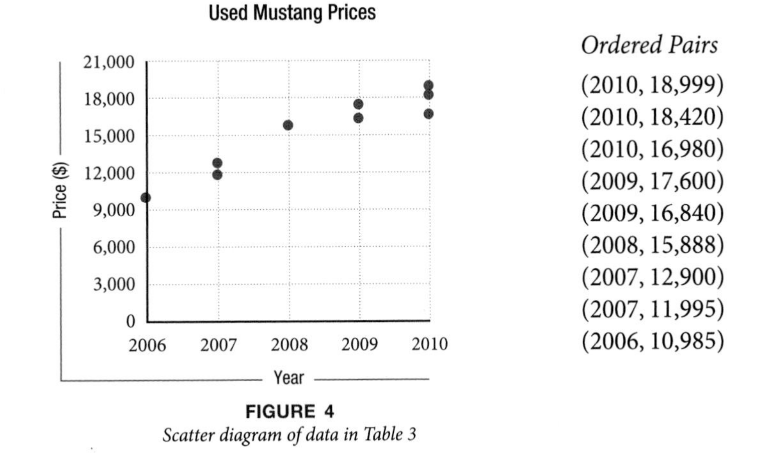

FIGURE 4
Scatter diagram of data in Table 3

SOLUTION In Table 3, the year 2010 is paired with three different prices: $18,999, $18,420, and $16,980. That is enough to disqualify the data from belonging to a function. For a set of paired data to be considered a function, each number in the domain must be paired with exactly one number in the range. ■

Still, there is a relationship between the first coordinates and second coordinates in the used car data. It is not a function relationship, but it is a relationship. To classify all relationships specified by ordered pairs, whether they are functions or not, we include the following two definitions.

(def DEFINITION *Relation*

A *relation* is a rule that pairs each element in one set, called the *domain*, with one or more elements from a second set, called the *range*.

(def ALTERNATE DEFINITION *Relation*

A *relation* is a set of ordered pairs. The set of all first coordinates is the *domain* of the relation. The set of all second coordinates is the *range* of the relation.

Here are some facts that will help clarify the distinction between relations and functions:

1. Any rule that assigns numbers from one set to numbers in another set is a relation. If that rule makes the assignment so no input has more than one output, then it is also a function.
2. Any set of ordered pairs is a relation. If none of the first coordinates of those ordered pairs is repeated, the set of ordered pairs is also a function.
3. Every function is a relation.
4. Not every relation is a function.

EXAMPLE 5 Sketch the graph of $x = y^2$.

SOLUTION Without going into much detail, we graph the equation $x = y^2$ by finding a number of ordered pairs that satisfy the equation, plotting these points, then drawing a smooth curve that connects them. A table of values for x and y that satisfy the equation follows, along with the graph of $x = y^2$ shown in Figure 5.

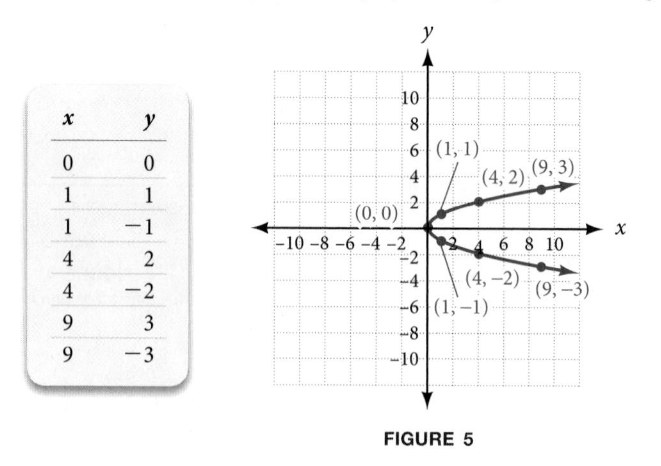

x	y
0	0
1	1
1	−1
4	2
4	−2
9	3
9	−3

FIGURE 5

As you can see from looking at the table and the graph in Figure 5, several ordered pairs whose graphs lie on the curve have repeated first coordinates, for instance

(1, 1) and (1, −1), (4, 2) and (4, −2), as well as (9, 3) and (9, −3). The graph is therefore not the graph of a function. ■

Vertical Line Test

Look back at the scatter diagram for used Mustang prices shown in Figure 4. Notice that some of the points on the diagram lie above and below each other along vertical lines. This is an indication that the data do not constitute a function. Two data points that lie on the same vertical line must have come from two ordered pairs with the same first coordinates.

Now, look at the graph shown in Figure 5. The reason this graph is the graph of a relation, but not of a function, is that some points on the graph have the same first coordinates, for example, the points (4, 2) and (4, −2). Furthermore, any time two points on a graph have the same first coordinates, those points must lie on a vertical line. [To convince yourself, connect the points (4, 2) and (4, −2) with a straight line. You will see that it must be a vertical line.] This allows us to write the following test that uses the graph to determine whether a relation is also a function.

[Δ≠Σ **RULE** *Vertical Line Test*

If a vertical line crosses the graph of a relation in more than one place, the relation cannot be a function. If no vertical line can be found that crosses a graph in more than one place, then the graph represents a function.

If we look back to the graph of $h = 32t - 16t^2$ as shown in Figure 3, we see that no vertical line can be found that crosses this graph in more than one place. The graph shown in Figure 3 is therefore the graph of a function.

EXAMPLE 6 Graph the equation $y = \frac{1}{x}$.

SOLUTION Since this is the first time we have graphed an equation of this form, we will make a table of values for x and y that satisfy the equation. Before we do, let's make some generalizations about the graph (Figure 6).

First, notice that since y is equal to 1 divided by x, y will be positive when x is positive. (The quotient of two positive numbers is a positive number.) Likewise, when x is negative, y will be negative. In other words, x and y always will have the same sign.

Next, notice that the expression $\frac{1}{x}$ will be undefined when x is 0, meaning that there is no value of y corresponding to $x = 0$. Because of this, the graph will not cross the y-axis. Further, the graph will not cross the x-axis either. If we try to find the x-intercept by letting $y = 0$, we have

$$0 = \frac{1}{x}$$

But there is no value of x to divide into 1 to obtain 0. Therefore, since there is no solution to this equation, our graph will not cross the x-axis.

To summarize, we can expect to find the graph in Quadrants I and III only, and the graph will cross neither axis.

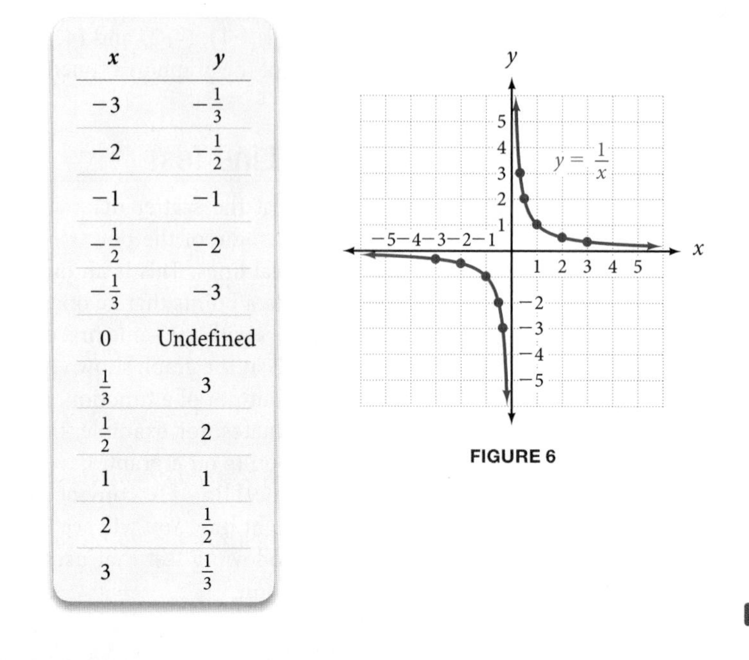

x	y
-3	$-\frac{1}{3}$
-2	$-\frac{1}{2}$
-1	-1
$-\frac{1}{2}$	-2
$-\frac{1}{3}$	-3
0	Undefined
$\frac{1}{3}$	3
$\frac{1}{2}$	2
1	1
2	$\frac{1}{2}$
3	$\frac{1}{3}$

FIGURE 6

EXAMPLE 7 Graph $y = \sqrt{x}$ and $y = \sqrt[3]{x}$.

SOLUTION The graphs are shown in Figures 7 and 8. Notice that the graph of $y = \sqrt{x}$ appears in the first quadrant only, because in the equation $y = \sqrt{x}$, x and y cannot be negative.

The graph of $y = \sqrt[3]{x}$ appears in Quadrants I and III because the cube root of a positive number is also a positive number, and the cube root of a negative number is a negative number. That is, when x is positive, y will be positive, and when x is negative, y will be negative.

The graphs of both equations will contain the origin, because $y = 0$ when $x = 0$ in both equations.

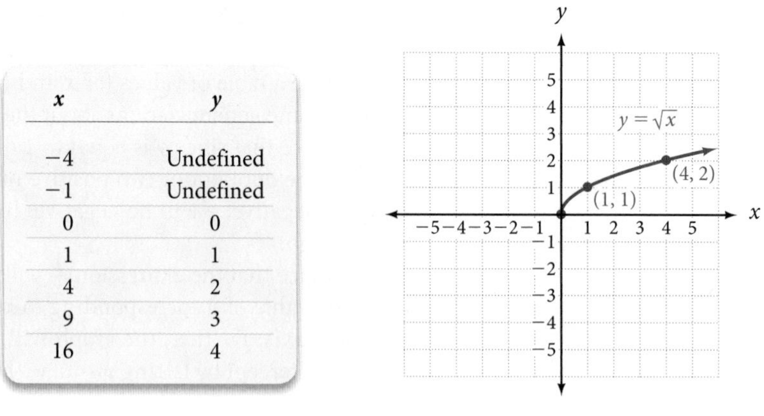

x	y
-4	Undefined
-1	Undefined
0	0
1	1
4	2
9	3
16	4

FIGURE 7

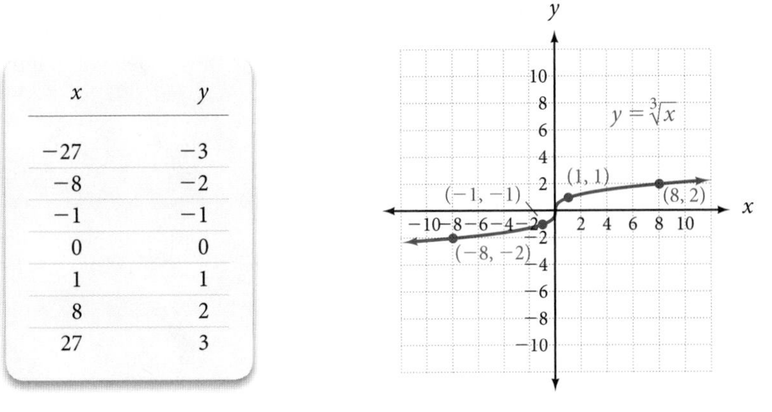

x	y
−27	−3
−8	−2
−1	−1
0	0
1	1
8	2
27	3

FIGURE 8

Modeling: Advertising Expenses

EXAMPLE 8 The manager at Albert's Appliances has \$3,000 to spend on advertising for the next fiscal quarter. A 30-second spot on television costs \$150 per broadcast, and a 30-second radio ad costs \$50.

a. The manager decides to buy x television ads and y radio ads. Write an equation relating x and y.

b. Make a table of values showing several choices for x and y.

c. Plot the points from your table, and graph the equation.

SOLUTION

a. Each television ad costs \$150, so x ads will cost \$150x. Similarly, y radio ads will cost \$50y. The manager has \$3,000 to spend, so the sum of the costs must be \$3,000. Thus,

$$150x + 50y = 3{,}000$$

b. Choose some values of x and solve the equation for the corresponding value of y. For example, if $x = 10$, then

$$150(10) + 50y = 3{,}000$$

$$1{,}500 + 50y = 3{,}000$$

$$50y = 1{,}500$$

$$y = 30$$

If the manager buys 10 television ads, she can also buy 30 radio ads. You can verify the other entries in the table.

x	y
8	36
10	30
12	24
14	18

c. Plot the points from the table. All the solutions lie on a straight line, as shown below in Figure 9.

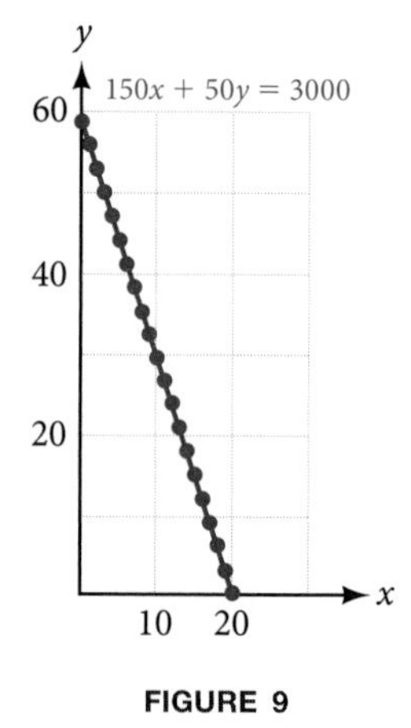

FIGURE 9

Using Technology 3.2

Many times, graphs of functions can be constructed more accurately and efficiently using technology. Here are some examples using Wolfram|Alpha.

EXAMPLE 9 Use Wolfram|Alpha to construct the graphs of the functions $f(x) = x^2$ and $f(x) = x^2 + 2$, on the same coordinate system.

SOLUTION Online, go to www.wolframalpha.com. Here you will see:

In the entry field, enter

graph f(x)=x^2, f(x)=x^2+2

Click the = sign or press the Return or Enter key. You will see the graph of both functions on the same coordinate system. This is what you will see:

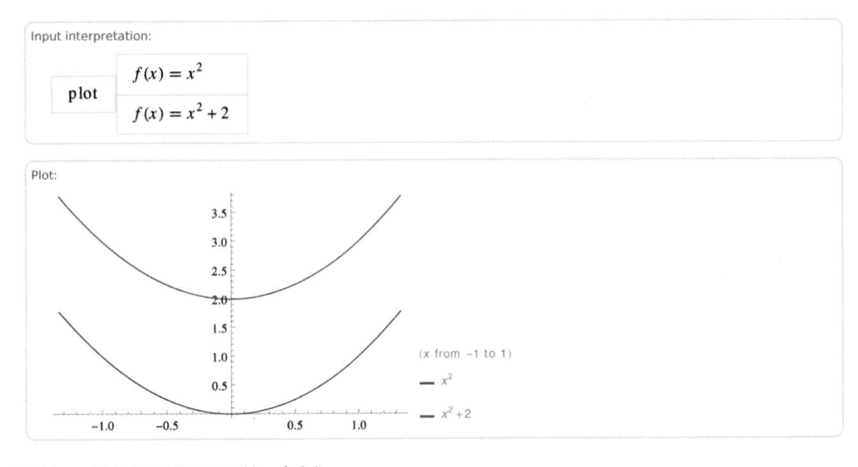

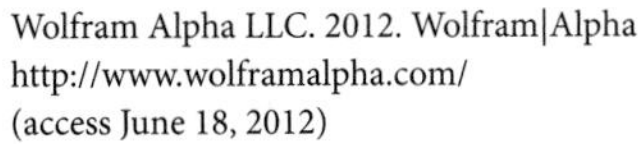
Wolfram Alpha LLC. 2012. Wolfram|Alpha
http://www.wolframalpha.com/
(access June 18, 2012)

If you want to specify the domain, or see the graph for certain values of x, you can specify that in the command line. The next example uses the same functions from Example 9, shown from $x = -6$ to $x = 6$.

EXAMPLE 10 Use Wolfram|Alpha to graph both the functions $f(x) = x^2$ and $f(x) = x^2 + 2$, from $x = -6$ to $x = 6$.

SOLUTION Online, go to www.wolframalpha.com. Here you will see, after you enter your instructions.

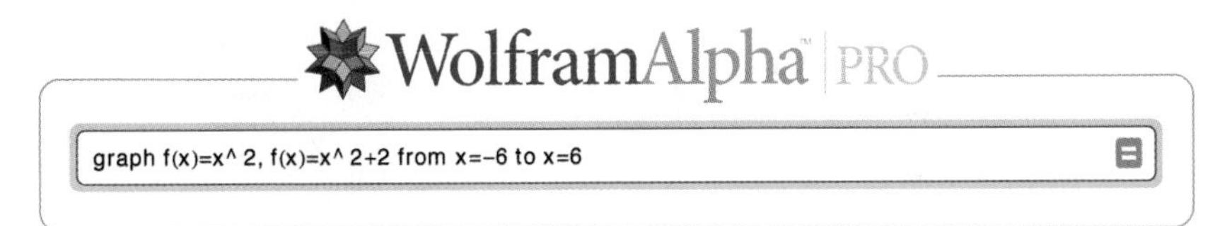

Click the = sign or press the Return or Enter key. Here is what you will see:

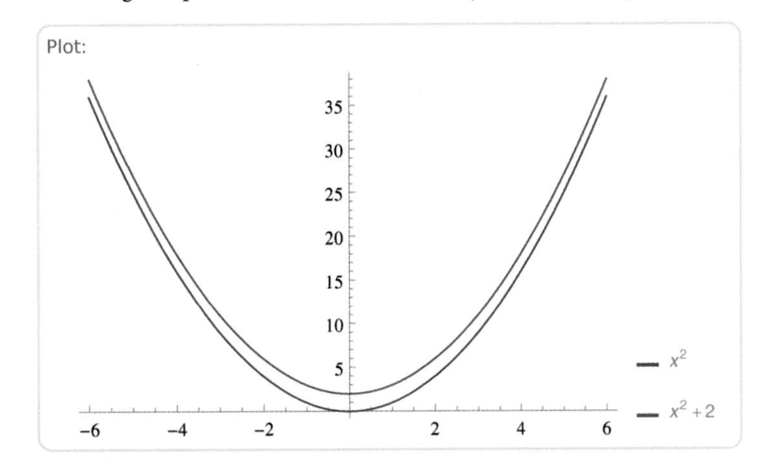

Wolfram Alpha LLC. 2012. Wolfram|Alpha
http://www.wolframalpha.com/
(access June 18, 2012)

EXAMPLE 11 We know the domain for the function $f(x) = \dfrac{x+3}{x-2}$ is all real numbers except 2. Use Wolfram|Alpha to find the domain.

SOLUTION Enter the following command line at Wolfram|Alpha

find the domain for f(x)=(x+3)/(x−2)

Notice how Wolfram|Alpha also specifies the domain.

Result:

$\{x \in \mathbb{R} : x \neq 2\}$

(assuming a function from reals to reals)

Wolfram Alpha LLC. 2012. Wolfram|Alpha
http://www.wolframalpha.com/
(access June 18, 2012)

It is telling us that the domain is all x such that x is a real number and x is not equal to 2. Notice that Wolfram|Alpha also returns the range for the function, though we are not going to show it here. ■

GETTING READY FOR CLASS

After reading through the preceding section, respond in your own words and in complete sentences.

A. What is a function?
B. What is the vertical line test?
C. Is every line the graph of a function? Explain.
D. Which variable is usually associated with the domain of a function?

Problem Set 3.2

For each of the following relations, give the domain and range, and indicate which are also functions.

1. $(1, 2), (3, 4), (5, 6), (7, 8)$
2. $(2, 1), (4, 3), (6, 5), (8, 7)$
3. $(2, 5), (3, 4), (1, 4), (0, 6)$
4. $(0, 4), (1, 6), (2, 4), (1, 5)$
5. $(a, 3), (b, 4), (c, 3), (d, 5)$
6. $(a, 5), (b, 5), (c, 4), (d, 5)$
7. $(a, 1), (a, 2), (a, 3), (a, 4)$
8. $(a, 1), (b, 1), (c, 1), (d, 1)$

State whether each of the following graphs represents a function.

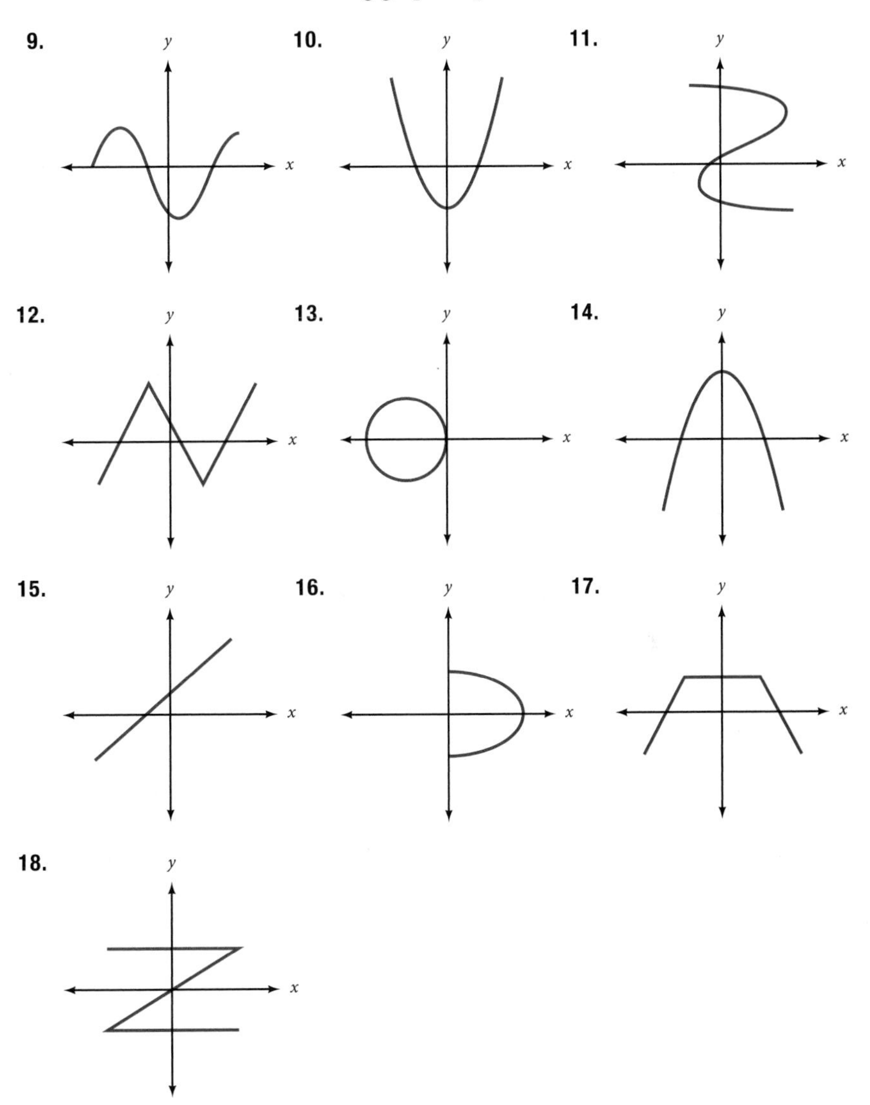

Determine the domain and range of the following functions. Assume the *entire* function is shown.

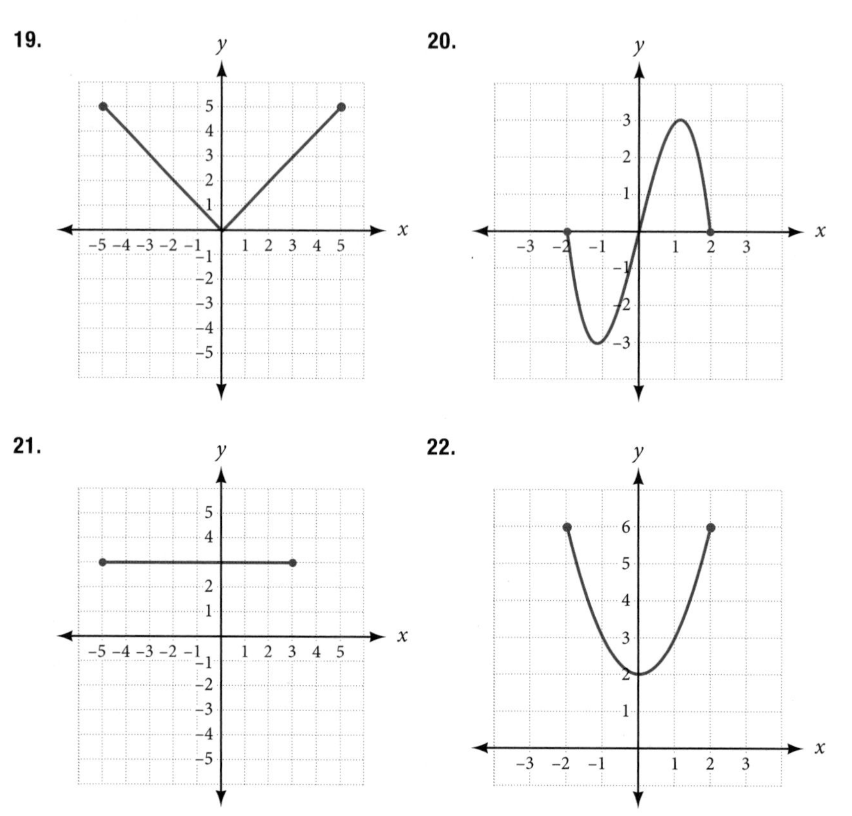

Graph each of the following relations. In each case, use the graph to find the domain and range, and indicate whether the graph is the graph of a function.

23. $y = x^2 - 1$ **24.** $y = x^2 + 1$ **25.** $y = x^2 + 4$ **26.** $y = x^2 - 9$

27. $x = y^2 - 1$ **28.** $x = y^2 + 1$ **29.** $y = (x + 2)^2$ **30.** $y = (x - 3)^2$

31. $x = (y + 1)^2$ **32.** $x = 3 - y^2$

Graph each of the following equations.

33. $y = \dfrac{-4}{x}$ **34.** $y = \dfrac{4}{x}$ **35.** $y = \dfrac{8}{x}$ **36.** $y = \dfrac{-8}{x}$

Graph each equation.

37. $y = \sqrt{x} - 2$ **38.** $y = \sqrt{x} + 2$ **39.** $y = \sqrt{x - 2}$ **40.** $y = \sqrt{x + 2}$

41. $y = \sqrt[3]{x} + 3$ **42.** $y = \sqrt[3]{x} - 3$ **43.** $y = \sqrt[3]{x + 3}$ **44.** $y = \sqrt[3]{x - 3}$

Modeling Practice

45. Hourly Pay Suppose you have a job that pays \$9.50 per hour and you work anywhere from 10 to 40 hours per week.

a. Write an equation, with a restriction on the variable x, that gives the amount of money, y, you will earn for working x hours in one week.

b. Use the function rule you have written in Part a to complete Table 4.

TABLE 4 Weekly Wages

Hours Worked	Function Rule	Gross Pay (\$)
x		y
10		
20		
30		
40		

c. Construct a line graph from the information in Table 4.

d. State the domain and range of this function.

e. What is the minimum amount you can earn in a week with this job? What is the maximum amount?

46. Weekly Wages The ad shown here was in the local newspaper. Suppose you are hired for the job described in the ad.

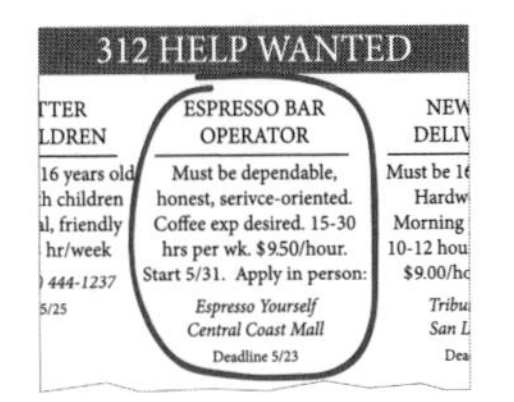

312 HELP WANTED

ESPRESSO BAR
OPERATOR

Must be dependable,
honest, serivce-oriented.
Coffee exp desired. 15-30
hrs per wk. \$9.50/hour.
Start 5/31. Apply in person:

Espresso Yourself
Central Coast Mall

Deadline 5/23

a. If x is the number of hours you work per week and y is your weekly gross pay, write the equation for y. (Be sure to include any restrictions on the variable x that are given in the ad.)

b. Use the function rule you have written in Part a to complete Table 5.

TABLE 5 Weekly Wages

Hours Worked	Function Rule	Gross Pay (\$)
x		y
15		
20		
25		
30		

c. Construct a line graph from the information in Table 5.

d. State the domain and range of this function.

e. What is the minimum amount you can earn in a week with this job? What is the maximum amount?

47. Pendulum Clock The length of time (T) in seconds it takes the pendulum of a clock to swing through one complete cycle is given by a square root function.

$$T = 2\pi\sqrt{\frac{L}{32}}$$

where L is the length, in feet, of the pendulum, and π is approximately $\frac{22}{7}$. How long must the pendulum be if one complete cycle takes 2 seconds?

48. Pollution A long straight river, 100 meters wide, is flowing at 1 meter per second. A pollutant is entering the river at a constant rate from one of its banks. As the pollutant disperses in the water, it forms a plume that is modeled by the function $y = \sqrt{x}$. Use this information to answer the following questions.

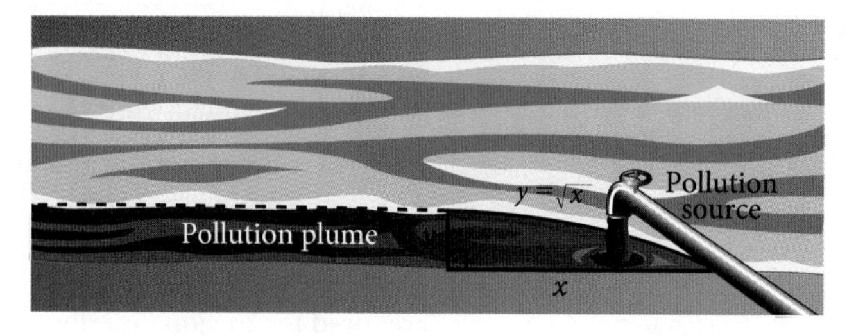

a. How wide is the plume 25 meters down river from the source of the pollution?

b. How wide is the plume 100 meters down river from the source of the pollution?

c. How far down river from the source of the pollution does the plume reach halfway across the river?

d. How far down the river from the source of the pollution does the plume reach the other side of the river?

49. Camera Phones The chart shows the estimated number of camera phones and non-camera phones sold from 2004 to 2010. Using the chart, list all the values in the domain and range for the total phones sales.

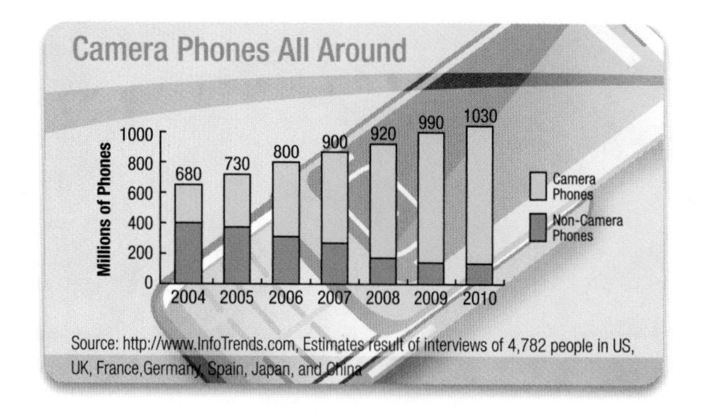

50. Light Bulbs The chart shows a comparison of power usage between incandescent and energy efficient light bulbs. Use the chart to state the domain and range of the function for an energy efficient bulb.

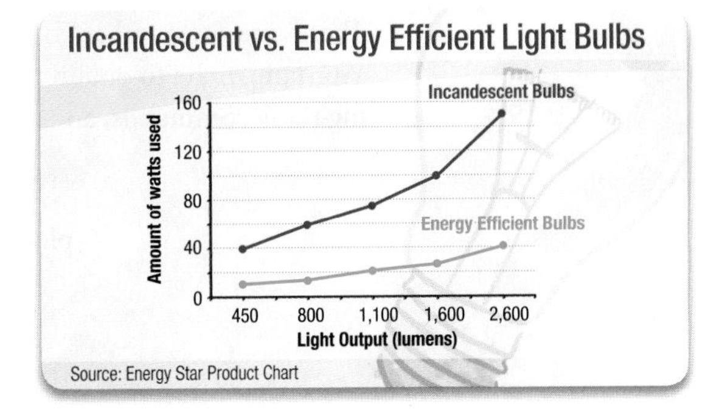

51. Wages Match each of the following statements to the appropriate graph indicated by labels I–IV.

a. Sarah works 25 hours to earn \$250.

b. Justin works 35 hours to earn \$560.

c. Rosemary works 30 hours to earn \$360.

d. Marcus works 40 hours to earn \$320.

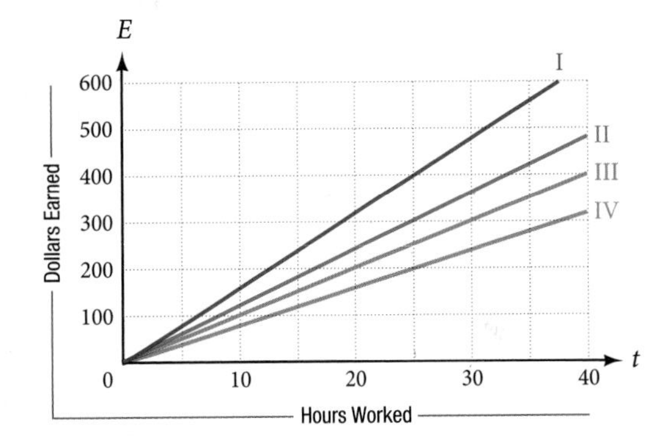

52. Find an equation for each of the functions shown in the graph above. Show dollars earned, E, as a function of hours worked, t. Then, indicate the domain and range of each function.

a. Graph I: $E =$ Domain $= \{t \mid \quad\}$ Range $= \{E \mid \quad\}$

b. Graph II: $E =$ Domain $= \{t \mid \quad\}$ Range $= \{E \mid \quad\}$

c. Graph III: $E =$ Domain $= \{t \mid \quad\}$ Range $= \{E \mid \quad\}$

d. Graph IV: $E =$ Domain $= \{t \mid \quad\}$ Range $= \{E \mid \quad\}$

Using Technology Exercises

Use Wolfram|Alpha for the following problems.

53. There are a number of different commands that you can use with Wolfram|Alpha to obtain the results in Example 9. Try the example again using these commands, and any others that you think of:

graph f(x)=x^2, f(x)=x^2+2

plot f(x)=x^2, f(x)=x^2+2

graph y=x^2, y=x^2+2

plot x^2, x^2+2

54. On the same coordinate system, construct the graphs of

$$f(x) = 2x, f(x) = 2^x, \text{ and } f(x) = x^2$$

55. On the same coordinate system, construct the graphs of

$$f(x) = x^2 - 2, f(x) = x^2, \text{ and } f(x) = x^2 + 2$$

56. On the same coordinate system, construct the graphs of

$$y = x, \quad y = \sqrt{x}, \quad \text{and } y = x^2$$

57. On the same coordinate system, construct the graphs of

$$y = x, \quad y = |x|, \quad \text{and } y = \sqrt{x^2}$$

58. On the same coordinate system, construct the graphs of

$$y = x, \quad y = |x|, \quad \text{and } y = \sqrt[3]{x^3}$$

59. Find the domain for each function.

a. $y = \sqrt{x}$ **b.** $y = \sqrt{x} - 2$ **c.** $y = \sqrt{x} + 2$

60. Find the domain for each function.

a. $y = \frac{1}{x}$ **b.** $y = \frac{1}{x - 3}$ **c.** $y = \frac{1}{x + 3}$

On your graphing calculator, graph each equation and build a table as indicated.

61. $y = 64t - 16t^2$ TblStart $= 0$ ΔTbl $= 1$

62. $y = \frac{1}{2}x - 4$ TblStart $= -5$ ΔTbl $= 1$

63. $y = \frac{12}{x}$ TblStart $= 0.5$ ΔTbl $= 0.5$

64. $y = 2^x$ TblStart $= -2$ ΔTbl $= 1$

Getting Ready for the Next Section

Simplify. Round to the nearest whole number if necessary.

65. $4(3.14)(9)$

66. $\frac{4}{3}(3.14) \cdot 3^3$

67. $4(-2) - 1$

68. $3(3)^2 + 2(3) - 1$

69. If $s = \frac{60}{t}$, find s when

a. $t = 10$

b. $t = 8$

70. If $y = 3x^2 + 2x - 1$, find y when

a. $x = 0$

b. $x = -2$

71. Find the value of $x^2 + 2$ for

a. $x = 5$

b. $x = -2$

72. Find the value of $125 \cdot 2^t$ for

a. $t = 0$

b. $t = 1$

For the equation $y = x^2 - 3$:

73. Find y if x is 2.

74. Find y if x is -2.

75. Find y if x is 0.

76. Find y if x is -4.

Function Notation and More Graphing

3.3

312 HELP WAN
YARD PERSON
NE
Full time 40 hrs. with weekend work required. Cleaning & loading trucks. $8.50/hr.
Valid CDL with clean record & drug screen required.
Submit current MVR to KCI, 225 Suburban Rd., SLO 93405 (805) 555-3304.
BABY SITTER

INTRODUCTION Let's return to the discussion that introduced us to functions. If a job pays \$8.50 per hour for working from 0 to 40 hours a week, then the amount of money, y, earned in one week is a function of the number of hours worked, x. The exact relationship between x and y is written

$$y = 8.5x \quad \text{for} \quad 0 \le x \le 40$$

Because the amount of money earned, y, depends on the number of hours worked, x, we call y the *dependent variable* and x the *independent variable*. Furthermore, if we let f represent all the ordered pairs produced by the equation, then we can write

$$f = \{(x, y) \mid y = 8.5x \quad \text{and} \quad 0 \le x \le 40\}$$

Once we have named a function with a letter, we can use an alternative notation to represent the dependent variable y. The alternative notation for y is $f(x)$. It is read "f of x" and can be used instead of the variable y when working with functions. The notation y and the notation $f(x)$ are equivalent. That is,

$$y = 8.5x \Leftrightarrow f(x) = 8.5x$$

When we use the notation $f(x)$ we are using *function notation*. The benefit of using function notation is that we can write more information with fewer symbols than we can by using just the variable y. For example, asking how much money a person will make for working 20 hours is simply a matter of asking for $f(20)$. Without function notation, we would have to say, "Find the value of y that corresponds to a value of $x = 20$." To illustrate further, using the variable y, we can say "y is 170 when x is 20." Using the notation $f(x)$, we simply say "$f(20) = 170$." Each expression indicates that you will earn \$170 for working 20 hours.

EXAMPLE 1 If $f(x) = 8.5x$, find $f(0)$, $f(10)$, and $f(20)$.

VIDEO EXAMPLES
SECTION 3.3

SOLUTION To find $f(0)$, we substitute 0 for x in the expression $8.5x$ and simplify. We find $f(10)$ and $f(20)$ in a similar manner—by substitution.

If $\quad f(x) = 8.5x$

then $\quad f(0) = 8.5(0) = 0$

$$f(10) = 8.5(10) = 85$$

$$f(20) = 8.5(20) = 170$$

If we changed the example in the discussion that opened this section so the hourly wage was \$9.50 per hour, we would have a new equation to work with, namely,

$$y = 9.5x \quad \text{for} \quad 0 \le x \le 40$$

Suppose we name this new function with the letter g. Then

$$g = \{(x, y) \mid y = 9.5x \quad \text{and} \quad 0 \le x \le 40\}$$

and

$$g(x) = 9.5x$$

If we want to talk about both functions in the same discussion, having two different letters, f and g, makes it easy to distinguish between them. For example, since $f(x) = 8.5x$ and $g(x) = 9.5x$, asking how much money a person makes for working 20 hours is simply a matter of asking for $f(20)$ or $g(20)$, avoiding any confusion over which hourly wage we are talking about.

The diagrams shown in Figure 1 further illustrate the similarities and differences between the two functions we have been discussing.

Note The symbol $\in$ means "is a member of."

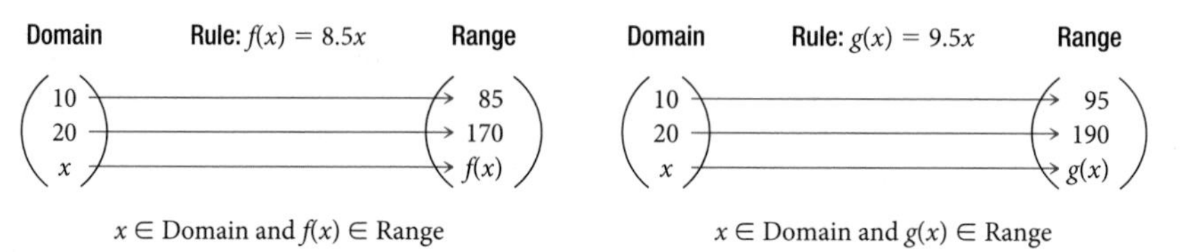

FIGURE 1 *Function maps*

EXAMPLE 2 If $f(x) = 3x^2 + 2x - 1$, find $f(0)$, $f(3)$, and $f(-2)$.

SOLUTION Since $f(x) = 3x^2 + 2x - 1$, we have

$$f(0) = 3(0)^2 + 2(0) - 1 = 0 - 1 = -1$$
$$f(3) = 3(3)^2 + 2(3) - 1 = 27 + 6 - 1 = 32$$
$$f(-2) = 3(-2)^2 + 2(-2) - 1 = 12 - 4 - 1 = 7$$

Note Some students like to think of functions as machines. Values of x are put into the machine, which transforms them into values of $f(x)$, which are then output by the machine.

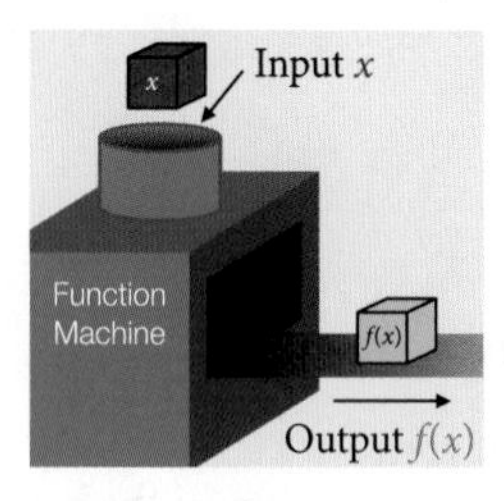

In Example 2, the function f is defined by the equation $f(x) = 3x^2 + 2x - 1$. We could just as easily have said $y = 3x^2 + 2x - 1$. That is, $y = f(x)$. Saying $f(-2) = 7$ is exactly the same as saying y is 7 when x is -2.

EXAMPLE 3 If $f(x) = 4x - 1$ and $g(x) = x^2 + 2$, then

$$f(5) = 4(5) - 1 = 19 \quad \text{and} \quad g(5) = 5^2 + 2 = 27$$
$$f(-2) = 4(-2) - 1 = -9 \quad \text{and} \quad g(-2) = (-2)^2 + 2 = 6$$
$$f(0) = 4(0) - 1 = -1 \quad \text{and} \quad g(0) = 0^2 + 2 = 2$$
$$f(z) = 4z - 1 \quad \text{and} \quad g(z) = z^2 + 2$$
$$f(a) = 4a - 1 \quad \text{and} \quad g(a) = a^2 + 2$$

$$\begin{aligned} f(a + 3) &= 4(a + 3) - 1 \\ &= 4a + 12 - 1 \\ &= 4a + 11 \end{aligned} \qquad \begin{aligned} g(a + 3) &= (a + 3)^2 + 2 \\ &= (a^2 + 6a + 9) + 2 \\ &= a^2 + 6a + 11 \end{aligned}$$

TECHNOLOGY NOTE *More About Example 3*

Most graphing calculators can use tables to evaluate functions. To work Example 3 using a graphing calculator table, set Y1 equal to 4X − 1 and Y2 equal to $X^2 + 2$. Then set the independent variable in the table to Ask instead of Auto. Go to your table and input 5, −2, and 0. Under Y1 in the table, you will find $f(5), f(-2)$, and $f(0)$. Under Y2, you will find $g(5), g(-2)$, and $g(0)$.

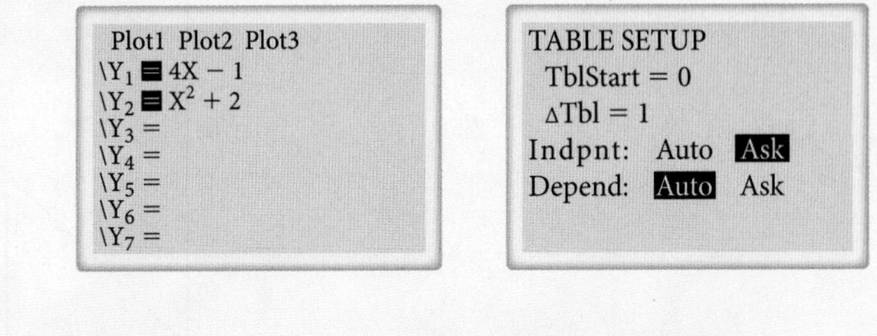

The table will look like this:

X	Y_1	Y_2
5	19	27
−2	−9	6
0	−1	2

Although the calculator asks us for a starting value and a table increment, these values don't matter because we are inputting the X values ourselves.

EXAMPLE 4 If the function f is given by

$$f = \{(-2, 0), (3, -1), (2, 4), (7, 5)\}$$

then $f(-2) = 0, f(3) = -1, f(2) = 4$, and $f(7) = 5$. ■

We can generalize the discussion at the end of Example 4 this way:

$$(a, b) \in f \qquad \text{if and only if} \qquad f(a) = b$$

EXAMPLE 5 If $f(x) = 2x^2$ and $g(x) = 3x - 1$, find

a. $f[g(2)]$ **b.** $g[f(2)]$

SOLUTION The expression $f[g(2)]$ is read "f of g of 2."

a. Because $g(2) = 3(2) - 1 = 5$,

$$f[g(2)] = f(5) = 2(5)^2 = 50$$

b. Because $f(2) = 2(2)^2 = 8$,

$$g[f(2)] = g(8) = 3(8) - 1 = 23$$

■

Function Notation and Graphs

We can visualize the relationship between x and $f(x)$ on the graph of the function. Figure 2 shows the graph of $f(x) = 8.5x$ along with two additional line segments. The horizontal line segment corresponds to $x = 20$, and the vertical line segment corresponds to $f(20)$. (Note that the domain is restricted to $0 \le x \le 40$.)

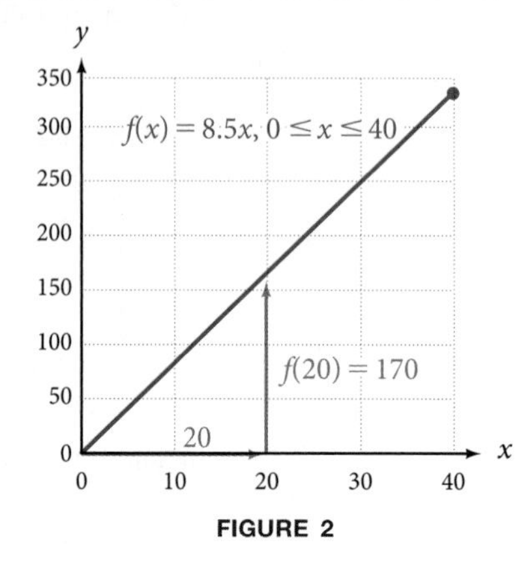

FIGURE 2

We can also use functions and function notation to talk about numbers in many of the charts and graphs we see in the media. For example, consider the chart on gasoline prices shown below. Let's let x represent one of the years in the chart.

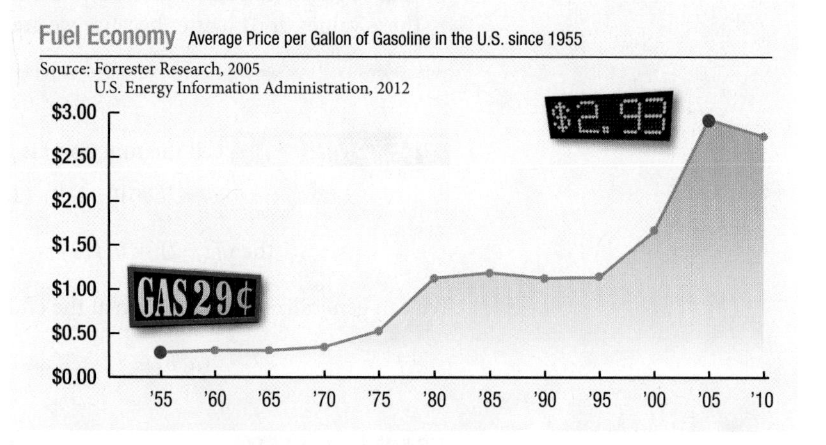

If the function f pairs each year in the chart with the average price of regular gasoline for that year, then each statement below is true:

$$f(1955) = \$0.29$$

The domain of $f =$
$\{1955, 1960, 1965, 1970, 1975, 1980, 1985, 1990, 1995, 2000, 2005, 2010\}$

In general, when we refer to the function f we are referring to the domain, the range, and the rule that takes elements in the domain and outputs elements in the range. When we talk about $f(x)$ we are talking about the rule itself, or an element in the range, or the variable y.

The function f

Domain of f	$y = f(x)$	Range of f
Inputs	*Rule*	*Outputs*

Modeling: Using Function Notation

The remaining examples in this section show a variety of ways to use and interpret function notation.

EXAMPLE 6 If it takes Lorena t minutes to run a mile, then her average speed s, in miles per hour, is given by the formula

$$s(t) = \frac{60}{t} \qquad \text{for} \qquad t > 0$$

© Andrew Rich/iStockPhoto

Find $s(10)$ and $s(8)$, and then explain what they mean.

SOLUTION To find $s(10)$, we substitute 10 for t in the equation and simplify:

$$s(10) = \frac{60}{10} = 6$$

In words: When Lorena runs a mile in 10 minutes, her average speed is 6 miles per hour.

We calculate $s(8)$ by substituting 8 for t in the equation. Doing so gives us

$$s(8) = \frac{60}{8} = 7.5$$

In words: Running a mile in 8 minutes is running at a rate of 7.5 miles per hour.

EXAMPLE 7 A painting is purchased as an investment for \$125. If its value increases continuously so that it doubles every 5 years, then its value is given by the function

$$V(t) = 125 \cdot 2^{t/5} \qquad \text{for} \qquad t \geq 0$$

© XYZ Textbooks

where t is the number of years since the painting was purchased, and V is its value (in dollars) at time t. Find $V(5)$ and $V(10)$, and explain what they mean.

SOLUTION The expression $V(5)$ is the value of the painting when $t = 5$ (5 years after it is purchased). We calculate $V(5)$ by substituting 5 for t in the equation $V(t) = 125 \cdot 2^{t/5}$. Here is our work:

$$V(5) = 125 \cdot 2^{5/5} = 125 \cdot 2^1 = 125 \cdot 2 = 250$$

In words: After 5 years, the painting is worth \$250.

The expression $V(10)$ is the value of the painting after 10 years. To find this number, we substitute 10 for t in the equation:

$$V(10) = 125 \cdot 2^{10/5} = 125 \cdot 2^2 = 125 \cdot 4 = 500$$

In words: The value of the painting 10 years after it is purchased is \$500.

Note Notice how important the units are in this problem. The numerical part of each answer is the same. It is the units that distinguish them.

EXAMPLE 8 A balloon has the shape of a sphere with a radius of 3 inches. Use the following formulas to find the volume and surface area of the balloon.

$$V(r) = \frac{4}{3}\pi r^3 \qquad S(r) = 4\pi r^2$$

SOLUTION As you can see, we have used function notation to write the formulas for volume and surface area, because each quantity is a function of the radius. To find these quantities when the radius is 3 inches, we evaluate $V(3)$ and $S(3)$:

$$V(3) = \frac{4}{3}\pi \cdot 3^3 = \frac{4}{3}\pi \cdot 27$$

$= 36\pi$ cubic inches, or 113 cubic inches (to the nearest whole number)

$$S(3) = 4\pi \cdot 3^2$$

$= 36\pi$ square inches, or 113 square inches (to the nearest whole number)

The fact that $V(3) = 36\pi$ means that the ordered pair $(3, 36\pi)$ belongs to the function V. Likewise, the fact that $S(3) = 36\pi$ tells us that the ordered pair $(3, 36\pi)$ is a member of function S.

TECHNOLOGY NOTE *More About Example 8*

If we look at Example 8, we see that when the radius of a sphere is 3, the numerical values of the volume and surface area are equal. How unusual is this? Are there other values of r for which $V(r)$ and $S(r)$ are equal? We can answer this question by looking at the graphs of both V and S.

To graph the function $V(r) = \frac{4}{3}\pi r^3$, set Y1 = $4\pi X^3/3$. To graph $S(r) = 4\pi r^2$, set Y2 = $4\pi X^2$. Graph the two functions in each of the following windows:

Window 1: X from -4 to 4, Y from -2 to 10

Window 2: X from 0 to 4, Y from 0 to 50

Window 3: X from 0 to 4, Y from 0 to 150

Then use the TRACE and ZOOM features of your calculator to locate the point in the first quadrant where the two graphs intersect. How do the coordinates of this point compare with the results in Example 8?

Greatest Integer Function (A Step Function)

The simplest step function is the one that outputs the greatest integer not greater than the number that is input. The equation for this function is

$$y = [[x]]$$

where [[x]] stands for the greatest integer not greater than x. We call this function the ***greatest integer function***. Here is a table and a graph (Figure 3) for this function, for positive values of x.

TABLE 1 The Greatest Integer Function Rule

Input		Output
x	(greatest integer not greater than x)	y
0	greatest integer not greater than 0	0
0.25	greatest integer not greater than 0.25	0
0.99	greatest integer not greater than 0.99	0
1	greatest integer not greater than 1	1
1.5	greatest integer not greater than 1.5	1
1.75	greatest integer not greater than 1.75	1
2	greatest integer not greater than 2	2
2.1	greatest integer not greater than 2.1	2
2.75	greatest integer not greater than 2.75	2
3	greatest integer not greater than 3	3

Once you see the graph of this step function, you know right away where the name "step function" comes from. The graph of the function $y = [[x]]$ extended to include negative numbers appears in Figure 4.

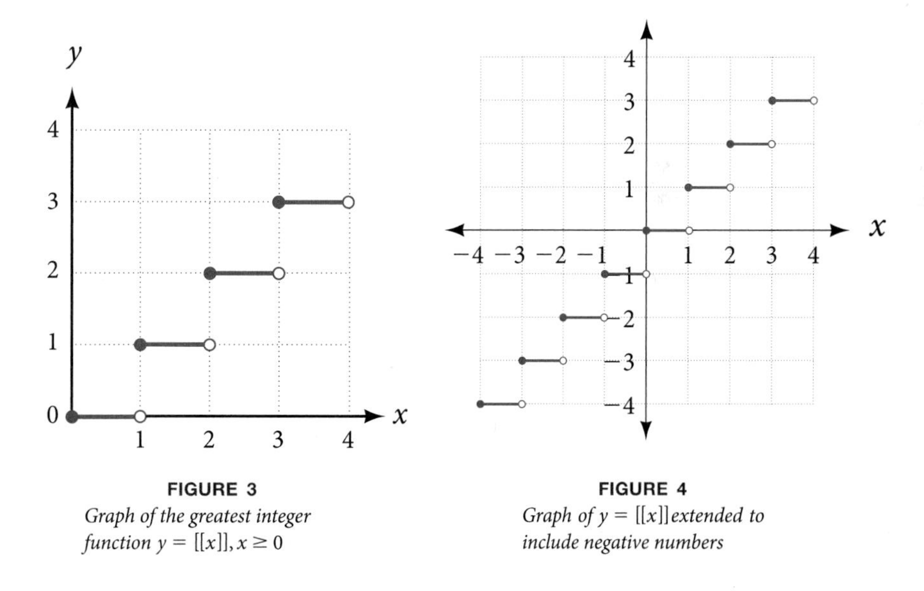

FIGURE 3
Graph of the greatest integer function $y = [[x]], x \geq 0$

FIGURE 4
Graph of $y = [[x]]$ *extended to include negative numbers*

Modeling: The Postage Stamp Function (A Step Function)

The postage stamp function is an example of a step function. If you mail a first class large envelope that weighs 1 ounce or less, you pay 90 cents. If your package weighs over 1 ounce, but not more than 2 ounces, then you pay 90 cents plus 20 cents, which is 110 cents. The charges for other weights are given in the table.

TABLE 2 The Postage Stamp Function	
Input x	Output $C(x)$
$0 < x \le 1$	90
$1 < x \le 2$	110
$2 < x \le 3$	130
$3 < x \le 4$	150
$4 < x \le 5$	170
$5 < x \le 6$	190

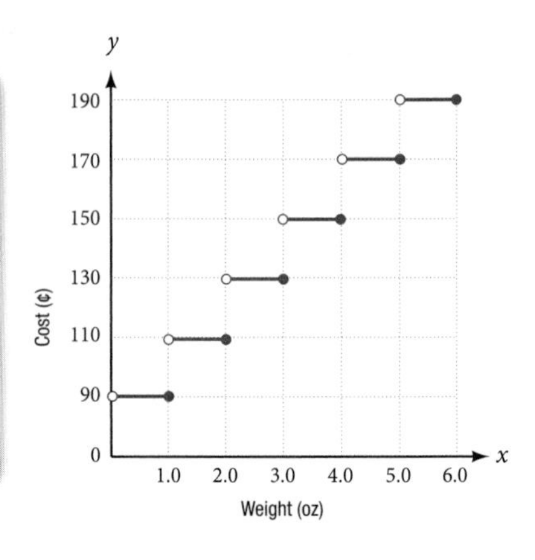

FIGURE 5
Graph of the postage stamp function

Using function notation, we can write the following expressions:

$$C(1) = 90¢$$

$$C(1.5) = 110¢$$

$$C(2.75) = 130¢$$

$$C(3) = 130¢$$

$$C(3.25) = 150¢$$

The graph of the postage stamp function appears in Figure 5.

Using Technology 3.3

One of the equations you will solve frequently in this course and in other mathematics courses you may take, especially calculus, is the equation $f(x) = 0$. This equation asks for the input values of x that return an output value of 0. Here are some examples using Wolfram|Alpha.

EXAMPLE 9 If $f(x) = x^2 - 4$, use Wolfram|Alpha to find x so that $f(x) = 0$.

SOLUTION We could do some of the work ourselves and simply say solve $x^2 - 4 = 0$, but with Wolfram|Alpha, that is unnecessary. Enter the following into the Wolfram|Alpha command line:

```
if f(x)=x^2-4, solve f(x)=0
```

Notice that Wolfram|Alpha solves this problem by finding the intersection of the graph and of $y = x^2$ and $y = 4$. We know this because part of the information that is returned to us is this graph:

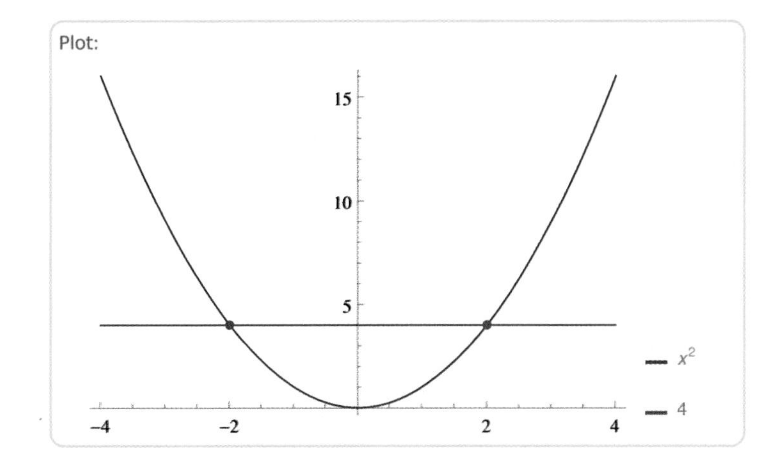

Wolfram Alpha LLC. 2012. Wolfram|Alpha
http://www.wolframalpha.com/
(access June 18, 2012)

GETTING READY FOR CLASS

After reading through the preceding section, respond in your own words and in complete sentences.

A. Explain what you are calculating when you find $f(2)$ for a given function f.

B. If $s(t) = \frac{60}{t}$, how do you find $s(10)$?

C. If $f(2) = 3$ for a function f, what is the relationship between the numbers 2 and 3 and the graph of f?

D. If $f(6) = 0$ for a particular function f, then you can immediately graph one of the intercepts. Explain.

Problem Set 3.3

Let $f(x) = 2x - 5$ and $g(x) = x^2 + 3x + 4$. Evaluate the following.

1. $f(2)$ **2.** $f(3)$ **3.** $f(-3)$ **4.** $g(-2)$

5. $g(-1)$ **6.** $f(-4)$ **7.** $g(-3)$ **8.** $g(2)$

9. $g(a)$ **10.** $f(a)$ **11.** $f(a + 6)$ **12.** $g(a + 6)$

Let $f(x) = 3x^2 - 4x + 1$ and $g(x) = 2x - 1$. Evaluate the following.

13. $f(0)$ **14.** $g(0)$ **15.** $g(-4)$ **16.** $f(1)$

17. $f(-1)$ **18.** $g(-1)$ **19.** $g\left(\frac{1}{2}\right)$ **20.** $g\left(\frac{1}{4}\right)$

21. $f(a)$ **22.** $g(a)$ **23.** $f(a + 2)$ **24.** $g(a + 2)$

If $f = \{(1, 4), (-2, 0), \left(3, \frac{1}{2}\right), (\pi, 0)\}$ and $g = \{(1, 1), (-2, 2), \left(\frac{1}{2}, 0\right)\}$, find each of the following values of f and g.

25. $f(1)$ **26.** $g(1)$ **27.** $g\left(\frac{1}{2}\right)$

28. $f(3)$ **29.** $g(-2)$ **30.** $f(\pi)$

Let $f(x) = x^2 - 2x$ and $g(x) = 5x - 4$. Evaluate the following.

31. $f(-4)$ **32.** $g(-3)$ **33.** $f(-2) + g(-1)$

34. $f(-1) + g(-2)$ **35.** $2f(x) - 3g(x)$ **36.** $f(x) - g(x^2)$

37. $f[g(3)]$ **38.** $g[f(3)]$

Let $f(x) = \dfrac{1}{x + 3}$ and $g(x) = \dfrac{1}{x} + 1$. Evaluate the following.

39. $f\left(\frac{1}{3}\right)$ **40.** $g\left(\frac{1}{3}\right)$ **41.** $f\left(-\frac{1}{2}\right)$

42. $g\left(-\frac{1}{2}\right)$ **43.** $f(-3)$ **44.** $g(0)$

45. For the function $f(x) = x^2 - 4$, evaluate each of the following expressions.

a. $f(a) - 3$ **b.** $f(a - 3)$ **c.** $f(x) + 2$

d. $f(x + 2)$ **e.** $f(a + b)$ **f.** $f(x + h)$

46. For the function $f(x) = 3x^2$, evaluate each of the following expressions.

a. $f(a) - 2$ **b.** $f(a - 2)$ **c.** $f(x) + 5$

d. $f(x + 5)$ **e.** $f(a + b)$ **f.** $f(x + h)$

47. Graph the function $f(x) = \frac{1}{2}x + 2$. Then draw and label the line segments that represent $x = 4$ and $f(4)$.

48. Graph the function $f(x) = -\frac{1}{2}x + 6$. Then draw and label the line segments that represent $x = 4$ and $f(4)$.

49. For the function $f(x) = \frac{1}{2}x + 2$, find the value of x for which $f(x) = x$.

50. For the function $f(x) = -\frac{1}{2}x + 6$, find the value of x for which $f(x) = x$.

51. Graph the function $f(x) = x^2$. Then draw and label the line segments that represent $x = 1$ and $f(1)$, $x = 2$ and $f(2)$ and, finally, $x = 3$ and $f(3)$.

52. Graph the function $f(x) = x^2 - 2$. Then draw and label the line segments that represent $x = 2$ and $f(2)$ and the line segments corresponding to $x = 3$ and $f(3)$.

Getting Ready for Calculus The problems below are representative of the type of problems you will need to be familiar with in order to be successful in calculus. They are not calculus problems, but are algebra problems that occur in the process of solving calculus problems. These particular problems are taken from *Applied Calculus* by Denny Burzynski and Guy Sanders, published by XYZ Textbooks.

53. If $f(x) = x^2 + 6x - 2$, find $f(5)$.

54. If $f(x) = \dfrac{x^4 - 5x^2 + 2x - 2}{(x^2 - 1)^2}$, find $f(0)$.

55. If $g(x) = \dfrac{-16.33x^2 + 483}{(7.2x^2 + 210)^2}$, find $g(4)$, to the nearest ten thousandth.

56. If $h(x) = \dfrac{x^4(21x - 5)}{(4x - 1)^{3/4}}$, find $h(-1)$.

57. If $f(x) = x^2 - 4x + 3$, find the values of x for which $f(x) = 0$.

58. If $g(t) = (t^2 - 4)(t^2 + 1)$, find the real values of t for which $g(t) = 0$.

59. If $h(t) = 4t^2 - 6t$, find the values of t for which $h(t) = 0$.

60. If $N(t) = 90t^2 - t^3, 0 \le t \le 70$,

a. state the domain for N.

b. find $N(15)$.

Modeling Practice

61. Investing in Art A painting is purchased as an investment for \$150. If its value increases continuously so that it doubles every 3 years, then its value is given by the function

$$V(t) = 150 \cdot 2^{t/3} \quad \text{for} \quad t \ge 0$$

where t is the number of years since the painting was purchased, and $V(t)$ is its value (in dollars) at time t. Find $V(3)$ and $V(6)$, and then explain what they mean.

62. Average Speed If it takes Minke t minutes to run a mile, then her average speed $s(t)$, in miles per hour, is given by the formula

$$s(t) = \frac{60}{t} \quad \text{for} \quad t > 0$$

Find $s(4)$ and $s(5)$, and then explain what they mean.

63. Antidepressant Sales Suppose x represents one of the years in the chart. Suppose further that we have three functions f, g, and h that do the following:

f pairs each year with the total sales of Zoloft in billions of dollars for that year.
g pairs each year with the total sales of Effexor in billions of dollars for that year.
h pairs each year with the total sales of Wellbutrin in billions of dollars for that year.

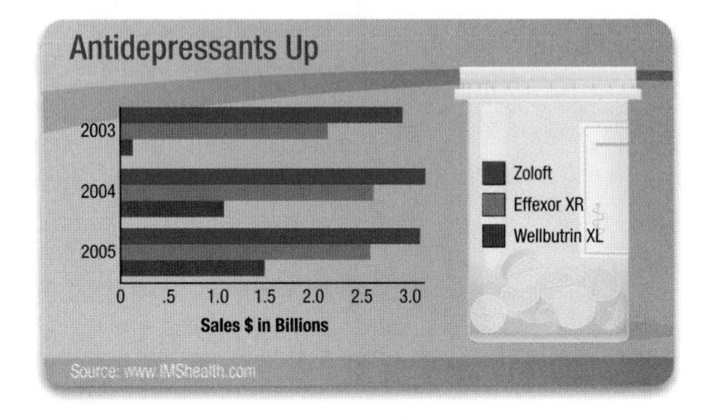

For each statement below, indicate whether the statement is true or false.

a. The domain of g is $\{2003, 2004, 2005\}$

b. The domain of g is $\{x \mid 2003 \leq x \leq 2005\}$

c. $f(2004) > g(2004)$

d. $h(2005) > 1.5$

e. $h(2005) > h(2004) > h(2003)$

64. Mobile Phone Sales Suppose x represents one of the years in the chart. Suppose further that we have three functions f, g, and h that do the following:

f pairs each year with the number of camera phones sold that year.
g pairs each year with the number of non-camera phones sold that year.
h is such that $h(x) = f(x) + g(x)$.

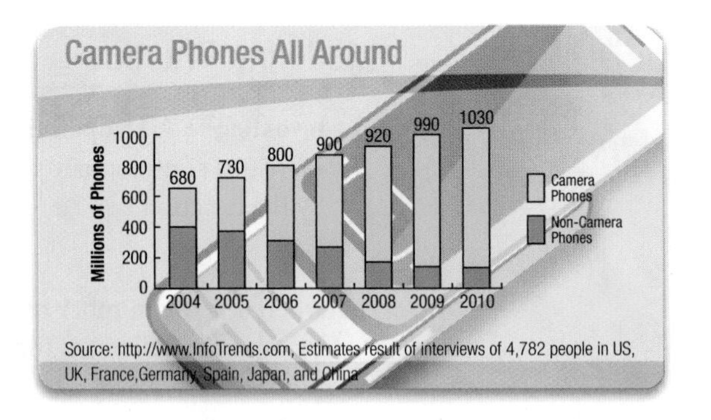

For each statement below, indicate whether the statement is true or false.

a. The domain of f is $\{2004, 2005, 2006, 2007, 2008, 2009, 2010\}$

b. $h(2005) = 741{,}000{,}000$

c. $f(2009) > g(2009)$

d. $f(2004) < f(2005)$

e. $h(2010) > h(2007) > h(2004)$

Straight-Line Depreciation Straight-line depreciation is an accounting method used to help spread the cost of new equipment over a number of years. It takes into account both the cost when new and the salvage value, which is the value of the equipment at the time it gets replaced.

65. Value of a Copy Machine The function $V(t) = -3{,}300t + 18{,}000$, where V is value and t is time in years, can be used to find the value of a large copy machine during the first 5 years of use.

a. What is the value of the copier after 3 years and 9 months?

b. What is the salvage value of this copier if it is replaced after 5 years?

c. State the domain of this function.

d. Sketch the graph of this function.

e. What is the range of this function?

f. After how many years will the copier be worth only $10,000?

66. Step Function Figure 6 shows the graph of the step function C that was used to calculate the first-class postage on a letter weighing x ounces in 2006. Use this graph to answer Parts a through d.

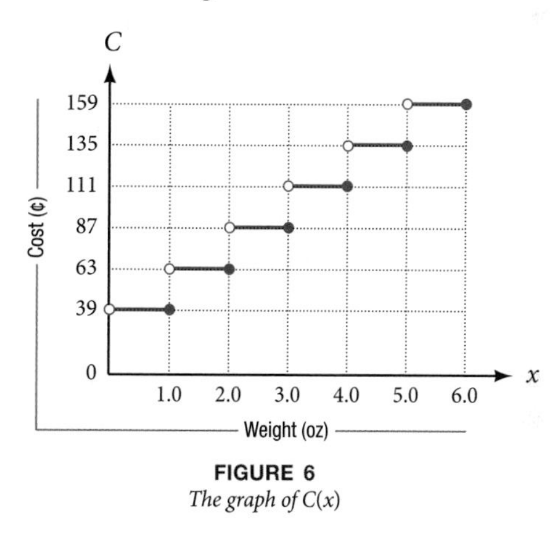

FIGURE 6
The graph of $C(x)$

a. Fill in the following table:

Weight (ounces)	0.6	1.0	1.1	2.5	3.0	4.8	5.0	5.3
Cost (cents)								

b. If a letter cost 87 cents to mail, how much does it weigh? State your answer in words. State your answer as an inequality.

c. If the entire function is shown in Figure 6, state the domain.

d. State the range of the function shown in Figure 6.

Using Technology Exercises

Use Wolfram|Alpha for the following problems.

67. If $f(x) = x^3 - 4x$, find x so that $f(x) = 0$.

68. If $f(x) = x^4 - 4x^2$, find x so that $f(x) = 0$.

69. If $f(x) = x^2 - 2$, find x so that $f(x) = 0$.

70. If $f(x) = x^2 - 2$, find x so that $f(x) = x$.

71. If $f(x) = x^2 - 2$, find x so that $f(x) = 2$.

Getting Ready for the Next Section

Graph each function.

72. $y = x$

73. $y = x^2$

74. $y = \frac{1}{x}$

75. $y = x^3$

76. $y = |x|$

77. $y = \sqrt{x}$

78. $y = x + 2$

79. $y = x^2 - 9$

80. $y = |x| - 3$

81. $y = 2\sqrt{x}$

SPOTLIGHT ON SUCCESS *Student Instructor Lauren*

There are a lot of word problems in algebra and many of them involve topics that I don't know much about. I am better off solving these problems if I know something about the subject. So, I try to find something I can relate to. For instance, an example may involve the amount of fuel used by a pilot in a jet airplane engine. In my mind, I'd change the subject to something more familiar, like the mileage I'd be getting in my car and the amount spent on fuel, driving from my hometown to my college. Changing these problems to more familiar topics makes math much more interesting and gives me a better chance of getting the problem right. It also helps me to understand how greatly math affects and influences me in my everyday life. We really do use math more than we would like to admit—budgeting our income, purchasing gasoline, planning a day of shopping with friends—almost everything we do is related to math. So the best advice I can give with word problems is to learn how to associate the problem with something familiar to you.

You should know that I have always enjoyed math. I like working out problems and love the challenges of solving equations like individual puzzles. Although there are more interesting subjects to me, and I don't plan on pursuing a career in math or teaching, I do think it's an important subject that will help you in any profession.

Transformations and Other Graphing Techniques

3.4

© Michael Bodmann/iStockPhoto

INTRODUCTION The function $E = f(h)$ shown in Figure 1a gives the amount of electrical power, in megawatts, drawn by a community from its local power plant during a 24-hour period in 2005. As the population increases, the amount of electrical power used by the population increases. The blue graph in Figure 1b shows the daily power consumption for the same community in 2012. The two graphs have the same shape. The blue graph is the red graph, just moved up 300 units. In mathematics, we say the blue graph is *translated* up 300 units from the red graph.

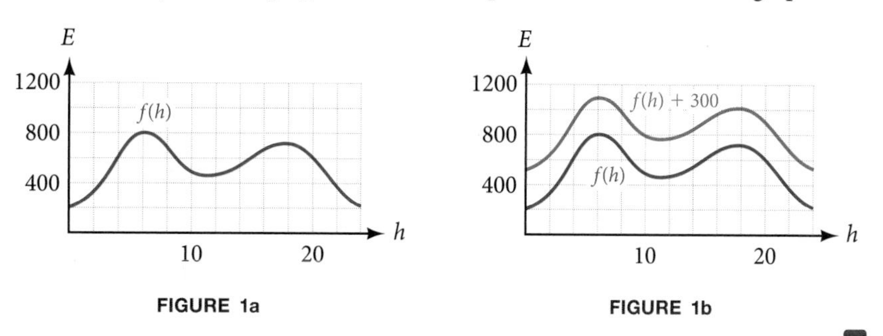

FIGURE 1a FIGURE 1b

Why Learn Graphing Techniques?

With all the graphing software available to us, you may be wondering why it is necessary to learn any graphing techniques at all. The answer is that we want to obtain a basic, intuitive, sense of what our graphs will look like before we do any graphing. It is also important to know how different numbers in the expression of a function affect the graph of that function. For example, you know intuitively what the graph of $y = x^2$ looks like. Do you also know how the number 2, in the functions below, affects the graph of $y = x^2$?

$$y = 2x^2 \qquad y = x^2 + 2 \qquad y = (x + 2)^2$$

If not, you will after you study this section. Knowing how these numbers affect the graph of the underlying function means that you have a more complete understanding of the relationship between a function and its graph.

We are not going to go into extreme detail with our techniques of graphing. We are going to do just enough to give you some good graphing tools that you can use in the next math class you take. To begin, let's list the six basic graphs that you should have memorized by this point in the course.

Six Basic Graphs

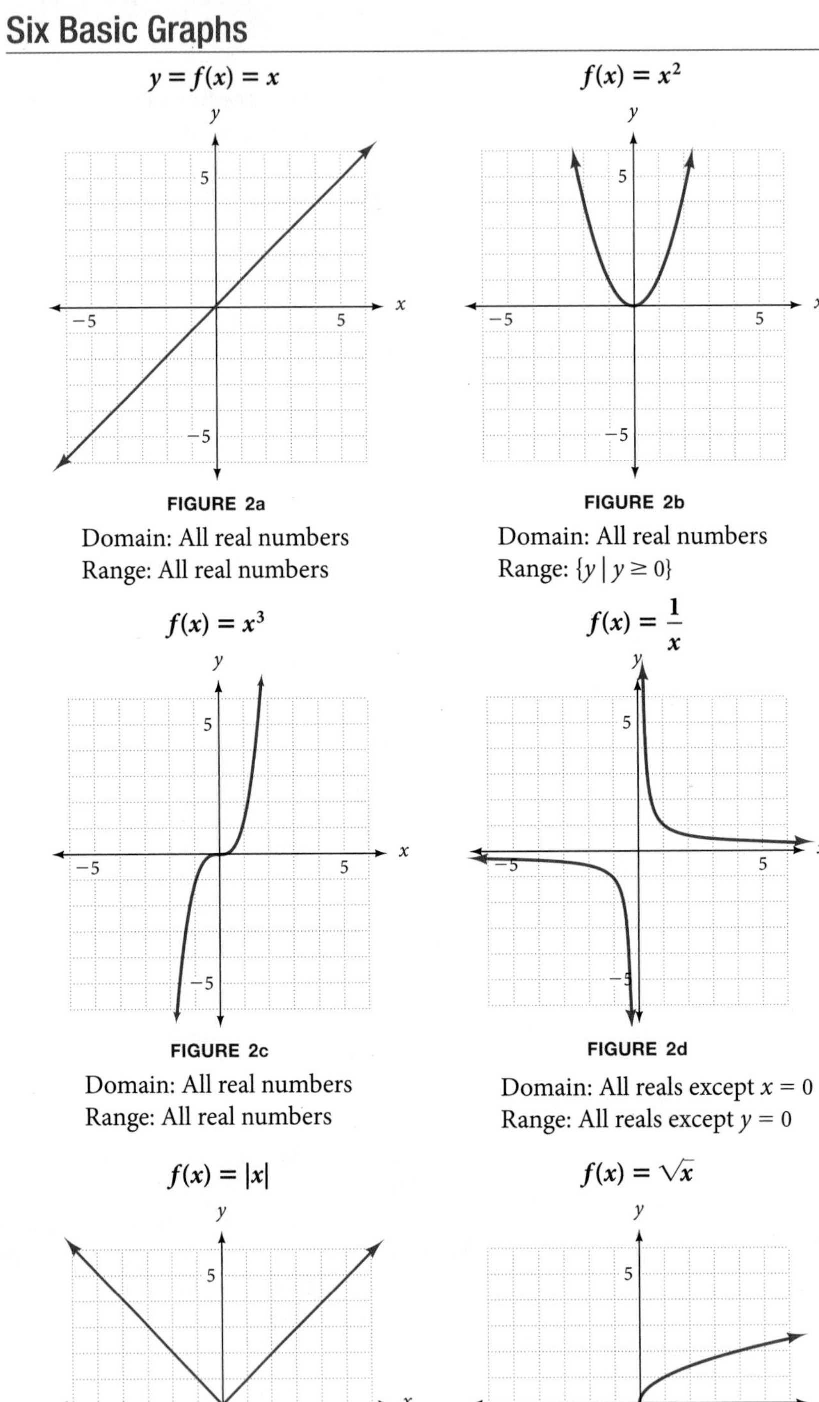

$y = f(x) = x$

FIGURE 2a

Domain: All real numbers
Range: All real numbers

$f(x) = x^2$

FIGURE 2b

Domain: All real numbers
Range: $\{y \mid y \geq 0\}$

$f(x) = x^3$

FIGURE 2c

Domain: All real numbers
Range: All real numbers

$f(x) = \dfrac{1}{x}$

FIGURE 2d

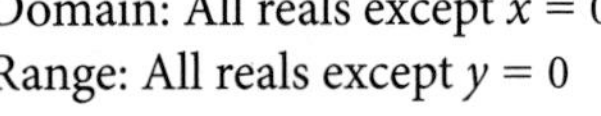

Domain: All reals except $x = 0$
Range: All reals except $y = 0$

$f(x) = |x|$

FIGURE 2e

Domain: All real numbers
Range: $\{y \mid y \geq 0\}$

$f(x) = \sqrt{x}$

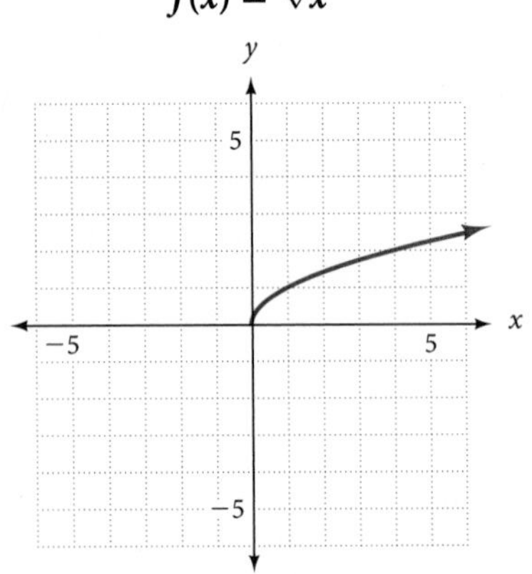

FIGURE 2f

Domain: $\{x \mid x \geq 0\}$
Range: $\{y \mid y \geq 0\}$

Translations

To begin our work with translations, recall the graph of $y = -\frac{1}{3}x$ from Section 3.1. We need it for comparison purposes in Example 1.

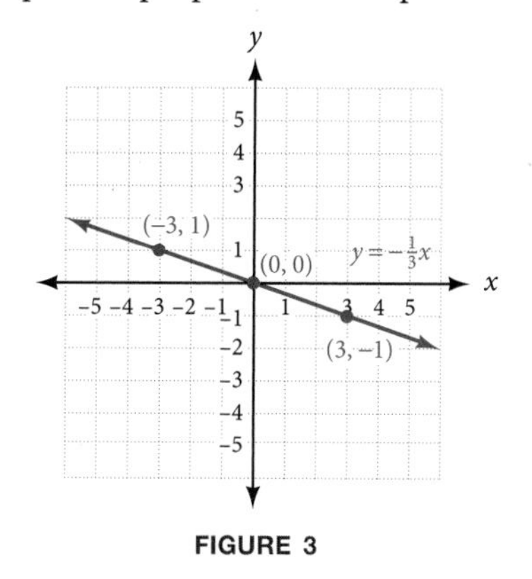

FIGURE 3

EXAMPLE 1 Graph the equation $y = -\frac{1}{3}x + 2$.

SOLUTION We need ordered pairs that are solutions to our equation. Noticing the similarity of this equation to the equation above, we choose the same values of x for our inputs.

Input x	Calculate Using the Equation	Output y	Form Ordered Pairs
-3	$y = -\frac{1}{3}(-3) + 2 = 1 + 2 =$	3	$(-3, 3)$
0	$y = -\frac{1}{3}(0) + 2 = 0 + 2 =$	2	$(0, 2)$
3	$y = -\frac{1}{3}(3) + 2 = -1 + 2 =$	1	$(3, 1)$

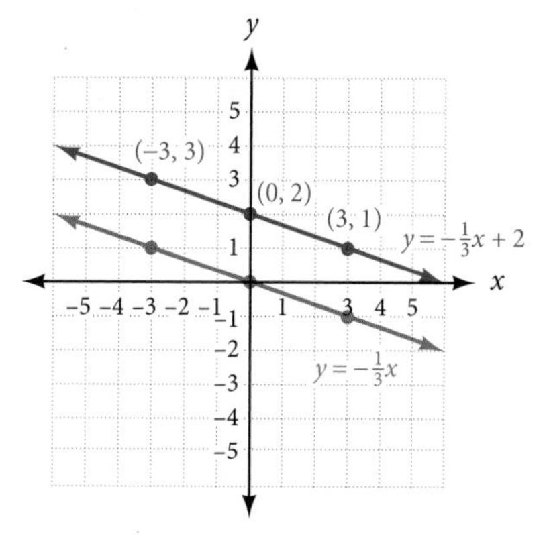

FIGURE 4

We can see that the graph of $y = -\frac{1}{3}x + 2$ looks just like the graph of $y = -\frac{1}{3}x$, but all points are moved up vertically 2 units.

EXAMPLE 2 Graph the equation $y = -\frac{1}{3}x - 4$.

SOLUTION We create a table from ordered pairs, then we graph the information in the table. However, even before we start, we are expecting the graph of $y = -\frac{1}{3}x - 4$ to be 4 units below the graph of $y = -\frac{1}{3}x$.

Input x	Calculate Using the Equation	Output y	Form Ordered Pairs
−3	$y = -\frac{1}{3}(-3) - 4 = 1 - 4 =$	−3	(−3, −3)
0	$y = -\frac{1}{3}(0) - 4 = 0 - 4 =$	−4	(0, −4)
3	$y = -\frac{1}{3}(3) - 4 = -1 - 4 =$	−5	(3, −5)

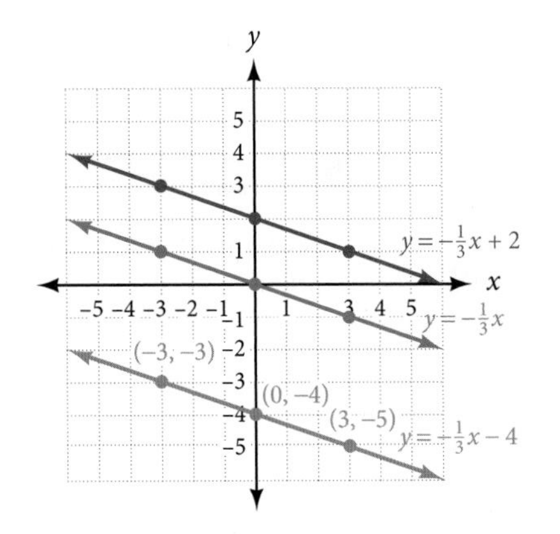

FIGURE 5

Vertical Translations

From our previous two examples we can generalize as follows.

RULE *Vertical Translations*

If $k > 0$, then

The Graph of	Is the Graph of $y = f(x)$ translated
$y = f(x) + k$	k units up
$y = f(x) - k$	k units down

EXAMPLE 3 Find an equation for the blue line.

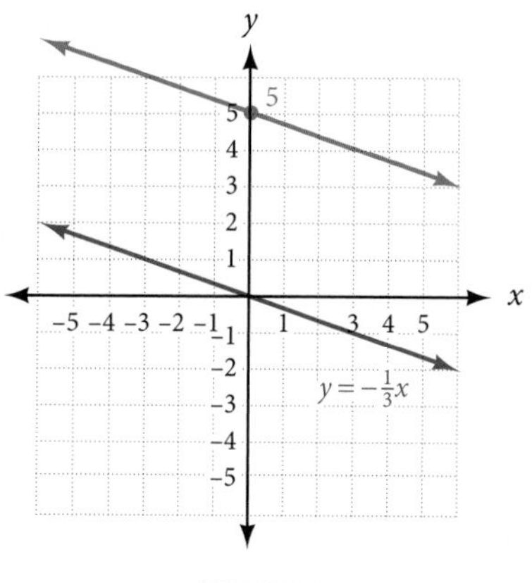

FIGURE 6

SOLUTION The blue line is parallel to the line $y = -\frac{1}{3}x$, but translated up 5 units from $y = -\frac{1}{3}x$. According to what we have done up to this point, the equation for the blue line is $y = -\frac{1}{3}x + 5$. ■

Below are a table and graph for the function $y = x^2$ that we considered previously. We need these for comparison purposes in Example 4.

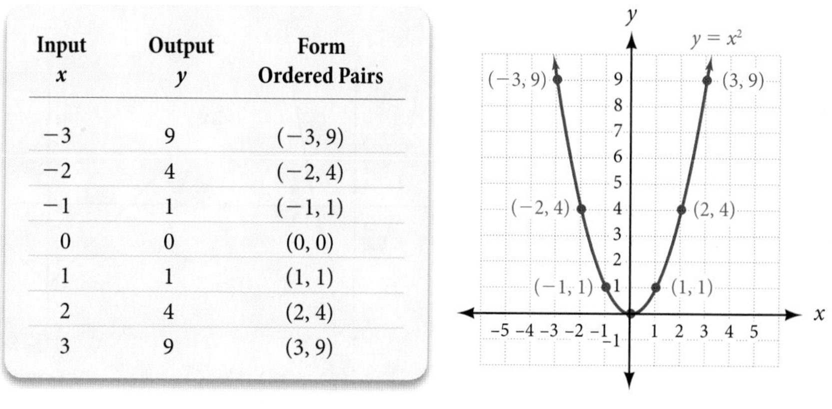

Input x	Output y	Form Ordered Pairs
−3	9	(−3, 9)
−2	4	(−2, 4)
−1	1	(−1, 1)
0	0	(0, 0)
1	1	(1, 1)
2	4	(2, 4)
3	9	(3, 9)

FIGURE 7

EXAMPLE 4 Graph $y = x^2 - 4$.

SOLUTION We make a table as we did in the previous example. If the vertical translation idea works for this type of equation, as it did with our straight lines, we expect this graph to be the graph of $y = x^2$ translated down 4 units.

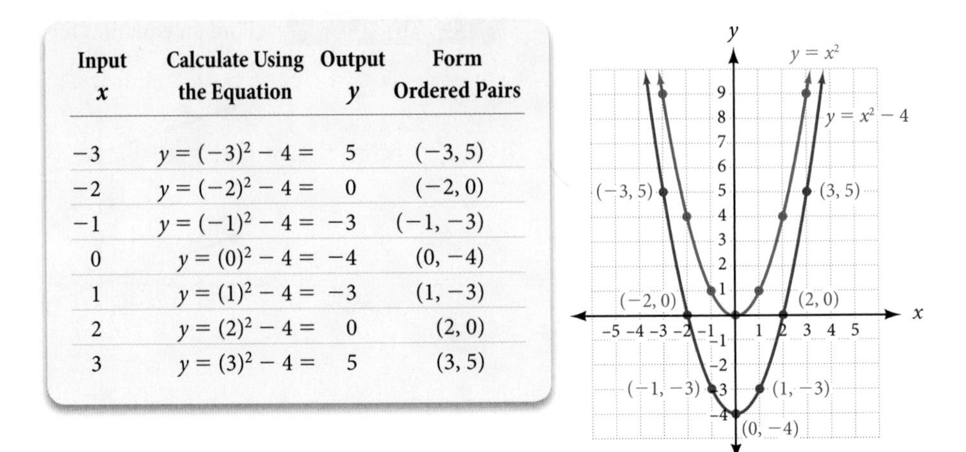

Input x	Calculate Using the Equation	Output y	Form Ordered Pairs
-3	$y = (-3)^2 - 4 =$	5	$(-3, 5)$
-2	$y = (-2)^2 - 4 =$	0	$(-2, 0)$
-1	$y = (-1)^2 - 4 =$	-3	$(-1, -3)$
0	$y = (0)^2 - 4 =$	-4	$(0, -4)$
1	$y = (1)^2 - 4 =$	-3	$(1, -3)$
2	$y = (2)^2 - 4 =$	0	$(2, 0)$
3	$y = (3)^2 - 4 =$	5	$(3, 5)$

FIGURE 8

As you can see, the graph of $y = x^2 - 4$ is the graph of $y = x^2$ translated down 4 units.

EXAMPLE 5 Match each relation with its graph, then indicate which relations are functions

a. $y = |x| - 4$ **b.** $y = x^2 - 4$ **c.** $y = 2x + 2$

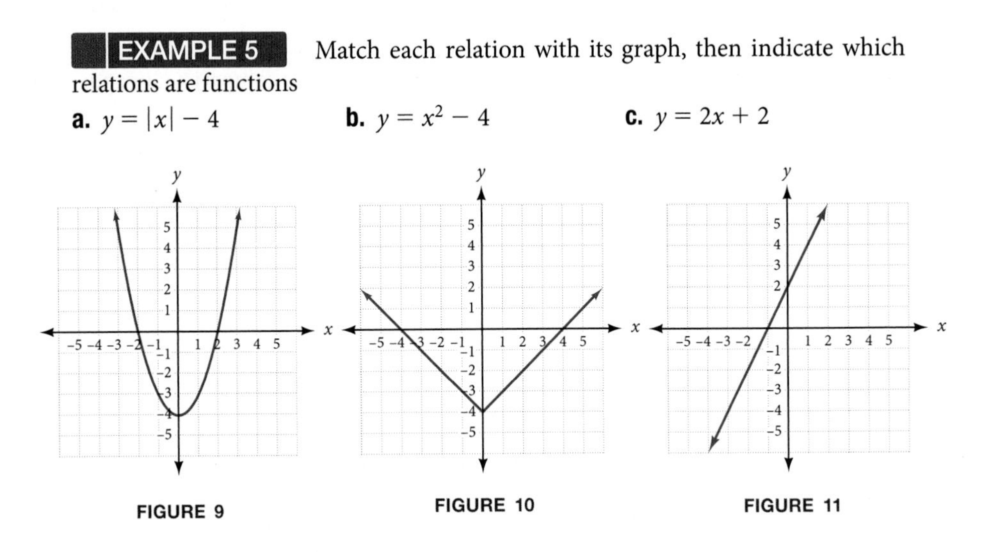

FIGURE 9 FIGURE 10 FIGURE 11

SOLUTION Using the basic graphs for a guide along with our knowledge of translations, we have the following:

a. Figure 10 **b.** Figure 9 **c.** Figure 11

And, since all of these graphs pass the vertical line test, all are functions.

If we look back to the introduction to this section, we see that the information in the two graphs can be explained by a vertical translation. The red graph shows power consumption each hour during a 24-hour period in 2005.

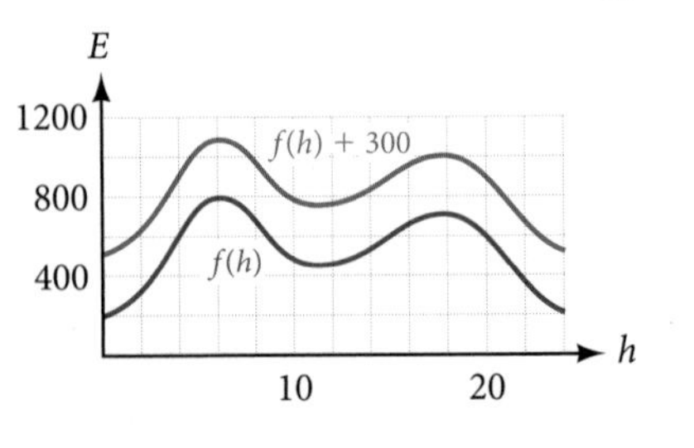

The blue graph is power consumption for the same 24-hour period, but in the year 2012. You can see that the blue graph is the red graph, translated up 300 units. The power consumption usage follows the same pattern in 2012 as it did in 2005, the upward translation of the graph can be attributed to the increase in population.

Horizontal Translations

Now consider the graphs of $f(x) = (x + 2)^2$ and $g(x) = (x - 2)^2$ shown in Figure 12. Compared with the graph of the basic function $y = x^2$, the graph of $f(x) = (x + 2)^2$ is shifted two units to the *left*, as shown by the arrows. Similarly, the graph of $g(x) = (x - 2)^2$ is shifted two units to the *right* compared to the graph of $y = x^2$. In general, we have the following principle:

$\Delta \neq \Sigma$ RULE ***Horizontal Translations***

If $h > 0$, then

The Graph of	Is the Graph of $y = f(x)$ translated
$y = f(x + h)$	h units to the left
$y = f(x - h)$	h units to the right

x	$y = x^2$	$f(x) = (x + 2)^2$
−3	9	1
−2	4	0
−1	1	1
0	0	4
1	1	9
2	4	16
3	9	25

x	$y = x^2$	$g(x) = (x - 2)^2$
−3	9	25
−2	4	16
−1	1	9
0	0	4
1	1	1
2	4	0
3	9	1

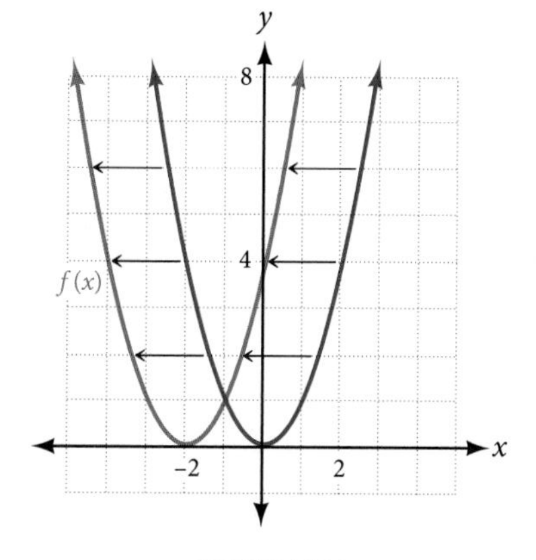

FIGURE 12a

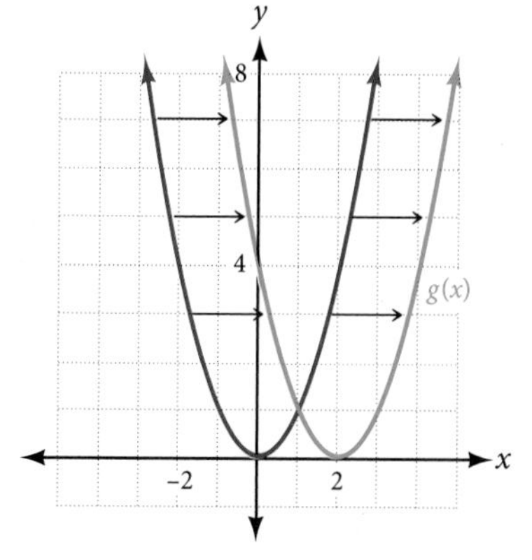

FIGURE 12b

EXAMPLE 6 State the domain and then graph each function.

a. $g(x) = \sqrt{x+1}$ **b.** $h(x) = \dfrac{1}{(x-3)^2}$

SOLUTIONS

a. Because the quantity under the radical must be positive, the domain will be all real numbers such that

$$x + 1 \geq 0$$
$$x \geq -1$$

Our domain is $\{x|x \geq -1\}$.

The table shows that each y-value for $g(x)$ occurs one unit to the left of the same y-value for the graph of $y = \sqrt{x}$. Consequently, each point on the graph of $y = g(x)$ is shifted one unit to the left of $y = \sqrt{x}$, as shown in Figure 13.

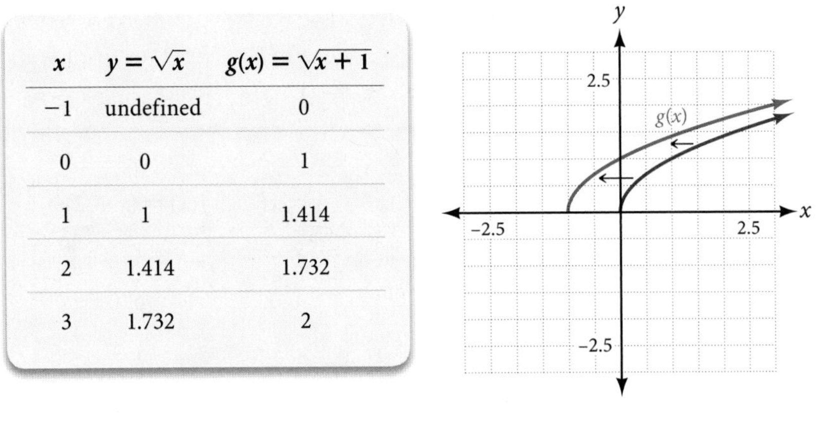

x	$y = \sqrt{x}$	$g(x) = \sqrt{x+1}$
−1	undefined	0
0	0	1
1	1	1.414
2	1.414	1.732
3	1.732	2

FIGURE 13

b. Because we cannot divide by 0, our domain will be $\{x|x \neq 3\}$ because 3 is the only number that causes the denominator to be 0. The table shows that each y-value for $h(x)$ occurs three units to the right of the same y-value for the graph of $y = \dfrac{1}{x^2}$. Consequently, each point on the graph of $y = h(x)$ is shifted three units to the right of $y = \dfrac{1}{x^2}$, as shown in Figure 14.

x	$y = \dfrac{1}{x^2}$	$h(x) = \dfrac{1}{(x-3)^2}$
−1	1	1/16
0	undefined	1/9
1	1	1/4
2	1/4	1
3	1/9	undefined
4	1/16	1

FIGURE 14

© David H. Lewis/iStockPhoto

Modeling: Glaucoma

EXAMPLE 7 The function $N = f(p)$ graphed in Figure 15 gives the number of people who have eye pressure level p from a sample of 100 people with healthy eyes, and the function g gives us the number of people with pressure level p in a sample of 100 glaucoma patients.

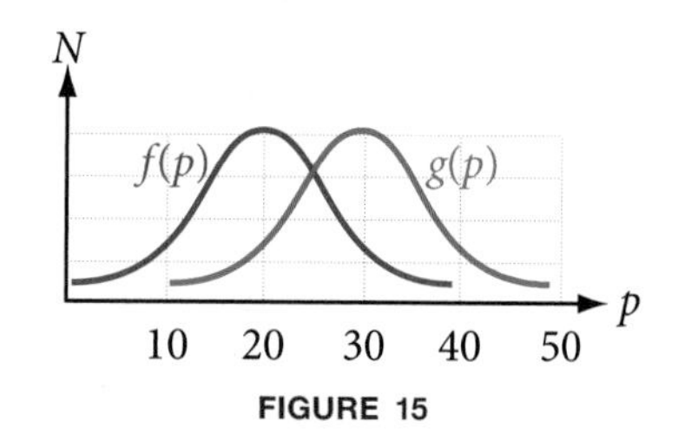

FIGURE 15

a. Write a formula for g as a transformation of f.

b. For what pressure readings could a doctor be fairly certain that a patient has glaucoma?

SOLUTIONS

a. The graph of g is translated 10 units to the right of f, so $g(p) = f(p - 10)$.

b. From the graph, it appears that pressure readings above 40 are a strong indication of glaucoma. Readings between 10 and 40 cannot conclusively distinguish healthy eyes from those with glaucoma. ∎

Our next example involves both a horizontal translation and a vertical translation of a graph.

EXAMPLE 8 Graph $y = |x - 3| + 4$.

SOLUTION Our graph will be a transformation of the basic graph $y = |x|$. We proceed according to the rule for order of operations which tells us to go within the first grouping symbol first and then work our way out. Here is a diagram showing how we approach this, along with the graphs that result:

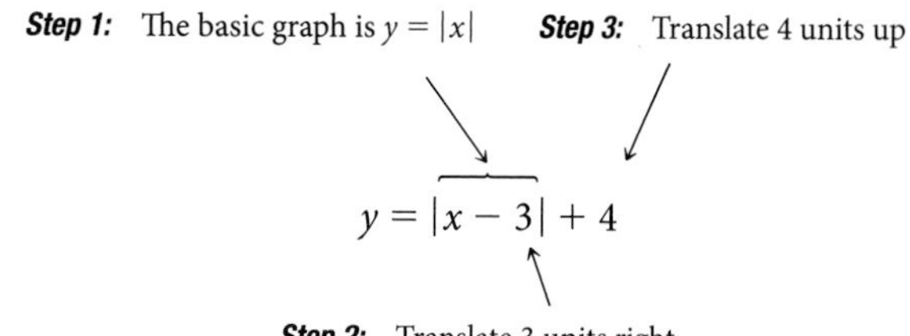

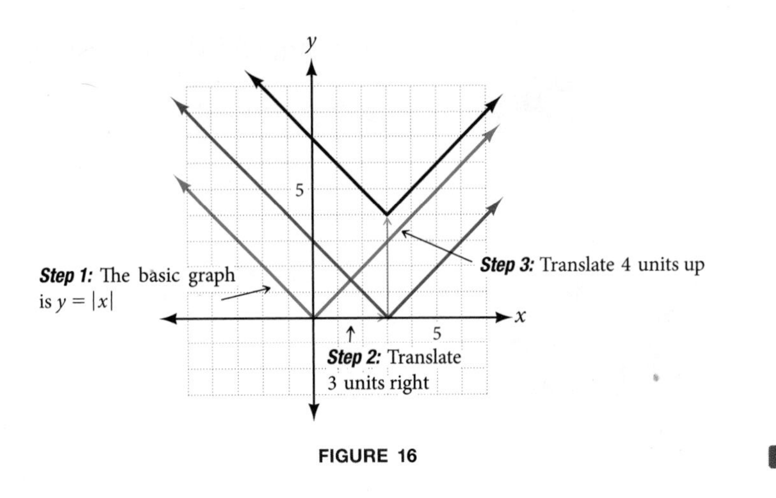

FIGURE 16

Scale Factors

We have seen that adding a constant to the expression defining a function results in a translation of its graph. What happens if we multiply the expression by a constant? Consider the graphs of the functions

$$f(x) = 2x^2,\ g(x) = \frac{1}{2}x^2, \text{ and } h(x) = -x^2$$

shown in Figure 17a, b, and c, and compare each to the graph of $y = x^2$.

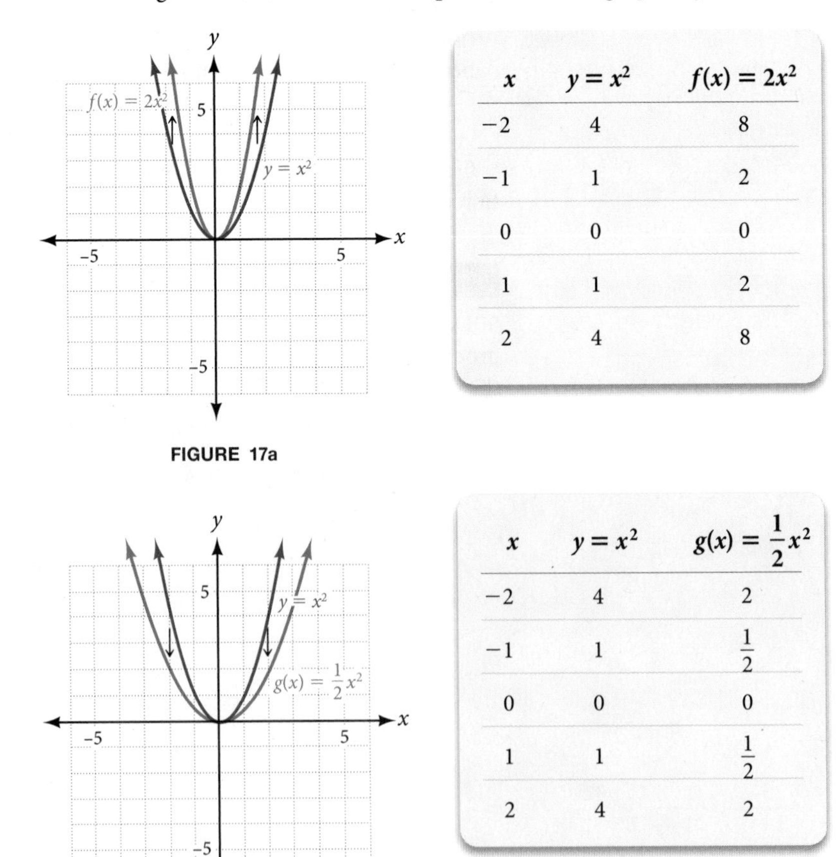

x	$y = x^2$	$f(x) = 2x^2$
−2	4	8
−1	1	2
0	0	0
1	1	2
2	4	8

FIGURE 17a

x	$y = x^2$	$g(x) = \frac{1}{2}x^2$
−2	4	2
−1	1	$\frac{1}{2}$
0	0	0
1	1	$\frac{1}{2}$
2	4	2

FIGURE 17b

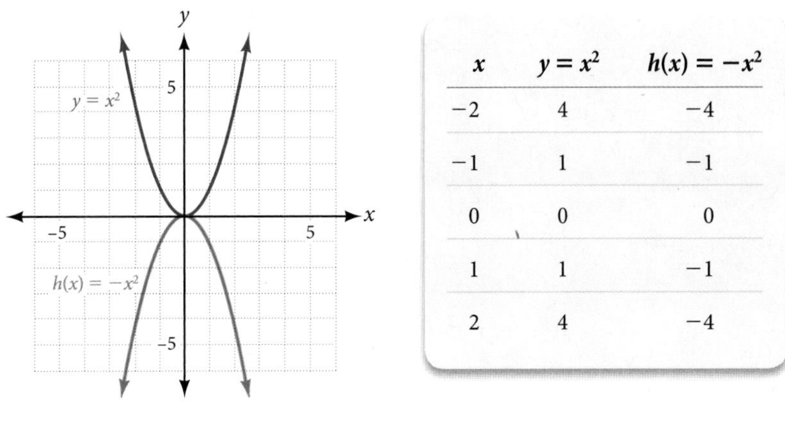

x	$y = x^2$	$h(x) = -x^2$
-2	4	-4
-1	1	-1
0	0	0
1	1	-1
2	4	-4

FIGURE 17c

Compared to the graph of $y = x^2$, the graph of $f(x) = 2x^2$ is expanded, or stretched, vertically by a factor of 2. The y-coordinate of each point on the graph has been doubled, as you can see in the table of values, so each point on the graph of f is twice as far from the x-axis as its counterpart on the basic graph $y = x^2$. The graph of $g(x) = \frac{1}{2}x^2$ is compressed vertically by a factor of $\frac{1}{2}$; each point is half as far from the x-axis as its counterpart on the graph of $y = x^2$. The graph of $h(x) = -x^2$ is flipped, or reflected, about the x-axis; the y-coordinate of each point on the graph of $y = x^2$ is replaced by its opposite.

In general, we have the following:

RULE *Scale Factors and Reflections*

Compared with the graph of $y = f(x)$, the graph of $y = af(x)$, where $a \neq 0$, is

1. stretched vertically by a factor of $|a|$ if $|a| > 1$,

2. compressed vertically by a factor of $|a|$ if $0 < |a| < 1$, and

3. reflected about the x-axis if $a < 0$.

The constant a is called the *scale factor* for the graph.

Modeling: Blood Alcohol Level

EXAMPLE 9 The function $A = f(t)$ graphed in Figure 18 gives a person's blood alcohol level t hours after drinking a martini. Sketch a graph of $g(t) = 2f(t)$ and explain what it tells you.

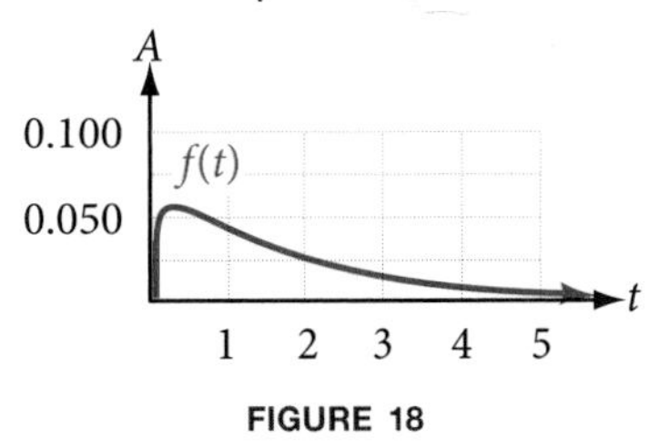

FIGURE 18

SOLUTION To sketch a graph of g, we stretch the graph of f vertically by a factor of 2, as shown in Figure 19. At each time t, the person's blood alcohol level is twice the value given by f. The function g could represent a person's blood alcohol level t hours after drinking two martinis.

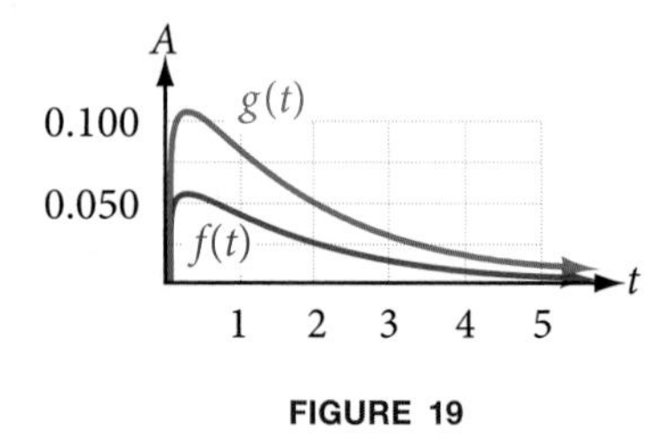

FIGURE 19

Here is an example that involves horizontal and vertical translation, accompanied by a scale factor of $\frac{1}{4}$.

EXAMPLE 10 Graph $f(x) = \frac{1}{4}(x + 4)^2 + 2$.

SOLUTION We identify the basic graph from the structure of the formula for $f(x)$. In this case, the basic graph is $y = x^2$. We will perform the translations separately, following the order of operations. Here is a diagram showing how we interpret the information in the equation, followed by Figure 20, which shows how we construct the graph.

Step 1: The basic graph is $y = x^2$

Step 3: Scale each y value by $\frac{1}{4}$ to flatten graph

Step 4: Translate 2 units up

$$y = \frac{1}{4}(x + 4)^2 + 2$$

Step 2: Translate 4 units left

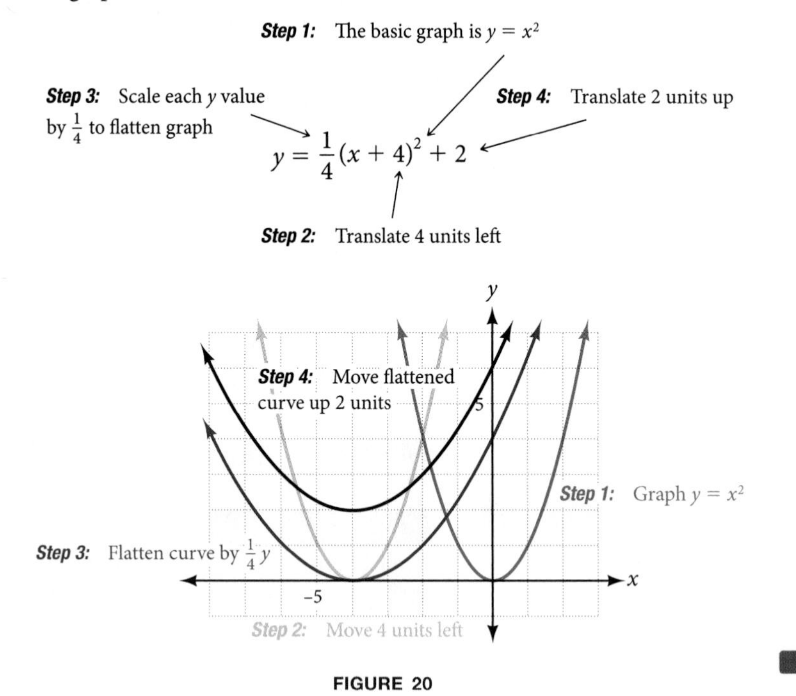

FIGURE 20

Using Technology 3.4

We can use a graphing calculator to confirm our work with translations. Let's do a simple example that translates our basic graph of $y = x$.

EXAMPLE 11 Use the graphing calculator to graph $y = x + b$ for $b = -3, -2, -1, 0, 1, 2,$ and 3.

SOLUTION

Method 1: Y-Variables List

To use the Y-variables list, enter each equation at one of the Y variables, set the graph window, then graph. The calculator will graph the equations in order, starting with Y1 and ending with Y7. Following is the Y-variables list, an appropriate window, and a sample of the type of graph obtained (Figure 21).

Y1 = X − 3
Y2 = X − 2
Y3 = X − 1
Y4 = X
Y5 = X + 1
Y6 = X + 2
Y7 = X + 3

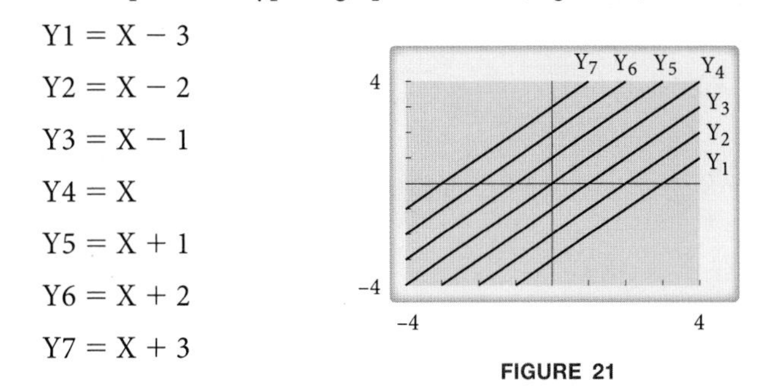

FIGURE 21

Window: X from −4 to 4, Y from −4 to 4

Method 2: Programming

The same result can be obtained by programming your calculator to graph $y = x + b$ for $b = -3, -2, -1, 0, 1, 2,$ and 3. Here is an outline of a program that will do this. Check the manual that came with your calculator to find the commands for your calculator.

Step 1: Clear screen
Step 2: Set window for X from −4 to 4 and Y from −4 to 4
Step 3: $-3 \rightarrow B$
Step 4: Label 1
Step 5: Graph Y = X + B
Step 6: $B + 1 \rightarrow B$
Step 7: If $B < 4$, Go to 1
Step 8: End

Method 3: Using Lists

On the TI-82/83 you can set Y1 as follows

$$Y1 = X + \{-3, -2, -1, 0, 1, 2, 3\}$$

When you press GRAPH, the calculator will graph each line from $y = x + (-3)$ to $y = x + 3$.

Each of the three methods will produce graphs similar to those in Figure 21. ■

GETTING READY FOR CLASS

Respond in your own words and in complete sentences.

A. Why do we learn graphing techniques when there are so many graphing programs available to us?

B. Without looking in the book, name and graph as many of the six basic functions as you can. Include the domain and range for each one.

C. What effect does 5 have on the graph of $y = 5x^2$?

D. What effect does 5 have on the graph of $y = x^2 - 5$?

Problem Set 3.4

For each problem below, the equation of the red curve is given. Find the equation for the blue graph.

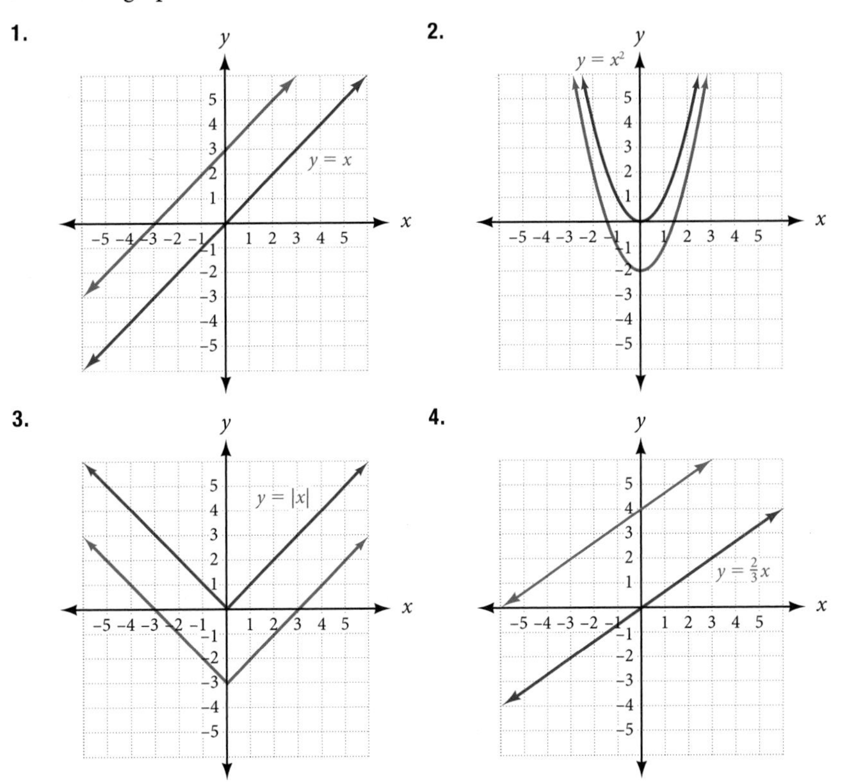

Graph each of the following. Use one coordinate system for each problem.

5. **a.** $y = 2x$
b. $y = 2x + 3$
c. $y = 2x - 5$

6. **a.** $y = \frac{1}{3}x$
b. $y = \frac{1}{3}x + 1$
c. $y = \frac{1}{3}x - 3$

7. **a.** $y = \frac{1}{2}x^2$
b. $y = \frac{1}{2}x^2 - 2$
c. $y = \frac{1}{2}x^2 + 2$

8. **a.** $y = 2x^2$
b. $y = 2x^2 - 8$
c. $y = 2x^2 + 1$

Identify each graph as a translation of a basic function, and write a formula for the graph.

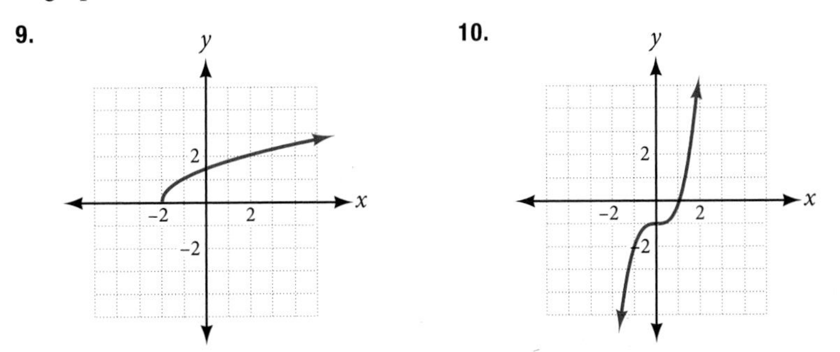

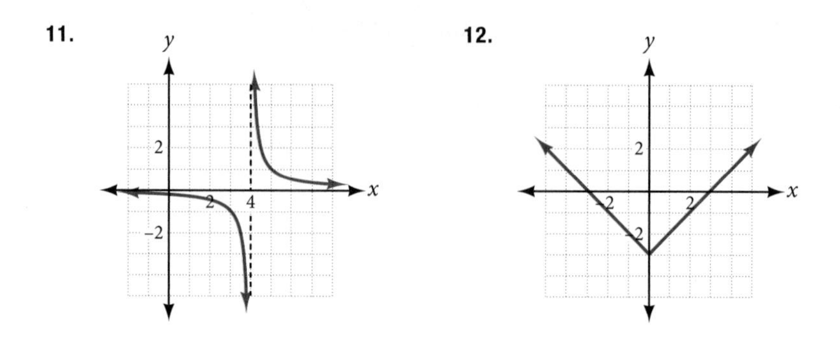

Sketch the basic graph and the graph of the given function on the same axes. Label the coordinates of two or three points on the graph of the given function.

13. $f(x) = |x| - 2$ **14.** $g(x) = (x + 1)^3$ **15.** $f(s) = s^2 + 3$

16. $G(t) = \sqrt{t - 2}$ **17.** $G(r) = (r + 2)^3$ **18.** $F(r) = \dfrac{1}{r - 4}$

19. $H(d) = \sqrt{d} - 3$ **20.** $h(v) = \dfrac{1}{v + 6}$ **21.** $f(x) = \dfrac{1}{3}|x|$

22. $H(x) = -3|x|$ **23.** $g(z) = \dfrac{2}{z}$ **24.** $G(v) = -3\sqrt{v}$

25. $f(s) = \dfrac{1}{8}s^3$ **26.** $H(x) = \dfrac{1}{3x}$

27. Match each graph with its equation.

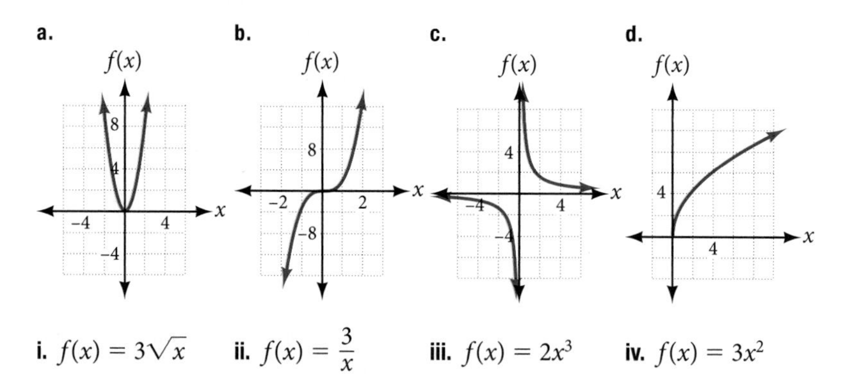

i. $f(x) = 3\sqrt{x}$ **ii.** $f(x) = \dfrac{3}{x}$ **iii.** $f(x) = 2x^3$ **iv.** $f(x) = 3x^2$

28. Match each graph with its equation.

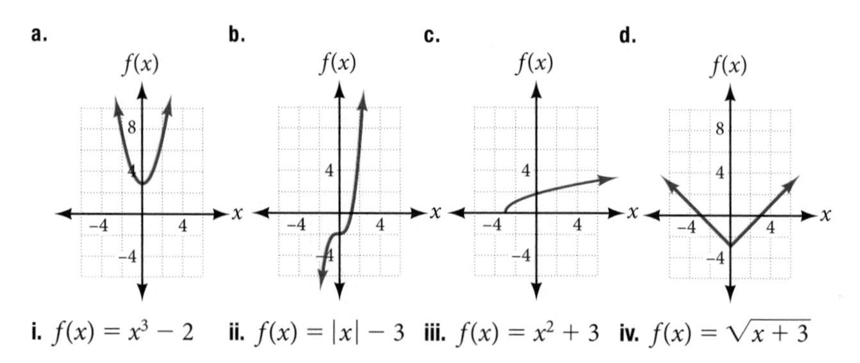

i. $f(x) = x^3 - 2$ **ii.** $f(x) = |x| - 3$ **iii.** $f(x) = x^2 + 3$ **iv.** $f(x) = \sqrt{x + 3}$

29. The graph of $y = g(t)$ is shown below, followed by three transformations of its graph. Write a formula for each transformation in terms of the original function, $y = g(t)$.

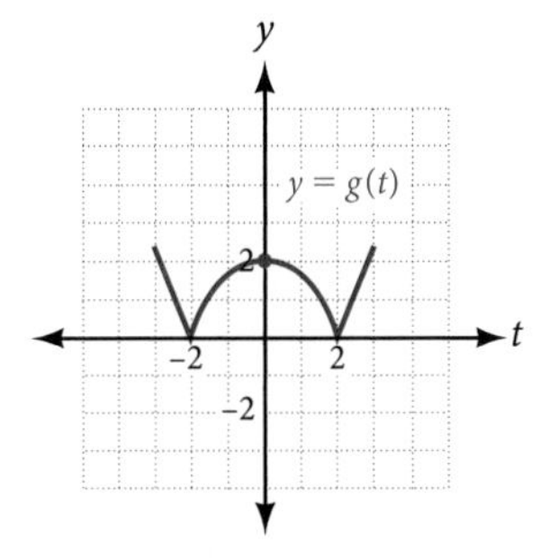

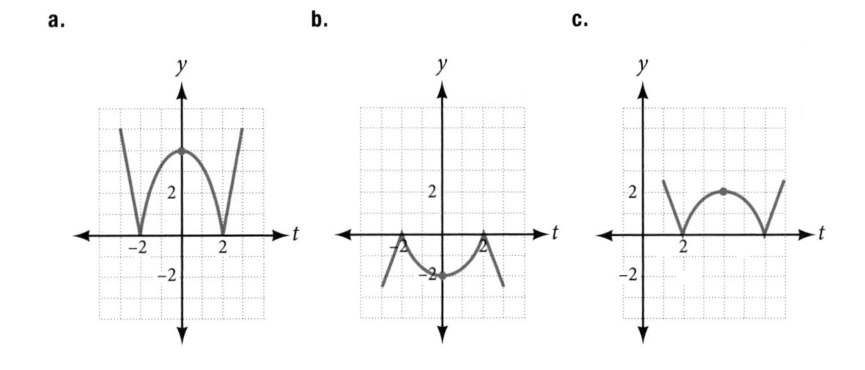

30. The graph of $T = h(v)$ is shown below followed by three transformations of its graph. Write a formula for each transformation in terms of the original function, $T = h(v)$.

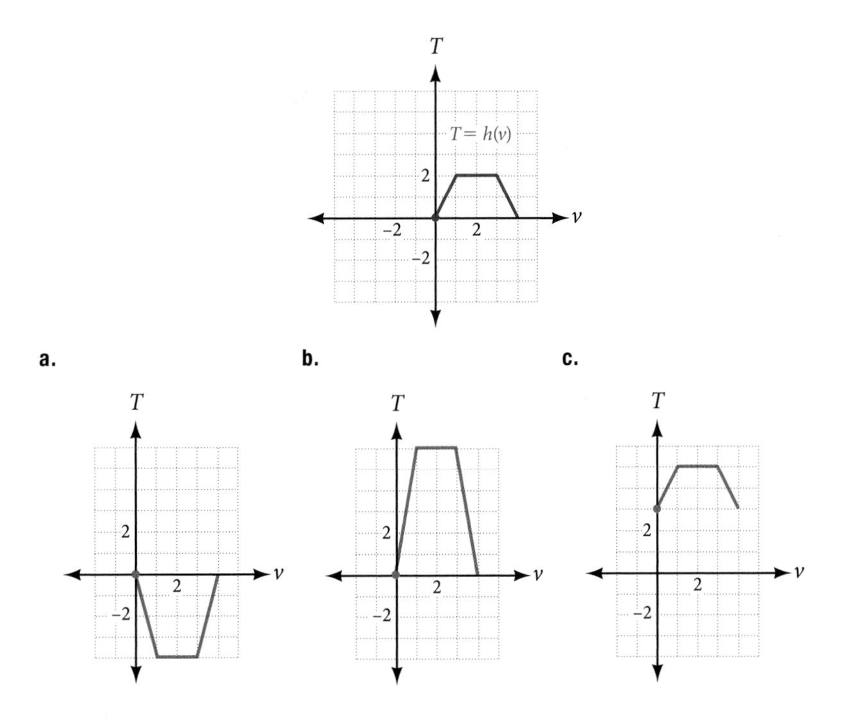

Sketch the basic graph and the graph of the given function by hand on the same axes. Label the coordinates of two or three points on the graph of the given function.

31. $f(x) = 2 + (x - 3)^2$

32. $f(x) = (x + 4)^2 + 1$

33. $g(z) = \dfrac{1}{z + 2} - 3$

34. $g(z) = \dfrac{1}{z - 1} + 1$

35. $F(u) = -3\sqrt{u + 4} + 4$

36. $F(u) = 4\sqrt{u - 3} - 5$

37. $G(t) = 2|t - 5| - 1$

38. $G(t) = 2 - |t + 4|$

Give an equation for each graph.

39.

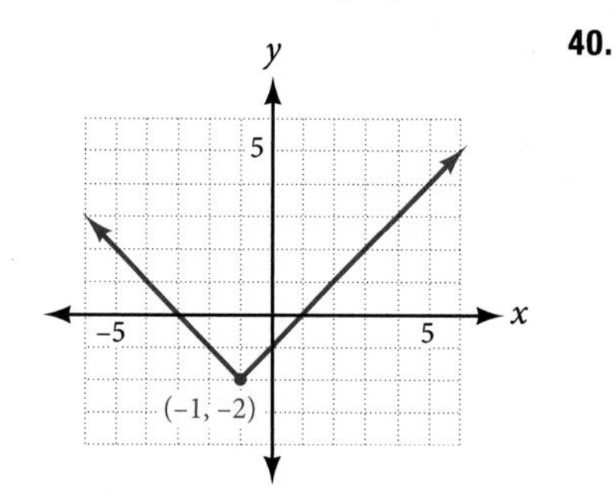

40.

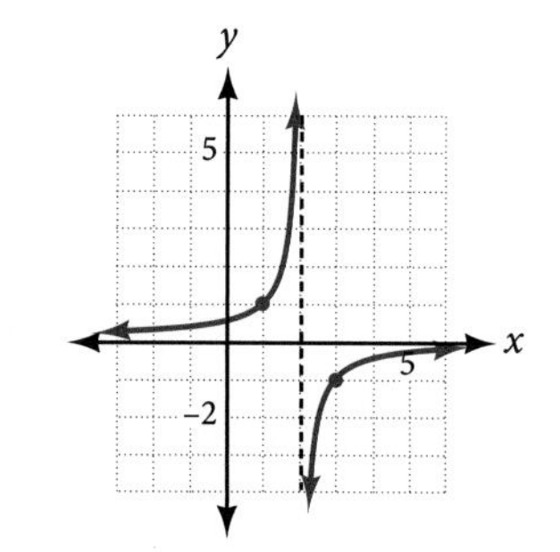

41.

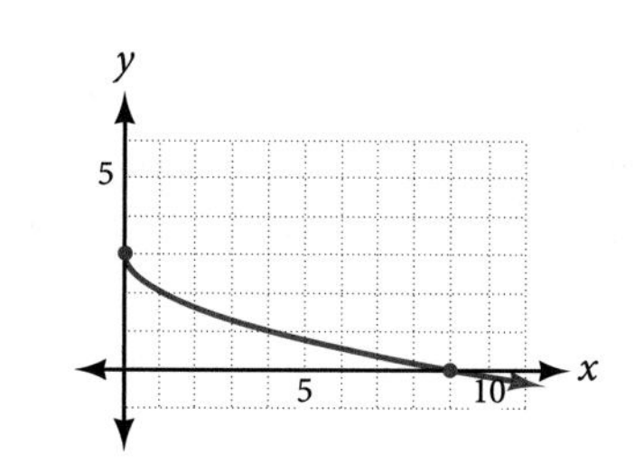

42.

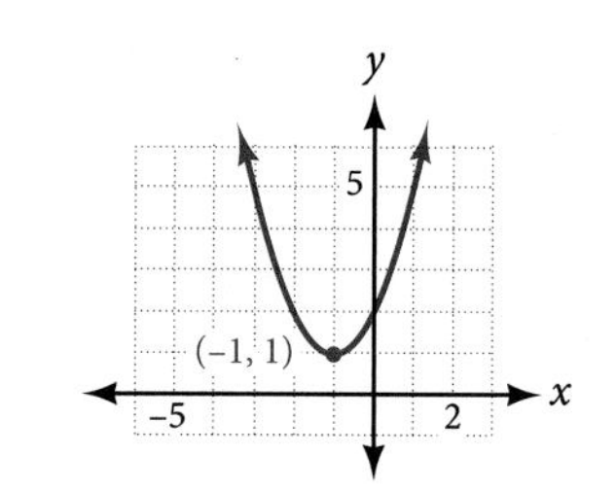

43.

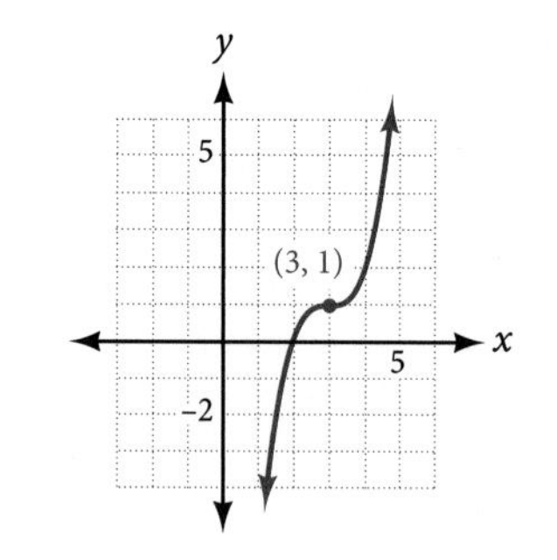

Using Technology Exercises

Use your graphing calculator to graph each of the following families of curves.

44. $y = 2x + b$ $\quad b = -3, -2, -1, 0, 1, 2, 3$

45. $y = 2(x + b)$ $\quad b = -3, -2, -1, 0, 1, 2, 3$

46. $y = x^2 + b$ $\quad b = -3, -2, -1, 0, 1, 2, 3$

47. $y = (x + b)^2$ $\quad b = -3, -2, -1, 0, 1, 2, 3$

Getting Ready for the Next Section

Multiply.

48. $x(35 - 0.1x)$

49. $0.6(M - 70)$

50. $(4x - 3)(x - 1)$

51. $(4x - 3)(4x^2 - 7x + 3)$

Simplify.

52. $(35x - 0.1x^2) - (8x + 500)$

53. $(4x - 3) + (4x^2 - 7x + 3)$

54. $(4x^2 + 3x + 2) - (2x^2 - 5x - 6)$

55. $(4x^2 + 3x + 2) + (2x^2 - 5x - 6)$

56. $4(2)^2 - 3(2)$

57. $4(-1)^2 - 7(-1)$

Algebra and Composition with Functions

INTRODUCTION People tend to drink more water on hot days than they do on cold days. Because most drinking water is sold in plastic bottles, as the temperature goes up, there is an increase in the number of plastic bottles recycled. We can model this situation with the following diagram.

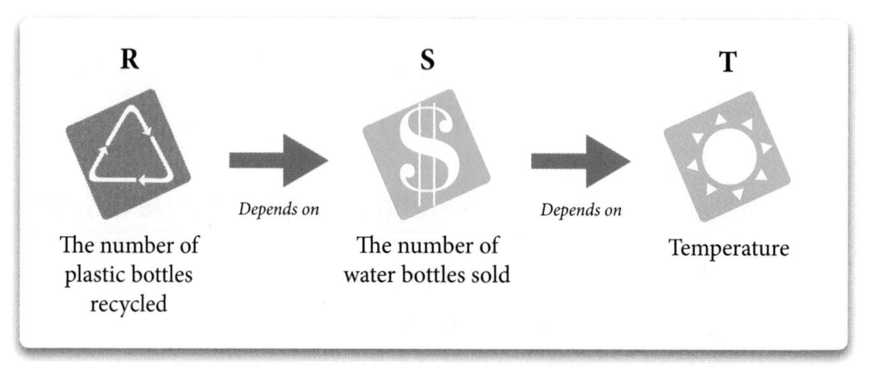

This situation involves a *composite function* because the number of plastic bottles recycled depends on the number of plastic bottles sold, which in turn, depends on the temperature. Composition of functions is one of the topics we will study in this section.

Algebra with Functions

If we are given two functions f and g with a common domain, we can define four other functions as follows.

DEFINITION

$(f+g)(x) = f(x) + g(x)$	The function $f + g$ is the sum of the functions f and g.
$(f-g)(x) = f(x) - g(x)$	The function $f - g$ is the difference of the functions f and g.
$(fg)(x) = f(x)g(x)$	The function fg is the product of the functions f and g.
$\left(\frac{f}{g}\right)(x) = \frac{f(x)}{g(x)}$	The function $\frac{f}{g}$ is the quotient of the functions f and g, where $g(x) \neq 0$.

Let's look at some examples of algebra with functions. In each example that follows, notice that combining functions with addition, subtraction, multiplication, and division produces another function.

VIDEO EXAMPLES

SECTION 3.5

EXAMPLE 1 If $f(x) = 4x^2 + 3x + 2$ and $g(x) = 2x^2 - 5x - 6$, write the formulas for the functions $f + g$, $f - g$, fg, and $\frac{f}{g}$.

SOLUTION The function $f + g$ is defined by

$$\begin{aligned}(f + g)(x) &= f(x) + g(x)\\ &= (4x^2 + 3x + 2) + (2x^2 - 5x - 6)\\ &= 6x^2 - 2x - 4\end{aligned}$$

The function $f - g$ is defined by

$$\begin{aligned}(f - g)(x) &= f(x) - g(x)\\ &= (4x^2 + 3x + 2) - (2x^2 - 5x - 6)\\ &= 4x^2 + 3x + 2 - 2x^2 + 5x + 6\\ &= 2x^2 + 8x + 8\end{aligned}$$

The function fg is defined by

$$\begin{aligned}(fg)(x) &= f(x)g(x)\\ &= (4x^2 + 3x + 2)(2x^2 - 5x - 6)\\ &= 8x^4 - 20x^3 - 24x^2 + 6x^3 - 15x^2 - 18x + 4x^2 - 10x - 12\\ &= 8x^4 - 14x^3 - 35x^2 - 28x - 12\end{aligned}$$

The function f / g is defined by

$$\begin{aligned}\left(\frac{f}{g}\right)(x) &= \frac{f(x)}{g(x)}\\ &= \frac{4x^2 + 3x + 2}{2x^2 - 5x - 6}\end{aligned}$$

■

EXAMPLE 2 Let $f(x) = 4x - 3$, $g(x) = 4x^2 - 7x + 3$, and $h(x) = x - 1$. Find $f + g$, fh, fg, and $\frac{g}{f}$.

SOLUTION The function $f + g$, the sum of functions f and g, is defined by

$$\begin{aligned}(f + g)(x) &= f(x) + g(x)\\ &= (4x - 3) + (4x^2 - 7x + 3)\\ &= 4x^2 - 3x\end{aligned}$$

The function fh, the product of functions f and h, is defined by

$$\begin{aligned}(fh)(x) &= f(x)h(x)\\ &= (4x - 3)(x - 1)\\ &= 4x^2 - 7x + 3\\ &= g(x)\end{aligned}$$

The function fg, the product of the functions f and g, is defined by

$$\begin{aligned}(fg)(x) &= f(x)g(x)\\ &= (4x - 3)(4x^2 - 7x + 3)\\ &= 16x^3 - 28x^2 + 12x - 12x^2 + 21x - 9\\ &= 16x^3 - 40x^2 + 33x - 9\end{aligned}$$

The function $\frac{g}{f}$, the quotient of the functions g and f, is defined by

$$\begin{aligned}\left(\frac{g}{f}\right)(x) &= \frac{g(x)}{f(x)}\\ &= \frac{4x^2 - 7x + 3}{4x - 3}\end{aligned}$$

Factoring the numerator, we can reduce to lowest terms:

$$\begin{aligned}&= \frac{(4x - 3)(x - 1)}{4x - 3}\\ &= x - 1\\ &= h(x)\end{aligned}$$

■

EXAMPLE 3 If f, g, and h are the same functions defined in Example 2, evaluate $(f + g)(2)$, $(fh)(-1)$, $(fg)(0)$, and $\left(\frac{g}{f}\right)(5)$.

SOLUTION We use the formulas for $f + g$, fh, fg, and $\frac{g}{f}$ found in Example 2:

$$\begin{aligned}(f + g)(2) &= 4(2)^2 - 3(2)\\ &= 16 - 6\\ &= 10\\ (fh)(-1) &= 4(-1)^2 - 7(-1) + 3\\ &= 4 + 7 + 3\\ &= 14\\ (fg)(0) &= 16(0)^3 - 40(0)^2 + 33(0) - 9\\ &= 0 - 0 + 0 - 9\\ &= -9\\ \left(\frac{g}{f}\right)(5) &= 5 - 1\\ &= 4\end{aligned}$$

■

Modeling: Revenue, Cost, and Profit

EXAMPLE 4 A company produces and sells copies of an accounting program for home computers. The price they charge for the program is related to the number of copies sold by the demand function

$$p(x) = 35 - 0.1x$$

a. Find the revenue function.

b. If the cost function is $C(x) = 8x + 500$, find the profit function.

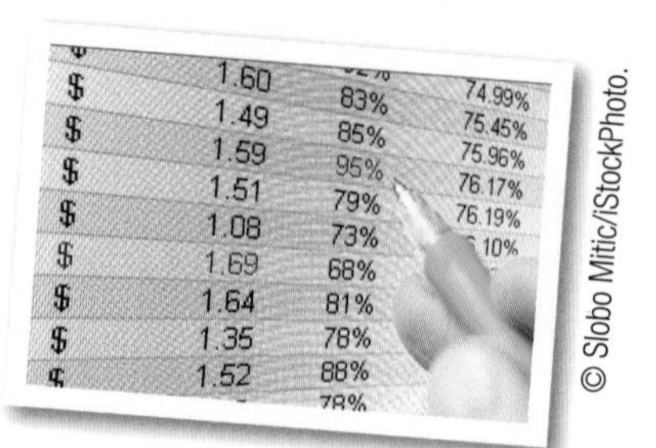

© Slobo Mitic/iStockPhoto.

SOLUTION

a. We find the revenue for this business by multiplying the number of items sold by the price per item. When we do so, we are forming a new function by combining two existing functions. That is, if $n(x) = x$ is the number of items sold and $p(x) = 35 - 0.1x$ is the price per item, then revenue is

$$R(x) = n(x) \cdot p(x) = x(35 - 0.1x) = 35x - 0.1x^2$$

In this case, the revenue function is the product of two functions. When we combine functions in this manner, we are applying our rules for algebra to functions.

b. To carry this situation further, we know the profit function is the difference between two functions. If the cost function for producing x copies of the accounting program is $C(x) = 8x + 500$, then the profit function is

$$P(x) = R(x) - C(x) = (35x - 0.1x^2) - (8x + 500) = -500 + 27x - 0.1x^2$$

The relationship between these last three functions is represented visually in Figure 1.

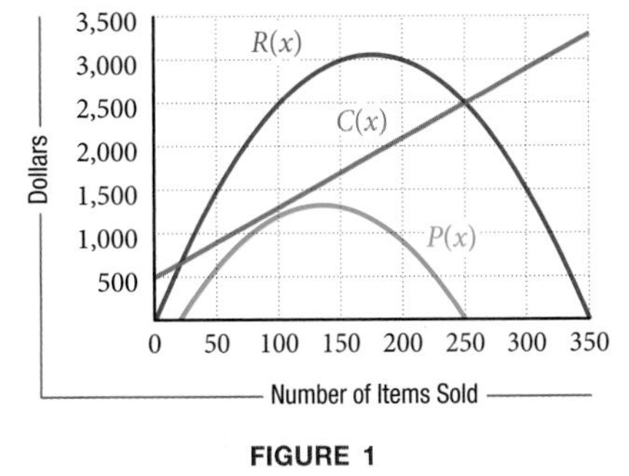

FIGURE 1

Composition of Functions and Training Heart Rate

In addition to the four operations used to combine functions shown so far in this section, there is a fifth way to combine two functions to obtain a new function. It is called *composition of functions.* To illustrate the concept, here is the definition of training heart rate: training heart rate, in beats per minute, is resting heart rate plus 60% of the difference between maximum heart rate and resting heart rate. If your resting heart rate is 70 beats per minute, then your training heart rate is a function of your maximum heart rate M.

$$T(M) = 70 + 0.6(M - 70) = 70 + 0.6M - 42 = 28 + 0.6M$$

But your maximum heart rate is found by subtracting your age in years from 220. So, if x represents your age in years, then your maximum heart rate is

$$M(x) = 220 - x$$

Therefore, if your resting heart rate is 70 beats per minute and your age in years is x, then your training heart rate can be written as a function of x.

$$T(x) = 28 + 0.6(220 - x)$$

This last line is the composition of functions T and M. We input x into function M, which outputs $M(x)$. Then, we input $M(x)$ into function T, which outputs $T(M(x))$, which is the training heart rate as a function of age x. Here is a diagram, called a function map, of the situation:

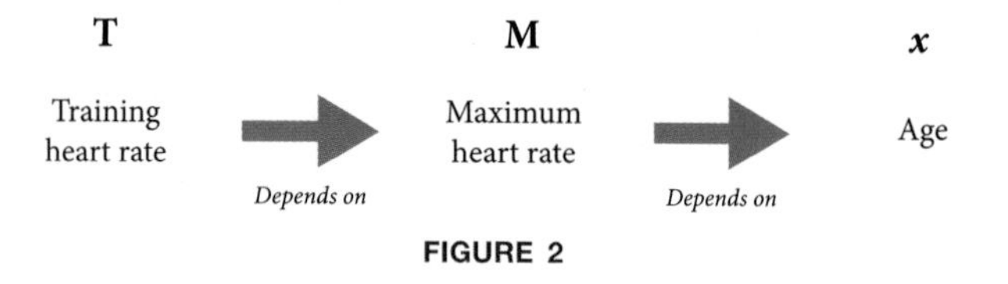

FIGURE 2

Now let's generalize the preceding ideas into a formal development of composition of functions. To find the composition of two functions f and g, we first require that the range of g have numbers in common with the domain of f. Then the composition of f with g, is defined this way:

$$(f \circ g)(x) = f(g(x))$$

To understand this new function, we begin with a number x, and we operate on it with g, giving us $g(x)$. Then we take $g(x)$ and operate on it with f, giving us $f(g(x))$. The only numbers we can use for the domain of the composition of f with g are numbers x in the domain of g, for which $g(x)$ is in the domain of f. The diagrams in Figure 3 illustrate the composition of f with g.

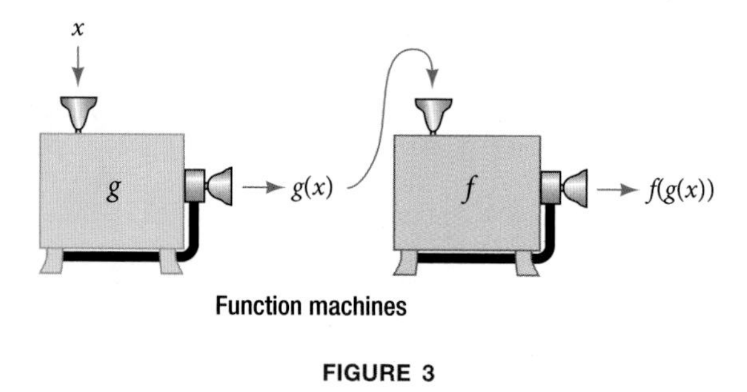

FIGURE 3

Composition of functions is not commutative. The composition of f with g, $f \circ g$, may therefore be different from the composition of g with f, $g \circ f$.

$$(g \circ f)(x) = g(f(x))$$

Again, the only numbers we can use for the domain of the composition of g with f are numbers in the domain of f, for which $f(x)$ is in the domain of g. The diagrams in Figure 4 illustrate the composition of g with f.

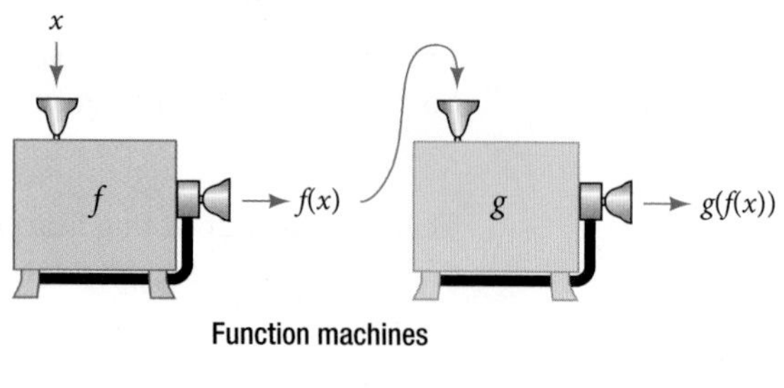

FIGURE 4

EXAMPLE 5 If $f(x) = x + 5$ and $g(x) = x^2 - 2x$, find $(f \circ g)(x)$ and $(g \circ f)(x)$.

SOLUTION The composition of f with g is

$$\begin{aligned}(f \circ g)(x) &= f(g(x)) \\ &= f(x^2 - 2x) \\ &= (x^2 - 2x) + 5 \\ &= x^2 - 2x + 5\end{aligned}$$

The composition of g with f is

$$\begin{aligned}(g \circ f)(x) &= g(f(x)) \\ &= g(x + 5) \\ &= (x + 5)^2 - 2(x + 5) \\ &= (x^2 + 10x + 25) - 2x - 10 \\ &= x^2 + 8x + 15\end{aligned}$$

Modeling: Temperature and Recycling

EXAMPLE 6 Suppose that, for temperatures greater than 10 °C, the average number of soft drinks sold each day, s (in hundreds), depends on the temperature, t, is given by

$$s(t) = 1{,}200 + 23t + 600(t - 10)^{1/3}$$

And, the number of soft drink aluminum cans recycled, r (in hundreds), depends on the number of soft drinks sold, s, as

$$r(s) = s^{3/4}$$

a. Give some reasonable numbers for the domain of the function s.

b. How many cans are sold when the temperature is 27 °C?

c. How many cans are recycled when the temperature is 27 °C?

d. Suppose we want to use the Fahrenheit temperature scale, instead of the Celsius scale. How would the problem change?

SOLUTION Here is a diagram of the situation.

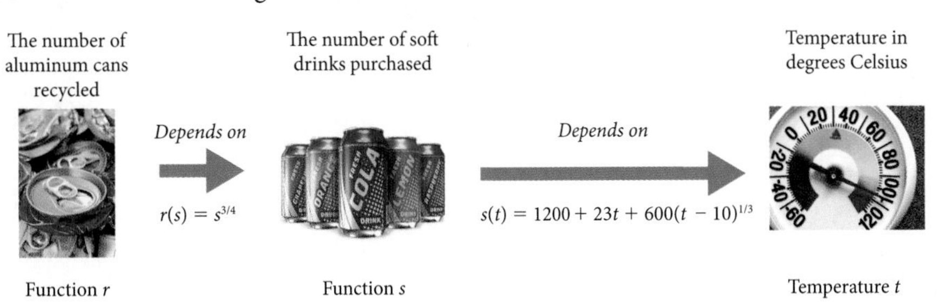

a. People tend to drink more soft drinks on hot days than on cold days. If we restrict the domain of this function so that t ranges from 20 °C (which is 68 °F) to 40 °C (104 °F), we should have a reasonable range of values to input for temperature.

b. To find the number of cans sold when the temperature is 27 °C, we simply substitute 27 for t and use our formulas.

$$\text{When } t = 27, \text{ then } s(27) = 1{,}200 + 23(27) + 600(27 - 10)^{1/3}$$
$$= 1{,}200 + 621 + 600(17)^{1/3}$$
$$\approx 1{,}200 + 621 + 600(2.57)$$
$$\approx 3{,}363$$

Since s represents hundreds of cans, the number of cans sold is 336,300.

c. To find the number of cans recycled, we substitute $s = 3{,}363$ into the formula for the number of cans recycled, giving us

$$r(3{,}363) = 3{,}363^{3/4} \approx 442$$

Since r is in hundreds of cans, the number of cans recycled when the temperature is 27 °C is 44,200.

d. Let's use the variable T to represent the temperature in degrees Fahrenheit. If we want to input our temperatures in degrees Fahrenheit, we must add a third function on to the end of the chain that will convert our degrees Fahrenheit (T) to degrees Celsius (t), in the beginning of the problem. Here is a diagram of this situation.

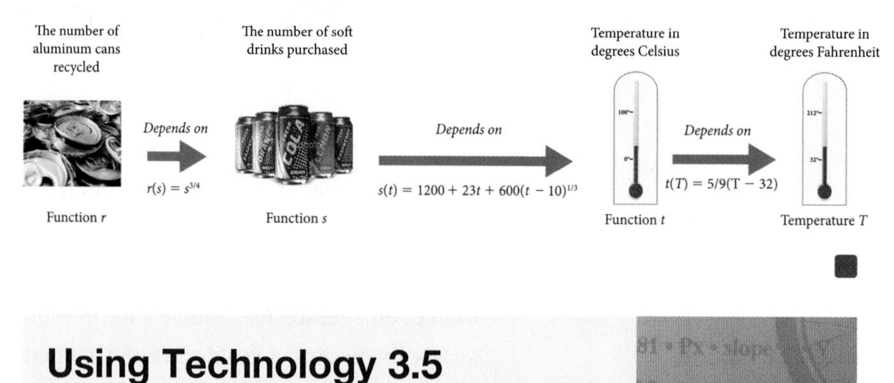

Using Technology 3.5

You can use your graphing calculator to evaluate a composite function.

EXAMPLE 7 Evaluate $(f \circ g)(2)$ for $f(x) = \sqrt{x - 3}$ and $g(x) = 2x + 3$.

SOLUTION Set up your Y-variables this way:

Y1 = √(X-3)

Y2 = 2X+3

In the computation window, enter

Y1(Y2(2))

The calculator responds with the value 2.

When $(g \circ f)(2)$ is evaluated, the calculator responds with an error. Can you get your calculator to respond with an error? Why does it do so?

GETTING READY FOR CLASS

Respond in your own words and in complete sentences.

A. How are profit, revenue, and cost related?
B. How do you find maximum heart rate?
C. For functions f and g, how do you find the composition of f with g?
D. For functions f and g, how do you find the composition of g with f?

Problem Set 3.5

Let $f(x) = 4x - 3$ and $g(x) = 2x + 5$. Write a formula for each of the following functions.

1. $f + g$ **2.** $f - g$ **3.** $g - f$ **4.** $g + f$

5. fg **6.** $\frac{f}{g}$ **7.** $\frac{g}{f}$ **8.** ff

If the functions f, g, and h are defined by $f(x) = 3x - 5$, $g(x) = x - 2$, and $h(x) = 3x^2 - 11x + 10$, write a formula for each of the following functions.

9. $g + f$ **10.** $f + h$ **11.** $g + h$ **12.** $f - g$

13. $g - f$ **14.** $h - g$ **15.** fg **16.** gf

17. fh **18.** gh **19.** $\frac{h}{f}$ **20.** $\frac{h}{g}$

21. $\frac{f}{h}$ **22.** $\frac{g}{h}$ **23.** $f + g + h$ **24.** $h - g + f$

25. $h + fg$ **26.** $h - fg$

Let $f(x) = 2x + 1$, $g(x) = 4x + 2$, and $h(x) = 4x^2 + 4x + 1$, and find the following.

27. $(f + g)(2)$ **28.** $(f - g)(-1)$ **29.** $(fg)(3)$ **30.** $\left(\frac{f}{g}\right)(-3)$

31. $\left(\frac{h}{g}\right)(1)$ **32.** $(hg)(1)$ **33.** $(fh)(0)$ **34.** $(h - g)(-4)$

35. $(f + g + h)(2)$ **36.** $(h - f + g)(0)$ **37.** $(h + fg)(3)$ **38.** $(h - fg)(5)$

39. Let $f(x) = x^2$ and $g(x) = x + 4$, and find

a. $(f \circ g)(5)$ **b.** $(g \circ f)(5)$ **c.** $(f \circ g)(x)$ **d.** $(g \circ f)(x)$

40. Let $f(x) = 3 - x$ and $g(x) = x^3 - 1$, and find

a. $(f \circ g)(0)$ **b.** $(g \circ f)(0)$ **c.** $(f \circ g)(x)$ **d.** $(g \circ f)(x)$

41. Let $f(x) = x^2 + 3x$ and $g(x) = 4x - 1$, and find

a. $(f \circ g)(0)$ **b.** $(g \circ f)(0)$ **c.** $(f \circ g)(x)$ **d.** $(g \circ f)(x)$

42. Let $f(x) = (x - 2)^2$ and $g(x) = x + 1$, and find

a. $(f \circ g)(-1)$ **b.** $(g \circ f)(-1)$ **c.** $(f \circ g)(x)$ **d.** $(g \circ f)(x)$

For each of the following pairs of functions f and g, show that $(f \circ g)(x) = (g \circ f)(x) = x$.

43. $f(x) = 5x - 4$ and $g(x) = \frac{x + 4}{5}$

44. $f(x) = \frac{x}{6} - 2$ and $g(x) = 6x + 12$

Getting Ready for Calculus The problems below are representative of the type of problems you will need to be familiar with in order to be successful in calculus. These particular problems are taken from *Applied Calculus* by Denny Burzynski and Guy Sanders, published by XYZ Textbooks.

45. If $u(x) = 5x^2 + 4$ and $v(x) = x^3 + 11$, find $u(x) \cdot v(x)$.

46. If $u(t) = t^2 - 4$, $v(t) = t^3 + 1$, and $d(t) = 2t$, find $h(t) = u(t) \cdot d(t) + v(t) \cdot d(t)$.

47. If $u(x) = x^3 + 5x + 4$, $v(x) = x^2 - 3$, $f(x) = 3x^2 + 2$, and $g(x) = 2x$, find $h(x) = v(x) \cdot f(x) - u(x) \cdot g(x)$.

48. If $f(x) = x^2 + 3x$ and $g(x) = 2x - 9$, find $(f \circ g)(x)$ and $(g \circ f)(x)$.

49. If $f(u) = u^2 + 3u$ and $u(x) = 2x - 9$, find $(f(u(x))$.

50. If $f(x) = x^6$ and $g(x) = 4x^3 + 5x + 2$, find $(f \circ g)(x)$ and $(g \circ f)(x)$.

Modeling Practice

51. Profit, Revenue, and Cost A company manufactures and sells DVDs. Here are the equations they use in connection with their business.

Number of DVDs sold each day: $n(x) = x$
Selling price for each DVD: $p(x) = 11.5 - 0.05x$
Daily fixed costs: $f(x) = 200$
Daily variable costs: $v(x) = 2x$
Find the following functions.

a. Revenue $= R(x) =$ the product of the number of DVDs sold each day and the selling price of each DVD.

b. Cost $= C(x) =$ the sum of the fixed costs and the variable costs.

c. Profit $= P(x) =$ the difference between revenue and cost.

d. Average cost $= \overline{C}(x) =$ the quotient of cost and the number of tapes sold each day.

52. Profit, Revenue, and Cost A company manufactures and sells CDs for home computers. Here are the equations they use in connection with their business.

Number of CDs sold each day: $n(x) = x$
Selling price for each CD: $p(x) = 3 - \frac{1}{300}x$
Daily fixed costs: $f(x) = 200$
Daily variable costs: $v(x) = 2x$
Find the following functions.

a. Revenue $= R(x) =$ the product of the number of CDs sold each day and the selling price of each diskette.

b. Cost $= C(x) =$ the sum of the fixed costs and the variable costs.

c. Profit $= P(x) =$ the difference between revenue and cost.

d. Average cost $= \overline{C}(x) =$ the quotient of cost and the number of CDs sold each day.

© Rob Cruse/iStockPhoto

53. Training Heart Rate Find the training heart rate function, $T(M)$, for a person with a resting heart rate of 62 beats per minute, then find the following.

a. Find the maximum heart rate function, $M(x)$, for a person x years of age.

b. What is the maximum heart rate for a 24-year-old person?

c. What is the training heart rate for a 24-year-old person with a resting heart rate of 62 beats per minute?

d. What is the training heart rate for a 36-year-old person with a resting heart rate of 62 beats per minute?

e. What is the training heart rate for a 48-year-old person with a resting heart rate of 62 beats per minute?

54. Training Heart Rate Find the training heart rate function, $T(M)$, for a person with a resting heart rate of 72 beats per minute, then find the following to the nearest whole number.

a. Find the maximum heart rate function, $M(x)$, for a person x years of age.

b. What is the maximum heart rate for a 20-year-old person?

c. What is the training heart rate for a 20-year-old person with a resting heart rate of 72 beats per minute?

d. What is the training heart rate for a 30-year-old person with a resting heart rate of 72 beats per minute?

e. What is the training heart rate for a 40-year-old person with a resting heart rate of 72 beats per minute?

Using Technology Exercises

Use your graphing calculator to evaluate each composite function.

55. If $f(x) = 5x^2 - x$ and $g(x) = \sqrt{x}$, evaluate both $f[g(16)]$ and $g[f(16)]$.

56. Carbon Monoxide Levels A study of a northwestern community indicates that the average daily level C of carbon monoxide in the air, in parts per million (ppm), is approximated by the function $C(p) = \sqrt{0.18p^2 + 9.6}$, $0 < p \leq 20$, where p represents the population of the community in thousands of people. (Enter this function using Y1.) The population of the community depends on the time t in years from now and is approximated by the function $p(t) = 1.8 + 0.04t^2$, $0 < t \leq 20$. (Enter this function using Y2.) Compute the level of carbon monoxide in this community's air 15 years from now in two ways. First, compute $p(15)$ and then substitute this result into $C(p)$. Second, compute directly using the method given above $C[p(15)]$. Compare the two results. If they are equal, write your conclusion in sentence form. If they are unequal, try again.

57. Carbon Monoxide Levels For the previous exercise, make a single change in the defining expression of Y1 so that when you turn off the graphing capability of Y2, the graph of the composite function $C[p(t)]$ appears in the window $0 < t \leq 25$ and $0 \leq C \leq 25$. Sketch your graph.

Maintaining Your Skills

The problems that follow review some of the more important skills you have learned in previous sections and chapters.

Solve the following equations.

58. $x - 5 = 7$

59. $3y = -4$

60. $5 - \frac{4}{7}a = -11$

61. $\frac{1}{5}x - \frac{1}{2} - \frac{1}{10}x + \frac{2}{5} = \frac{3}{10}x + \frac{1}{2}$

62. $5(x - 1) - 2(2x + 3) = 5x - 4$

63. $0.07 - 0.02(3x + 1) = -0.04x + 0.01$

Solve for the indicated variable.

64. $P = 2l + 2w$ for w

65. $A = \frac{1}{2}h(b + B)$ for B

Solve the following inequalities. Write the solution set using interval notation, then graph the solution set.

66. $-5t \leq 30$

67. $5 - \frac{3}{2}x > -1$

68. $1.6x - 2 < 0.8x + 2.8$

69. $3(2y + 4) \geq 5(y - 8)$

Solve the following equations.

70. $\left|\frac{1}{4}x - 1\right| = \frac{1}{2}$

71. $\left|\frac{2}{3}a + 4\right| = 6$

72. $|3 - 2x| + 5 = 2$

73. $5 = |3y + 6| - 4$

Chapter 3 Summary

Ordered Pairs [3.1]

A pair of numbers enclosed in parentheses and separated by a comma, such as $(-2, 1)$, is called an *ordered pair* of numbers. The first number in the pair is called the *x-coordinate* of the ordered pair; the second number is called the *y-coordinate*. For the ordered pair $(-2, 1)$, the x-coordinate is -2 and the y-coordinate is 1.

EXAMPLES

1. Graph the ordered pairs $(2, 5)$, $(-2, 5)$, $(-2, -5)$, and $(2, -5)$.

To graph the ordered pair $(2, 5)$, we start at the origin and move 2 units to the right, then 5 units up. We are now at the point whose coordinates are $(2, 5)$. We graph the other three ordered pairs in a similar manner (see Figure 4 in Section 3.1).

Rectangular Coordinate System [3.1]

A *rectangular coordinate system* is made by drawing two real number lines at right angles to each other. The two number lines, called *axes*, cross each other at 0. This point is called the *origin*. Positive directions are to the right and up. Negative directions are to the left and down. The two number lines divide the coordinate system into four *quadrants*, which we number I through IV in a counterclockwise direction.

2.

Graphing Equations [3.1]

We can plot a single point from an ordered pair, but to draw a line or a curve, we need more points. To graph an equation in two variables, we simply graph its solution set. That is, we draw a line or smooth curve through all the points whose coordinates satisfy the equation.

3. Plotting the ordered pairs $(-3, 1)$, $(0, 0)$, and $(3, -1)$ and drawing a straight line through their graphs, we have the graph of the equation $y = -\frac{1}{3}x$ as shown here.

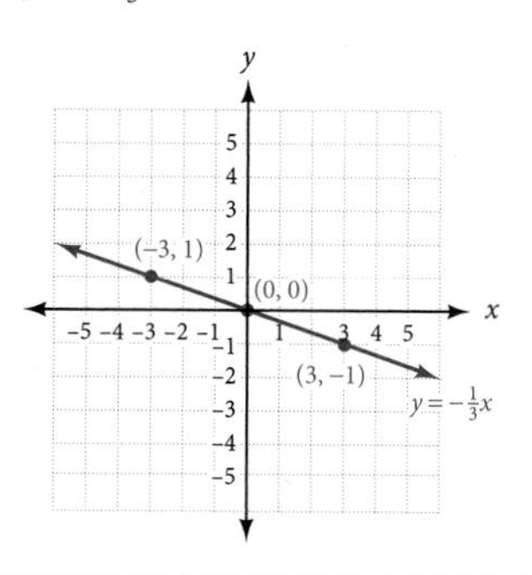

Intercepts [3.1]

The *x-intercept* of an equation is the *x-coordinate* of the point where the graph crosses the *x-axis*. The *y-intercept* is the *y-coordinate* of the point where the graph crosses the *y-axis*. We find the y-intercept by substituting $x = 0$ into the equation and solving for y. The x-intercept is found by letting $y = 0$ and solving for x.

4. To find the x-intercept for $3x + 2y = 6$, we let $y = 0$ and get

$$3x = 6$$
$$x = 2$$

In this case the x-intercept is 2, and the graph crosses the x-axis at $(2, 0)$.

Relations and Functions [3.2]

5. The relation

$$\{(8, 1), (6, 1), (-3, 0)\}$$

is also a function because no ordered pairs have the same first coordinates. The domain is $\{8, 6, -3\}$ and the range is $\{1, 0\}$.

A *function* is a rule that pairs each element in one set, called the *domain,* with exactly one element from a second set, called the *range.*

A *relation* is any set of ordered pairs. The set of all first coordinates is called the *domain* of the relation, and the set of all second coordinates is the *range* of the relation. A function is a relation in which no two different ordered pairs have the same first coordinates.

Vertical Line Test [3.2]

If a vertical line crosses the graph of a relation in more than one place, the relation cannot be a function. If no vertical line can be found that crosses a graph in more than one place, then the graph represents a function.

Function Notation [3.3]

6. If $f(x) = 5x - 3$ then

$$\begin{aligned} f(0) &= 5(0) - 3 \\ &= -3 \\ f(1) &= 5(1) - 3 \\ &= 2 \\ f(-2) &= 5(-2) - 3 \\ &= -13 \\ f(a) &= 5a - 3 \end{aligned}$$

The alternative notation for y is $f(x)$. It is read "f of x" and can be used instead of the variable y when working with functions. The notation y and the notation $f(x)$ are equivalent; that is, $y = f(x)$.

Greatest Integer Functions [3.3]

7.

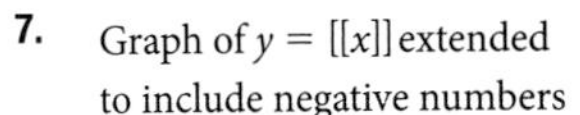

Graph of $y = [[x]]$ extended to include negative numbers

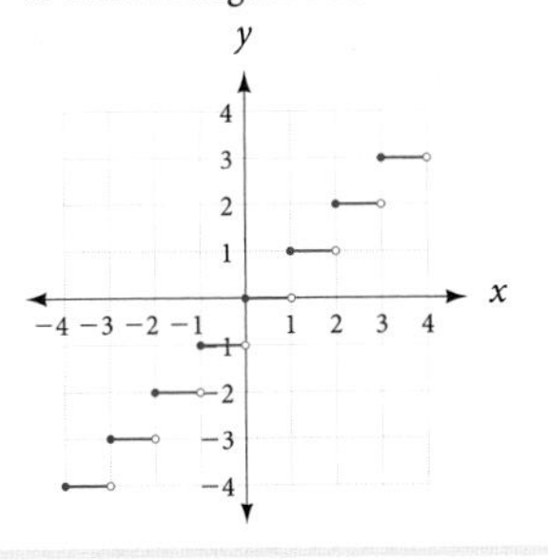

The simplest step function is the one that outputs the greatest integer not greater than the number that is input. The equation for this function is

$$y = [[x]]$$

where $[[x]]$ stands for the greatest integer not greater than x. We call this function the *greatest integer function.*

Six Basic Functions [3.4]

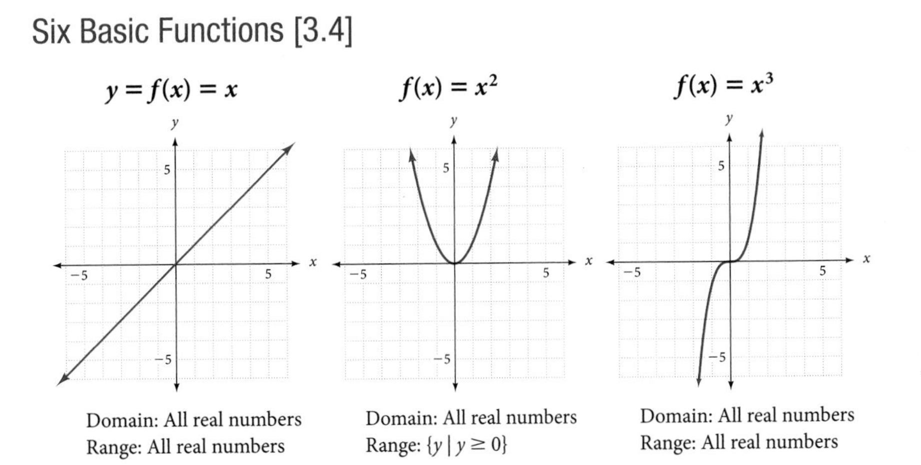

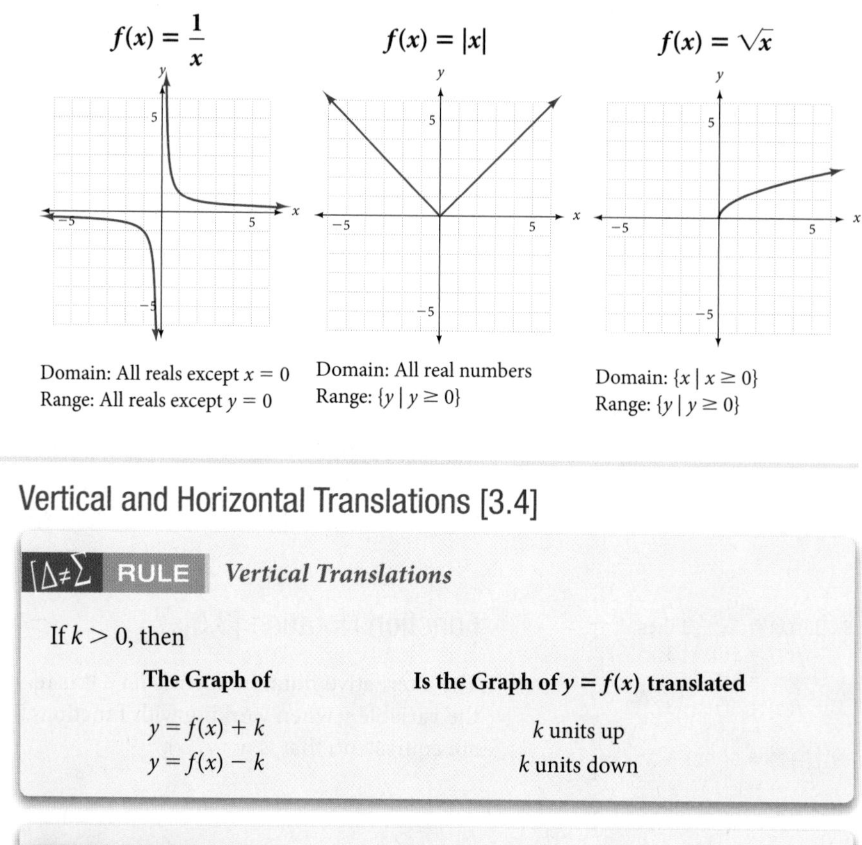

Domain: All reals except $x = 0$
Range: All reals except $y = 0$

Domain: All real numbers
Range: $\{y \mid y \geq 0\}$

Domain: $\{x \mid x \geq 0\}$
Range: $\{y \mid y \geq 0\}$

Vertical and Horizontal Translations [3.4]

RULE *Vertical Translations*

If $k > 0$, then

The Graph of	Is the Graph of $y = f(x)$ translated
$y = f(x) + k$	k units up
$y = f(x) - k$	k units down

RULE *Horizontal Translations*

If $h > 0$, then

The Graph of	Is the Graph of $y = f(x)$ translated
$y = f(x + h)$	h units to the left
$y = f(x - h)$	h units to the right

Algebra with Functions [3.5]

$(f + g)(x) = f(x) + g(x)$ The function $f + g$ is the sum of the functions f and g.

$(f - g)(x) = f(x) - g(x)$ The function $f - g$ is the difference of the f and g.

$(fg)(x) = f(x)g(x)$ The function fg is the product of the functions f and g.

$\left(\frac{f}{g}\right)(x) = \frac{f(x)}{g(x)}$ The function $\frac{f}{g}$ is the quotient of the functions f and g, where $g(x) \neq 0$.

Composition of Functions [3.5]

If f and g are two functions for which the range of each has numbers in common with the domain of the other, then we have the following definitions:

The composition of f with g: $(f \circ g)(x) = f[g(x)]$
The composition of g with f: $(g \circ f)(x) = g[f(x)]$

Chapter 3 Test

State the domain for each of the following functions. [3.2]

1. $f(x) = x^2 - 4$

2. $g(x) = \sqrt{x - 3}$

3. $h(x) = \dfrac{1}{x^2 - 9}$

4. $\{(-2, 0), (-3, 0), (2, 1)\}$

5. $h(t) = 4 - 4t - 16t^2, t \geq 0$

Find the intercepts for each of the following. [3.1]

6. $f(x) = \frac{2}{3}x - 4$

7. $g(x) = \sqrt{x - 3}$

8. $h(x) = x^4 - 9x^2$

9. $f(x) = x^2 + 4$

Graph each of the following. Indicate any that are not functions. [3.2, 3.3, 3.4]

10. $y = (x - 2)^2 - 4$

11. $y = \dfrac{1}{t - 2} - 4$

12. $x = y^2 - 4$

13. $g(x) = [[x]]$

14. $f(t) = \frac{2}{3}t - 2$

Let $f(x) = x - 2$, $g(x) = 3x + 4$, and $h(x) = 3x^2 - 2x - 8$, and find the following. [3.5]

15. $f(3) + g(2)$

16. $h(0) + g(0)$

17. $(f \circ g)(2)$

18. $(g \circ f)(2)$

19. Show that $(fg)(x) = h(x)$.

20. Find $\frac{h}{g}(2)$ and $f(2)$.

For the functions below, find x so that $f(x) = x$. [3.5]

21. $f(x) = 5x - 8$

22. $f(x) = x^2 - 2$

CLASSIFICATION OF FUNCTIONS

As we progress through the rest of the book, we will be working with a variety of different functions. In order to help organize our functions, we have the following general categories for the functions we will see throughout the book.

Polynomial Functions

$f(x)$ = a polynomial in the variable x

Subcategories of Polynomial Functions

For the definitions below, a, b, c, d, k, and m are real numbers.

Constant Function	$f(x) = k$	
Linear Function	$f(x) = mx + b$	
Quadratic Function	$f(x) = ax^2 + bx + c$	$a \neq 0$
Cubic Function	$f(x) = ax^3 + bx^2 + cx + d$	$a \neq 0$
Power Function	$f(x) = kx^p$	$a \neq 0$
Direct Variation	$f(x) = kx$	$k \neq 0$
Inverse Variation	$f(x) = \dfrac{k}{x}$	$k \neq 0$
Square Root Function	$f(x) = \sqrt{x}$	

Rational Functions

If $P(x)$ and $Q(x)$ are polynomials, and $Q(x) \neq 0$, then the function below is a rational function.

$$f(x) = \frac{P(x)}{Q(x)}$$

Exponential Functions

If a, b, and k are nonzero real numbers, then the function below is an exponential function.

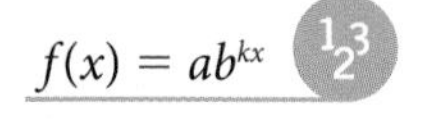

$$f(x) = ab^{kx}$$

Logarithmic Functions

If $b > 0$, $b \neq 1$, and $k \neq 0$, then the function below is a logarithmic function.

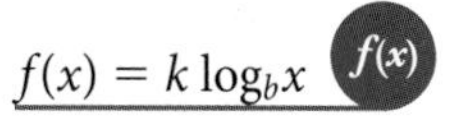

$$f(x) = k \log_b x$$

The examples below show some of the ways we will see the four modeling types as we progress through the book.

IN WORDS

A skydiver jumps from a plane. Like any object that falls toward earth, the distance the skydiver falls is directly proportional to the square of the time he has been falling, until he reaches his terminal velocity. If the skydiver falls 64 feet in the first 2 seconds of the jump, then how far will he have fallen after 3.5 seconds?

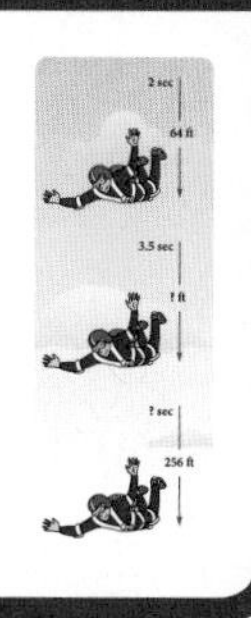

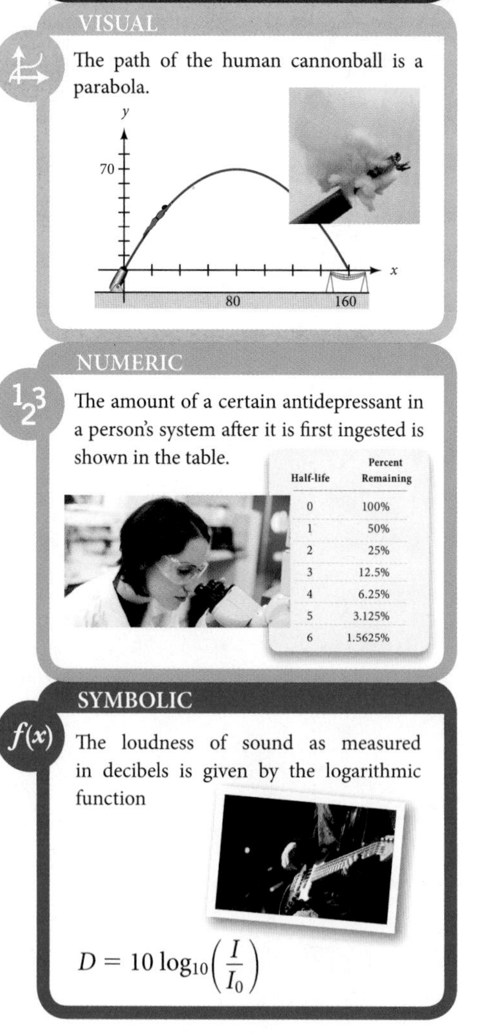

VISUAL

The path of the human cannonball is a parabola.

NUMERIC

The amount of a certain antidepressant in a person's system after it is first ingested is shown in the table.

Half-life	Percent Remaining
0	100%
1	50%
2	25%
3	12.5%
4	6.25%
5	3.125%
6	1.5625%

SYMBOLIC

The loudness of sound as measured in decibels is given by the logarithmic function

$$D = 10 \log_{10}\left(\frac{I}{I_0}\right)$$

4 Slope, Rates of Change, and Linear Functions

© alexey_ds/iStockphoto

Chapter Outline

A student is heating water in a chemistry lab. As the water heats, she records the temperature readings from two thermometers, one giving temperature in degrees Fahrenheit and the other in degrees Celsius. The table below shows some of the data she collects. The line graph gives a visual representation of the data in the table.

Degrees Celsius	Degrees Fahrenheit
0	32
25	77
50	122
75	167
100	212

The exact relationship between the Fahrenheit and Celsius temperature scales is given by the formula

$$F = \frac{9}{5}C + 32$$

We have three ways to describe the relationship between the two temperature scales: a table, a graph, and an equation. We can also describe this relationship in words by saying something like this: The relationship between the two temperature scales is linear. Water boils at 212 °F and 100 °C. Water freezes at 32 °F and 0 °C.

Note When you see this icon next to an example or problem in this chapter, you will know that we are using the topics in this chapter to model situations in the world around us.

Study Skills

The study skills for this chapter are concerned with getting ready to take an exam.

1. **Getting Ready to Take an Exam** Try to arrange your daily study habits so you have little studying to do the night before your next exam. The next two goals will help you achieve goal number 1.

2. **Review with the Exam in Mind** You should review material that will be covered on the next exam every day. Your review should consist of working problems. Preferably, the problems you work should be problems from your list of difficult problems.

3. **Continue to List Difficult Problems** You should continue to list and rework the problems that give you the most difficulty. It is this list that you will use to study for the next exam. Your goal is to go into the next exam knowing you can successfully work any problem from your list of hard problems.

4. **Pay Attention to Instructions** Taking a test is different from doing homework. When you take a test, the problems will be mixed up. When you do your homework, you usually work a number of similar problems. Sometimes students who do well on their homework become confused when they see the same problems on a test, because they have not paid attention to the instructions on their homework. For example, suppose you see the equation $y = 3x - 2$ on your next test. By itself, the equation is simply a statement. There isn't anything to do unless the equation is accompanied by instructions. Each of the following is a valid instruction with respect to the equation $y = 3x - 2$ and the result of applying the instructions will be different in each case:

   Find x when y is 10.
   Solve for x.
   Graph the equation.
   Find the intercepts.
   Find the slope.

   There are many things to do with the equation. If you train yourself to pay attention to the instructions that accompany a problem as you work through the assigned problems, you will not find yourself confused about what to do with a problem when you see it on a test.

The Slope of a Line

4.1

© Alex Nikada/iStockphoto

INTRODUCTION You get in an elevator on the ground floor of a high-rise building and ride it up, non-stop, to the 26th floor. As the elevator moves up you feel the sensation of being pushed down towards the floor, and as you pass the 12th floor your ears pop. Both of these sensations are your body's reaction to rates of change, which we study as slopes of lines.

For an intuitive introduction to the slope of a line, imagine that a highway sign tells us we are approaching a 6% downgrade. As we drive down this hill, each 100 feet we travel horizontally is accompanied by a 6-foot drop in elevation.

In mathematics we say the slope of the highway is $-0.06 = -\frac{6}{100} = -\frac{3}{50}$. The slope is the ratio of the vertical change to the accompanying horizontal change.

In defining the slope of a straight line, we want to associate a number with the line. First, we want the slope of a line to measure the "steepness" of the line. That is, in comparing two lines, the slope of the steeper line should have the larger numerical value. Second, we want a line that rises going from left to right to have a *positive* slope. We want a line that falls going from left to right to have a *negative* slope. (A line that neither rises nor falls going from left to right must, therefore, have 0 slope.) Geometrically, we can define the *slope* of a line as the ratio of the vertical change to the horizontal change encountered when moving from one point to another on the line. The vertical change is sometimes called the *rise*. The horizontal change is called the *run*.

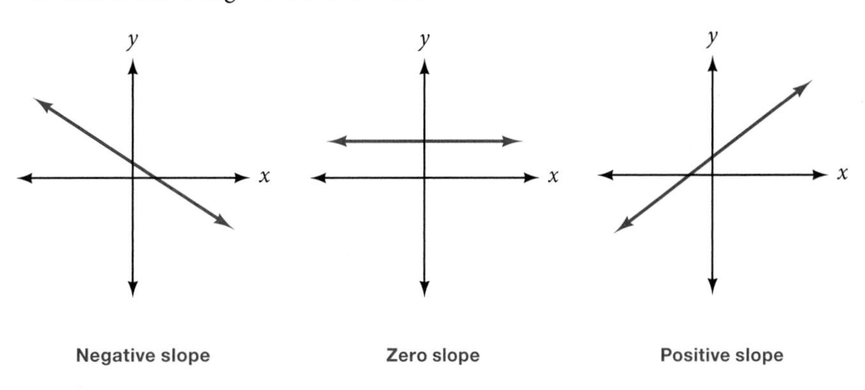

VIDEO EXAMPLES

SECTION 4.1

EXAMPLE 1 Find the slope of the line $y = 2x - 3$.

SOLUTION To use our geometric definition, we first graph $y = 2x - 3$ (Figure 1). We then pick any two convenient points and find the ratio of rise to run. By convenient points we mean points with integer coordinates. If we let $x = 2$ in the equation, then $y = 1$. Likewise, if we let $x = 4$, then y is 5.

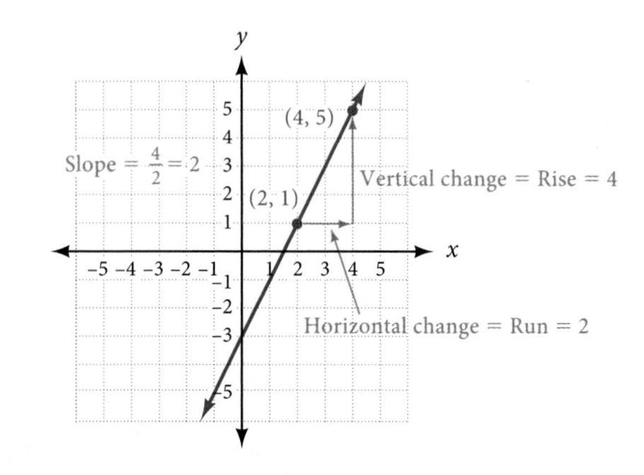

FIGURE 1

The ratio of vertical change to horizontal change is 4 to 2, giving us a slope of $\frac{4}{2} = 2$. Our line has a slope of 2.

Notice that we can measure the vertical change (rise) by subtracting the y-coordinates of the two points shown in Figure 1: $5 - 1 = 4$. The horizontal change (run) is the difference of the x-coordinates: $4 - 2 = 2$. This gives us a second way of defining the slope of a line.

DEFINITION *Slope*

The *slope* of the line between two points (x_1, y_1) and (x_2, y_2) is given by

$$\text{Slope} = m = \frac{\text{Rise}}{\text{Run}} = \frac{y_2 - y_1}{x_2 - x_1}$$

↗ Geometric Form ↖ Algebraic Form

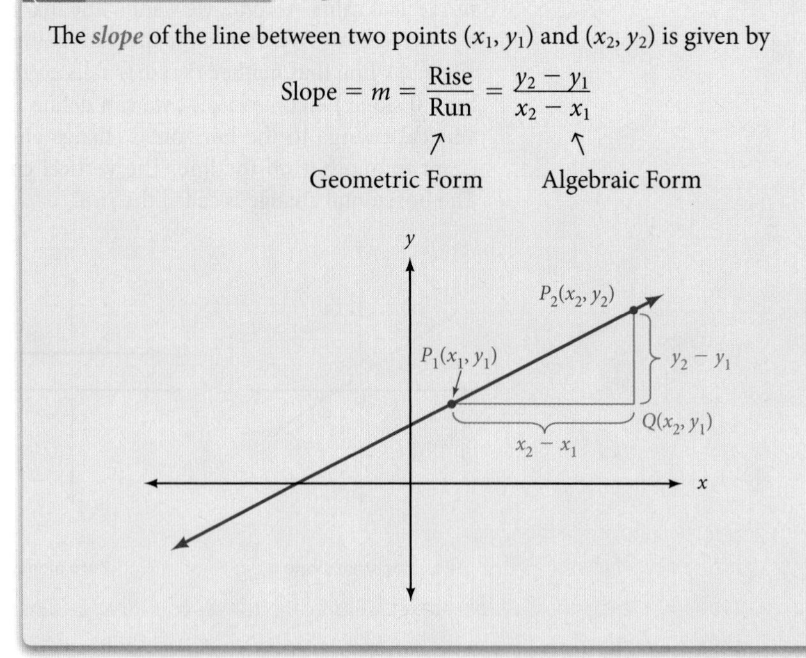

EXAMPLE 2 Find the slope of the line through $(-2, -3)$ and $(-5, 1)$.

SOLUTION

$$m = \frac{y_2 - y_1}{x_2 - x_1} = \frac{1 - (-3)}{-5 - (-2)} = \frac{4}{-3} = -\frac{4}{3}$$

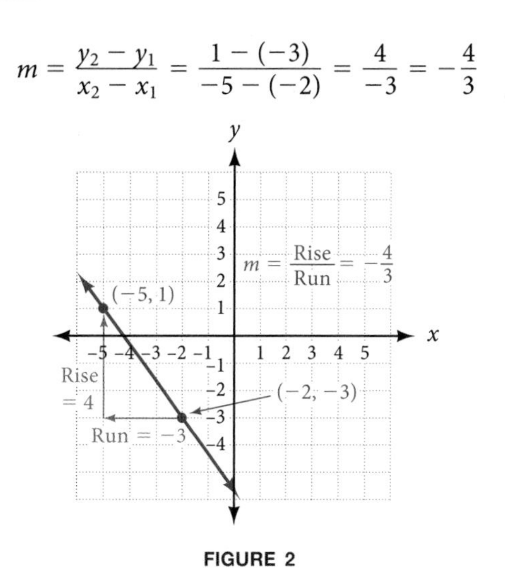

FIGURE 2

Looking at the graph of the line between the two points (Figure 2), we can see our geometric approach does not conflict with our algebraic approach.

We should note here that it does not matter which ordered pair we call (x_1, y_1) and which we call (x_2, y_2). If we were to reverse the order of subtraction of both the x- and y-coordinates in the preceding example, we would have

$$m = \frac{-3 - 1}{-2 - (-5)} = \frac{-4}{3} = -\frac{4}{3}$$

which is the same as our previous result.

Note The two most common mistakes students make when first working with the formula for the slope of a line are

1. Putting the difference of the x-coordinates over the difference of the y-coordinates.
2. Subtracting in one order in the numerator and then subtracting in the opposite order in the denominator. You would make this mistake in Example 2 if you wrote $1 - (-3)$ in the numerator and then $-2 - (-5)$ in the denominator.

EXAMPLE 3 Find the slope of the line containing $(3, -1)$ and $(3, 4)$.

SOLUTION Using the definition for slope, we have

$$m = \frac{-1 - 4}{3 - 3} = \frac{-5}{0}$$

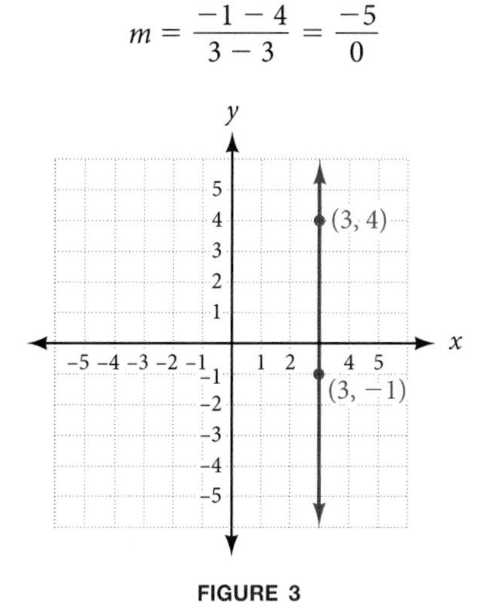

FIGURE 3

The expression $\frac{-5}{0}$ is undefined. That is, there is no real number to associate with it. In this case, we say the slope is *undefined*, or we say the line has *no slope*.

Note If slope is meant to represent the steepness of a line, then a vertical line would have the largest number possible for its slope. Some people would say the slope of a vertical line is infinite, or infinity. We have to be careful with terms such as these because we have used a ratio to define slope. Infinity is not a number, and we don't want to start using it to describe slope. However, on an intuitive level saying the slope is infinite is probably a more accurate description of what is happening than to say the slope is undefined, or the line has no slope.

The graph of our line is shown in Figure 3. Our line with undefined slope is a vertical line. All vertical lines have undefined slopes. All horizontal lines, as we mentioned earlier, have 0 slope.

Slopes of Parallel and Perpendicular Lines

In geometry, we call lines in the same plane that never intersect parallel. For two lines to be nonintersecting, they must rise or fall at the same rate. In other words, two lines are *parallel* if and only if they have the *same slope.*

Although it is not as obvious, it is also true that two nonvertical lines are *perpendicular* if and only if the *product of their slopes is* -1. This is the same as saying their slopes are negative reciprocals.

We can state these facts with symbols as follows: If line l_1 has slope m_1 and line l_2 has slope m_2, then

$$l_1 \text{ is parallel to } l_2 \Leftrightarrow m_1 = m_2$$

and

$$l_1 \text{ is perpendicular to } l_2 \Leftrightarrow m_1 \cdot m_2 = -1 \text{ or } \left(m_1 = \frac{-1}{m_2}\right)$$

For example, if a line has a slope of $\frac{2}{3}$, then any line parallel to it has a slope of $\frac{2}{3}$. Any line perpendicular to it has a slope of $-\frac{3}{2}$ (the negative reciprocal of $\frac{2}{3}$).

Although we cannot give a formal proof of the relationship between the slopes of perpendicular lines at this level of mathematics, we can offer some justification for the relationship. Figure 4 shows the graphs of two lines. One of the lines has a slope of $\frac{2}{3}$; the other has a slope of $-\frac{3}{2}$. As you can see, the lines are perpendicular.

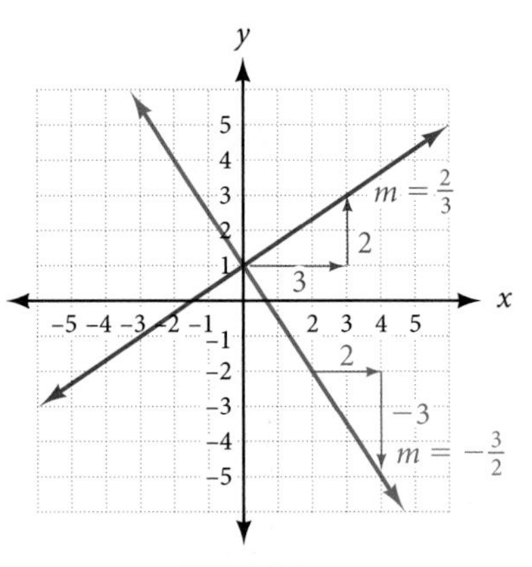

FIGURE 4

Modeling: Slope and Rate of Change

So far, the slopes we have worked with represent the ratio of the change in y to the corresponding change in x, or, on the graph of the line, the slope is the ratio of vertical change to horizontal change in moving from one point on the line to another. However, when our variables represent quantities from the world around us, slope can have additional interpretations.

EXAMPLE 4 On the chart below, find the slope of the line connecting the first point (1955, 0.29) with the highest point (2005, 2.93). Explain the significance of the result.

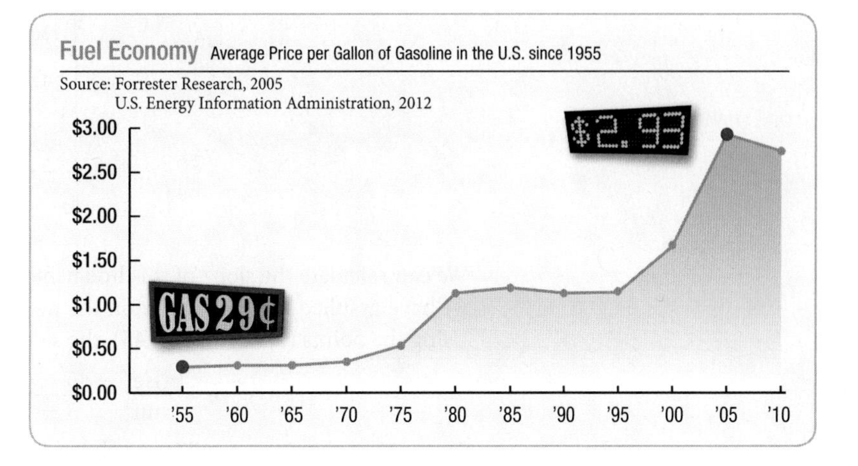

SOLUTION The slope of the line connecting the first point (1955, 0.29) with the highest point (2005, 2.93), is

$$m = \frac{2.93 - 0.29}{2005 - 1955} = \frac{2.64}{50} = 0.0528$$

The units are dollars/year. If we write this in terms of cents we have

$$m = 5.28 \text{ cents/year}$$

which is the average change in the price of a gallon of gasoline over a 50-year period of time.

Likewise, if we connect the points (2005, 2.93) and (2010, 2.74), the line that results has a slope of

$$m = \frac{2.74 - 2.93}{2010 - 2005} = \frac{-0.19}{5} = -0.038 \text{ dollars/year} = -3.8 \text{ cents/year}$$

which is the average change in the price of a gallon of gasoline over the most recent 5-year period. As you can imagine by looking at the chart, the line connecting the first point and highest point is very different from the line connecting the points from 2005 and 2010, and this is what we are seeing numerically with our slope calculations. If we were summarizing this information for an article in the newspaper, we could say, "Although the price of a gallon of gasoline increased 5.28 cents per year from 1955 to its peak in 2005, in the last 5 years the average annual rate has actually decreased at a rate of 3.8 cents per year." When describing rates of change, positive slopes translate into increases, while negative slopes are decreases.

Modeling: Slope and Average Speed

Previously we introduced the rate equation $d = rt$. Suppose that a boat is traveling at a constant speed of 15 miles per hour in still water. The following table shows the distance the boat will have traveled in the specified number of hours. The graph of this data shown in Figure 5. Notice that the points all lie along a line.

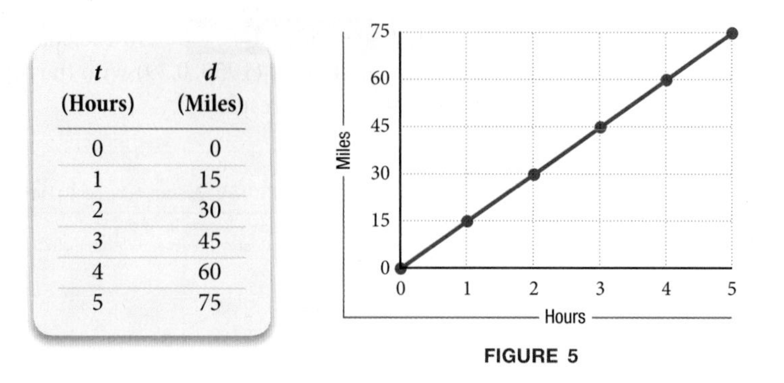

t (Hours)	d (Miles)
0	0
1	15
2	30
3	45
4	60
5	75

FIGURE 5

We can calculate the slope of this line using any two points from the table. Notice we have graphed the data with t on the horizontal axis and d on the vertical axis. Using the points (2, 30) and (3, 45), the slope will be

$$m = \frac{\text{rise}}{\text{run}} = \frac{45 - 30}{3 - 2} = \frac{15}{1} = 15$$

The units of the rise are miles and the units of the run are hours, so the slope will be in units of miles per hour. We see that the slope is simply the change in distance divided by the change in time, which is how we compute the average speed. Since the speed is constant, the slope of the line represents the speed of 15 miles per hour.

EXAMPLE 5 A car is traveling at a constant speed. A graph of the distance the car has traveled over time is shown below (Figure 6). Use the graph to find the speed of the car.

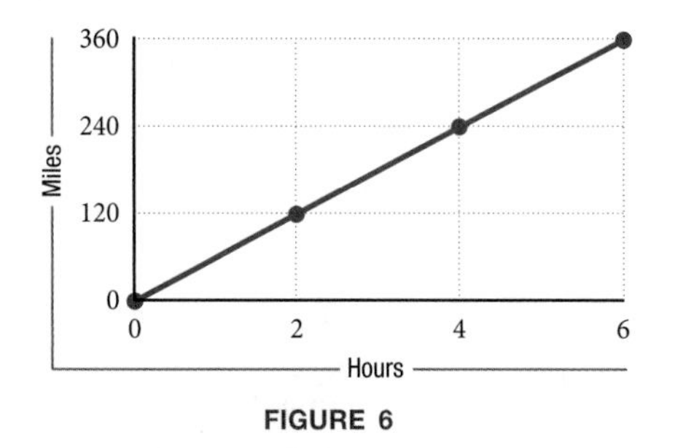

FIGURE 6

SOLUTION Using the second and third points, we see the rise is $240 - 120 = 120$ miles, and the run is $4 - 2 = 2$ hours. The speed is given by the slope, which is

$$\begin{aligned} m &= \frac{\text{rise}}{\text{run}} \\ &= \frac{120 \text{ miles}}{2 \text{ hours}} \\ &= 60 \text{ miles per hour} \end{aligned}$$

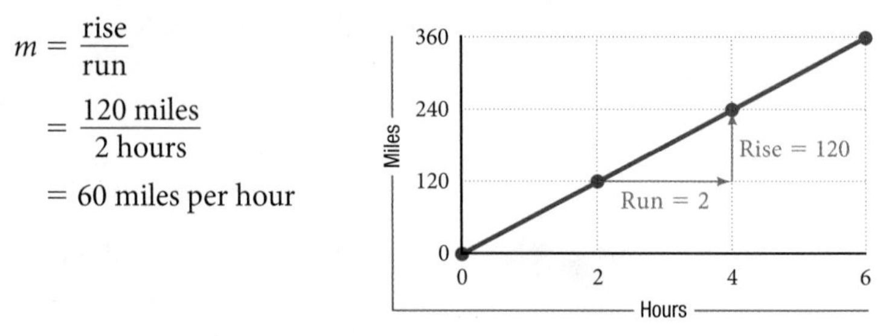

Modeling: Sensing Hypoglycemia

Hypoglycemia occurs when your glucose level in your blood drops too low. Glucose levels are measured in mg/dL (milligrams per deciliter). Generally, blood glucose levels below 70 mg/dL are considered hypoglycemic.

EXAMPLE 6 A patient tells her doctor that she is experiencing hypoglycemia during her workouts at the gym. Her doctor gives her a blood testing kit and has her repeat her workout, monitoring her blood sugar level every 5 minutes, and noting any symptoms she is experiencing. The results are shown in Figure 7.

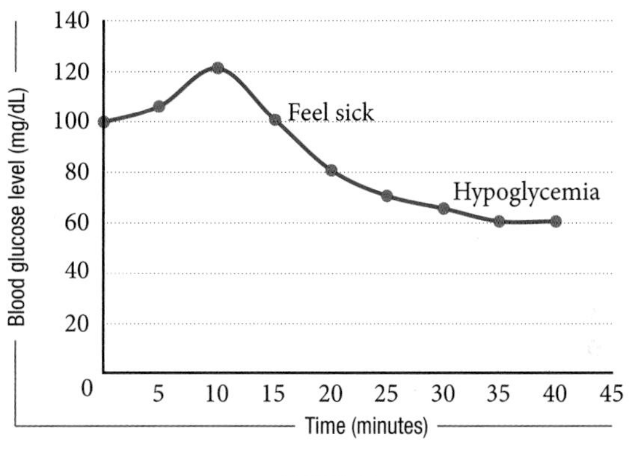

FIGURE 7

After looking at the graph, her doctor said, "The first time you felt poorly, you were sensing the rate of change in your blood sugar level, but you were probably not experiencing hypoglycemia until your sixth reading."

a. Is her first experience of feeling poorly because of the slope of a line, or a point on the graph?

b. Is she feeling poorly at her sixth reading because of the slope of a line, or a point on the graph?

SOLUTION

a. She starts to feel sick on her fourth reading. Generally, a blood glucose level of 100 mg/dL is not considered to be hypoglycemic, so she is feeling poorly because of the *rate* at which her blood sugar level is dropping. Mathematically, it is because of the slope of the line that passes through her third, fourth, and fifth readings.

b. At her sixth reading she is experiencing hypoglycemia because her blood glucose level is 70 mg/dL, which corresponds to a point on the graph.

Difference Quotients

One of the important applications of calculus is finding rates of change. As you may recall, the slope of a line represents a rate of change. In the diagram below, the slope of the blue line is the average rate of change of the red curve from point P to point Q. The slope of the line passing through the points P and Q is given by the formula

$$\text{Slope of line through } PQ = m = \frac{f(x) - f(a)}{x - a}$$

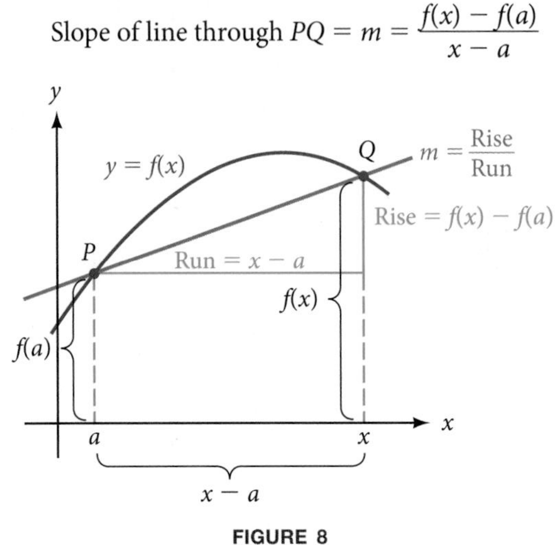

FIGURE 8

The expression $\frac{f(x) - f(a)}{x - a}$ is called a *difference quotient*. It represents the average rate of change of the function f from the point $(a, f(a))$ to the point $(x, f(x))$.

EXAMPLE 7 If $f(x) = 3x - 5$, find $\dfrac{f(x) - f(a)}{x - a}$.

SOLUTION

$$\begin{aligned}\frac{f(x) - f(a)}{x - a} &= \frac{(3x - 5) - (3a - 5)}{x - a} \\ &= \frac{3x - 3a}{x - a} \\ &= \frac{3(x - a)}{x - a} \\ &= 3\end{aligned}$$

EXAMPLE 8 If $f(x) = x^2 - 4$, find $\dfrac{f(x) - f(a)}{x - a}$ and simplify.

SOLUTION Because $f(x) = x^2 - 4$ and $f(a) = a^2 - 4$, we have

$$\begin{aligned}\frac{f(x) - f(a)}{x - a} &= \frac{(x^2 - 4) - (a^2 - 4)}{x - a} \\ &= \frac{x^2 - 4 - a^2 + 4}{x - a} \\ &= \frac{x^2 - a^2}{x - a} \\ &= \frac{(x + a)(x - a)}{x - a} \qquad \text{Factor and divide out common factor} \\ &= x + a\end{aligned}$$

There is a second form of the difference quotient. The diagram in Figure 9 is similar to the one in Figure 8. The main difference is in how we label the points. From Figure 9, we can see another difference quotient that gives us the slope of the line through the points P and Q.

$$\text{Slope of line through } PQ = m = \frac{f(x+h) - f(x)}{h}$$

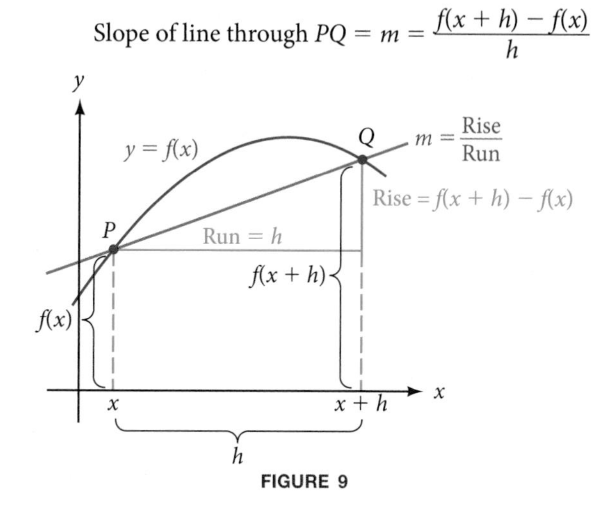

FIGURE 9

Examples 9 and 10 use the same functions used in Examples 7 and 8, but this time the new difference quotient is used.

EXAMPLE 9 If $f(x) = 3x - 5$, find $\dfrac{f(x+h) - f(x)}{h}$.

SOLUTION The expression $f(x + h)$ is given by

$$\begin{aligned} f(x+h) &= 3(x+h) - 5 \\ &= 3x + 3h - 5 \end{aligned}$$

Using this result gives us

$$\begin{aligned} \frac{f(x+h) - f(x)}{h} &= \frac{(3x + 3h - 5) - (3x - 5)}{h} \\ &= \frac{3h}{h} \\ &= 3 \end{aligned}$$

EXAMPLE 10 If $f(x) = x^2 - 4$, find $\dfrac{f(x+h) - f(x)}{h}$.

SOLUTION The expression $f(x + h)$ is given by

$$\begin{aligned} f(x+h) &= (x+h)^2 - 4 \\ &= x^2 + 2xh + h^2 - 4 \end{aligned}$$

Using this result gives us

$$\begin{aligned} \frac{f(x+h) - f(x)}{h} &= \frac{(x^2 + 2xh + h^2 - 4) - (x^2 - 4)}{h} \\ &= \frac{2xh + h^2}{h} \\ &= \frac{h(2x + h)}{h} \\ &= 2x + h \end{aligned}$$

Using Technology 4.1

You can use your graphing calculator to find the average rate of change of a function. The following entries illustrate one way of computing the average rate of change of the function $f(x) = (5x - 3)^{2/3}$ as x increases from 16 to 20.

EXAMPLE 11 If $f(x) = (5x - 3)^{2/3}$, find $\dfrac{f(20) - f(16)}{20 - 16}$.

SOLUTION Begin by putting the function in the Y-variables list.

Y1=(5X-3)^2/3

Now, in the computation window, enter

(Y1(20)-Y1(16))/(20-16)

The result is 0.7530075664.

GETTING READY FOR CLASS

After reading through the preceding section, respond in your own words and in complete sentences.

A. If you were looking at a graph that described the performance of a stock you had purchased, why would it be better if the slope of the line were positive, rather than negative?
B. Describe the behavior of a line with a negative slope.
C. Would you rather climb a hill with a slope of $\frac{1}{2}$ or a slope of 3? Explain why.
D. What is a difference quotient?

Problem Set 4.1

Find the slope of each of the following lines from the given graph.

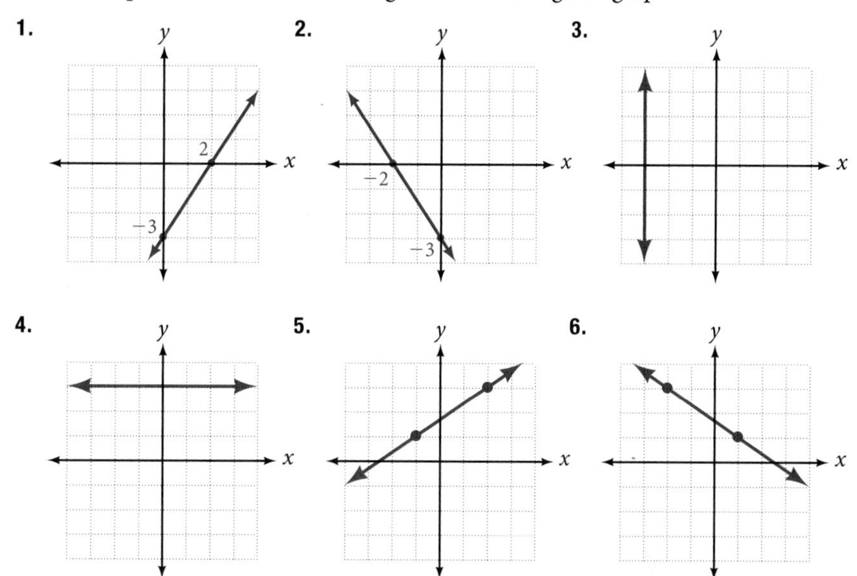

Find the slope of the line through each of the following pairs of points. Then, plot each pair of points, draw a line through them, and indicate the rise and run in the graph in the manner shown in Example 2.

7. $(2, 1), (4, 4)$

8. $(3, 1), (5, 4)$

9. $(1, 4), (5, 2)$

10. $(1, 3), (5, 2)$

11. $(1, -3), (4, 2)$

12. $(2, -3), (5, 2)$

13. $(2, -4)$ and $(5, -9)$

14. $(-3, 2)$ and $(-1, 6)$

15. $(-3, 5)$ and $(1, -1)$

16. $(-2, -1)$ and $(3, -5)$

17. $(-4, 6)$ and $(2, 6)$

18. $(2, -3)$ and $(2, 7)$

19. $(a, -3)$ and $(a, 5)$

20. $(x, 2y)$ and $(4x, 8y)$

Solve for the indicated variable if the line through the two given points has the given slope.

21. $(a, 3)$ and $(2, 6)$, $m = -1$

22. $(a, -2)$ and $(4, -6)$, $m = -3$

23. $(2, b)$ and $(-1, 4b)$, $m = -2$

24. $(-4, y)$ and $(-1, 6y)$, $m = 2$

25. $(2, 4)$ and (x, x^2), $m = 5$

26. $(3, 9)$ and (x, x^2), $m = -2$

27. $(1, 3)$ and $(x, 2x^2 + 1)$, $m = -6$

28. $(3, 7)$ and $(x, x^2 - 2)$, $m = -4$

For each of the equations in Problems 29–32, complete the table, and then use the results to find the slope of the graph of the equation.

29. $2x + 3y = 6$

x	y
0	
	0

30. $3x - 2y = 6$

x	y
0	
	0

31. $y = \frac{2}{3}x - 5$

x	y
0	
3	

32. $y = -\frac{3}{4}x + 2$

x	y
0	
4	

33. **Finding Slope from Intercepts** Graph the line that has an x-intercept of 3 and a y-intercept of -2. What is the slope of this line?
34. **Finding Slope from Intercepts** Graph the line with x-intercept -4 and y-intercept -2. What is the slope of this line?
35. **Parallel Lines** Find the slope of any line parallel to the line through $(2, 3)$ and $(-8, 1)$.
36. **Parallel Lines** Find the slope of any line parallel to the line through $(2, 5)$ and $(5, -3)$.
37. **Perpendicular Lines** Line l contains the points $(5, -6)$ and $(5, 2)$. Give the slope of any line perpendicular to l.
38. **Perpendicular Lines** Line l contains the points $(3, 4)$ and $(-3, 1)$. Give the slope of any line perpendicular to l.
39. **Parallel Lines** Line l contains the points $(-2, 1)$ and $(4, -5)$. Find the slope of any line parallel to l.
40. **Parallel Lines** Line l contains the points $(3, -4)$ and $(-2, -6)$. Find the slope of any line parallel to l.
41. **Perpendicular Lines** Line l contains the points $(-2, -5)$ and $(1, -3)$. Find the slope of any line perpendicular to l.
42. **Perpendicular Lines** Line l contains the points $(6, -3)$ and $(-2, 7)$. Find the slope of any line perpendicular to l.

43. Determine if each of the following tables could represent ordered pairs from an equation of a line.

a.

x	y
0	5
1	7
2	9
3	11

b.

x	y
-2	-5
0	-2
2	0
4	1

44. The following lines have slope 2, $\frac{1}{2}$, 0, and -1. Match each line to its slope value.

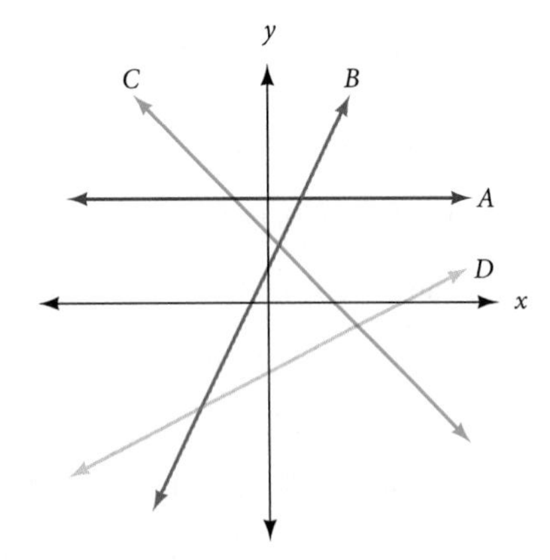

For Problems 45-54, evaluate

a. $\dfrac{f(x) - f(a)}{x - a}$ **b.** $\dfrac{f(x + h) - f(x)}{h}$

45. $f(x) = 4x$
46. $f(x) = -3x$

47. $f(x) = 5x + 3$

48. $f(x) = 6x - 5$

49. $f(x) = x^2$

50. $f(x) = 3x^2$

51. $f(x) = x^2 + 1$

52. $f(x) = x^2 - 3$

53. $f(x) = x^2 - 3x + 4$

54. $f(x) = x^2 + 4x - 7$

Modeling Practice

An object is traveling at a constant speed. The distance and time data are shown on the given graph. Use the graph to find the speed of the object.

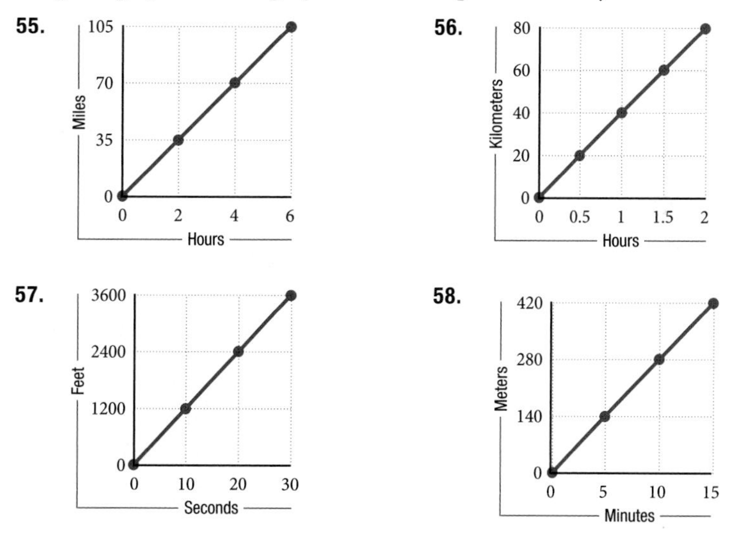

59. Heating a Block of Ice A block of ice with an initial temperature of −20 °C is heated at a steady rate. The graph shows how the temperature changes as the ice melts to become water and the water boils to become steam and water.

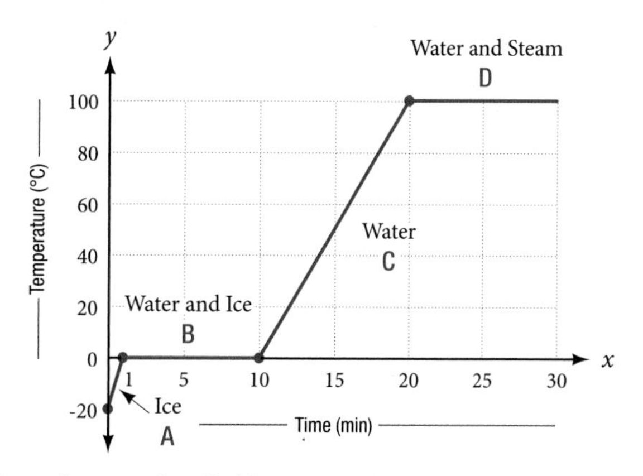

a. How long does it take all the ice to melt?

b. From the time the heat is applied to the block of ice, how long is it before the water boils?

c. Find the slope of the line segment labeled A. What units would you attach to this number?

d. Find the slope of the line segment labeled C. Be sure to attach units to your answer.

e. Is the temperature changing faster during the 1st minute or the 16th minute?

60. Slope of a Highway A sign at the top of the Cuesta Grade, outside of San Luis Obispo, reads "7% downgrade next 3 miles." The following diagram is a model of the Cuesta Grade that takes into account the information on that sign.

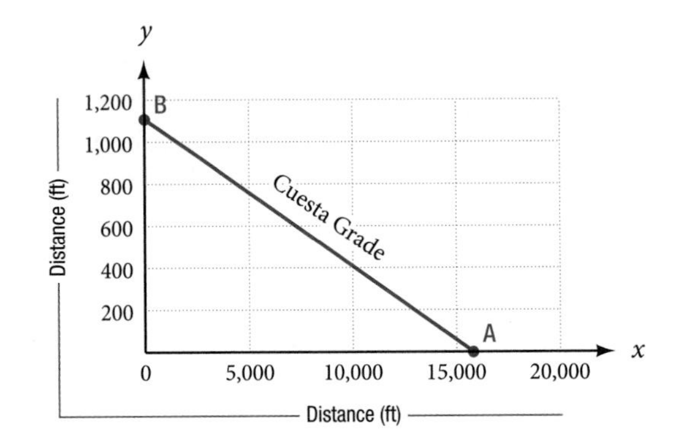

a. At point B, the graph crosses the y-axis at 1,106 feet. How far is it from the origin to point A?

b. What is the slope of the Cuesta Grade?

61. Solar Energy The graph below shows the annual shipments of solar thermal collectors in the United States. Using the graph below, find the slope of the line connecting the first (1997, 8,000) and last (2006, 20,000) endpoints and then explain in words what the slope represents.

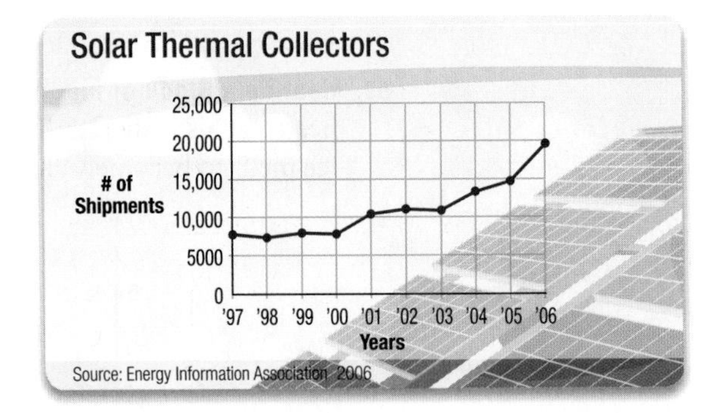

62. Age of New Mothers The graph shows the increase in average age of first time mothers in the U.S. since 1970. Find the slope of the line that connects the points (1975, 21.75) to (1990, 24.25). Round to the nearest hundredth. Explain in words what the slope represents.

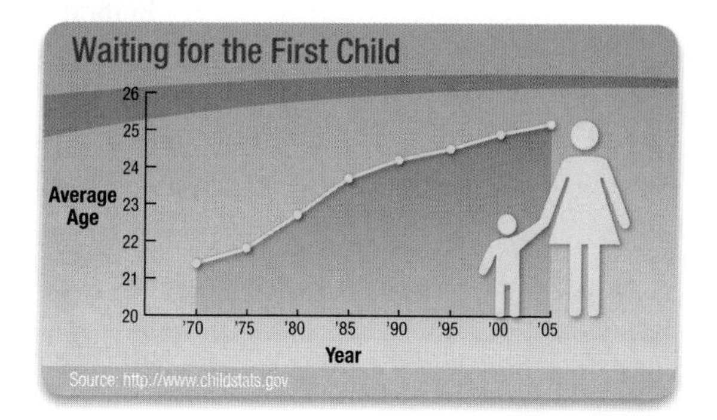

63. Light Bulbs The chart shows a comparison of power usage between incandescent and energy efficient light bulbs. Use the chart to work the following problems involving slope.

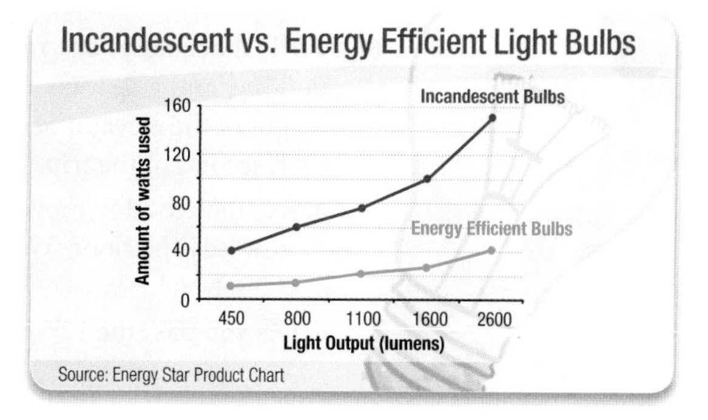

a. Find the slope of the line for the incandescent bulb from the two endpoints and then explain in words what the slope represents.

b. Find the slope of the line for the energy efficient bulb from the two endpoints and then explain in words what the slope represents.

c. Which light bulb is better? Why?

64. Horse Racing The graph shows the amount of money bet on horse racing from 1985 to 2005. Use the chart to work the following problems involving slope.

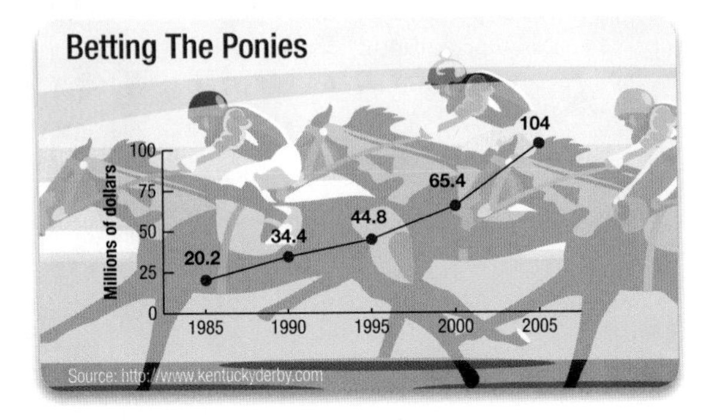

a. Find the slope of the line from 1985 to 1990, and then explain in words what the slope represents.

b. Find the slope of the line from 2000 to 2005, and then explain in words what the slope represents.

65. Vertical Acceleration You get in an elevator on the ground floor of a high-rise building and ride it up, non-stop, to the 26th floor. Figure 10 below shows the height of the elevator as it ascends to the 26th floor. Figure 11 shows its speed each second of the trip.

a. As the elevator moves up, you feel the sensation of being pushed down towards the floor. Which point on the graphs below corresponds to this sensation?

b. As you pass the 12th floor, your ears pop. Name the point that corresponds to this sensation.

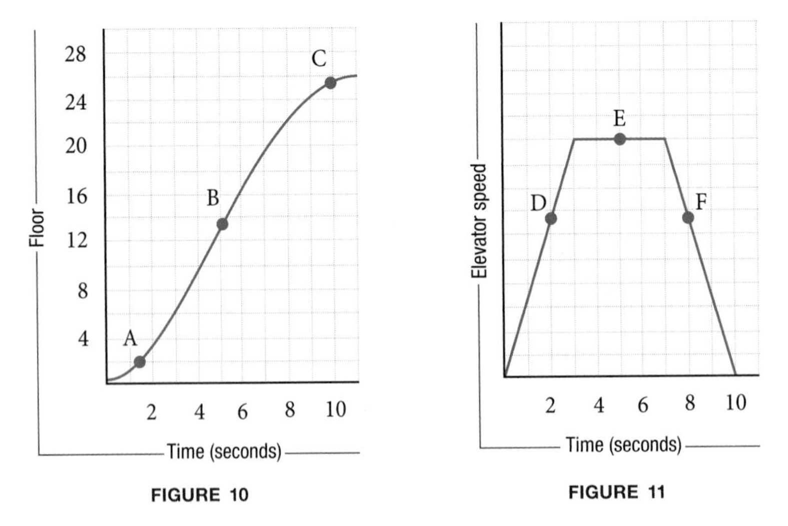

FIGURE 10

FIGURE 11

66. Physiology and Slope You fly from Lima to Cuzco, Peru. The elevation of Cuzco is around 11,000 feet above sea level. As the plane takes off in Lima, you feel yourself being pushed against the seat. As the plane descends into Cuzco, your ears pop. When you get out of the plane in Cuzco, your breathing is a little more difficult than usual. Match the sensation on the left with the quantity on the right.

Pushed back in your seat	Change in altitude
Ears pop	Change in velocity
Difficulty breathing	Altitude

Using Technology Exercises

Use your calculator to solve the following problems.

67. If $f(x) = 4x - 5$, find $\dfrac{f(x + 3) - f(x)}{3}$.

68. If $f(x) = 0.85x^2 - 1.82x - 5$, find $\dfrac{f(6) - f(0)}{6 - 0}$.

69. A company that manufactures a particular product has determined that the cost of producing x units of the product is described by the function $C(x) = 0.035x^2 + 1{,}855$. Find the average rate of change of the cost as the number of units produced increases from 200 to 500.

Getting Ready for the Next Section

Simplify.

70. $2\left(-\dfrac{1}{2}\right)$

71. $\dfrac{3 - (-1)}{-3 - 3}$

72. $-\dfrac{5 - (-3)}{2 - 6}$

73. $3\left(-\dfrac{2}{3}x + 1\right)$

Solve for y.

74. $\dfrac{y - b}{x - 0} = m$

75. $2x + 3y = 6$

76. $y - 3 = -2(x + 4)$

77. $y + 1 = -\dfrac{2}{3}(x - 3)$

78. If $y = -\dfrac{4}{3}x + 5$, find y when x is 0.

79. If $y = -\dfrac{4}{3}x + 5$, find y when x is 3.

Success Skills

If you have made it this far, then you have the study skills necessary to be successful in this course. Success skills are more general in nature and will help you with all your classes and ensure your success in college as well.

Let's start with a question:

Question: What quality is most important for success in any college course?

Answer: Independence. You want to become an independent learner.

We all know people like this. They are generally happy. They don't worry about getting the right instructor, or whether or not things work out every time. They have a confidence that comes from knowing that they are responsible for their success or failure in the goals they set for themselves.

Here are some of the qualities of an independent learner:

- Intends to succeed.
- Doesn't let setbacks deter them.
- Knows their resources.
  - Instructor's office hours
  - Math lab
  - Student Solutions Manual
  - Group study
  - Internet
- Doesn't mistake activity for achievement.
- Has a positive attitude.

There are other traits as well. The first step in becoming an independent learner is doing a little self-evaluation and then making of list of traits that you would like to acquire. What skills do you have that align with those of an independent learner? What attributes do you have that keep you from being an independent learner? What qualities would you like to obtain that you don't have now?

Linear Functions and Equations of Lines

INTRODUCTION The table and illustrations below show some corresponding temperatures on the Fahrenheit and Celsius temperature scales. For example, water freezes at 32 °F and 0 °C, and boils at 212 °F and 100 °C.

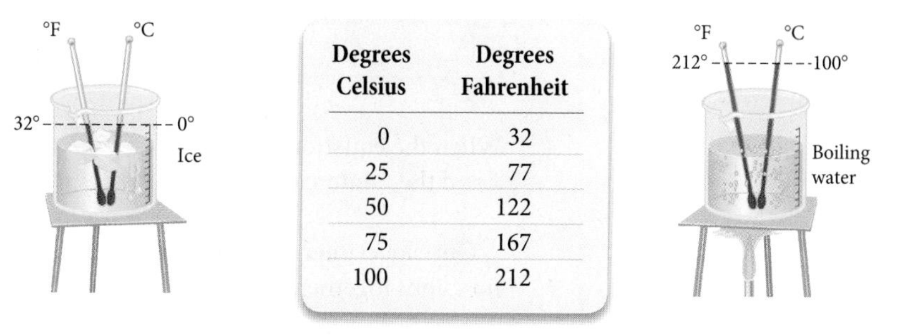

Degrees Celsius	Degrees Fahrenheit
0	32
25	77
50	122
75	167
100	212

If we plot all the points in the table using the x-axis for temperatures on the Celsius scale and the y-axis for temperatures on the Fahrenheit scale, we see that they line up in a straight line (Figure 1).

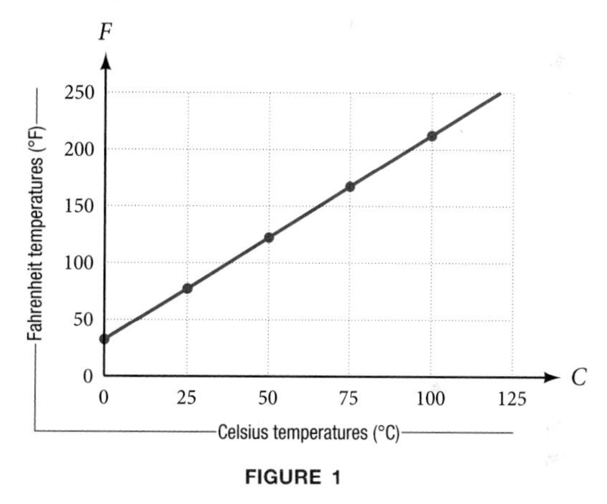

FIGURE 1

This means that a linear equation in two variables will give a perfect description of the relationship between the two scales. That equation is

$$F = \frac{9}{5}C + 32$$

The techniques we use to find the equation of a line from a set of points is what this section is all about.

Suppose line l has slope m and y-intercept b. What is the equation of l? Because the y-intercept is b, we know the point $(0, b)$ is on the line. If (x, y) is any other point on l, then using the definition for slope, we have

$$\frac{y - b}{x - 0} = m \qquad \text{Definition of Slope}$$

$$y - b = mx \qquad \text{Multiply both sides by } x$$

$$y = mx + b \qquad \text{Add } b \text{ to both sides}$$

This last equation is known as the ***slope-intercept form*** of the equation of a straight line.

PROPERTY *Slope-Intercept Form of the Equation of a Line*

The equation of any line with slope m and y-intercept b is given by

$$y = mx + b$$

(↗ Slope; ↑ y-intercept)

When the equation is in this form, the *slope* of the line is always the *coefficient* of x and the y-intercept is always the *constant term*.

Our slope-intercept form of the equation can also be used to classify all *linear functions* together.

DEFINITION *Linear Function*

A *Linear Function* is any function that can be put in the form

$$y = f(x) = mx + b$$

where m and b are real numbers.

VIDEO EXAMPLES

SECTION 4.2

EXAMPLE 1 Find the equation of the line with slope $-\frac{4}{3}$ and y-intercept 5. Then graph the line.

SOLUTION Substituting $m = -\frac{4}{3}$ and $b = 5$ into the equation $y = mx + b$, we have

$$y = -\frac{4}{3}x + 5$$

Finding the equation from the slope and y-intercept is just that easy. If the slope is m and the y-intercept is b, then the equation is always $y = mx + b$. Now, let's graph the line.

Because the y-intercept is 5, the graph goes through the point (0, 5). To find a second point on the graph, we start at (0, 5) and move 4 units down (that's a rise of -4) and 3 units to the right (a run of 3). The point we end up at is (3, 1). Drawing a line that passes through (0, 5) and (3, 1), we have the graph of our equation. (Note that we could also let the rise = 4 and the run = -3 and obtain the same graph.) The graph is shown in Figure 2.

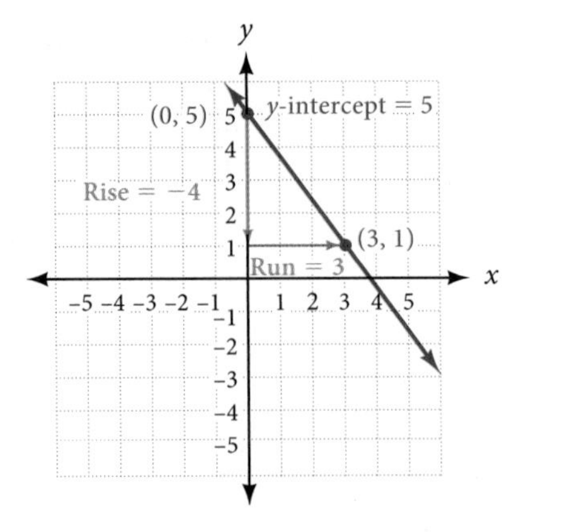

FIGURE 2

EXAMPLE 2 Give the slope and y-intercept for the line $2x - 3y = 5$.

SOLUTION To use the slope-intercept form, we must solve the equation for y in terms of x:

$$2x - 3y = 5$$

$$-3y = -2x + 5 \qquad \text{Add } -2x \text{ to both sides}$$

$$y = \frac{2}{3}x - \frac{5}{3} \qquad \text{Divide by } -3$$

The last equation has the form $y = mx + b$. The slope must be $m = \frac{2}{3}$ and the y-intercept is $b = -\frac{5}{3}$.

EXAMPLE 3 Graph the linear function $f(x) = -\frac{2}{3}x + 2$ using the slope and y-intercept.

SOLUTION The slope is $m = -\frac{2}{3}$ and the y-intercept is $b = 2$. Therefore, the point $(0, 2)$ is on the graph, and the ratio of rise to run going from $(0, 2)$ to any other point on the line is $-\frac{2}{3}$. If we start at $(0, 2)$ and move 2 units up (that's a rise of 2) and 3 units to the left (a run of -3), we will be at another point on the graph. (We could also go down 2 units and right 3 units and still be assured of ending up at another point on the line because $\frac{2}{-3}$ is the same as $\frac{-2}{3}$.)

Note As we mentioned earlier, the rectangular coordinate system is the tool we use to connect algebra and geometry. Example 3 illustrates this connection, as do many other examples in this chapter. In Example 3, Descartes's rectangular coordinate system allows us to associate the equation $y = -\frac{2}{3}x + 2$ (an algebraic concept) with the straight line (a geometric concept) shown in Figure 3.

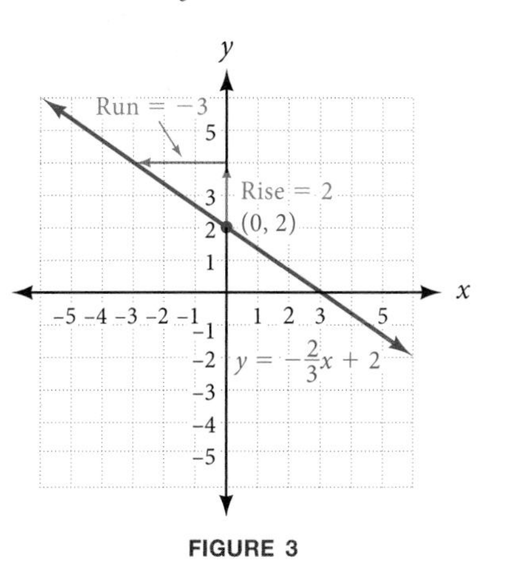

FIGURE 3

A second useful form of the equation of a line is the point-slope form.

Let line l contain the point (x_1, y_1) and have slope m. If (x, y) is any other point on l, then by the definition of slope we have

$$\frac{y - y_1}{x - x_1} = m$$

Multiplying both sides by $(x - x_1)$ gives us

$$(x - x_1) \cdot \frac{y - y_1}{x - x_1} = m(x - x_1)$$

$$y - y_1 = m(x - x_1)$$

This last equation is known as the ***point-slope form*** of the equation of a line.

PROPERTY ***Point-Slope Form of the Equation of a Line***

The equation of the line through (x_1, y_1) with slope m is given by

$$y - y_1 = m(x - x_1)$$

This form of the equation of a line is used to find the equation of a line, either given one point on the line and the slope, or given two points on the line.

EXAMPLE 4 Find the equation of the line with slope -2 that contains the point $(-4, 3)$. Write the answer in slope-intercept form.

SOLUTION

Using	$(x_1, y_1) = (-4, 3)$ and $m = -2$	
in	$y - y_1 = m(x - x_1)$	Point-slope form
gives us	$y - 3 = -2(x + 4)$	Note: $x - (-4) = x + 4$
	$y - 3 = -2x - 8$	Multiply out right side
	$y = -2x - 5$	Add 3 to each side

Figure 4 is the graph of the line that contains $(-4, 3)$ and has a slope of -2. Notice that the y-intercept on the graph matches that of the equation we found.

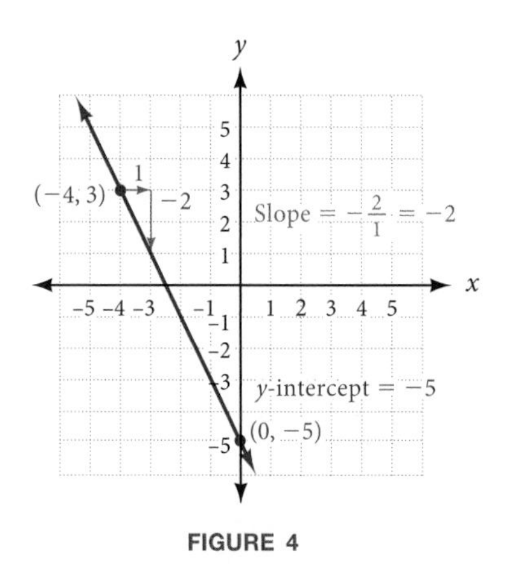

FIGURE 4

EXAMPLE 5 Find the equation of the line that passes through the points $(-3, 3)$ and $(3, -1)$.

SOLUTION We begin by finding the slope of the line:

$$m = \frac{3 - (-1)}{-3 - 3} = \frac{4}{-6} = -\frac{2}{3}$$

Note We could have used the point $(-3, 3)$ instead of $(3, -1)$ and obtained the same equation. That is, using $(x_1, y_1) = (-3, 3)$ and $m = -\frac{2}{3}$ in $y - y_1 = m(x - x_1)$ gives us

$$y - 3 = -\frac{2}{3}(x + 3)$$
$$y - 3 = -\frac{2}{3}x - 2$$
$$y = -\frac{2}{3}x + 1$$

which is the same result we obtained using $(3, -1)$.

Using $(x_1, y_1) = (3, -1)$ and $m = -\frac{2}{3}$ in $y - y_1 = m(x - x_1)$ yields

$$y + 1 = -\frac{2}{3}(x - 3)$$
$$y + 1 = -\frac{2}{3}x + 2 \quad \text{Multiply out right side}$$
$$y = -\frac{2}{3}x + 1 \quad \text{Add } -1 \text{ to each side}$$

Figure 5 shows the graph of the line that passes through the points $(-3, 3)$ and $(3, -1)$. As you can see, the slope and y-intercept are $-\frac{2}{3}$ and 1, respectively.

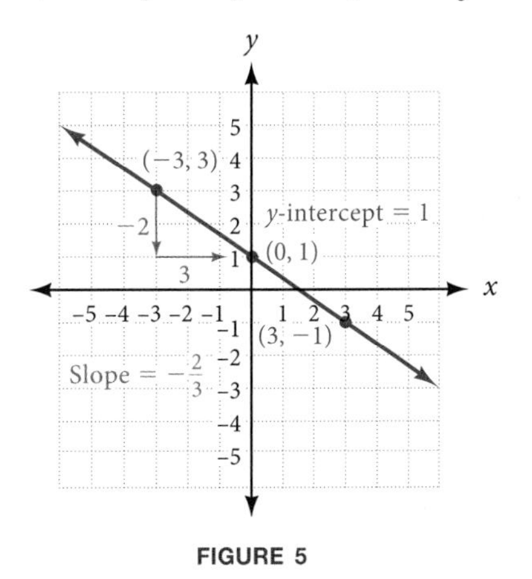

FIGURE 5

The last form of the equation of a line that we will consider in this section is called the *standard form*. It is used mainly to write equations in a form that is free of fractions and is easy to compare with other equations.

PROPERTY *Standard Form for the Equation of a Line*

If a, b, and c are integers, then the equation of a line is in standard form when it has the form

$$ax + by = c$$

and a and b are not both 0.

If we were to write the equation

$$y = -\frac{2}{3}x + 1$$

in standard form, we would first multiply both sides by 3 to obtain

$$3y = -2x + 3$$

Then we would add $2x$ to each side, yielding

$$2x + 3y = 3$$

which is a linear equation in standard form.

EXAMPLE 6 Give the equation of the line through $(-1, 4)$ whose graph is perpendicular to the graph of $2x - y = -3$. Write the answer in standard form.

SOLUTION To find the slope of $2x - y = -3$, we solve for y:

$$2x - y = -3$$
$$y = 2x + 3$$

The slope of this line is 2. The line we are interested in is perpendicular to the line with slope 2 and must, therefore, have a slope of $-\frac{1}{2}$.

Using $(x_1, y_1) = (-1, 4)$ and $m = -\frac{1}{2}$, we have

$$y - y_1 = m(x - x_1)$$
$$y - 4 = -\frac{1}{2}(x + 1)$$

Because we want our answer in standard form, we multiply each side by 2.

$$2y - 8 = -1(x + 1)$$
$$2y - 8 = -x - 1$$
$$x + 2y - 8 = -1$$
$$x + 2y = 7$$

The last equation is in standard form.

TECHNOLOGY NOTE *Graphing Calculators*

One advantage of using a graphing calculator to graph lines is that a calculator does not care whether the equation has been simplified or not. To illustrate, in Example 5 we found that the equation of the line with slope $-\frac{2}{3}$ that passes through the point $(3, -1)$ is

$$y + 1 = -\frac{2}{3}(x - 3)$$

Normally, to graph this equation we would simplify it first. With a graphing calculator, we add -1 to each side and enter the equation this way:

$$\text{Y1} = -(2/3)(\text{X} - 3) - 1$$

No simplification is necessary. We can graph the equation in this form, and the graph will be the same as that of the simplified form of the equation, which is $y = -\frac{2}{3}x + 1$. To convince yourself that this is true, graph both the simplified form for the equation and the unsimplified form in the same window. As you will see, the two graphs coincide.

Modeling: Manufacturing

EXAMPLE 7 It cost a bicycle company $9,000 to make 50 touring bikes in its first month of operation and $15,000 to make 125 bikes during its second month.

a. Find a linear equation for the company's monthly production cost, C, in terms of the number of bikes made, x.

b. State the slope and vertical intercept of your line, including units. What do they tell you about the problem?

SOLUTION

a. We first find two data points, (x, C), from the information given.

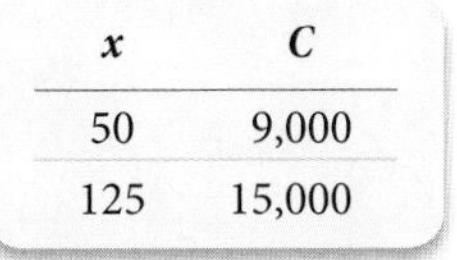

x	C
50	9,000
125	15,000

Step 1: Compute the slope of the line.

$$m = \frac{C_2 - C_1}{x_2 - x_1} = \frac{15,000 - 9,000}{125 - 50} = 80$$

FIGURE 6

Step 2: Apply the point-slope formula, using (50, 9,000) for (x_1, y_1).

$$C - C_1 = m\,(x - x_1)$$

$$C - 9,000 = 80(x - 50)$$

$$= 9,000 + 80x - 4,000$$

$$= 80x + 5,000$$

b. The slope is 80 dollars per bike, and it tells us the cost of producing each bike. The vertical intercept is 5,000, and it tells us that the bicycle company's fixed costs (before production begins) are $5,000.

The following summary reminds us that all horizontal lines have equations of the form $y = b$, and slopes of 0. Since they cross the y-axis at b, the y-intercept is b; there is no x-intercept. Vertical lines have no slope, and equations of the form $x = a$. Each will have an x-intercept at a, and no y-intercept. Finally, equations of the form $y = mx$ have graphs that pass through the origin. The slope is always m and both the x-intercept and the y-intercept are 0.

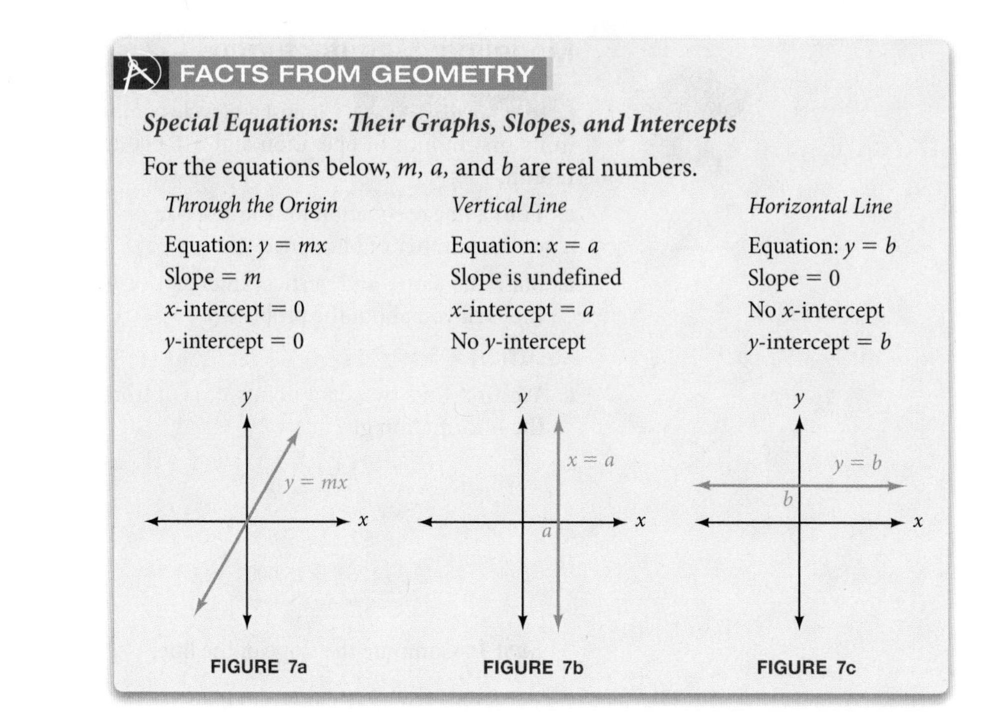

FACTS FROM GEOMETRY

Special Equations: Their Graphs, Slopes, and Intercepts

For the equations below, m, a, and b are real numbers.

Through the Origin	*Vertical Line*	*Horizontal Line*
Equation: $y = mx$	Equation: $x = a$	Equation: $y = b$
Slope $= m$	Slope is undefined	Slope $= 0$
x-intercept $= 0$	x-intercept $= a$	No x-intercept
y-intercept $= 0$	No y-intercept	y-intercept $= b$

FIGURE 7a **FIGURE 7b** **FIGURE 7c**

Piecewise Defined Functions

A function may be defined by different formulas on different portions of the x-axis. Such a function is said to be defined *piecewise*.

EXAMPLE 8 Graph the function defined by

$$f(x) = \begin{cases} x + 1 & \text{if } x \leq 1 \\ 3 & \text{if } x > 1 \end{cases}$$

SOLUTION Think of the coordinate system as divided into two regions by the vertical line $x = 1$, as shown in Figure 8. In the left-hand region ($x \leq 1$), we graph the line $y = x + 1$. Notice that the point $(1, 2)$ is included in the graph. We indicate this with a solid dot at the point $(1, 2)$. In the right-hand region ($x > 1$), we graph the horizontal line $y = 3$. The point $(1, 3)$ is *not* part of the graph. We indicate this with an open circle at the point $(1, 3)$.

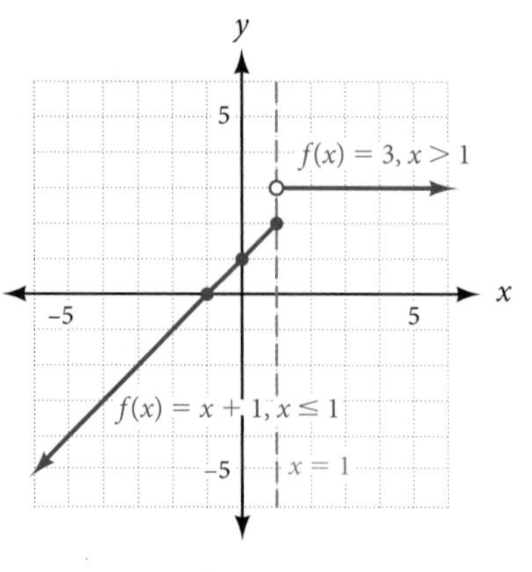

FIGURE 8

The absolute value function $f(x) = |x|$ is an example of a function that is defined piecewise.

$$f(x) = |x| = \begin{cases} x & \text{if } x \geq 0 \\ -x & \text{if } x < 0 \end{cases}$$

To sketch the absolute value function, we graph the line $y = x$ in the first quadrant and the line $y = -x$ in the second quadrant.

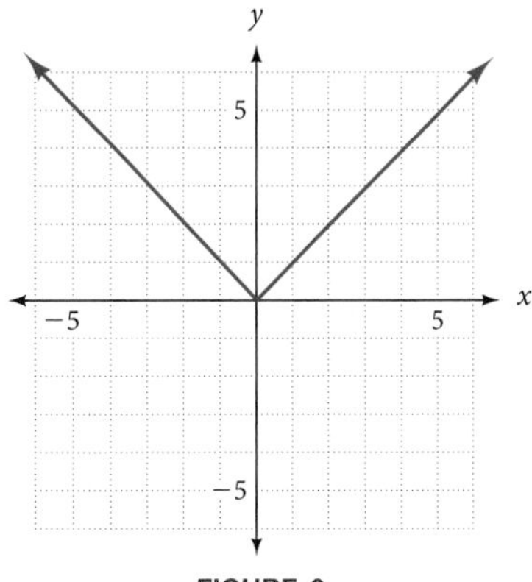

FIGURE 9

EXAMPLE 9

a. Write a piecewise definition for $g(x) = |x - 3|$.

b. Sketch a graph of $g(x) = |x - 3|$.

SOLUTION

a. In the definition for $|x|$, we replace x by $x - 3$ to get

$$g(x) = |x - 3| = \begin{cases} x - 3 & \text{if } x - 3 \geq 0 \\ -(x - 3) & \text{if } x - 3 < 0 \end{cases}$$

We can simplify this expression to

$$g(x) = |x - 3| = \begin{cases} x - 3 & \text{if } x \geq 3 \\ -x + 3 & \text{if } x < 3 \end{cases}$$

b. In the first region, $x \geq 3$, we graph the line $y = x - 3$. Because $x = 3$ is included in this region, the endpoint of this portion of the graph, $(3, 0)$, is included, too. In the second region, $x < 3$, we graph the line $y = -x + 3$. Note that the two pieces of the graph meet at the point $(3, 0)$, as show in Figure 10.

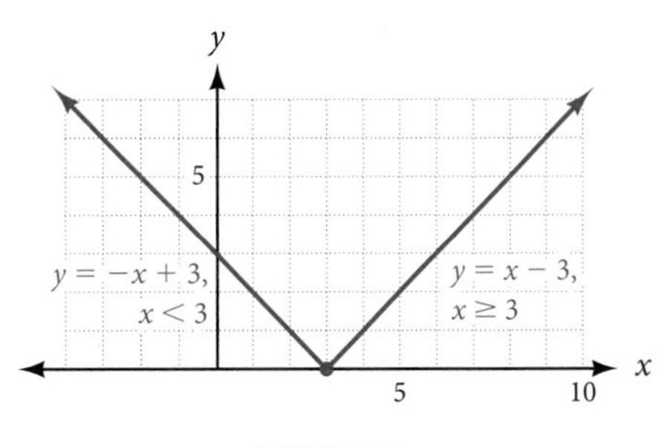

FIGURE 10

Using Technology 4.2

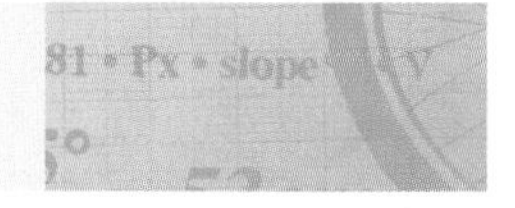

Your calculator can construct graphs of functions that are defined piecewise. The following entry shows how to construct the graph of $f(x) = \begin{cases} x^2 - 4 & \text{if } x \le 2 \\ -x + 2 & \text{if } x > 2 \end{cases}$.

EXAMPLE 10 Graph the following function.

$$f(x) = \begin{cases} x^2 - 4 & \text{if } x \le 2 \\ -x + 2 & \text{if } x > 2 \end{cases}$$

SOLUTION Here is how we enter the function into our Y-variables list.

Y1=(X^2-4)(X≤2)+(-X+2)(X>2)

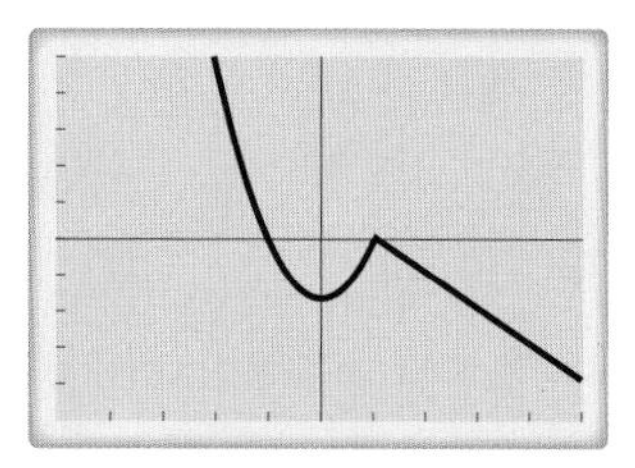

GETTING READY FOR CLASS

After reading through the preceding section, respond in your own words and in complete sentences.

A. How would you graph the line $y = \frac{1}{2}x + 3$?
B. What is the slope-intercept form of the equation of a line?
C. Describe how you would find the equation of a line if you knew the slope and the y-intercept of the line.
D. If you had the graph of a line, how would you use it to find the equation of the line?

Problem Set 4.2

Give the equation of the line with the following slope and y-intercept.

1. $m = -4, b = -3$ **2.** $m = -6, b = \frac{4}{3}$ **3.** $m = -\frac{2}{3}, b = 0$

4. $m = 0, b = \frac{3}{4}$ **5.** $m = -\frac{2}{3}, b = \frac{1}{4}$ **6.** $m = \frac{5}{12}, b = -\frac{3}{2}$

Find the slope of a line **a.** parallel and **b.** perpendicular to the given line.

7. $y = 3x - 4$ **8.** $y = -4x + 1$ **9.** $3x + y = -2$

10. $2x - y = -4$ **11.** $2x + 5y = -11$ **12.** $3x - 5y = -4$

Give the slope and y-intercept for each of the following equations. Sketch the graph using the slope and y-intercept. Give the slope of any line perpendicular to the given line.

13. $y = 3x - 2$ **14.** $y = 2x + 3$ **15.** $2x - 3y = 12$

16. $3x - 2y = 12$ **17.** $4x + 5y = 20$ **18.** $5x - 4y = 20$

For each of the following lines, name the slope and y-intercept. Then write the equation of the line in slope-intercept form.

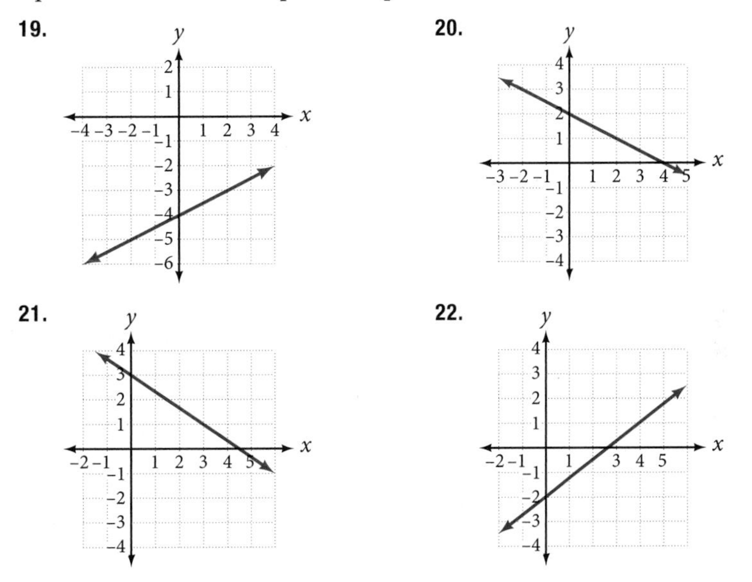

For each of the following problems, the slope and one point on the line are given. In each case, find the equation of that line. (Write the equation for each line in slope-intercept form.)

23. $(-2, -5); m = 2$ **24.** $(-1, -5); m = 2$ **25.** $(-4, 1); m = -\frac{1}{2}$

26. $(-2, 1); m = -\frac{1}{2}$ **27.** $\left(-\frac{1}{3}, 2\right); m = -3$ **28.** $\left(-\frac{2}{3}, 5\right); m = -3$

29. $(-4, 2); m = \frac{2}{3}$ **30.** $(3, -4); m = -\frac{1}{3}$ **31.** $(-5, -2); m = -\frac{1}{4}$

32. $(-4, -3); m = \frac{1}{6}$

Find the equation of the line that passes through each pair of points. Write your answers in standard form.

33. $(3, -2), (-2, 1)$ **34.** $(-4, 1), (-2, -5)$ **35.** $\left(-2, \frac{1}{2}\right), \left(-4, \frac{1}{3}\right)$

36. $(-6, -2), (-3, -6)$ **37.** $\left(\frac{1}{3}, -\frac{1}{5}\right), \left(-\frac{1}{3}, -1\right)$ **38.** $\left(-\frac{1}{2}, -\frac{1}{2}\right), \left(\frac{1}{2}, \frac{1}{10}\right)$

For each of the following lines, name the coordinates of any two points on the line. Then use those two points to find the equation of the line.

39. **40.** **41.** **42.**

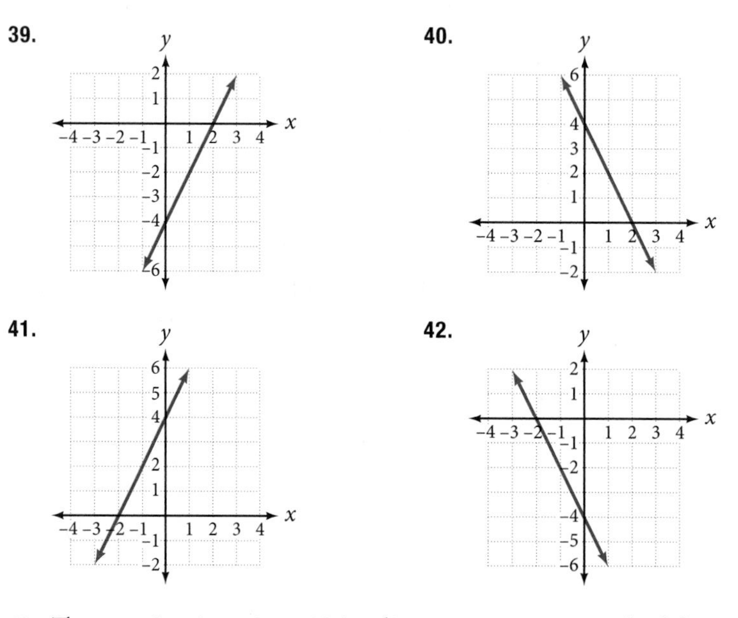

43. The equation $3x - 2y = 10$ is a linear equation in standard form. From this equation, answer the following:
a. Find the x and y intercepts.
b. Find a solution to this equation other than the intercepts in Part a.
c. Write this equation in slope-intercept form.
d. Is the point $(2, 2)$ a solution to the equation?

44. The equation $4x + 3y = 8$ is a linear equation in standard form. From this equation, answer the following:
a. Find the x and y intercepts.
b. Find a solution to this equation other than the intercepts in Part a.
c. Write this equation in slope-intercept form.
d. Is the point $(-3, 2)$ a solution to the equation?

Paying Attention to Instructions The next two problems are intended to give you practice reading, and paying attention to, the instructions that accompany the problems you are working.

45. Work each problem according to the instructions given:

a. Solve: $-2x + 1 = -3$ **b.** Find x when y is 0: $-2x + y = -3$

c. Find y when x is 0: $-2x + y = -3$ **d.** Graph: $-2x + y = -3$

e. Solve for y: $-2x + y = -3$

46. Work each problem according to the instructions given:

a. Solve: $\frac{x}{3} + \frac{1}{4} = 1$ **b.** Find x when y is 0: $\frac{x}{3} + \frac{y}{4} = 1$

c. Find y when x is 0: $\frac{x}{3} + \frac{y}{4} = 1$ **d.** Graph: $\frac{x}{3} + \frac{y}{4} = 1$

e. Solve for y: $\frac{x}{3} + \frac{y}{4} = 1$

47. Graph each of the following lines. In each case, name the slope, the x-intercept, and the y-intercept.

a. $y = \frac{1}{2}x$ **b.** $x = 3$ **c.** $y = -2$

48. Graph each of the following lines. In each case, name the slope, the x-intercept, and the y-intercept.

a. $y = -2x$ **b.** $x = 2$ **c.** $y = -4$

49. Find the equation of the line parallel to the graph of $3x - y = 5$ that contains the point $(-1, 4)$.

50. Find the equation of the line parallel to the graph of $2x - 4y = 5$ that contains the point $(0, 3)$.

51. Line l is perpendicular to the graph of the equation $2x - 5y = 10$ and contains the point $(-4, -3)$. Find the equation for l.

52. Line l is perpendicular to the graph of the equation $-3x - 5y = 2$ and contains the point $(2, -6)$. Find the equation for l.

53. Give the equation of the line perpendicular to the graph of $y = -4x + 2$ that has an x-intercept of -1.

54. Write the equation of the line parallel to the graph of $7x - 2y = 14$ that has an x-intercept of 5.

55. Find the equation of the line with x-intercept 3 and y-intercept 2.

56. Find the equation of the line with x-intercept 2 and y-intercept 3.

For Problems 57-68, graph the following piecewise defined functions. Indicate whether the endpoints of each piece are included on the graph.

57. $f(x) = \begin{cases} -2 & \text{if } x \leq 1 \\ x - 3 & \text{if } x > 1 \end{cases}$ **58.** $h(x) = \begin{cases} -x + 2 & \text{if } x \leq -1 \\ 3 & \text{if } x > -1 \end{cases}$

59. $G(t) = \begin{cases} 3t + 9 & \text{if } t < -2 \\ -3 - \frac{1}{2}t & \text{if } t \geq -2 \end{cases}$ **60.** $F(s) = \begin{cases} \frac{1}{3}s + 3 & \text{if } s < 3 \\ 2s - 3 & \text{if } s \geq 3 \end{cases}$

61. $H(t) = \begin{cases} t^2 & \text{if } t \leq 1 \\ \frac{1}{2}t + \frac{1}{2} & \text{if } t > 1 \end{cases}$ **62.** $g(t) = \begin{cases} \frac{3}{2}t + 7 & \text{if } t \leq -2 \\ t^2 & \text{if } t > -2 \end{cases}$

63. $k(x) = \begin{cases} |x| & \text{if } x \leq 2 \\ \sqrt{x} & \text{if } x > 2 \end{cases}$

64. $S(x) = \begin{cases} \frac{1}{x} & \text{if } x < 1 \\ |x| & \text{if } x \geq 1 \end{cases}$

65. $D(x) = \begin{cases} |x| & \text{if } x < -1 \\ x^3 & \text{if } x \geq -1 \end{cases}$

66. $m(x) = \begin{cases} x^2 & \text{if } x \leq \frac{1}{2} \\ |x| & \text{if } x > \frac{1}{2} \end{cases}$

67. $P(t) = \begin{cases} t^3 + 1 & \text{if } t \leq 1 \\ t^2 & \text{if } t > 1 \end{cases}$

68. $Q(t) = \begin{cases} t^2 & \text{if } t \leq -1 \\ \sqrt[3]{t} & \text{if } t > -1 \end{cases}$

Write a piecewise definition for the function and sketch its graph.

69. $f(x) = |2x - 8|$

70. $g(x) = |3x + 6|$

71. $g(t) = \left|1 + \frac{t}{3}\right|$

72. $f(t) = \left|\frac{1}{2}t - 3\right|$

73. $F(x) = |x^3|$

74. $G(x) = \left|1 + \frac{1}{x}\right|$

Modeling Practice

75. **Deriving the Temperature Equation** The table below resembles the table from the introduction to this section. The rows of the table give us ordered pairs (C, F).

Degrees Celsius	Degrees Fahrenheit
C	F
0	32
25	77
50	122
75	167
100	212

a. Use any two of the ordered pairs from the table to derive the equation $F = \frac{9}{5}C + 32$.

b. Use the equation from Part a to find the Fahrenheit temperature that corresponds to a Celsius temperature of 30 °.

76. **Maximum Heart Rate** The table below gives the maximum heart rate for adults 30, 40, 50, and 60 years old. Each row of the table gives us an ordered pair (A, M).

Age (years)	Maximum Heart Rate (beats per minute)
A	M
30	190
40	180
50	170
60	160

a. Use any two of the ordered pairs from the table to derive the equation $M = 220 - A$, which gives the maximum heart rate M for an adult whose age is A.

b. Use the equation from Part a to find the maximum heart rate for a 25-year-old adult.

© Dario Egidi/iStockphoto

77. **Textbook Cost** To produce this textbook, suppose the publisher spent \$125,000 for typesetting and \$6.50 per book for printing and binding. The total cost to produce and print n books can be written as

$$C = 125{,}000 + 6.5n$$

a. Suppose the number of books printed in the first printing is 10,000. What is the total cost?

b. If the average cost is the total cost divided by the number of books printed, find the average cost of producing 10,000 textbooks.

c. Find the cost to produce one more textbook when you have already produced 10,000 textbooks.

78. **Exercise Heart Rate** In an aerobics class, the instructor indicates that her students' exercise heart rate is 60% of their maximum heart rate, where maximum heart rate is 220 minus their age.

a. Determine the equation that gives exercise heart rate E in terms of age A.

b. Use the equation to find the exercise heart rate of a 22-year-old student.

c. Sketch the graph of the equation for students from 18 to 80 years of age.

79. **Solar Energy** The graph shows the annual number of solar thermal collector shipments in the United States. Find an equation for the line segment that connects the points (7, 13,750) and (8, 15,000), where x is the number of years past 1997. Write your answer in slope-intercept form.

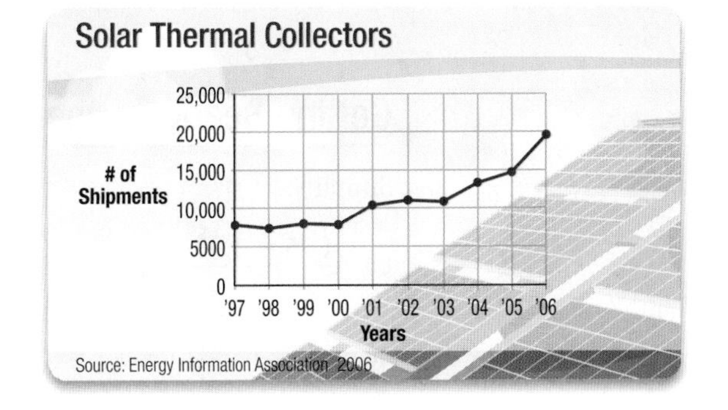

80. Ice, Water, Steam Here is the graph that shows how the temperature changes as the ice melts to become water and the water boils to become steam and water, that we encountered previously. Write a piecewise defined function for this graph.

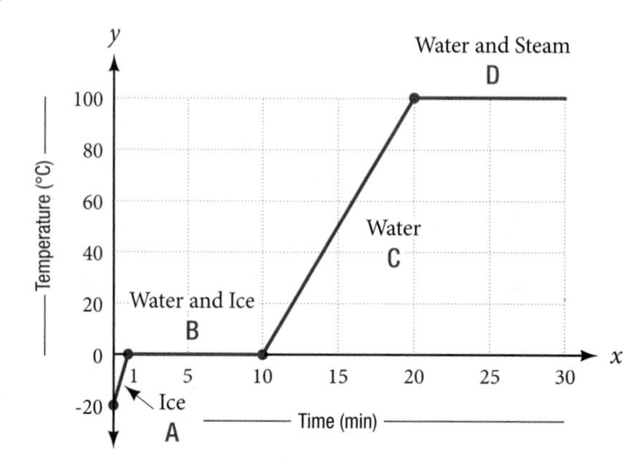

Using Technology Exercises

Use your graphing calculator to construct the graphs of each piecewise defined function.

81. $f(x) = \begin{cases} 3 - x^2 & \text{if } x \leq 1 \\ x^3 - 4x & \text{if } x > 1 \end{cases}$

82. $f(x) = \begin{cases} x^2 - 0.9x + 3.5 & \text{if } x \leq 3 \\ 2.8x - 4.5 & \text{if } x > 3 \end{cases}$

83. $f(x) = \begin{cases} 2.6 - 1.1x^2 & \text{if } |x| < 6 \\ 2.6 & \text{if } |x| > 6 \end{cases}$

84. $f(x) = \begin{cases} x^3 - 8x & \text{if } -3 \leq x < 3 \\ 10 & \text{if } 3 \leq x < 7 \\ \dfrac{10}{x - 7} & \text{if } x \geq 7 \end{cases}$

Getting Ready for the Next Section

Simplify.

85. $\dfrac{1.25 - 1.00}{5}$

86. $\dfrac{2.10 - 1.60}{5}$

87. $\dfrac{12 - 32}{16 - 8}$

88. $-2.5(T - 16) + 12$

89. Find the value of $-2.5x + 52$ when x is 9.

90. Find the equation of the line through (16, 12) if the slope is -2.5.

Linear Regression

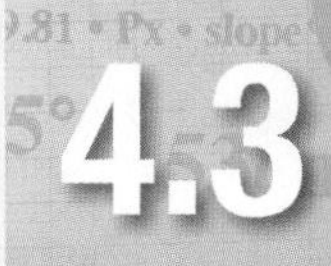

INTRODUCTION In 1929, the astronomer Edwin Hubble (shown in the picture) announced his discovery that the other galaxies in the universe are moving away from us at velocities that increase with distance.

Figure 1 shows a plot of velocity versus distance where each dot represents a galaxy. The fact that the dots all lie approximately on a straight line is the basis of "Hubble's law".

The line in Figure 1 was found using the least-squares method of curve fitting, which we will examine in the Using Technology section.

Recessional velocity

Distance

FIGURE 1

Finding a straight line that approximates data that suggests a linear relationship is called *linear regression*, and it is what we will study in this section.

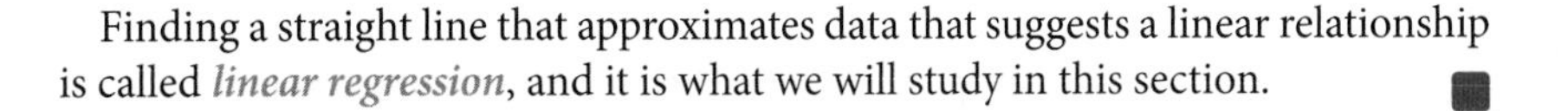

Lines of Best Fit

In most cases, a mathematical model is not a perfect description of reality. Many factors can affect empirical data, including measurement error, environmental conditions, and the influence of related variables. Nonetheless, we can often find an equation that approximates the data in a useful way. The graphs of these equations are called *lines of best fit*, or *regression lines*.

For our first example, we take one of the graphs we have used previously and we find a line of best fit that approximates all the data in the graph.

VIDEO EXAMPLES

SECTION 4.3

EXAMPLE 1 The graph below shows the annual average price for gasoline in 5-year increments from 1955 to 2010.

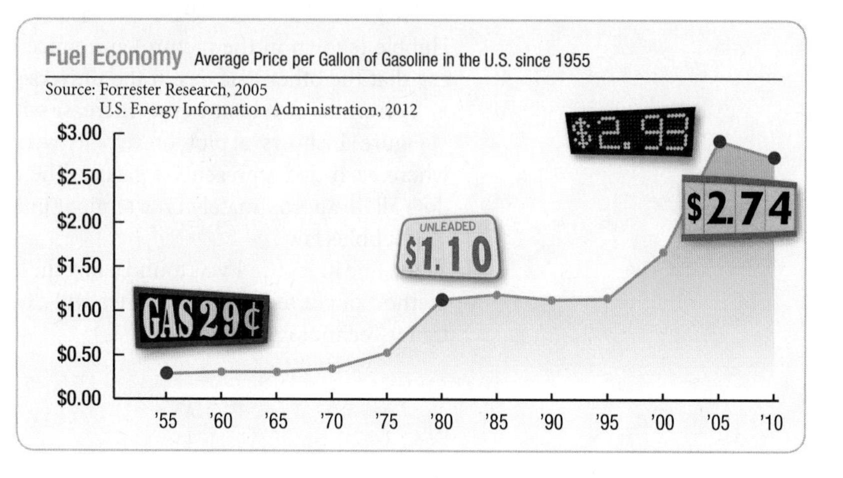

a. Form a line of best fit by connecting the point associated with the year 1980 with the last point on the graph.

b. Find the equation of the line of best fit.

c. Use the line of best fit to predict the average price of gasoline in the year 2015.

SOLUTION We start by drawing our line of best fit by connecting the point at the year 1980 and the last point on the graph.

a.

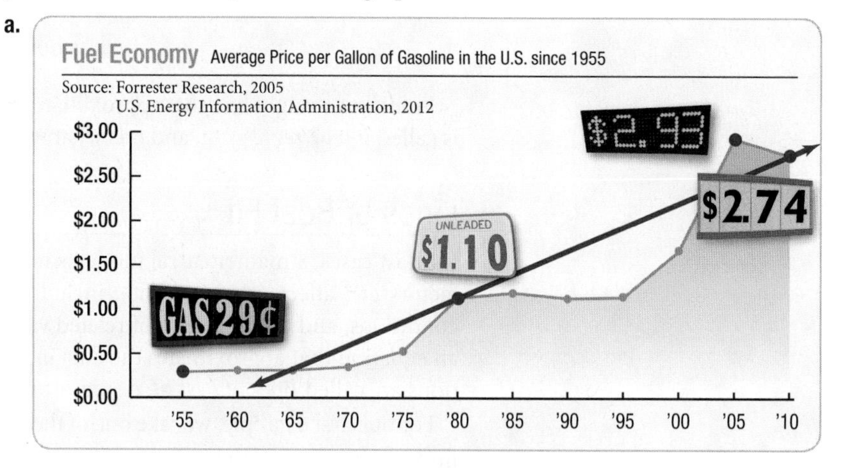

b. If we let x represent the year and y represent the price of gasoline, then the coordinates of our first point are (1980, 1.10) and the coordinates of our last point are (2010, 2.74). First we find the slope, then we use the point slope form of the equation of a line to find the equation of our line of best fit.

$$\text{Slope} = m = \frac{2.74 - 1.10}{2010 - 1980} = \frac{1.64}{30} \approx 0.055$$

Using (2010, 2.74) in

$$y - y_1 = m(x - x_1)$$

we have

$$y - 2.74 = 0.055(x - 2010)$$
$$y - 2.74 = 0.055x - 110.55$$
$$y = 0.055x - 107.81$$

c. To estimate the average price of gasoline in the year 2015, we substitute 2015 for x in our equation.

In the year 2015:

$$y = 0.055(2015) - 107.81$$
$$= 110.825 - 107.81$$
$$= \$3.015$$

Using our line of best fit, we predict the price of gasoline will be approximately \$3.02 per gallon in the year 2015.

Could we have used a different line to approximate the data in Example 1? Yes. There are many ways to approximate the data in Example 1. We are looking for a line that approximates all the data we have. If we had connected the first point on our graph with the point at the year 2005, all the data points would have been below our line, in which case our line would not have been a good representation of all the data. In general, we try to find a line that has as many points above it as it has points below it. The line we chose has more points below it than above it, yet it appears to be a good fit.

EXAMPLE 2 The table shows the minimum hourly wage in the U.S. at five-year intervals. (Source: U.S. Department of Labor)

Year	1960	1965	1970	1975	1980	1985	1990	1995	2000	2005	2010
Minimum Wage (\$)	1.00	1.25	1.60	2.10	3.10	3.35	3.80	4.25	5.15	5.15	7.25

a. Let t represent the number of years after 1960, and plot the data. Is the data linear?

b. Draw a line that "fits" the data.

SOLUTION

a. The graph shown is called a *scatter-plot*. The points do not line up in a straight line exactly. However, the data points do appear to lie close to an imaginary line.

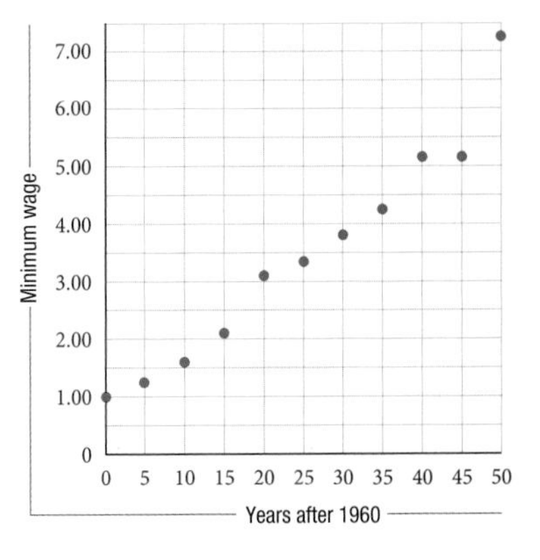

FIGURE 2

b. We would like to draw a line that comes as close as possible to all the data points, even though it may not pass precisely through any of them. In particular, we try to adjust the line so that we have the same number of points above the line and below the line. One possible solution is shown in the figure at right. The line shown here is a regression line. In this case, our regression line passes through (5, 1.25) and (25, 3.35). The slope is

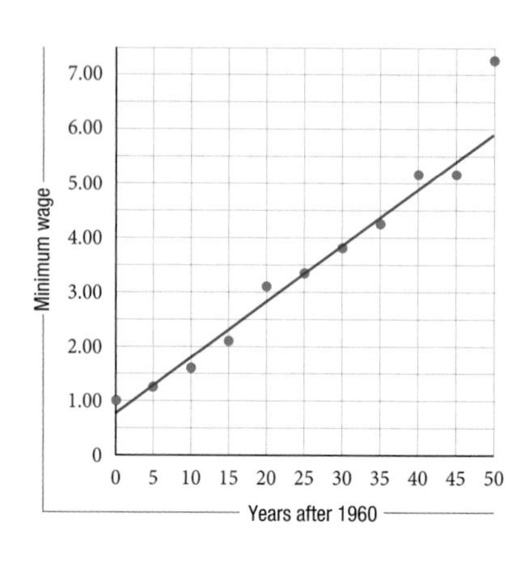

FIGURE 3

$$m = \frac{3.35 - 1.25}{25 - 5} = \frac{2.1}{20} = 0.105$$

Using our point-slope formula we have

$$y - 1.25 = 0.105(x - 5)$$

$$y = 0.105x + 0.725$$

EXAMPLE 3 A consumer group wants to test the gas mileage of a new model SUV. They test-drive six vehicles under similar conditions and record the distance each drove on various amounts of gasoline.

Gasoline used (gal)	9.6	11.3	8.8	5.2	10.3	6.7
Miles driven	155.8	183.6	139.6	80.4	167.1	99.7

a. Are the data linear?

b. Draw a line that fits the data.

c. What does the slope of the line tell us about the data?

SOLUTION

a. No, the data are not strictly linear. If we compute the slopes between successive data points, the values are not constant. We can see from an accurate plot of the data, shown in Figure 4a, that the points lie close to, but not precisely on, a straight line.

b. We would like to draw a line that comes as close as possible to all the data points, even though it may not pass precisely through any of them. In particular, we try to adjust the line so that we have the same number of data points above the line and below the line. One possible solution is shown in Figure 4b.

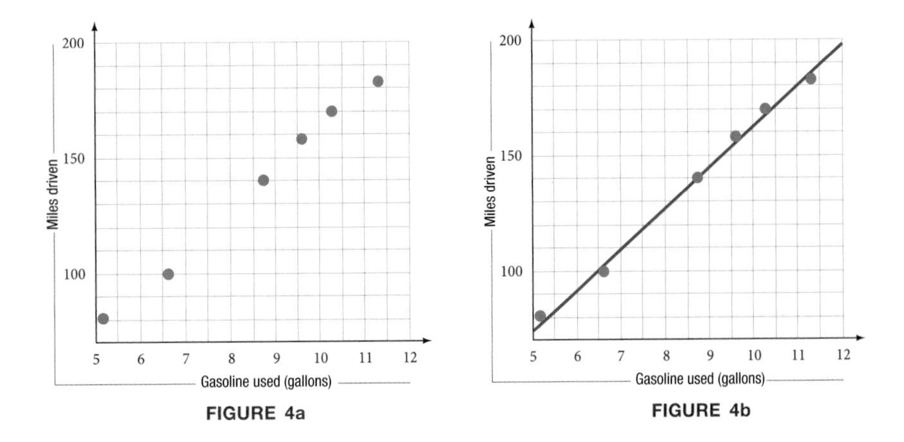

FIGURE 4a

FIGURE 4b

c. To compute the slope of the line of best fit, we first choose two points on the line. Our line appears to pass through one of the data points, (8.8, 139.6). Look for a second point on the line whose coordinates are easy to read, perhaps (6.5, 100). The slope is

$$m = \frac{139.6 - 100}{8.8 - 6.5} = 17.2 \text{ miles per gallon}$$

According to our data, the SUV gets about 17.2 miles to the gallon.

Interpolation and Extrapolation

A regression line is a type of linear model. We can use it to analyze data and to make predictions.

Interpolation: Using our regression line to estimate values between data points is called *interpolation*. For example, if we want to estimate a minimum wage in the year 1972, we would substitute 12 for x in the equation from Example 2 and obtain

$$y = 0.105(12) + 0.725 = 1.985$$

Our estimate of minimum wage for 1972 is $1.99 an hour.

Extrapolation: When we make predictions that go beyond the known data, we are *extrapolating*. The process is called *extrapolation*. In Example 1, when we predicted the price of gasoline to be $3.02 per gallon in the year 2015, we were extrapolating.

EXAMPLE 4 An outdoor snack bar collected the following data showing the number of cups of cocoa, C, they sold when the high temperature for the day was T °Celsius.

©Andreas Arnold/iStockphoto

Temperature (°C), T	2	4	5	8	10	11	12	15	16	18
Cups of cocoa, C	45	42	42	35	25	25	17	16	15	6

a. Make a scatterplot of the data, and draw a regression line.

b. Read values from your line for the number of cups of cocoa that will be sold when the temperature is 8 °C and when the temperature is 16 °C.

c. Find an equation for the regression line.

d. Use your equation to predict the number of cups of cocoa that will be sold when the temperature is 9 °C, and when the temperature is 24 °C.

SOLUTION

a. The scatterplot and a regression line are shown in the figure.

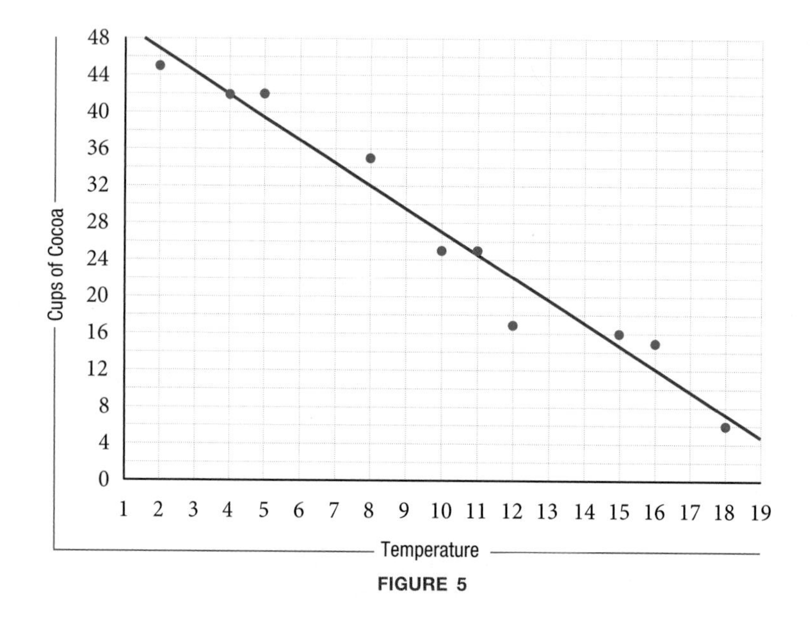

FIGURE 5

The regression line need not pass through any of the data points, but it should be as close as possible. We try to draw the regression line so that there are an equal number of data points above and below the line.

b. The points (8, 32) and (16, 12) appear to lie on the regression line. According to this model, the snack bar will sell 32 cups of cocoa when the temperature is 8 °C, and 12 cups when it is 16 °C. These values are close to the actual data, but not exact.

c. To find an equation for the regression line, we use two points on the line—not data points! We will use (8, 32) and (16, 12). First we compute the slope.

$$m = \frac{C_2 - C_1}{T_2 - T_1} = \frac{12 - 32}{16 - 8} = -2.5$$

Next, we apply the point-slope formula. We'll use the point (16, 12).

$$C - 12 = -2.5(T - 16)$$

$$C = -2.5T + 52$$

d. When $T = 9$,

$$C = -2.5(9) + 52 = 29.5$$

We predict that the snack bar will sell 29 or 30 cups of cocoa when the temperature is 9 °C. When $T = 24$,

$$C = -2.5(24) + 52 = -8$$

Because the snack bar cannot sell -8 cups of cocoa, this prediction is not useful. (What is the Fahrenheit equivalent of 24 °C?)

As we mentioned previously, using a regression line to estimate values between known data points is called *interpolation*. If the data points lie fairly close to the regression line, then interpolation will usually give a fairly accurate estimate. In Example 4, the estimate of 29 or 30 cups of cocoa at 9 °C seems reasonable in the context of the data.

Likewise, extrapolation can often give useful information, but if we try to extrapolate too far beyond our data, we may get unreasonable results. The conditions that produced the data may no longer hold, as in Example 4, or other unexpected conditions may arise to alter the situation.

Using Technology 4.3

Estimating a line of best fit is a subjective process. Rather than base their estimates on such a line, statisticians often use the *least squares regression line*. This regression line minimizes the sum of the squares of all the vertical distances between the data points and the corresponding points on the line (see Figure 6). Many calculators are programmed to find the least squares regression line, using an algorithm that depends only on the data, not on the appearance of the graph.

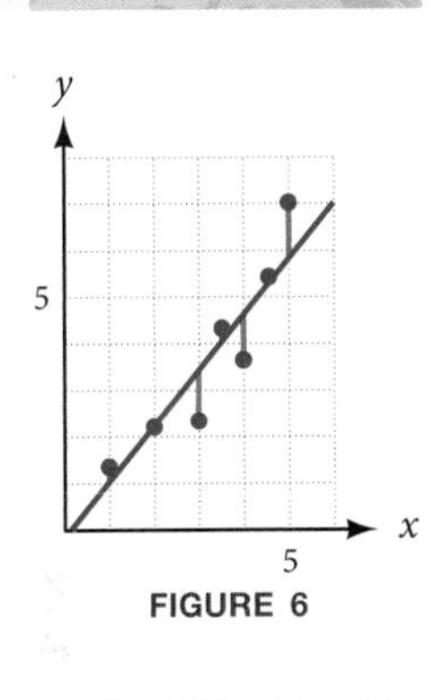

FIGURE 6

You can use a graphing calculator to make a scatterplot, find a regression line, and graph the regression line with the data points. On the TI-83 calculator, we use the statistics mode, which you can access by pressing [STAT]. You will see a display that looks like Figure 7. Choose [1] to Edit (enter or alter) data.

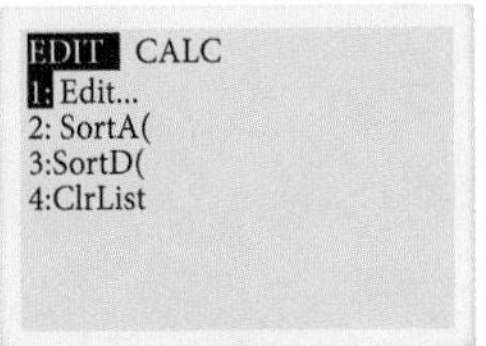

FIGURE 7

Now follow the instructions for using your calculator's statistics features.

EXAMPLE 5 Find the equation of the least squares regression line for the following data:

$$(10, 12), (11, 14), (12, 14), (12, 16), (14, 20)$$

Then plot the data points and the least squares regression line on the same axes.

SOLUTION We must first enter the data. Press [STAT] [ENTER] then select EDIT. You should see a screen like Figure 8a. If there are data in column L1 or L2, clear them out: Use the [▲] key to select L1, press [CLEAR], then do the same for L2. Enter the

x-coordinates of the data points in the L1 column and enter the y-coordinates in the L2 column, as shown in Figure 8b.

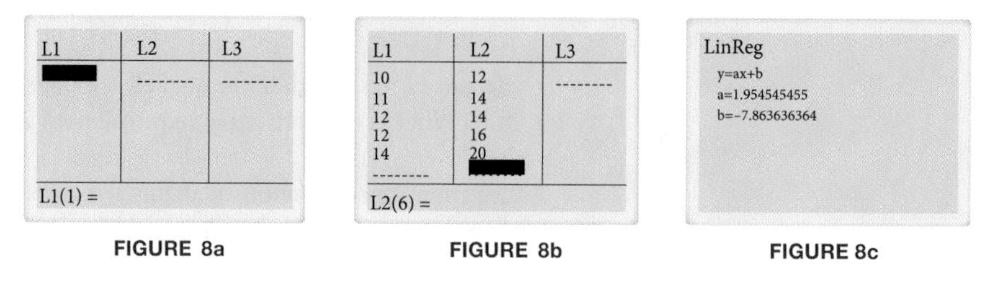

FIGURE 8a FIGURE 8b FIGURE 8c

Now we are ready to find the regression equation for our data. Press [STAT] [▸] 4 to select linear regression, or LinReg ($ax + b$), then press [ENTER]. The calculator will display the equation $y = ax + b$ and the values for a and b, as shown in Figure 8c. You should find that your regression line is approximately $y = 1.95x - 7.86$.

To plot our points and line, we first need to clear out any old definitions in the [Y=] list. Position the cursor after Y1 = and copy in the regression equation as follows: Press [VARS] 5 [▸] [▸] [ENTER]. To draw a scatterplot, press [2nd] [Y=] [1] and set the Plot1 menu as shown in Figure 9a. Finally, press [ZOOM] 9 to see the scatterplot of the data and the regression line. The graph is shown in Figure 9b.

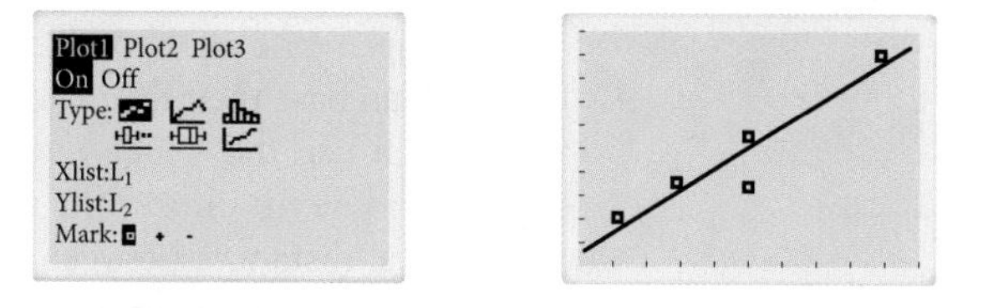

FIGURE 9a FIGURE 9b

GETTING READY FOR CLASS

After reading through the preceding section, respond in your own words and in complete sentences.

A. What is a scatterplot?
B. What is a regression line?
C. When are you using data and extrapolating?
D. What is the least squares regression line?

Problem Set 4.3

For each graph below, draw a line of best fit that you think approximates all the data in the graph.

1. Annual shipments of solar thermal collectors in the United States

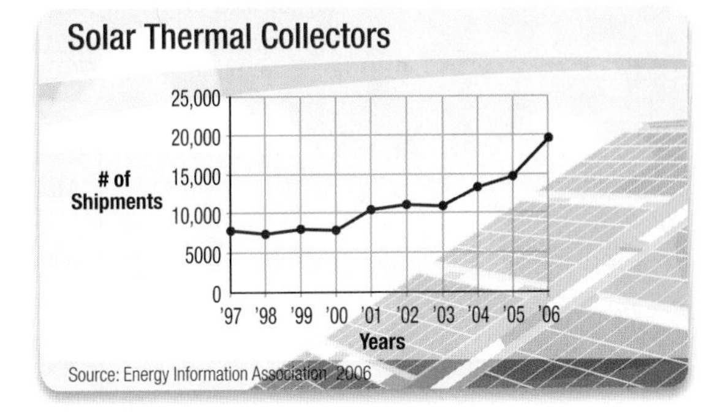

2. Average age of first-time mothers in the U.S. since 1970.

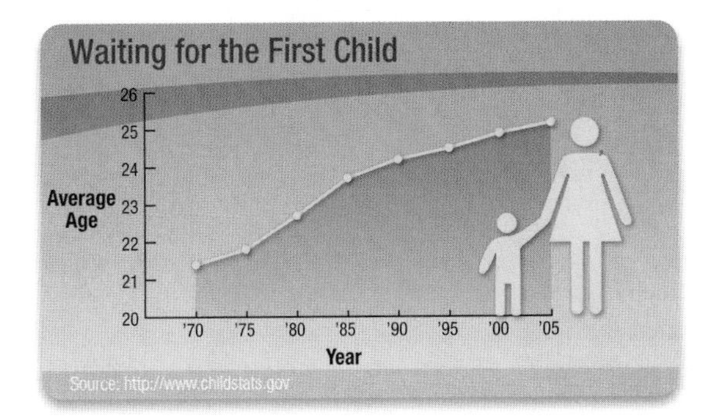

3. Manatee Accidents The number of manatees killed by watercraft in Florida waters has been increasing since 1975. Data are given at 5-year intervals in the table. (Source: Florida Fish and Wildlife Conservation Commission)

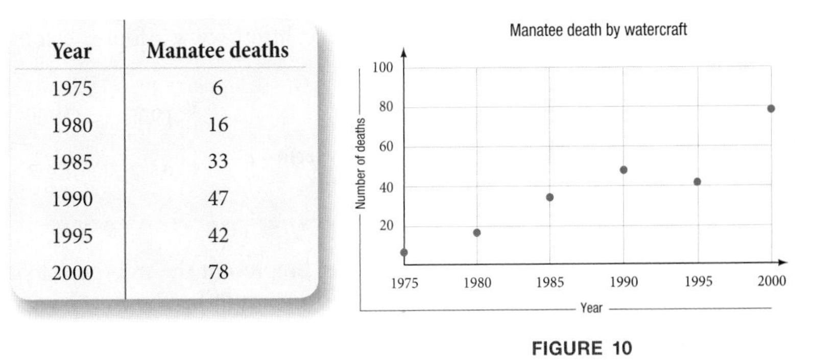

Year	Manatee deaths
1975	6
1980	16
1985	33
1990	47
1995	42
2000	78

FIGURE 10

a. Draw a regression line through the data points shown in Figure 10.

b. Find an equation for the regression line, using $t = 0$ in 1975.

c. Use the regression equation to estimate the number of manatees killed by watercraft in 1998.

4. **Brain Metabolism** Human brains consume a large amount of energy, about 16 times as much as muscle tissue per unit weight. In fact, brain metabolism accounts for about 25% of an adult human's energy needs, as compared to about 5% for other mammals. As hominid species evolved, their brains required larger and larger amounts of energy, as shown in Figure 11. (Source: *Scientific American*, December 2002)

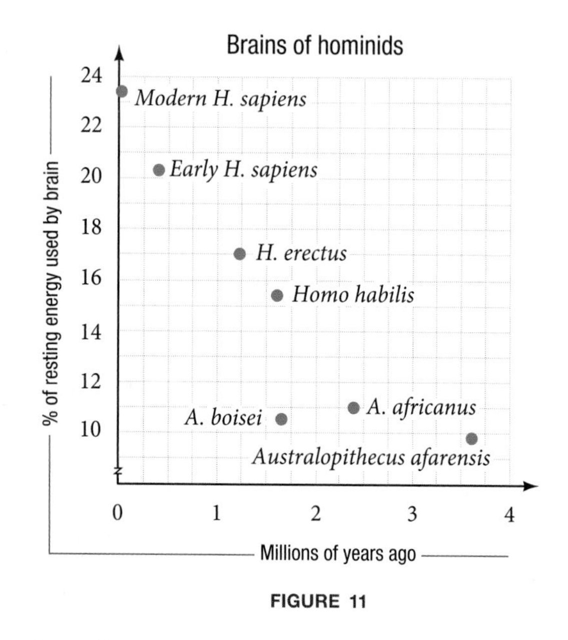

FIGURE 11

a. Draw a line of best fit through the data points.
b. Estimate the amount of energy used by the brain of a hominid species that lived three million years ago.

5. **Life Expectancy** Life expectancy in the United States has been rising since the nineteenth century. The table shows the U.S. life expectancy in selected years. (Source: http://www.infoplease.com)

Year	1950	1960	1970	1980	1990	2000
Life expectancy at birth	68.2	69.7	70.8	73.7	75.4	77

a. Let t represent the number of years after 1950, and plot the data. Draw a line of best fit for the data points.
b. Find an equation for your regression line.
c. Estimate the life expectancy of someone born in 1987.
d. Predict the life expectancy of someone born in 2010.

6. **Earning Potential** "The earnings gap between high-school and college graduates continues to widen, the Census Bureau says. On average, college graduates now earn just over $51,000 a year, almost twice as much as high-school graduates. And those with no high-school diploma have actually seen their earnings drop in recent years." The table shows the unemployment rate and the median weekly earnings for employees with different levels of education. (Source: Morning Edition, National Public Radio, March 28, 2005)

	Years of education	Weekly earnings ($)
Some high school, no diploma	10	396
High-school graduate	12	554
Some college, no degree	13	622
Associate's degree	14	672
Bachelor's degree	16	900
Master's degree	18	1,064
Professional degree	20	1,307

   a. Plot years of education on the horizontal axis and weekly earnings on the vertical axis.
   b. Find an equation for the regression line.
   c. State the slope of the regression line, including units, and explain what it means in the context of the data.
   d. Do you think this model is useful for extrapolation or interpolation? For example, what weekly earnings does the model predict for someone with 15 years of education? For 25 years? Do you think these predictions are valid? Why or why not?

7. **Radiation and Mutations** High-frequency radiation is harmful to living things because it can cause changes in their genetic material. The data below, collected by C. P. Oliver in 1930, show the frequency of genetic transmutations induced in fruit flies by doses of X-rays, measured in roentgens. (Source: C. P. Oliver, 1930)

© bncc369/iStockphoto

Dosages (roentgens)	285	570	1,640	3,280	6,560
Percent of mutated genes	1.18	2.99	4.56	9.63	15.85

   a. Plot the data and draw a line of best fit through the data points.
   b. Find the equation of your regression line.
   c. Use the regression equation to predict the percent of mutations that might result from exposure to 5,000 roentgens of radiation.

8. **Area Codes** With Americans' increased use of faxes, pagers, and cell phones, new area codes are being created at a steady rate. The table shows the number of area codes in the US each year. (Source: USA Today, NeuStar, Inc.)

Year	1997	1998	1999	2000	2001	2002	2003
Number of Area Codes	151	186	204	226	239	262	274

a. Let t represent the number of years after 1997, and plot the data. Draw a line of best fit for the data points.
b. Find an equation for your regression line.
c. How many area codes do you predict for 2010?

9. **Carbon Levels** The table shows the amount of carbon released into the atmosphere annually from burning fossil fuels, in billions of tons, at 5-year intervals from 1950 to 1995. (Source: www.worldwatch.org)

Year	50	55	60	65	70	75	80	85	90	95
Carbon emissions	1.6	2.0	2.5	3.1	4.0	4.5	5.2	5.3	5.9	6.2

a. Let t represent the number of years after 1950, and plot the data.
b. Draw a line of best fit for the data points.
c. Find an equation for your regression line.
d. Use interpolation to estimate the amount of carbon released in 1992.

10. **Chemistry** Six students are trying to identify an unknown chemical compound by heating the substance and measuring the density of the gas that evaporates. (Density = mass/volume.) The students record the mass lost by the solid substance and the volume of the gas that evaporated from it. They know that the mass lost by the solid must be the same as the mass of the gas that evaporated. (Source: Hunt and Sykes, 1984)

Student	A	B	C	D	E	F
Volume of gas (cm^3)	48	60	24	81	76	54
Loss in mass (mg)	64	81	32	107	88	72

a. Plot the data with volume on the horizontal axis. Which student made an error in the experiment?
b. Ignoring the incorrect data point, draw a line of best fit through the other points.
c. Find an equation of the form $y = kx$ for the data. Why should you expect the regression line to pass through the origin?
d. Use your equation to calculate the mass of 1,000 cm^3 (one liter) of the gas.

e. Here are the densities of some gases at room temperature:

Hydrogen 83 mg/liter
Nitrogen 1,160 mg/liter
Oxygen 1,330 mg/liter
Carbon dioxide 1,830 mg/liter

Which of these might have been the gas that evaporated from the unknown substance? (Hint: Use your answer to Part d to calculate the density of the gas. 1 cm^3 = 1 milliliter.)

Using Technology Exercises

11. Predators/Prey Birds' nests are always in danger from predators. If there are other nests close by, the chances of predators finding the nest increase. The table shows the probability of a nest being found by predators and the distance to the nearest neighboring nest. (Source: Perrins, 1979)

Distance to nearest neighbor (meters)	20	40	60	80	100
Probability of predators (%)	47	34	32	17	1.5

a. Plot the data and the least squares regression line.

b. Use the regression line to estimate the probability of predators finding a nest if its nearest neighbor is 50 meters away.

c. If the probability of predators finding a nest is 10%, how far away is its nearest neighbor?

d. What is the probability of predators finding a nest if its nearest neighbor is 120 meters away? Is your answer reasonable?

12. Bicycle Speed A trained cyclist pedals faster as he increases his cycling speed, even with a multiple-gear bicycle. The table shows the pedal frequency, p (in revolutions per minute), and the cycling speed, c (in kilometers per hour), of one cyclist. (Source: Pugh, 1974)

© Ljupco/iStockphoto

Speed (km/hr)	8.8	12.5	16.2	24.4	31.9	35.0
Pedal frequency (rpm)	44.5	50.7	60.6	77.9	81.9	95.3

a. Plot the data and the least squares regression line.

b. Estimate the cyclist's pedal frequency at a speed of 20 kilometers per hour.

c. Estimate the cyclist's speed when he is pedaling at 70 revolutions per minute.

d. Does your regression line give a reasonable prediction for the pedaling frequency when the cyclist is not moving? Explain.

13. Use your graphing calculator, or any technology you choose, to find the least squares regression line for the data in the graph. Plot the points and the regression line on graph paper so it looks similar to the graph below (if x is defined in years since 1980).

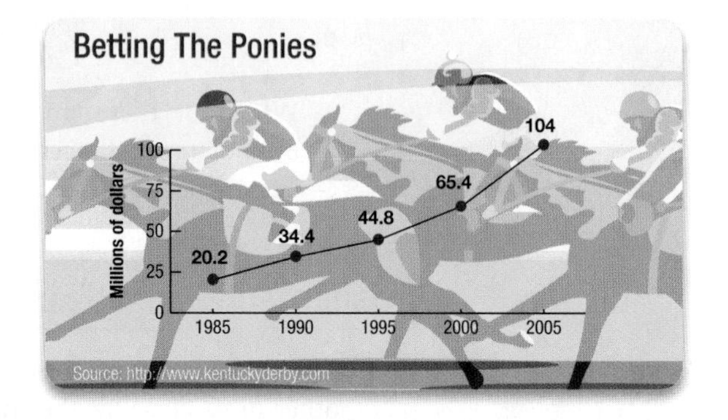

14. Below is the table that shows the relationship between the two temperature scales. We already know this relationship is linear. For comparison, use your graphing calculator, or any technology you choose, to find the least squares regression line for the data in the table. Compare your results with the equation that gives the linear relationship between the two scales.

Degrees Celsius	Degrees Fahrenheit
0	32
25	77
50	122
75	167
100	212

Getting Ready for the Next Section

Simplify.

15. $16(3.5)^2$

16. $\dfrac{2{,}400}{100}$

17. $\dfrac{180}{45}$

18. $4(2)(4)^2$

19. $\dfrac{0.0005(200)}{(0.25)^2}$

20. $\dfrac{0.2(0.5)^2}{100}$

21. If $y = Kx$, find K if $x = 5$ and $y = 15$.

22. If $d = Kt^2$, find K if $t = 2$ and $d = 64$.

23. If $P = \frac{K}{V}$, find K if $P = 48$ and $V = 50$.

24. If $y = Kxz^2$, find K if $x = 5$, $z = 3$, and $y = 180$.

Variation

INTRODUCTION If you are a runner and you average t minutes for every mile you run during one of your workouts, then your speed s in miles per hour is given by the equation and graph shown here. The graph (Figure 1) is shown in the first quadrant only because both t and s are positive.

$$s = \frac{60}{t}$$

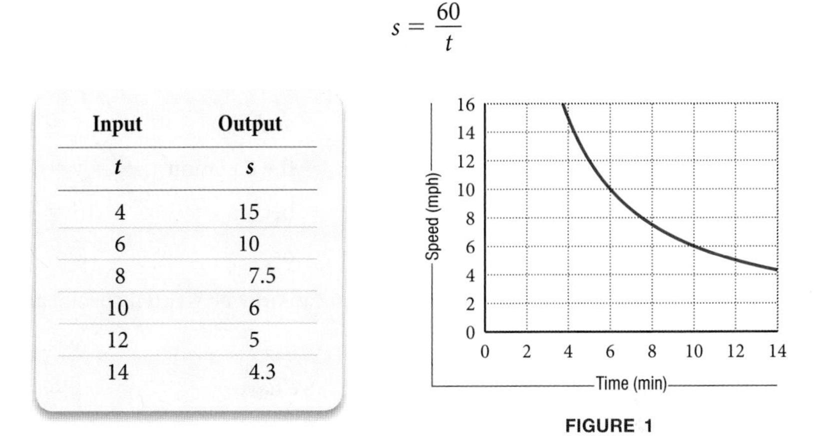

Input t	Output s
4	15
6	10
8	7.5
10	6
12	5
14	4.3

FIGURE 1

You know intuitively that as your average time per mile t increases, your speed s decreases. Likewise, lowering your time per mile will increase your speed. The table and Figure 1 also show this to be true: Increasing t decreases s, and decreasing t increases s. Quantities that are connected in this way are said to *vary inversely* with each other. Inverse variation is one of the topics we will study in this section. There are two main types of variation: *direct variation* and *inverse variation*. Variation problems are most common in the sciences, particularly in chemistry and physics. ■

Direct Variation

When we say the variable y *varies directly* with the variable x, we mean that the relationship can be written in symbols as $y = Kx$, where K is a nonzero constant called the *constant of variation* (or *proportionality constant*).

Another way of saying y varies directly with x is to say y is *directly proportional* to x.

Study the following list. It gives the mathematical equivalent of some direct variation statements.

Verbal Phrase	Algebraic Equation
y varies directly with x.	$y = Kx$
s varies directly with the square of t.	$s = Kt^2$
y is directly proportional to the cube of z.	$y = Kz^3$
u is directly proportional to the square root of v.	$u = K\sqrt{v}$

VIDEO EXAMPLES

SECTION 4.4

EXAMPLE 1 y varies directly with x. If y is 15 when x is 5, find y when x is 7.

SOLUTION The first sentence gives us the general relationship between x and y. The equation equivalent to the statement "y varies directly with x" is

$$y = Kx$$

The first part of the second sentence in our example gives us the information necessary to evaluate the constant K:

When	$y = 15$
and	$x = 5$
the equation	$y = Kx$
becomes	$15 = K \cdot 5$
or	$K = 3$

The equation can now be written specifically as

$$y = 3x$$

Letting $x = 7$, we have

$$y = 3 \cdot 7$$
$$y = 21$$

Modeling: Skydiving

EXAMPLE 2 A skydiver jumps from a plane. Like any object that falls toward earth, the distance the skydiver falls is directly proportional to the square of the time he has been falling, until he reaches his terminal velocity. If the skydiver falls 64 feet in the first 2 seconds of the jump, then

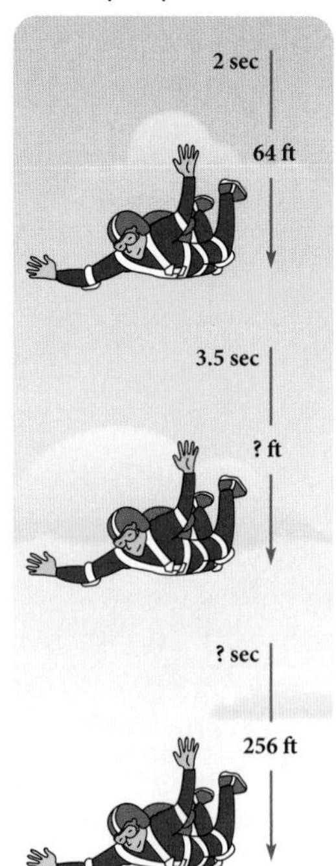

a. How far will he have fallen after 3.5 seconds?
b. Graph the relationship between distance and time.
c. How long will it take him to fall 256 feet?

SOLUTION We let t represent the time the skydiver has been falling, then we can let $d(t)$ represent the distance he has fallen.

a. Since $d(t)$ is directly proportional to the square of t, we have the general function that describes this situation:

$$d(t) = Kt^2$$

Next, we use the fact that $d(2) = 64$ to find K.

$$64 = K(2)^2$$
$$K = 16$$

The specific equation that describes this situation is

$$d(t) = 16t^2$$

To find how far a skydiver will fall after 3.5 seconds, we find $d(3.5)$,

$$d(3.5) = 16(3.5)^2$$

$$d(3.5) = 196$$

A skydiver will fall 196 feet after 3.5 seconds.

b. To graph this equation, we use a table:

Input	Output
t	$d(t)$
0	0
1	16
2	64
3	144
4	256
5	400

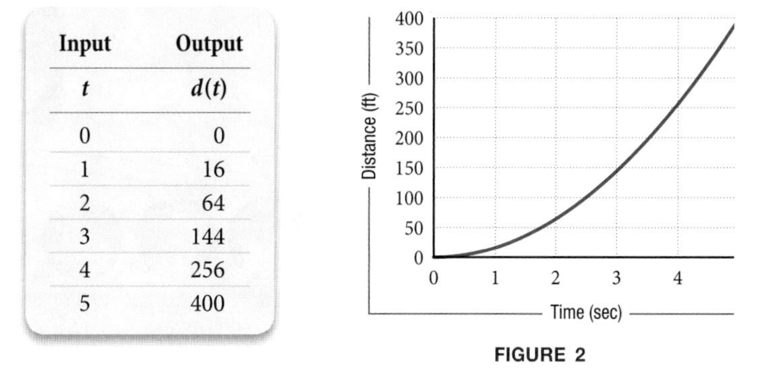

FIGURE 2

c. From the table or the graph (Figure 2), we see that it will take 4 seconds for the skydiver to fall 256 feet.

Inverse Variation

Running

From the introduction to this section, we know that the relationship between the number of minutes t it takes a person to run a mile and his or her average speed in miles per hour s can be described with the following equation and table, and with Figure 3.

$$s = \frac{60}{t}$$

Input	Output
t	s
4	15
6	10
8	7.5
10	6
12	5
14	4.3

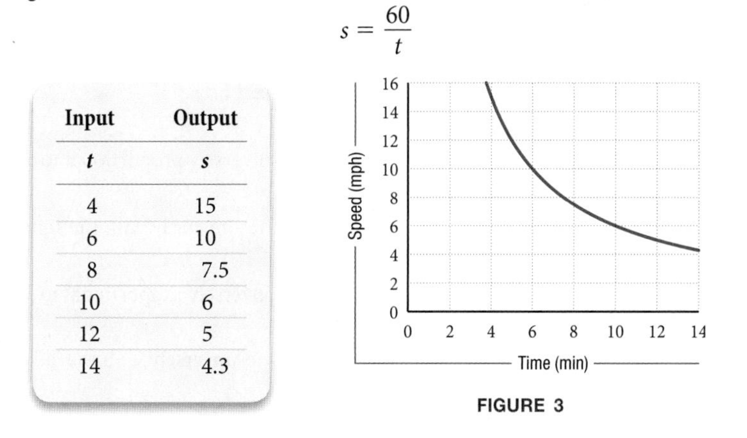

FIGURE 3

If t decreases, then s will increase, and if t increases, then s will decrease. The variable s is *inversely proportional* to the variable t. In this case, the *constant of proportionality* is 60.

Photography

If you are familiar with the terminology and mechanics associated with photography, you know that the f-stop for a particular lens will increase as the aperture (the maximum diameter of the opening of the lens) decreases. In mathematics, we say that f-stop and aperture vary inversely with each other. The following diagram illustrates this relationship.

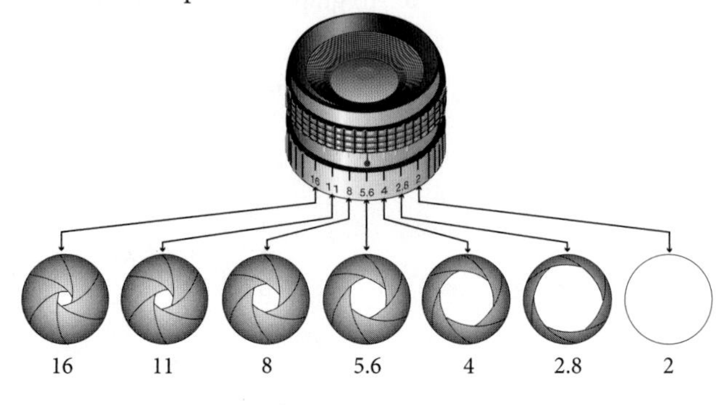

If f is the f-stop and d is the aperture, then their relationship can be written

$$f = \frac{K}{d}$$

In this case, K is the constant of proportionality. (Those of you familiar with photography know that K is also the focal length of the camera lens.)

In General

We generalize this discussion of inverse variation as follows: If y varies inversely with x, then

$$y = K\frac{1}{x} \qquad \text{or} \qquad y = \frac{K}{x}$$

We can also say y is inversely proportional to x. The constant K is again called the constant of variation or proportionality constant.

Verbal Phrase	**Algebraic Equation**
y is inversely proportional to x.	$y = \frac{K}{x}$
s varies inversely with the square of t.	$s = \frac{K}{t^2}$
y is inversely proportional to x^4.	$y = \frac{K}{x^4}$
z varies inversely with the cube root of t.	$z = \frac{K}{\sqrt[3]{t}}$

Modeling: Volume and Pressure

EXAMPLE 3 The volume of a gas is inversely proportional to the pressure of the gas on its container. If a pressure of 48 pounds per square inch corresponds to a volume of 50 cubic feet, what pressure is needed to produce a volume of 100 cubic feet?

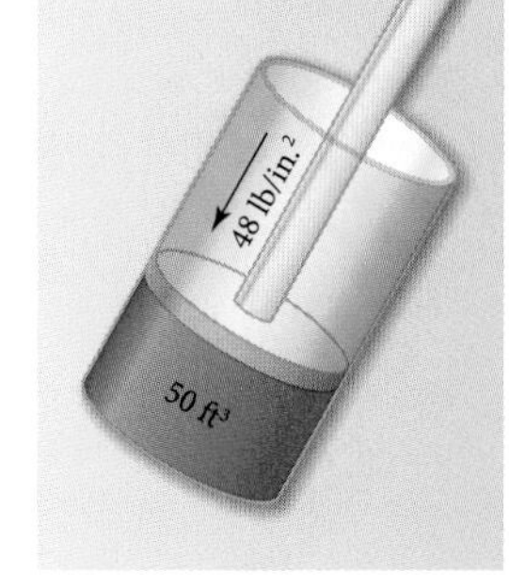

SOLUTION We can represent volume with V and pressure with P:

$$V = \frac{K}{P}$$

Using $P = 48$ and $V = 50$, we have

$$50 = \frac{K}{48}$$

$$K = 50(48)$$

$$K = 2{,}400$$

The equation that describes the relationship between P and V is

$$V = \frac{2{,}400}{P}$$

Here is a graph of this relationship.

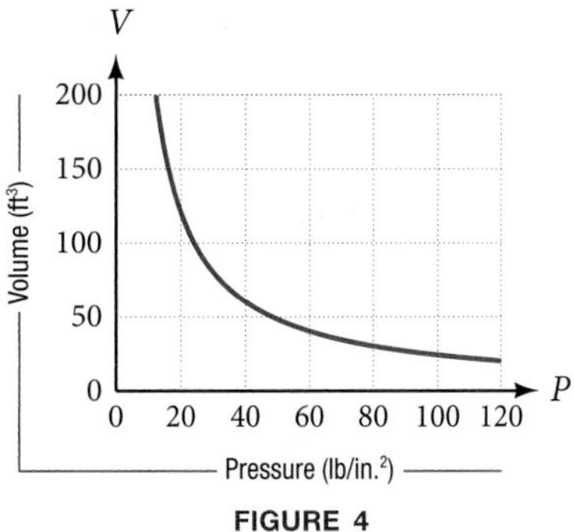

FIGURE 4

Note The relationship between pressure and volume as given in this example is known as Boyle's law and applies to situations such as those encountered in a piston-cylinder arrangement. It was Robert Boyle (1627–1691) who, in 1662, published the results of some of his experiments that showed, among other things, that the volume of a gas decreases as the pressure increases. This is an example of inverse variation.

Substituting $V = 100$ into our last equation, we get

$$100 = \frac{2{,}400}{P}$$

$$100P = 2{,}400$$

$$P = \frac{2{,}400}{100}$$

$$P = 24$$

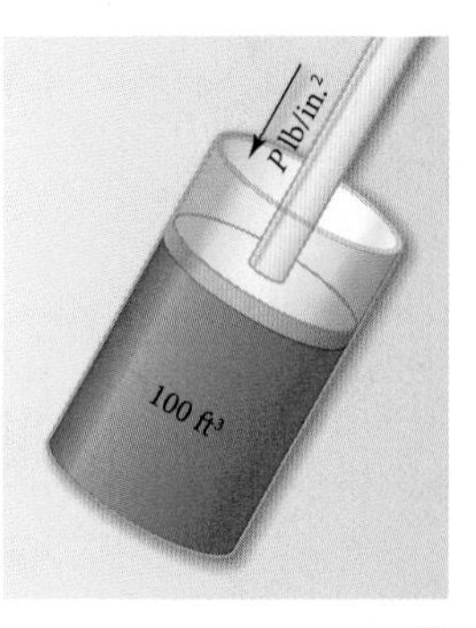

A volume of 100 cubic feet is produced by a pressure of 24 pounds per square inch.

Joint Variation and Other Variation Combinations

Many times relationships among different quantities are described in terms of more than two variables. If the variable y varies directly with *two* other variables, say x and z, then we say y varies *jointly* with x and z. In addition to *joint variation*, there are many other combinations of direct and inverse variation involving more than two variables. The following table is a list of some variation statements and their equivalent mathematical forms:

Verbal Phrase	Algebraic Equation
y varies jointly with x and z.	$y = Kxz$
z varies jointly with r and the square of s.	$z = Krs^2$
V is directly proportional to T and inversely proportional to P.	$V = \frac{KT}{P}$
F varies jointly with m_1 and m_2 and inversely with the square of r.	$F = \frac{Km_1m_2}{r^2}$

EXAMPLE 4 y varies jointly with x and the square of z. When x is 5 and z is 3, y is 180. Find y when x is 2 and z is 4.

SOLUTION The general equation is given by

$$y = Kxz^2$$

Substituting $x = 5$, $z = 3$, and $y = 180$, we have

$$180 = K(5)(3)^2$$

$$180 = 45K$$

$$K = 4$$

The specific equation is

$$y = 4xz^2$$

When $x = 2$ and $z = 4$, the last equation becomes

$$y = 4(2)(4)^2$$

$$y = 128$$

Modeling: Electricity

EXAMPLE 5 In electricity, the resistance of a cable is directly proportional to its length and inversely proportional to the square of the diameter. If a 100-foot cable 0.5 inch in diameter has a resistance of 0.2 ohm, what will be the resistance of a cable made from the same material if it is 200 feet long with a diameter of 0.25 inch?

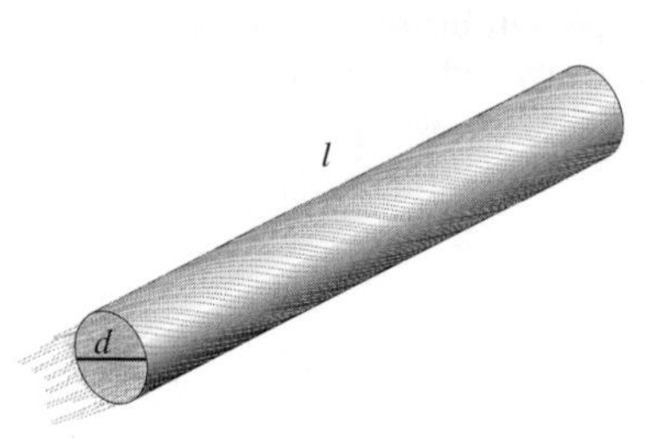

SOLUTION Let R = resistance, l = length, and d = diameter. The equation is

$$R = \frac{Kl}{d^2}$$

When $R = 0.2$, $l = 100$, and $d = 0.5$, the equation becomes

$$0.2 = \frac{K(100)}{(0.5)^2}$$

or

$$K = 0.0005$$

Using this value of K in our original equation, the result is

$$R = \frac{0.0005l}{d^2}$$

When $l = 200$ and $d = 0.25$, the equation becomes

$$R = \frac{0.0005(200)}{(0.25)^2}$$

$$R = 1.6 \text{ ohms}$$

GETTING READY FOR CLASS

After reading through the preceding section, respond in your own words and in complete sentences.

A. Give an example of a direct variation statement, and then translate it into symbols.

B. Translate the equation $y = \frac{K}{x}$ into words.

C. For the inverse variation equation $y = \frac{3}{x}$, what happens to the values of y as x gets larger?

D. How are direct variation statements and linear equations in two variables related?

Problem Set 4.4

For the following problems, y varies directly with x.

1. If y is 10 when x is 2, find y when x is 6.
2. If y is -32 when x is 4, find x when y is -40.

For the following problems, r is inversely proportional to s.

3. If r is -3 when s is 4, find r when s is 2.
4. If r is 8 when s is 3, find s when r is 48.

For the following problems, d varies directly with the square of r.

5. If $d = 10$ when $r = 5$, find d when $r = 10$.
6. If $d = 12$ when $r = 6$, find d when $r = 9$.

For the following problems, y varies inversely with the square of x.

7. If $y = 45$ when $x = 3$, find y when x is 5.
8. If $y = 12$ when $x = 2$, find y when x is 6.

For the following problems, z varies jointly with x and the square of y.

9. If z is 54 when x and y are 3, find z when $x = 2$ and $y = 4$.
10. If z is 27 when $x = 6$ and $y = 3$, find x when $z = 50$ and $y = 4$.

For the following problems, I varies inversely with the cube of w.

11. If $I = 32$ when $w = \frac{1}{2}$, find I when $w = \frac{1}{3}$.
12. If $I = \frac{1}{25}$ when $w = 5$, find I when $w = 10$.

For the following problems, z varies jointly with y and the square of x.

13. If $z = 72$ when $x = 3$ and $y = 2$, find z when $x = 5$ and $y = 3$.
14. If $z = 240$ when $x = 4$ and $y = 5$, find z when $x = 6$ and $y = 3$.
15. If $x = 1$ when $z = 25$ and $y = 5$, find x when $z = 160$ and $y = 8$.
16. If $x = 4$ when $z = 96$ and $y = 2$, find x when $z = 108$ and $y = 1$.

For the following problems, F varies directly with m and inversely with the square of d.

17. If $F = 150$ when $m = 240$ and $d = 8$, find F when $m = 360$ and $d = 3$.
18. If $F = 72$ when $m = 50$ and $d = 5$, find F when $m = 80$ and $d = 6$.
19. If $d = 5$ when $F = 24$ and $m = 20$, find d when $F = 18.75$ and $m = 40$.
20. If $d = 4$ when $F = 75$ and $m = 20$, find d when $F = 200$ and $m = 120$.

Modeling Practice

21. **Length of a Spring** The length a spring stretches is directly proportional to the force applied. If a force of 5 pounds stretches a spring 3 inches, how much force is necessary to stretch the same spring 10 inches?

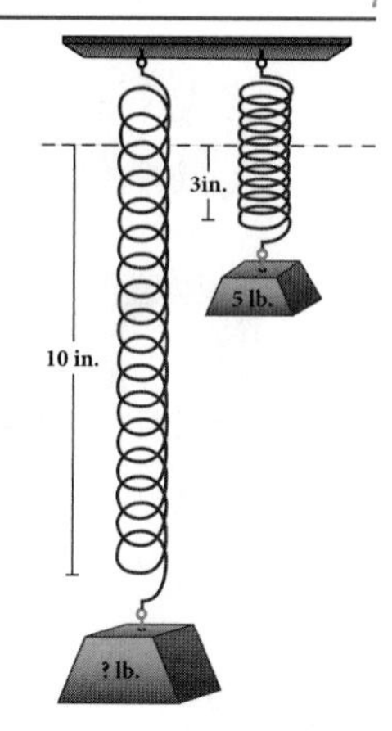

22. **Weight and Surface Area** The weight of a certain material varies directly with the surface area of that material. If 8 square feet weighs half a pound, how much will 10 square feet weigh?

23. **Pressure and Temperature** The temperature of a gas varies directly with its pressure. A temperature of 200 K produces a pressure of 50 pounds per square inch.
    a. Find the equation that relates pressure and temperature.
    b. Graph the equation from Part a in the first quadrant only.
    c. What pressure will the gas have at 280 K?

24. **Circumference and Diameter** The circumference of a wheel is directly proportional to its diameter. A wheel has a circumference of 8.5 feet and a diameter of 2.7 feet.
    a. Find the equation that relates circumference and diameter.
    b. Graph the equation from Part a in the first quadrant only.
    c. What is the circumference of a wheel that has a diameter of 11.3 feet?

25. **Volume and Pressure** The volume of a gas is inversely proportional to the pressure. If a pressure of 36 pounds per square inch corresponds to a volume of 25 cubic feet, what pressure is needed to produce a volume of 75 cubic feet?

26. **Wavelength and Frequency** The frequency of an electromagnetic wave varies inversely with the wavelength. If a wavelength of 200 meters has a frequency of 800 kilocycles per second, what frequency will be associated with a wavelength of 500 meters?

27. **f-Stop and Aperture Diameter** The relative aperture, or *f*-stop, for a camera lens is inversely proportional to the diameter of the aperture. An *f*-stop of 2 corresponds to an aperture diameter of 40 millimeters for the lens on an automatic camera.
    a. Find the equation that relates *f*-stop and diameter.
    b. Graph the equation from Part a in the first quadrant only.
    c. What is the *f*-stop of this camera when the aperture diameter is 10 millimeters?

© Mirjana Jovic/iStockphoto

28. **f-Stop and Aperture Diameter** The relative aperture, or *f*-stop, for a camera lens is inversely proportional to the diameter of the aperture. An *f*-stop of 2.8 corresponds to an aperture diameter of 75 millimeters for a certain telephoto lens.
    a. Find the equation that relates *f*-stop and diameter.
    b. Graph the equation from Part a in the first quadrant only.
    c. What aperture diameter corresponds to an *f*-stop of 5.6?

29. **Surface Area of a Cylinder** The surface area of a hollow cylinder varies jointly with the height and radius of the cylinder. If a cylinder with radius 3 inches and height 5 inches has a surface area of 94 square inches, what is the surface area of a cylinder with radius 2 inches and height 8 inches?

30. **Capacity of a Cylinder** The capacity of a cylinder varies jointly with its height and the square of its radius. If a cylinder with a radius of 3 centimeters and a height of 6 centimeters has a capacity of 169.56 cubic centimeters, what will be the capacity of a cylinder with radius 4 centimeters and height 9 centimeters?

31. **Electrical Resistance** The resistance of a wire varies directly with its length and inversely with the square of its diameter. If 100 feet of wire with diameter 0.01 inch has a resistance of 10 ohms, what is the resistance of 60 feet of the same type of wire if its diameter is 0.02 inch?

32. **Volume and Temperature** The volume of a gas varies directly with its temperature and inversely with the pressure. If the volume of a certain gas is 30 cubic feet at a temperature of 300 K and a pressure of 20 pounds per square inch, what is the volume of the same gas at 340 K when the pressure is 30 pounds per square inch?

33. **Period of a Pendulum** The time it takes for a pendulum to complete one period varies directly with the square root of the length of the pendulum. A 100-centimeter pendulum takes 2.1 seconds to complete one period.
    a. Find the equation that relates period and pendulum length.
    b. Graph the equation from Part a in quadrant I only.
    c. How long does it take to complete one period if the pendulum hangs 225 centimeters?

Maintaining Your Skills

For each of the following straight lines, identify the x-intercept, y-intercept, and slope, and sketch the graph.

34. $2x + y = 6$

35. $y = \frac{3}{2}x + 4$

36. $x = -2$

Find the equation for each line.

37. Give the equation of the line through $(-1, 3)$ that has slope $m = 2$.

38. Give the equation of the line through $(-3, 2)$ and $(4, -1)$.

39. Line l contains the point $(5, -3)$ and has a graph parallel to the graph of $2x - 5y = 10$. Find the equation for l.

40. Give the equation of the vertical line through $(4, -7)$.

State the domain and range for the following relations, and indicate which relations are also functions.

41. $\{(-2, 0), (-3, 0), (-2, 1)\}$

42. $y = x^2 - 9$

Let $f(x) = x - 2$, $g(x) = 3x + 4$, and $h(x) = 3x^2 - 2x - 8$, and find the following.

43. $f(3) + g(2)$

44. $h(0) + g(0)$

45. $f[g(2)]$

46. $g[f(2)]$

SPOTLIGHT ON SUCCESS *Student Instructor CJ*

We are what we repeatedly do.
Excellence, then, is not an act, but a habit.
—Aristotle

Something that has worked for me in college, in addition to completing the assigned homework, is working on some extra problems from each section. Working on these extra problems is a great habit to get into because it helps further your understanding of the material, and you see the many different types of problems that can arise. If you have completed every problem that your book offers, and you still don't feel confident that you have a full grasp of the material, look for more problems. Many problems can be found online or in other books. Your professors may even have some problems that they would suggest doing for extra practice. The biggest benefit to working all the problems in the course's assigned textbook is that often teachers will choose problems either straight from the book or ones similar to problems that were not assigned for tests. Doing this will ensure that you do your best in all your classes.

Chapter 4 Summary

Linear Equations in Two Variables [4.1]

EXAMPLES

1. The equation $3x + 2y = 6$ is an example of a linear equation in two variables.

A *linear equation in two variables* is any equation that can be put in *standard form* $ax + by = c$ where a and b are not both zero. The graph of every linear equation is a straight line.

The Slope of a Line [4.1]

2. The slope of the line through $(1, -1)$ and $(6, 9)$ is

$$m = \frac{9 - (-1)}{6 - 1} = \frac{10}{5} = 2$$

The *slope* of the line containing points (x_1, y_1) and (x_2, y_2) is given by

$$\text{Slope} = m = \frac{\text{Rise}}{\text{Run}} = \frac{y_2 - y_1}{x_2 - x_1}$$

Horizontal lines have 0 slope, and vertical lines have no slope.
Parallel lines have equal slopes, and perpendicular lines have slopes that are negative reciprocals.

Difference Quotients [4.1]

3. If $f(x) = x^2$, then

$$\frac{f(x) - f(a)}{x - a} = \frac{x^2 - a^2}{x - a}$$

$$= \frac{(x + a)(x - a)}{x - a}$$

$$= x + a$$

Each of the following expressions is called a *difference quotient*.

$$\frac{f(x) - f(a)}{x - a} \qquad \frac{f(x + h) - f(x)}{h}$$

The Slope-Intercept Form of a Line [4.2]

4. The equation of the line with slope 5 and y-intercept 3 is

$$y = 5x + 3$$

The equation of a line with slope m and y-intercept b is given by

$$y = mx + b$$

The Point-Slope Form of a Line [4.2]

5. The equation of the line through $(3, 2)$ with slope -4 is

$$y - 2 = -4(x - 3)$$

which can be simplified to

$$y = -4x + 14$$

The equation of the line through (x_1, y_1) that has slope m can be written as

$$y - y_1 = m(x - x_1)$$

Piecewise-Defined Functions [4.2]

6. The following is a piecewise-defined function.

$$f(x) = \begin{cases} x & \text{if } x \le 0 \\ x^2 & \text{if } x > 0 \end{cases}$$

A function that is defined by more than one formula for different values in its domain is a piecewise function.

Linear Regression [4.3]

7.

When we approximate a linear pattern with a straight line, that line is called the *line of best fit* or *regression line*. If we use that line to predict values that lie within our original data, we are using *interpolation*. If we use the regression line to make predictions that lie outside our original data we are using *extrapolation*. We can use our graphing calculators to find regression lines that are found using the *least squares* method.

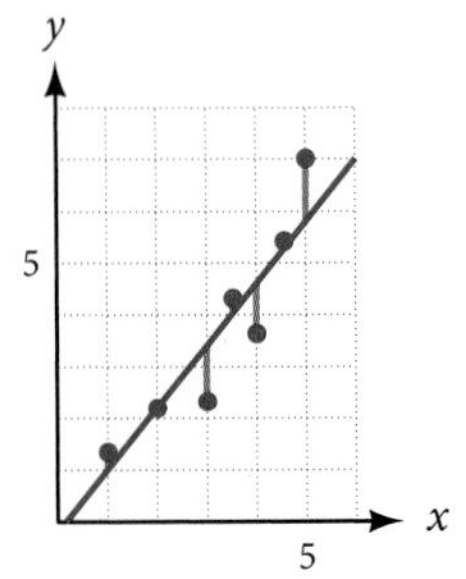

Variation [4.4]

8. If y varies directly with x, then

$$y = Kx$$

Then if y is 18 when x is 6,

$$18 = K \cdot 6$$

or

$$K = 3$$

So the equation can be written more specifically as

$$y = 3x$$

If we want to know what y is when x is 4, we simply substitute:

$$y = 3 \cdot 4$$
$$y = 12$$

If y *varies directly* with x (y is directly proportional to x), then

$$y = Kx$$

If y *varies inversely* with x (y is inversely proportional to x), then

$$y = \frac{K}{x}$$

If z *varies jointly* with x and y (z is directly proportional to both x and y), then

$$z = Kxy$$

In each case, K is called the *constant of variation*.

Chapter 4 Test

For each of the following straight lines, identify the x-intercept, y-intercept, and slope, and sketch the graph. [4.1–4.2]

1. $2x + y = 6$ **2.** $y = -2x - 3$ **3.** $y = \frac{3}{2}x + 4$ **4.** $x = -2$

Find the following difference quotients. [4.1]

5. If $f(x) = \frac{1}{2}x + 1$, find $\frac{f(x) - f(a)}{x - a}$.

6. If $f(x) = x^2 + 1$, find $\frac{f(x + h) - f(x)}{h}$.

7. If $f(x) = 3x - 2$, find $\frac{f(5) - f(2)}{5 - 2}$.

8. If $f(x) = 3x^2$, find $\frac{f(1) - f(-1)}{1 - (-1)}$.

Find the equation for each line. [4.2]

9. Give the equation of the line through $(-1, 3)$ that has slope $m = 2$.

10. Give the equation of the line through $(-3, 2)$ and $(4, -1)$.

11. Line l contains the point $(5, -3)$ and has a graph parallel to the graph of $2x - 5y = 10$. Find the equation for l.

12. Line l contains the point $(-1, -2)$ and has a graph perpendicular to the graph of $y = 3x - 1$. Find the equation for l.

13. Give the equation of the vertical line through $(4, -7)$.

Graph each function. [4.2]

14. $f(x) = \begin{cases} x & \text{if } x \le 0 \\ 1 & \text{if } x > 0 \end{cases}$

15. $f(x) = \begin{cases} x - 2 & \text{if } x \le 2 \\ (x - 2)^2 & \text{if } x > 2 \end{cases}$

© Irina Kozhemyakina/iStockphoto

16. Turtle Population Decline Zoologists in Costa Rica have been counting the number of leatherback turtles that nest in Playa Grande each year. The results are displayed in the table and scatter plot below. [4.3]

Year	Nestings
1988	1,367
1989	1,340
1990	665
1991	770
1992	909
1993	180
1994	506
1995	421
1996	125
1997	195
1998	117
1999	110
2000	88

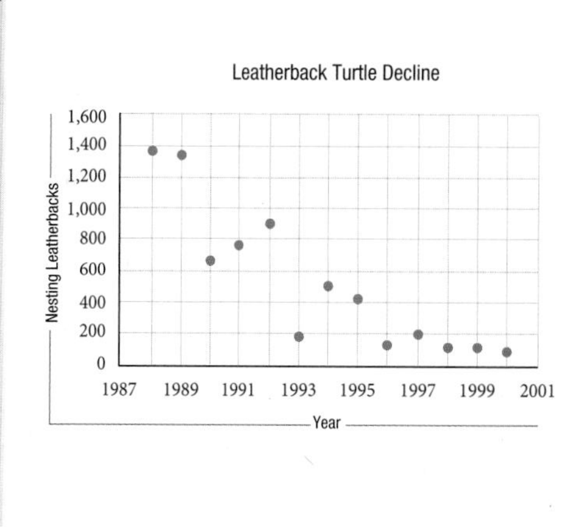

a. Draw a line that connects the first point and last points on your scatter diagram.

b. Find the slope of this line.

c. Find the equation of this line. Let $x = 0$ for 1987.

d. Do you think this line gives a good representation of the trend shown in the data?

e. Draw a line on your scatter diagram that gives a better representation of the data in your scatter diagram.

17. Mobile Homes The number of mobile homes in the United States has been increasing since 1960. The data in the table are given in millions of mobile homes. (Source: *USA Today*, U.S. Census Bureau)

Year	Number of mobile homes
1960	0.8
1970	2.1
1980	4.7
1990	7.4
2000	8.8

a. Let t represent the number of years after 1960 and plot the data. Draw a line of best fit for the data points.

b. Find an equation for your regression line.

Solve the following variation problems. [4.4]

18. Direct Variation Quantity y varies directly with the square of x. If y is 50 when x is 5, find y when x is 3.

19. Joint Variation Quantity z varies jointly with x and the cube of y. If z is 15 when x is 5 and y is 2, find z when x is 2 and y is 3.

20. Maximum Load The maximum load (L) a horizontal beam can safely hold varies jointly with the width (w) and the square of the depth (d) and inversely with the length (l). If a 10-foot beam with width 3 feet and depth 4 feet will safely hold up to 800 pounds, how many pounds will a 12-foot beam with width 3 feet and depth 4 feet hold?

Polynomials and Rational Functions

5

Chapter Outline

© Карина Чекарева/iStockPhoto

If you have ever put yourself on a weight loss diet, you know that you lose more weight at the beginning of the diet than you do later. If we let $W(x)$ represent a person's weight after x weeks on the diet, then the rational function

$$W(x) = \frac{80(2x + 15)}{x + 6}$$

is a mathematical model of the person's weekly progress on a diet intended to take them from 200 pounds to about 160 pounds. Rational functions are good models for quantities that fall off rapidly to begin with, and then level off over time. The table shows some values for this function, along with the graph of this function.

Note When you see this icon next to an example or problem in this chapter, you will know that we are using the topics in this chapter to model situations in the world around us.

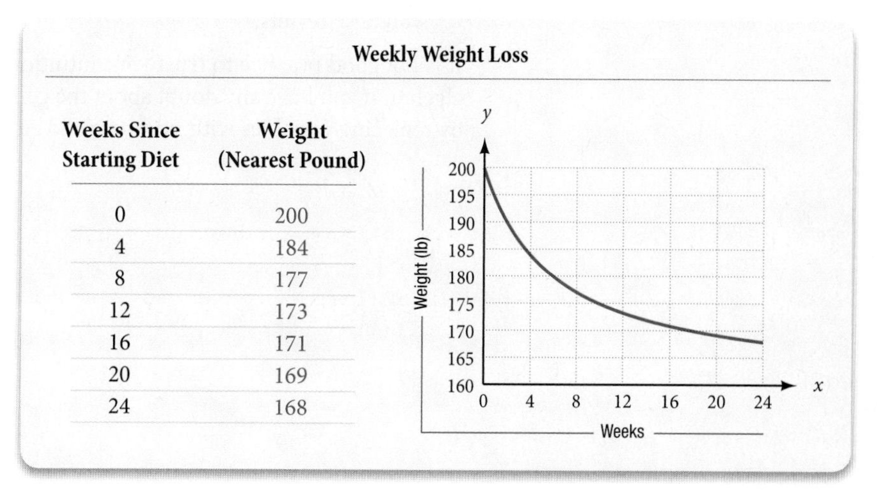

Weekly Weight Loss

Weeks Since Starting Diet	Weight (Nearest Pound)
0	200
4	184
8	177
12	173
16	171
20	169
24	168

As you progress through this chapter, you will acquire an intuitive feel for these types of functions, and as a result, you will see why they are good models for situations such as dieting.

Study Skills

The study skills for this chapter cover the way you approach new situations in mathematics. The first study skill is a point of view you hold about your natural instincts for what does and doesn't work in mathematics. The second study skill gives you a way of testing your instincts.

1. **Don't Let Your Intuition Fool You** As you become more experienced and more successful in mathematics you will be able to trust your mathematical intuition. For now, though, it can get in the way of your success. For example, if you ask some students to "subtract 3 from -5" they will answer -2 or 2. Both answers are incorrect, even though they may seem intuitively true. Likewise, some students will expand $(a + b)^2$ and arrive at $a^2 + b^2$, which is incorrect. In both cases, intuition leads directly to the wrong answer.

2. **Test Properties of Which You Are Unsure** From time to time, you will be in a situation where you would like to apply a property or rule, but you are not sure it is true. You can always test a property or statement by substituting numbers for variables. For instance, I always have students that rewrite $(x + 3)^2$ as $x^2 + 9$, thinking that the two expressions are equivalent. The fact that the two expressions are not equivalent becomes obvious when we substitute 10 for x in each one.

   When $x = 10$, the expression $(x + 3)^2$ is $(10 + 3)^2 = 13^2 = 169$

   When $x = 10$, the expression $x^2 + 9 = 10^2 + 9 = 100 + 9 = 109$

   When you test the equivalence of expressions by substituting numbers for the variable, make it easy on yourself by choosing numbers that are easy to work with, such as 10. Don't try to verify the equivalence of expressions by substituting 0, 1, or 2 for the variable, as using these numbers will occasionally give you false results.

It is not good practice to trust your intuition or instincts in every new situation in algebra. If you have any doubt about the generalizations you are making, test them by replacing variables with numbers and simplifying.

5.1 Quadratic Functions

© Bettmann/Corbis/iStockPhoto

INTRODUCTION If you have been to the circus or the county fair recently, you may have witnessed one of the more spectacular acts, the human cannonball. The human cannonball shown in the photograph will reach a height of 70 feet, and travel a distance of 160 feet, before landing in a safety net. In this chapter, we use this information to derive the equation

$$f(x) = -\frac{7}{640}(x - 80)^2 + 70 \quad \text{for } 0 \leq x \leq 160$$

which describes the path flown by this particular cannonball. This function is a *polynomial* function, and, in particular, it is a *quadratic* function. The table and graph below were constructed from this function.

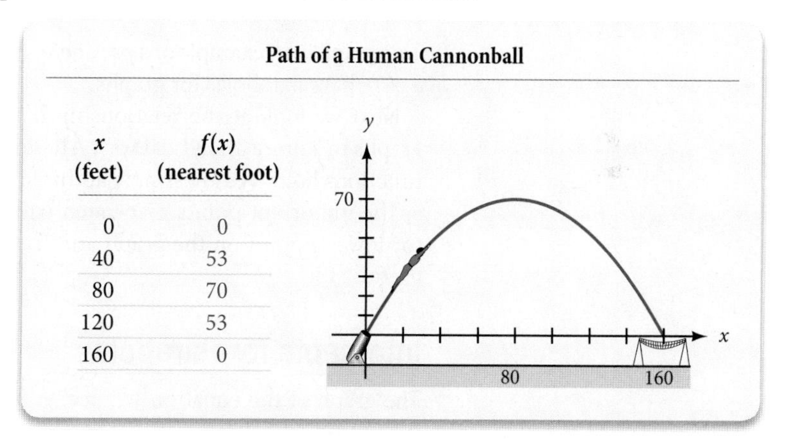

Path of a Human Cannonball

x (feet)	$f(x)$ (nearest foot)
0	0
40	53
80	70
120	53
160	0

All objects that are projected into the air, whether they are basketballs, bullets, arrows, or coins, follow parabolic paths like the one shown in the graph. And, these parabolic graphs are characteristic of quadratic functions. Studying the material in this chapter will give you a more mathematical hold on the world around you.

Quadratic Functions

Recall that the solution set to the equation

$$y = x^2 - 3$$

consists of ordered pairs. One method of graphing the solution set is to find a number of ordered pairs that satisfy the equation and to graph them. We can obtain some ordered pairs that are solutions to $y = x^2 - 3$ by use of a table. We could also use our translation techniques and move the basic graph of $y = x^2$ down 3 units. (See the table and graph on the next page.)

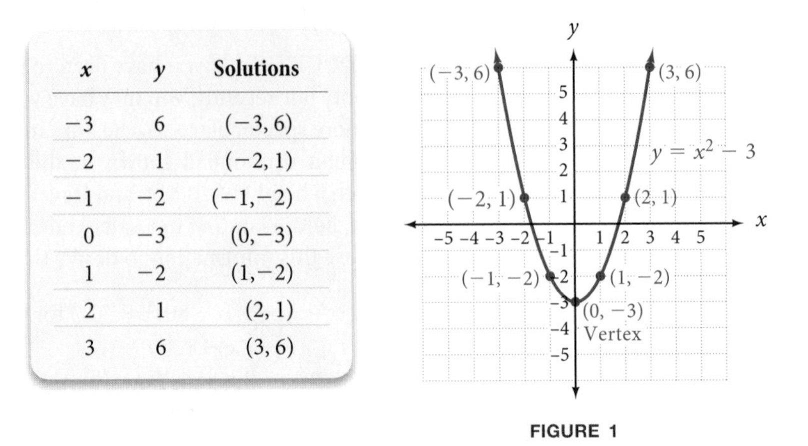

x	y	Solutions
-3	6	$(-3, 6)$
-2	1	$(-2, 1)$
-1	-2	$(-1, -2)$
0	-3	$(0, -3)$
1	-2	$(1, -2)$
2	1	$(2, 1)$
3	6	$(3, 6)$

FIGURE 1

This graph is an example of a *parabola*. All equations of the form $y = ax^2 + bx + c$, $a \neq 0$, have parabolas for graphs.

Next we look at the relationship between our quadratic functions and their graphs in a more analytical way. All our previous methods of graphing quadratic functions hold. We are simply adding new tools to our graphing procedures.

The important points associated with the graph of a parabola are the highest (or lowest) point on the graph and the x-intercepts. The y-intercept can also be useful.

Intercepts for Parabolas

The graph of the equation $y = ax^2 + bx + c$ crosses the y-axis at $y = c$, because substituting $x = 0$ into $y = ax^2 + bx + c$ yields $y = c$.

Because the graph crosses the x-axis when $y = 0$, the x-intercepts are those values of x that are solutions to the quadratic equation $0 = ax^2 + bx + c$.

The Vertex of a Parabola

The highest or lowest point on a parabola is called the *vertex*. The x-coordinate of the vertex for the graph of $y = ax^2 + bx + c$ is

$$x = \frac{-b}{2a}$$

To see this, we must transform the right side of $y = ax^2 + bx + c$ into an expression that contains x in just one of its terms. This is accomplished by completing the square on the first two terms. Here is what it looks like:

$$\begin{aligned}
y &= ax^2 + bx + c \\
&= a\left(x^2 + \frac{b}{a}x\right) + c \\
&= a\left[x^2 + \frac{b}{a}x + \left(\frac{b}{2a}\right)^2\right] + c - a\left(\frac{b}{2a}\right)^2 \\
&= a\left(x + \frac{b}{2a}\right)^2 + \frac{4ac - b^2}{4a}
\end{aligned}$$

It may not look like it, but this last line indicates that the vertex of the graph of $y = ax^2 + bx + c$ has an x-coordinate of $\frac{-b}{2a}$. Because a, b, and c are constants, the only quantity that is varying in the last expression is the x in $\left(x + \frac{b}{2a}\right)^2$. The quantity $\left(x + \frac{b}{2a}\right)^2$ will never be negative, thus the smallest it will ever be is 0, and that will happen when $x = \frac{-b}{2a}$.

We can use the vertex point along with the x and y-intercepts to sketch the graph of any equation of the form $y = ax^2 + bx + c$. Here is a summary of the preceding information.

Note In the box at the right, we show the quadratic formula as a way of finding the x-intercepts, because it always works. If the equation is easily factorable, however, you may wish to find the intercepts by factoring.

$[\Delta \neq \Sigma]$ *Graphing Parabolas I*

The graph of $y = ax^2 + bx + c$, $a \neq 0$, will be a parabola with

1. A y-intercept at $y = c$
2. x-intercepts (if they exist) at

$$x = \frac{-b \pm \sqrt{b^2 - 4ac}}{2a}$$

3. A vertex when $x = \dfrac{-b}{2a}$

VIDEO EXAMPLES

SECTION 5.1

EXAMPLE 1 Sketch the graph of $y = x^2 - 6x + 5$.

SOLUTION Since $c = 5$, the graph has the y-intercept at $y = 5$. To find the x-intercepts, we let $y = 0$ and solve for x:

$$0 = x^2 - 6x + 5$$

$$0 = (x - 5)(x - 1)$$

$$x = 5 \quad \text{or} \quad x = 1$$

To find the coordinates of the vertex, we first find the x-coordinate of the vertex:

$$x = \frac{-b}{2a} = \frac{-(-6)}{2(1)} = 3$$

The x-coordinate of the vertex is 3. To find the y-coordinate, we substitute 3 for x in our original equation:

$$y = 3^2 - 6(3) + 5 = 9 - 18 + 5 = -4$$

The graph crosses the x-axis at 1 and 5 and has its vertex at $(3, -4)$. It also crosses the y-axis at 5. Plotting these points and connecting them with a smooth curve, we have the graph shown in Figure 2.

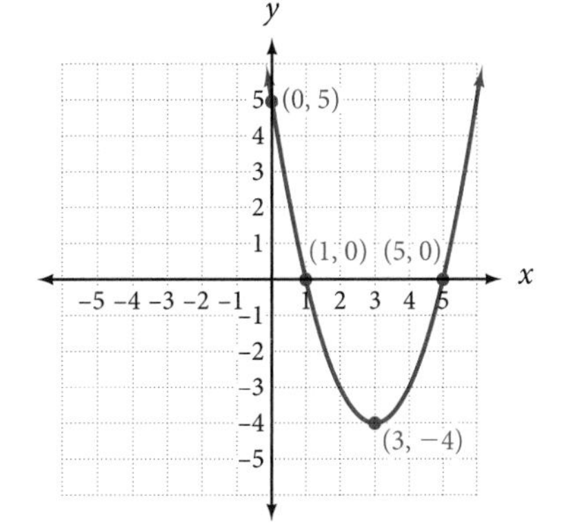

FIGURE 2

The graph is a parabola that opens up, so we say the graph is *concave up*. The vertex is the lowest point on the graph. ■

Finding the Vertex by Completing the Square

Note When we complete the square on a quadratic function, we put it in a form that allows us to use our translation techniques to find the graph.

Another way to locate the vertex of the parabola in Example 1 is by completing the square on the first two terms on the right side of the equation $y = x^2 - 6x + 5$. In this case, we would do so by adding 9 to and subtracting 9 from the right side of the equation. This amounts to adding 0 to the equation, so we know we haven't changed its solutions. This is what it looks like:

$$\begin{aligned} y &= (x^2 - 6x \qquad) + 5 \\ &= (x^2 - 6x + 9) + 5 - 9 \\ &= (x - 3)^2 - 4 \end{aligned}$$

You may have to look at this last equation awhile to see this, but when $x = 3$, then $y = (x - 3)^2 - 4 = 0^2 - 4 = -4$ is the smallest y will ever be. That is why the vertex is at $(3, -4)$. As a matter of fact, this is the same kind of reasoning we used when we derived the formula $x = \frac{-b}{2a}$ for the x-coordinate of the vertex.

EXAMPLE 2 Graph $y = -x^2 - 2x + 3$.

SOLUTION To find the y-intercept, we set $x = 0$ which yields $y = 3$. To find the x-intercepts, we let $y = 0$.

$$\begin{aligned} 0 &= -x^2 - 2x + 3 \\ 0 &= x^2 + 2x - 3 && \text{Multiply each side by } -1 \\ 0 &= (x + 3)(x - 1) \\ x &= -3 \quad \text{or} \quad x = 1 \end{aligned}$$

The x-coordinate of the vertex is given by

$$x = \frac{-b}{2a} = \frac{-(-2)}{2(-1)} = \frac{2}{-2} = -1$$

To find the y-coordinate of the vertex, we substitute -1 for x in our original equation to get

$$y = -(-1)^2 - 2(-1) + 3 = -1 + 2 + 3 = 4$$

Our parabola has a y-intercept at 3, x-intercepts at -3 and 1, and a vertex at $(-1, 4)$. Figure 3 shows the graph.

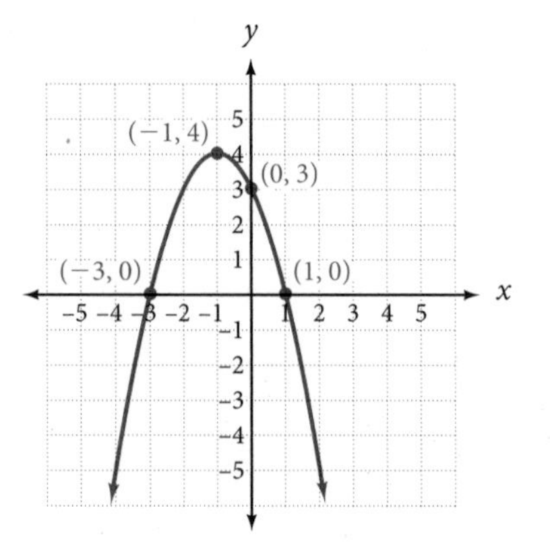

FIGURE 3

We say the graph is *concave down* because it opens downward. Again, we could have obtained the coordinates of the vertex by completing the square on the first two terms on the right side of our equation. To do so, we must first factor -1 from the first two terms. (Remember, the leading coefficient must be 1 to complete the square.) When we complete the square, we add 1 inside the parentheses, which actually decreases the right side of the equation by -1 because everything in the parentheses is multiplied by -1. To make up for it, we add 1 outside the parentheses.

$$\begin{aligned} y &= -1(x^2 + 2x \quad\ ) + 3 \\ &= -1(x^2 + 2x + 1) + 3 + 1 \\ &= -1(x + 1)^2 + 4 \end{aligned}$$

The last line tells us that the *largest* value of y will be 4, and that will occur when $x = -1$. ■

EXAMPLE 3 Graph $y = 3x^2 - 6x + 1$.

SOLUTION We set $x = 0$ to find the y-intercept at $(0, 1)$. To find the x-intercepts, we let $y = 0$ and solve for x:

$$0 = 3x^2 - 6x + 1$$

Because the right side of this equation does not factor, we can look at the discriminant to see what kind of solutions are possible. The discriminant for this equation is

$$b^2 - 4ac = 36 - 4(3)(1) = 24$$

Note The expression under the radical in the quadratic formula is called the *discriminant*. The discriminant indicates the number and type of solutions to a quadratic equation when the original equation has integer coefficients.

Because the discriminant is a positive number but not a perfect square, the equation will have irrational solutions. This means that the x-intercepts are irrational numbers and will have to be approximated with decimals using the quadratic formula. Rather than use the quadratic formula, we will find some other points on the graph, but first let's find the vertex.

Here are both methods of finding the vertex:

Using the formula that gives us the x-coordinate of the vertex, we have:

$$x = \frac{-b}{2a} = \frac{-(-6)}{2(3)} = 1$$

Substituting 1 for x in the equation gives us the y-coordinate of the vertex:

$$y = 3 \cdot 1^2 - 6 \cdot 1 + 1 = -2$$

To complete the square on the right side of the equation, we factor 3 from the first two terms, add 1 inside the parentheses, and add -3 outside the parentheses (this amounts to adding 0 to the right side):

$$\begin{aligned} y &= 3(x^2 - 2x \quad\ ) + 1 \\ &= 3(x^2 - 2x + 1) + 1 - 3 \\ &= 3(x - 1)^2 - 2 \end{aligned}$$

In either case, the vertex is $(1, -2)$.

Finally, since we already have the y-intercept, let's find a point on the opposite side of the vertex from it.

When $x = 2$

$$\begin{aligned} y &= 3(2)^2 - 6(2) + 1 \\ &= 12 - 12 + 1 \\ &= 1 \end{aligned}$$

The point just found is (2, 1). Plotting this point along with the vertex $(1, -2)$ and y-intercept (0, 1), we have the graph shown in Figure 4.

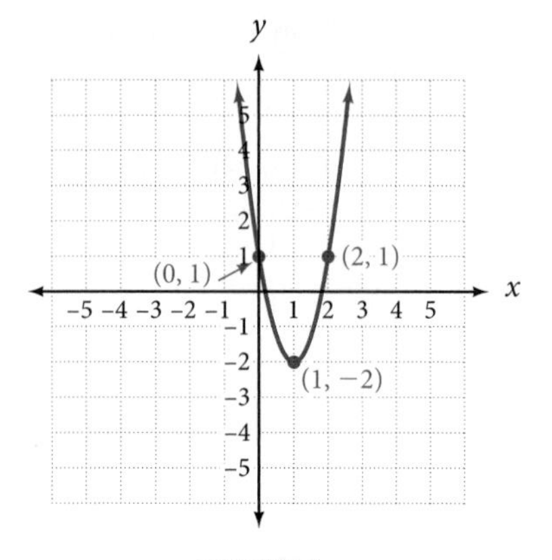

FIGURE 4

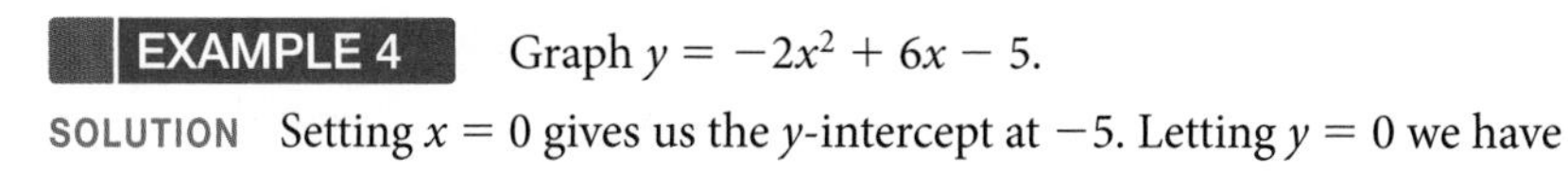

EXAMPLE 4 Graph $y = -2x^2 + 6x - 5$.

SOLUTION Setting $x = 0$ gives us the y-intercept at -5. Letting $y = 0$ we have

$$0 = -2x^2 + 6x - 5$$

Again, the right side of this equation does not factor. The discriminant is $b^2 - 4ac = 36 - 4(-2)(-5) = -4$ which indicates that the solutions are complex numbers. This means that our original equation does not have x-intercepts. The graph does not cross the x-axis.

Let's find the vertex.

Using our formula for the x-coordinate of the vertex, we have:

$$x = \frac{-b}{2a} = \frac{-6}{2(-2)} = \frac{6}{4} = \frac{3}{2}$$

To find the y-coordinate, we let $x = \frac{3}{2}$

$$\begin{aligned} y &= -2\left(\frac{3}{2}\right)^2 + 6\left(\frac{3}{2}\right) - 5 \\ &= \frac{-18}{4} + \frac{18}{2} - 5 \\ &= \frac{-18 + 36 - 20}{4} \\ &= -\frac{1}{2} \end{aligned}$$

Finding the vertex by completing the square is a more complicated matter. To make the coefficient of x^2 a 1, we must factor -2 from the first two terms. To complete the square inside the parentheses, we add $\frac{9}{4}$. Since each term inside the parentheses is multiplied by -2, we add $\frac{9}{2}$ outside the parentheses so that the net result is the same as adding 0 to the right side:

$$\begin{aligned} y &= -2(x^2 - 3x \qquad) - 5 \\ &= -2\left(x^2 - 3x + \frac{9}{4}\right) - 5 + \frac{9}{2} \\ &= -2\left(x - \frac{3}{2}\right)^2 - \frac{1}{2} \end{aligned}$$

The vertex is $\left(\frac{3}{2}, -\frac{1}{2}\right)$. If we let $x = 3$, we will have a point that is the same distance from the vertex as the y-intercept.

When $x = 3$

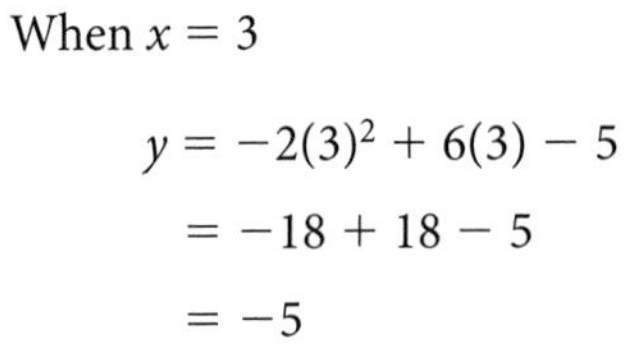

$$\begin{aligned} y &= -2(3)^2 + 6(3) - 5 \\ &= -18 + 18 - 5 \\ &= -5 \end{aligned}$$

The additional point on the graph is $(3, -5)$. Figure 5 shows the graph.

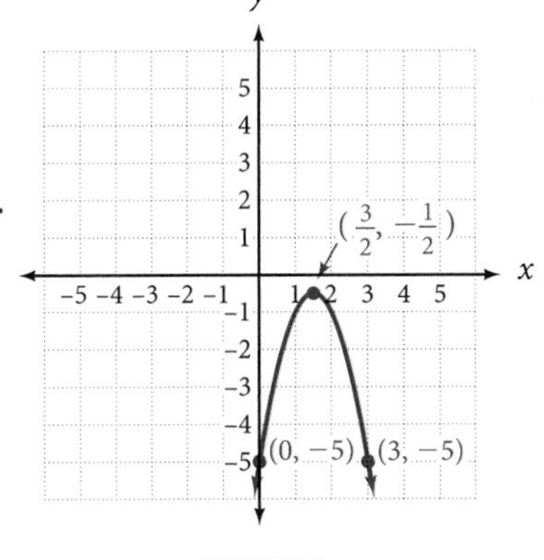

FIGURE 5

The graph is concave down. The vertex is the highest point on the graph. ■

By looking at the equations and graphs in Examples 1 through 4, we can conclude that the graph of $y = ax^2 + bx + c$ will be concave up when a is positive, and concave down when a is negative. Taking this even further, if $a > 0$ then the vertex is the lowest point on the graph, and if $a < 0$ the vertex is the highest point on the graph. Finally, if we complete the square on x in the equation we can rewrite the equation of our parabola as $y = a(x - h)^2 + k$. When the equation is in this form, called *vertex form*, the vertex is at the point (h, k). Here is a summary:

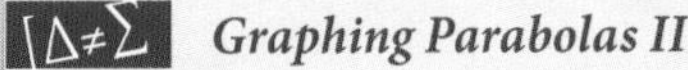

[Δ≠Σ] ***Graphing Parabolas II***

The graph of

$$y = a(x - h)^2 + k, a \neq 0$$

will be a parabola with a vertex at (h, k). The vertex will be the highest point on the graph when $a < 0$, and the lowest point on the graph when $a > 0$.

Modeling: Business Applications

The fact that the vertex of a parabola is the highest or lowest point on the graph is particularly useful in applications where we wish to find a maximum or minimum value.

EXAMPLE 5 A company selling copies of an accounting program for home computers finds that it will make a weekly profit of P dollars from selling x copies of the program, according to the equation

$$P(x) = -0.1x^2 + 27x - 500$$

How many copies of the program should it sell to make the largest possible profit, and what is the largest possible profit?

SOLUTION Because the coefficient of x^2 is negative, we know the graph of this parabola will be concave down, meaning that the vertex is the highest point of the curve. We find the vertex by first finding its x-coordinate:

$$x = \frac{-b}{2a} = \frac{-27}{2(-0.1)} = \frac{27}{0.2} = 135$$

This represents the number of programs the company needs to sell each week to make a maximum profit. To find the maximum profit, we substitute 135 for x in the original equation. (A calculator is helpful for these kinds of calculations.)

$$\begin{aligned} P(135) &= -0.1(135)^2 + 27(135) - 500 \\ &= -0.1(18{,}225) + 3{,}645 - 500 \\ &= -1{,}822.5 + 3{,}645 - 500 \\ &= 1{,}322.5 \end{aligned}$$

The maximum weekly profit is \$1,322.50 and is obtained by selling 135 programs a week. ■

© zhang bo/iStockPhoto

EXAMPLE 6 An art supply store finds that they can sell x sketch pads each week at p dollars each, according to the equation $x = 900 - 300p$. Graph the revenue equation $R = xp$. Then use the graph to find the price p that will bring in the maximum revenue. Finally, find the maximum revenue.

SOLUTION As it stands, the revenue equation contains three variables. Because we are asked to find the value of p that gives us the maximum value of R, we rewrite the equation using just the variables R and p. Because $x = 900 - 300p$, we have

$$R = xp = (900 - 300p)p$$

The graph of this equation is shown in Figure 6. The graph appears in the first quadrant only, because R and p are both positive quantities.

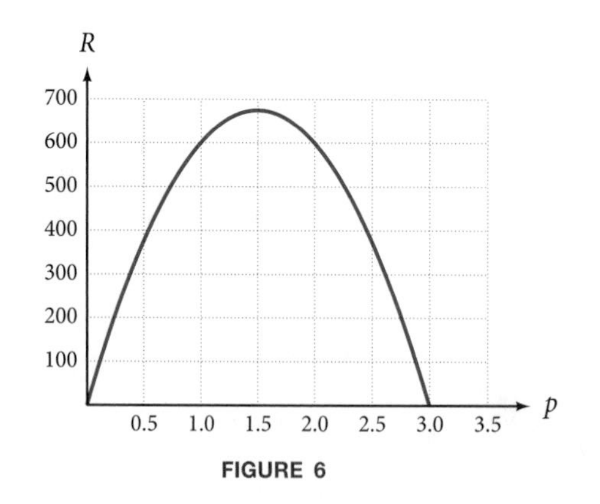

FIGURE 6

From the graph, we see that the maximum value of R occurs when $p = \$1.50$. We can calculate the maximum value of R from the equation:

When	$p = 1.5$
the equation	$R = (900 - 300p)p$

becomes $\quad R = (900 - 300 \cdot 1.5)1.5$

$$= (900 - 450)1.5$$

$$= 450 \cdot 1.5$$

$$= 675$$

The maximum revenue is $675. It is obtained by setting the price of each sketch pad at $p = \$1.50$. ■

TECHNOLOGY NOTE *Graphing Calculators*

If you have been using a graphing calculator for some of the material in this course, you are well aware that your calculator can draw all the graphs in this section very easily. It is important, however, that you be able to recognize and sketch the graph of any parabola by hand. It is a skill that all successful college algebra students should possess, even if they are proficient in the use of a graphing calculator. My suggestion is that you work the problems in this section and problem set without your calculator. Then use your calculator to check your results.

Modeling: Finding the Equation from the Graph

EXAMPLE 7 At the 1997 Washington County Fair in Oregon, David Smith, Jr., The Bullet, was shot from a cannon. As a human cannonball, he reached a height of 70 feet before landing in a net 160 feet from the cannon. Sketch the graph of his path, and then find the equation of the graph.

SOLUTION We assume that the path taken by the human cannonball is a parabola. If the origin of the coordinate system is at the opening of the cannon, then the net that catches him will be at 160 on the x-axis. Figure 7 shows the graph.

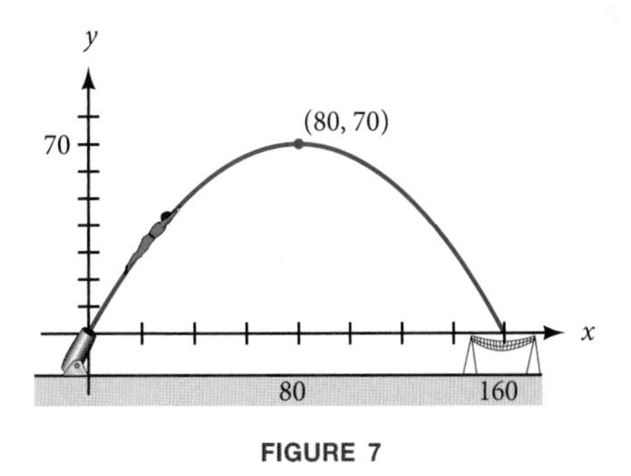

FIGURE 7

Because the curve is a parabola, we know the equation will have the form

$$y = a(x - h)^2 + k$$

Because the vertex of the parabola is at (80, 70), we can fill in two of the three constants in our equation, giving us

$$y = a(x - 80)^2 + 70$$

To find a, we note that the landing point will be (160, 0). Substituting the coordinates of this point into the equation, we solve for a:

$$0 = a(160 - 80)^2 + 70$$
$$0 = a(80)^2 + 70$$
$$0 = 6{,}400a + 70$$
$$a = -\frac{70}{6{,}400} = -\frac{7}{640}$$

The equation that describes the path of the human cannonball is

$$y = -\frac{7}{640}(x - 80)^2 + 70 \quad \text{for } 0 \le x \le 160$$

Using Technology 5.1

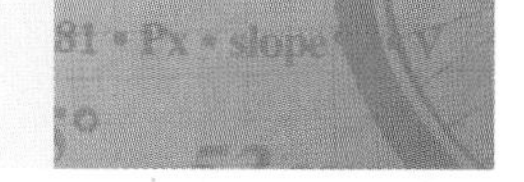

Let's use Wolfram|Alpha to draw the graph of a parabola through three points that do not lie in a straight line.

EXAMPLE 8 Use Wolfram|Alpha to graph the parabola from the human cannonball in Example 7.

SOLUTION The path of the human cannonball in Example 7 passes through the points (0, 0), (80, 70) and (160, 0). Here is what we enter into Wolfram|Alpha, and the results that are given back:

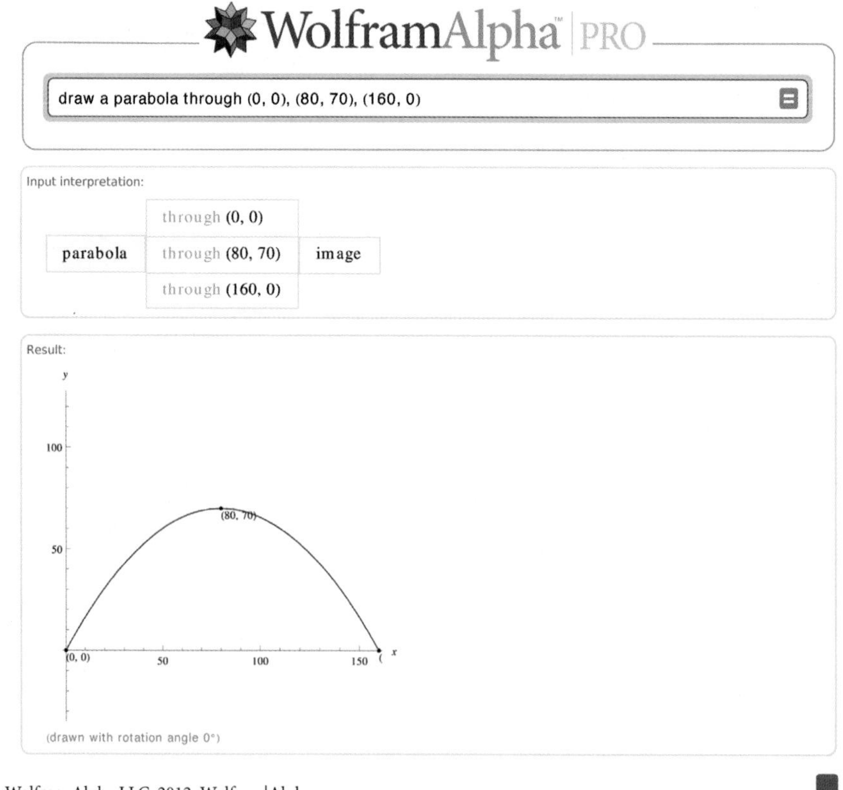

Wolfram Alpha LLC. 2012. Wolfram|Alpha
http://www.wolframalpha.com/
(access June 18, 2012)

EXAMPLE 9 Use Wolfram|Alpha to find the equation of the parabola from the human cannonball in Example 7.

SOLUTION We simply change the question to this:

Find the equation of the parabola through (0, 0), (80, 70), (160, 0).

Here are the results:

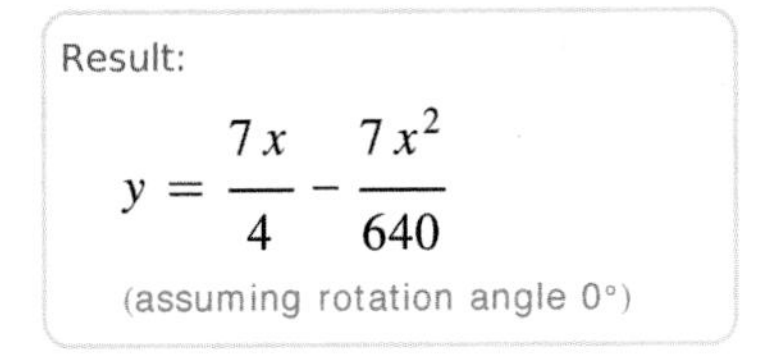
Result:

$$y = \frac{7x}{4} - \frac{7x^2}{640}$$

(assuming rotation angle 0°)

Wolfram Alpha LLC. 2012. Wolfram|Alpha
http://www.wolframalpha.com/
(access June 18, 2012)

To see if this is the same result we obtained, let's manipulate our result from Example 7.

$$\begin{aligned} y &= -\frac{7}{640}(x - 80)^2 + 70 \\ &= -\frac{7}{640}(x^2 - 160x + 6{,}400) + 70 \\ &= -\frac{7}{640}x^2 + \frac{7}{640}160x - \frac{7}{640}6{,}400 + 70 \\ &= -\frac{7}{640}x^2 + \frac{7}{4}x - 70 + 70 \\ &= -\frac{7}{640}x^2 + \frac{7}{4}x \end{aligned}$$

Our equation is equivalent to the one returned by Wolfram|Alpha. ■

GETTING READY FOR CLASS

After reading through the preceding section, respond in your own words and in complete sentences.

A. What is a parabola?
B. What part of the equation of a parabola determines whether the graph is concave up or concave down?
C. Suppose $f(x) = ax^2 + bx + c$ is the equation of a parabola. Explain how $f(4) = 1$ relates to the graph of the parabola.
D. A line can be graphed with two points. How many points are necessary to get a reasonable sketch of a parabola? Explain.

Problem Set 5.1

For each of the following equations, give the x-intercepts, y-intercept and the coordinates of the vertex, and sketch the graph.

1. $y = x^2 + 2x - 3$
2. $y = x^2 + 4x - 5$
3. $y = x^2 - 1$
4. $y = x^2 - 4$
5. $y = -x^2 + 9$
6. $y = -x^2 + 1$
7. $y = 2x^2 - 4x - 6$
8. $y = 2x^2 + 4x - 6$
9. $y = x^2 - 2x - 4$
10. $y = x^2 - 2x - 2$

Graph each parabola. Label the vertex and any intercepts that exist.

11. $y = 2(x - 1)^2 + 3$
12. $y = 2(x + 1)^2 - 3$
13. $f(x) = -(x + 2)^2 + 4$
14. $f(x) = -(x - 3)^2 + 1$
15. $g(x) = \frac{1}{2}(x - 2)^2 - 4$
16. $g(x) = \frac{1}{3}(x - 3)^2 - 3$
17. $f(x) = -2(x - 4)^2 - 1$
18. $f(x) = -4(x - 1)^2 + 4$

Find the vertex and any two convenient points to sketch the graphs of the following equations.

19. $y = x^2 - 4x - 4$
20. $y = x^2 - 2x + 3$
21. $y = -x^2 + 4x - 2$
22. $f(x) = x^2 + 1$
23. $f(x) = x^2 + 4$
24. $y = -x^2 - 3$
25. $y = -x^2 - 2$
26. $g(x) = 2x^2 + 4x + 3$

For each of the following equations, find the coordinates of the vertex, and indicate whether the vertex is the highest point on the graph or the lowest point on the graph. (Do not graph.)

27. $y = x^2 - 6x + 5$
28. $y = -x^2 + 6x - 5$
29. $y = -x^2 + 2x + 8$
30. $y = x^2 - 2x - 8$
31. $y = 12 + 4x - x^2$
32. $y = -12 - 4x + x^2$
33. $y = -x^2 - 8x$
34. $y = x^2 + 8x$

Modeling Practice

35. **Maximum Profit** A company earns a weekly profit of P dollars by selling x items, according to the equation $P(x) = -0.5x^2 + 40x - 300$. Find the number of items the company must sell each week to obtain the largest possible profit. Then, find the largest possible profit.

36. **Maximum Profit** A company earns a weekly profit of P dollars by selling x items, according to the equation $P(x) = -0.5x^2 + 100x - 1{,}600$. Find the number of items the company must sell each week to obtain the largest possible profit. Then, find the largest possible profit.

37. **Maximum Profit** A company finds that it can make a profit of P dollars each month by selling x patterns, according to the formula $P(x) = -0.002x^2 + 3.5x - 800$. How many patterns must it sell each month to have a maximum profit? What is the maximum profit?

38. **Maximum Profit** A company selling picture frames finds that it can make a profit of P dollars each month by selling x frames, according to the formula $P(x) = -0.002x^2 + 5.5x - 1{,}200$. How many frames must it sell each month to have a maximum profit? What is the maximum profit?

39. Maximum Height Chaudra is tossing a softball into the air with an underhand motion. The distance of the ball above her hand at any time is given by the function

$$h(t) = 32t - 16t^2 \quad \text{for} \quad 0 \le t \le 2$$

where $h(t)$ is the height of the ball (in feet) and t is the time (in seconds). Find the times at which the ball is in her hand and the maximum height of the ball.

40. Maximum Height Hali is tossing a quarter into the air with an underhand motion. The distance of the quarter above her hand at any time is given by the function

$$h(t) = 16t - 16t^2 \quad \text{for} \quad 0 \le t \le 1$$

where $h(t)$ is the height of the quarter (in feet) and t is the time (in seconds). Find the times at which the quarter is in her hand and the maximum height of the quarter.

41. Maximum Height A ball is projected into the air with an upward velocity of 64 feet per second. The equation that gives the height h (in feet) of the ball at any time t (in seconds) is $h(t) = 64t - 16t^2$. Find the maximum height attained by the ball.

42. Maximum Area Justin wants to fence three sides of a rectangular exercise yard for his dog. The fourth side of the exercise yard will be a side of the house. He has 80 feet of fencing available. Find the dimensions of the exercise yard that will enclose the maximum area.

43. Maximum Area Repeat Problem 42, assuming that Justin has 60 feet of fencing available.

44. Maximum Revenue The relationship between the number x of pencil sharpeners a company sells each week and the price p of each sharpener is given by the equation $x = 1{,}800 - 100p$. Graph the revenue equation $R = xp$ and use the graph to find the price p that will bring in the maximum revenue. Then find the maximum revenue.

Finding the Equation from the Graph For each of the following problems, the graph is a parabola. In each case, find an equation in the form $y = a(x - h)^2 + k$ that describes the graph.

45. **46.**

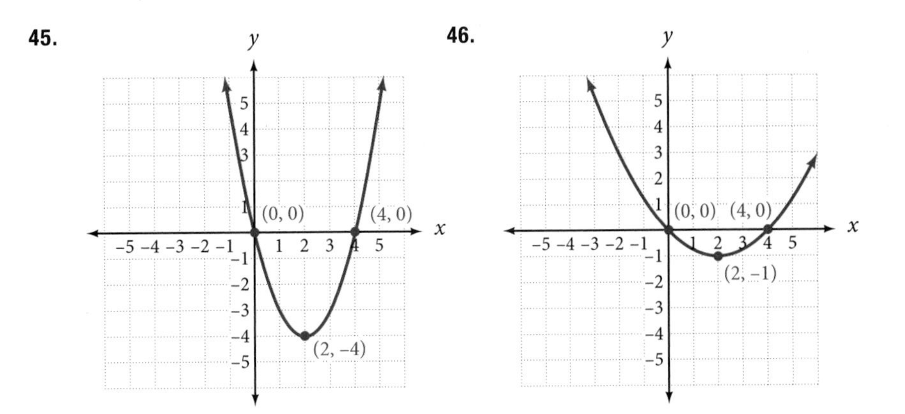

47. Human Cannonball A human cannonball is shot from a cannon at the county fair. He reaches a height of 60 feet before landing in a net 180 feet from the cannon. Sketch the graph of his path, and then find the equation of the graph.

48. Human Cannonball Referring to Problem 47, find the height reached by the human cannonball after he has traveled 30 feet horizontally, and after he has traveled 150 feet horizontally.

49. Comparing Expressions, Equations, and Functions Four problems follow. The solution to Problem 3 is shown in Figure 10. Solve the other three problems, and then explain how the solutions to the four problems are related.

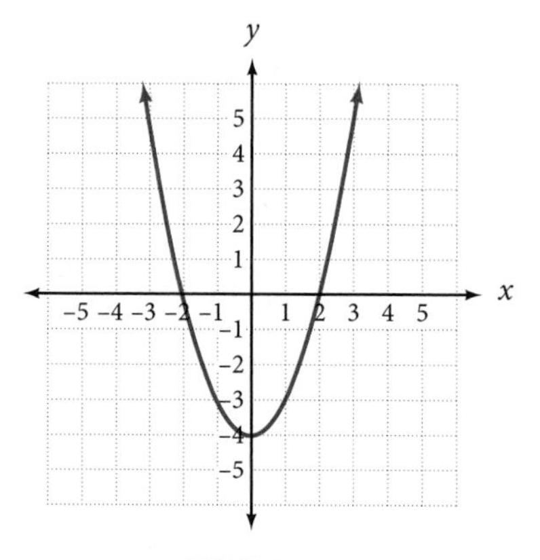

FIGURE 10

Problem 1 Factor the expression $x^2 - 4$.
Problem 2 Solve the equation $x^2 - 4 = 0$.
Problem 3 Graph the function $y = x^2 - 4$.
Problem 4 If $f(x) = x^2 - 4$, find the values of x for which $f(x) = 0$.

50. Comparing Expressions, Equations, and Functions Four problems are shown here. The solution to Problem 3 is shown in Figure 11. Solve the other three problems, and then explain how the solutions to the four problems are related.

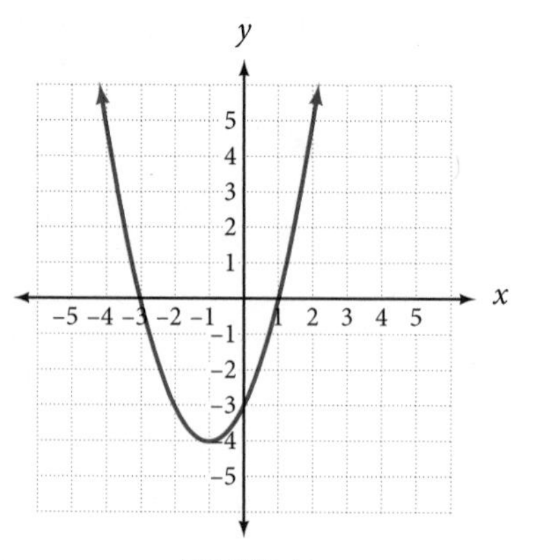

FIGURE 11

Problem 1 Factor the expression $x^2 + 2x - 3$.
Problem 2 Solve the equation $x^2 + 2x - 3 = 0$.
Problem 3 Graph the function $y = x^2 + 2x - 3$.
Problem 4 If $f(x) = x^2 + 2x - 3$, find the values of x for which $f(x) = 0$.

© Bettmann/Corbis/iStockPhoto

51. Human Cannonball Suppose the equation gives the height $h(x)$ of the human cannonball after he has traveled x feet horizontally. What is $h(30)$, and what does it represent?

$$h(x) = -\frac{7}{640}(x - 80)^2 + 70$$

52. Interpreting Graphs The graph below shows the different paths taken by the human cannonball when his velocity out of the cannon is 50 miles/hour, and his cannon is inclined at varying angles.

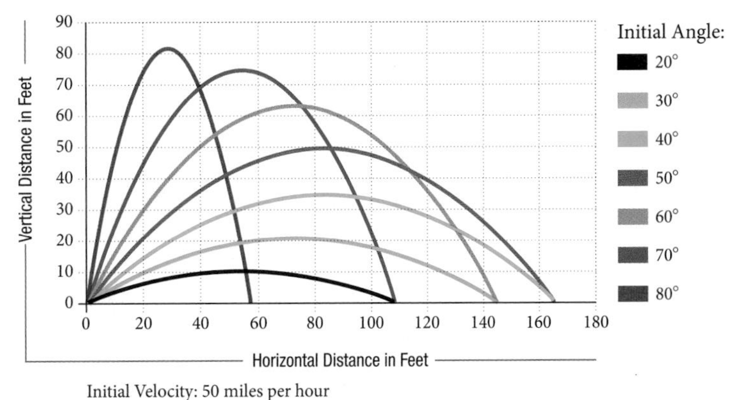

a. If his landing net is placed 107 feet from the cannon, at what angle should the cannon be inclined so that he lands in the net?

b. Approximately where do you think he would land if the cannon was inclined at 45°?

c. If the cannon was inclined at 45°, approximately what height do you think he would attain?

d. Do you think there is another angle for which he would travel the same distance he travels at 80°? Give an estimate of that angle.

e. The fact that every landing point can come from two different paths makes us think that the equations that give us the landing points must be what type of equations?

Using Technology Exercises

Use Wolfram|Alpha to find the equation of the parabola through the given set of points. Rewrite the equation, if necessary, in the form you are used to seeing it. Be sure that your equation and the result from Wolfram|Alpha are equivalent.

53. $(0, 0), (2, -4), (4, 0)$

54. $(0, 0), (2, 4), (4, 0)$

55. $(0, 4), (2, 8), (4, 4)$

56. $(0, -4), (2, -8), (4, -4)$

Getting Ready for the Next Section

Divide.

57. $\dfrac{10x^2}{5x^2}$

58. $\dfrac{-15x^4}{5x^2}$

59. $\dfrac{4x^4y^3}{-2x^2y}$

60. $\dfrac{10a^4b^2}{4a^2b^2}$

61. $4{,}628 \div 25$

62. $7{,}546 \div 35$

Multiply.

63. $2x^2(2x - 4)$

64. $3x^2(x - 2)$

65. $(2x - 4)(2x^2 + 4x + 5)$

66. $(x - 2)(3x^2 + 6x + 15)$

Subtract.

67. $(2x^2 - 7x + 9) - (2x^2 - 4x)$

68. $(x^2 - 6xy - 7y^2) - (x^2 + xy)$

Factor.

69. $x^2 - a^2$

70. $x^2 - 1$

71. $x^2 - 6xy - 7y^2$

72. $2x^2 - 5xy + 3y^2$

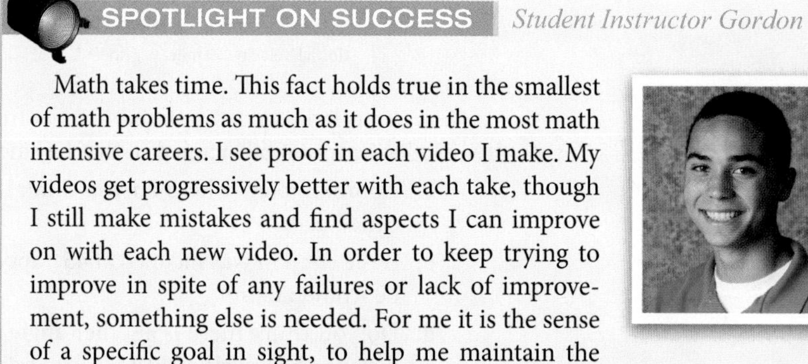

SPOTLIGHT ON SUCCESS *Student Instructor Gordon*

Math takes time. This fact holds true in the smallest of math problems as much as it does in the most math intensive careers. I see proof in each video I make. My videos get progressively better with each take, though I still make mistakes and find aspects I can improve on with each new video. In order to keep trying to improve in spite of any failures or lack of improvement, something else is needed. For me it is the sense of a specific goal in sight, to help me maintain the desire to put in continued time and effort.

When I decided on the number one university I wanted to attend, I wrote the name of that school in bold block letters on my door, written to remind myself daily of my ultimate goal. Stuck in the back of my head, this end result pushed me little by little to succeed and meet all of the requirements for the university I had in mind. And now I can say I'm at my dream school bringing with me that skill.

I recognize that others may have much more difficult circumstances than my own to endure, with the goal of improving or escaping those circumstances, and I deeply respect that. But that fact demonstrates to me how easy but effective it is, in comparison, to "stay with the problems longer" with a goal in mind of something much more easily realized, like a good grade on a test. I've learned to set goals, small or big, and to stick with them until they are realized.

Division with Polynomials

5.2

INTRODUCTION First Bank of San Luis Obispo charges \$2.00 per month and \$0.15 per check for a regular checking account. So, if you write x checks in one month, the total monthly cost of the checking account will be $C(x) = 2.00 + 0.15x$. From this formula, we see that the more checks we write in a month, the more we pay for the account. But it is also true that the more checks we write in a month, the lower the cost per check. To find the cost per check, we use the *average cost* function. To find the average cost function, we divide the total cost by the number of checks written.

© Randy Mayes/iStockPhoto

$$\text{Average Cost} = \overline{C}(x) = \frac{C(x)}{x} = \frac{2.00 + 0.15x}{x}$$

This last expression gives us the average cost per check for each of the x checks written. To work with this last expression, we need to know something about division with polynomials, and that is what we will cover in this section. ■

In the past we have worked with the following types of functions:

Linear functions, which have the form

$$f(x) = ax + b \qquad a \neq 0$$

Quadratic functions, which have the form

$$f(x) = ax^2 + bx + c \qquad a \neq 0$$

Cubic functions, which have the form

$$f(x) = ax^3 + bx^2 + cx + d \qquad a \neq 0$$

Each of these functions is a special case of a more general classification of function—the *polynomial function.*

DEFINITION *Polynomial functions*

A *polynomial function* of degree n in x is any function that can be written in the form

$$P(x) = a_nx^n + a_{n-1}x^{n-1} + a_{n-2}x^{n-2} + \cdots + a_2x^2 + a_1x + a_0$$

where $a_n \neq 0$, n is a nonnegative integer, and each coefficient is a real number.

Note Although our definition does not allow for complex coefficients, if you go on in mathematics you may find a more general definition for polynomial functions that does allow the coefficients to be complex numbers.

Before we begin our work with polynomial functions, we need to review division with polynomials and introduce a new tool called *synthetic division.*

We begin this section by considering division of a polynomial by a monomial. This is the simplest kind of polynomial division. The rest of the section is devoted to division of a polynomial by a polynomial. This kind of division is similar to long division with whole numbers.

Dividing a Polynomial by a Monomial

To divide a polynomial by a monomial, we use the definition of division and apply the distributive property. The following example illustrates the procedure.

VIDEO EXAMPLES

SECTION 5.2

EXAMPLE 1 Divide $\dfrac{10x^5 - 15x^4 + 20x^3}{5x^2}$.

SOLUTION

$= (10x^5 - 15x^4 + 20x^3) \cdot \dfrac{1}{5x^2}$ — Dividing by $5x^2$ is the same as multiplying by $\frac{1}{5x^2}$

$= 10x^5 \cdot \dfrac{1}{5x^2} - 15x^4 \cdot \dfrac{1}{5x^2} + 20x^3 \cdot \dfrac{1}{5x^2}$ — Distributive property

$= \dfrac{10x^5}{5x^2} - \dfrac{15x^4}{5x^2} + \dfrac{20x^3}{5x^2}$ — Multiplying by $\frac{1}{5x^2}$ is the same as dividing by $5x^2$

$= 2x^3 - 3x^2 + 4x$ ■

Notice that division of a polynomial by a monomial is accomplished by dividing each term of the polynomial by the monomial. The first two steps are usually not shown in a problem like this. They are part of Example 1 to justify distributing $5x^2$ under all three terms of the polynomial $10x^5 - 15x^4 + 20x^3$.

Here are some more examples of this kind of division.

EXAMPLE 2 Divide $\dfrac{8x^3y^5 - 16x^2y^2 + 4x^4y^3}{-2x^2y}$. Write the result with positive exponents.

SOLUTION

$$\frac{8x^3y^5 - 16x^2y^2 + 4x^4y^3}{-2x^2y} = \frac{8x^3y^5}{-2x^2y} + \frac{-16x^2y^2}{-2x^2y} + \frac{4x^4y^3}{-2x^2y}$$

$$= -4xy^4 + 8y - 2x^2y^2$$ ■

EXAMPLE 3 Divide $\dfrac{10a^4b^2 + 8ab^3 - 12a^3b + 6ab}{4a^2b^2}$. Write the result with positive exponents.

SOLUTION

$$\frac{10a^4b^2 + 8ab^3 - 12a^3b + 6ab}{4a^2b^2} = \frac{10a^4b^2}{4a^2b^2} + \frac{8ab^3}{4a^2b^2} - \frac{12a^3b}{4a^2b^2} + \frac{6ab}{4a^2b^2}$$

$$= \frac{5a^2}{2} + \frac{2b}{a} - \frac{3a}{b} + \frac{3}{2ab}$$ ■

Notice in Example 3 that the result is not a polynomial because of the last three terms. If we were to write each as a product, some of the variables would have negative exponents. For example, the second term would be

$$\frac{2b}{a} = 2a^{-1}b$$

Recall that, by definition, a polynomial cannot have any terms with negative exponents. The divisor in each of the preceding examples was a monomial. We now want to turn our attention to division of polynomials in which the divisor has two or more terms.

Dividing a Polynomial by a Polynomial

EXAMPLE 4 Divide: $\dfrac{x^2 - 6xy - 7y^2}{x + y}$

SOLUTION In this case, we can factor the numerator and perform division by simply dividing out common factors, just as we did in the previous section:

$$\frac{x^2 - 6xy - 7y^2}{x + y} = \frac{(x + y)(x - 7y)}{x + y}$$

$$= x - 7y$$

Long Division

For the type of division shown in Example 4, the denominator must be a factor of the numerator. When the denominator is not a factor of the numerator, or in the case where we can't factor the numerator, the method used in Example 4 won't work. We need to develop a new method for these cases. Because this new method is very similar to ***long division*** with whole numbers, we will review the method of long division here.

EXAMPLE 5 Divide $25\overline{)4{,}628}$.

SOLUTION

$$\begin{array}{r} 1 \\ 25\overline{)4{,}628} \\ 25 \\ \hline 21 \end{array}$$

Estimate 25 into 46

Multiply $1 \times 25 = 25$

Subtract $46 - 25 = 21$

$$\begin{array}{r} 1 \\ 25\overline{)4{,}628} \\ 25\!\downarrow \\ \hline 212 \end{array}$$

Bring down the 2

These are the four basic steps in long division: estimate, multiply, subtract, and bring down the next term. To complete the problem, we simply perform the same four steps:

$$\begin{array}{r} 18 \\ 25\overline{)4{,}628} \\ 25 \\ \hline 212 \\ 200\!\downarrow \\ \hline 128 \end{array}$$

8 is the estimate

Multiply to get 200

Subtract to get 12, then bring down the 8

One more time:

$$\begin{array}{r} 185 \\ 25\overline{)4,628} \\ \underline{25} \\ 212 \\ \underline{200}\downarrow \\ 128 \\ \underline{125} \\ 3 \end{array}$$

5 is the estimate

Multiply to get 125

Subtract to get 3

Because 3 is less than 25 and we have no more terms to bring down, we have our answer:

$$\frac{4,628}{25} = 185 + \frac{3}{25}$$

To check our answer, we multiply 185 by 25 and then add 3 to the result:

$$25(185) + 3 = 4,625 + 3 = 4,628$$

■

Long division with polynomials is similar to long division with whole numbers. Both use the same four basic steps: estimate, multiply, subtract, and bring down the next term. We use long division with polynomials when the denominator has two or more terms and is not a factor of the numerator (or we are not sure if it is) as in the next example.

EXAMPLE 6 Divide $\dfrac{2x^2 - 7x + 9}{x - 2}$.

SOLUTION

$$\begin{array}{r} 2x\phantom{{}-7x+9} \\ x-2\overline{)\,2x^2 - 7x + 9} \\ \underline{\overset{-}{\cancel{+}}\,2x^2 \overset{+}{\cancel{-}}\,4x}\phantom{{}+9} \\ -3x\phantom{{}+9} \end{array}$$

Estimate $2x^2 \div x = 2x$

Multiply $2x(x - 2) = 2x^2 - 4x$

Subtract $(2x^2 - 7x) - (2x^2 - 4x) = -3x$

$$\begin{array}{r} 2x\phantom{{}-7x+9} \\ x-2\overline{)\,2x^2 - 7x + 9} \\ \underline{\overset{-}{\cancel{+}}\,2x^2 \overset{+}{\cancel{-}}\,4x}\downarrow \\ -3x + 9 \end{array}$$

Bring down the 9

Notice we change the signs on $2x^2 - 4x$ and add in the subtraction step. Subtracting a polynomial is equivalent to adding its opposite.

We repeat the four steps.

$$\begin{array}{r} 2x - 3\phantom{{}+9} \\ x-2\overline{)\,2x^2 - 7x + 9} \\ \underline{\overset{-}{\cancel{+}}\,2x^2 \overset{+}{\cancel{-}}\,4x}\phantom{{}+9} \\ -3x + 9 \\ \underline{\overset{+}{\cancel{-}}3x \overset{-}{\cancel{+}}\,6} \\ 3 \end{array}$$

-3 is the estimate: $-3x \div x = -3$

Multiply $-3(x - 2) = -3x + 6$

Subtract $(-3x + 9) - (-3x + 6) = 3$

Because we have no other term to bring down, we have our answer:

$$\frac{2x^2 - 7x + 9}{x - 2} = 2x - 3 + \frac{3}{x - 2}$$

To check, we multiply $(2x - 3)(x - 2)$ to get $2x^2 - 7x + 6$; then, adding the remainder 3 to this result, we have $2x^2 - 7x + 9$. ■

In setting up a division problem involving two polynomials, we must remember two things: (1) Both polynomials should be in decreasing powers of the variable, and (2) neither should skip any powers from the highest power down to the constant term. If there are any missing terms, they can be filled in using a coefficient of 0.

EXAMPLE 7 Divide $2x - 4\overline{)4x^3 - 6x - 11}$.

SOLUTION Because the trinomial is missing a term in x^2, we can fill it in with $0x^2$:

$$4x^3 - 6x - 11 = 4x^3 + 0x^2 - 6x - 11$$

Adding $0x^2$ does not change our original problem.

$$\begin{array}{rrrrr}
 & 2x^2 & +\ 4x & +\ 5 & \\
2x - 4\,) & 4x^3 & +\ 0x^2 & -\ 6x & -\ 11 \\
 & - & + & & \\
 & \cancel{+}4x^3 & \cancel{-}\ 8x^2 & \downarrow & \\
\hline
 & & +\ 8x^2 & -\ 6x & \\
 & & - & + & \\
 & & \cancel{+}\ 8x^2 & \cancel{-}\ 16x & \downarrow \\
\hline
 & & & +\ 10x & -\ 11 \\
 & & & - & + \\
 & & & \cancel{+}\ 10x & \cancel{-}\ 20 \\
\hline
 & & & & +\ 9
\end{array}$$

Note Adding the $0x^2$ term gives us a column in which to write $-8x^2$.

We can write our result that way.

$$\frac{4x^3 - 6x - 11}{2x - 4} = 2x^2 + 4x + 5 + \frac{9}{2x - 4}$$

To check this result, we multiply $2x - 4$ and $2x^2 + 4x + 5$:

$$\begin{array}{lrrrr}
 & 2x^2 & +\ 4x & +\ 5 & \\
\times & & 2x & -\ 4 & \\
\hline
 4x^3 & +\ 8x^2 & +\ 10x & & \\
 + & -\ 8x^2 & -\ 16x & -\ 20 & \\
\hline
 4x^3 & & -\ 6x & -\ 20 &
\end{array}$$

Adding 9 (the remainder) to this result gives us the polynomial $4x^3 - 6x - 11$. Our answer checks. ■

For our next example, let's do Example 4 again, but this time use long division.

EXAMPLE 8 Divide $\dfrac{x^2 - 6xy - 7y^2}{x + y}$.

SOLUTION

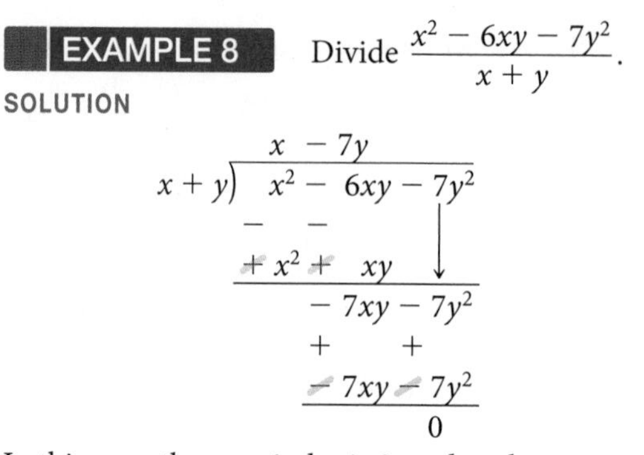

In this case, the remainder is 0, and we have

$$\frac{x^2 - 6xy - 7y^2}{x + y} = x - 7y$$

which is easy to check because

$$(x + y)(x - 7y) = x^2 - 6xy - 7y^2$$

■

EXAMPLE 9 Factor $x^3 + 9x^2 + 26x + 24$ completely if $x + 2$ is one of its factors.

SOLUTION Because $x + 2$ is one of the factors of the polynomial we are trying to factor, it must divide that polynomial evenly—that is, without a remainder. Therefore, we begin by dividing the polynomial by $x + 2$:

$$\begin{array}{r|l}
 & x^2 + 7x + 12 \\
x + 2 & x^3 + 9x^2 + 26x + 24 \\
 & \underline{-x^3 - 2x^2} \\
 & \quad 7x^2 + 26x \\
 & \quad \underline{-7x^2 - 14x} \\
 & \qquad 12x + 24 \\
 & \qquad \underline{-12x - 24} \\
 & \qquad\qquad 0
\end{array}$$

Now we know that the polynomial we are trying to factor is equal to the product of $x + 2$ and $x^2 + 7x + 12$. To factor completely, we simply factor $x^2 + 7x + 12$:

$$\begin{aligned} x^3 + 9x^2 + 26x + 24 &= (x + 2)(x^2 + 7x + 12) \\ &= (x + 2)(x + 3)(x + 4) \end{aligned}$$

■

Synthetic Division

Synthetic division is a short form of long division with polynomials. We will consider synthetic division only for those cases in which the divisor is of the form $x + k$, where k is a constant.

Let's begin by looking over an example of long division with polynomials.

$$\begin{array}{r}
3x^2 - 2x + 4 \\
x+3\overline{)3x^3 + 7x^2 - 2x - 4} \\
\underline{3x^3 + 9x^2} \phantom{- 2x - 4} \\
-2x^2 - 2x \phantom{- 4} \\
\underline{-2x^2 - 6x} \phantom{- 4} \\
4x - 4 \\
\underline{4x + 12} \\
-16
\end{array}$$

We can rewrite the problem without showing the variable since the variable is written in descending powers and similar terms are in alignment. It looks like this:

$$\begin{array}{rrrrr}
 & 3 & -2 & 4 & \\
1+3\,) & 3 & 7 & -2 & -4 \\
 & (3) & +9 & & \\
 & & -2 & (-2) & \\
 & & (-2) & -6 & \\
 & & & 4 & (-4) \\
 & & & (4) & 12 \\
 & & & & -16
\end{array}$$

We have used parentheses to enclose the numbers that are repetitions of the numbers above them. We can compress the problem by eliminating all repetitions except the first one:

$$\begin{array}{rrrrr}
 & & 3 & -2 & 4 \\
1+3\,) & 3 & 7 & -2 & -4 \\
 & & 9 & -6 & 12 \\
\hline
 & 3 & -2 & 4 & -16
\end{array}$$

The top line is the same as the first three terms of the bottom line, so we eliminate the top line. Also, the 1 that was the coefficient of x in the original problem can be eliminated since we will consider only division problems where the divisor is of the form $x + k$. The following is the most compact form of the original division problem:

$$\begin{array}{rrrrr}
+3\,) & 3 & 7 & -2 & -4 \\
 & & 9 & -6 & 12 \\
\hline
 & 3 & -2 & 4 & -16
\end{array}$$

If we check over the problem, we find that the first term in the bottom row is exactly the same as the first term in the top row—and it always will be in problems of this type. Also, the last three terms in the bottom row come from multiplication by $+3$ and then subtraction. We can get an equivalent result by multiplying by -3 and adding. The problem would then look like this:

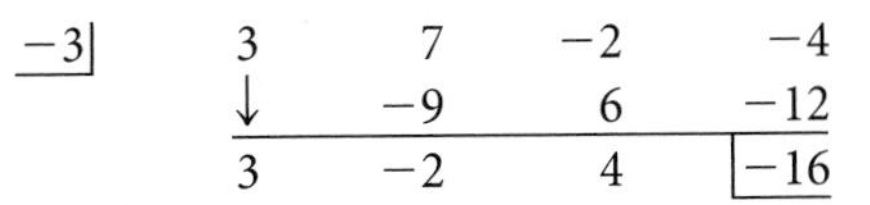

We have used the brackets ⌋ ⌊ to separate the divisor and the remainder. This last expression is synthetic division. It is an easy process to remember. Simply change the sign of the constant term in the divisor, then bring down the first term of the dividend. The process is then just a series of multiplications and additions, as indicated in the following diagram by the arrows:

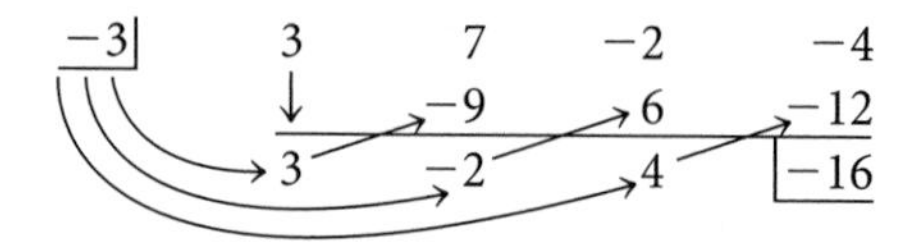

The last term of the bottom row is always the remainder.

Here are some additional examples of synthetic division with polynomials.

EXAMPLE 10 Divide $x^4 - 2x^3 + 4x^2 - 6x + 2$ by $x - 2$.

SOLUTION We change the sign of the constant term in the divisor to get $+2$ and then complete the procedure:

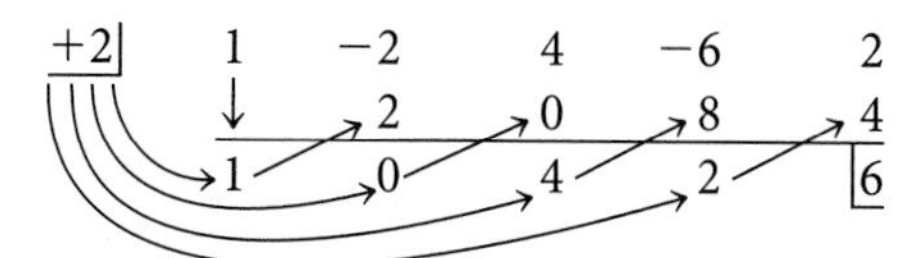

From the last line we have the answer:

$$1x^3 + 0x^2 + 4x + 2 + \frac{6}{x-2}$$

Further simplifying the result, we get:

$$x^3 + 4x + 2 + \frac{6}{x-2}$$

EXAMPLE 11 Divide $\dfrac{3x^3 - 4x + 5}{x + 4}$.

SOLUTION Since we cannot skip any powers of the variable in the polynomial $3x^3 - 4x + 5$, we rewrite it as $3x^3 + 0x^2 - 4x + 5$ and proceed as we did in Example 10:

$$\begin{array}{r|rrrr} -4 & 3 & 0 & -4 & 5 \\ & \downarrow & -12 & 48 & -176 \\ \hline & 3 & -12 & 44 & \boxed{-171} \end{array}$$

From the synthetic division, we have

$$\frac{3x^3 - 4x + 5}{x + 4} = 3x^2 - 12x + 44 + \frac{-171}{x + 4}$$

If we multiply both sides of the equation above by $x + 4$, we have

$$3x^3 - 4x + 5 = (x + 4)(3x^2 - 12x + 44) + (-171)$$

which has the form

$$\text{dividend} = (\text{divisor})(\text{quotient}) + (\text{remainder})$$

In this chapter our divisors will be limited to binomials of the form $x - r$. Since the degree of the remainder is always less than the degree of the divisor, our remainders will be real numbers. We summarize this information with the following theorem. The theorem is called an ***existence theorem*** because it guarantees the existence of the quotient and remainder whenever a polynomial is divided by a binomial.

THEOREM ***Existence Theorem for Division***

If $P(x)$ is a polynomial and r is a real number, then there exists a unique polynomial $q(x)$ and a unique real number R such that

$$P(x) = (x - r)q(x) + R$$

The polynomial $q(x)$ is called the *quotient*, $x - r$ is called the *divisor*, and R is called the *remainder*. The degree of $q(x)$ is 1 less than the degree of $P(x)$.

EXAMPLE 12 Identify the quotient and remainder when

$$x^3 - 4x^2 + 2x - 5$$

is divided by $x - 3$.

SOLUTION Using synthetic division we have

$$\begin{array}{r|rrrr} 3 & 1 & -4 & 2 & -5 \\ & \downarrow & 3 & -3 & -3 \\ \hline & 1 & -1 & -1 & \boxed{-8} \end{array}$$

The quotient is $x^2 - x - 1$ and the remainder is -8. Writing the results in the form

$$\text{dividend} = (\text{divisor})(\text{quotient}) + (\text{remainder})$$

we have

$$x^3 - 4x^2 + 2x - 5 = (x - 3)(x^2 - x - 1) + (-8)$$

EXAMPLE 13 Find the remainders when $P(x) = 2x^3 - x^2 + 3x - 4$ is divided by $x - 2, x - 1, x + \frac{1}{2}, x + 3$.

SOLUTION Here are the four synthetic divisions:

Division by $x - 2$:

$$\begin{array}{r|rrrr} 2 & 2 & -1 & 3 & -4 \\ & \downarrow & 4 & 6 & 18 \\ \hline & 2 & 3 & 9 & \boxed{14} \end{array}$$

Division by $x - 1$:

$$\begin{array}{r|rrrr} 1 & 2 & -1 & 3 & -4 \\ & \downarrow & 2 & 1 & 4 \\ \hline & 2 & 1 & 4 & \boxed{0} \end{array}$$

Division by $x + \frac{1}{2}$:

$$\begin{array}{r|rrrr} -\frac{1}{2} & 2 & -1 & 3 & -4 \\ & \downarrow & -1 & 1 & -2 \\ \hline & 2 & -2 & 4 & \boxed{-6} \end{array}$$

Division by $x + 3$:

$$\begin{array}{r|rrrr} -3 & 2 & -1 & 3 & -4 \\ & \downarrow & -6 & 21 & -72 \\ \hline & 2 & -7 & 24 & \boxed{-76} \end{array}$$

As we progress through this chapter you will find that there are occasions when you will do this kind of repeated synthetic division. On those occasions, you can save yourself some time in writing by making a table similar to the one below. The table is a condensed version of the four synthetic divisions on the previous page. The table is created by doing the second line of each synthetic division mentally and writing only the third line.

		2	−1	3	−4	
Division by $x - 2$:	2	2	3	9	14	remainder
Division by $x - 1$:	1	2	1	4	0	remainder
Division by $x + \frac{1}{2}$:	$-\frac{1}{2}$	2	−2	4	−6	remainder
Division by $x + 3$:	−3	2	−7	24	−76	remainder

■

EXAMPLE 14 Divide $\dfrac{x^3 - 1}{x - 1}$.

SOLUTION Writing the numerator as $x^3 + 0x^2 + 0x - 1$ and using synthetic division, we have

$$\begin{array}{r|rrrr} +1 & 1 & 0 & 0 & -1 \\ & \downarrow & 1 & 1 & 1 \\ \hline & 1 & 1 & 1 & \boxed{0} \end{array}$$

which indicates

$$\frac{x^3 - 1}{x - 1} = x^2 + x + 1$$

■

Modeling: Average Cost

EXAMPLE 15 First Bank of San Luis Obispo charges \$2.00 per month and \$0.15 per check for a regular checking account. As we mentioned in the introduction to this section, the total monthly cost of this account is $C(x) = 2.00 + 0.15x$. To find the average cost of each of the x checks, we divide the total cost by the number of checks written. That is,

$$\overline{C}(x) = \frac{C(x)}{x}$$

© bluestocking/iStockPhoto

a. Find the formula for the average cost function, $\overline{C}(x)$.

b. Use the average cost function to fill in the following table.

x	1	5	10	15	20
$\overline{C}(x)$					

c. What happens to the average cost as more checks are written?

d. Assume that you write at least 1 check a month, but never more than 20 checks per month, and graph both $y = C(x)$ and $y = \overline{C}(x)$ on the same set of axes.

SOLUTION

a. The average cost function is

$$\overline{C}(x) = \frac{C(x)}{x}$$

$$= \frac{2.00 + 0.15x}{x}$$

$$= 0.15 + \frac{2.00}{x}$$

b. Using the formula from Part a we have

x	1	5	10	15	20
$\overline{C}(x)$	2.15	0.55	0.35	0.28	0.25

c. The average cost decreases as more checks are written.

d.

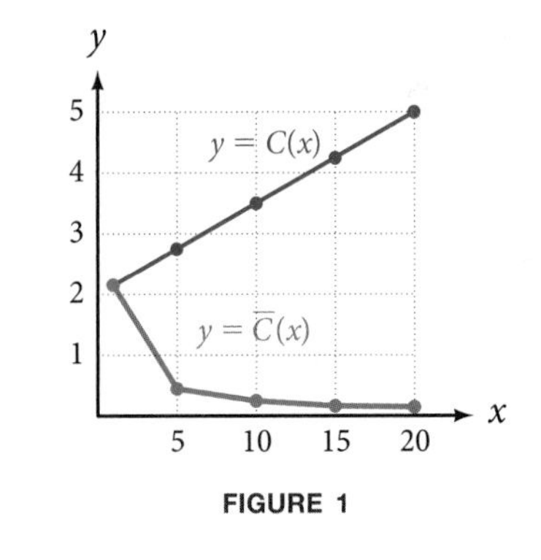

FIGURE 1

Using Technology 5.2

Let's see what happens when we use Wolfram|Alpha to do long division with polynomials.

EXAMPLE 16 Use Wolfram|Alpha to repeat Example 7 by asking it to divide $4x^3 - 6x - 11$ by $2x - 4$.

SOLUTION Here is how we enter the information.

You will notice lots of information displayed on the screen. We pick out the correct answer, which is

$$2x^2 + 4x + \frac{9}{2(x-2)} + 5$$

Wolfram Alpha LLC. 2012. Wolfram|Alpha
http://www.wolframalpha.com/
(access June 18, 2012)

Do you see that it matches the answer we produced in Example 7?

EXAMPLE 17 Use Wolfram|Alpha to repeat Example 9 by asking it to divide $x^3 + 9x^2 + 26x + 24$ by $x + 2$.

SOLUTION Here is how we enter the information

Can you pick out the correct answers? Do you know what some of the other pieces of information represent?

GETTING READY FOR CLASS

After reading through the preceding section, respond in your own words and in complete sentences.

A. What are the four steps used in long division with polynomials?
B. What does it mean to have a remainder of 0?
C. When must long division be performed, and when can factoring be used to divide polynomials?
D. When can you use synthetic division?

Problem Set 5.2

Find the following quotients.

1. $\dfrac{6x^3 + 12x^2 - 9x}{3x}$
2. $\dfrac{12x^5 - 18x^4 - 6x^3}{6x^3}$
3. $\dfrac{6y^4 - 3y^3 + 18y^2}{9y^2}$
4. $\dfrac{-9x^5 + 10x^3 - 12x}{-6x^4}$
5. $\dfrac{a^2b + ab^2}{ab}$
6. $\dfrac{9x^4y^4 + 18x^3y^4 - 27x^2y^4}{-9xy^3}$

Divide by factoring numerators and then dividing out common factors.

7. $\dfrac{x^2 - x - 6}{x + 2}$
8. $\dfrac{2a^2 + 3a - 9}{2a - 3}$
9. $\dfrac{5x^2 - 26xy - 24y^2}{5x + 4y}$
10. $\dfrac{x^3 + 8}{x + 2}$
11. $\dfrac{y^4 - 81}{y - 3}$
12. $\dfrac{x^3 + 2x^2 - 25x - 50}{x + 5}$

Divide using the long division method.

13. $\dfrac{x^2 + 4x - 8}{x - 3}$
14. $\dfrac{8x^2 - 26x - 9}{2x - 7}$
15. $\dfrac{3x^3 - 5x^2 + 2x - 1}{x - 2}$
16. $\dfrac{3y^3 - 19y^2 + 17y + 4}{3y - 4}$
17. $\dfrac{6x^3 + 7x^2 - x + 3}{3x^2 - x + 1}$
18. $\dfrac{9y^3 - 6y^2 + 8}{3y - 3}$
19. $\dfrac{a^4 + a^3 - 1}{a + 2}$
20. $\dfrac{y^4 - 81}{y - 3}$

Use synthetic division to find the following quotients.

21. $\dfrac{x^2 - 5x + 6}{x + 2}$
22. $\dfrac{x^2 + 8x - 12}{x - 3}$
23. $\dfrac{3x^2 - 4x + 1}{x - 1}$
24. $\dfrac{4x^2 - 2x - 6}{x + 1}$
25. $\dfrac{x^3 + 2x^2 + 3x + 4}{x - 2}$
26. $\dfrac{x^3 - 2x^2 - 3x - 4}{x - 2}$
27. $\dfrac{3x^3 - x^2 + 2x + 5}{x - 3}$
28. $\dfrac{2x^3 - 5x^2 + x + 2}{x - 2}$

29. $\dfrac{2x^3 + x - 3}{x - 1}$

30. $\dfrac{3x^3 - 2x + 1}{x - 5}$

31. $\dfrac{x^4 + 2x^2 + 1}{x + 4}$

32. $\dfrac{x^4 - 3x^2 + 1}{x - 4}$

33. $\dfrac{x^5 - 2x^4 + x^3 - 3x^2 - x + 1}{x - 2}$

34. $\dfrac{2x^5 - 3x^4 + x^3 - x^2 + 2x + 1}{x + 2}$

35. $\dfrac{x^2 + x + 1}{x - 1}$

36. $\dfrac{x^2 + x + 1}{x + 1}$

37. $\dfrac{x^4 - 1}{x + 1}$

38. $\dfrac{x^4 + 1}{x - 1}$

39. $\dfrac{x^3 - 1}{x - 1}$

40. $\dfrac{x^3 - 1}{x + 1}$

Use synthetic division to identify the quotient and remainder when $P(x)$ is divided by $f(x)$.

41. $P(x) = x^2 + 3x - 4, f(x) = x - 5$

42. $P(x) = 3x^2 + 5x - 2, f(x) = x - 1$

43. $P(x) = x^3 - 2x^2 + 5x + 4, f(x) = x + 3$

44. $P(x) = x^3 - 5x^2 + 2x - 3, f(x) = x + 4$

45. $P(x) = 5x^3 - 2x^2 + 4, f(x) = x - 3$

46. $P(x) = 7x^3 - 4x^2 + 1, f(x) = x - 2$

Use a table like the one in Example 13 to work the following problems.

47. Divide $2x^3 - 3x^2 + 5x - 6$ by $x - 3$, $x - \frac{1}{2}$, $x + 1$, and $x + 2$.

48. Divide $3x^3 + 5x^2 + 5x - 6$ by $x - 3$, $x - \frac{1}{2}$, $x + 1$, and $x + 2$.

49. Find the remainders when $2x^3 - 9x^2 + 3$ is divided by $x - 3$, $x - \frac{1}{2}$, $x + 1$, and $x + 2$.

50. Find the remainders when $4x^3 - 3x^2 - 2$ is divided by $x - 2$, $x - 3$, $x + \frac{1}{4}$, and $x + 3$.

51. Factor $x^3 + 6x^2 + 11x + 6$ completely if one of its factors is $x + 3$.

52. Factor $x^3 + 10x^2 + 29x + 20$ completely if one of its factors is $x + 4$.

53. Factor $x^3 + 5x^2 - 2x - 24$ completely if one of its factors is $x + 3$.

54. Factor $x^3 + 3x^2 - 10x - 24$ completely if one of its factors is $x + 2$.

55. Find $P(-2)$ if $P(x) = x^2 - 5x + 6$. Compare it with the remainder in Problem 21.

56. Find $P(3)$ if $P(x) = x^2 + 4x - 8$. Compare it with the remainder in Problem 13.

Modeling Practice

© Randy Mayes/iStockPhoto

57. Checking Account Founders Bank of San Luis Obispo charges \$1.50 per month and \$0.12 per check for a regular checking account. The total monthly cost of this account is $C(x) = 1.50 + 0.12x$. To find the average cost of each of the x checks, we divide the total cost by the number of checks written. That is,

$$\overline{C}(x) = \frac{C(x)}{x}$$

a. Use the total cost function to fill in the following table.

x	1	5	10	15	20
$C(x)$					

b. Find the formula for the average cost function, $\overline{C}(x)$.

c. Use the average cost function to fill in the following table.

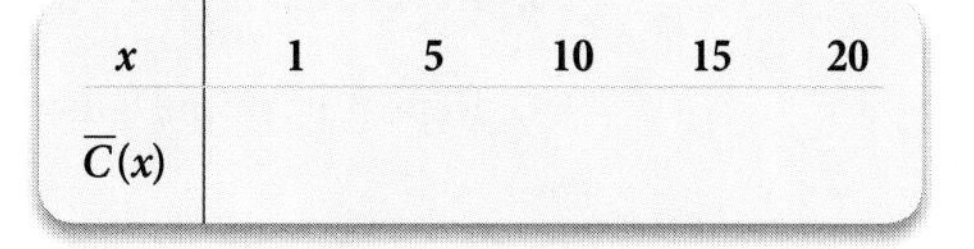

x	1	5	10	15	20
$\overline{C}(x)$					

d. What happens to the average cost as more checks are written?

e. Assume that you write at least 1 check a month, but never more than 20 checks per month, and graph both $y = C(x)$ and $y = \overline{C}(x)$ on the same set of axes.

f. Give the domain and range of each of the functions you graphed in Part e.

58. Average Cost A company that manufactures computer diskettes uses the function $C(x) = 200 + 2x$ to represent the daily cost of producing x diskettes.

a. Find the average cost function, $\overline{C}(x)$.

b. Use the average cost function to fill in the following table:

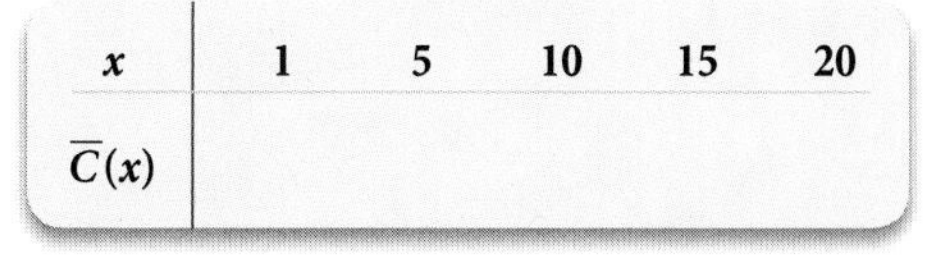

x	1	5	10	15	20
$\overline{C}(x)$					

c. What happens to the average cost as more items are produced?

d. Graph the function $y = \overline{C}(x)$ for $x > 0$.

e. What is the domain of this function?

f. What is the range of this function?

59. Average Cost For long distance service, a particular phone company charges a monthly fee of \$4.95 plus \$0.07 per minute of calling time used. The relationship between the number of minutes of calling time used, m, and the amount of the monthly phone bill $T(m)$ is given by the function $T(m) = 4.95 + 0.07m$.

a. Find the total cost when 100, 400, and 500 minutes of calling time is used in 1 month.

b. Find a formula for the average cost per minute function $\overline{T}(m)$.

c. Find the average cost per minute of calling time used when 100, 400, and 500 minutes are used in 1 month.

© Dan Fletcher/iStockPhoto

60. Average Cost A company manufactures electric pencil sharpeners. Each month they have fixed costs of \$40,000 and variable costs of \$8.50 per sharpener. Therefore, the total monthly cost to manufacture x sharpeners is given by the function $C(x) = 40{,}000 + 8.5x$.

a. Find the total cost to manufacture 1,000, 5,000, and 10,000 sharpeners a month.

b. Write an expression for the average cost per sharpener function $\overline{C}(x)$.

c. Find the average cost per sharpener to manufacture 1,000, 5,000, and 10,000 sharpeners per month.

Using Technology Exercises

61. Repeat Example 6 using Wolfram|Alpha.

62. Repeat Example 3 using Wolfram|Alpha.

63. Repeat Example 14 using Wolfram|Alpha.

64. Repeat Example 8 using Wolfram|Alpha.

Getting Ready for the Next Section

65. If $P(x) = 2x^3 - 3x^2 + 5x - 7$, find $P(5)$.

66. If $C(x) = 0.0002x^3 - 0.03x^2 + 4x + 800$, find $C(100)$ and $C(200)$.

67. Divide $4x^3 - 16x^2 - 9x + 36$ by $x - 4$.

68. Divide $x^3 - 5x + 2$ by $x - 2$.

69. Solve for x: $x^2 + 2x - 1 = 0$.

70. Solve for x: $6x^2 - 7x - 5 = 0$.

Zeros of a Polynomial Function

In Section 5.2 we found that dividing $P(x)$ by $x - r$ yields a polynomial $q(x)$ and a real number R such that

$$P(x) = (x - r)q(x) + R$$

Look at what happens when we find $P(r)$ by replacing x with r in this equation:

$$\begin{aligned} P(r) &= (r - r)q(r) + R \\ &= 0q(r) + R \\ &= 0 + R \\ &= R \end{aligned}$$

The remainder R is the same as $P(r)$. We state this fact formally as a theorem.

[Δ≠Σ THEOREM ***Remainder Theorem for Division***

If the polynomial $P(x)$ is divided by $x - r$, the remainder is $P(r)$.

The remainder theorem gives us an alternate way to calculate function values for polynomials. For instance, if we want to find $P(5)$ for some polynomial $P(x)$, we can divide $P(x)$ by $x - 5$. The remainder from that division process will be $P(5)$.

VIDEO EXAMPLES

SECTION 5.3

EXAMPLE 1 If $P(x) = 2x^3 - 3x^2 + 5x - 7$, find $P(5)$ by substitution and also by using the remainder theorem.

SOLUTION To find $P(5)$ by substitution we substitute 5 for x in the expression $P(x) = 2x^3 - 3x^2 + 5x - 7$.

$$\begin{aligned} P(5) &= 2 \cdot 5^3 - 3 \cdot 5^2 + 5 \cdot 5 - 7 \\ &= 2(125) - 3(25) + 25 - 7 \\ &= 250 - 75 + 25 - 7 \\ &= 193 \end{aligned}$$

To find $P(5)$ using the remainder theorem we look for the remainder when $P(x)$ is divided by $x - 5$. Using synthetic division to divide $P(x)$ by $x - 5$, we have

$$\begin{array}{r|rrrr} 5 & 2 & -3 & 5 & -7 \\ & & 10 & 35 & 200 \\ \hline & 2 & 7 & 40 & \boxed{193} \end{array}$$

As you can see, the remainder 193 is the same as $P(5)$. ■

Modeling: Business Application

EXAMPLE 2 A small company can manufacture between 50 and 300 prerecorded DVDs each day with a daily cost of

$$C(x) = 0.0002x^3 - 0.03x^2 + 4x + 800$$

Find the daily cost to manufacture 100 DVDs a day and the cost to manufacture 200 DVDs a day.

© Kazuhiko Yoshino/iStockPhoto

SOLUTION We must find $C(100)$ and $C(200)$. To do so, we could simply substitute the numbers 100 and 200 for x in the cost equation and then simplify the result. With a calculator, the work would not be difficult. On the other hand, we can use our remainder theorem and synthetic division to obtain the same result.

$$\begin{array}{r|rrrr} 100 & 0.0002 & -0.03 & 4 & 800 \\ & & 0.02 & -1 & 300 \\ \hline & 0.0002 & -0.01 & 3 & \boxed{1{,}100} \end{array}$$

The remainder is 1,100, which indicates that $C(100) = 1{,}100$. The cost to produce 100 DVDs a day is $1,100.

$$\begin{array}{r|rrrr} 200 & 0.0002 & -0.03 & 4 & 800 \\ & & 0.04 & 2 & 1{,}200 \\ \hline & 0.0002 & 0.01 & 6 & \boxed{2{,}000} \end{array}$$

Since the remainder is 2,000, we have $C(200) = 2{,}000$, telling us that the cost to produce 200 DVDs a day is $2,000. ■

The Factor Theorem

Next we develop the relationships that exist between the linear factors of a polynomial and values of x that make $P(x)$ equal to 0.

From Section 5.2 we know that dividing a polynomial $P(x)$ by $x - r$ will result in a polynomial $q(x)$ and a constant R such that

$$P(x) = (x - r)q(x) + R$$

Using the remainder theorem we can replace R with $P(r)$ giving us

$$P(x) = (x - r)q(x) + P(r)$$

Now, if, $P(r) = 0$ this last expression becomes

$$P(x) = (x - r)q(x)$$

which means that $x - r$ is a factor of $P(x)$.

Likewise, if $x - r$ is a factor of $P(x)$ to begin with, then dividing $P(x)$ by $x - r$ will result in a remainder of 0. Since the remainder when dividing by $x - r$ is $P(r)$, it follows that $P(r) = 0$.

This line of reasoning gives rise to our next theorem.

[Δ≠Σ THEOREM *Factor Theorem for Polynomials*

The binomial $x - r$ is a factor of the polynomial $P(x)$ if and only if $P(r) = 0$.

EXAMPLE 3 Find a polynomial $P(x)$ such that $P(2) = 0$, $P(1) = 0$, and $P(-2) = 0$.

SOLUTION By the factor theorem we have the following:

Since $P(2) = 0$, $x - 2$ is a factor of $P(x)$;

Since $P(1) = 0$, $x - 1$ is a factor of $P(x)$;

Since $P(-2) = 0$, $x + 2$ is a factor of $P(x)$.

Therefore, one expression for $P(x)$ is

$$\begin{aligned} P(x) &= (x-2)(x-1)(x+2) \\ &= (x^2-4)(x-1) \\ &= x^3 - x^2 - 4x + 4 \end{aligned}$$

The polynomial $P(x) = x^3 - x^2 - 4x + 4$ is a polynomial for which

$$P(2) = P(1) = P(-2) = 0$$

EXAMPLE 4 Is $x - 4$ a factor of $P(x) = 4x^3 - 16x^2 - 9x + 36$?

SOLUTION From the factor theorem, we know that $x - 4$ is a factor of $P(x)$ if $P(4) = 0$. The remainder theorem tells us that $P(4)$ is the remainder when $P(x)$ is divided by $x - 4$. Therefore, if we divide $P(x)$ by $x - 4$ and find that the remainder is 0, then $x - 4$ is a factor of $P(x)$. On the other hand, if the remainder is not 0, then $x - 4$ is not a factor of $P(x)$.

Using synthetic division to divide by $x - 4$ we have

$$\begin{array}{r|rrrr} 4 & 4 & -16 & -9 & 36 \\ & & 16 & 0 & -36 \\ \hline & 4 & 0 & -9 & \boxed{0} \end{array}$$

Since the remainder is 0, $x - 4$ is a factor of $P(x)$.

For a polynomial $P(x)$, a number r for which $P(r) = 0$ is a special number. Not only does r lead us to a factor of $P(x)$ (as the factor theorem indicates), it is also a solution to the polynomial equation $P(x) = 0$, as we shall see in a moment. Because of this, we give r a special name, as reflected in the next definition.

(dĕf DEFINITION *Zeros*

If $P(x)$ is a polynomial and $P(r) = 0$, then r is a *zero* of $P(x)$.

We could restate the factor theorem, using this definition, by saying

$$x - r \text{ is a factor of } P(x) \Leftrightarrow r \text{ is a zero of } P(x)$$

Zeros, Factors, and Solutions

Saying that r is a zero of $P(x)$ is the same as saying that $x = r$ is a solution to the polynomial equation $P(x) = 0$. Here's why: If r is a zero of $P(x)$, then, by definition, $P(r) = 0$, meaning that $x = r$ is a solution to the equation $P(x) = 0$. [Substituting r for x in $P(x) = 0$ gives us $P(r) = 0$, which must be true since r is a zero of $P(x)$.] On the other hand, if r is a solution to $P(x) = 0$, then $P(r) = 0$, meaning r is a zero of $P(x)$.

To help clarify the relationship between a zero and a factor of a polynomial and a solution to a polynomial equation, consider the following three closely-related quantities:

The polynomial $x^2 - 2x - 3$

The polynomial function $P(x) = x^2 - 2x - 3$

The polynomial equation $x^2 - 2x - 3 = 0$

Because the polynomial $x^2 - 2x - 3$ has a factor of $x - 3$, $P(3) = 0$, meaning that 3 is a zero of the polynomial function $P(x) = x^2 - 2x - 3$. In addition, since $P(3) = 0$, $x = 3$ is a solution to the polynomial equation $x^2 - 2x - 3 = 0$.

In general, we say: polynomials have factors, polynomials and polynomial functions have zeros, and polynomial equations have solutions.

EXAMPLE 5 Solve $x^3 - 5x + 2 = 0$ completely if $x = 2$ is one solution.

SOLUTION Since $x = 2$ is a solution, $x - 2$ is a factor. Dividing by $x - 2$ with synthetic division we have

$$\begin{array}{r|rrrr} 2 & 1 & 0 & -5 & 2 \\ & & 2 & 4 & -2 \\ \hline & 1 & 2 & -1 & \boxed{0} \end{array}$$

Using this result, we have

$$x^3 - 5x + 2 = 0$$
$$(x - 2)(x^2 + 2x - 1) = 0$$

This is as far as we can factor. We have one solution of $x = 2$. The other two solutions come from the quadratic formula.

$$\begin{aligned} x &= \frac{-2 \pm \sqrt{4 - 4(1)(-1)}}{2(1)} \\ &= \frac{-2 \pm 2\sqrt{2}}{2} \\ &= -1 \pm \sqrt{2} \end{aligned}$$

■

EXAMPLE 6 Find all zeros for $P(x) = 6x^3 - 25x^2 + 16x + 15$ if 3 is one of the zeros.

SOLUTION We know from the discussion above that if 3 is a zero of $P(x)$, then $x - 3$ is one of the factors of $P(x)$. Dividing $P(x)$ by $x - 3$ with synthetic division we have

$$\begin{array}{r|rrrr} 3 & 6 & -25 & 16 & 15 \\ & & 18 & -21 & -15 \\ \hline & 6 & -7 & -5 & \boxed{0} \end{array}$$

which tells us that

$$6x^3 - 25x^2 + 16x + 15 = (x - 3)(6x^2 - 7x - 5)$$

Since zeros of $P(x)$ and solutions to $P(x) = 0$ are the same, we can find the zeros we are looking for by solving the equation $P(x) = 0$.

$$6x^3 - 25x^2 + 16x + 15 = 0$$

$$(x - 3)(6x^2 - 7x - 5) = 0$$

$$(x - 3)(2x + 1)(3x - 5) = 0$$

$$x = 3 \quad \text{or} \quad x = -\frac{1}{2} \quad \text{or} \quad x = \frac{5}{3}$$

The zeros of $P(x)$ are 3, $-\frac{1}{2}$, and $\frac{5}{3}$. ■

Modeling: Profit Functions

EXAMPLE 7 The cost to a company to produce x units of a product is given by the cost function $C(x) = 0.02x^3 - 0.5x^2 + 10x + 150$. The revenue the company expects to realize from the sale of x units of the product is given by the revenue function $R(x) = -1.1x^2 + 41.5x$. If $P(x)$ represents the profit realized on the sale of x units of the product, compute and interpret $P(20)$.

© Patrick Heagney/iStockPhoto

SOLUTION As you know, profit is defined to be the difference between the revenue and the cost. That is,

$$P(x) = R(x) - C(x)$$

In this case,

$$\begin{aligned} P(x) &= -1.1x^2 + 41.5x - (0.02x^3 - 0.5x^2 + 10x + 150) \\ &= -1.1x^2 + 41.5x - 0.02x^3 + 0.5x^2 - 10x - 150 \\ &= -0.02x^3 - 0.6x^2 + 31.5x - 150 \end{aligned}$$

Substituting 20 for x in $P(x)$, we get $P(20) = 80$. This means that when the company produces and sells 20 units of its product, it realizes a profit of \$80. ■

GETTING READY FOR CLASS

After reading through the preceding section, respond in your own words and in complete sentences.

A. If the polynomial $P(x)$ is divided by $x - r$, what is the remainder?
B. If a polynomial $P(x)$ is divided by $x - 5$, and the remainder is 193, what is $P(5)$?
C. If $x - 2$ is a factor of polynomial $P(x)$, then what number is $P(2)$?
D. How are profit, revenue, and cost related?

Problem Set 5.3

1. If $P(x) = 5x^3 - 2x^2 + 4$ find $P(3)$ by substitution and by synthetic division.
2. If $P(x) = 7x^3 - 4x^2 + 1$ find $P(2)$ by substitution and by synthetic division.
3. If $P(x) = x^2 + 3x - 4$ find $P(5)$ by substitution and by synthetic division.
4. If $P(x) = 3x^2 + 5x - 2$ find $P(6)$ by substitution and by synthetic division.
5. If $P(x) = x^3 - 2x^2 + 4$ find $P(-3)$.
6. If $P(x) = x^3 - 5x^2 + 2x - 3$ if $P(-4)$.
7. Find $P(4)$ if $P(x) = 2x^4 - 3x^3 + 4x^2 - 5x + 4$.
8. Find $P(2)$ if $P(x) = x^4 + 2x^3 - 3x^2 - 8x + 1$.

Use synthetic division and the factor theorem to answer questions 9 through 16.

9. Is $x - 5$ a factor of $2x^3 - 8x^2 - 7x - 15$?
10. Is $x + 5$ a factor of $2x^3 - 8x^2 - 7x - 15$?
11. Is $x - 2$ a factor of $3x^3 + 6x^2 + 2x + 4$?
12. Is $x + 2$ a factor of $3x^3 + 6x^2 + 2x + 4$?
13. Is $x - 1$ a factor of $2x^4 - 3x^3 + 4x^2 - 5x + 4$?
14. Is $x + 1$ a factor of $2x^4 - 3x^3 + 4x^2 - 5x + 4$?
15. Is $x + 2$ a factor of $2x^4 + 3x^3 - 3x^2 - 8x - 4$?
16. Is $x - 2$ a factor of $2x^4 + 3x^3 - 3x^2 - 8x - 4$?
17. Factor $x^3 - 6x^2 + 9x - 4$ completely if $x - 4$ is one factor.
18. Factor $x^3 - 3x^2 - 13x + 15$ completely if $x - 5$ is one factor.
19. Find all solutions to $12x^3 - 47x^2 + 28x + 15 = 0$ if $x = 3$ is one solution.
20. Find all solutions to $8x^3 - 26x^2 + 17x + 6 = 0$ if $x = 2$ is one solution.
21. Find all solutions to $2x^3 - 11x^2 + 17x - 6 = 0$ if $x = \frac{1}{2}$ is one solution.
22. Find all solutions to $3x^3 - 13x^2 + 13x - 3 = 0$ if $x = \frac{1}{3}$ is one solution.
23. Find all solutions to $x^3 - 3x - 2 = 0$ if $x = 2$ is one solution.
24. Find all solutions to $x^3 + 3x^2 - 4 = 0$ if $x = 1$ is one solution.
25. Find all solutions to $x^3 + 4x^2 + x + 4 = 0$ if $x = -4$ is one solution.
26. Find all solutions to $x^3 + x^2 + 4x + 4 = 0$ if $x = -1$ is one solution.
27. Solve $x^3 - 6x + 4 = 0$ if $x = 2$ is one of the solutions.
28. Solve $2x^3 - 5x + 6 = 0$ if $x = -2$ is one of the solutions.

29. Use the factor theorem to see if $x + 1$ is a factor of $P(x) = x^3 - 1$. Do not use synthetic division.

30. Use the factor theorem to see if $x - 1$ is a factor of $P(x) = x^3 - 1$. Do not use synthetic division.

Getting Ready for Calculus The problems below are representative of the type of problems you will need to be familiar with in order to be successful in calculus. These particular problems are taken from *Applied Calculus* by Denny Burzynski and Guy Sanders, published by XYZ Textbooks.

31. Find $f(20) - f(0)$ if $f(x) = 1200x - x^3$.

32. Find $f(52.70) - f(0)$ if $f(x) = 13.89x$.

33. Find $f(5) - f(2)$ if $f(x) = \frac{x^3}{3} - \frac{5x^2}{2} + 4x$.

34. Find $f(4) - f(2)$ if $f(x) = -\frac{x^3}{3} + \frac{5x^2}{2} - 4x$.

Modeling Practice

35. A company can manufacture x items for a total cost of

$$C(x) = -0.01x^2 + 5x + 300$$

dollars. How much does it cost to manufacture 200 items and how much does it cost to manufacture 300 items?

36. A company can manufacture x items for a total cost of

$$C(x) = -0.01x^2 + 7x + 400$$

dollars. How much does it cost to manufacture 200 items and how much does it cost to manufacture 500 items?

37. A company can manufacture between 100 and 500 calculators per day with a daily cost of

$$C(x) = 0.0001x^3 - 0.04x^2 + 5x + 300$$

dollars. Find the daily cost to manufacture 200 calculators and the cost to manufacture 300 calculators a day.

© Daniel Laflor/iStockPhoto

38. A company finds that the daily cost in dollars to manufacture between 100 and 500 bottles of sunscreen per day is given by the equation

$$C(x) = 0.0001x^3 - 0.04x^2 + 3x + 2000$$

Find the daily cost to manufacture 200 and 300 bottles of sunscreen.

39. The revenue realized by a company on the sale of x units of one of its products is approximated by

$$R(x) = \frac{1}{3}x^3 - \frac{7}{2}x^2 - 300x + 200$$

Compute and interpret $R(85)$.

40. The cost to a company to produce x units of a product is given by the cost function $C(x) = -0.035x^2 + 40x + 25$. Compute and interpret $C(600)$.

41. Data indicates that the cost to a company of shipping x units of product is approximately $C(x) = 0.0001x^3 - 0.002x^2 + 0.03x$. Compute and interpret $C(350)$.

42. The revenue realized by a company on the sale of x units of its product is given by the revenue function $R(x) = x^3 + 4x^2 + 160x$. Compute and interpret $R(70)$.

Getting Ready for the Next Section

43. Solve $x^3 - 4x = 0$.

44. Factor $x^3 - x^2 - 9x + 9$.

45. Solve $x^2 - 9 = 0$.

46. If $f(x) = x^3 - x^2 - 9x + 9$, find

a. $f(-4)$ **b.** $f(0)$ **c.** $f(2)$

47. Solve $x^2 - 3 = 0$.

48. Factor $x^4 - 10x^2 + 9 = 0$.

Graphing Polynomial Functions

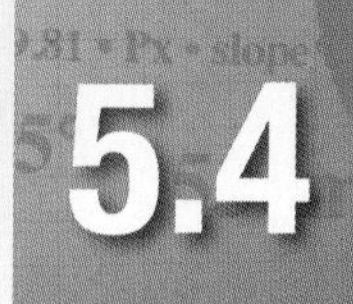

The linear, quadratic, and cubic functions we have graphed previously are all examples of polynomial functions. In this section we will extend the work we have done previously with graphing polynomial functions to include more complicated third- and fourth-degree polynomial functions.

To graph the polynomial functions in this section we will use a combination of three items we have covered previously: x-intercepts, a sign chart, and a table of ordered pairs that satisfy the equation. In addition, we will assume that the graph of a polynomial function with integer coefficients is a smooth continuous curve with no breaks or sharp points. Here is our first example.

VIDEO EXAMPLES

SECTION 5.4

EXAMPLE 1 Graph $y = x^3 - 4x$.

SOLUTION We start by finding the x-intercepts. Recall that we do so by letting $y = 0$ and solving for x.

x-intercepts

$0 = x^3 - 4x$	Let y equal 0
$0 = x(x^2 - 4)$	Begin factoring
$0 = x(x + 2)(x - 2)$	Factor completely
$x = 0$ or $x + 2 = 0$ or $x - 2 = 0$	Set factors to 0
$x = 0$ or $x = -2$ or $x = 2$	The x-intercepts

The x-intercepts are -2, 0, and 2. (Note that these numbers are also zeros of the polynomial function $P(x) = x^3 - 4x$.)

A *sign chart* will show us where the factors of y are positive and where they are negative, which in turn will tell us where y is positive and where y is negative. The points on the graph for which y is positive are points that lie above the x-axis. Likewise, the points on the graph for which y is negative are points that lie below the x-axis. Here is the sign chart for our equation.

SIGN CHART

	$x < -2$	$-2 < x < 0$	$0 < x < 2$	$x > 2$	
Sign of x	$- - - -$	$- - - -$	$+ + + +$	$+ + + +$	
Sign of $x + 2$	$- - - -$	$+ + + +$	$+ + + +$	$+ + + +$	
Sign of $x - 2$	$- - - -$	$- - - -$	$- - - -$	$+ + + +$	
$y = x(x + 2)(x - 2)$ is	negative	positive	negative	positive	
The graph is	below	above	below	above	the x-axis

(Boundaries on the number line: -2, 0, 2)

To be able to make a more accurate sketch of the graph we need some additional points. The table next to Figure 1 shows some additional points found by letting $x = -3, -1, 1$, and 3 and finding the corresponding values of y. This is called a *table of ordered pairs*.

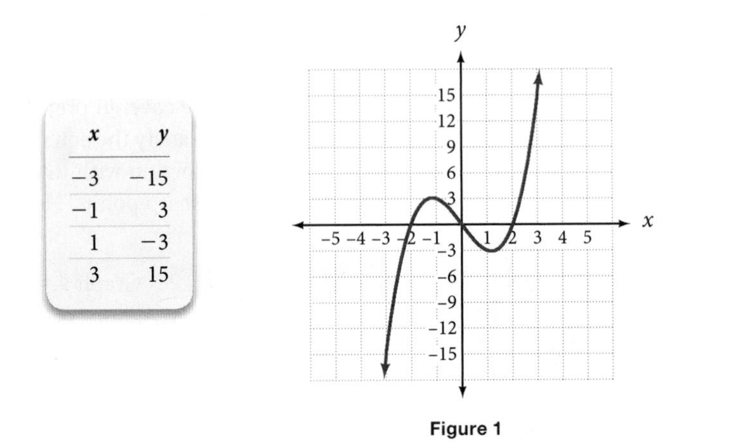

x	y
-3	-15
-1	3
1	-3
3	15

Figure 1

There are three things to note about the graph shown in Figure 1:

1. We have labeled the axes differently in order to plot more points. In this case units on the y-axis are multiples of 3, while the x-axis is labeled as usual.
2. The graph has two turning points. That is, there are two points at which the graph changes its vertical direction from going up to going down or from going down to going up.
3. The turning points are not actually at the points $(-1, 3)$ and $(1, -3)$, although it may look like they are. In fact, the turning points have x-coordinates of $\frac{-2\sqrt{3}}{3}$ and $\frac{2\sqrt{3}}{3}$. However, to show that this is true we would need some of the tools developed in calculus. For our purposes, when we graph polynomial functions of degree larger than 2, we are interested in the general shape of the graph only; we do not need to show the exact location of the turning points.

EXAMPLE 2 Graph $f(x) = x^3 - x^2 - 9x + 9$.

SOLUTION The x-intercepts and sign chart both come from the factors of y.

x-intercepts

$$x^3 - x^2 - 9x + 9 = 0$$

$$x^2(x - 1) - 9(x - 1) = 0$$

$$(x - 1)(x^2 - 9) = 0 \quad \text{Factor by grouping}$$

$$(x - 1)(x + 3)(x - 3) = 0 \quad \text{Factor completely}$$

$$x - 1 = 0 \quad x + 3 = 0 \quad x - 3 = 0$$

$$x = 1 \quad x = -3 \quad x = 3 \quad \text{The x-intercepts}$$

SIGN CHART

Sign of $x - 1$	$- - - -$	$- - - -$	$+ + + +$	$+ + + +$	
Sign of $x + 3$	$- - - -$	$+ + + +$	$+ + + +$	$+ + + +$	
Sign of $x - 3$	$- - - -$	$- - - -$	$- - - -$	$+ + + +$	
y is	negative	positive	negative	positive	
	-3	1	3		
The graph is	below	above	below	above	the x-axis

Recall from Section 5.2 that the remainder theorem tells us that the remainder when $P(x)$ is divided by $x - r$ is $P(r)$. As far as ordered pairs that satisfy the equation $y = P(x)$ are concerned, when $x = r$, the associated y value is $y = P(r)$, the remainder when $P(x)$ is divided by $x - r$. Below is a condensed table of six synthetic divisions that give us six ordered pairs that are solutions to our equation.

	1	**−1**	**−9**	**9**	**ordered pair**
−4	1	−5	11	−35	(−4, −35)
−2	1	−3	−3	15	(−2, 15)
−1	1	−2	−7	16	(−1, 16)
0	1	−1	−9	9	(0, 9)
2	1	1	−7	−5	(2, −5)
4	1	3	3	21	(4, 21)

The graph is shown in Figure 2. Note that we have labeled the units on the y-axis in multiples of 5 this time, in order to see more of the graph. Note also that the graph has two turning points, just like the graph in Figure 1.

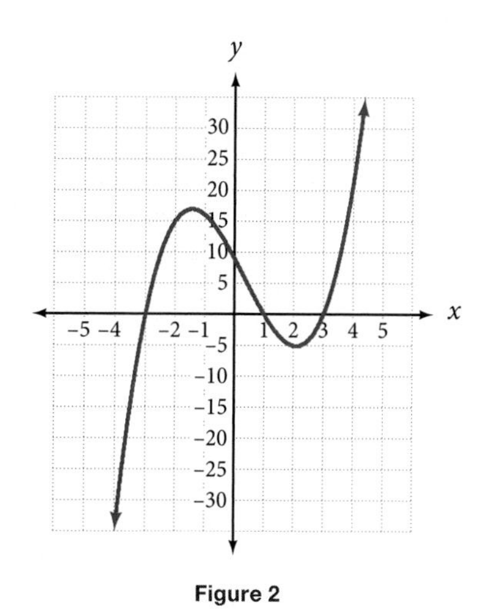

Figure 2

EXAMPLE 3 Graph $y = x^4 - 3x^2$.

SOLUTION We follow the same procedure shown in the first two examples.

x-intercepts

$$x^4 - 3x^2 = 0$$
$$x^2(x^2 - 3) = 0$$
$$x^2(x + \sqrt{3})(x - \sqrt{3}) = 0$$
$$x^2 = 0 \qquad x + \sqrt{3} = 0 \qquad x - \sqrt{3} = 0$$
$$x = 0 \qquad x = -\sqrt{3} \qquad x = \sqrt{3}$$

Note Notice that $x^2 - 3$ factors as the difference of two squares into $(x - \sqrt{3})(x + \sqrt{3})$. If you need to convince yourself that this is the correct factorization, do so by multiplying the factors. You will obtain $x^2 - 3$.

The x-intercepts are 0 and $\pm\sqrt{3}$. We can approximate $\sqrt{3}$ with 1.7 when graphing. In the sign chart, note that the factor x^2 is positive for all nonzero values of x.

SIGN CHART

Sign of x^2	+ + + +	+ + + +	+ + + +	+ + + +	
Sign of $x + \sqrt{3}$	− − − −	+ + + +	+ + + +	+ + + +	
Sign of $x - \sqrt{3}$	− − − −	− − − −	− − − −	+ + + +	
$y = x^2(x + \sqrt{3})(x - \sqrt{3})$ is	positive	negative	negative	positive	
	$-\sqrt{3}$	0	$\sqrt{3}$		
The graph is	above	below	below	above	the x-axis

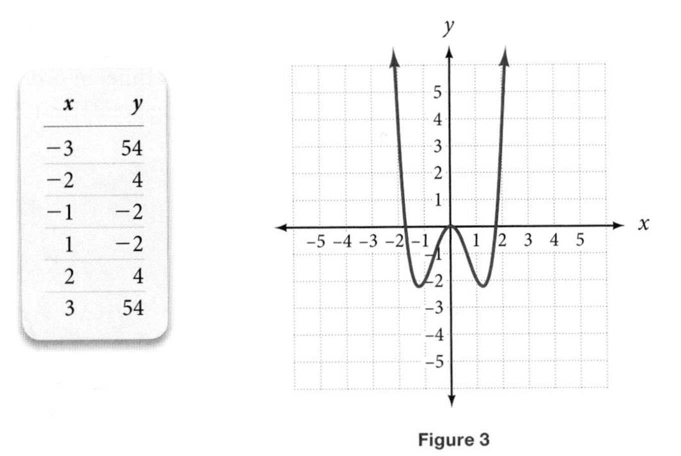

x	y
−3	54
−2	4
−1	−2
1	−2
2	4
3	54

Figure 3

The graph is shown in Figure 3 along with a table of values for x and y. Note that for $x = \pm 3$, y is 54, meaning that the graph increases rapidly for x less than $-\sqrt{3}$ and x greater than $\sqrt{3}$. We should note also that the smallest values of y do not occur at -1 and 1, although it may look that way in Figure 3. The smallest values of y actually occur when $x = -\sqrt{1.5}$ and $x = \sqrt{1.5}$, but again, to show that this is true we would need some of the tools developed in calculus. Remember, for our purposes, we are interested only in the general shape of the graph. And finally, note that this graph has 3 turning points. ■

If we look at the degree of the equations in Examples 1, 2, and 3 and the number of turning points in their associated graphs, we find that in each case, the number of turning points is one less than the degree of the equation. In general, we can state the following.

$[\Delta \neq \Sigma$ *Turning Points*

The graph of a polynomial function of degree n will have at most $n - 1$ turning points.

For example, the graph of a 5th-degree polynomial function will have 4 or fewer turning points.

EXAMPLE 4 Graph $g(x) = x^4 - 10x^2 + 9$.

SOLUTION First, find the x-intercepts:

x-intercepts

$$x^4 - 10x^2 + 9 = 0$$

$$(x^2 - 9)(x^2 - 1) = 0$$

$$(x + 3)(x - 3)\ (x + 1)\ (x - 1) = 0$$

$$x + 3 = 0 \qquad x - 3 = 0 \qquad x + 1 = 0 \qquad x - 1 = 0$$

$$x = -3 \qquad x = 3 \qquad x = -1 \qquad x = 1$$

SIGN CHART

Sign of $x + 1$	$- - - -$	$- - - -$	$+ + + +$	$+ + + +$	$+ + + +$
Sign of $x - 1$	$- - - -$	$- - - -$	$- - - -$	$+ + + +$	$+ + + +$
Sign of $x + 3$	$- - - -$	$+ + + +$	$+ + + +$	$+ + + +$	$+ + + +$
Sign of $x - 3$	$- - - -$	$- - - -$	$- - - -$	$- - - -$	$+ + + +$
y is	positive	negative	positive	negative	positive
	-3	-1	1	3	
The graph is	above	below	above	below	above the x-axis

As was the case in Example 2, we use the remainder theorem and synthetic division to create a table of ordered pairs that satisfy the equation.

	1	**0**	**−10**	**0**	**9**	**ordered pair**
−4	1	−4	6	−24	105	(−4, 105)
−2	1	−2	−6	12	−15	(−2, −15)
0	1	0	−10	0	9	(0, 9)
2	1	2	−6	−12	−15	(2, −15)
4	1	4	6	24	105	(4, 105)

The graph is shown in Figure 4.

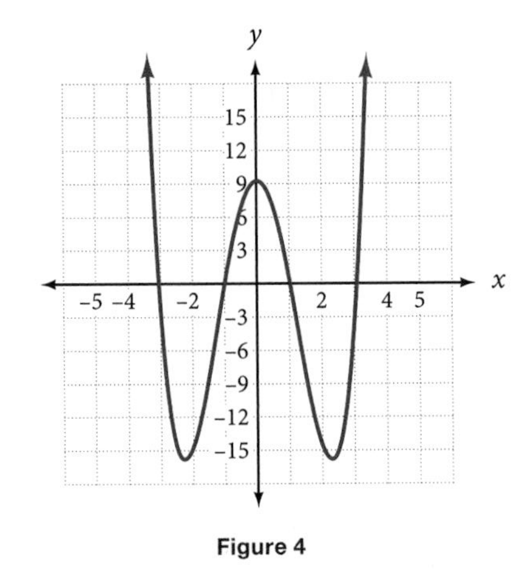

Figure 4

Modeling: Online Advertising Revenue

EXAMPLE 5 Social networking and video-sharing sites, such as Facebook, dating sites, and YouTube, often include rich media and display advertising placed around user-generated content. Suppose that, for a particular social networking company, the annual revenue from rich media advertisements, in millions of dollars, for the years 2007 through 2012 can be approximated with the model $R(x) = -x^4 + 11x^3 - 39x^2 + 45x$, where x is the number of years from the beginning of 2007. (Source: eMarketer April 2008)

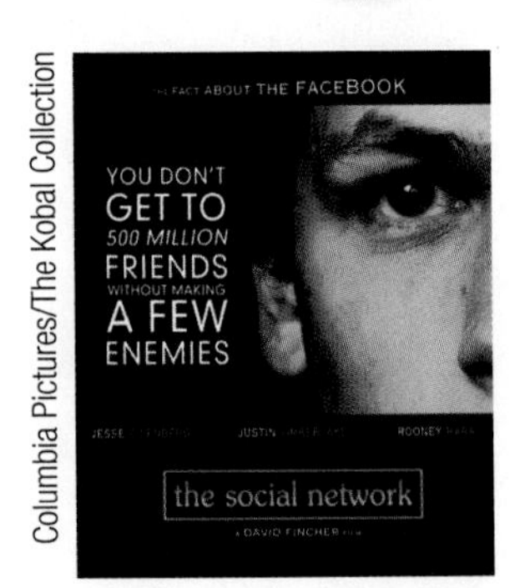

Columbia Pictures/The Kobal Collection

a. Find the company's revenue from rich media advertisements at the beginning of the years 2007, 2008, 2009, 2010, 2011, and 2012. Summarize the results in a table.

b. If the company adheres to its present business model, can it expect its revenues to rise after 2012?

c. Construct the graph of this function.

d. Does the information contained in the graph correspond with the answer to Part b?

SOLUTION

a. You can find the revenues by direct computation:

In 2007, $x = 2007 - 2007 = 0$, and $R(0) = -0^4 + 11 \cdot 0^3 - 39 \cdot 0^2 + 45 \cdot 0 = 0$

In 2008, $x = 2008 - 2007 = 1$, and $R(1) = -1^4 + 11 \cdot 1^3 - 39 \cdot 1^2 + 45 \cdot 1 = 16$

In 2009, $x = 2009 - 2007 = 2$, and $R(2) = -2^4 + 11 \cdot 2^3 - 39 \cdot 2^2 + 45 \cdot 2 = 6$

In 2010, $x = 2010 - 2007 = 3$, and $R(3) = -3^4 + 11 \cdot 3^3 - 39 \cdot 3^2 + 45 \cdot 3 = 0$

In 2011, $x = 2011 - 2007 = 4$, and $R(4) = -4^4 + 11 \cdot 4^3 - 39 \cdot 4^2 + 45 \cdot 4 = 4$

In 2012, $x = 2012 - 2007 = 5$, and $R(5) = -5^4 + 11 \cdot 5^3 - 39 \cdot 5^2 + 45 \cdot 5 = 0$

Year	2007	2008	2009	2010	2011	2012
x	0	1	2	3	4	5
Revenue	0	16	6	0	4	0

b. The table shows that the company experienced three revenue turning points.

One in 2008, when the revenue went up from 0 to 16 then back down to 6.
One in 2010, when the revenue went down from 6 to 0 then back up to 4.
One in 2011, when the revenue went up from 0 to 4 then back down to 0.

The function $R(x) = -x^4 + 11x^3 - 39x^2 + 45x$ is 4th degree and can have at most $4 - 1 = 3$ turning points. Since all three have been accounted for, there cannot be any others. Therefore, revenues cannot be expected to rise after 2012.

c. To graph the function, it would be convenient to factor the expression $-x^4 + 11x^3 - 39x^2 + 45x$. First, notice $-x$ can be factored out, giving $-x(x^3 - 11x^2 + 39x - 45)$. Now you need to factor $x^3 - 11x^2 + 39x - 45$. It may look like a daunting task, but notice you have some valuable information. You know that the function has a zero at $x = 5$. (In 2012, when $x = 5$, the revenue is zero.) Since $x = 5$ is a zero, $x - 5$ is a factor. Use synthetic division to divide $x^3 - 11x^2 + 39x - 45$ by $x - 5$.

$$\begin{array}{r|rrrr} 5 & 1 & -11 & 39 & -45 \\ & & 5 & -30 & 45 \\ \hline & 1 & -6 & 9 & \boxed{0} \end{array}$$

Using this result we have

$$\begin{aligned} -x^4 + 11x^3 - 39x^2 + 45x &= -x(x-5)(x^2 - 6x + 9) \\ &= -x(x-5)(x-3)^2 \end{aligned}$$

The x-intercepts and the sign chart both come directly from the factors of $R(x)$.

$$-x^4 + 11x^3 - 39x^2 + 45x = 0$$

$$-x(x-5)(x-3)^2 = 0$$

$$x = 0 \quad x = 3 \quad x = 5$$

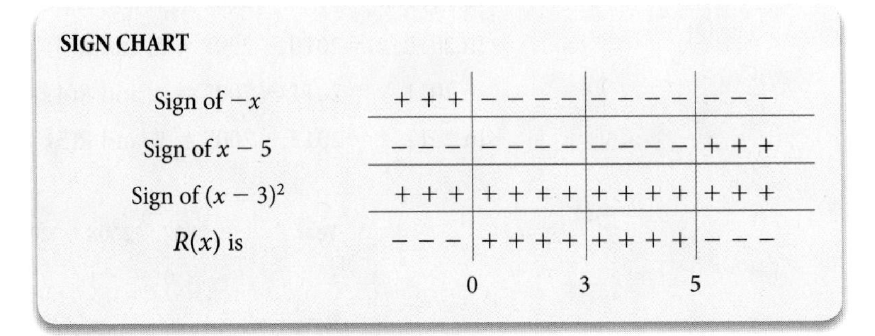

SIGN CHART

		0		3		5	
Sign of $-x$	+ + +		− − − −		− − − −		− − −
Sign of $x - 5$	− − −		− − − −		− − − −		+ + +
Sign of $(x-3)^2$	+ + +		+ + + +		+ + + +		+ + +
$R(x)$ is	− − −		+ + + +		+ + + +		− − −

Using the information we have compiled, we sketch the graph in Figure 5.

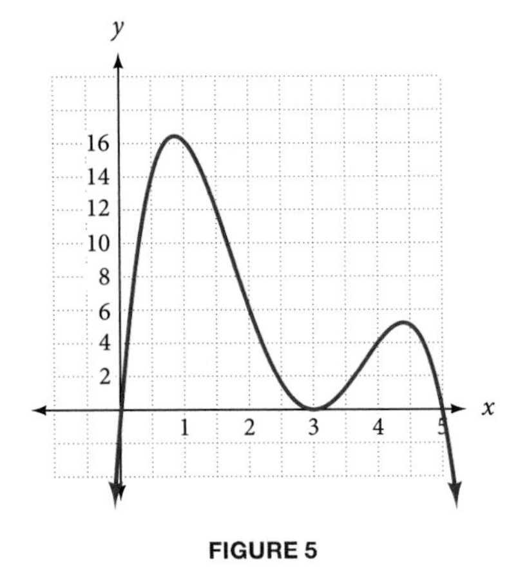

FIGURE 5

d. The graph corresponds nicely with the behavior determined from Part b. ■

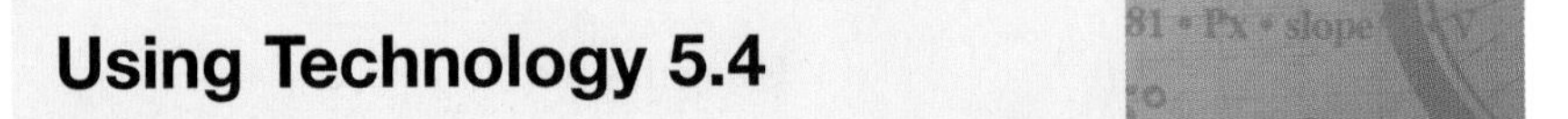

Using Technology 5.4

You can use your graphing calculator to convert tables of data into linear, quadratic, cubic, quartic, and exponential functions that model the phenomena you are studying. The process is called *regression analysis* and, using the statistic mode of your calculator, you are able to perform the analysis and create an equation simply by inputting the data and pressing the appropriate keys. You can study regression analysis more fully in an introductory statistics course.

Modeling: Advertising Spending

© AIMSTOCK/iStockPhoto

EXAMPLE 6 Advertisers spend huge amounts of money advertising their products on online social networking sites. To analyze the relationship between the amount of money advertisers spend on social networking sites and number of years they have been doing so, a researcher collected the following data. In the table, x represents the number of years from the year 2008, and the amounts spent on advertising are in billions of dollars. (Source: eMarketer)

Year	2008	2009	2010	2011	2012
x	0	1	2	3	4
Amount	1.2	1.3	1.3	1.4	1.5

a. Use the data set to construct a cubic model (function) that relates the amount advertisers spend to promote their products on social media networks to the number of years since 2008.

b. Use your model to predict the number of dollars advertisers will spend in the year 2015.

SOLUTION

a. Here is how to construct a function from a table of data values.

Step 1: ***Clear previous data out of your lists:***
Press [STAT] → ClrList [L1],[L2]
Step 2: ***Enter your data into two lists:***
Press [STAT] [ENTER] and enter the x-values into L1 and the y-values into L2.

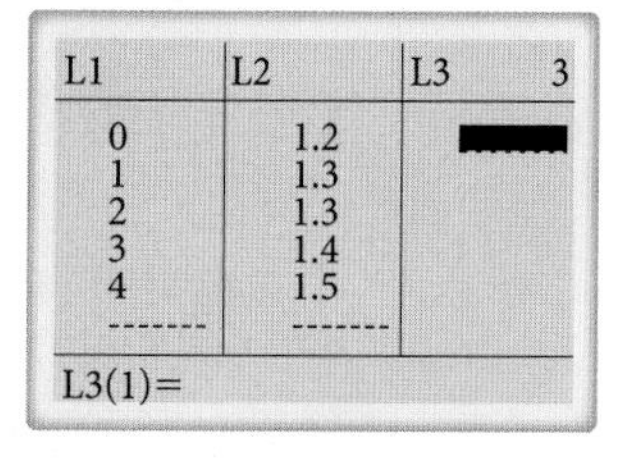

FIGURE 6

Step 3: ***Construct a graph of the ordered pairs:***
Press [2nd] [Y=] [ENTER] [ENTER] to activate the statistical plot feature. Scroll down to Type and select the first type. Then scroll down to XList and YList and enter L1 and L2, respectively (Figure 7a). Finally, scroll to Mark and select the first mark. Now press [WINDOW] and enter the smallest of your x-values in Xmin, and the largest value of x in Xmax. Then enter the smallest of your y-values into Ymin and the largest of your y-values in Ymax (Figure 7b). Press the [GRAPH] key. The calculator displays a scatter diagram in Figure 7c.

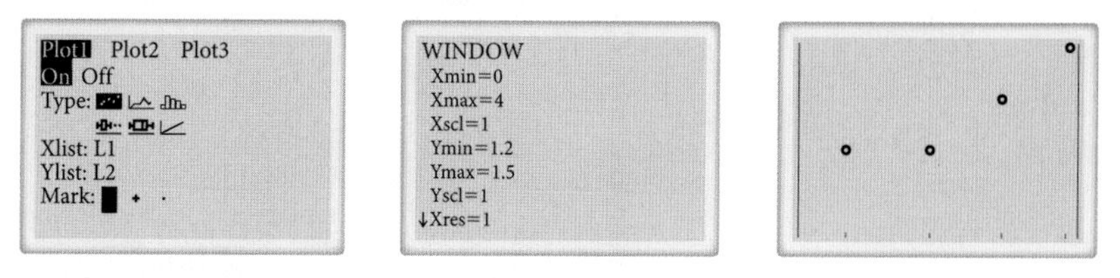

FIGURE 7a FIGURE 7b FIGURE 7c

Step 4: Construct the function:
Press STAT. Right arrow over to CALC. Under the CALC options, choose item 6, CubicReg (Figure 8a). Enter L1, L2, Y1 to make the data in those lists available to the calculator's regression algorithm, and to put the regression equation into Y1. Press ENTER and the regression equation appears on your screen (Figure 8b).

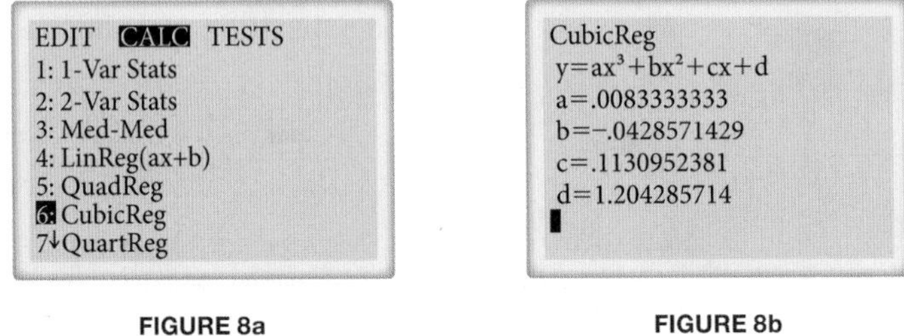

FIGURE 8a **FIGURE 8b**

Our cubic regression equation is

$$y = 0.008x^3 - 0.043x^2 + 0.113x + 1.204$$

Step 5: If we press GRAPH, the graph of our curve appears with our original scatter plot (Figure 9a). If we press ZOOM, then arrow down to 9:ZoomStat, we have the screen shown in Figure 9b.

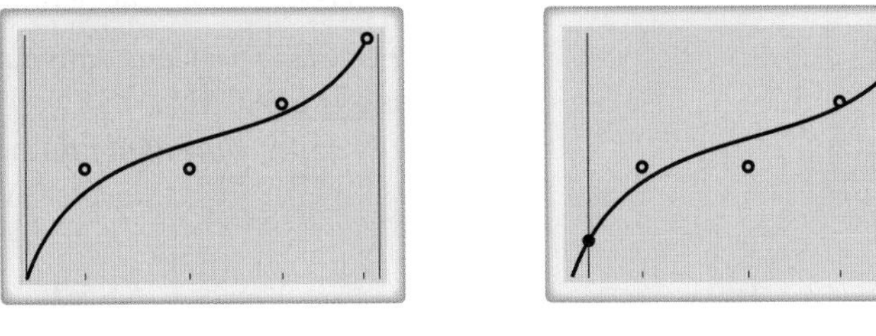

FIGURE 9a **FIGURE 9b**

b. To predict the amount of money advertisers will spend in the year 2015, substitute $x = 7$ into the regression equation.

$$y = 0.008(7)^3 - 0.043(7)^2 + 0.113(7) + 1.204$$
$$= 2.63$$

If advertisers keep spending at their current rate, they will spend approximately \$2.63 billion in the year 2015 for social networking advertising. ■

GETTING READY FOR CLASS

After reading through the preceding section, respond in your own words and in complete sentences.

A. If you have a polynomial function, how do you find the intercepts for its graph?
B. How is a sign chart helpful in graphing?
C. If a polynomial function has degree 4, how many turning points does the graph have?
D. What is the y-intercept for the graph of $y = x^3 - 9x$?

Problem Set 5.4

Use the method developed in this section to sketch the graph of each of the following.

1. $y = x^3 - x^2$ **2.** $y = x^3 - 4x^2$ **3.** $y = \frac{1}{4}x^3 - x^2$

4. $y = \frac{1}{4}x^3 - 2x^2$ **5.** $y = x^3 - 4x$ **6.** $y = \frac{1}{3}x^3 - 2x^2$

7. $y = x^3 - x$ **8.** $y = x^3 - 9x$ **9.** $y = x^3 - 3x$

10. $y = 4x^3 - 9x$ **11.** $y = x^3 + 3x^2 - x - 3$ **12.** $y = x^3 - 2x^2 - x - 2$

13. $y = x^3 + x^2 - 9x - 9$ **14.** $y = x^3 + 4x^2 - x - 4$ **15.** $y = x^4 - 4x^2$

16. $y = x^4 - 9x^2$ **17.** $y = x^4 - 2x^2$ **18.** $y = x^4 - 5x^2$

19. $y = x^4 - 5x^2 + 4$ **20.** $y = x^4 - 4x^2 + 3$ **21.** $y = x^4 - 11x^2 + 18$

22. $y = x^4 - 13x^2 + 36$

Modeling Practice

23. Advertising Many companies advertise on niche social networks that are devoted to specific hobbies or interests. Suppose that, for a particular niche social network, the annual percent of all niche specific advertising dollars for the years 2010 through 2012 can be approximated with the model $P(x) = x^3 - 6x^2 + 9x$, where x is the number of years from the beginning of 2010.

a. Find the percent of all niche specific advertising this collection of social network companies garnered in the years 2010, 2011, and 2012 and project anticipated percents for 2013 and 2014. Summarize the results in a table.

b. If the company adheres to its present business model, can it expect its revenues to rise after 2014 without declining?

c. Construct the graph of this function.

24. Text Message Advertising Some companies place advertisements in text messages. They spend money on such things as display banners, links or icons placed on wireless application protocols (WAP), mobile HTML sites or embedded in mobile applications such as maps or games or videos. Suppose that through the years 2008-2015, a company uses the text message advertising spending plan $S(x) = -x^3 + 9x^2 - 15x + 9$, where $S(x)$ is in millions of dollars and x is the number of years from the year 2008.

a. Find the amount of money this company spent or plans to spend each year on text message advertising for the years 2008 through 2015. Summarize the results in a table.

b. If the company adheres to this spending plan, can it expect to increase its spending after the year 2015?

c. Construct the graph of this function.

© Siniša Botaš/iStockPhoto

25. Band Revenue Over the years 2006 through 2012, a touring back-up band saw its annual revenues modeled by the function

$$R(x) = x^4 - 16x^3 + 78x^2 - 112x + 49$$

where $R(x)$ is in thousands of dollars and x is the number of years since 2006. (Source: GrabStats.com)

a. Find the annual revenues for this band for the years 2006 through 2012. Summarize the results in a table.

b. If the band's income continues according to this model, can it expect to increase its revenues after the year 2013?

c. Construct the graph of this function.

26. Water Quality An engineering company, studying the quality of water in a particular city, has determined that under a new set of requirements, x years from now, the amount A (in parts per million) of a particular particle in the city's water can be modeled by the function $A(x) = -x^5 - 2x^4 + 2x^3 + 8x^2 + 7x + 2$. It is known that 1 year ago, there were zero of these particles in the city's water and, under the new requirements, that 2 years from now there will again be 0 of these particles in the water.

a. Use synthetic division, and the fact that you know two zeros, to help you factor this function into two linear factors, one appearing to the first power and the other to the 4th power.

b. Make a table of values for $x = -1, 0, 0.5, 1, 1.5$, and 2.

c. Graph this function for $x = -1$ to $x = 2$.

d. The graph shows only one turning point. Based on your knowledge of the relationship between the factored form and the number of zeros of a polynomial function, could this curve turn around again after $x = 2$ and increase back through $x = 0$?

Using Technology Exercises

27. Wireless Users The weekly number of digital media and wireless internet users in the U.S. has increased dramatically since 2005. The table displays the number of users for each year from 2005 through 2010. (Source: Bridge Ratings)

Year	2005	2006	2007	2008	2009	2010
Number of users (in millions)	5.7	10.1	19.2	28.8	51.8	93.3

a. Use this data set to construct a quadratic model (function) that relates the year to the weekly number N of U.S. users of digital media and wireless internet since the year 2005. Let x represent the number of years since 2005 so that $x = 0$ in 2005. Round all numbers to two decimal places.

b. Use your model to predict the weekly number N of U.S. users of digital media and wireless internet in the year 2011.

c. Use your model to predict the weekly number N of U.S. users of digital media and wireless internet in the year 2015. The population of the United States in the year 2012 is about 300 million people. What does this say about the model?

© Grady Reese/iStockPhoto

28. Cell Phone Subscriptions The number of cell phone subscribers in the U.S. for each year through the years of 1985 through 2010 is given in the table. Let x represent the number of years from 1985 (so that $x = 0$ represents 1985) and $N(x)$ represent the number of subscribers. (Source: CTIA—The Wireless Association)

Year	1985	1986	1987	1988	1989	1990	1991	1992
Number	0.3	0.7	1.2	3.5	5.3	7.6	11.0	16.0
Year	**1993**	**1994**	**1995**	**1996**	**1997**	**1998**	**1999**	**2000**
Number	20	24.1	33.8	44.0	55.3	69.2	86.0	109.5
Year	**2001**	**2002**	**2003**	**2004**	**2005**	**2006**	**2007**	**2008**
Number	128.3	140.8	158.7	182.1	207.9	233.0	250.0	262.7
Year	**2009**	**2010**						
Number	276.6	300.5						

a. Use this data set to construct a cubic model (function) that relates the year to the annual number of U.S. cell phone subscribers for the years 1985 through 2010. Round all numbers to three decimal places.

b. Use your model to predict the annual number of U.S. cell phone subscribers in the year 2000. Does the predicted value underestimate or overestimate the actual value? By how much?

c. The population of the United States in the year 2012 is about 300 million people. How is it that the number of cell phone subscribers in the U.S. in the year 2010 is 300 million?

Getting Ready for the Next Section

29. If $f(x) = \dfrac{x-4}{x-2}$, find

a. $f(-2)$

b. $f(2)$

30. Reduce to lowest terms: $\dfrac{x^2-9}{x-3}$.

31. Find the intercepts for $y = x + 3$.

32. For what value of x is $\dfrac{6}{x-2}$ undefined?

33. For the expression $\dfrac{x-4}{x-2}$,

a. What value of x makes the expression 0?

b. What value of x makes the expression undefined?

Success Skills

Don't complain about anything, ever.

Do you complain to your classmates about your teacher? If you do, it could be getting in the way of your success in the class.

I have students that tell me that they like the way I teach and that they are enjoying my class. I have other students, in the same class, that complain to each other about me. They say I don't explain things well enough. Are the complaining students giving themselves a reason for not doing well in the class? I think so. They are shifting the responsibility for their success from themselves to me. It's not their fault they are not doing well, it's mine. When these students are alone, trying to do homework, they start thinking about how unfair everything is and they lose their motivation to study. Without intending to, they have set themselves up to fail by making their complaints more important than their progress in the class.

What happens when you stop complaining? You put yourself back in charge of your success. When there is no one to blame if things don't go well, you are more likely to do well. I have had students tell me that, once they stopped complaining about a class, the teacher became a better teacher and they started to actually enjoy going to class.

If you find yourself complaining to your friends about a class or a teacher, make a decision to stop. When other people start complaining to each other about the class or the teacher, walk away; don't participate in the complaining session. Try it for a day, or a week, or for the rest of the term. It may be difficult to do at first, but I'm sure you will like the results, and if you don't, you can always go back to complaining.

Graphing Rational Functions

Rational Functions

The function $r(t) = \frac{785}{t}$ is called a *rational function* because the right side, $\frac{785}{t}$, is a rational expression (the numerator, 785, is a polynomial of degree 0 and the denominator, t, is a polynomial of degree 1). We can extend our knowledge of rational expressions to functions with the following definition:

DEFINITION *Rational Function*

A *rational function* is any function that can be written in the form

$$f(x) = \frac{P(x)}{Q(x)}$$

where $P(x)$ and $Q(x)$ are polynomials and $Q(x) \neq 0$.

VIDEO EXAMPLES

SECTION 5.5

EXAMPLE 1 For the rational function $f(x) = \dfrac{x-4}{x-2}$ find $f(0), f(-4)$, $f(4), f(-2)$, and $f(2)$.

SOLUTION To find these function values, we substitute the given value of x into the rational expression, and then simplify if possible.

$$f(0) = \frac{0-4}{0-2} = \frac{-4}{-2} = 2 \qquad f(-2) = \frac{-2-4}{-2-2} = \frac{-6}{-4} = \frac{3}{2}$$

$$f(-4) = \frac{-4-4}{-4-2} = \frac{-8}{-6} = \frac{4}{3} \qquad f(2) = \frac{2-4}{2-2} = \frac{-2}{0} \quad \text{Undefined}$$

$$f(4) = \frac{4-4}{4-2} = \frac{0}{2} = 0$$

Because the rational function in Example 1 is not defined when x is 2, the domain of that function does not include 2. We have more to say about the domain of a rational function next.

The Domain of a Rational Function

We can limit the domain of a rational function by specifying the values the variable can assume, as long as they don't make the function undefined. For example, if we want to evaluate $r(t) = \frac{785}{t}$ for t from 10 to 30, we can state the domain as $\{t \mid 10 \leq t \leq 30\}$. If the domain of a rational function is not specified, it is assumed to be all real numbers for which the function is defined. That is, the domain of the rational function

$$f(x) = \frac{P(x)}{Q(x)}$$

is all x for which $Q(x)$ is nonzero. For example:

$$\text{The domain for } f(x) = \frac{x-4}{x-2}, \text{ is } \{x \mid x \neq 2\}$$

$$\text{The domain for } g(x) = \frac{x^2 + 5}{x + 1}, \text{ is } \{x \mid x \neq -1\}$$

$$\text{The domain for } h(x) = \frac{x}{x^2 - 9}, \text{ is } \{x \mid x \neq -3, x \neq 3\}$$

Notice that, for these functions, $f(2)$, $g(-1)$, $h(-3)$, and $h(3)$ are all undefined, and that is why the domains are written as shown.

EXAMPLE 2 Graph the equation $y = \dfrac{x^2 - 9}{x - 3}$. How is this graph different from the graph of $y = x + 3$?

SOLUTION We know from the discussion on factoring and reducing to lowest terms that

$$y = \frac{x^2 - 9}{x - 3} = \frac{(x + 3)(x - 3)}{x - 3} = x + 3$$

This relationship is true for all x except $x = 3$ because the rational expressions with $x - 3$ in the denominator are undefined when x is 3. However, for all other values of x, the expressions

$$\frac{x^2 - 9}{x - 3} \quad \text{and} \quad x + 3$$

are equal. Therefore, the graphs of

$$y = \frac{x^2 - 9}{x - 3} \quad \text{and} \quad y = x + 3$$

will be the same except when x is 3. In the first equation, there is no value of y to correspond to $x = 3$. In the second equation, $y = x + 3$ so y is 6 when x is 3.

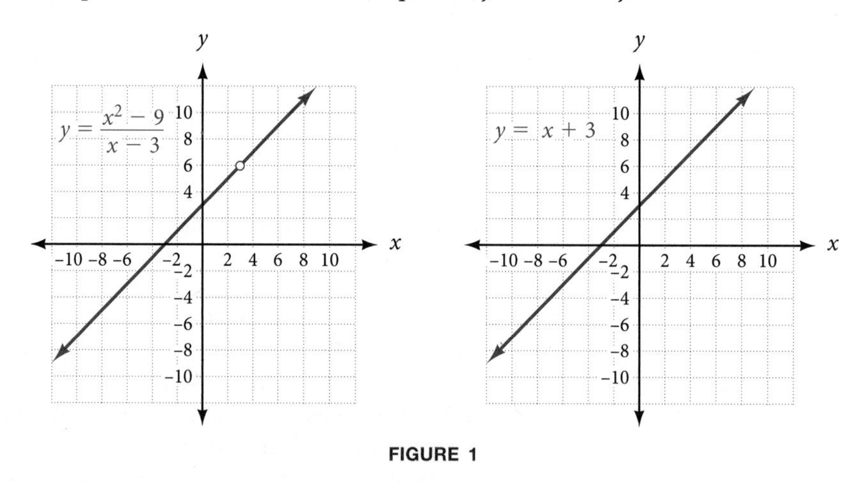

FIGURE 1

Now you can see the difference in the graphs of the two equations. To show that there is no y value for $x = 3$ in the graph on the left in Figure 1, we draw an open circle at that point on the line. We say that there is a *hole* at that point.

Notice that the two graphs shown in Figure 1 are both graphs of functions. Suppose we use function notation to designate them as follows:

$$f(x) = \frac{x^2 - 9}{x - 3} \qquad \text{and} \qquad g(x) = x + 3$$

The two functions, f and g, are equivalent except when $x = 3$, because $f(3)$ is undefined, while $g(3) = 6$. The domain of the function f is all real numbers except $x = 3$, while the domain for g is all real numbers, with no restrictions.

Graphing Rational Functions

In the introduction to this chapter, we looked at the graph of a rational function that illustrated a person's weekly weight loss on a diet. Our next example continues our investigation of these types of graphs.

EXAMPLE 3 Graph the rational function $f(x) = \dfrac{6}{x - 2}$.

SOLUTION To find the y-intercept, we let x equal 0.

$$\text{When } x = 0: \qquad y = \frac{6}{0 - 2} = \frac{6}{-2} = -3 \qquad y\text{-intercept}$$

The graph will not cross the x-axis. If it did, we would have a solution to the equation

$$0 = \frac{6}{x - 2}$$

which has no solution because there is no number to divide 6 by to obtain 0.

The graph of our equation is shown in Figure 2 along with a table giving values of x and y that satisfy the equation. Notice that y is undefined when x is 2. This means that the graph will not cross the vertical line $x = 2$. (If it did, there would be a value of y for $x = 2$.) The line $x = 2$ is called a *vertical asymptote* of the graph. The graph will get very close to the vertical asymptote, but will never touch or cross it.

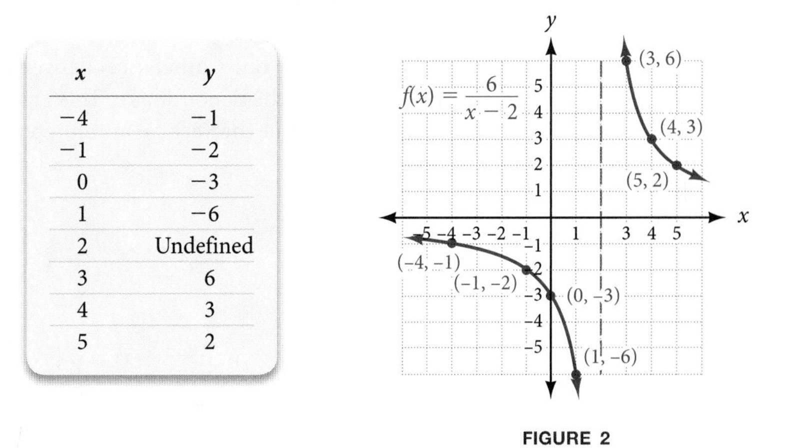

x	y
−4	−1
−1	−2
0	−3
1	−6
2	Undefined
3	6
4	3
5	2

FIGURE 2

If you were to graph $y = \frac{6}{x}$ on the coordinate system in Figure 2, you would see that the graph of $y = \frac{6}{x - 2}$ is the graph of $y = \frac{6}{x}$ with all points shifted 2 units to the right.

TECHNOLOGY NOTE *More About Example 3*

We know the graph of $f(x) = \frac{6}{x-2}$ will not cross the vertical asymptote $x = 2$ because replacing x with 2 in the equation gives us an undefined expression, meaning there is no value of y to associate with $x = 2$. We can use a graphing calculator to explore the behavior of this function when x gets closer and closer to 2 by using the table function on the calculator. To see how the function behaves as x gets close to 2, we let X take on values of 1.9, 1.99, and 1.999. Then we move to the other side of 2 and let X become 2.1, 2.01, and 2.001. Here is the setup:

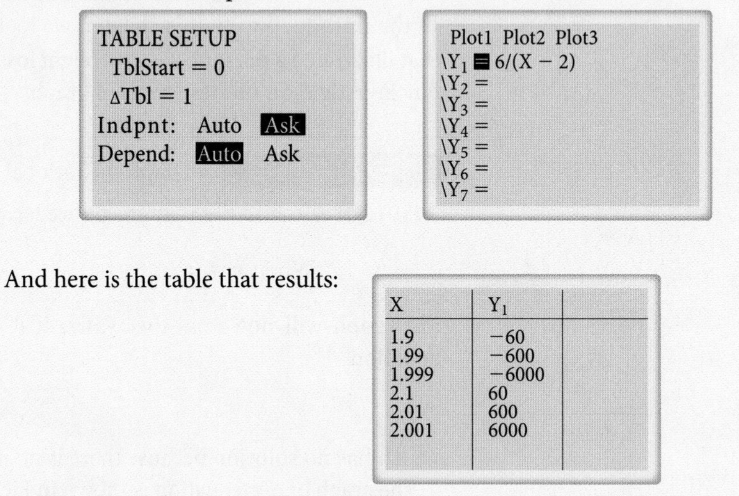

And here is the table that results:

X	Y_1
1.9	−60
1.99	−600
1.999	−6000
2.1	60
2.01	600
2.001	6000

As you can see, the values in the table support the shape of the curve in Figure 2 around the vertical asymptote.

EXAMPLE 4 Graph: $g(x) = \dfrac{6}{x+2}$

SOLUTION The only difference between this equation and the equation in Example 3 is in the denominator. This graph will have the same shape as the graph in Example 3, but the vertical asymptote will be $x = -2$ instead of $x = 2$. Figure 3 shows the graph.

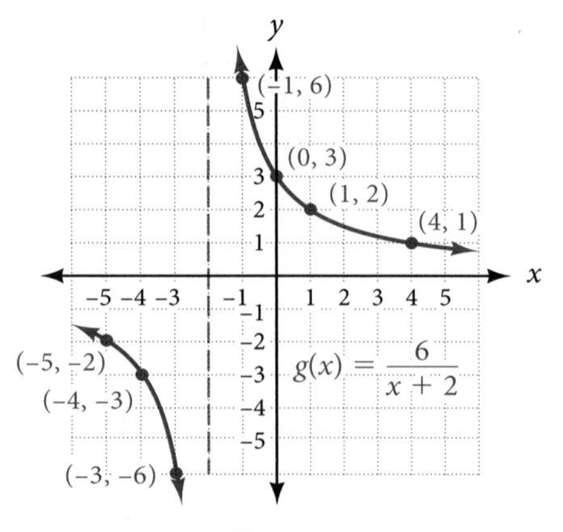

Figure 3

Notice that the graphs shown in Figures 2 and 3 are both graphs of functions because no vertical line will cross either graph in more than one place. Notice the similarities and differences in our two functions,

$$f(x) = \frac{6}{x - 2} \quad \text{and} \quad g(x) = \frac{6}{x + 2}$$

and their graphs. The vertical asymptotes shown in Figures 2 and 3 correspond to the fact that both $f(2)$ and $g(-2)$ are undefined. The domain for the function f is all real numbers except $x = 2$, while the domain for g is all real numbers except $x = -2$.

More About Graphing Rational Functions

We continue our investigation of the graphs of rational functions by considering the graph of a rational function with binomials in the numerator and denominator.

EXAMPLE 5 Graph the rational function $y = \dfrac{x - 4}{x - 2}$.

SOLUTION In addition to making a table to find some points on the graph, we can analyze the graph as follows:

1. The graph will have a y-intercept of 2, because when $x = 0$, $y = \dfrac{-4}{-2} = 2$.
2. To find the x-intercept, we let $y = 0$ to get

$$0 = \frac{x - 4}{x - 2}$$

   The only way this expression can be 0 is if the numerator is 0, which happens when $x = 4$. (If you want to solve this equation, multiply both sides by $x - 2$. You will get the same solution, $x = 4$.)
3. The graph will have a *vertical asymptote* at $x = 2$, because $x = 2$ will make the denominator of the function 0, meaning y is undefined when x is 2.
4. The graph will have a *horizontal asymptote* at $y = 1$ because for very large values of x, $\frac{x-4}{x-2}$ is very close to 1. The larger x is, the closer $\frac{x-4}{x-2}$ is to 1. The same is true for very small values of x, such as $-1{,}000$ and $-10{,}000$. We will have more to say about horizontal asymptotes after we finish this example.

Putting this information together with the ordered pairs in the table next to the figure, we have the graph shown in Figure 4.

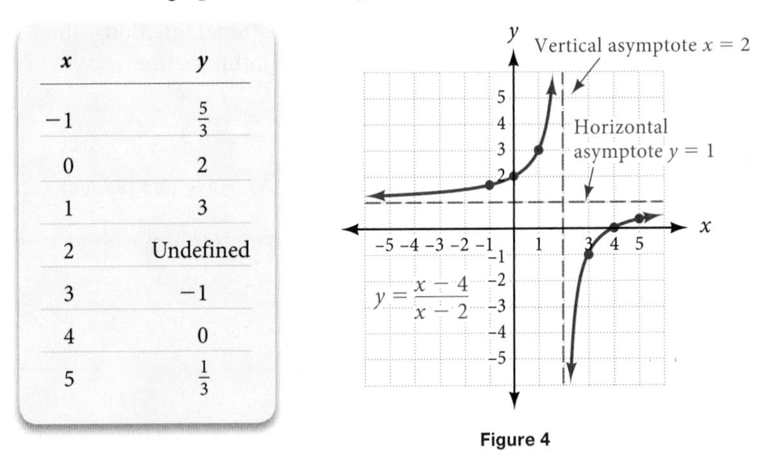

x	y
-1	$\frac{5}{3}$
0	2
1	3
2	Undefined
3	-1
4	0
5	$\frac{1}{3}$

Figure 4

TECHNOLOGY NOTE *More About Example 5*

In Figure 4, the horizontal asymptote is at $y = 1$. To show that the graph approaches this line as x becomes very large, we use the table function on our graphing calculator, with X taking values of 100, 1,000, and 10,000. To show that the graph approaches the line $y = 1$ on the left side of the coordinate system, we let X become -100, $-1{,}000$, and $-10{,}000$.

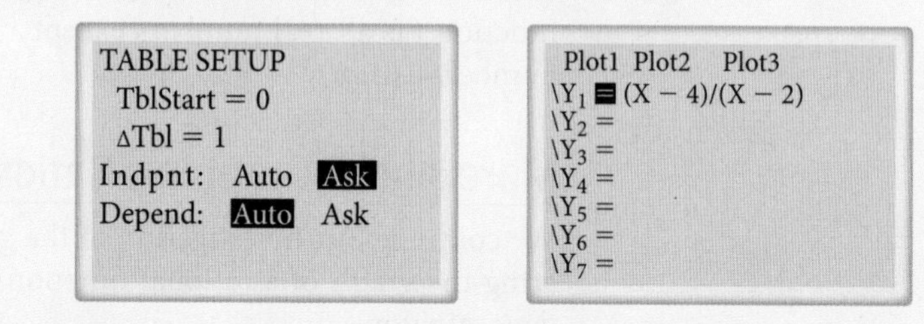

Here is the resulting table:

X	Y_1
100	.97959
1000	.998
10000	.9998
−100	1.0196
−1000	1.002
−10000	1.0002

As you can see, as x becomes very large in the positive direction, the graph approaches the line from below. As x becomes very small in the negative direction, the graph approaches the line from above.

Here is more information on horizontal asymptotes.

Horizontal Asymptotes

A ***horizontal asymptote*** for the graph of a rational function is the horizontal line that the graph approaches as x takes on extreme values. The horizontal asymptote for the rational function $f(x) = \frac{P(x)}{Q(x)}$ will occur at

1. the x-axis, when the degree of $P(x)$ is less than the degree of $Q(x)$, or
2. the line $y = \frac{a}{b}$, where a is the leading coefficient for $P(x)$, and b is the leading coefficient for $Q(x)$, when the degree of $P(x)$ and the degree of $Q(x)$ are equal.

Here are two rational functions which have horizontal asymptotes (their graphs are shown later in this section):

$y = \dfrac{x+2}{x-3}$	$y = \dfrac{x}{x^2-9}$
Degree of $P(x)$ = Degree of $Q(x)$	Degree of $P(x)$ < Degree of $Q(x)$
Horizontal asymptote at $y = \frac{1}{1} = 1$	Horizontal asymptote at the x-axis.

Previously we used tables in our graphing calculators to give some justification for why the horizontal asymptotes appear where they do. We can use long division for further justification. Consider the function

$$y = \frac{x+2}{x+3}$$

If we divide the numerator by the denominator on the right side of the equation, we have

$$y = 1 - \frac{1}{x+3} \quad \text{because} \quad \begin{array}{r} 1 \\ x+3 \overline{)x+2} \\ - \quad - \\ \underline{\cancel{+}x \cancel{+} 3} \\ -1 \end{array}$$

Now as x becomes very large and takes on values of 100, 1,000, and 10,000, the fraction $\frac{1}{x+3}$ becomes very small. In fact, we can make it as small as we want, by making x larger and larger. As we do this we are subtracting less and less from 1 in the expression:

$$1 - \frac{1}{x+3}$$

We say $1 - \frac{1}{x+3}$ goes toward 1 as x goes to ∞.

We can use the same type of reasoning as x goes further and further in the negative direction. That is,

$$1 - \frac{1}{x+3} \text{ goes toward 1 as } x \text{ goes toward } -\infty.$$

For another example, we can do long division on the right side of the function

$$y = \frac{2x-5}{x+1}$$

giving us

$$y = \frac{2x-5}{x+1} = 2 - \frac{7}{x+1}$$

The horizontal asymptote is 2 because

$$2 - \frac{7}{x+1} \text{ approaches 2 as } x \text{ goes to } \infty \text{ or } -\infty.$$

For a third example, if we reasoned the same way with the function:

$$y = \frac{2x+1}{3x-4}$$

we would find the horizontal asymptote at $y = \frac{2}{3}$.

The following steps can be used as a guide to help you sketch the graph of a rational function:

1. Find all intercepts.
2. Find the vertical asymptote(s).
3. Determine the horizontal asymptote.
4. Fill in gaps with the table values.
5. Draw a smooth curve through the points you have found.

EXAMPLE 6 Graph the rational function below, showing all the intercepts and asymptotes.

$$y = \frac{x+2}{x-3}$$

SOLUTION Following our guidelines for graphing, we have

x-intercept: $x = -2$ $\quad$ y-intercept: $y = -\frac{2}{3}$

Vertical Asymptote: $x = 3$ $\quad$ Horizontal Asymptote: $y = 1$

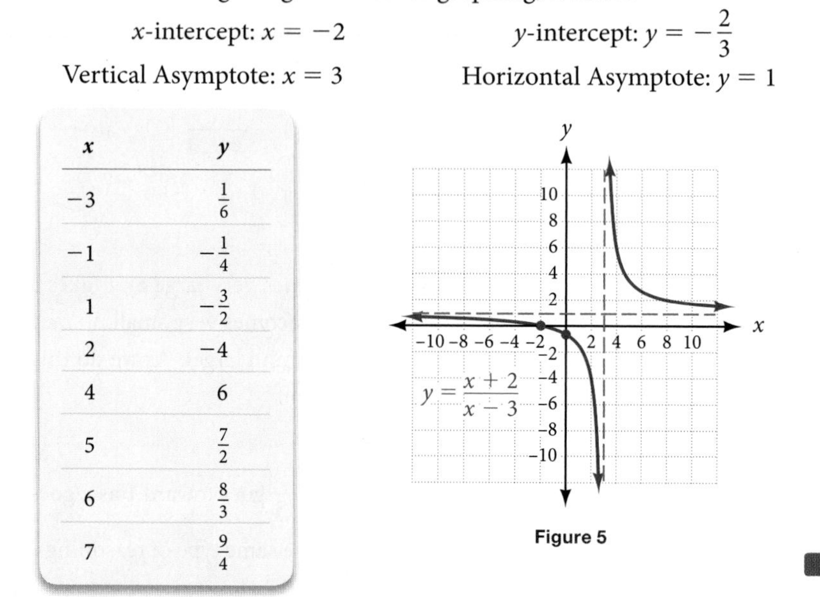

x	y
-3	$\frac{1}{6}$
-1	$-\frac{1}{4}$
1	$-\frac{3}{2}$
2	-4
4	6
5	$\frac{7}{2}$
6	$\frac{8}{3}$
7	$\frac{9}{4}$

Figure 5

EXAMPLE 7 Graph the rational function below, showing all the intercepts and asymptotes.

$$y = \frac{x}{x^2 - 9}$$

SOLUTION Following our guidelines for graphing, we have

x-intercept: $x = 0$ $\quad$ y-intercept: $y = 0$

Vertical Asymptotes: $x = -3, x = 3$ $\quad$ Horizontal Asymptote: x-axis

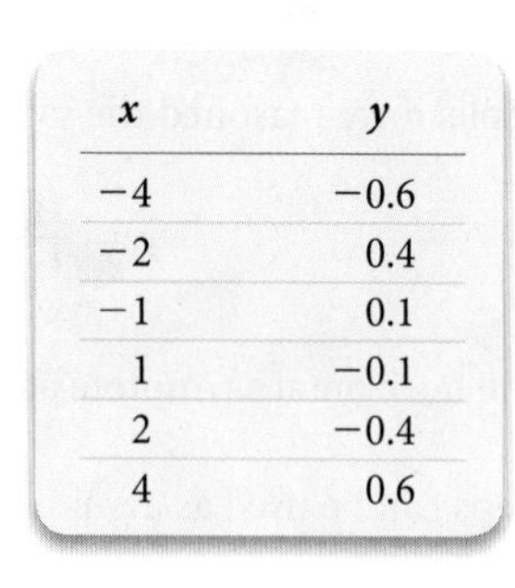

x	y
-4	-0.6
-2	0.4
-1	0.1
1	-0.1
2	-0.4
4	0.6

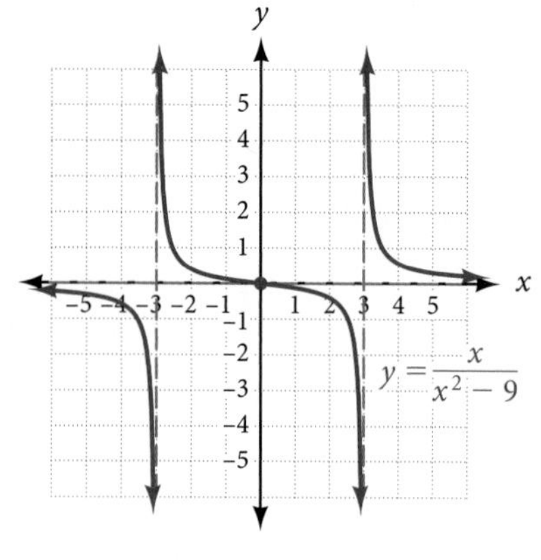

Figure 6

Modeling: Ground Speed and Wind Speed

Our next example involves a pedal-powered airplane traveling with the wind in one direction, and then against the wind on the return trip. This is a common situation with airplanes and also with watercraft traveling with, and against, a water current.

© Bob Rhine/NASA Dryden Flight Research Center

EXAMPLE 8 Francine is planning a 60-mile training flight through the desert on her cycle-plane, a pedal-driven aircraft. If there is no wind, she can pedal at an average speed of 15 miles per hour, so she can complete the flight in 4 hours.

a. If there is a headwind of x miles per hour, it will take Francine longer to fly 60 miles. Express the time it will take for Francine to complete the training flight as a function of x.

b. Make a table of values for the function.

c. Graph the function and explain what it tells you about the time Francine should allot for the ride.

SOLUTION

a. If there is a headwind of x miles per hour, Francine's ground speed will be $15 - x$ miles per hour. Using the fact that $time = \frac{distance}{rate}$, we find that the time needed for the flight will be

$$t = f(x) = \frac{60}{15 - x}$$

b. We evaluate the function for several values of x, as shown in the table.

x	0	3	5	7	9	10
t	4	5	6	7.5	10	12

For example, if the headwind is 5 miles per hour, then

$$t = \frac{60}{15 - 5} = \frac{60}{10} = 6$$

Francine's effective speed is only 10 miles per hour, and it will take her 6 hours to fly the 60 miles. The table shows that as the speed of the headwind increases, the time required for the flight increases also.

c. To graph our function, we use the information in the table, and we note the following:

1. the graph will appear above the x-axis only because t cannot be negative.
2. the graph will have a vertical asymptote at $x = 15$.
3. the graph will have a horizontal asymptote at the x-axis.

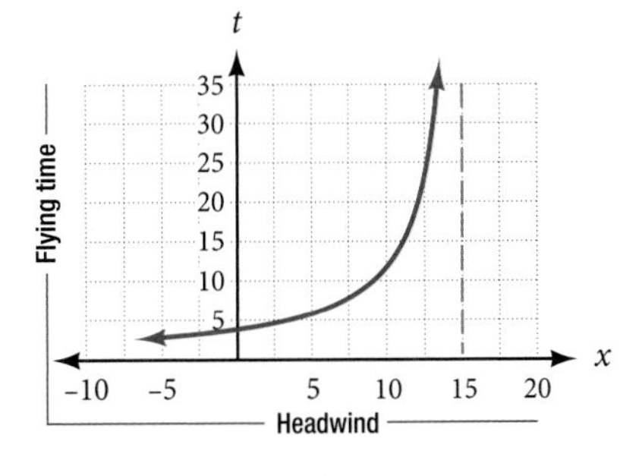

FIGURE 7

The graph of the function is shown on the previous page. As the speed of the wind gets close to 15 miles per hour, Francine's flying time becomes extremely large. In theory, if the wind speed were exactly 15 miles per hour Francine would never complete her flight.

What about negative values for x? If we interpret a negative headwind as a tailwind, Francine's flying time should decrease for negative x-values. For example, if $x = -5$, there is a tailwind of 5 miles per hour, so Francine's effective speed is 20 miles per hour, and she can complete the flight in 3 hours. As the tailwind gets stronger (that is, as we move further to the left in the x-direction), Francine's flying time continues to decrease, and the graph approaches the x-axis. ■

Modeling: Average Cost

EXAMPLE 9 MathTV decides to sell T-shirts to use for advertising. They make an initial investment of \$100 to pay for the design of the T-shirt and to set up the printing process. After that the T-shirts cost \$5 each for labor and materials.

a. Express MathTV's average cost per T-shirt as a function of the number of T-shirts they produce.

b. Make a table of values for the function.

c. Graph the function and explain what it tells you about the cost of the T-shirts.

SOLUTION

a. If MathTV produces x T-shirts, their total costs will be $100 + 5x$ dollars. To find the average cost per T-shirt, we divide the total cost by the number of T-shirts produced, to get

$$\text{Cost Function } C(x) = 5x + 100$$

$$\text{Average Cost Function } \overline{C}(x) = \frac{5x + 100}{x}$$

b. Evaluate the function for several values of x, as shown in the table below.

x	1	2	4	5	10	20
$\overline{C}(x)$	105	55	30	25	15	10

If MathTV makes only one T-shirt, its cost is \$105. But if they make more T-shirts, the cost of the original \$100 investment is distributed among them. For example, the average cost per T-shirt for two T-shirts is

$$\frac{5(2) + 100}{2} = 55$$

and the average cost for five T-shirts is

$$\frac{5(5) + 100}{5} = 25$$

c. To graph the average cost function, we use the information in the table, and we note the following:

1. the graph will appear in the first quadrant only because x cannot be negative.

2. there is a vertical asymptote at $x = 0$ (the y-axis).
3. there is a horizontal asymptote at $\overline{C}(x) = 5$.

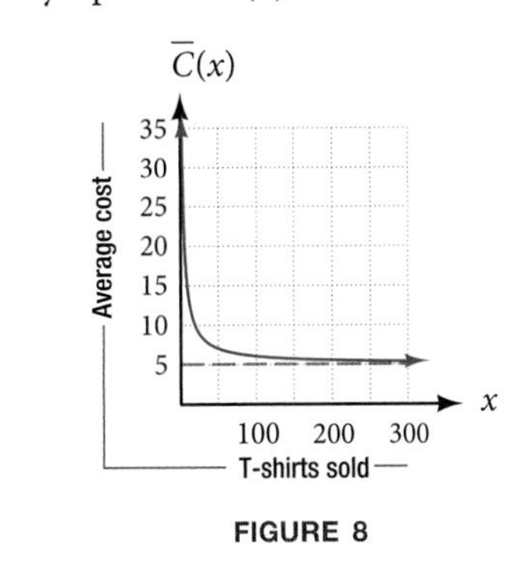

FIGURE 8

The graph is shown in Figure 8. The point (5, 25) indicates that if MathTV makes five T-shirts the cost per shirt is $25. The graph shows that as the number of T-shirts increases, the average cost per shirt continues to decrease, but not as rapidly as at first. Eventually the average cost levels off and approaches $5 per T-shirt. For example, if MathTV produces 400 T-shirts, the average cost per shirt is

$$\frac{5(400) + 100}{400} = 5.25$$

Using Technology 5.5

You can use your graphing calculator to construct the graph of a function. If that function has vertical asymptotes, however, you will need to be careful when you describe its behavior.

EXAMPLE 10 The function below has a vertical asymptote at $x = 3$.

$$f(x) = \frac{2x + 1}{x - 3}$$

a. Graph the function in both the Connected mode and in the Dot mode. Explain the differences in the graphs.

b. Name the horizontal asymptote.

c. Name the intercepts.

SOLUTION

a. Your calculator may construct the graph so that it appears that the asymptote is part of the curve by connecting it to the branches of the curve. This connection happens when your calculator is set in the Connected mode (Figure 9a). The calculator is being instructed to connect the *turn-on* pixels. The problem can be fixed by placing your calculator in Dot mode (Figure 9b). However, the Dot mode gives the curve a very discontinuous appearance. It is best if you can determine the existence of vertical asymptotes by analyzing the function.

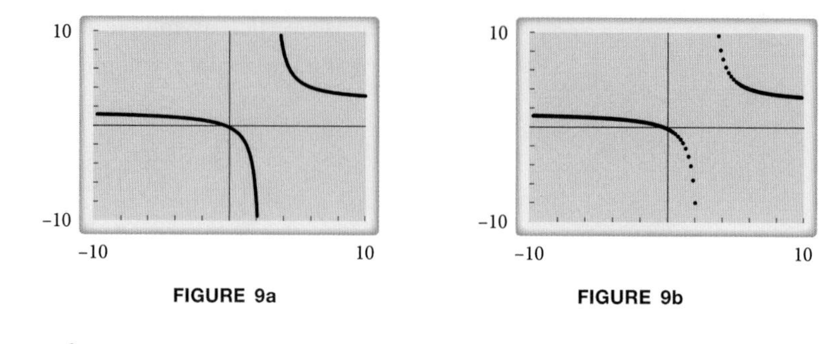

FIGURE 9a **FIGURE 9b**

b. $y = 2$

c. x-intercept is $-\frac{1}{2}$, y-intercept is $-\frac{1}{3}$.

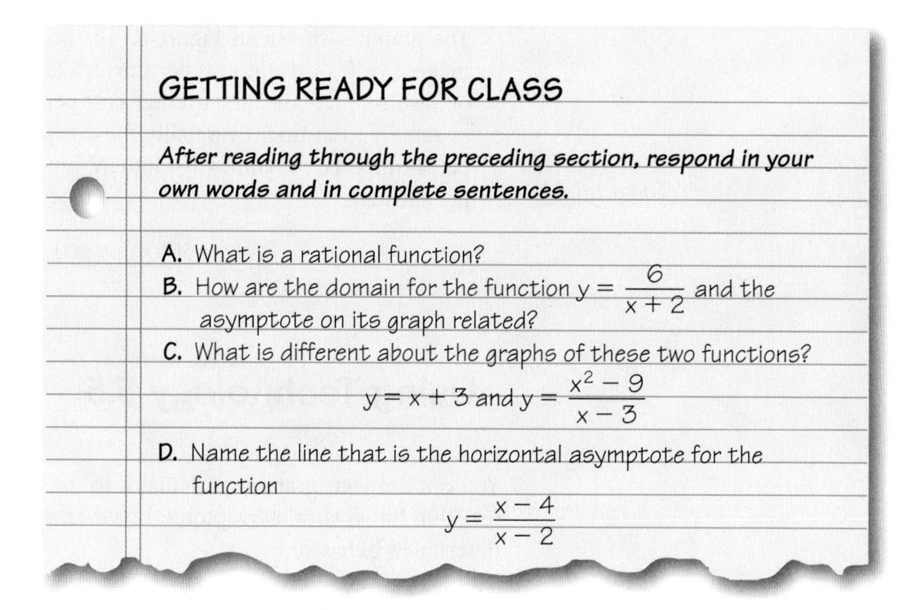

GETTING READY FOR CLASS

After reading through the preceding section, respond in your own words and in complete sentences.

A. What is a rational function?

B. How are the domain for the function $y = \frac{6}{x+2}$ and the asymptote on its graph related?

C. What is different about the graphs of these two functions?

$$y = x + 3 \text{ and } y = \frac{x^2 - 9}{x - 3}$$

D. Name the line that is the horizontal asymptote for the function

$$y = \frac{x - 4}{x - 2}$$

Problem Set 5.5

1. If $g(x) = \frac{x+3}{x-1}$, find $g(0)$, $g(-3)$, $g(3)$, $g(-1)$, and $g(1)$, if possible.

2. If $g(x) = \frac{x-2}{x-1}$, find $g(0)$, $g(-2)$, $g(2)$, $g(-1)$, and $g(1)$, if possible.

3. If $h(t) = \frac{t-3}{t+1}$, find $h(0)$, $h(-3)$, $h(3)$, $h(-1)$, and $h(1)$, if possible.

4. If $h(t) = \frac{t-2}{t+1}$, find $h(0)$, $h(-2)$, $h(2)$, $h(-1)$, and $h(1)$, if possible.

State the domain for each rational function.

5. $f(x) = \frac{x-3}{x-1}$

6. $f(x) = \frac{x+4}{x-2}$

7. $g(x) = \frac{x^2-4}{x-2}$

8. $g(x) = \frac{x^2-9}{x-3}$

9. $h(t) = \frac{t-4}{t^2-16}$

10. $h(t) = \frac{t-5}{t^2-25}$

Let $f(x) = \frac{x^2-4}{x-2}$ and $g(x) = x + 2$ and evaluate the following expressions, if possible.

11. $f(0)$ and $g(0)$

12. $f(1)$ and $g(1)$

13. $f(2)$ and $g(2)$

14. $f(3)$ and $g(3)$

Let $f(x) = \frac{x^2-1}{x-1}$ and $g(x) = x + 1$ and evaluate the following expressions, if possible.

15. $f(0)$ and $g(0)$

16. $f(1)$ and $g(1)$

17. $f(2)$ and $g(2)$

18. $f(3)$ and $g(3)$

19. Graph the equation $y = \frac{x^2-4}{x-2}$. Then explain how this graph is different from the graph of $y = x + 2$.

20. Graph the equation $y = \frac{x^2-1}{x-1}$. Then explain how this graph is different from the graph of $y = x + 1$.

The graphs of two rational functions are given in Figures 10 and 11. Use the graphs to find the following.

21. a. $f(2)$ b. $f(-1)$ c. $f(0)$ d. $g(3)$

22. a. $g(6)$ b. $g(-1)$ c. $f(g(6))$ d. $g(f(-2))$

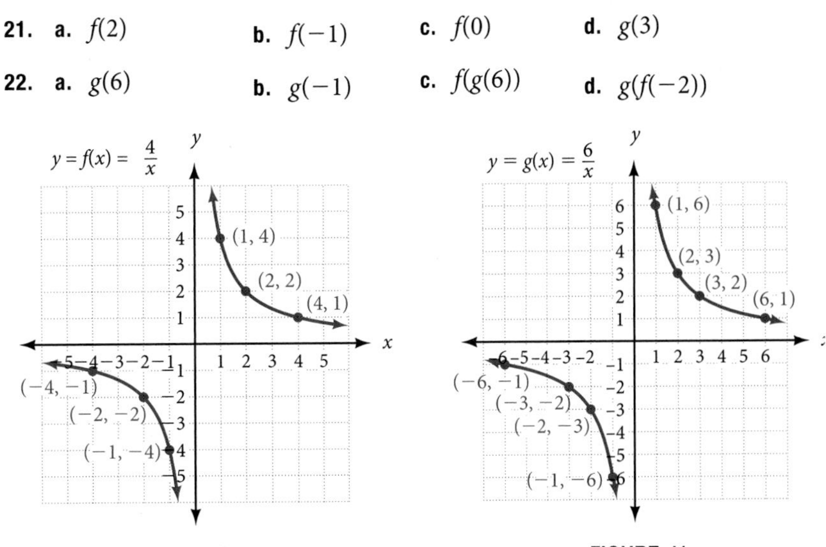

FIGURE 10

FIGURE 11

For each rational function, find all the intercepts and asymptotes. Then use any of the other tools you have to help sketch the graph.

23. $y = \dfrac{x-3}{x+2}$

24. $y = \dfrac{x}{x+3}$

25. $y = \dfrac{x}{x^2-4}$

26. $y = \dfrac{-x}{x^2-4}$

27. $y = \dfrac{x^2}{x^2-4}$

28. $y = \dfrac{-x^2}{x^2-4}$

Getting Ready for Calculus The problems below are representative of the type of problems you will need to be familiar with in order to be successful in calculus. These particular problems are taken from *Applied Calculus* by Denny Burzynski and Guy Sanders, published by XYZ Textbooks.

29. If $g(x) = \dfrac{x^4 - 5x^2 + 2x - 2}{(x^2-1)^2}$, find $g(2)$.

30. If $f(x) = \dfrac{2x-5}{x-3}$, find $f(4)$.

31. If $g(x) = \dfrac{1}{(x-3)^2}$, find $g(4)$.

32. If $h(x) = \dfrac{(x+3)(x-6)}{(x-1)^2}$, find $h(3)$.

Find the intercepts and vertical asymptotes for each function.

33. $y = \dfrac{5x+2}{3x-4}$

34. $f(x) = \dfrac{7x+3}{7x-3}$

35. $y = \dfrac{2(t+8)(t-6)}{(t+1)^2}$

36. $f(x) = \dfrac{t(t-4)}{(t-2)^2}$

Modeling Practice

37. Diet The following rational function is the one we mentioned in the introduction to this chapter. The quantity $W(x)$ is the weight (in pounds) of the person after x weeks of dieting. Use the function to fill in the table (round your answers to the nearest tenth). Then compare your results with the graph in the chapter introduction.

$$W(x) = \frac{80(2x+15)}{x+6}$$

Weeks x	0	1	4	12	24
Weight $W(x)$ (lb)					

38. Drag Racing The following rational function gives the speed $V(x)$, in miles per hour, of a dragster at each second x during a quarter-mile race.

$$V(x) = \frac{340x}{x + 3}$$

Use the function to fill in the table. Round answers to the nearest whole number.

Time x (sec)	**0**	**1**	**2**	**3**	**4**	**5**	**6**
Speed $V(x)$ (mi/hr)							

For Problems 39 and 40, use 3.14 as an approximation for π. Round answers to the nearest tenth.

39. Average Speed A person riding a Ferris wheel with a diameter of 65 feet travels once around the wheel in 30 seconds. What is the average speed of the rider in feet per second?

40. Average Speed A person riding a Ferris wheel with a diameter of 102 feet travels once around the wheel in 3.5 minutes. What is the average speed of the rider in feet per minute?

The abbreviation "rpm" stands for revolutions per minute. If a point on a circle rotates at 300 rpm, then it rotates through one complete revolution 300 times every minute. The length of time it takes to rotate once around the circle is $\frac{1}{300}$ minute. Use 3.14 as an approximation for π.

41. Average Speed A $3\frac{1}{2}$ inch diskette, when placed in the disk drive of a computer, rotates at 300 rpm (1 revolution takes $\frac{1}{300}$ minute). Find the average speed of a point 2 inches from the center of the diskette. Then find the average speed of a point 1.5 inches from the center of the diskette.

42. Average Speed A 5-inch fixed disk in a computer rotates at 3,600 rpm. Find the average speed of a point 2 inches from the center of the disk. Then find the average speed of a point 1.5 inches from the center.

43. Average Speed The Ferris wheel in Problem 39 has a circumference of 204 feet (to the nearest foot). If a ride on the wheel takes from 20 to 50 seconds, then the relationship between the average speed of a rider and the amount of time it takes to complete one revolution is given by the function

$$r(t) = \frac{204}{t} \qquad 20 \le t \le 50$$

where $r(t)$ is in feet per second and t is in seconds.

a. State the domain for this function.

b. Graph the function.

44. **Average Speed** The Ferris wheel in Problem 40 has a circumference of 320 feet (to the nearest foot). If a ride on the wheel takes from 3 to 5 minutes, then the relationship between the average speed of a rider and the amount of time it takes to complete one revolution is given by the function

$$r(t) = \frac{320}{t} \qquad 3 \le t \le 5$$

where $r(t)$ is in feet per minute and t is in minutes.

a. State the domain for this function.

b. Graph the function.

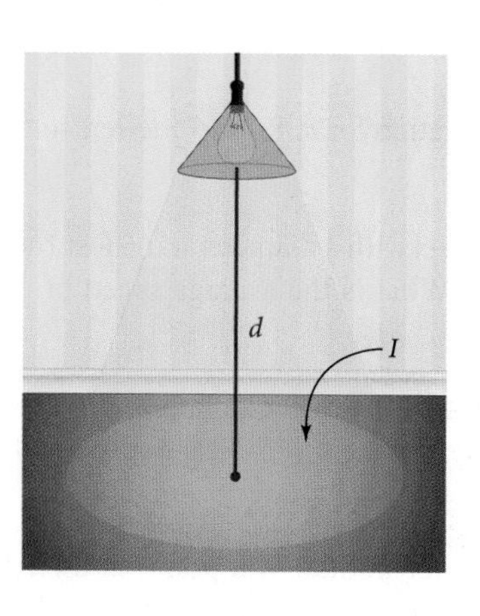

45. **Intensity of Light** The relationship between the intensity of light that falls on a surface from a 100-watt light bulb and the distance from that surface is given by the rational function

$$I(d) = \frac{120}{d^2} \qquad \text{for} \qquad 10 \le d \le 15$$

where $I(d)$ is the intensity of light (in lumens per square foot) and d is the distance (in feet) from the light bulb to the surface.

a. State the domain for this function.

b. Graph this function.

46. **Average Speed** If it takes Maria t minutes to run a mile, then her average speed $s(t)$ is given by the rational function

$$s(t) = \frac{60}{t} \quad \text{for} \quad 6 \le t \le 12$$

where $s(t)$ is in miles per hour and t is in minutes.

a. State the domain for this function.

b. Graph this function.

47. The following is a graph of a person's weight loss (in pounds) after x weeks of dieting:

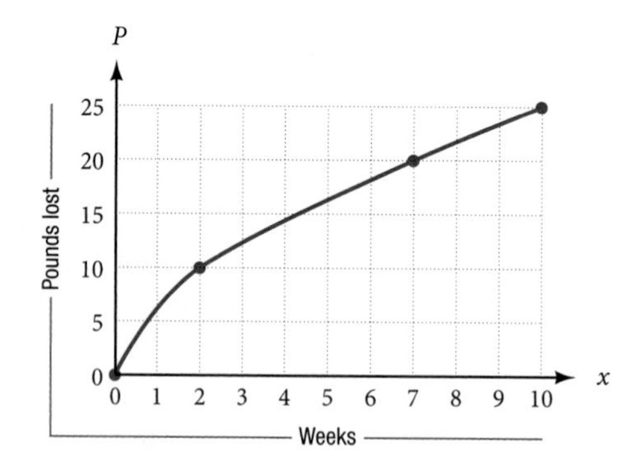

a. Find the ratio of weight lost after 2 weeks to weight lost after 10 weeks.

b. Find the ratio of weight lost after 7 weeks to weight lost after 10 weeks.

48. **Traveling Time** Suppose you travel 220 miles on the freeway to get to a relative's house. If the speed limit has been changed from 55 miles per hour to 70 miles per hour, how much time will you save with the new speed limit?

49. Marathon If Guillermo can run a marathon at an average rate r, then the time $t(r)$ it will take him to complete the race is given by the function

$$t(r) = \frac{24}{r} \quad \text{for} \quad 10 \le r \le 15$$

where $t(r)$ is in hours and r is in miles per hour.

a. State the domain of this function.

b. Graph this function.

c. Guillermo's sister Consuela runs 2 miles per hour slower than Guillermo. If Guillermo can run at an average rate r, find an expression for the average rate at which Consuela can run.

d. Use your answer to Part c to write the equation of the function that would give the time $T(r)$ for Consuela to run a marathon.

e. Graph the function $T(r)$ from Part d.

50. Century If Gary can ride a century (100 miles) in t hours, then his average speed for the ride is given by the function

$$s(t) = \frac{100}{t} \quad \text{for} \quad 4 \le t \le 10$$

where $s(t)$ is in miles per hour.

a. State the domain of this function.

b. Graph this function.

c. Gary has a friend Todd who can complete a century in one hour less than the time it takes Gary. If Gary can complete the ride in time t, find an expression for the time in which Todd can complete the ride.

d. Use your answer to Part c to write the equation of the function that would give the average speed $S(t)$ for Todd on a century ride.

e. Graph the function $S(t)$ from Part d.

Using Technology Exercises

Use your graphing calculator to construct the graph of each function. Then list

a. the asymptotes.

b. the intercepts.

51. $f(x) = \dfrac{x + 1}{x - 3}$

52. $f(x) = \dfrac{3x + 1}{x - 3}$

53. $f(x) = \dfrac{2x + 1}{3x - 3}$

54. $f(x) = \dfrac{4}{(x + 2)(x - 2)}$

55. **Deer Population** A state's forestry service is attempting to establish a herd of deer as part of a forestry restoration program. Their plan is to initially place 150 deer in various locations in the forest so that the population of deer t years from now will be approximated by the function

$$N(t) = \frac{150 + 38t}{1 + 0.037t}$$

Create a table that displays the time and the population for the first 30 years in 3-year increments. Can you see a trend in the growth of the population? Does the increase in population seem to be speeding up or slowing down? How can you tell?

Getting Ready for the Next Section

56. Use synthetic division to find the remainder when $2x^3 - 3x^2 - 5x + 6$ is divided by

a. $x - 6$ **b.** $x - 3$ **c.** $x - 2$

57. Use synthetic division to find the remainder when $2x^3 - 7x^2 + 4x + 3$ is divided by

a. $x - 3$ **b.** $x - \frac{3}{2}$

58. Multiply $(2x - 3)(x^2 - 2x - 1)$.

59. If $P(x) = 7x^4 - 2x^3 + 3x^2 + 4x - 5$, find $P(-x)$.

Solving Polynomial Equations

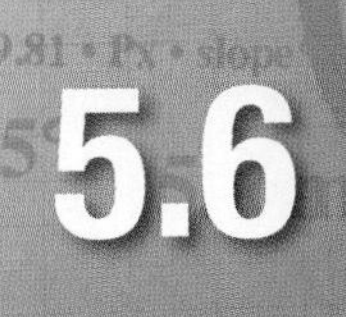

In this section we will concern ourselves with solutions to equations of the form $P(x) = 0$ where $P(x)$ is a polynomial of degree n, with real numbers for coefficients. We will list some facts (in the form of theorems) about the solutions to these equations.

Our first theorem tells us how many solutions we can expect from a polynomial equation in which the coefficients are real numbers. We state it without proof.

> **Number of Solutions Theorem**
>
> If $P(x)$ is an nth-degree polynomial, then the polynomial equation $P(x) = 0$ has exactly n solutions, some of which may be repeated solutions. All the solutions will be complex numbers and some of them may also be real numbers. (Remember, the real numbers are a subset of the complex numbers.)

This theorem tells us that an nth-degree polynomial equation will have at most n distinct solutions. For example, the third-degree equation

$$(x + 3)(x - 4)(x - 5) = 0$$

has three solutions, -3, 4, and 5. (We know the equation is third-degree because if we multiply the factors on the left side, the highest power of x in the result is 3.)

On the other hand, the sixth-degree equation

$$(x + 3)^2(x - 4)(x - 5)^3 = 0$$

has a total of six solutions: -3, -3, 4, 5, 5, and 5, only three of which are unique. We indicate the number of times a solution is repeated by using the word multiplicity. For the equation above, we say that -3 is a solution of *multiplicity* 2 because it occurs twice. Likewise, we say 5 is a solution of multiplicity 3 because it occurs three times. Note that the multiplicity of a solution is given by the exponent on the linear factor that leads to that solution.

As our next theorem indicates, we can generalize even further about the solutions to a polynomial equation by counting what are called the variations in sign of the polynomials $P(x)$ and $P(-x)$. Before we state the theorem, we should give some explanation on how to count the variations in sign for $P(x)$ and $P(-x)$.

We count the number of variations in sign of a polynomial $P(x)$ (written in descending powers of x) by counting the number of times the signs of the coefficients change as we move left to right across the polynomial. For example, the polynomial below has 3 variations in sign.

$$\underbrace{7x^4 - 2x^3}_{1}\underbrace{\;+ 3x^2}_{2} \underbrace{+ 4x - 5}_{3}$$

If a polynomial contains a coefficient of 0, we simply ignore that term when counting variations in sign. For example, the coefficients of x^3 and x are both 0 in the polynomial

$$\underbrace{x^4 - x^2}_{1} - 6$$

The polynomial has exactly 1 variation in sign.

To find the number of variations in sign for the polynomial $P(-x)$, we first replace each x with $-x$ and simplify. Then we count the number of variations in sign in the resulting polynomial. For example,

$$\text{If} \quad P(x) = 7x^4 - 2x^3 + 3x^2 + 4x - 5$$
$$\text{then } P(-x) = 7(-x)^4 - 2(-x)^3 + 3(-x)^2 + 4(-x) - 5$$
$$= 7x^4 + 2x^3 + \underbrace{3x^2 - 4x}_{1} - 5$$

which has 1 variation in sign. In general, replacing x with $-x$ will leave each term with an even exponent unchanged, but will change the sign on each term with an odd exponent.

Here is our theorem.

Descartes' Rule of Signs

Let $P(x) = 0$ be a polynomial equation with real coefficients.

1. The number of positive real solutions is either equal to the number of variations in sign of $P(x)$, or is less than the number of variations in sign of $P(x)$ by a positive even number.
2. The number of negative real solutions is either equal to the number of variations in sign of $P(-x)$, or is less than the number of variations in sign of $P(-x)$ by a positive even number.

VIDEO EXAMPLES

SECTION 5.6

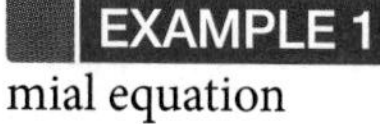

EXAMPLE 1 What can we say about the real solutions to the polynomial equation

$$4x^5 - 3x^4 + 2x^3 - 2x^2 + 3x - 4 = 0$$

SOLUTION Here are the variations in sign for $P(x)$.

$$P(x) = \underbrace{4x^5 - 3x^4}_{1} \underbrace{\;+ 2x^3}_{3} \ldots$$

$$P(x) = 4x^5 - 3x^4 + 2x^3 - 2x^2 + 3x - 4 = 0$$
$$\quad 1 \quad 2 \quad 3 \quad 4 \quad 5$$

With 5 variations in sign for $P(x)$, the equation $P(x) = 0$ will have either 5 positive real solutions, 3 positive real solutions, or 1 positive real solution, because 3 and 1 are each smaller than 5 by an even number.

The number of negative real solutions is given by the variations in sign of

$$P(-x) = 4(-x)^5 - 3(-x)^4 + 2(-x)^3 - 2(-x)^2 + 3(-x) - 4$$
$$= -4x^5 - 3x^4 - 2x^3 - 2x^2 - 3x - 4$$

Since there are 0 variations in sign for $P(-x)$, the equation $P(x) = 0$ has no negative real solutions. ■

EXAMPLE 2 What are the possibilities for real solutions to the polynomial equation $x^6 + 10x^4 + 9x^2 = 0$?

SOLUTION In this case $P(x) = x^6 + 10x^4 + 9x^2 = 0$ has 0 variations in sign, so there are no positive real solutions. To see if there are any possibilities for negative real solutions we look at $P(-x)$.

$$P(-x) = (-x)^6 + 10(-x)^4 + 9(-x)^2 = x^6 + 10x^4 + 9x^2$$

which also has 0 variations in sign, meaning there are no negative real solutions either.

The only possible real solution is $x = 0$, which is in fact a solution.

Since the only real solution is $x = 0$ (multiplicity 2), the other solutions are all complex numbers of the form $a + bi$ where $b \neq 0$, which, you may recall, are called imaginary numbers. (If you solve the original equation by factoring, you will find the solutions are 0, i, $-i$, $3i$, and $-3i$.) ■

EXAMPLE 3 State the number of possible real and imaginary solutions to $2x^3 + 3x^2 - 2x - 3 = 0$.

SOLUTION Here $P(x) = 2x^3 + 3x^2 - 2x - 3$ has 1 variation in sign, meaning we have exactly 1 positive real solution to the equation.

The polynomial

$$\begin{aligned} P(-x) &= 2(-x)^3 + 3(-x)^2 - 2(-x) - 3 \\ &= -2x^3 + 3x^2 + 2x - 3 \end{aligned}$$

has 2 variations in sign. The number of negative real solutions will be 2 or 0.

Since the equation has degree 3, there are 3 solutions total.

If we put all the information we have together, we have two different combinations of types of solutions for this equation, as shown in the table below. Each row of the table gives one of the possible combinations of solutions for the equation. Remember that imaginary solutions are solutions of the form $a + bi$ where $b \neq 0$; that is, complex numbers for which the coefficient of i is not 0.

Real Solutions		Imaginary Solutions
Positive	Negative	
1	2	0
1	0	2

■

EXAMPLE 4 What are the possible combinations of solutions to the equation $2x^4 - 7x^3 + 4x^2 + 7x - 6 = 0$?

SOLUTION The number of variations in sign of $P(x)$ is 3, giving us 3 or 1 for the number of positive real solutions.

Here is $P(-x)$:

$$\begin{aligned} P(-x) &= 2(-x)^4 - 7(-x)^3 + 4(-x)^2 + 7(-x) - 6 \\ &= 2x^4 + 7x^3 + 4x^2 - 7x - 6 \end{aligned}$$

There is exactly 1 variation in sign in $P(-x)$, so there is exactly 1 negative real solution to the equation.

Since there are 4 solutions total (this is a 4th-degree equation), we have the following possible combinations of solutions.

Real Solutions Positive	Real Solutions Negative	Imaginary Solutions
3	1	0
1	1	2

Our next theorem tells us how to list all the rational numbers that may be solutions to a polynomial equation. We state it without proof.

 ***Rational Solution Theorem***

If the rational number $\frac{p}{q}$ (in lowest terms) is a solution to the polynomial equation

$$a_nx^n + a_{n-1}x^{n-1} + \cdots + a_1x + a_0 = 0 \qquad a_n \neq 0$$

where the coefficients a_0 through a_n are integers, then p is a factor of the constant term a_0 and q is a factor of the leading coefficient a_n.

To illustrate this theorem, suppose we are looking for solutions to the equation

$$2x^3 - 3x^2 - 5x + 6 = 0$$

If this equation has a rational number p/q for one of its solutions, then the numerator p of that solution must be a factor of the constant term 6 and the denominator q must be a factor of the leading coefficient 2. The factors of 6 are ±1, ±2, ±3, and ±6, and p must be one of these numbers. The factors of 2 are ±1 and ±2, and q must be one of these numbers. Therefore, if there are rational solutions to this equation, they are among the numbers

$$\pm\frac{1}{2}, \pm1, \pm\frac{3}{2}, \pm2, \pm3, \text{ and } \pm6$$

To test these numbers we use synthetic division and the factor theorem: If the remainder from synthetic division is 0, we have a solution.

Try $x = 6$

$$\begin{array}{r|rrrr} 6 & 2 & -3 & -5 & 6 \\ & & 12 & 54 & 294 \\ \hline & 2 & 9 & 49 & \boxed{300} \end{array}$$

which shows $x = 6$ is not a solution

Try $x = 3$

$$\begin{array}{r|rrrr} 3 & 2 & -3 & -5 & 6 \\ & & 6 & 9 & 12 \\ \hline & 2 & 3 & 4 & \boxed{18} \end{array}$$

$x = 3$ is not a solution

Try $x = 2$

$$\begin{array}{r|rrrr} 2 & 2 & -3 & -5 & 6 \\ & & 4 & 2 & -6 \\ \hline & 2 & 1 & -3 & \boxed{0} \end{array}$$

$x = 2$ is a solution

The last line in this synthetic division tells us that $x = 2$ is a solution and the original equation factors as follows:

$$(x - 2)(2x^2 + x - 3) = 0$$
$$(x - 2)(2x + 3)(x - 1) = 0$$

These three factors give us solutions $x = 2$, $x = -\frac{3}{2}$, and $x = 1$.

As the next two examples indicate, we can use the information on the number of possible positive and negative real solutions, and the rational solution theorem to solve polynomial equations.

EXAMPLE 5 Find all solutions to $2x^3 - 7x^2 + 4x + 3 = 0$.

SOLUTION There are 2 variations in sign for $P(x)$, so we may have 2 positive real solutions or no positive real solutions.

Since $P(-x) = -2x^3 - 7x^2 - 4x + 3$ has 1 variation in sign, we will have 1 negative real solution.

Now we list the possible rational solutions p/q:

possible values for p: $\pm 1, \pm 3$
possible values for q: $\pm 1, \pm 2$
possible rational solutions p/q: $\pm \frac{1}{2}, \pm 1, \pm \frac{3}{2}, \pm 3$

$$\begin{array}{r|rrrr} 3 & 2 & -7 & 4 & 3 \\ & & 6 & -3 & 3 \\ \hline & 2 & -1 & 1 & \boxed{6} \end{array} \quad x = 3 \text{ is not a solution}$$

$$\begin{array}{r|rrrr} \frac{3}{2} & 2 & -7 & 4 & 3 \\ & & 3 & -6 & -3 \\ \hline & 2 & -4 & -2 & \boxed{0} \end{array} \quad x = \frac{3}{2} \text{ is a solution}$$

Now that we have one solution, we can use the factor theorem to write our equation as

$$\left(x - \frac{3}{2}\right)(2x^2 - 4x - 2) = 0$$

To find the solutions that come from the second factor, we use the quadratic formula to obtain

$$x = 1 \pm \sqrt{2}$$

The three solutions are $\frac{3}{2}$, $1 + \sqrt{2}$, and $1 - \sqrt{2}$. The first two are positive and the last one is negative. ■

EXAMPLE 6 Solve $x^5 + x^4 - 2x^3 - 2x^2 + x + 1 = 0$.

SOLUTION We will have 5 solutions total. Since there are 2 variations in sign for $P(x)$, we can expect either 2 or 0 positive real solutions. To see the possible number of negative real solutions we look at $P(-x)$, which is

$$P(-x) = -x^5 + x^4 + 2x^3 - 2x^2 - x + 1 = 0$$

Since $P(-x)$ has 3 variations in sign, we can expect either 3 or 1 negative real solutions.

For rational solutions of the form $x = p/q$, we have the following:

possible values for p: ± 1
possible values for q: ± 1
possible rational solutions $x = p/q$: ± 1

Let's see if $x = 1$ is a solution:

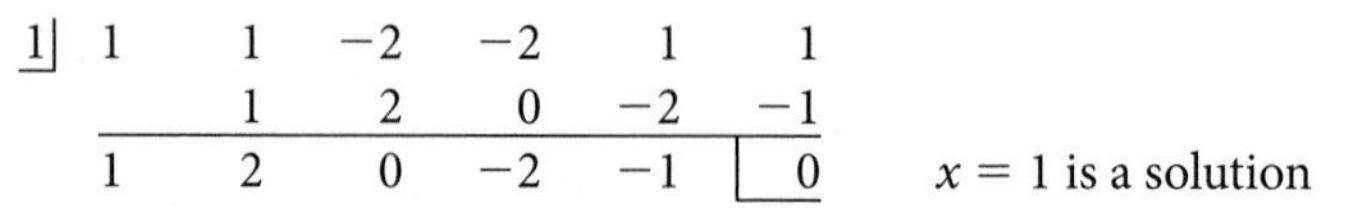

1	1	1	−2	−2	1	1	
		1	2	0	−2	−1	
	1	2	0	−2	−1	0	$x = 1$ is a solution

Using this result and the factor theorem we can write our equation as

$$(x - 1)(x^4 + 2x^3 - 2x - 1) = 0$$

Since the second factor has exactly one variation in sign, exactly one of the solutions that comes from setting the factor equal to 0 will be positive. The only possible positive rational solution is $x = 1$.

1	1	2	0	−2	−1	
		1	3	3	1	
	1	3	3	1	0	$x = 1$ is a solution

Our original equation can now be written as

$$(x - 1)(x - 1)(x^3 + 3x^2 + 3x + 1) = 0$$

There are no variations in sign for the last factor, so it will not have any positive real solutions. (Notice that this just confirms what we already knew from the variations in sign above—that we would have at most two positive real solutions.) The only possible rational solution will be -1.

−1	1	3	3	1	
		−1	−2	−1	
	1	2	1	0	$x = -1$ is a solution

We use this last result to write our equation as

$$(x - 1)(x - 1)(x + 1)(x^2 + 2x + 1) = 0$$

which we factor completely and write as

$$(x - 1)(x - 1)(x + 1)(x + 1)(x + 1) = 0$$
$$(x - 1)^2(x + 1)^3 = 0$$

The number 1 is a solution of multiplicity 2 and -1 is a solution of multiplicity 3. This accounts for all five solutions. ■

Problem Set 5.6

For each of the following equations, give

a. the number of solutions

b. the possible number of positive real solutions

c. the possible number of negative real solutions

Do not solve the equations.

1. $3x^4 - 2x^3 + x^2 - 2x + 3 = 0$

2. $-2x^4 - 3x^3 + 4x^2 - 3x + 2 = 0$

3. $x^4 + x^2 + 1 = 0$

4. $x^6 + x^4 + x^2 = 0$

5. $x^4 + x^2 - 1 = 0$

6. $x^6 - x^4 + x^2 = 0$

7. $7x^3 - 6x^2 + 5x - 4 = 0$

8. $-5x^3 + 6x^2 - 5x + 4 = 0$

9. $x^5 + 3x^4 - 5x^3 - 15x^2 + 4x + 12 = 0$

10. $x^5 - 3x^4 - 5x^3 + 15x^2 + 4x - 12 = 0$

Solve each of the following equations completely. If a solution occurs more than once, give its multiplicity.

11. $x^3 - 3x^2 + 4x - 12 = 0$

12. $x^3 - 2x^2 + 9x - 18 = 0$

13. $2x^4 - 3x^3 - 3x - 2 = 0$

14. $3x^4 - 9x^3 + 9x - 3 = 0$

15. $x^5 + 2x^4 - 2x^3 - 4x^2 + x + 2 = 0$

16. $x^5 - 2x^4 - 2x^3 + 4x^2 + x - 2 = 0$

17. $x^5 + 3x^4 - 5x^3 - 15x^2 + 4x + 12 = 0$

18. $x^5 - 3x^4 - 5x^3 + 15x^2 + 4x - 12 = 0$

19. $4x^3 + 8x^2 - 11x + 3 = 0$

20. $9x^3 - 51x^2 + 31x - 5 = 0$

21. $x^3 - 9x^2 + 25x - 21 = 0$

22. $x^3 - 8x^2 - 5x + 40 = 0$

23. $x^4 - 2x^3 + x^2 + 8x - 20 = 0$

24. $x^4 - 2x^3 - 4x^2 + 18x - 45 = 0$

25. $4x^4 - 8x^3 + 3x^2 + 8x - 7 = 0$

26. $4x^4 - 8x^3 - 9x^2 + 32x - 28 = 0$

Graph each of the following equations. In each case, you will need to use the methods of this section to find the x-intercepts.

27. $y = x^3 - 6x^2 + 3x + 10$

28. $y = x^3 + 4x^2 - 7x - 10$

29. $y = x^4 - 8x^3 + 14x^2 + 8x - 15$

30. $y = x^4 + 6x^3 + 7x^2 - 6x - 8$

Maintaining Your Skills

Perform the indicated operations.

31. $\frac{2a + 10}{a^3} \cdot \frac{a^2}{3a + 15}$

32. $\frac{4a + 8}{a^2 - a - 6} \div \frac{a^2 + 7a + 12}{a^2 - 9}$

33. $(x^2 - 9)\left(\frac{x + 2}{x + 3}\right)$

34. $\frac{1}{x + 4} + \frac{8}{x^2 - 16}$

35. $\frac{2x - 7}{x - 2} - \frac{x - 5}{x - 2}$

36. $2 + \frac{25}{5x - 1}$

Simplify each expression.

37. $\frac{\frac{1}{x} - \frac{1}{3}}{\frac{1}{x} + \frac{1}{3}}$

38. $\frac{1 - \frac{9}{x^2}}{1 - \frac{1}{x} - \frac{6}{x^2}}$

Solve each equation.

39. $\frac{x}{x - 3} + \frac{3}{2} = \frac{3}{x - 3}$

40. $1 - \frac{3}{x} = \frac{-2}{x^2}$

Chapter 5 Summary

EXAMPLES

1. The graph of the quadratic function $y = x^2 - 6x + 5$.

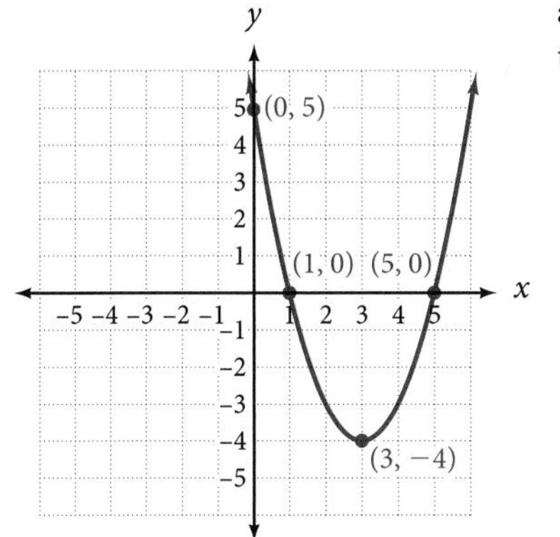

Quadratic Functions [5.1]

All objects that are projected into the air, whether they are basketballs, bullets, arrows, or coins, follow parabolic paths. And, these parabolic graphs are characteristic of quadratic functions.

2. The x-coordinate of the vertex for the graph above is found this way:

$$x = \frac{-b}{2a} = \frac{-(-6)}{2(1)} = 3$$

Vertex of a Parabola [5.1]

The highest or lowest point on a parabola is called the *vertex*. The x-coordinate of the vertex for the graph of $y = ax^2 + bx + c$ is

$$x = \frac{-b}{2a}$$

3. Dividing $3x^3 - 4x + 5$ by $x + 4$ with synthetic division:

$$\begin{array}{r|rrrr} -4 & 3 & 0 & -4 & 5 \\ & \downarrow & -12 & 48 & -176 \\ \hline & 3 & -12 & 44 & \boxed{-171} \end{array}$$

Synthetic Division [5.2]

Synthetic division is a short form of long division with polynomials. We will consider synthetic division only for those cases in which the divisor is of the form $x + k$, where k is a constant.

4. The function

$$P(x) = 2x^4 - 3x^2 + 5$$

is a polynomial function of degree 4. The leading coefficient is 2 and the constant term is 5.

Polynomial Functions [5.2]

A polynomial function of degree n in x is any function that can be written in the form

$$P(x) = a_nx^n + a_{n-1}x^{n-1} + \cdots + a_1x + a_0$$

where $a_n \neq 0$, n is a nonnegative integer, and each coefficient is a real number.

5. When $x^3 - 4x^2 + 2x - 5$ is divided by $x - 3$, the quotient is $x^2 - x - 1$ and the remainder is -8. Writing the results in the form dividend = (divisor)(quotient) + (remainder) we have

$$x^3 - 4x^2 + 2x - 5$$
$$= (x - 3)(x^2 - x - 1) + (-8)$$

Existence Theorem for Division [5.2]

If $P(x)$ is a polynomial and r is a real number, then there exists a unique polynomial $q(x)$ and a unique real number R such that

$$P(x) = (x - r)q(x) + R$$

The polynomial $q(x)$ is called the quotient, $x - r$ is called the divisor, and R is called the remainder. The degree of $q(x)$ is 1 less than the degree of $P(x)$.

Remainder Theorem for Division [5.3]

6. If $P(x) = x^3 - 4x^2 + 2x - 5$ as in Example 5, then the remainder theorem tells us that $P(3) = -8$.

If the polynomial $P(x)$ is divided by $x - r$, the remainder is $P(r)$.

Factor Theorem for Polynomials [5.3]

7. The binomial $x - 2$ is a factor of $P(x) = x^5 - 32$ because

$$P(2) = 2^5 - 32 = 32 - 32 = 0$$

The binomial $x - r$ is a factor of the polynomial $P(x)$ if and only if $P(r) = 0$.

Zeros, Factors, and Solutions [5.3]

8. From Example 7 above, 2 is a zero of $P(x) = x^5 - 32$ and $x = 2$ is a solution to the equation $x^5 - 32 = 0$.

If $P(x)$is a polynomial and $P(r) = 0$, then r is a zero of $P(x)$.

In general, we say: polynomials have factors, polynomials and polynomial functions have zeros, and polynomial equations have solutions.

Graphing Polynomial Functions [5.4]

9. The graph of the polynomial function $y = x^3 - 4x$.

To graph a polynomial function we use the x-intercepts, a sign chart, and a table of ordered pairs that satisfy the equation. We assume that the graph is smooth continuous curve with no breaks or sharp points. Further, if the polynomial function has degree n, then the graph will have at most $n - 1$ turning points.

Rational Functions [5.5]

10. Each of the following is a rational function.

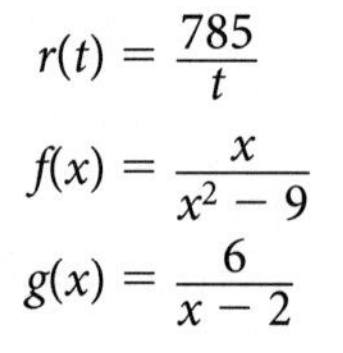

$$r(t) = \frac{785}{t}$$

$$f(x) = \frac{x}{x^2 - 9}$$

$$g(x) = \frac{6}{x - 2}$$

A *rational function* is any function that can be written in the form

$$f(x) = \frac{P(x)}{Q(x)}$$

where $P(x)$ and $Q(x)$ are polynomials and $Q(x) \neq 0$.

Graphing Rational Functions [5.5]

11.

Check for intercepts and asymptotes, and then use a table to find points on the graph.

Number of Solutions Theorem [5.6]

12. The equation

$$7x^4 - 2x^3 + 3x^2 + 4x - 5 = 0$$

will have 4 solutions.

If $P(x)$ is an nth-degree polynomial, then the polynomial equation $P(x) = 0$ has exactly n solutions, some of which may be repeated solutions. All the solutions will be complex numbers and some of them may also be real numbers.

Descartes' Rule of Signs [5.6]

13. The equation

$$P(x) = 7x^4 + 2x^3 + 3x^2 - 4x - 5$$

has only 1 variation in sign.

$$P(-x) = 7(-x)^4 + 2(-x)^3 + 3(-x)^2 - 4(-x) - 5$$
$$= 7x^4 - 2x^3 + 3x^2 + 4x - 5$$

has 3 variations in sign.

The equation

$$7x^4 - 2x^3 + 3x^2 + 4x - 5 = 0$$

will have either 3 or 1 negative real solutions, and exactly 1 positive real solution.

Let $P(x) = 0$ be a polynomial equation with real coefficients.

1. The number of positive real solutions is either equal to the number of variations in sign of $P(x)$, or is less than the number of variations in sign of $P(x)$ by a positive even number.
2. The number of negative real solutions is either equal to the number of variations in sign of $P(-x)$, or is less than the number of variations in sign of $P(-x)$ by a positive even number.

Rational Solution Theorem [5.6]

14. To list the possible rational solutions for the equation $2x^3 - 7x^2 + 4x + 3 = 0$ we look at the possible values for p and the possible values for q.

possible values for p: $\pm 1, \pm 3$
possible values for q: $\pm 1, \pm 2$
possible rational solutions p/q: $\pm 1/2, \pm 1, \pm 3/2, \pm 3$

If the rational number p/q (in lowest terms) is a solution to the polynomial equation

$$a_n x^n + a_{n-1}x^{n-1} + \cdots + a_1 x + a_0 = 0 \qquad a_n \neq 0$$

where the coefficients a_0 through a_n are integers, then p is a factor of the constant term a_0 and q is a factor of the leading coefficient a_n.

Chapter 5 Test

For each of the following equations, give the x-intercepts and the coordinates of the vertex, and sketch the graph. [5.1]

1. $y = x^2 - 2x - 3$

2. $y = -x^2 - 4x + 5$

Find the vertex and any two convenient points to sketch the graphs of the following equations. [5.1]

3. $y = -x^2 + 2x - 5$

4. $g(x) = 3x^2 + 4x + 1$

Identify the quotient and remainder when $P(x)$ is divided by $f(x)$. [5.2]

5. $P(x) = 8x^4 - 2x^3 + x^2 - 3x + 1, f(x) = x - \frac{1}{2}$

6. $P(x) = 16x^4 + 3x^2 - 3x + 4, f(x) = x + \frac{1}{4}$

7. Is $x + 2$ a factor of $P(x) = x^5 - 32$?

8. Use synthetic division to find $P(4)$ if $P(x) = 3x^3 - 4x^2 + 2x - 5$. [5.2]

9. Factor $x^3 - 7x^2 + 16x - 12$ completely if $x - 3$ is one of its factors. [5.3]

10. Find all solutions to $4x^3 - 8x^2 - 37x + 20 = 0$ if $x = \frac{1}{2}$ is one solution. [5.3]

Graph each of the following polynomial functions. [5.4]

11. $y = x^3 - 2x$

12. $y = x^3 - x^2 - 5x + 5$

13. $y = x^4 - 3x^2$

14. $y = x^4 - 8x^2 + 7$

Graph each rational function, Label all intercepts and asymptotes, when they exist. [5.5]

15. $y = \frac{4}{x}$

16. $y = \frac{-4}{x - 2}$

17. $y = \frac{x + 4}{x - 2}$

18. $y = \frac{x}{x^2 - 1}$

For Problems 19 and 20, follow the instructions. Do not solve the equations. [5.6]

19. How many positive real solutions can we expect from the equation

$$4x^4 - 8x^3 + 19x^2 + 2x - 5 = 0$$

20. How many negative real solutions can we expect from the equation

$$4x^4 - 8x^3 + 19x^2 + 2x - 5 = 0$$

Solve each of the following equations. If a solution occurs more than once, give its multiplicity. [5.6]

21. $2x^4 - 5x^3 - 2x^2 + 11x - 6 = 0$

22. $2x^4 - 9x^3 + 15x^2 - 11x + 3 = 0$

23. $4x^4 - 8x^3 + 19x^2 + 2x - 5 = 0$

© Srebrina Yaneva/iStockPhoto

24. **Wildlife** The eider duck, one of the world's fastest flying birds, can exceed an airspeed of 65 miles per hour. A flock of eider ducks is migrating south at an average airspeed of 50 miles per hour against a moderate headwind. Their next feeding grounds are 150 miles away.

    a. Express the ducks' travel time t as a function of the wind speed, v.

    b. Complete the table showing the travel time for various wind speeds.

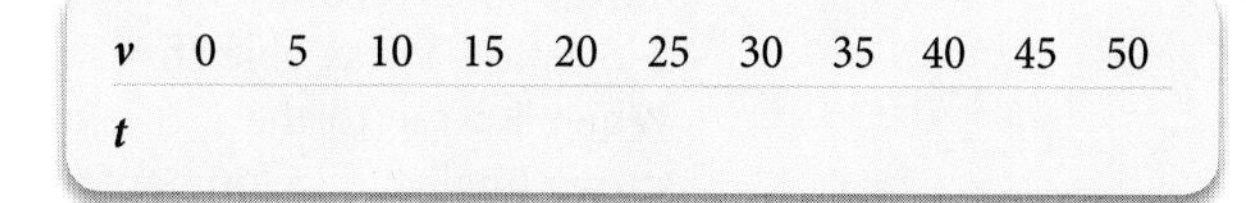

v	0	5	10	15	20	25	30	35	40	45	50
t											

    What happens to the travel time as the headwind increases?

    c. Use the table to choose an appropriate scale and graph your function $t(v)$. Label the scales on the axes.

    d. Estimate the wind speed if the travel time was 12 hours. Illustrate your result on a graph.

    e. Give the equations of any horizontal or vertical asymptotes. What does the vertical asymptote signify in the context of the problem? [5.5]

25. **Solution Concentration** Delbert prepares a 25% glucose solution by mixing 2 mL of glucose with 8 mL of water. If he adds x mL of glucose to the solution, its concentration is given by

$$C(x) = \frac{2 + x}{8 + x}$$

    a. How many mL of glucose should Delbert add to increase the concentration to 50%?

    b. Graph the function for $0 \leq x \leq 100$.

    c. What is the horizontal asymptote of the graph? What does it tell you about the solution? [5.5]

26. **Maximum Height** An arrow is shot straight up into the air with an initial velocity of 128 feet per second. If h represents the height (in feet) of the arrow at any time t (in seconds), then the equation that gives h in terms of t is $h(t) = 128t - 16t^2$. Find the maximum height attained by the arrow. [5.1]

27. **Maximum Revenue** The relationship between the number of calculators x a company sells each day and the price p of each calculator is given by the equation $x = 1{,}700 - 100p$. Graph the revenue equation $R = xp$ and use the graph to find the price p that will bring in the maximum revenue. Then find the maximum revenue. [5.1]

SPOTLIGHT ON SUCCESS *Instructor Edwin*

You never fail until you stop trying.
—Albert Einstein

Coming to the United States at the age of 10 and not knowing how to speak English was a very difficult hurdle to overcome. However, with hard work and dedication I was able to rise above those obstacles. When I came to the U.S. our school did not have a strong English development program as it was known at that time, English as a Second Language (ESL). The approach back then was "sink or swim." When my self-esteem was low, my mom and my three older sisters were always there for me and they would always encourage me to do well. My mom was a single parent, and her number one priority was that we would receive a good education. My mother's perseverance is what has made me the person I am today. At a young age I was able to see that she had overcome more than what my situation was, and I would always tell myself, "if Mom can do it, I could also do it." Not only did she not have an education, but she also saved us from a civil war that was happening in my home country of El Salvador.

When things in school got hard, I would always reflect on all the hard work, sacrifice and effort of my mother. I would just tell myself that I should not have any excuses and that I needed to keep going. If my mother, who worked as a housekeeper, could send all four of her kids to college doesn't motivate you, I don't know what does. It definitely motivated me. The day everything began to change for me was when I was in eighth grade. I was sitting in my biology class not paying attention to the teacher because I was really focusing on a piece of paper on the wall. It said, "You never fail until you stop trying." I read it over and over, trying to digest what the quote meant. With my limited English I was doing my best to translate what it meant in my native language. It finally clicked! I was able to figure out what those seven words meant. I memorized the quote and began to apply it to my academics and to real-life situations. I began to really focus in my studies. I wanted to do well in school, and most important I wanted to improve my English. To this day I always reflect to that quote when I feel I can't do something.

I was able to finish junior high successfully. Going to high school was a lot easier and I ended up with very good grades and eventually I was accepted to an excellent college. I was never the smartest student on campus, but I always did well because I never quit. I earned my college degree and now I teach at a dual immersion elementary school. I have that same quote in my classroom and I constantly remind my students to never stop trying.

Exponential and Logarithmic Functions

6

Chapter Outline

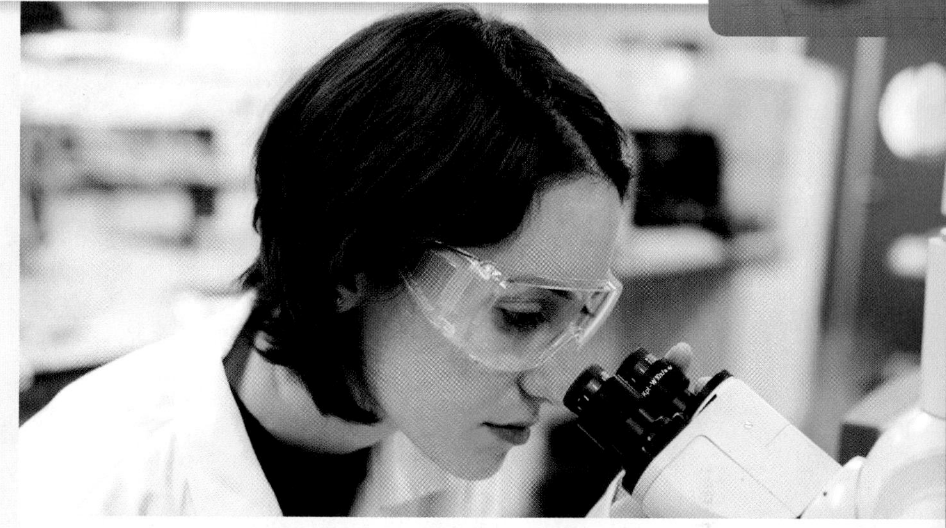

All living things contain large amounts of carbon, most of it in the most common form, carbon-12. However, one part in a trillion is in the form of carbon-14, which is unstable and undergoes radioactive decay. The half-life of carbon-14 is 5,730 years, which means that every 5,730 years a sample of carbon-14 will decrease to half its original amount. When a living organism dies, the carbon-14 begins to decay. By measuring the amount left, the age of the organism can be determined with surprising accuracy. Objects that have been dated this way include the Dead Sea Scrolls, the "Iceman" found frozen in the Alps in 1991, and campsites of prehistoric humans.

Table 1 and Figure 1 illustrate how a decaying substance such as carbon-14 decreases by half each half-life.

Note When you see this icon next to an example or problem in this chapter, you will know that we are using the topics in this chapter to model situations in the world around us.

Half-life	Percent Remaining
0	100%
1	50%
2	25%
3	12.5%
4	6.25%
5	3.125%
6	1.5625%

TABLE 1

FIGURE 1

Carbon dating, as this technique is known, is effective for dating organisms that died in the last 50,000 years or so (about nine half-lives of carbon-14). As the graph shows, after this the amount of carbon-14 remaining is so small, and changes so slowly, that the method is no longer accurate.

Study Skills

Never mistake activity for achievement.

— John Wooden, legendary UCLA basketball coach

You may think that this John Wooden quote has to do with being productive and efficient, or using your time wisely, but it is really about being honest with yourself. I have had students come to me after failing a test saying, "I can't understand why I got such a low grade after I put so much time in studying." One student even had help from a tutor and felt she understood everything that we covered. After asking her a few questions, it became clear that she spent all her time studying with a tutor and the tutor was doing most of the work. The tutor can work all the homework problems, but the student cannot. She has mistaken activity for achievement.

Can you think of situations in your life when you are mistaking activity for achievement?

How would you describe someone who is mistaking activity for achievement in the way they study for their math class?

Which of the following best describes the idea behind the John Wooden quote?

- Always be efficient.
- Don't kid yourself.
- Take responsibility for your own success.
- Study with purpose.

Exponential Functions

INTRODUCTION To obtain an intuitive idea of how exponential functions behave, we can consider the heights attained by a bouncing ball. When a ball used in the game of racquetball is dropped from any height, the first bounce will reach a height that is $\frac{2}{3}$ of the original height. The second bounce will reach $\frac{2}{3}$ of the height of the first bounce, and so on, as shown in Figure 1.

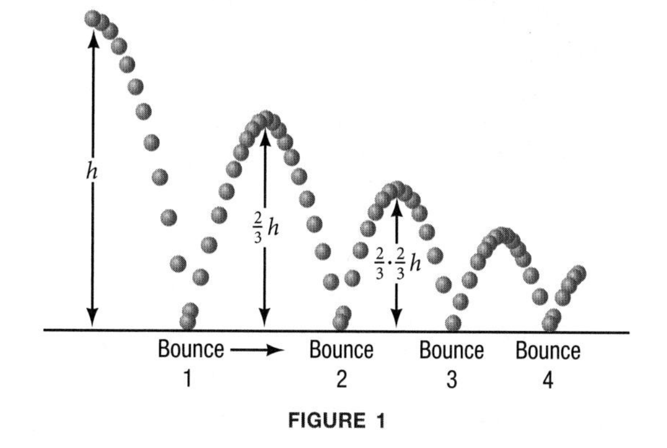

FIGURE 1

If the ball is initially dropped from a height of 1 meter, then during the first bounce it will reach a height of $\frac{2}{3}$ meter. The height of the second bounce will reach $\frac{2}{3}$ of the height reached on the first bounce. The maximum height of any bounce is $\frac{2}{3}$ of the height of the previous bounce.

Initial height: $h = 1$

Bounce 1: $h = \frac{2}{3}(1) = \frac{2}{3}$

Bounce 2: $h = \frac{2}{3}\left(\frac{2}{3}\right) = \left(\frac{2}{3}\right)^2$

Bounce 3: $h = \frac{2}{3}\left(\frac{2}{3}\right)^2 = \left(\frac{2}{3}\right)^3$

Bounce 4: $h = \frac{2}{3}\left(\frac{2}{3}\right)^3 = \left(\frac{2}{3}\right)^4$

. .
. .
. .

Bounce n: $h = \frac{2}{3}\left(\frac{2}{3}\right)^{n-1} = \left(\frac{2}{3}\right)^n$ ■

This last equation is exponential in form. We classify all exponential functions together with the following definition.

DEFINITION *Exponential function*

An *exponential function* is any function that can be written in the form

$$f(x) = a \cdot b^{kx}$$

where b is a positive real number other than 1, and both a and k are nonzero.

Each of the following is an exponential function:

$$f(x) = 2^x \qquad y = 3^x \qquad f(x) = \left(\frac{1}{4}\right)^x$$

The first step in becoming familiar with exponential functions is to find some values for specific exponential functions.

EXAMPLE 1 If the exponential functions f and g are defined by

$$f(x) = 2^x \quad \text{and} \quad g(x) = 3^x$$

then

$$f(0) = 2^0 = 1 \qquad g(0) = 3^0 = 1$$
$$f(1) = 2^1 = 2 \qquad g(1) = 3^1 = 3$$
$$f(2) = 2^2 = 4 \qquad g(2) = 3^2 = 9$$
$$f(3) = 2^3 = 8 \qquad g(3) = 3^3 = 27$$
$$f(-2) = 2^{-2} = \frac{1}{2^2} = \frac{1}{4} \qquad g(-2) = 3^{-2} = \frac{1}{3^2} = \frac{1}{9}$$
$$f(-3) = 2^{-3} = \frac{1}{2^3} = \frac{1}{8} \qquad g(-3) = 3^{-3} = \frac{1}{3^3} = \frac{1}{27}$$

Modeling: Radioactive Decay

In the introduction to this chapter, we discussed the half-life of radioactive carbon-14. Iodine-131 is another radioactive substance that undergoes decay at a known rate. It is used in thyroid testing. The half-life of iodine-131 is 8 days, which means that every 8 days a sample of iodine-131 will decrease to half of its original amount. If we start with A_0 micrograms of iodine-131, then after t days the sample will contain

$$A(t) = A_0 \cdot 2^{-t/8}$$

micrograms of iodine-131.

© ZoneCreative/iStockPhoto

EXAMPLE 2 A patient is administered a 1,200-microgram dose of iodine-131. How much iodine-131 will be in the patient's system after 10 days, and after 16 days?

SOLUTION The initial amount of iodine-131 is $A_0 = 1{,}200$, so the function that gives the amount left in the patient's system after t days is

$$A(t) = 1{,}200 \cdot 2^{-t/8}$$

After 10 days, the amount left in the patient's system is

$$A(10) = 1{,}200 \cdot 2^{-10/8} = 1{,}200 \cdot 2^{-1.25} \approx 504.5 \text{ micrograms}$$

After 16 days, the amount left in the patient's system is

$$A(16) = 1{,}200 \cdot 2^{-16/8} = 1{,}200 \cdot 2^{-2} = 300 \text{ micrograms}$$

Note Recall that the symbol $\approx$ is read "is approximately equal to".

Graphing Exponential Functions

EXAMPLE 3 Sketch the graph of the exponential function $y = 2^x$.

SOLUTION Using the results of Example 1, we produce the following table. Graphing the ordered pairs given in the table and connecting them with a smooth curve, we have the graph of $y = 2^x$ shown in Figure 2.

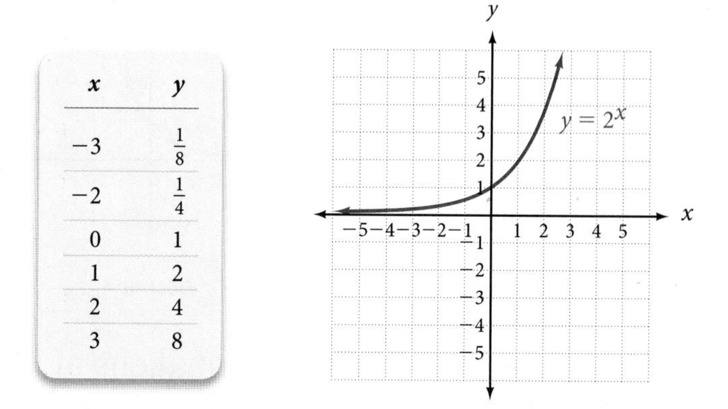

x	y
-3	$\frac{1}{8}$
-2	$\frac{1}{4}$
0	1
1	2
2	4
3	8

FIGURE 2

Notice that the graph does not cross the x-axis. It *approaches* the x-axis—in fact, we can get it as close to the x-axis as we want without it actually intersecting the x-axis. For the graph of $y = 2^x$ to intersect the x-axis, we would have to find a value of x that would make $2^x = 0$. Because no such value of x exists, the graph of $y = 2^x$ cannot intersect the x-axis, and the x-axis is a horizontal asymptote of $y = 2^x$. ■

EXAMPLE 4 Sketch the graph of $y = \left(\frac{1}{3}\right)^x$.

SOLUTION The table beside Figure 3 gives some ordered pairs that satisfy the equation. Using the ordered pairs from the table, we have the graph shown in Figure 3.

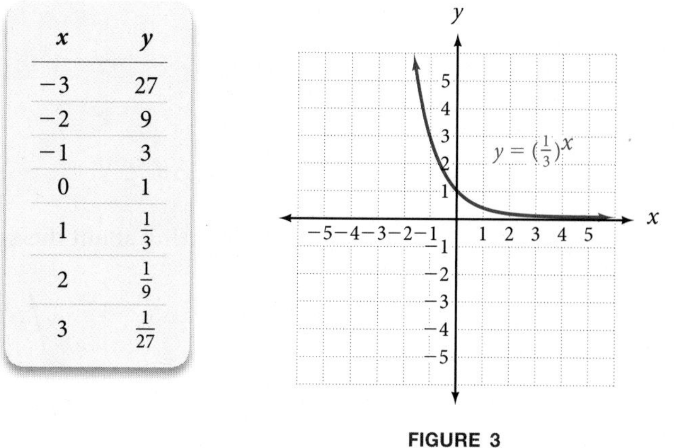

x	y
-3	27
-2	9
-1	3
0	1
1	$\frac{1}{3}$
2	$\frac{1}{9}$
3	$\frac{1}{27}$

FIGURE 3

■

The graphs of all basic exponential functions have two things in common:

1. Each crosses the y-axis at $(0, 1)$ because $b^0 = 1$; and
2. none can cross the x-axis because $b^x = 0$ is impossible due to the restrictions on b.

Figures 4 and 5 show some families of exponential curves to help you become more familiar with them on an intuitive level.

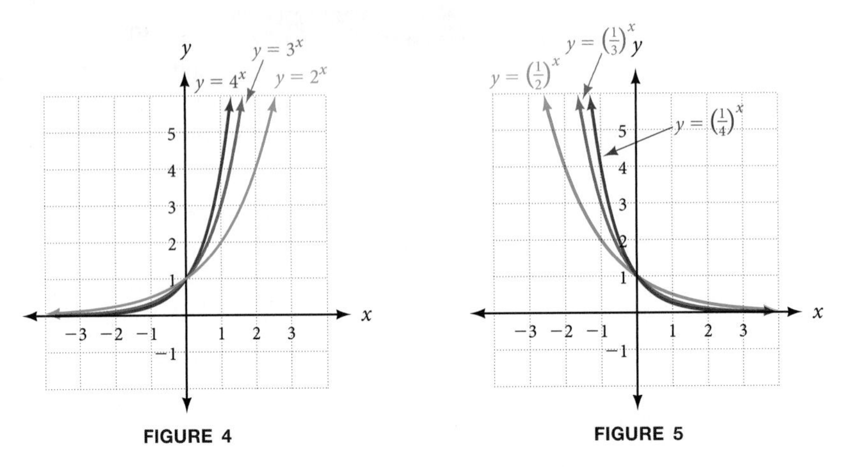

FIGURE 4

FIGURE 5

Transformations of the Graphs of Exponential Functions

All of the rules we have for translating, scaling, and reflecting of graphs will hold for the exponential functions as well. Below are some variations on the graph of $y = 2^x$.

Translating 3 units right

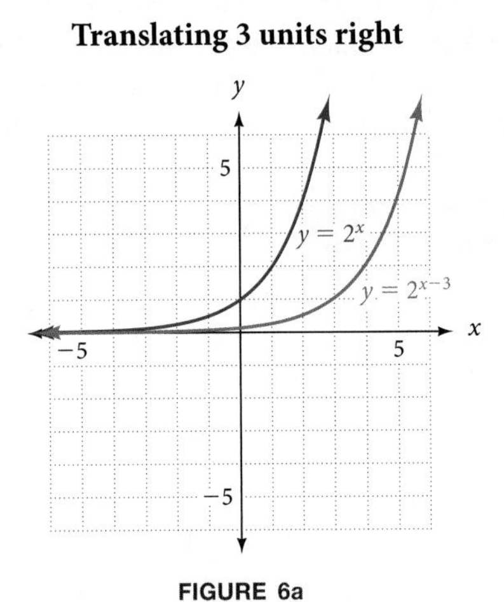

FIGURE 6a

Translating 3 units down

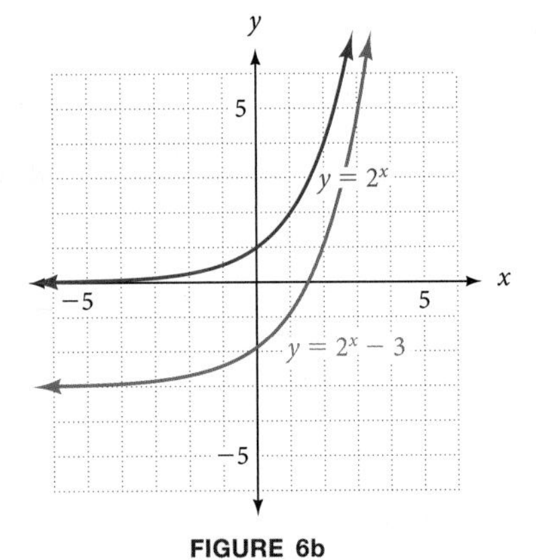

FIGURE 6b

Reflecting about the x-axis

FIGURE 6c

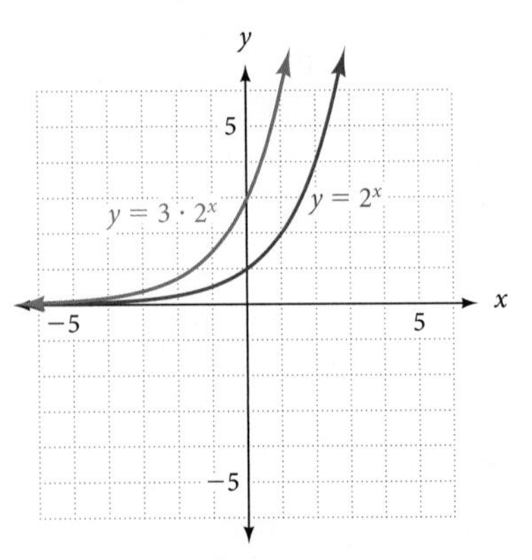

FIGURE 6d

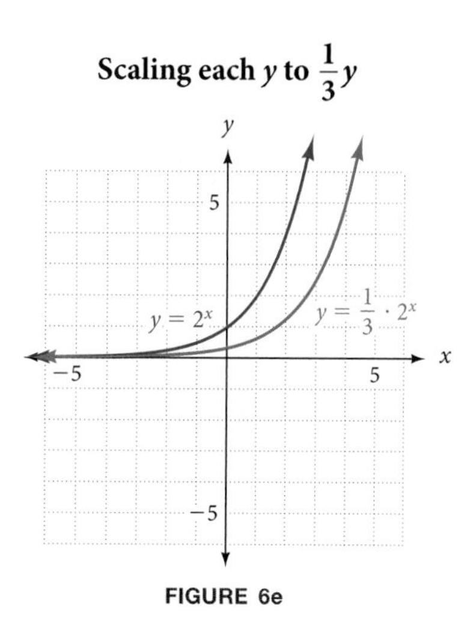

FIGURE 6e

Modeling: Medications and Half-Life

Many medications are modeled using half-lives, in the same way that radioactive materials are modeled. For example, a popular antidepressant has a half-life of 5 days. If a patient is taking that medication, and the concentration in the patient's system is relatively constant, then, if the patient stops taking the medication, the concentration of the medication in the patient's system will decrease according to the half-life. For example, suppose an antidepressant has a half-life of 5 days, and its concentration in a patient's system is 80 ng/mL (nanograms/milliliter) when the patient stops taking the antidepressant. To find the concentration after a half-life passes, we multiply the previous concentration by $\frac{1}{2}$. Here is the information displayed in a table.

Days Since Discontinuing	Concentration (ng/mL)
0	80
5	40
10	20
15	10
20	5

The information in the table is shown visually in Figures 7 and 8. The diagram in Figure 7 is called a *scatter diagram*. If we connect the points in the scatter diagram with a smooth curve, we have the diagram in Figure 8.

Note Medications with short half-lives are eliminated from a patient's system more quickly than medications with longer half-lives. In some cases, eliminating a medication from a patient's system too quickly can cause unpleasant side-effects, sometimes called *withdrawal.* Doctors will tell patients to taper off certain medications and not others, depending on the half-life of the medication.

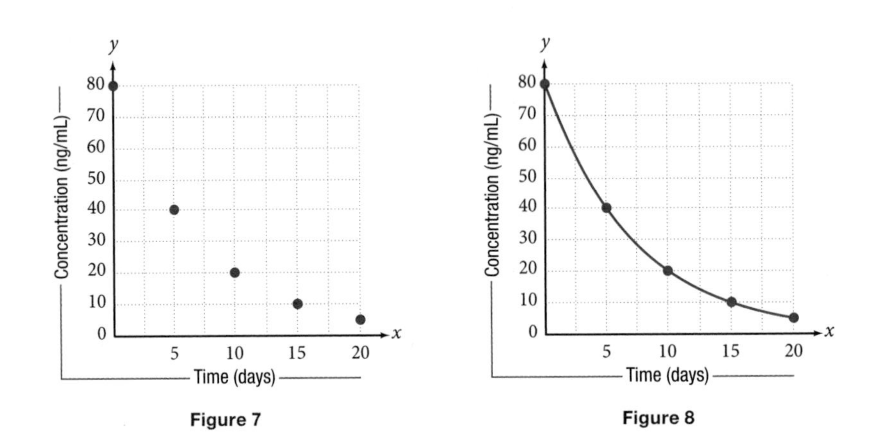

Figure 7 Figure 8

The graph in Figure 8 looks like the graph of an exponential function, and, in fact, it is. The equation for the function is

$$y = 80 \cdot 2^{-t/5}$$

where t is measured in days. As you can see, the initial concentration appears as the coefficient, and the half-life is part of the exponent. We can generalize this relationship as follows:

Half-Life and Exponential Functions

If the initial amount of a substance is A_0, and the substance decays with a half-life of k hours, then the amount of the substance present t hours later is

$$A = A_0 \cdot 2^{-t/k}$$

If the half-life is given in days, then t is in days; if the half-life is given in years, then t is in years; and so on.

Modeling: Compound Interest

Among the many applications of exponential functions are the applications having to do with interest-bearing accounts. Here are the details.

If P dollars are deposited in an account with annual interest rate r, compounded n times per year, then the amount of money in the account after t years is given by the formula

$$A(t) = P\left(1 + \frac{r}{n}\right)^{nt}$$

EXAMPLE 5 Suppose you deposit \$500 in an account with an annual interest rate of 8% compounded quarterly. Find an equation that gives the amount of money in the account after t years. Then find

a. The amount of money in the account after 5 years.

b. The number of years it will take for the account to contain \$1,000.

SOLUTION First, we note that $P = 500$ and $r = 0.08$. Interest that is compounded quarterly is compounded four times a year, giving us $n = 4$. Substituting these numbers into the preceding formula, we have our function

$$A(t) = 500\left(1 + \frac{0.08}{4}\right)^{4t} = 500(1.02)^{4t}$$

a. To find the amount after 5 years, we let $t = 5$:

$$A(5) = 500(1.02)^{4 \cdot 5} = 500(1.02)^{20} \approx \$742.97$$

Our answer is found on a calculator, and then rounded to the nearest cent.

b. To see how long it will take for this account to total \$1,000, we graph the equation $Y1 = 500(1.02)^{4X}$ on a graphing calculator, and then look to see where it intersects the line $Y2 = 1{,}000$. The two graphs are shown in Figure 9.

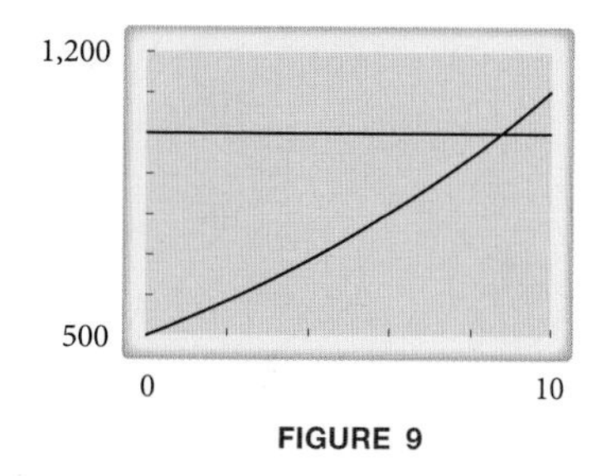

FIGURE 9

Using ZOOM and TRACE, or the Intersect function on the graphing calculator, we find that the two curves intersect at $X \approx 8.75$ and $Y = 1{,}000$. This means that our account will contain \$1,000 after the money has been on deposit for about 8.75 years.

The Natural Exponential Function

A commonly occurring exponential function is based on a special number we denote with the letter e. The *number e* is a number like π. It is irrational and occurs in many formulas that describe the world around us. Like π, it can be approximated with a decimal number. Whereas π is approximately 3.1416, e is approximately 2.7183. (If you have a calculator with a key labeled e^x, you can use it to find e^1 to find a more accurate approximation to e.) We can give a more precise definition of the number e by using some notation from calculus.

$$e = \lim_{x \to \infty}\left(1 + \frac{1}{x}\right)^x$$

This notation indicates that e is the number that is approached by the expression

$$\left(1 + \frac{1}{x}\right)^x$$

as x becomes larger and larger. To visualize this process, we can build a table.

x	$\left(1 + \frac{1}{x}\right)^x$
1	$\left(1 + \frac{1}{1}\right)^1 = 2$
10	$\left(1 + \frac{1}{10}\right)^{10} = (1.1)^{10} = 2.6$
10^2	$\left(1 + \frac{1}{100}\right)^{10^2} = (1.01)^{10^2} = 2.70$
10^3	$\left(1 + \frac{1}{1{,}000}\right)^{10^3} = (1.001)^{10^3} = 2.717$
10^4	$\left(1 + \frac{1}{10{,}000}\right)^{10^4} = (1.0001)^{10^4} = 2.7181$
10^5	$\left(1 + \frac{1}{100{,}000}\right)^{10^5} = (1.00001)^{10^5} = 2.71827$
10^6	$\left(1 + \frac{1}{1{,}000{,}000}\right)^{10^6} = (1.000001)^{10^6} = 2.718280$
10^7	$\left(1 + \frac{1}{10{,}000{,}000}\right)^{10^7} = (1.0000001)^{10^7} = 2.7182817$
$\downarrow$ ∞	$\downarrow$ e

By making x large enough, we can approximate e to as many decimal places as we like. The number e to 9 decimal places is 2.718281828. For the work we are going to do with the number e, we only need to know that it is an irrational number that is approximately 2.7183.

Here are a table and graph (Figure 10) for the natural exponential function

$$y = f(x) = e^x$$

x	$f(x) = e^x$
-2	$f(-2) = e^{-2} = \frac{1}{e^2} \approx 0.135$
-1	$f(-1) = e^{-1} = \frac{1}{e} \approx 0.368$
0	$f(0) = e^0 = 1$
1	$f(1) = e^1 = e \approx 2.72$
2	$f(2) = e^2 \approx 7.39$
3	$f(3) = e^3 \approx 20.09$

FIGURE 10

One common application of natural exponential functions is with interest-bearing accounts. In Example 5, we worked with the formula

$$A(t) = P\left(1 + \frac{r}{n}\right)^{nt}$$

that gives the amount of money in an account if P dollars are deposited for t years at annual interest rate r, compounded n times per year. In Example 5, the number of compounding periods was four. What would happen if we let the number of compounding periods become larger and larger, so that we compounded the interest every day, then every hour, then every second, and so on? If we take this as far as it can go, we end up compounding the interest every moment. When this happens, we have an account with interest that is compounded continuously, and the amount of money in such an account depends on the number e. Here are the details.

Modeling: Continuously Compounded Interest

If P dollars are deposited in an account with annual interest rate r, compounded continuously, then the amount of money in the account after t years is given by the formula

$$A(t) = Pe^{rt}$$

EXAMPLE 6 Suppose you deposit \$500 in an account with an annual interest rate of 8% compounded continuously. Find an equation that gives the amount of money in the account after t years. Then find the amount of money in the account after 5 years.

SOLUTION Because the interest is compounded continuously, we use the formula $A(t) = Pe^{rt}$. Substituting $P = 500$ and $r = 0.08$ into this formula, we have

$$A(t) = 500e^{0.08t}$$

After 5 years, this account will contain

$$A(5) = 500e^{0.08 \cdot 5} = 500e^{0.4} \approx \$745.91$$

to the nearest cent. Compare this result with the answer to Example 5a.

Using Technology 6.1

Let's use Wolfram|Alpha to confirm some of the graphs of our exponential functions from this section.

EXAMPLE 7 Use Wolfram|Alpha to graph the functions below on the same coordinate system.

$$y = 2^x \text{ and } y = \left(\frac{1}{2}\right)^x$$

SOLUTION Online, go to www.wolframalpha.com. Enter the information below, into the box on the website, exactly as it is shown.

Graph y=2^x, y=(1/2)^x

Here is what you will see:

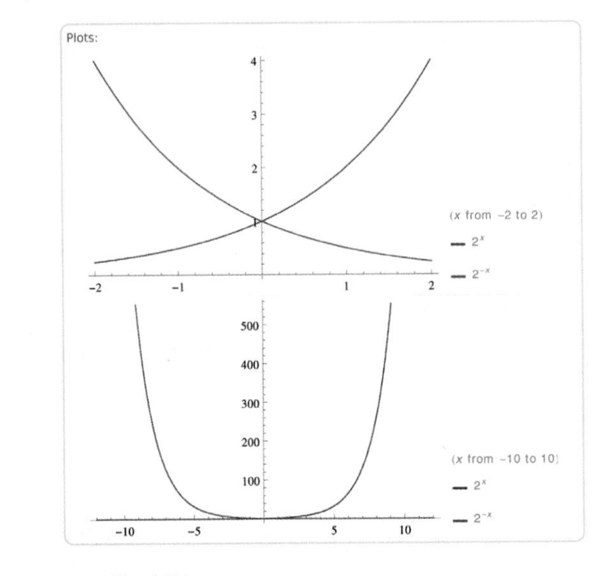

Wolfram Alpha LLC. 2012. Wolfram|Alpha
http://www.wolframalpha.com/
(access June 18, 2012)

Wolfram|Alpha simplifies $y = \left(\frac{1}{2}\right)^x$ to $y = 2^{-x}$. These two equations are equivalent. ■

Our next example involves a logistic function. Many things in the world around us can be modeled with logistic functions.

EXAMPLE 8 A study by the United Nations and published in *The New York Times* in November, 1995, produced data about the world's population. From the data, the logistic function

$$P(t) = \frac{11.5}{1 + 12.8e^{-0.0266t}}$$

was constructed. The variable t represents the number of years since 1900, and the variable P represents the world's population in billions of people.

a. Create a table of values that show the world's population in the years 1900, 1920, 1940, 1960, 1980, 2000, 2020, 2040, 2060, 2080, and 2100.

b. Construct a graph of the function $P(t)$ from $t = 0$ to 200 (1900 to 2100).

c. Evaluate the function at $t = 112$ (2012).

SOLUTION Online, go to www.wolframalpha.com.

a. Pay close attention to capital letters and the usage of the [] {} () symbols and make sure your variables match; that is, use both lowercase t's or uppercase T's. In the insert field, type

Table[11.5/(1+12.8e^(-0.0266t))], {t,0,200,20}

This code will create a table of population values for the years 1900 to 2100 in intervals of 20 years. Here is that table:

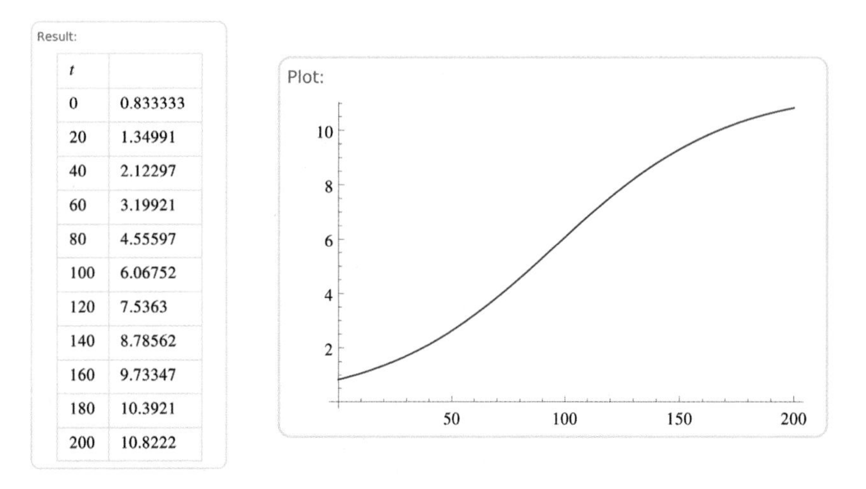

Result:

t	
0	0.833333
20	1.34991
40	2.12297
60	3.19921
80	4.55597
100	6.06752
120	7.5363
140	8.78562
160	9.73347
180	10.3921
200	10.8222

Wolfram Alpha LLC. 2012. Wolfram|Alpha
http://www.wolframalpha.com/
(access June 18, 2012)

b. To construct the graph of the function $P(t)$ from $t = 0$ to 200 (1900 to 2100) above, enter the following at the Wolfram|Alpha site:

graph 11.5/(1+12.8e^(-0.0266t)) from t=0 to t=200

c. To evaluate the function at $t = 112$ (the year 2012), we enter the following at the Wolfram|Alpha site:

evaluate 11.5/(1+12.8e^(-0.0266t)) at t = 112

The result is 7.67333, meaning that our formula gives the world population in the year 2012 as approximately 7.7 billion people. Notice also that Wolfram|Alpha also gives us the graph (the same one produced in Part b above) and the point where $t = 112$. ■

GETTING READY FOR CLASS

After reading through the preceding section, respond in your own words and in complete sentences.

A. What is an exponential function?

B. In an exponential function, explain why the base b cannot equal 1. (What kind of function would you get if the base was equal to 1?)

C. Explain continuously compounded interest.

D. What characteristics do the graphs of $y = 2^x$ and $y = \left(\frac{1}{2}\right)^x$ have in common?

Problem Set 6.1

Let $f(x) = 3^x$ and $g(x) = \left(\frac{1}{2}\right)^x$, and evaluate each of the following.

1. $g(0)$ **2.** $f(0)$ **3.** $g(-1)$ **4.** $g(-4)$

5. $f(-3)$ **6.** $f(-1)$ **7.** $f(2) + g(-2)$ **8.** $f(2) - g(-2)$

Let $f(x) = 4^x$ and $g(x) = \left(\frac{1}{3}\right)^x$. Evaluate each of the following.

9. $f(-1) + g(1)$ **10.** $f(2) + g(-2)$ **11.** $\dfrac{f(-2)}{g(1)}$ **12.** $f(3) - f(2)$

Graph each of the following functions.

13. $y = 4^x$ **14.** $y = 2^{-x}$ **15.** $y = 3^{-x}$ **16.** $y = \left(\frac{1}{3}\right)^{-x}$

17. $y = e^x$ **18.** $y = e^{-x}$ **19.** $y = \left(\frac{1}{3}\right)^x$ **20.** $y = \left(\frac{1}{2}\right)^{-x}$

For Problems 21-24, graph Parts a-c on the same coordinate system.

21. **a.** $y = 2^x$ **b.** $y = 2^x - 2$ **c.** $y = 2^x + 2$

22. **a.** $y = 3^x$ **b.** $y = 3^{x-2}$ **c.** $y = 3^{x+2}$

23. **a.** $y = 2^x$ **b.** $y = \frac{1}{3} \cdot 2^x$ **c.** $y = -2^x$

24. **a.** $y = 3^x$ **b.** $y = 2 \cdot 3^x$ **c.** $y = -2 \cdot 3^x$

25. Graph the following functions on the same coordinate system for positive values of x only.

$$y = 2x, y = x^2, y = 2^x$$

26. **Reading Graphs** The graphs of two exponential functions are given in Figures 11 and 12. Use the graphs to find the following:

a. $f(0)$ **b.** $f(-1)$ **c.** $f(1)$ **d.** $g(0)$

e. $g(1)$ **f.** $g(-1)$ **g.** $f(g(0))$ **h.** $g(f(0))$

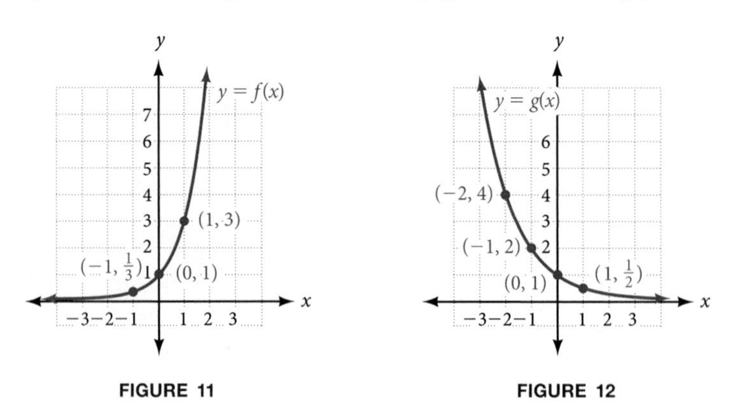

FIGURE 11 FIGURE 12

Getting Ready for Calculus Problems 27-35 are problems you will see again if you go on to a calculus class.

Graph each function.

27. $y = 3^x$ **28.** $y = \left(\frac{1}{3}\right)^x$ **29.** $y = \left(\frac{2}{3}\right)^x$ **30.** $y = \left(\frac{3}{2}\right)^x$

31. If $f(x) = \left(1 + \frac{1}{x}\right)^x$, find each of the following.

a. $f(1)$ **b.** $f(10)$ **c.** $f(100)$ **d.** $f(1{,}000)$

32. If $f(t) = e^{2t}$, find each of the following.

a. $f(-2)$ **b.** $f(0)$ **c.** $f(1)$ **d.** $f(2)$

33. If $f(t) = Ae^{kt}$, find $f(0)$.

34. If $N(p) = 35e^{-0.11(p-1)}$, find $N(6)$.

35. If $A = 2000\left(1 + \frac{0.08}{k}\right)^{15k}$, find A when $k = 4$.

Modeling Practice

36. Bouncing Ball Suppose the ball mentioned in the introduction to this section is dropped from a height of 6 feet above the ground. Find an exponential equation that gives the height h the ball will attain during the nth bounce. How high will it bounce on the fifth bounce?

37. Bouncing Ball A golf ball is manufactured so that if it is dropped from A feet above the ground onto a hard surface, the maximum height of each bounce will be one half of the height of the previous bounce. Find an exponential equation that gives the height h the ball will attain during the nth bounce. If the ball is dropped from 10 feet above the ground onto a hard surface, how high will it bounce on the eighth bounce?

38. Half-Life The half-life of a drug is 4 hours. A patient has been taking the drug on a regular basis for a few months and then discontinues taking it. The concentration of the drug in a patient's system is 60 ng/mL when the patient stops taking the medication.

a. Complete the table

Hours Since Discontinuing	Concentration (ng/mL)
0	60
4	
8	
12	
16	

b. Use the data from the table to construct a graph.
c. Write the equation for the exponential function described in Parts a and b.

39. Half-Life The half-life of a drug is 8 hours. A patient has been taking the drug on a regular basis for a few months and then discontinues taking it. The concentration of the drug in a patient's system is 120 ng/mL when the patient stops taking the medication.
a. Complete the table

Hours Since Discontinuing	Concentration (ng/mL)
0	120
8	
16	
24	
32	

b. Use the data from the table to construct a graph.
c. Write the equation for the exponential function described in Parts a and b.

40. Half-Life The half-lives of two antidepressants are given in the following table. A person on antidepressant 1 tells his doctor that he begins to feel sick if he misses his morning dose, while a patient taking antidepressant 2 tells his doctor that he does not notice a difference if he misses a day of taking the medication. Explain these situations in terms of the half-life of the medication.

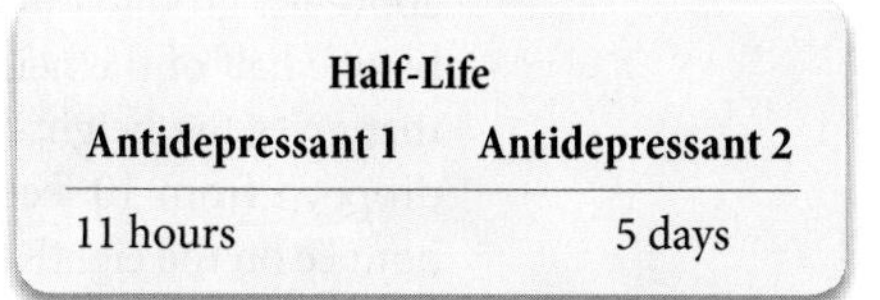

Half-Life	
Antidepressant 1	Antidepressant 2
11 hours	5 days

41. Half-Life Two patients are taking the antidepressants mentioned in the previous problem. Both decide to stop their current medications and take another medication. The physician tells the patient on antidepressant 2 to simply stop taking it, but instructs the patient on antidepressant 1 to take half the normal dose for 3 days, then one-fourth the normal dose for another 3 days before stopping the medication all together. Use half-life to explain why the doctor uses two different methods of taking the patients off their medications.

42. Exponential Decay Twinkies on the shelf of a convenience store lose their fresh tastiness over time. We say that the taste quality is 1 when the Twinkies are first put on the shelf at the store, and that the quality of tastiness declines according to the function $Q(t) = 0.85^t$ (t in days). Graph this function on a graphing calculator, and determine when the taste quality will be one half of its original value.

43. Exponential Growth Automobiles built before 1993 use Freon in their air conditioners. The federal government now prohibits the manufacture of Freon. Because the supply of Freon is decreasing, the price per pound is increasing exponentially. Current estimates put the formula for the price per pound of Freon at $p(t) = 1.89(1.25)^t$, where t is the number of years since 1990. Find the price of Freon in 1995 and 1990. How much did Freon cost in the year 2010?

44. Compound Interest Suppose you deposit $1,200 in an account with an annual interest rate of 6% compounded quarterly.

a. Find an equation that gives the amount of money in the account after t years.
b. Find the amount of money in the account after 8 years.
c. How many years will it take for the account to contain $2,400?
d. If the interest were compounded continuously, how much money would the account contain after 8 years?

45. Compound Interest Suppose you deposit $500 in an account with an annual interest rate of 8% compounded monthly.

a. Find an equation that gives the amount of money in the account after t years.
b. Find the amount of money in the account after 5 years.
c. How many years will it take for the account to contain $1,000?
d. If the interest were compounded continuously, how much money would the account contain after 5 years?

Declining-Balance Depreciation The declining-balance method of depreciation is an accounting method businesses use to deduct most of the cost of new equipment during the first few years of purchase. Unlike other methods, the declining-balance formula does not consider salvage value.

© Eduardo Luzzatti Buyé/ iStockPhoto

46. Value of a Crane The function

$$V(t) = 450{,}000(1 - 0.30)^t$$

where V is value and t is time in years, can be used to find the value of a crane for the first 6 years of use.

a. What is the value of the crane after 3 years and 6 months?
b. State the domain of this function.
c. Sketch the graph of this function.
d. State the range of this function.
e. After how many years will the crane be worth only $85,000?

47. Value of a Printing Press The function $V(t) = 375{,}000(1 - 0.25)^t$, where V is value and t is time in years, can be used to find the value of a printing press during the first 7 years of use.

a. What is the value of the printing press after 4 years and 9 months?
b. State the domain of this function.
c. Sketch the graph of this function.
d. State the range of this function.
e. After how many years will the printing press be worth only $65,000?

48. Bacteria Growth Suppose it takes 12 hours for a certain strain of bacteria to reproduce by dividing in half. If 50 bacteria are present to begin with, then the total number present after x days will be $f(x) = 50 \cdot 4^x$. Find the total number present after 1 day, 2 days, and 3 days.

49. Bacteria Growth Suppose it takes 1 day for a certain strain of bacteria to reproduce by dividing in half. If 100 bacteria are present to begin with, then the total number present after x days will be $f(x) = 100 \cdot 2^x$. Find the total number present after 1 day, 2 days, 3 days, and 4 days. Estimate how many days must elapse before over 100,000 bacteria are present.

50. Value of a Painting A painting is purchased as an investment for \$150. If the painting's value doubles every 3 years, then its value is given by the function

$$V(t) = 150 \cdot 2^{t/3} \quad \text{for } t \geq 0$$

where t is the number of years since it was purchased, and $V(t)$ is its value (in dollars) at that time. Graph this function.

© XYZ Textbooks

51. Value of a Painting A painting is purchased as an investment for \$125. If the painting's value doubles every 5 years, then its value is given by the function

$$V(t) = 125 \cdot 2^{t/5} \quad \text{for } t \geq 0$$

where t is the number of years since it was purchased, and $V(t)$ is its value (in dollars) at that time. Graph this function.

52. Cost Increase The cost of a can of Coca Cola in 1960 was \$0.10. The exponential function that models the cost of a Coca Cola by year is given below, where t is the number of years since 1960.

$$C(t) = 0.10e^{0.0576t}$$

a. What was the expected cost of a can of Coca Cola in 1985?
b. What was the expected cost of a can of Coca Cola in 2000?
c. What was the expected cost of a can of Coca Cola in 2010?
d. What is the expected cost of a can of Coca Cola in 2050?

53. Value of a Car As a car ages, its value decreases. The value of a particular car with an original purchase price of \$25,600 is modeled by the following function, where c is the value at time t (Kelly Blue Book).

$$c(t) = 25{,}600(1 - 0.22)^t$$

a. What is the value of the car when it is 3 years old?
b. What is the total depreciation amount after 4 years?

54. Bacteria Decay You are conducting a biology experiment and begin with 5,000,000 cells, but some of those cells are dying each minute. The rate of death of the cells is modeled by the function $A(t) = A_0 \cdot e^{-0.598t}$, where A_0 is the original number of cells, t is time in minutes, and A is the number of cells remaining after t minutes.
a. How may cells remain after 5 minutes?
b. How many cells remain after 10 minutes?
c. How many cells remain after 20 minutes?

© Jason Lugo/iStockPhoto

55. Drag Racing Table 1 gives the speed of a dragster every second during one race at the 1993 Winternationals. Figure 13 is a line graph constructed from the data in Table 1. The graph of the function $s(t) = 250(1 - 1.5^{-t})$ contains the first point and the last point shown in Figure 13. That is, both (0, 0) and (6, 228.1) satisfy the function. Graph the function to see how close it comes to the other points in Figure 13.

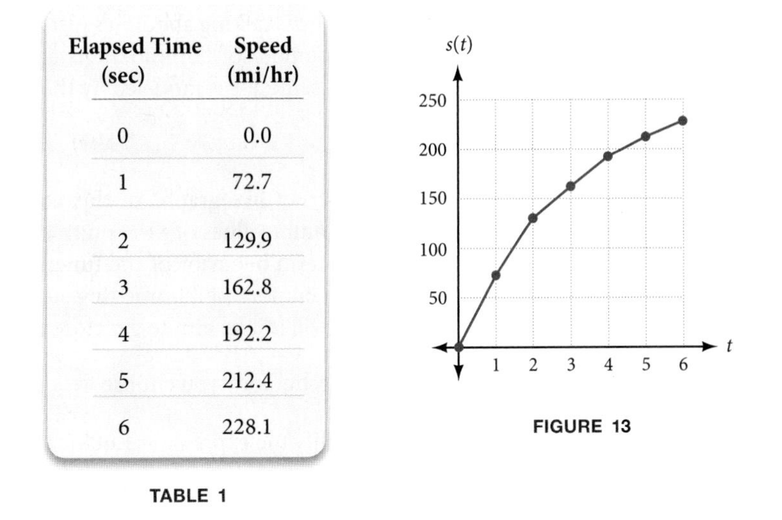

Elapsed Time (sec)	Speed (mi/hr)
0	0.0
1	72.7
2	129.9
3	162.8
4	192.2
5	212.4
6	228.1

TABLE 1

FIGURE 13

Using Technology Exercises

Use Wolfram|Alpha to graph the following groups of equations on the same coordinate system.

56. $f(x) = 3^x$, $g(x) = x^3$, and $h(x) = 3x$

57. $f(x) = 2 \cdot 3^x$, $g(x) = 3 \cdot 2^x$, and $h(x) = 6^x$

58. $f(x) = e^x$, $g(x) = e^{-x}$, and $h(x) = \dfrac{1}{e^x}$

59. $f(x) = 2^x$, $g(x) = -2^x$, and $h(x) = 1 - 2^x$

60. On a graphing calculator, graph the family of curves $y = b^x$, $b = 2, 4, 6, 8$.

61. On a graphing calculator, graph the family of curves $y = b^x$, $b = \frac{1}{2}, \frac{1}{4}, \frac{1}{6}, \frac{1}{8}$.

62. Spreading a Rumor A person starts and passes along a rumor and t hours later N people have heard it. Suppose N and t are related by the logistic function

$$N(t) = \frac{600}{1 + 400e^{-0.8t}}$$

Construct the graph of this function and use it to describe its behavior over a 24-hour period. Your description should include a statement about the maximum number of people who could hear the rumor. Does this number appear in the function's expression? Try changing this number and see what happens to your graph.

63. Math Professor A professor of mathematics is well known by his students to make statements during the course of the semester that confuse them and leave them flummoxed. The professor, of course, claims to be baffled and befuddled by their comments. Some examples of the professor's flummoxing statements are "I started out with nothing and still have most of it left." "For this problem, choose the heavier of the numbers -5 and $+5$." One of his students even commented "The professor is a very smart fellow and really knows what he is talking about. It's just that we don't know what he is talking about." The number of flummoxing statements the professor makes during the 16-week semester is modeled by the logistic equation

$$N(t) = \frac{200}{1 + 200e^{-0.85t}}$$

Construct five graphs of this equation, starting each at $t = 0$ and using as maximum values of t the numbers 4, 8, 10, 12 and 16 so that you can see the long-term behavior of the function. Based on your graph, what is the maximum number of flummoxing statements the professor can make? About how long will it take him to get close to that number?

Use any technology you choose to work the following problems.

64. Simplify the expression $1{,}600 \cdot 2^{-t/8}$ for the following values of t.

a. $t = 0$ **b.** $t = 1$ **c.** $t = 8$ **d.** $t = 4$

65. Simplify the expression $450{,}000\,(1 - 0.30)^t$ for the following values of t.

a. $t = 1$ **b.** $t = 2$ **c.** $t = 3$ **d.** $t = 4$

66. Graph each of the following on the same coordinate system.

a. $y = e^x - 3$ **b.** $y = e^{x-3}$

67. Graph each of the following on the same coordinate system.

a. $y = 3^x$ **b.** $y = 2 \cdot 3^x$ **c.** $y = -3^x$ **d.** $y = 3^{-x}$

68. Solve each equation for x:

a. $e^x(2x - 3) = 0$ **b.** $3xe^x + e^x = 0$

Getting Ready for the Next Section

Solve each equation for y.

69. $x = 2y - 3$

70. $x = \frac{y + 7}{5}$

71. $x = y^2 - 3$

72. $x = (y + 4)^3$

73. $x = \frac{y - 4}{y - 2}$

74. $x = \frac{y + 5}{y - 3}$

75. $x = \sqrt{y - 3}$

76. $x = \sqrt{y} + 5$

The Inverse of a Function

INTRODUCTION The following diagram (Figure 1) shows the route Justin takes to school. He leaves his home and drives 3 miles east, and then turns left and drives 2 miles north. When he leaves school to drive home, he drives the same two segments, but in the reverse order and the opposite direction; that is, he drives 2 miles south, turns right, and drives 3 miles west. When he arrives home from school, he is right where he started. His route home "undoes" his route to school, leaving him where he began.

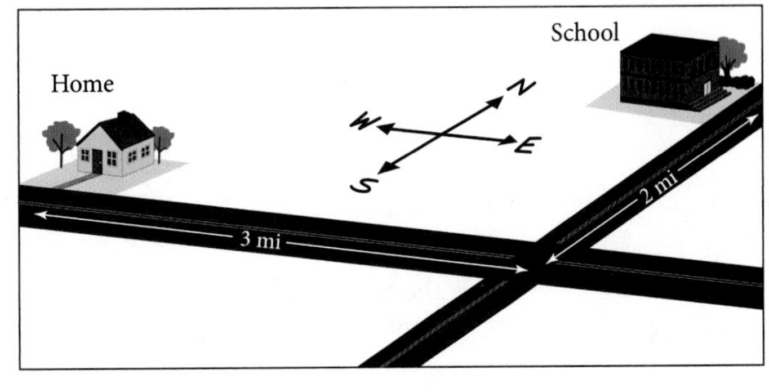

FIGURE 1

As you will see, the relationship between a function and its inverse is similar to the relationship between Justin's route from home to school and his route from school to home.

Finding the Inverse of a Function

Suppose the function f is given by

$$f = \{(1, 4), (2, 5), (3, 6)\}$$

The inverse of f is obtained by reversing the order of the coordinates in each ordered pair in f. Here is a diagram of the situation:

$$\text{The function } f = \{(1, 4), (2, 5), (3, 6)\}$$

$$\text{Its inverse } g = \{(4, 1), (5, 2), (6, 3)\}$$

$$\text{Domain of } f = \{1, 2, 3\} = \text{Range of } g$$

$$\text{Range of } f = \{4, 5, 6\} = \text{Domain of } g$$

As you can see, the domain of f is now the range of g, and the range of f is now the domain of g. Every function (or relation) has an inverse that is obtained from the original function by interchanging the components of each ordered pair.

Suppose a function f is defined with an equation instead of a list of ordered pairs. We can obtain the equation of the inverse of f by interchanging the role of x and y in the equation for f.

VIDEO EXAMPLES

SECTION 6.2

EXAMPLE 1 If the function f is defined by $f(x) = 2x - 3$, find the equation that represents the inverse of f.

SOLUTION Because the inverse of f is obtained by interchanging the components of all the ordered pairs belonging to f, and each ordered pair in f satisfies the equation $y = 2x - 3$, we simply exchange x and y in the equation $y = 2x - 3$ to get the formula for the inverse of f:

$$\text{The function: } y = 2x - 3$$

$$\text{Its inverse: } x = 2y - 3$$

We now solve this equation for y in terms of x:

$$x + 3 = 2y \qquad \text{Add 3 to each side}$$

$$\frac{x+3}{2} = y \qquad \text{Divide each side by 2}$$

$$y = \frac{x+3}{2} \qquad \text{Exchange sides}$$

The last line gives the equation that defines the inverse of f. Let's compare the graphs of f and its inverse as given here. (See Figure 2.)

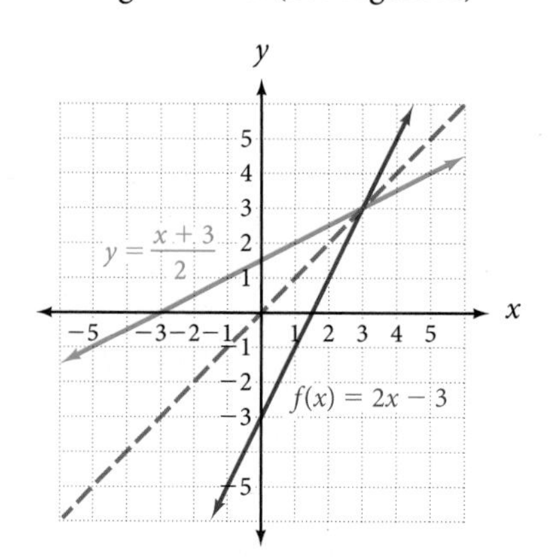

FIGURE 2

The graphs of f and its inverse have symmetry about the line $y = x$. This is a reasonable result since the one function was obtained from the other by interchanging x and y in the equation. The ordered pairs (a, b) and (b, a) always have symmetry about the line $y = x$. ■

EXAMPLE 2 Graph the function $f(x) = x^2 - 2$ and its inverse. Give the equation for the inverse.

SOLUTION We can obtain the graph of the inverse of $f(x) = x^2 - 2$ by graphing $y = x^2 - 2$ by the usual methods, and then reflecting the graph about the line $y = x$.

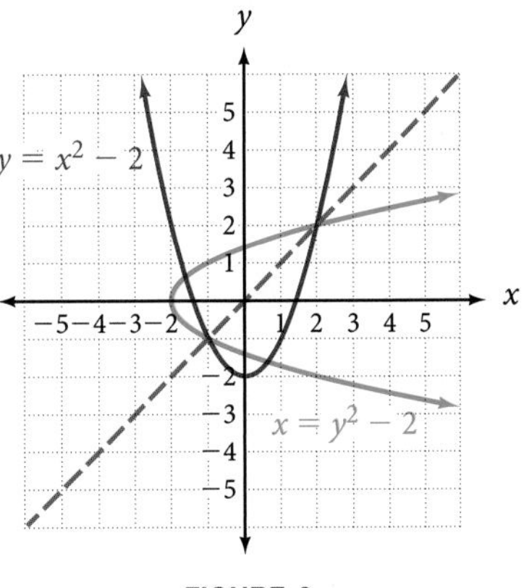

FIGURE 3

The equation that corresponds to the inverse of $y = x^2 - 2$ is obtained by interchanging x and y to get $x = y^2 - 2$.

We can solve the equation $x = y^2 - 2$ for y in terms of x as follows:

$$x = y^2 - 2$$
$$x + 2 = y^2$$
$$y = \pm\sqrt{x + 2}$$

■

Comparing the graphs from Examples 1 and 2, we observe that the inverse of a function is not always a function. In Example 1, both f and its inverse have graphs that are nonvertical straight lines and therefore both represent functions. In Example 2, the inverse of function f is not a function, since a vertical line crosses it in more than one place.

One-to-One Functions

We can distinguish between those functions with inverses that are also functions and those functions with inverses that are not functions with the following definition.

> **DEFINITION** *One-to-one functions*
>
> A function is a *one-to-one function* if every element in the range comes from exactly one element in the domain.

This definition indicates that a one-to-one function will yield a set of ordered pairs in which no two different ordered pairs have the same second coordinates. For example, the function

$$f = \{(2, 3), (-1, 3), (5, 8)\}$$

Domain f **Range**

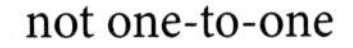

not one-to-one

is not one-to-one because the element 3 in the range comes from both 2 and -1 in the domain. On the other hand, the function

$$g = \{(5, 7), (3, -1), (4, 2)\}$$

Domain g **Range**

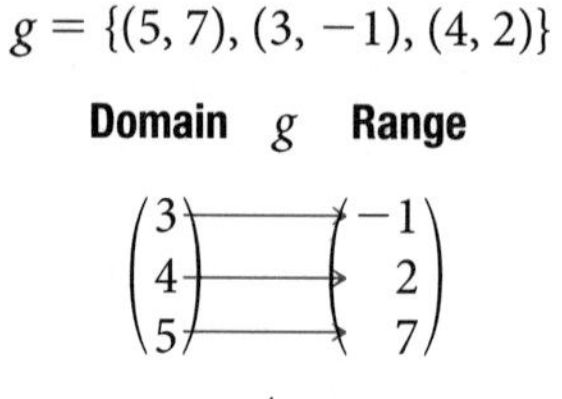

one-to-one

is a one-to-one function because every element in the range comes from only one element in the domain.

Horizontal Line Test

If we have the graph of a function, we can determine if the function is one-to-one with the following test. If a horizontal line crosses the graph of a function in more than one place, then the function is not a one-to-one function because the points at which the horizontal line crosses the graph will be points with the same y-coordinates, but different x-coordinates. Therefore, the function will have an element in the range (the y-coordinate) that comes from more than one element in the domain (the x-coordinates).

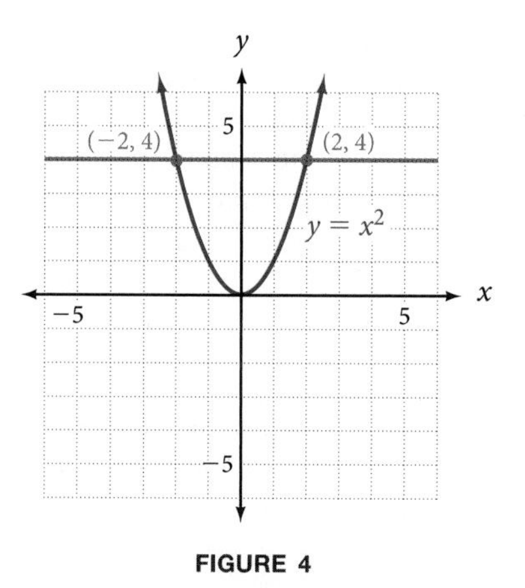

FIGURE 4

Of the functions we have covered previously, all the linear functions and exponential functions are one-to-one functions because no horizontal lines can be found that will cross their graphs in more than one place.

Functions Whose Inverses Are Also Functions

Because one-to-one functions do not repeat second coordinates, when we reverse the order of the ordered pairs in a one-to-one function, we obtain a relation in which no two ordered pairs have the same first coordinate—by definition, this relation must be a function. In other words, every one-to-one function has an inverse that is itself a function. Because of this, we can use function notation to represent that inverse.

Inverse Function Notation

If $y = f(x)$ is a one-to-one function, then the inverse of f is also a function and can be denoted by $y = f^{-1}(x)$.

Note The notation f^{-1} does not represent the reciprocal of f. That is, the -1 in this notation is not an exponent. The notation f^{-1} is defined as representing the inverse function for a one-to-one function.

To illustrate, in Example 1 we found that the inverse of $f(x) = 2x - 3$ was the function $y = \frac{x+3}{2}$. We can write this inverse function with inverse function notation as

$$f^{-1}(x) = \frac{x+3}{2}$$

On the other hand, the inverse of the function in Example 2 is not itself a function, so we do not use the notation $f^{-1}(x)$ to represent it.

EXAMPLE 3 Find the inverse of $g(x) = \dfrac{x-4}{x-2}$.

SOLUTION To find the inverse for g, we begin by replacing $g(x)$ with y to obtain

$$y = \frac{x-4}{x-2} \qquad \text{The original function}$$

To find an equation for the inverse, we exchange x and y.

$$x = \frac{y-4}{y-2} \qquad \text{The inverse of the original function}$$

To solve for y, we first multiply each side by $y - 2$ to obtain

$$x(y-2) = y - 4$$

$$xy - 2x = y - 4 \qquad \text{Distributive property}$$

$$xy - y = 2x - 4 \qquad \text{Collect all terms containing } y \text{ on the left side}$$

$$y(x-1) = 2x - 4 \qquad \text{Factor } y \text{ from each term on the left side}$$

$$y = \frac{2x-4}{x-1} \qquad \text{Divide each side by } x - 1$$

Because our original function is one-to-one, as verified by the graph in Figure 5a, its inverse is also a function. Therefore, we can use inverse function notation to write the inverse function. Here are the graphs of g and g^{-1}.

$$g(x) = \frac{x-4}{x-2} \qquad\qquad g^{-1}(x) = \frac{2x-4}{x-1}$$

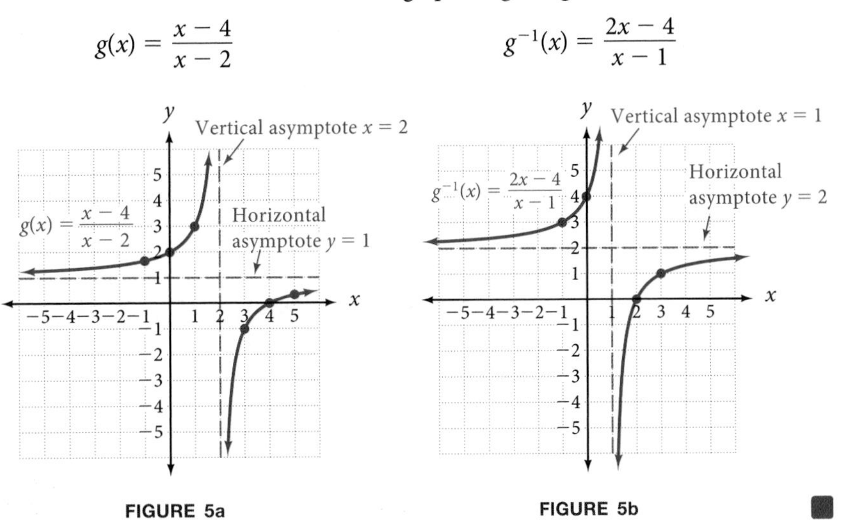

FIGURE 5a

FIGURE 5b

EXAMPLE 4 Graph the function $y = 2^x$ and its inverse $x = 2^y$.

SOLUTION We graphed $y = 2^x$ in the preceding section. We simply reflect its graph about the line $y = x$ to obtain the graph of its inverse $x = 2^y$.

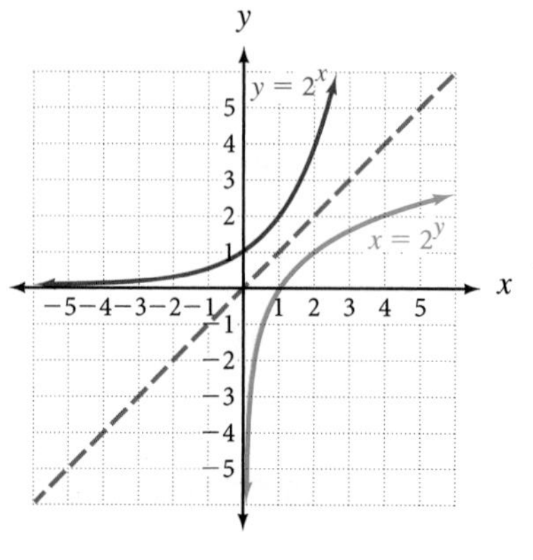

FIGURE 6

As you can see from the graph, $x = 2^y$ is a function. We do not have the mathematical tools yet to solve this equation for y, however. Therefore, we are unable to use the inverse function notation to represent this function. In the next section, we will give a definition that solves this problem. For now, we simply leave the equation as $x = 2^y$. ■

Functions, Relations, and Inverses—A Summary

Here is a summary of some of the things we know about functions, relations, and their inverses:

1. Every function is a relation, but not every relation is a function.
2. Every function has an inverse, but only one-to-one functions have inverses that are also functions.
3. The domain of a function is the range of its inverse, and the range of a function is the domain of its inverse.
4. If $y = f(x)$ is a one-to-one function, then we can use the notation $y = f^{-1}(x)$ to represent its inverse function.
5. The graph of a function and its inverse have symmetry about the line $y = x$.
6. If (a, b) belongs to the function f, then the point (b, a) belongs to its inverse.

Some interesting things happen when we use Wolfram|Alpha to find the inverse of a function. Let's use Wolfram|Alpha to work Example 1.

Using Technology 6.2

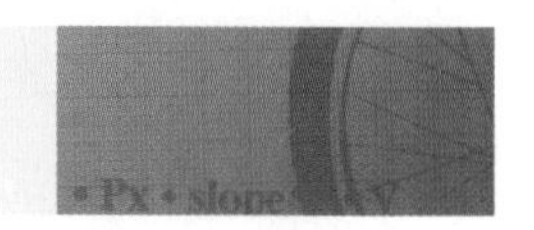

EXAMPLE 5 Use Wolfram|Alpha to find the inverse function for $f(x) = 2x - 3$.

SOLUTION Enter the information below, into the box on the Wolfram|Alpha website, exactly as it is shown.

find the inverse function for f(x)=2x-3

Here is what you will see:

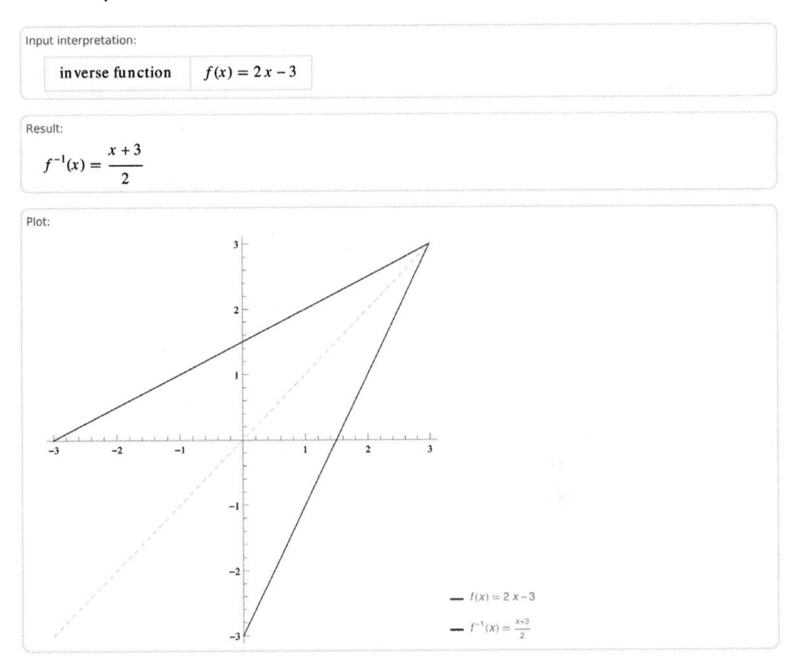

Wolfram Alpha LLC. 2012. Wolfram|Alpha
http://www.wolframalpha.com/
(access June 18, 2012)

Notice that we get the same results that we found in Example 1 of this section. In the Problem Set that follows you will be asked to work the same problem again, but with y instead of $f(x)$. The results may surprise you.

GETTING READY FOR CLASS

After reading through the preceding section, respond in your own words and in complete sentences.

A. What is the inverse of a function?

B. What is the relationship between the graph of a function and the graph of its inverse?

C. Explain why only one-to-one functions have inverses that are also functions.

D. Describe the vertical line test, and explain the difference between the vertical line test and the horizontal line test.

Problem Set 6.2

For each of the following one-to-one functions, find the equation of the inverse. Write the inverse using the notation $f^{-1}(x)$.

1. $f(x) = 3x - 1$
2. $f(x) = 2x - 5$
3. $f(x) = x^3$
4. $f(x) = x^3 - 2$
5. $f(x) = \dfrac{x - 3}{x - 1}$
6. $f(x) = \dfrac{x - 2}{x - 3}$
7. $f(x) = \dfrac{x - 3}{4}$
8. $f(x) = \dfrac{x + 7}{2}$
9. $f(x) = \dfrac{1}{2}x - 3$
10. $f(x) = \dfrac{1}{3}x + 1$
11. $f(x) = \dfrac{2}{3}x - 3$
12. $f(x) = -\dfrac{1}{2}x + 4$
13. $f(x) = x^3 - 4$
14. $f(x) = -3x^3 + 2$
15. $f(x) = \dfrac{4x - 3}{2x + 1}$
16. $f(x) = \dfrac{3x - 5}{4x + 3}$
17. $f(x) = \dfrac{2x + 1}{3x + 1}$
18. $f(x) = \dfrac{3x + 2}{5x + 1}$

For each of the following relations, sketch the graph of the relation and its inverse, and write an equation for the inverse. If the inverse is also a function, write the equation using inverse function notation.

19. $y = 2x - 1$
20. $y = 3x + 1$
21. $y = x^2 - 3$
22. $y = x^2 + 1$
23. $y = x^2 - 2x - 3$
24. $y = x^2 + 2x - 3$
25. $y = 3^x$
26. $y = \left(\dfrac{1}{2}\right)^x$
27. $y = 4$
28. $y = -2$
29. $y = \dfrac{1}{2}x^3$
30. $y = x^3 - 2$
31. $y = \dfrac{1}{2}x + 2$
32. $y = \dfrac{1}{3}x - 1$
33. $y = \sqrt{x + 2}$
34. $y = \sqrt{x} + 2$

35. Determine if the following functions are one-to-one.

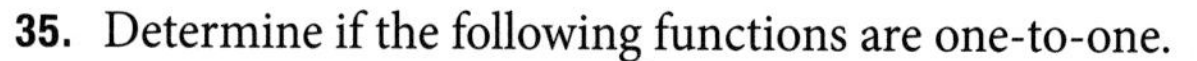

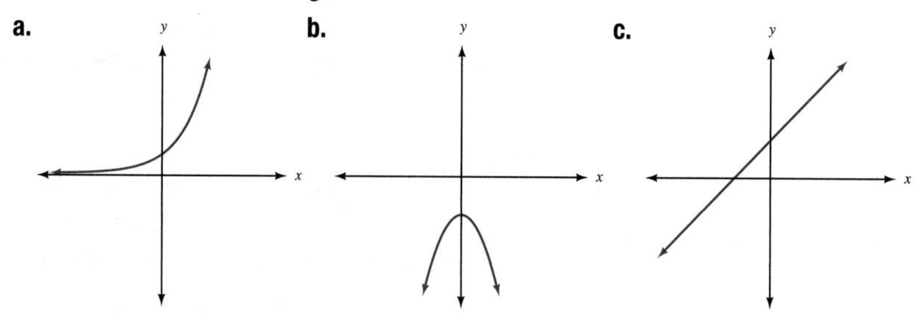

36. Could the following tables of values represent ordered pairs from one-to-one functions? Explain your answer.

a.

x	y
−2	5
−1	4
0	3
1	4
2	5

b.

x	y
1.5	0.1
2.0	0.2
2.5	0.3
3.0	0.4
3.5	0.5

37. If $f(x) = 3x - 2$, then $f^{-1}(x) = \dfrac{x+2}{3}$. Use these two functions to find

a. $f(2)$ **b.** $f^{-1}(2)$ **c.** $f[f^{-1}(2)]$ **d.** $f^{-1}[f(2)]$

38. If $f(x) = \dfrac{1}{2}x + 5$, then $f^{-1}(x) = 2x - 10$. Use these two functions to find

a. $f(-4)$ **b.** $f^{-1}(-4)$ **c.** $f[f^{-1}(-4)]$ **d.** $f^{-1}[f(-4)]$

39. Let $f(x) = \dfrac{1}{x}$, and find $f^{-1}(x)$.

40. Let $f(x) = \dfrac{a}{x}$, and find $f^{-1}(x)$. (a is a real number constant.)

Inverse Functions in Words Inverses may also be found by *inverse reasoning*. For example, to find the inverse of $f(x) = 3x + 2$, first list, in order, the operations done to variable x:

Step 1: Multiply by 3.
Step 2: Add 2.

Then, to find the inverse, simply apply the inverse operations, in reverse order, to the variable x. That is:

Step 3: Subtract 2.
Step 4: Divide by 3.

The inverse function then becomes $f^{-1}(x) = \frac{x-2}{3}$.

41. Use this method of "inverse reasoning" to find the inverse of the function $f(x) = \frac{x}{7} - 2$.

42. Inverse Functions in Words Use *inverse reasoning* to find the following inverses:

a. $f(x) = 2x + 7$ **b.** $f(x) = \sqrt{x} - 9$

c. $f(x) = x^3 - 4$ **d.** $f(x) = \sqrt{x^3 - 4}$

43. Reading Tables Evaluate each of the following functions using the functions defined by Tables 1 and 2.

a. $f[g(-3)]$ **b.** $g[f(-6)]$ **c.** $g[f(2)]$

d. $f[g(3)]$ **e.** $f[g(-2)]$ **f.** $g[f(3)]$

g. What can you conclude about the relationship between functions f and g?

TABLE 1

x	$f(x)$
−6	3
2	−3
3	−2
6	4

TABLE 2

x	$g(x)$
−3	2
−2	3
3	−6
4	6

44. Reading Tables Use the functions defined in Tables 1 and 2 in Problem 43 to answer the following questions.

a. What are the domain and range of f?

b. What are the domain and range of g?

c. How are the domain and range of f related to the domain and range of g?

d. Is f a one-to-one function?

e. Is g a one-to-one function?

Modeling Practice

45. Social Security A function that models the billions of dollars of Social Security payment (as shown in the chart) per year is $s(t) = 16t + 249.4$, where t is time in years since 1990 (U.S. Census Bureau).

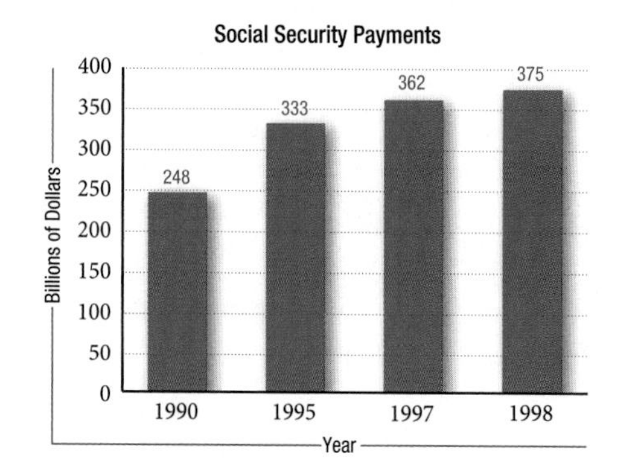

a. Use the model to estimate the amount of Social Security payments to be paid in 2005.

b. Write the inverse of the function.

c. Using the inverse function, estimate the year in which payments will reach \$507 billion.

46. Families The following chart shows the census data for the percentage of one-parent families.

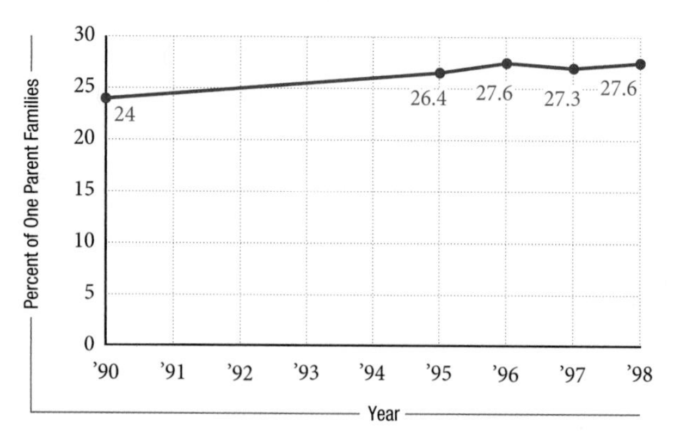

A function that closely approximates this data is $f(x) = 0.417x + 24$, when x is the time in years since 1990 (U.S. Census Bureau).

a. Use the function to predict the percentage of families with one parent in the year 2010.

b. Determine the inverse of the function, and estimate the year in which approximately 29% of the families are one-parent families.

47. Speed The fastest type of plane, a rocket plane, can travel at a speed of 4,520 miles per hour. The function $f(m) = \frac{22m}{15}$ converts miles per hour, m, to feet per second (*World Book Encyclopedia*).

a. Use the function to convert the speed of the rocket plane to feet per second.

b. Write the inverse of the function.

c. Using the inverse function, convert 2 feet per second to miles per hour.

NASA

48. Speed A Lockheed SR-71A airplane set a world record (as reported by Air Force Armament Museum in 1996) with an absolute speed record of 2,193.167 miles per hour. The function $s(h) = 0.4468424h$ converts miles per hour, h, to meters per second, s.

a. What is the absolute speed of the Lockheed SR-71A in meters per second?

b. What is the inverse of this function?

c. Using the inverse function, determine the speed of an airplane in miles per hour that flies 150 meters per second.

Using Technology Exercises

Use Wolfram|Alpha to answer the following questions.

49. Find the inverse of $y = 2x - 3$. (Be sure to use y instead of $f(x)$.)

50. Repeat Example 2 using Wolfram|Alpha. Note their result. We limit the use of $f^{-1}(x)$ to functions. They use it for an inverse relation as well.

51. Repeat Example 3 using Wolfram|Alpha.

52. Repeat Example 4 using Wolfram|Alpha.

Getting Ready for the Next Section

Simplify.

53. 3^{-2}

54. 2^3

Solve.

55. $2 = 3x$

56. $3 = 5x$

57. $4 = x^3$

58. $12 = x^2$

Fill in the boxes to make each statement true.

59. $8 = 2^{\square}$

60. $27 = 3^{\square}$

61. $10{,}000 = 10^{\square}$

62. $1{,}000 = 10^{\square}$

63. $81 = 3^{\square}$

64. $81 = 9^{\square}$

65. $6 = 6^{\square}$

66. $1 = 5^{\square}$

Logarithms Are Exponents

6.3

INTRODUCTION In March 2011, the world learned that an earthquake had occurred in Japan, causing massive destruction and loss of life. News stations reported the strength of the quake by indicating that it measured 9.0 on the Richter scale. For comparison, Table 1 gives the Richter magnitude of a number of other earthquakes.

Although the size of the numbers in the table do not seem to be very different, the intensity of the earthquakes they measure can be very different. For example, the 2011 Japan earthquake was more than 10 times stronger than the 2008 earthquake in China. The reason behind this is that the Richter scale is a *logarithmic scale*.

Year	Earthquake	Richter Magnitude
1985	Michoacan, Mexico	8.0
1994	Northridge, Calif.	6.7
2004	Indonesia	9.1
2008	Sichuan, China	7.9
2010	Haiti	7.0
2011	Honshu, Japan	9.0

TABLE 1

In this section, we start our work with logarithms, which will give you an understanding of the Richter scale. Let's begin.

As you know from your work in the previous sections, equations of the form

$$y = b^x \quad b > 0, b \neq 1$$

are called exponential functions. Because the equation of the inverse of a function can be obtained by exchanging x and y in the equation of the original function, the inverse of an exponential function must have the form

$$x = b^y \quad b > 0, b \neq 1$$

Now, this last equation is actually the equation of a logarithmic function, as the following definition indicates:

DEFINITION

The expression $y = \log_b x$ is read "y is the logarithm to the base b of x" and is equivalent to the expression

$$x = b^y \qquad b > 0, b \neq 1$$

In words, we say "y is the number we raise b to in order to get x."

Notation When an expression is in the form $x = b^y$, it is said to be in *exponential form*. On the other hand, if an expression is in the form $y = \log_b x$, it is said to be in *logarithmic form*.

Here are some equivalent statements written in both forms.

Exponential Form		Logarithmic Form
$8 = 2^3$	$\Leftrightarrow$	$\log_2 8 = 3$
$25 = 5^2$	$\Leftrightarrow$	$\log_5 25 = 2$
$0.1 = 10^{-1}$	$\Leftrightarrow$	$\log_{10} 0.1 = -1$
$\frac{1}{8} = 2^{-3}$	$\Leftrightarrow$	$\log_2 \frac{1}{8} = -3$
$r = z^s$	$\Leftrightarrow$	$\log_z r = s$

VIDEO EXAMPLES

SECTION 6.3

EXAMPLE 1 Solve for x: $\log_3 x = -2$

SOLUTION In exponential form, the equation looks like this:

$$x = 3^{-2}$$

or

$$x = \frac{1}{9}$$

The solution is $\frac{1}{9}$.

EXAMPLE 2 Solve $\log_x 4 = 3$.

SOLUTION Again, we use the definition of logarithms to write the expression in exponential form:

$$4 = x^3$$

Taking the cube root of both sides, we have

$$\sqrt[3]{4} = \sqrt[3]{x^3}$$

$$x = \sqrt[3]{4}$$

The solution set is $\{\sqrt[3]{4}\}$.

EXAMPLE 3 Solve $\log_8 4 = x$.

SOLUTION We write the expression again in exponential form:

$$4 = 8^x$$

Because both 4 and 8 can be written as powers of 2, we write them in terms of powers of 2:

$$2^2 = (2^3)^x$$

$$2^2 = 2^{3x}$$

The only way the left and right sides of this last line can be equal is if the exponents are equal—that is, if

$$2 = 3x$$

or

$$x = \frac{2}{3}$$

The solution is $\frac{2}{3}$. We check as follows:

$$\log_8 4 = \frac{2}{3} \Leftrightarrow 4 = 8^{2/3}$$
$$4 \stackrel{?}{=} (\sqrt[3]{8})^2$$
$$4 \stackrel{?}{=} 2^2$$
$$4 = 4$$

The solution checks when used in the original equation. ■

Graphing Logarithmic Functions

Graphing logarithmic functions can be done using the graphs of exponential functions and the fact that the graphs of inverse functions have symmetry about the line $y = x$. Here's an example to illustrate.

EXAMPLE 4 Graph the equation $y = \log_2 x$.

SOLUTION The equation $y = \log_2 x$ is, by definition, equivalent to the exponential equation

$$x = 2^y$$

which is the equation of the inverse of the function

$$y = 2^x$$

The graph of $y = 2^x$ was given in Figure 2 of Section 6.1. We simply reflect the graph of $y = 2^x$ about the line $y = x$ to get the graph of $x = 2^y$, which is also the graph of $y = \log_2 x$. (See Figure 1.)

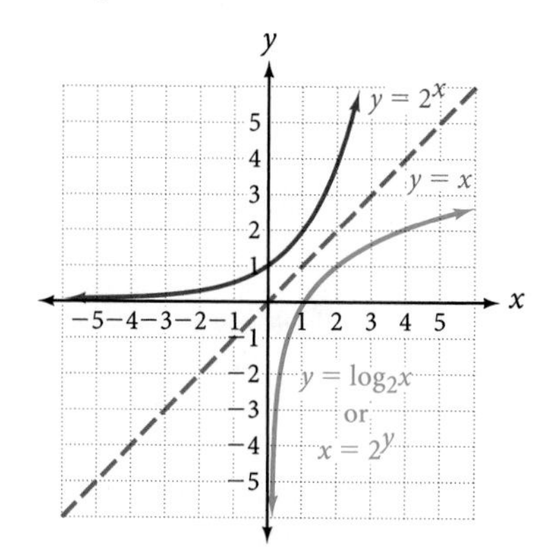

FIGURE 1

It is apparent from the graph that $y = \log_2 x$ is a function, because no vertical line will cross its graph in more than one place. The same is true for all logarithmic equations of the form $y = \log_b x$, where b is a positive number other than 1. Note also that the graph of $y = \log_b x$ will always appear to the right of the y-axis, meaning that x will always be positive in the expression $y = \log_b x$. ■

Inverse Functions

As you can see from Example 4, the functions $y = \log_2 x$ and $y = 2^x$ are inverse functions. We can generalize this observation to a wider group of functions

$$f(x) = \log_b x \quad \text{and} \quad f^{-1}(x) = b^x \quad b > 0, b \neq 1$$

are inverse functions.

Two Special Identities

Note If $f(x) = \log_b x$ and $f^{-1}(x) = b^x$ are inverse functions, then $f(f^{-1}(x)) = x$ gives us $\log_b b^x = x$.

If b is a positive real number other than 1, then each of the following is a consequence of the definition of a logarithm:

$$(1)\ b^{\log_b x} = x \qquad \text{and} \qquad (2)\ \log_b b^x = x$$

The justifications for these identities are similar. Let's consider only the first one. Consider the expression

$$y = \log_b x$$

By definition, it is equivalent to

$$x = b^y$$

Substituting $\log_b x$ for y in the last line gives us

$$x = b^{\log_b x}$$

The next examples in this section show how these two special properties can be used to simplify expressions involving logarithms.

EXAMPLE 5 Simplify the following logarithmic expressions.

a. $\log_2 8$ **b.** $\log_{10} 10{,}000$ **c.** $\log_b b$

d. $\log_b 1$ **e.** $\log_4 (\log_5 5)$

SOLUTION

a. Substitute 2^3 for 8:

$$\begin{aligned}\log_2 8 &= \log_2 2^3 \\ &= 3\end{aligned}$$

b. 10,000 can be written as 10^4:

$$\begin{aligned}\log_{10} 10{,}000 &= \log_{10} 10^4 \\ &= 4\end{aligned}$$

c. Because $b^1 = b$, we have

$$\begin{aligned}\log_b b &= \log_b b^1 \\ &= 1\end{aligned}$$

d. Because $1 = b^0$, we have

$$\begin{aligned}\log_b 1 &= \log_b b^0 \\ &= 0\end{aligned}$$

e. Because $\log_5 5 = 1$,

$$\begin{aligned}\log_4(\log_5 5) &= \log_4 1 \\ &= 0\end{aligned}$$

Modeling: Earthquakes

As we mentioned in the introduction to this section, one application of logarithms is in measuring the magnitude of an earthquake. If an earthquake has a shock wave T times greater than the smallest shock wave that can be measured on a seismograph, then the magnitude M of the earthquake, as measured on the Richter scale, is given by the formula

$$M = \log_{10} T$$

(When we talk about the size of a shock wave, we are talking about its amplitude. The amplitude of a wave is half the difference between its highest point and its lowest point.)

To illustrate the discussion, an earthquake that produces a shock wave that is 10,000 times greater than the smallest shock wave measurable on a seismograph will have a magnitude M on the Richter scale of

$$M = \log_{10} 10{,}000 = 4$$

© Claudia Dewald/iStockPhoto

EXAMPLE 6 If an earthquake has a magnitude of $M = 5$ on the Richter scale, what can you say about the size of its shock wave?

SOLUTION To answer this question, we put $M = 5$ into the formula $M = \log_{10} T$ to obtain

$$5 = \log_{10} T$$

Writing this expression in exponential form, we have

$$T = 10^5 = 100{,}000$$

We can say that an earthquake that measures 5 on the Richter scale has a shock wave 100,000 times greater than the smallest shock wave measurable on a seismograph.

From Example 6 and the discussion that preceded it, we find that an earthquake of magnitude 5 has a shock wave that is 10 times greater than an earthquake of magnitude 4, because 100,000 is 10 times 10,000.

GETTING READY FOR CLASS

After reading through the preceding section, respond in your own words and in complete sentences.

A. What is a logarithm?
B. What is the relationship between $y = 2^x$ and $y = \log_2 x$? How are their graphs related?
C. Will the graph of $y = \log_b x$ ever appear in the second or third quadrants? Explain why or why not.
D. Explain why $\log_2 0 = x$ has no solution for x.

Problem Set 6.3

Write each of the following expressions in logarithmic form.

1. $2^4 = 16$ **2.** $3^2 = 9$ **3.** $125 = 5^3$ **4.** $16 = 4^2$

5. $0.01 = 10^{-2}$ **6.** $0.001 = 10^{-3}$ **7.** $2^{-5} = \frac{1}{32}$ **8.** $4^{-2} = \frac{1}{16}$

9. $\left(\frac{1}{2}\right)^{-3} = 8$ **10.** $\left(\frac{1}{3}\right)^{-2} = 9$ **11.** $27 = 3^3$ **12.** $81 = 3^4$

Write each of the following expressions in exponential form.

13. $\log_{10} 100 = 2$ **14.** $\log_2 8 = 3$ **15.** $\log_2 64 = 6$

16. $\log_2 32 = 5$ **17.** $\log_8 1 = 0$ **18.** $\log_9 9 = 1$

19. $\log_{10} 0.001 = -3$ **20.** $\log_{10} 0.0001 = -4$ **21.** $\log_6 36 = 2$

22. $\log_7 49 = 2$ **23.** $\log_5 \frac{1}{25} = -2$ **24.** $\log_3 \frac{1}{81} = -4$

Solve each of the following equations for x.

25. $\log_3 x = 2$ **26.** $\log_4 x = 3$ **27.** $\log_5 x = -3$ **28.** $\log_2 x = -4$

29. $\log_2 16 = x$ **30.** $\log_3 27 = x$ **31.** $\log_8 2 = x$ **32.** $\log_{25} 5 = x$

33. $\log_x 4 = 2$ **34.** $\log_x 16 = 4$ **35.** $\log_x 5 = 3$ **36.** $\log_x 8 = 2$

37. $\log_5 25 = x$ **38.** $\log_5 x = -2$ **39.** $\log_x 36 = 2$ **40.** $\log_x \frac{1}{25} = 2$

41. $\log_8 4 = x$ **42.** $\log_{16} 8 = x$ **43.** $\log_9 \frac{1}{3} = x$ **44.** $\log_{27} 9 = x$

45. $\log_8 x = -2$ **46.** $\log_{36} \frac{1}{6} = x$

Sketch the graph of each of the following logarithmic equations.

47. $y = \log_3 x$ **48.** $y = \log_{1/2} x$ **49.** $y = \log_{1/3} x$ **50.** $y = \log_4 x$

51. $y = \log_5 x$ **52.** $y = \log_{1/5} x$ **53.** $y = \log_{10} x$ **54.** $y = \log_{1/4} x$

Each of the following graphs has an equation of the form $y = b^x$ or $y = \log_b x$. Find the equation for each graph.

55. **56.**

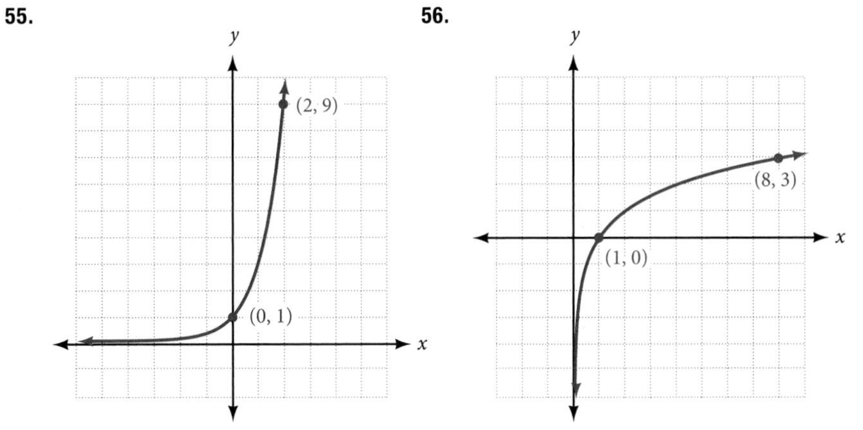

57. **58.**

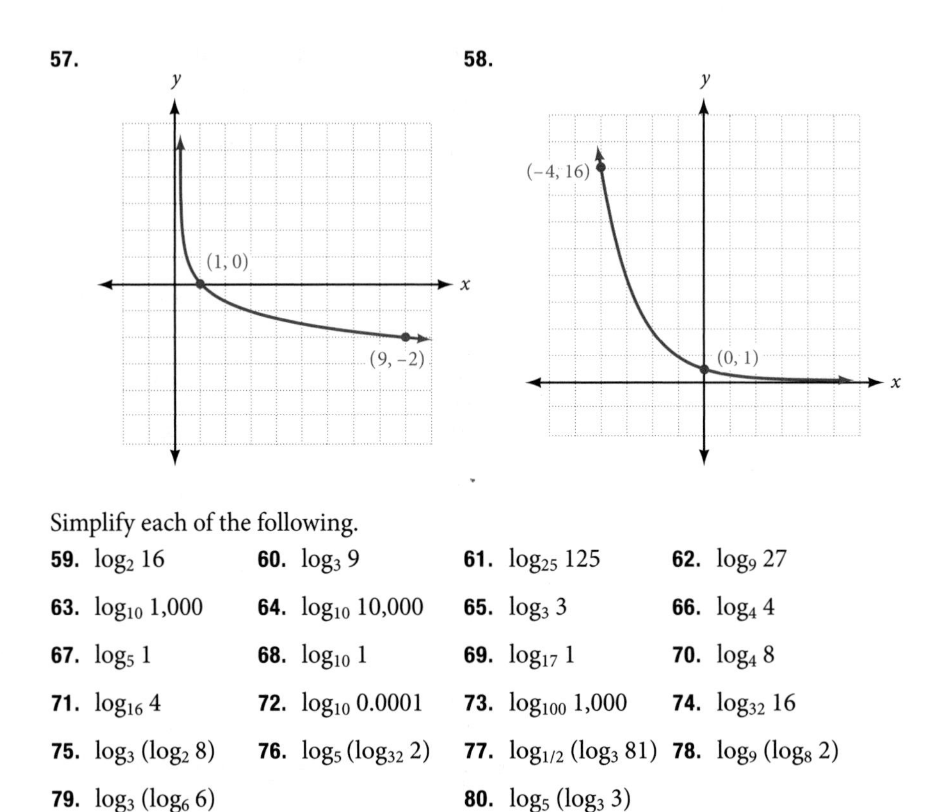

Simplify each of the following.

59. $\log_2 16$ **60.** $\log_3 9$ **61.** $\log_{25} 125$ **62.** $\log_9 27$

63. $\log_{10} 1{,}000$ **64.** $\log_{10} 10{,}000$ **65.** $\log_3 3$ **66.** $\log_4 4$

67. $\log_5 1$ **68.** $\log_{10} 1$ **69.** $\log_{17} 1$ **70.** $\log_4 8$

71. $\log_{16} 4$ **72.** $\log_{10} 0.0001$ **73.** $\log_{100} 1{,}000$ **74.** $\log_{32} 16$

75. $\log_3 (\log_2 8)$ **76.** $\log_5 (\log_{32} 2)$ **77.** $\log_{1/2} (\log_3 81)$ **78.** $\log_9 (\log_8 2)$

79. $\log_3 (\log_6 6)$ **80.** $\log_5 (\log_3 3)$

81. $\log_4 [\log_2(\log_2 16)]$ **82.** $\log_4 [\log_3(\log_2 8)]$

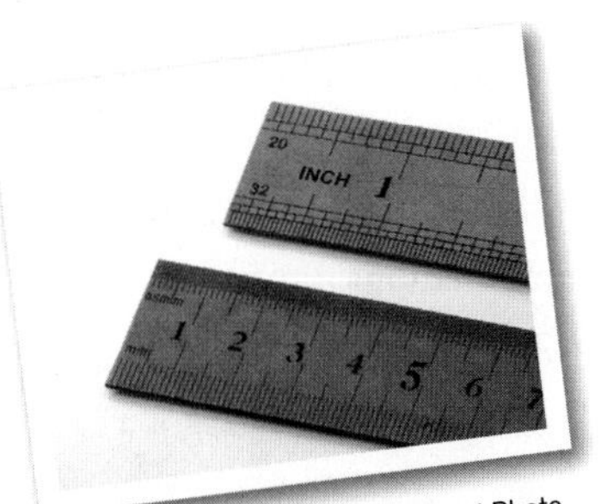

© Arno Staub/iStockPhoto

83. Metric System The metric system uses logical and systematic prefixes for multiplication. For instance, to multiply a unit by 100, the prefix "hecto" is applied, so a hectometer is equal to 100 meters. For each of the prefixes in the following table find the logarithm, base 10, of the multiplying factor.

Prefix	Multiplying Factor	$\log_{10}$ (Multiplying Factor)
Nano	0.000 000 001	
Micro	0.000 001	
Deci	0.1	
Giga	1,000,000,000	
Peta	1,000,000,000,000,000	

84. Domain and Range Use the graphs of $y = 2^x$ and $y = \log_2 x$ shown in Figure 1 of this section to find the domain and range for each function. Explain how the domain and range found for $y = 2^x$ relate to the domain and range found for $y = \log_2 x$.

Modeling Practice

85. Magnitude of an Earthquake Find the magnitude M of an earthquake with a shock wave that measures $T = 100$ on a seismograph.

86. Magnitude of an Earthquake Find the magnitude M of an earthquake with a shock wave that measures $T = 100{,}000$ on a seismograph.

87. Shock Wave If an earthquake has a magnitude of 8 on the Richter scale, how many times greater is its shock wave than the smallest shock wave measurable on a seismograph?

88. Shock Wave If the 2011 Japan earthquake had a magnitude of 9.0 on the Richter scale, how many times greater was its shock wave than the smallest shock wave measurable on a seismograph?

Earthquake The table below categorizes earthquakes by the magnitude and identifies the average annual occurrence.

Earthquakes

Descriptor	Magnitude	Average Annual Occurrence
Great	≥8.0	1
Major	7–7.9	18
Strong	6–6.9	120
Moderate	5–5.9	800
Light	4–4.9	6,200
Minor	3–3.9	49,000
Very Minor	2–2.9	1,000 per day
Very Minor	1–1.9	8,000 per day

Source: *USGS National Earthquake Information.*

89. What is the average number of earthquakes that occur per year when the number of times the associated shock wave is greater than the smallest measurable shock wave, T, is 1,000,000?

90. What is the average number of earthquakes that occur per year when $T = 1{,}000{,}000$ or greater?

Getting Ready for the Next Section

Simplify.

91. $8^{2/3}$

92. $27^{2/3}$

Solve.

93. $(x + 2)(x) = 2^3$

94. $(x + 3)(x) = 2^2$

95. $\dfrac{x - 2}{x + 1} = 9$

96. $\dfrac{x + 1}{x - 4} = 25$

Write in exponential form.

97. $\log_2 [(x + 2)(x)] = 3$

98. $\log_4 [x(x - 6)] = 2$

99. $\log_3\left(\dfrac{x - 2}{x + 1}\right) = 4$

100. $\log_3\left(\dfrac{x - 1}{x - 4}\right) = 2$

SPOTLIGHT ON SUCCESS *Student Instructor Stefanie*

Never confuse a single defeat with a final defeat.
—F. Scott Fitzgerald

The idea that has worked best for my success in college, and more specifically in my math courses, is to stay positive and be resilient. I have learned that a 'bad' grade doesn't make me a failure; if anything it makes me strive to do better. That is why I never let a bad grade on a test or even in a class get in the way of my overall success.

By sticking with this positive attitude, I have been able to achieve my goals. My grades have never represented how well I know the material. This is because I have struggled with test anxiety and it has consistently lowered my test scores in a number of courses. However, I have not let it defeat me. When I applied to graduate school, I did not meet the grade requirements for my top two schools, but that did not stop me from applying.

One school asked that I convince them that my knowledge of mathematics was more than my grades indicated. If I had let my grades stand in the way of my goals, I wouldn't have been accepted to both of my top two schools, and wouldn't be attending one of them in the fall, on my way to becoming a mathematics teacher.

Properties of Logarithms

© Sergey Sverdelov/iStockPhoto

INTRODUCTION If we search for a definition of the word *decibel*, we find the following: A unit used to express relative difference in power or intensity, usually between two acoustic or electric signals, equal to ten times the common logarithm of the ratio of the two levels.

Decibels	Comparable to
10	A light whisper
20	Quiet conversation
30	Normal conversation
40	Light traffic
50	Typewriter, loud conversation
60	Noisy office
70	Normal traffic, quiet train
80	Rock music, subway
90	Heavy traffic, thunder
100	Jet plane at takeoff

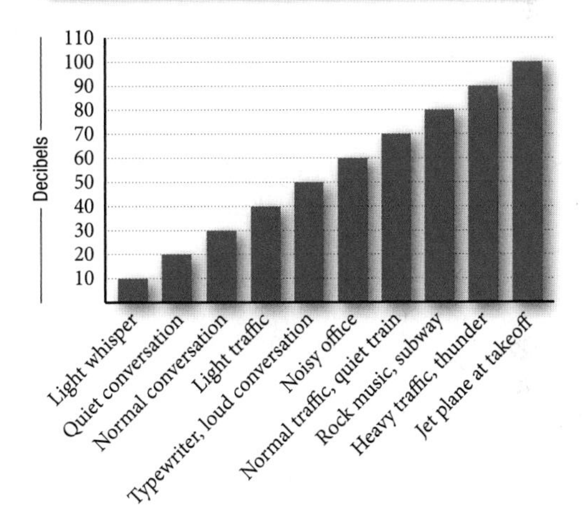

The precise definition for a *decibel* is

$$D = 10 \log_{10}\left(\frac{I}{I_0}\right)$$

where I is the intensity of the sound being measured, and I_0 is the intensity of the least audible sound. (Sound intensity is related to the amplitude of the sound wave that models the sound and is given in units of watts per meter^2.) In this section, we will see that the preceding formula can also be written as

$$D = 10(\log_{10} I - \log_{10} I_0)$$ ■

The rules we use to rewrite expressions containing logarithms are called the *properties of logarithms*. We will introduce three of them in this section.

For the following three properties, x, y, and b are all positive real numbers, $b \neq 1$, and r is any real number.

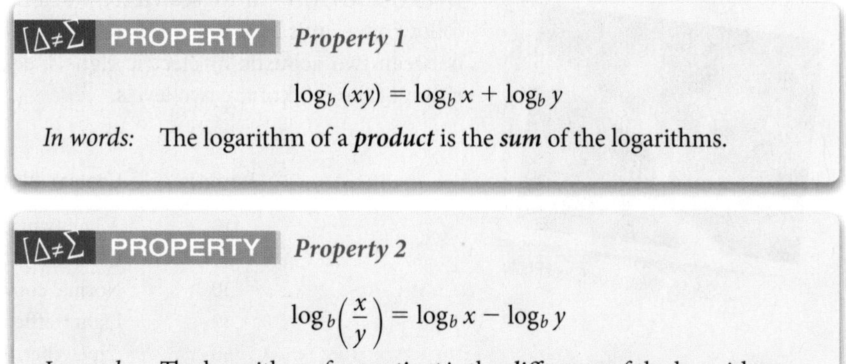

PROPERTY *Property 1*

$$\log_b (xy) = \log_b x + \log_b y$$

In words: The logarithm of a ***product*** is the ***sum*** of the logarithms.

PROPERTY *Property 2*

$$\log_b\left(\frac{x}{y}\right) = \log_b x - \log_b y$$

In words: The logarithm of a ***quotient*** is the ***difference*** of the logarithms.

PROPERTY *Property 3*

$$\log_b x^r = r \log_b x$$

In words: The logarithm of a number raised to a ***power*** is the ***product*** of the power and the logarithm of the number.

Proof of Property 1 To prove Property 1, we simply apply the first identity for logarithms given in the preceding section:

$$b^{\log_b xy} = xy = (b^{\log_b x})(b^{\log_b y}) = b^{\log_b x + \log_b y}$$

Because the first and last expressions are equal and the bases are the same, the exponents $\log_b xy$ and $\log_b x + \log_b y$ must be equal. Therefore,

$$\log_b xy = \log_b x + \log_b y$$

The proofs of Properties 2 and 3 proceed in much the same manner, so we will omit them here. The examples that follow show how the three properties can be used.

VIDEO EXAMPLES

SECTION 6.4

EXAMPLE 1 Expand, using the properties of logarithms: $\log_5 \dfrac{3xy}{z}$

SOLUTION Applying Property 2, we can write the quotient of $3xy$ and z in terms of a difference:

$$\log_5 \frac{3xy}{z} = \log_5 3xy - \log_5 z$$

Applying Property 1 to the product $3xy$, we write it in terms of addition:

$$\log_5 \frac{3xy}{z} = \log_5 3 + \log_5 x + \log_5 y - \log_5 z$$

■

EXAMPLE 2 Expand, using the properties of logarithms:

$$\log_2 \frac{x^4}{\sqrt{y} \cdot z^3}$$

SOLUTION We write $\sqrt{y}$ as $y^{1/2}$ and apply the properties:

$$\log_2 \frac{x^4}{\sqrt{y} \cdot z^3} = \log_2 \frac{x^4}{y^{1/2}z^3} \qquad \sqrt{y} = y^{1/2}$$

$$= \log_2 x^4 - \log_2 (y^{1/2} \cdot z^3) \qquad \text{Property 2}$$

$$= \log_2 x^4 - (\log_2 y^{1/2} + \log_2 z^3) \qquad \text{Property 1}$$

$$= \log_2 x^4 - \log_2 y^{1/2} - \log_2 z^3 \qquad \text{Remove parentheses and distribute } -1$$

$$= 4 \log_2 x - \frac{1}{2}\log_2 y - 3 \log_2 z \qquad \text{Property 3}$$

We can also use the three properties to write an expression in expanded form as just one logarithm.

EXAMPLE 3 Write as a single logarithm:

$$2 \log_{10} a + 3 \log_{10} b - \frac{1}{3}\log_{10} c$$

SOLUTION We begin by applying Property 3:

$$2 \log_{10} a + 3 \log_{10} b - \frac{1}{3}\log_{10} c = \log_{10} a^2 + \log_{10} b^3 - \log_{10} c^{1/3} \qquad \text{Property 3}$$

$$= \log_{10} (a^2 \cdot b^3) - \log_{10} c^{1/3} \qquad \text{Property 1}$$

$$= \log_{10} \frac{a^2b^3}{c^{1/3}} \qquad \text{Property 2}$$

$$= \log_{10} \frac{a^2b^3}{\sqrt[3]{c}} \qquad c^{1/3} = \sqrt[3]{c}$$

The properties of logarithms along with the definition of logarithms are useful in solving equations that involve logarithms.

EXAMPLE 4 Solve for x: $\log_2(x + 2) + \log_2 x = 3$

SOLUTION Applying Property 1 to the left side of the equation allows us to write it as a single logarithm:

$$\log_2(x + 2) + \log_2 x = 3$$

$$\log_2[(x + 2)(x)] = 3$$

The last line can be written in exponential form using the definition of logarithms:

$$(x + 2)(x) = 2^3$$

Solve as usual:

$$x^2 + 2x = 8$$

$$x^2 + 2x - 8 = 0$$

$$(x + 4)(x - 2) = 0$$

$$x + 4 = 0 \quad \text{or} \quad x - 2 = 0$$

$$x = -4 \quad \text{or} \quad x = 2$$

In the previous section, we noted the fact that x in the expression $y = \log_b x$ cannot be a negative number. Because substitution of $x = -4$ into the original equation gives

$$\log_2(-2) + \log_2(-4) = 3$$

which contains logarithms of negative numbers, we cannot use -4 as a solution. The solution set is 2.

GETTING READY FOR CLASS

After reading through the preceding section, respond in your own words and in complete sentences.

A. Explain why the following statement is false: "The logarithm of a product is the product of the logarithms."

B. Explain why the following statement is false: "The logarithm of a quotient is the quotient of the logarithms."

C. Explain the difference between $\log_b m + \log_b n$ and $\log_b(m + n)$. Are they equivalent?

D. Explain the difference between $\log_b(mn)$ and $(\log_b m)(\log_b n)$. Are they equivalent?

Problem Set 6.4

Use the three properties of logarithms given in this section to expand each expression as much as possible.

1. $\log_3 4x$ **2.** $\log_2 5x$ **3.** $\log_6 \frac{5}{x}$ **4.** $\log_3 \frac{x}{5}$

5. $\log_2 y^5$ **6.** $\log_7 y^3$ **7.** $\log_9 \sqrt[3]{z}$ **8.** $\log_8 \sqrt{z}$

9. $\log_6 x^2y^4$ **10.** $\log_{10} x^2y^4$ **11.** $\log_5 \sqrt{x} \cdot y^4$ **12.** $\log_8 \sqrt[3]{xy^6}$

13. $\log_b \frac{xy}{z}$ **14.** $\log_b \frac{3x}{y}$ **15.** $\log_{10} \frac{4}{xy}$ **16.** $\log_{10} \frac{5}{4y}$

17. $\log_{10} \frac{x^2y}{\sqrt{z}}$ **18.** $\log_{10} \frac{\sqrt{x} \cdot y}{z^3}$ **19.** $\log_{10} \frac{x^3\sqrt{y}}{z^4}$ **20.** $\log_{10} \frac{x^4\sqrt[3]{y}}{\sqrt{z}}$

21. $\log_b \sqrt[3]{\frac{x^2y}{z^4}}$ **22.** $\log_b \sqrt[4]{\frac{x^4y^3}{z^5}}$ **23.** $\log_3 \sqrt[3]{\frac{x^2y}{z^6}}$ **24.** $\log_8 \sqrt[4]{\frac{x^5y^6}{z^3}}$

25. $\log_a \frac{4x^5}{9a^2}$ **26.** $\log_b \frac{16b^2}{25y^3}$

Write each expression as a single logarithm.

27. $\log_b x + \log_b z$ **28.** $\log_b x - \log_b z$

29. $2 \log_3 x - 3 \log_3 y$ **30.** $4 \log_2 x + 5 \log_2 y$

31. $\frac{1}{2} \log_{10} x + \frac{1}{3} \log_{10} y$ **32.** $\frac{1}{3} \log_{10} x - \frac{1}{4} \log_{10} y$

33. $3 \log_2 x + \frac{1}{2} \log_2 y - \log_2 z$ **34.** $2 \log_3 x + 3 \log_3 y - \log_3 z$

35. $\frac{1}{2} \log_2 x - 3 \log_2 y - 4 \log_2 z$ **36.** $3 \log_{10} x - \log_{10} y - \log_{10} z$

37. $\frac{3}{2} \log_{10} x - \frac{3}{4} \log_{10} y - \frac{4}{5} \log_{10} z$ **38.** $3 \log_{10} x - \frac{4}{3} \log_{10} y - 5 \log_{10} z$

39. $\frac{1}{2} \log_5 x + \frac{2}{3} \log_5 y - 4 \log_5 z$ **40.** $\frac{1}{4} \log_7 x + 5 \log_7 y - \frac{1}{3} \log_7 z$

41. $\log_3(x^2 - 16) - 2 \log_3(x + 4)$ **42.** $\log_4(x^2 - x - 6) - \log_4(x^2 - 9)$

Solve each of the following equations.

43. $\log_2 x + \log_2 3 = 1$ **44.** $\log_3 x + \log_3 3 = 1$

45. $\log_3 x - \log_3 2 = 2$ **46.** $\log_3 x + \log_3 2 = 2$

47. $\log_3 x + \log_3(x - 2) = 1$ **48.** $\log_6 x + \log_6(x - 1) = 1$

49. $\log_3(x + 3) - \log_3(x - 1) = 1$ **50.** $\log_4(x - 2) - \log_4(x + 1) = 1$

51. $\log_2 x + \log_2(x - 2) = 3$ **52.** $\log_4 x + \log_4(x + 6) = 2$

53. $\log_8 x + \log_8(x - 3) = \frac{2}{3}$ **54.** $\log_{27} x + \log_{27}(x + 8) = \frac{2}{3}$

55. $\log_3(x + 2) - \log_3 x = 1$ **56.** $\log_2(x + 3) - \log_2(x - 3) = 2$

57. $\log_2(x + 1) + \log_2(x + 2) = 1$ **58.** $\log_3 x + \log_3(x + 6) = 3$

59. $\log_9\sqrt{x} + \log_9\sqrt{2x + 3} = \frac{1}{2}$ **60.** $\log_8\sqrt{x} + \log_8\sqrt{5x + 2} = \frac{2}{3}$

61. $4 \log_3 x - \log_3 x^2 = 6$

62. $9 \log_4 x - \log_4 x^3 = 12$

63. $\log_5 \sqrt{x} + \log_5 \sqrt{6x + 5} = 1$

64. $\log_2 \sqrt{x} + \log_2 \sqrt{6x + 5} = 1$

65. Decibel Formula Use the properties of logarithms to rewrite the decibel formula $D = 10 \log_{10}\left(\frac{I}{I_0}\right)$ as

$$D = 10(\log_{10} I - \log_{10} I_0)$$

66. Decibel Formula In the decibel formula $D = 10 \log_{10}\left(\frac{I}{I_0}\right)$, the threshold of hearing, I_0, is

$$I_0 = 10^{-12} \text{ watts/meter}^2$$

Substitute 10^{-12} for I_0 in the decibel formula, then show that it simplifies to

$$D = 10(\log_{10} I + 12)$$

67. Finding Logarithms If $\log_{10} 8 = 0.903$ and $\log_{10} 5 = 0.699$, find the following without using a calculator.

a. $\log_{10} 40$

b. $\log_{10} 320$

c. $\log_{10} 1{,}600$

68. Matching Match each expression in the first column with an equivalent expression in the second column:

a. $\log_2 (ab)$	**i.** b
b. $\log_2 \left(\frac{a}{b}\right)$	**ii.** 2
c. $\log_5 a^b$	**iii.** $\log_2 a + \log_2 b$
d. $\log_a b^a$	**iv.** $\log_2 a - \log_2 b$
e. $\log_a a^b$	**v.** $a \log_a b$
f. $\log_3 9$	**vi.** $b \log_5 a$

Getting Ready for Calculus Problems 69-72 are problems you will see again if you go on to a calculus class.

69. Expand $\log_2 \dfrac{x^3y^2}{z^4}$.

70. Expand $\log_{10} \dfrac{8x}{x + 4}$.

Write as a single logarithm.

71. $\log_3 x + 6\log_3 y$

72. $\log_5 a - \log_5 b - \log_5 c$

Modeling Practice

73. Henderson-Hasselbalch Formula Doctors use the Henderson-Hasselbalch formula to calculate the pH of a person's blood. pH is a measure of the acidity and/or the alkalinity of a solution. This formula is represented as

$$\text{pH} = 6.1 + \log_{10}\left(\frac{x}{y}\right)$$

where x is the base concentration and y is the acidic concentration. Rewrite the Henderson-Hasselbalch formula so that the logarithm of a quotient is not involved.

74. Henderson-Hasselbalch Formula Refer to the information in the preceding problem about the Henderson-Hasselbalch formula. If most people have a blood pH of 7.4, use the Henderson-Hasselbalch formula to find the ratio of $\frac{x}{y}$ for an average person.

75. Food Processing The formula $M = 0.21(\log_{10} a - \log_{10} b)$ is used in the food processing industry to find the number of minutes M of heat processing a certain food should undergo at 250 °F to reduce the probability of survival of *Clostridium botulinum* spores. The letter a represents the number of spores per can before heating, and b represents the number of spores per can after heating. Find M if $a = 1$ and $b = 10^{-12}$. Then find M using the same values for a and b in the formula $M = 0.21 \log_{10} \frac{a}{b}$.

76. Acoustic Powers The formula $N = \log_{10} \frac{P_1}{P_2}$ is used in radio electronics to find the ratio of the acoustic powers of two electric circuits in terms of their electric powers. Find N if P_1 is 100 and P_2 is 1. Then use the same two values of P_1 and P_2 to find N in the formula $N = \log_{10} P_1 - \log_{10} P_2$.

Getting Ready for the Next Section

Simplify.

77. 5^0

78. 4^1

79. $\log_3 3$

80. $\log_5 5$

81. $\log_b b^4$

82. $\log_a a^k$

Use a calculator to find each of the following. Write your answer in scientific notation with the first number in each answer rounded to the nearest tenth.

83. $10^{-5.6}$

84. $10^{-4.1}$

Divide and round to the nearest whole number.

85. $\dfrac{2.00 \times 10^8}{3.96 \times 10^6}$

86. $\dfrac{3.25 \times 10^{12}}{1.72 \times 10^{10}}$

Common Logarithms and Natural Logarithms

INTRODUCTION Acid rain was first discovered in the 1960s by Gene Likens and his research team who studied the damage caused by acid rain to Hubbard Brook in New Hampshire. Acid rain is rain with a pH of below 5.6. As you will see as you work your way through this section, pH is defined in terms of common logarithms — one of the topics we present in this section. So, when you are finished with this section, you will have a more detailed knowledge of pH and acid rain.

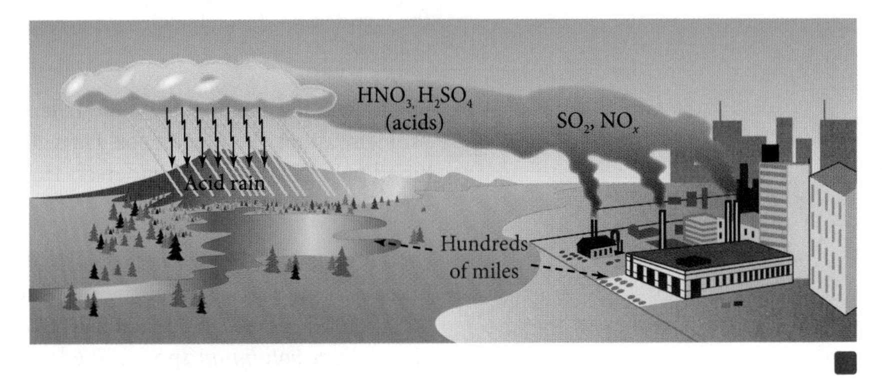

Two kinds of logarithms occur more frequently than other logarithms. Logarithms with a base of 10 are very common because our number system is a base-10 number system. For this reason, we call base-10 logarithms *common logarithms*.

DEFINITION *Common logarithms*

A *common logarithm* is a logarithm with a base of 10. Because common logarithms are used so frequently, it is customary, in order to save time, to omit notating the base. That is,

$$\log_{10} x = \log x$$

When the base is not shown, it is assumed to be 10.

Common Logarithms

Common logarithms of powers of 10 are simple to evaluate. We need only recognize that $\log 10 = \log_{10} 10 = 1$ and apply the third property of logarithms: $\log_b x^r = r \log_b x$.

$$\begin{aligned}
\log 1{,}000 &= \log 10^3 = 3 \log 10 = 3(1) = 3 \\
\log 100 &= \log 10^2 = 2 \log 10 = 2(1) = 2 \\
\log 10 &= \log 10^1 = 1 \log 10 = 1(1) = 1 \\
\log 1 &= \log 10^0 = 0 \log 10 = 0(1) = 0 \\
\log 0.1 &= \log 10^{-1} = -1 \log 10 = -1(1) = -1 \\
\log 0.01 &= \log 10^{-2} = -2 \log 10 = -2(1) = -2 \\
\log 0.001 &= \log 10^{-3} = -3 \log 10 = -3(1) = -3
\end{aligned}$$

To find common logarithms of numbers that are not powers of 10, we use a calculator with a LOG key.

Check the following logarithms to be sure you know how to use your calculator. (These answers have been rounded to the nearest ten-thousandth.)

$$\log 7.02 \approx 0.8463$$
$$\log 1.39 \approx 0.1430$$
$$\log 6.00 \approx 0.7782$$
$$\log 9.99 \approx 0.9996$$

VIDEO EXAMPLES

SECTION 6.5

EXAMPLE 1 Use a calculator to find log 2,760.

SOLUTION

$$\log 2{,}760 \approx 3.4409$$

To work this problem on a scientific calculator, we simply enter the number 2,760 and press the key labeled [LOG]. On a graphing calculator we press the [LOG] key first, then 2,760.

The 3 in the answer is called the *characteristic*, and the decimal part of the logarithm is called the *mantissa*.

EXAMPLE 2 Find log 0.0391.

SOLUTION $\log 0.0391 \approx -1.4078$

EXAMPLE 3 Find log 0.00523.

SOLUTION $\log 0.00523 \approx -2.2815$

EXAMPLE 4 Find x if $\log x = 3.8774$.

SOLUTION We are looking for the number whose logarithm is 3.8774. On a scientific calculator, we enter 3.8774 and press the key labeled [10^x]. On a graphing calculator we press [10^x] first, then 3.8774. The result is 7,540 to four significant digits. Here's why:

$$\begin{aligned} \text{If} \quad \log x &= 3.8774 \\ \text{then} \quad x &= 10^{3.8774} \\ &\approx 7{,}540 \end{aligned}$$

The number 7,540 is called the *antilogarithm* or just *antilog* of 3.8774. That is, 7,540 is the number whose logarithm is 3.8774.

EXAMPLE 5 Find x if $\log x = -2.4179$.

SOLUTION Using the [10^x] key, the result is 0.00382.

$$\begin{aligned} \text{If} \quad \log x &= -2.4179 \\ \text{then} \quad x &= 10^{-2.4179} \\ &\approx 0.00382 \end{aligned}$$

The antilog of -2.4179 is 0.00382. That is, the logarithm of 0.00382 is -2.4179.

Modeling: Earthquakes

In Section 6.3, we found that the magnitude M of an earthquake that produces a shock wave T times larger than the smallest shock wave that can be measured on a seismograph is given by the formula $M = \log_{10} T$. We can rewrite this formula using our shorthand notation for common logarithms as $M = \log T$.

EXAMPLE 6 The San Francisco earthquake of 1906 is estimated to have measured 8.3 on the Richter scale. The San Fernando earthquake of 1971 measured 6.6 on the Richter scale. Find T for each earthquake, and then give some indication of how much stronger the 1906 earthquake was than the 1971 earthquake.

Richter Scale

8.3

SOLUTION For the 1906 earthquake:

$$\text{If } \log T = 8.3, \text{ then } T = 2.00 \times 10^8.$$

For the 1971 earthquake:

$$\text{If } \log T = 6.6, \text{ then } T = 3.98 \times 10^6.$$

Dividing the two values of T and rounding our answer to the nearest whole number, we have

$$\frac{2.00 \times 10^8}{3.98 \times 10^6} \approx 50$$

The shock wave for the 1906 earthquake was approximately 50 times larger than the shock wave for the 1971 earthquake. ■

Modeling: Acid/Base Solutions

In chemistry, the pH of a solution is the measure of the acidity of the solution. The definition for pH involves common logarithms. Here it is:

$$\text{pH} = -\log[\text{H}^+]$$

where $[\text{H}^+]$ is the concentration of the hydrogen ion in moles per liter. The range for pH is from 0 to 14. Pure water, a neutral solution, has a pH of 7. An acidic solution, such as vinegar, will have a pH less than 7, and an alkaline solution, such as ammonia, has a pH above 7.

Note Notice that "clean rain" has been defined as that with a pH of 5.6 and above. Since pure water has a pH of 7, we can see that normal rain is actually slightly acidic.

THE ACID SCALE

Increasingly alkaline ← NEUTRAL → Increasingly acidic

ACID RAIN

14 13 12 11 10 9 8 7 6 5 4 3 2 1 0

Lye
Ammonia
Milk of magnesia
Seawater, baking soda
Lake Ontario
Blood
NEUTRAL
Milk
Mean pH of Adirondack Lakes, 1930
Upper limit at which some fish affected
"Clean" rain
Mean pH of Adirondack Lakes, 1975
Average pH of Killarney Lakes, 1971
Tomato juice
Average pH of rainfall, Toronto, Feb. 1979
Apple juice
Vinegar
Lemon juice
Most acidic rainfall recorded in U.S.
Battery acid

EXAMPLE 7 Normal rainwater has a pH of 5.6. What is the concentration of the hydrogen ion in normal rainwater?

SOLUTION Substituting 5.6 for pH in the formula $pH = -\log[H^+]$, we have

$$\begin{aligned} 5.6 &= -\log[H^+] && \text{Substitution} \\ \log[H^+] &= -5.6 && \text{Isolate the logarithm} \\ [H^+] &= 10^{-5.6} && \text{Write in exponential form} \\ &\approx 2.5 \times 10^{-6} \text{ mole per liter} && \text{Answer in scientific notation} \end{aligned}$$

EXAMPLE 8 The concentration of the hydrogen ion in a sample of acid rain known to kill fish is 3.2×10^{-5} mole per liter. Find the pH of this acid rain to the nearest tenth.

SOLUTION Substituting 3.2×10^{-5} for $[H^+]$ in the formula $pH = -\log[H^+]$, we have

$$\begin{aligned} pH &= -\log[3.2 \times 10^{-5}] && \text{Substitution} \\ &\approx -(-4.5) && \text{Evaluate the logarithm} \\ &\approx 4.5 && \text{Simplify} \end{aligned}$$

Natural Logarithms

(dēf′ **DEFINITION** *Natural logarithms*

A *natural logarithm* is a logarithm with a base of e. The natural logarithm of x is denoted by $\ln x$. That is,

$$\ln x = \log_e x$$

We can assume that all our properties of exponents and logarithms hold for expressions with a base of e, because e is a real number. Here are some examples intended to make you more familiar with the number e and natural logarithms.

EXAMPLE 9 Simplify each of the following expressions.

a. $e^0 = 1$

b. $e^1 = e$

c. $\ln e = 1$ — In exponential form, $e^1 = e$

d. $\ln 1 = 0$ — In exponential form, $e^0 = 1$

e. $\ln e^3 = 3$

f. $\ln e^{-4} = -4$

g. $\ln e^t = t$

EXAMPLE 10 Use the properties of logarithms to expand the expression $\ln Ae^{5t}$.

SOLUTION Because the properties of logarithms hold for natural logarithms, we have

$$\ln Ae^{5t} = \ln A + \ln e^{5t}$$
$$= \ln A + 5t \ln e$$
$$= \ln A + 5t \qquad \text{Because } \ln e = 1$$

EXAMPLE 11 If $\ln 2 = 0.6931$ and $\ln 3 = 1.0986$, find

a. $\ln 6$ **b.** $\ln 0.5$ **c.** $\ln 8$

SOLUTION

a. Writing 6 as $2 \cdot 3$ and applying Property 1 for logarithms, we have

$$\ln 6 = \ln 2 \cdot 3$$
$$= \ln 2 + \ln 3$$
$$= 0.6931 + 1.0986$$
$$= 1.7917$$

b. Writing 0.5 as $\frac{1}{2}$ and applying Property 2 for logarithms gives us

$$\ln 0.5 = \ln \frac{1}{2}$$
$$= \ln 1 - \ln 2$$
$$= 0 - 0.6931$$
$$= -0.6931$$

c. Writing 8 as 2^3 and applying Property 3 for logarithms, we have

$$\ln 8 = \ln 2^3$$
$$= 3 \ln 2$$
$$= 3(0.6931)$$
$$= 2.0793$$

GETTING READY FOR CLASS

After reading through the preceding section, respond in your own words and in complete sentences.

A. What is a common logarithm?
B. What is a natural logarithm?
C. What real-life situation can be modeled with common logarithms?
D. Find $\ln e$, and explain how you arrived at your answer.

Problem Set 6.5

Find the following logarithms. Give answer to 4 decimal places.

1. log 378 **2.** log 426 **3.** log 37.8 **4.** log 42,600

5. log 3,780 **6.** log 0.4260 **7.** log 0.0378 **8.** log 0.0426

9. log 37,800 **10.** log 4,900 **11.** log 600 **12.** log 900

13. log 2,010 **14.** log 10,200 **15.** log 0.00971 **16.** log 0.0312

17. log 0.0314 **18.** log 0.00052 **19.** log 0.399 **20.** log 0.111

Find x in the following equations. Express answers to 3 significant digits where appropriate.

21. $\log x = 2.8802$ **22.** $\log x = 4.8802$ **23.** $\log x = -2.1198$

24. $\log x = -3.1198$ **25.** $\log x = 3.1553$ **26.** $\log x = 5.5911$

27. $\log x = -5.3497$ **28.** $\log x = -1.5670$ **29.** $\log x = -7.0372$

30. $\log x = -4.2000$ **31.** $\log x = 10$ **32.** $\log x = -1$

33. $\log x = -10$ **34.** $\log x = 1$ **35.** $\log x = 20$

36. $\log x = -20$ **37.** $\log x = -2$ **38.** $\log x = 4$

39. $\log x = \log_2 8$ **40.** $\log x = \log_3 9$ **41.** $\ln x = -1$

42. $\ln x = 4$ **43.** $\log x = 2 \log 5$ **44.** $\log x = -\log 4$

45. $\ln x = -3 \ln 2$ **46.** $\ln x = 5 \ln 3$

Simplify each of the following expressions without using a calculator.

47. $\ln e$ **48.** $\ln 1$ **49.** $\ln e^5$ **50.** $\ln e^{-3}$

51. $\ln e^x$ **52.** $\ln e^y$ **53.** $\log 10{,}000$ **54.** $\log 0.001$

55. $\ln \frac{1}{e^3}$ **56.** $\ln \sqrt{e}$ **57.** $\log \sqrt{1000}$ **58.** $\log \sqrt[3]{10{,}000}$

Use the properties of logarithms to expand each of the following expressions.

59. $\ln 10e^{3t}$ **60.** $\ln 10e^{4t}$ **61.** $\ln Ae^{-2t}$

62. $\ln Ae^{-3t}$ **63.** $\log [100(1.01)^{3t}]$ **64.** $\log \left[\frac{1}{10} (1.5)^{t+2} \right]$

65. $\ln (Pe^{rt})$ **66.** $\ln \left(\frac{1}{2} e^{-kt} \right)$ **67.** $-\log (4.2 \times 10^{-3})$

68. $-\log (5.7 \times 10^{-10})$

If $\ln 2 = 0.6931$, $\ln 3 = 1.0986$, and $\ln 5 = 1.6094$, find each of the following.

69. $\ln 15$ **70.** $\ln 10$ **71.** $\ln \frac{1}{3}$ **72.** $\ln \frac{1}{5}$

73. $\ln 9$ **74.** $\ln 25$ **75.** $\ln 16$ **76.** $\ln 81$

Getting Ready for Calculus Problems 77-84 are problems you will see again if you go on to a calculus class.

Expand each expression.

77. $\ln 5x$ **78.** $\ln\left(\frac{3x}{x+9}\right)$ **79.** $\ln x^4$ **80.** $\ln(x^3y^7)$

Write as a single logarithm.

81. $3\ln x + 2\ln y$ **82.** $\ln a - \ln b - \ln c$

83. $5\ln x + 6\ln y - 7\ln z$ **84.** $4\ln a - 3\ln b - 5\ln c$

Modeling Practice

85. Atomic Bomb Tests The formula for determining the magnitude, M, of an earthquake on the Richter scale is $M = \log_{10} T$, where T is the number of times the shock wave is greater than the smallest measurable shock wave. The Bikini Atoll in the Pacific Ocean was used as a location for atomic bomb tests by the United States government in the 1950s. One such test resulted in an earthquake measurement of 5.0 on the Richter scale. Compare the 1906 San Francisco earthquake of estimated magnitude 8.3 on the Richter scale to this atomic bomb test. Use the shock wave T for purposes of comparison.

Photograph by U.S. Navy

86. Atomic Bomb Tests Today's nuclear weapons are 1,000 times more powerful than the atomic bombs tested in the Bikini Atoll mentioned in Problem 85. Use the shock wave T to determine the Richter scale measurement of a nuclear test today.

87. University Enrollment The percentage of students enrolled in a university who are between the ages of 25 and 34 can be modeled by the formula $s = 5\ln x$, where s is the percentage of students and x is the number of years since 1989. Predict the year in which approximately 15% of students enrolled in a university are between the ages of 25 and 34.

88. Memory A class of students take a test on the mathematics concept of solving quadratic equations. That class agrees to take a similar form of the test each month for the next 6 months to test their memory of the topic since instruction. The function of the average score earned each month on the test is $m(x) = 75 - 5\ln(x+1)$, where x represents time in months. Complete the table to indicate the average score earned by the class at each month.

Time, x	Score, m
0	
1	
2	
3	
4	
5	
6	

89. pH Find the pH of orange juice if the concentration of the hydrogen ion in the juice is $[H^+] = 6.50 \times 10^{-4}$.

90. pH Find the pH of milk if the concentration of the hydrogen ion in milk is $[H^+] = 1.88 \times 10^{-6}$.

91. pH Find the concentration of hydrogen ions in a glass of wine if the pH is 4.75.

92. pH Find the concentration of hydrogen ions in a bottle of vinegar if the pH is 5.75.

The Richter Scale Find the relative size T of the shock wave of earthquakes with the following magnitudes, as measured on the Richter scale.

93. 5.5 **94.** 6.6 **95.** 8.3 **96.** 8.7

97. Earthquake The chart below is a partial listing of earthquakes that were recorded in Canada during the year 2000. Complete the chart by computing the magnitude on the Richter scale, M, or the number of times the associated shock wave is larger than the smallest measurable shock wave, T.

Location	Date	Magnitude M	Shock wave T
Moresby Island	Jan. 23	4.3	
Vancouver Island	Apr. 30		1.26×10^5
Quebec City	June 29	3.2	
Mould Bay	Nov. 13	5.2	
St. Lawrence	Dec. 14		5.01×10^3

Source: *National Resources Canada, National Earthquake Hazards Program.*

98. Earthquake On January 26, 2001, an earthquake with a magnitude of 7.7 on the Richter scale hit southern India. By what factor was this earthquake's shock wave greater than the smallest measurable shock wave? (Source: *National Earthquake Information Center*)

Depreciation The annual rate of depreciation r on a car that is purchased for P dollars and is worth W dollars t years later can be found from the formula

$$\log(1 - r) = \frac{1}{t} \log \frac{W}{P}$$

99. Find the annual rate of depreciation on a car that is purchased for $9,000 and sold 5 years later for $4,500.

100. Find the annual rate of depreciation on a car that is purchased for $9,000 and sold 4 years later for $3,000.

Two cars depreciate in value according to the following depreciation tables. In each case, find the annual rate of depreciation.

101.

Age in Years	Value in Dollars
New	7,550
5	5,750

102.

Age in Years	Value in Dollars
New	7,550
3	5,750

Getting Ready for the Next Section

Solve.

103. $5(2x + 1) = 12$

104. $4(3x - 2) = 21$

Use a calculator to evaluate; give answers to 4 decimal places.

105. $\frac{100{,}000}{32{,}000}$

106. $\frac{1.4982}{6.5681} + 3$

107. $\frac{1}{2}\left(\frac{-0.6931}{1.4289} + 3\right)$

108. $1 + \frac{0.04}{52}$

Use the power rule to rewrite the following logarithms.

109. $\log 1.05^t$

110. $\log 1.033^t$

Use identities to simplify.

111. $\ln e^{0.05t}$

112. $\ln e^{-0.000121t}$

Exponential Equations and Change of Base

6.6

For items involved in exponential growth, the time it takes for a quantity to double is called the *doubling time*. For example, if you invest \$5,000 in an account that pays 5% annual interest, compounded quarterly, you may want to know how long it will take for your money to double in value. You can find this doubling time if you can solve the equation

$$10{,}000 = 5{,}000(1.0125)^{4t}$$

As you will see as you progress through this section, logarithms are the key to solving equations of this type.

Logarithms are very important in solving equations in which the variable appears as an exponent. The equation

$$5^x = 12$$

is an example of one such equation. Equations of this form are called *exponential equations*. Because the quantities 5^x and 12 are equal, so are their common logarithms. We begin our solution by taking the logarithm of both sides:

$$\log 5^x = \log 12$$

We now apply Property 3 for logarithms, $\log x^r = r \log x$, to turn x from an exponent into a coefficient:

$$x \log 5 = \log 12$$

Dividing both sides by log 5 gives us

$$x = \frac{\log 12}{\log 5}$$

If we want a decimal approximation to the solution, we can find log 12 and log 5 on a calculator and divide:

$$x \approx \frac{1.0792}{0.6990}$$

$$\approx 1.5439$$

The complete problem looks like this:

$$\begin{aligned} 5^x &= 12 \\ \log 5^x &= \log 12 \\ x \log 5 &= \log 12 \\ x &= \frac{\log 12}{\log 5} \\ &\approx \frac{1.0792}{0.6990} \\ &\approx 1.5439 \end{aligned}$$

Here is another example of solving an exponential equation using logarithms.

VIDEO EXAMPLES

SECTION 6.6

EXAMPLE 1 Solve for x: $25^{2x+1} = 15$

SOLUTION Taking the logarithm of both sides and then writing the exponent $(2x + 1)$ as a coefficient, we proceed as follows:

$$25^{2x+1} = 15$$

$$\log 25^{2x+1} = \log 15 \qquad \text{Take the log of both sides}$$

$$(2x + 1)\log 25 = \log 15 \qquad \text{Property 3}$$

$$2x + 1 = \frac{\log 15}{\log 25} \qquad \text{Divide by log 25}$$

$$2x = \frac{\log 15}{\log 25} - 1 \qquad \text{Add } -1 \text{ to both sides}$$

$$x = \frac{1}{2}\left(\frac{\log 15}{\log 25} - 1\right) \qquad \text{Multiply both sides by } \tfrac{1}{2}$$

Using a calculator, we can write a decimal approximation to the answer:

$$x \approx \frac{1}{2}\left(\frac{1.1761}{1.3979} - 1\right)$$

$$\approx \frac{1}{2}(0.8413 - 1)$$

$$\approx \frac{1}{2}(-0.1587)$$

$$\approx -0.0794$$

■

Modeling: Investments

If you invest P dollars in an account with an annual interest rate r that is compounded n times a year, then t years later the amount of money in that account will be

$$A = P\left(1 + \frac{r}{n}\right)^{nt}$$

EXAMPLE 2 How long does it take for $5,000 to double if it is deposited in an account that yields 5% interest compounded once a year?

SOLUTION Substituting $P = 5{,}000$, $r = 0.05$, $n = 1$, and $A = 10{,}000$ into our formula, we have

$$10{,}000 = 5{,}000(1 + 0.05)^t$$

$$10{,}000 = 5{,}000(1.05)^t$$

$$2 = (1.05)^t \qquad \text{Divide by 5,000}$$

This is an exponential equation. We solve by taking the logarithm of both sides:

$$\log 2 = \log(1.05)^t$$

$$= t \log 1.05$$

Dividing both sides by log 1.05, we have

$$t = \frac{\log 2}{\log 1.05}$$
$$\approx 14.2$$

It takes a little over 14 years for \$5,000 to double if it earns 5% interest per year, compounded once a year. ■

There is a fourth property of logarithms we have not yet considered. This last property allows us to change from one base to another and is therefore called the *change-of-base property*.

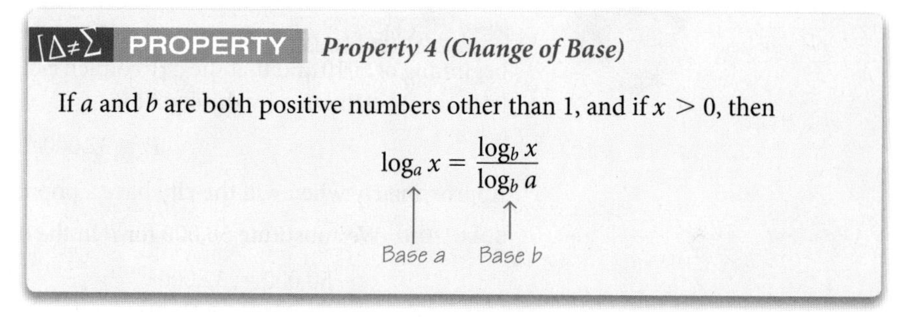

PROPERTY *Property 4 (Change of Base)*

If a and b are both positive numbers other than 1, and if $x > 0$, then

$$\log_a x = \frac{\log_b x}{\log_b a}$$

The logarithm on the left side has a base of a, and both logarithms on the right side have a base of b. This allows us to change from base a to any other base b that is a positive number other than 1. Here is a proof of Property 4 for logarithms.

Proof We begin by writing the identity

$$a^{\log_a x} = x$$

Taking the logarithm base b of both sides and writing the exponent $\log_a x$ as a coefficient, we have

$$\log_b a^{\log_a x} = \log_b x$$
$$\log_a x \log_b a = \log_b x$$

Dividing both sides by $\log_b a$, we have the desired result:

$$\frac{\log_a x \log_b a}{\log_b a} = \frac{\log_b x}{\log_b a}$$

$$\log_a x = \frac{\log_b x}{\log_b a}$$

We can use this property to find logarithms we could not otherwise compute on our calculators—that is, logarithms with bases other than 10 or e. The next example illustrates the use of this property.

EXAMPLE 3 Find $\log_8 24$.

SOLUTION Because we do not have base-8 logarithms on our calculators, we can change this expression to an equivalent expression that contains only base-10 logarithms:

$$\log_8 24 = \frac{\log 24}{\log 8} \qquad \text{Property 4}$$

Don't be confused. We did not just drop the base, we changed to base 10. We could have written the last line like this:

$$\log_8 24 = \frac{\log_{10} 24}{\log_{10} 8}$$

From our calculators, we write

$$\log_8 24 \approx \frac{1.3802}{0.9031}$$
$$\approx 1.5283$$

Modeling: Population Growth

EXAMPLE 4 Suppose that the population in a small city is 32,000 in the beginning of 2010 and that the city council assumes that the population size t years later can be estimated by the equation

$$P = 32{,}000e^{0.05t}$$

Approximately when will the city have a population of 50,000?

SOLUTION We substitute 50,000 for P in the equation and solve for t:

$$50{,}000 = 32{,}000e^{0.05t}$$
$$1.5625 = e^{0.05t} \qquad \frac{50{,}000}{32{,}000} = 1.5625$$

To solve this equation for t, we can take the natural logarithm of each side:

$$\ln 1.5625 = \ln e^{0.05t}$$
$$= 0.05t \ln e \qquad \text{Property 3 for logarithms}$$
$$= 0.05t \qquad \text{Because } \ln e = 1$$
$$t = \frac{\ln 1.5625}{0.05} \qquad \text{Divide each side by 0.05}$$
$$\approx 8.93 \text{ years}$$

We can estimate that the population will reach 50,000 toward the end of 2018.

Using Technology 6.6

We can evaluate many logarithmic expressions on a graphing calculator by using the fact that logarithmic functions and exponential functions are inverses.

EXAMPLE 5 Evaluate the logarithmic expression $\log_3 7$ from the graph of an exponential function.

SOLUTION First, we let $\log_3 7 = x$. Next, we write this expression in exponential form as $3^x = 7$. We can solve this equation graphically by finding the intersection of the graphs $Y1 = 3^x$ and $Y2 = 7$, as shown in Figure 1.

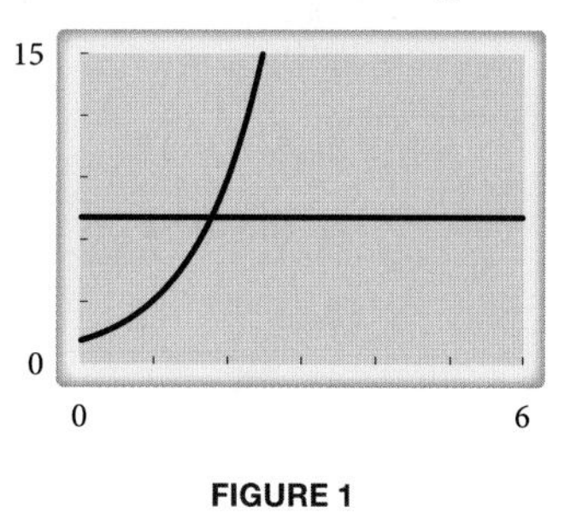

FIGURE 1

Using the calculator, we find the two graphs intersect at (1.77, 7). Therefore, $\log_3 7 = 1.77$ to the nearest hundredth. We can check our work by evaluating the expression $3^{1.77}$ on our calculator with the key strokes

3 [^] 1.77 [ENTER]

The result is 6.99 to the nearest hundredth, which seems reasonable since 1.77 is accurate to the nearest hundredth. To get a result closer to 7, we would need to find the intersection of the two graphs more accurately. ■

GETTING READY FOR CLASS

After reading through the preceding section, respond in your own words and in complete sentences.

A. What is an exponential equation?
B. How do logarithms help you solve exponential equations?
C. What is the change-of-base property?
D. Write an application modeled by the equation
$A = 10{,}000\left(1 + \frac{0.08}{2}\right)^{2 \cdot 5}$.

Problem Set 6.6

Solve each exponential equation. Use a calculator to write the answer in decimal form.

1. $3^x = 5$
2. $4^x = 3$
3. $5^x = 3$
4. $3^x = 4$
5. $5^{-x} = 12$
6. $7^{-x} = 8$
7. $12^{-x} = 5$
8. $8^{-x} = 7$
9. $8^{x+1} = 4$
10. $9^{x+1} = 3$
11. $4^{x-1} = 4$
12. $3^{x-1} = 9$
13. $3^{2x+1} = 2$
14. $2^{2x+1} = 3$
15. $3^{1-2x} = 2$
16. $2^{1-2x} = 3$
17. $15^{3x-4} = 10$
18. $10^{3x-4} = 15$
19. $6^{5-2x} = 4$
20. $9^{7-3x} = 5$
21. $3^{-4x} = 81$
22. $2^{5x} = \frac{1}{16}$
23. $5^{3x-2} = 15$
24. $7^{4x+3} = 200$
25. $100e^{3t} = 250$
26. $150e^{0.065t} = 400$
27. $1{,}200\left(1 + \frac{0.072}{4}\right)^{4t} = 25{,}000$
28. $2{,}700\left(1 + \frac{0.086}{12}\right)^{12t} = 10{,}000$
29. $50e^{-0.0742t} = 32$
30. $19e^{-0.000243t} = 12$

Use the change-of-base property and a calculator to find a decimal approximation to each of the following logarithms.

31. $\log_8 16$
32. $\log_9 27$
33. $\log_{16} 8$
34. $\log_{27} 9$
35. $\log_7 15$
36. $\log_3 12$
37. $\log_{15} 7$
38. $\log_{12} 3$
39. $\log_8 240$
40. $\log_6 180$
41. $\log_4 321$
42. $\log_5 462$

Find a decimal approximation to each of the following natural logarithms.

43. $\ln 345$
44. $\ln 3{,}450$
45. $\ln 0.345$
46. $\ln 0.0345$
47. $\ln 10$
48. $\ln 100$
49. $\ln 45{,}000$
50. $\ln 450{,}000$

Modeling Practice

51. **Compound Interest** How long will it take for \$500 to double if it is invested at 6% annual interest compounded 2 times a year?

52. **Compound Interest** How long will it take for \$500 to double if it is invested at 6% annual interest compounded 12 times a year?

53. **Compound Interest** How long will it take for \$1,000 to triple if it is invested at 12% annual interest compounded 6 times a year?

54. **Compound Interest** How long will it take for \$1,000 to become \$4,000 if it is invested at 12% annual interest compounded 6 times a year?

55. **Doubling Time** How long does it take for an amount of money P to double itself if it is invested at 8% interest compounded 4 times a year?

56. **Tripling Time** How long does it take for an amount of money P to triple itself if it is invested at 8% interest compounded 4 times a year?

57. **Tripling Time** If a \$25 investment is worth \$75 today, how long ago must that \$25 have been invested at 6% interest compounded twice a year?

58. **Doubling Time** If a \$25 investment is worth \$50 today, how long ago must that \$25 have been invested at 6% interest compounded twice a year?

Recall from Section 6.1 that if P dollars are invested in an account with annual interest rate r, compounded continuously, then the amount of money in the account after t years is given by the formula

$$A(t) = Pe^{rt}$$

59. **Continuously Compounded Interest** Repeat Problem 51 if the interest is compounded continuously.

60. **Continuously Compounded Interest** Repeat Problem 54 if the interest is compounded continuously.

61. **Continuously Compounded Interest** How long will it take \$500 to triple if it is invested at 6% annual interest, compounded continuously?

62. **Continuously Compounded Interest** How long will it take \$500 to triple if it is invested at 12% annual interest, compounded continuously?

63. **Continuously Compounded Interest** How long will it take for \$1,000 to be worth \$2,500 at 8% interest, compounded continuously?

64. **Continuously Compounded Interest** How long will it take for \$1,000 to be worth \$5,000 at 8% interest, compounded continuously?

65. **Exponential Growth** Suppose that the population in a small city is 32,000 at the beginning of 2005 and that the city council assumes that the population size t years later can be estimated by the equation

$$P(t) = 32{,}000e^{0.05t}$$

Approximately when will the city have a population of 64,000?

66. **Exponential Growth** Suppose the population of a city is given by the equation

$$P(t) = 100{,}000e^{0.05t}$$

where t is the number of years from the present time. How large is the population now? (*Now* corresponds to a certain value of t. Once you realize what that value of t is, the problem becomes very simple.)

67. **Airline Travel** The number of airline passengers in 1990 was 466 million. The number of passengers traveling by airplane each year has increased exponentially according to the model $P(t) = 466 \cdot 1.035^t$, where t is the number of years since 1990 (U.S. Census Bureau). In what year was it predicted that 900 million passengers would travel by airline?

68. **Bankruptcy Model** In 1997, there were a total of 1,316,999 bankruptcies filed under the Bankruptcy Reform Act. The model for the number of bankruptcies filed is $B(t) = 0.798 \cdot 1.164^t$, where t is the number of years since 1994 and B is the number of bankruptcies filed in terms of millions (Administrative Office of the U.S. Courts, *Statistical Tables for the Federal Judiciary*). In what year is it predicted that 12 million bankruptcies would be filed?

69. Health Care In 1990, \$699 billion was spent on health care expenditures. The amount of money, E, in billions spent on health care expenditures can be estimated using the function $E(t) = 78.16(1.11)^t$, where t is time in years since 1970 (U.S. Census Bureau). In what year was it estimated that \$800 billion would be spent on health care expenditures?

70. Value of a Car As a car ages, its value decreases. The value of a particular car with an original purchase price of \$25,600 is modeled by the function $c(t) = 25{,}600(1 - 0.22)^t$, where c is the value at time t (Kelly Blue Book). How old is the car when its value is \$10,000?

71. Compound Interest In 1986, the average cost of attending a public university through graduation was \$16,552 (U.S. Department of Education, National Center for Educational Statistics). If John's parents deposited that amount in an account in 1986 at an interest rate of 7% compounded semi-annually, how long would it take for the money to double?

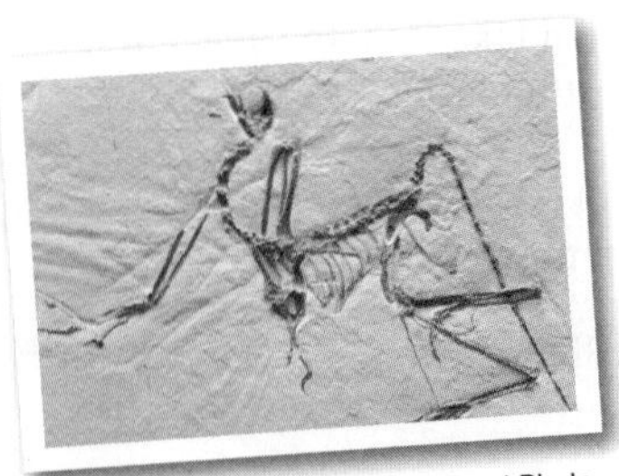
© Wicki58/iStockPhoto

72. Carbon Dating Scientists use carbon-14 dating to find the age of fossils and other artifacts. The amount of carbon-14 in an organism will yield information concerning its age. A formula used in carbon-14 dating is $A(t) = A_0 \cdot 2^{-t/5{,}730}$, where A_0 is the amount of carbon originally in the organism, t is time in years, and A is the amount of carbon remaining after t years. Determine the number of years since an organism died if it originally contained 1,000 grams of carbon-14 and it currently contains 600 grams of carbon-14.

73. Cost Increase The cost of a can of Coca Cola in 1960 was \$0.10. The function that models the cost of a Coca Cola by year is $C(t) = 0.10e^{0.0576t}$, where t is the number of years since 1960. In what year was it expected that a can of Coca Cola would cost \$1.00?

74. Online Banking Use The number of households using online banking services increased from 754,000 in 1995 to 12,980,000 in 2000. The formula $H(t) = 0.76e^{0.55t}$ models the number of households, H, in millions when time is t years since 1995 according to the Home Banking Report. In what year was it estimated that 50,000,000 households would use online banking services?

Maintaining Your Skills

The following problems review material we covered in Section 5.1.

Find the vertex for each of the following parabolas, and then indicate if it is the highest or lowest point on the graph.

75. $y = 2x^2 + 8x - 15$

76. $y = 3x^2 - 9x - 10$

77. $y = 12x - 4x^2$

78. $y = 18x - 6x^2$

79. Maximum Height An object is projected into the air with an initial upward velocity of 64 feet per second. Its height h at any time t is given by the formula $h = 64t - 16t^2$. Find the time at which the object reaches its maximum height. Then, find the maximum height.

80. Maximum Height An object is projected into the air with an initial upward velocity of 64 feet per second from the top of a building 40 feet high. If the height h of the object t seconds after it is projected into the air is $h = 40 + 64t - 16t^2$, find the time at which the object reaches its maximum height. Then, find the maximum height it attains.

Chapter 6 Summary

EXAMPLES

Exponential Functions [6.1]

1. For the exponential function $f(x) = 2^x$,

$$f(0) = 2^0 = 1$$
$$f(1) = 2^1 = 2$$
$$f(2) = 2^2 = 4$$
$$f(3) = 2^3 = 8$$

Any function of the form

$$f(x) = b^x$$

where $b > 0$ and $b \neq 1$, is an *exponential function.*

One-to-One Functions [6.2]

2. The function $f(x) = x^2$ is not one-to-one because 9, which is in the range, comes from both 3 and -3 in the domain.

A function is a *one-to-one function* if every element in the range comes from exactly one element in the domain.

Inverse Functions [6.2]

3. The inverse of $f(x) = 2x - 3$ is

$$f^{-1}(x) = \frac{x + 3}{2}$$

The *inverse* of a function is obtained by reversing the order of the coordinates of the ordered pairs belonging to the function. Only one-to-one functions have inverses that are also functions.

Definition of Logarithms [6.3]

4. The definition allows us to write expressions like

$$y = \log_3 27$$

equivalently in exponential form as

$$3^y = 27$$

which makes it apparent that y is 3.

If b is a positive number not equal to 1, then the expression

$$y = \log_b x$$

is equivalent to $x = b^y$; that is, in the expression $y = \log_b x$, y is the number to which we raise b in order to get x. Expressions written in the form $y = \log_b x$ are said to be in *logarithmic form*. Expressions like $x = b^y$ are in *exponential form*.

Two Special Identities [6.3]

5. Examples of the two special identities are

$$5^{\log_5 12} = 12$$

and

$$\log_8 8^3 = 3$$

For $b > 0$, $b \neq 1$, the following two expressions hold for all positive real numbers x:

(1) $b^{\log_b x} = x$

(2) $\log_b b^x = x$

Properties of Logarithms [6.4]

6. We can rewrite the expression

$$\log_{10} \frac{45^6}{273}$$

using the properties of logarithms, as

$$6 \log_{10} 45 - \log_{10} 273$$

If x, y, and b are positive real numbers, $b \neq 1$, and r is any real number, then:

1. $\log_b (xy) = \log_b x + \log_b y$
2. $\log_b \left(\frac{x}{y}\right) = \log_b x - \log_b y$
3. $\log_b x^r = r \log_b x$

Common Logarithms [6.5]

7. $\log_{10} 10{,}000 = \log 10{,}000$
$= \log 10^4$
$= 4$

Common logarithms are logarithms with a base of 10. To save time in writing, we omit the base when working with common logarithms; that is,

$$\log x = \log_{10} x$$

Natural Logarithms [6.5]

8. $\ln e = 1$
$\ln 1 = 0$

Natural logarithms, written *ln x*, are logarithms with a base of e, where the number e is an irrational number (like the number π). A decimal approximation for e is 2.7183. All the properties of exponents and logarithms hold when the base is e.

Change of Base [6.6]

9. $\log_6 475 = \dfrac{\log 475}{\log 6}$
$\approx \dfrac{2.6767}{0.7782}$
≈ 3.44

If x, a, and b are positive real numbers, $a \neq 1$ and $b \neq 1$, then

$$\log_a x = \frac{\log_b x}{\log_b a}$$

COMMON MISTAKE

The most common mistakes that occur with logarithms come from trying to apply the three properties of logarithms to situations in which they don't apply. For example, a very common mistake looks like this:

$$\frac{\log 3}{\log 2} = \log 3 - \log 2 \qquad \text{Mistake}$$

This is not a property of logarithms. To write the equation $\log 3 - \log 2$, we would have to start with

$$\log \frac{3}{2} \qquad \textit{NOT} \qquad \frac{\log 3}{\log 2}$$

There is a difference.

Chapter 6 Test

Graph each exponential function. [6.1]

1. $f(x) = 2^x$

2. $g(x) = 3^{-x}$

Sketch the graph of each function and its inverse. Find $f^{-1}(x)$ for Problem 3. [6.2]

3. $f(x) = 2x - 3$

4. $f(x) = x^2 - 4$

Solve for x. [6.3]

5. $\log_4 x = 3$

6. $\log_x 5 = 2$

Graph each of the following. [6.3]

7. $y = \log_2 x$

8. $y = \log_{1/2} x$

Evaluate each of the following. [6.3, 6.5, 6.6]

9. $\log_8 4$

10. $\log_7 21$

11. $\log 23{,}400$

12. $\log 0.0123$

13. $\ln 46.2$

14. $\ln 0.0462$

Use the properties of logarithms to expand each expression. [6.4]

15. $\log_2 \frac{8x^2}{y}$

16. $\log \frac{\sqrt{x}}{(y^4)\sqrt[5]{z}}$

Write each expression as a single logarithm. [6.4]

17. $2 \log_3 x - \frac{1}{2} \log_3 y$

18. $\frac{1}{3} \log x - \log y - 2 \log z$

Use a calculator to find x. [6.5]

19. $\log x = 4.8476$

20. $\log x = -2.6478$

Solve for x. [6.4, 6.6]

21. $5 = 3^x$

22. $4^{2x-1} = 8$

23. $\log_5 x - \log_5 3 = 1$

24. $\log_2 x + \log_2(x - 7) = 3$

25. pH Find the pH of a solution in which $[H^+] = 6.6 \times 10^{-7}$. [6.5]

26. Compound Interest If $400 is deposited in an account that earns 10% annual interest compounded twice a year, how much money will be in the account after 5 years? [6.1]

27. Compound Interest How long will it take $600 to become $1,800 if the $600 is deposited in an account that earns 8% annual interest compounded 4 times a year? [6.6]

28. Depreciation If a car depreciates in value 20% per year for the first 5 years after it is purchased for P_0 dollars, then its value after t years will be $V(t) = P_0(1 - r)^t$ for $0 \le t \le 5$. To the nearest dollar, find the value of a car 4 years after it is purchased for $18,000. [6.1]

7 Systems of Equations

Chapter Outline

© Neustock/iStockphoto

Note When you see this icon next to an example or problem in this chapter, you will know that we are using the topics in this chapter to model situations in the world around us.

Suppose you decide to buy a prepaid cellular phone and are trying to decide between two service plans. Plan A has no monthly charge, but calls are charged \$0.17 for each minute, or fraction of a minute, that you use the phone. Plan B is \$0.50 per month plus \$0.10 for each minute, or fraction of a minute. The monthly cost $C(x)$ for each plan can be represented with a linear equation in two variables:

$$\text{Plan A: } C(x) = 0.17x$$

$$\text{Plan B: } C(x) = 0.10x + 0.50$$

To compare the two plans, we use the table and graph shown below.

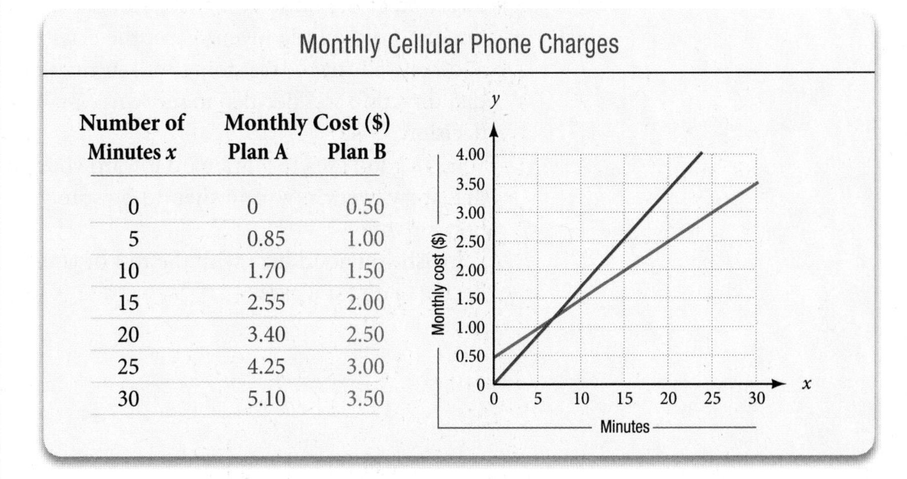

Monthly Cellular Phone Charges

Number of Minutes x	Monthly Cost (\$) Plan A	Plan B
0	0	0.50
5	0.85	1.00
10	1.70	1.50
15	2.55	2.00
20	3.40	2.50
25	4.25	3.00
30	5.10	3.50

The point of intersection of the two lines in the graph is the point at which the monthly costs of the two plans are equal. In this chapter, we will develop methods of finding that point of intersection.

Study Skills

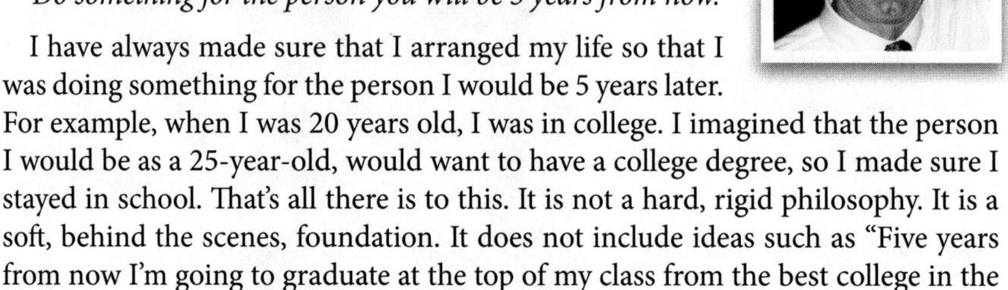

Dear Student,

Now that you are close to finishing this course, I want to pass on a couple of things that have helped me a great deal with my career. I'll introduce each one with a quote:

Do something for the person you will be 5 years from now.

I have always made sure that I arranged my life so that I was doing something for the person I would be 5 years later. For example, when I was 20 years old, I was in college. I imagined that the person I would be as a 25-year-old, would want to have a college degree, so I made sure I stayed in school. That's all there is to this. It is not a hard, rigid philosophy. It is a soft, behind the scenes, foundation. It does not include ideas such as "Five years from now I'm going to graduate at the top of my class from the best college in the country." Instead, you think, "five years from now I will have a college degree, or I will still be in school working towards it."

This philosophy led to a community college teaching job, writing textbooks, doing videos with the textbooks, then to MathTV and the book you are reading right now. Along the way there were many other options and directions that I didn't take, but all the choices I made were due to keeping the person I would be in 5 years in mind.

It's easier to ride a horse in the direction it is going.

I started my college career thinking that I would become a dentist. I enrolled in all the courses that were required for dental school. When I completed the courses, I applied to a number of dental schools, but wasn't accepted. I kept going to school, and applied again the next year, again, without success. My life was not going in the direction of dental school, even though I had worked hard to put it in that direction. So I did a little inventory of the classes I had taken and the grades I earned, and realized that I was doing well in mathematics. My life was actually going in that direction so I decided to see where mathematics would take me. It was a good decision.

It is a good idea to work hard toward your goals, but it is also a good idea to take inventory every now and then to be sure you are headed in the direction that is best for you.

I wish you good luck with the rest of your college years, and with whatever you decide to do for a career.

Pat McKeague
Fall 2012

Systems of Linear Equations in Two Variables

Previously, we found the graph of an equation of the form $ax + by = c$ to be a straight line. Because the graph is a straight line, the equation is said to be a linear equation. Two linear equations considered together form a *linear system* of equations. For example,

$$3x - 2y = 6$$
$$2x + 4y = 20$$

is a linear system. The solution set to the system is the set of all ordered pairs that satisfy both equations. If we graph each equation on the same set of axes, we can see the solution set (see Figure 1).

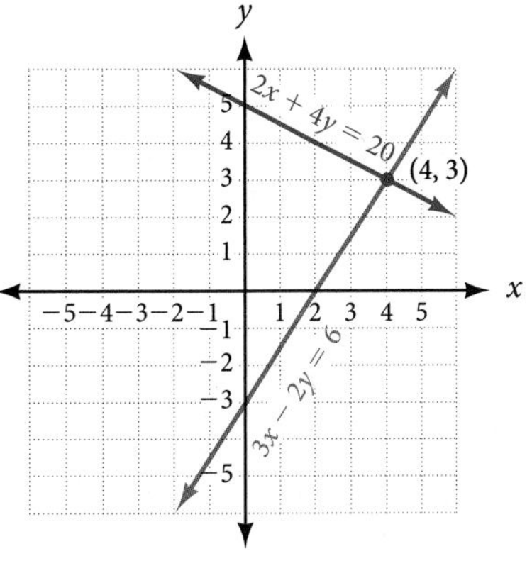

FIGURE 1

The point (4, 3) lies on both lines and therefore must satisfy both equations. It is obvious from the graph that it is the only point that does so. The solution set for the system is {(4, 3)}.

More generally, if $a_1x + b_1y = c_1$ and $a_2x + b_2y = c_2$ are linear equations, then the solution set for the system

$$a_1x + b_1y = c_1$$
$$a_2x + b_2y = c_2$$

can be illustrated through one of the graphs in Figure 2.

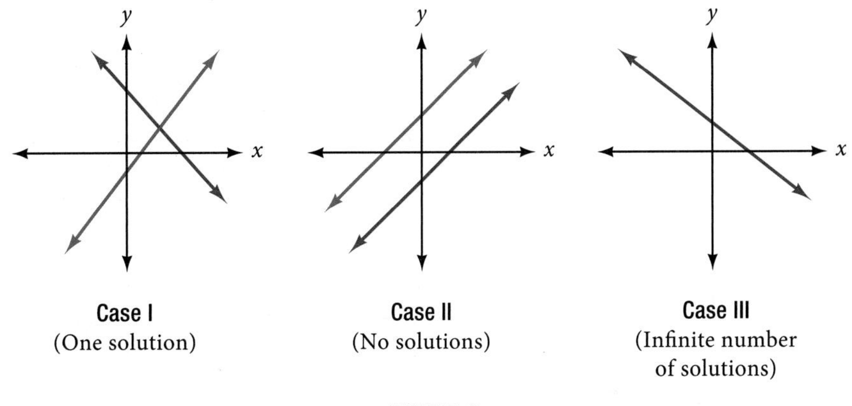

Case I (One solution) | **Case II** (No solutions) | **Case III** (Infinite number of solutions)

FIGURE 2

Case I The two lines intersect at one and only one point. The coordinates of the point give the solution to the system. This is what usually happens.

Case II The lines are parallel and therefore have no points in common. The solution set to the system is the empty set, $\varnothing$. In this case, we say the system of equations is *inconsistent.*

Case III The lines coincide. That is, their graphs represent the same line. The solution set consists of all ordered pairs that satisfy either equation. In this case, the equations are said to be *dependent.*

Note A system of equations is *consistent* if it has at least one solution. It is *inconsistent* if it has no solution. Two equations are *dependent* if one is a multiple of the other. Otherwise, they are *independent.*

In the beginning of this section, we found the solution set for the system

$$3x - 2y = 6$$
$$2x + 4y = 20$$

by graphing each equation and then reading the solution set from the graph. Solving a system of linear equations by graphing is the least accurate method. If the coordinates of the point of intersection are not integers, it can be difficult to read the solution set from the graph. There is another method of solving a linear system that does not depend on the graph. It is called the *addition method.*

The Addition Method

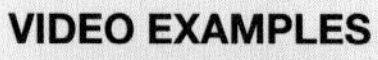

SECTION 7.1

Solve the system.

$$4x + 3y = 10$$
$$2x + y = 4$$

SOLUTION If we multiply the bottom equation by -3, the coefficients of y in the resulting equation and the top equation will be opposites:

$$4x + 3y = 10 \xrightarrow{\text{No Change}} 4x + 3y = 10$$
$$2x + y = 4 \xrightarrow[\text{Multiply by } -3]{} -6x - 3y = -12$$

Adding the left and right sides of the resulting equations, we have

$$\begin{aligned} 4x + 3y &= 10 \\ -6x - 3y &= -12 \\ \hline -2x \quad &= -2 \end{aligned}$$

The result is a linear equation in one variable. We have eliminated the variable y from the equations by addition. (It is for this reason we call this method of solving a linear system the *addition method.*) Solving $-2x = -2$ for x, we have

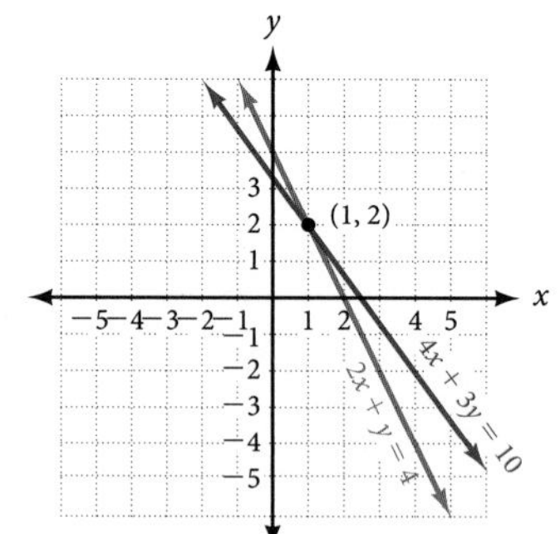

FIGURE 3 *A visual representation of the solution to the system in Example 1*

$$x = 1$$

This is the x-coordinate of the solution to our system. To find the y-coordinate, we substitute $x = 1$ into any of the equations containing both the variables x and y.

Let's try the second equation in our original system:

$$2(1) + y = 4$$
$$2 + y = 4$$
$$y = 2$$

This is the y-coordinate of the solution to our system. The ordered pair (1, 2) is the solution to the system.

> **Note** If we had put $x = 1$ into the first equation in our system, we would have obtained $y = 2$ also:
>
> $$4(1) + 3y = 10$$
> $$3y = 6$$
> $$y = 2$$

Checking Solutions

We can check our solution by substituting it into both of our equations.

Substituting $x = 1$ and $y = 2$ into $4x + 3y = 10$, we have

$$4(1) + 3(2) \stackrel{?}{=} 10$$
$$4 + 6 \stackrel{?}{=} 10$$
$$10 = 10 \quad \text{A true statement}$$

Substituting $x = 1$ and $y = 2$ into $2x + y = 4$, we have

$$2(1) + 2 \stackrel{?}{=} 4$$
$$2 + 2 \stackrel{?}{=} 4$$
$$4 = 4 \quad \text{A true statement}$$

Our solution satisfies both equations; therefore, it is a solution to our system of equations. ■

EXAMPLE 2 Solve the system.

$$3x - 5y = -2$$
$$2x - 3y = 1$$

SOLUTION We can eliminate either variable. Let's decide to eliminate the variable x. We can do so by multiplying the top equation by 2 and the bottom equation by -3, and then adding the left and right sides of the resulting equations:

$$3x - 5y = -2 \xrightarrow{\text{Multiply by 2}} 6x - 10y = -4$$
$$2x - 3y = 1 \xrightarrow[\text{Multiply by } -3]{} -6x + 9y = -3$$
$$-y = -7$$
$$y = 7$$

The y-coordinate of the solution to the system is 7. Substituting this value of y into any of the equations with both x- and y-variables gives $x = 11$. The solution to the system is (11, 7). It is the only ordered pair that satisfies both equations.

Checking (11, 7) in each equation looks like this:

Substituting $x = 11$ and $y = 7$ into $3x - 5y = -2$, we have

$$3(11) - 5(7) \stackrel{?}{=} -2$$
$$33 - 35 \stackrel{?}{=} -2$$
$$-2 = -2 \quad \text{A true statement}$$

Substituting $x = 11$ and $y = 7$ into $2x - 3y = 1$, we have

$$2(11) - 3(7) \stackrel{?}{=} 1$$
$$22 - 21 \stackrel{?}{=} 1$$
$$1 = 1 \quad \text{A true statement}$$

Our solution satisfies both equations; therefore, (11, 7) is a solution to our system. ■

EXAMPLE 3 Solve the system.

$$2x - 3y = 4$$
$$4x + 5y = 3$$

SOLUTION We can eliminate x by multiplying the top equation by -2 and adding it to the bottom equation:

$$\begin{array}{rcl} 2x - 3y = 4 & \xrightarrow{\text{Multiply by } -2} & -4x + 6y = -8 \\ 4x + 5y = 3 & \xrightarrow[\text{No Change}]{} & 4x + 5y = 3 \\ & & \hline 11y = -5 \\ & & y = -\frac{5}{11} \end{array}$$

The y-coordinate of our solution is $-\frac{5}{11}$. If we were to substitute this value of y back into either of our original equations, we would find the arithmetic necessary to solve for x cumbersome. For this reason, it is probably best to go back to the original system and solve it a second time—for x instead of y. Here is how we do that:

$$\begin{array}{rcl} 2x - 3y = 4 & \xrightarrow{\text{Multiply by } 5} & 10x - 15y = 20 \\ 4x + 5y = 3 & \xrightarrow[\text{Multiply by } 3]{} & 12x + 15y = 9 \\ & & \hline 22x \quad = 29 \\ & & x = \frac{29}{22} \end{array}$$

The solution to our system is $\left(\frac{29}{22}, -\frac{5}{11}\right)$. ■

The main idea in solving a system of linear equations by the addition method is to use the multiplication property of equality on one or both of the original equations, if necessary, to make the coefficients of either variable opposites. The following box shows some steps to follow when solving a system of linear equations by the addition method.

HOW TO *Solve a System of Linear Equations by the Addition Method*

Step 1: Decide which variable to eliminate. (In some cases, one variable will be easier to eliminate than the other. With some practice, you will notice which one it is.)

Step 2: Use the multiplication property of equality on each equation separately to make the coefficients of the variable that is to be eliminated opposites.

Step 3: Add the respective left and right sides of the system together.

Step 4: Solve for the remaining variable.

Step 5: Substitute the value of the variable from Step 4 into an equation containing both variables and solve for the other variable. (Or repeat Steps 2–4 to eliminate the other variable.)

Step 6: Check your solution in both equations, if necessary.

EXAMPLE 4 Solve the system.

$$\begin{aligned} 5x - 2y &= 5 \\ -10x + 4y &= 15 \end{aligned}$$

SOLUTION We can eliminate y by multiplying the first equation by 2 and adding the result to the second equation:

$$\begin{aligned} 5x - 2y &= 5 \xrightarrow{\text{Multiply by 2}} & 10x - 4y &= 10 \\ -10x + 4y &= 15 \xrightarrow[\text{No Change}]{} & -10x + 4y &= 15 \\ & & 0 &= 25 \end{aligned}$$

The result is the false statement $0 = 25$, which indicates there is no solution to the system. If we were to graph the two lines, we would find that they are parallel. In a case like this, we say the system is *inconsistent*. Whenever both variables have been eliminated and the resulting statement is false, the solution set for the system will be the empty set, $\varnothing$.

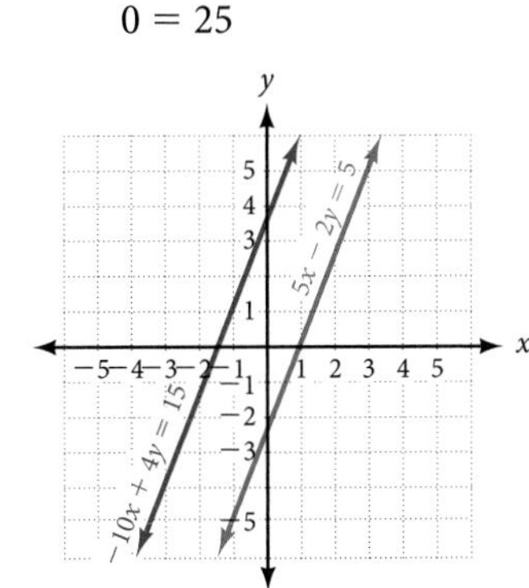

FIGURE 4 *A visual representation of the situation in Example 4 — the two lines are parallel*

EXAMPLE 5 Solve the system.

$$\begin{aligned} 4x + 3y &= 2 \\ 8x + 6y &= 4 \end{aligned}$$

SOLUTION Multiplying the top equation by -2 and adding, we can eliminate the variable x:

$$\begin{aligned} 4x + 3y &= 2 \xrightarrow{\text{Multiply by } -2} & -8x - 6y &= -4 \\ 8x + 6y &= 4 \xrightarrow[\text{No Change}]{} & 8x + 6y &= 4 \\ & & 0 &= 0 \end{aligned}$$

Both variables have been eliminated and the resulting statement $0 = 0$ is true. In this case, the lines coincide; they are *dependent*. The solution set consists of all ordered pairs that satisfy either equation. We can write the solution set as $\{(x, y) \mid 4x + 3y = 2\}$ or $\{(x, y) \mid 8x + 6y = 4\}$.

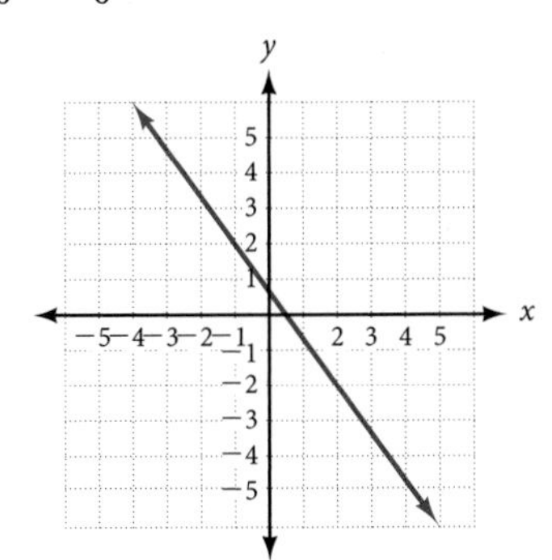

FIGURE 5 *A visual representation of the situation in Example 5 — both equations produce the same graph*

Special Cases

The previous two examples illustrate the two special cases in which the graphs of the equations in the system either coincide or are parallel. In both cases the left-hand sides of the equations were multiples of each other. In the case of the dependent equations, the right-hand sides were also multiples. We can generalize these observations for the system

$$a_1x + b_1y = c_1$$
$$a_2x + b_2y = c_2$$

Inconsistent System

What happens	*Geometric Interpretation*	*Algebraic Interpretation*
Both variables are eliminated, and the resulting statement is false.	The lines are parallel, and there is no solution to the system.	$\frac{a_1}{a_2} = \frac{b_1}{b_2} \neq \frac{c_1}{c_2}$

Dependent Equations

What happens	*Geometric Interpretation*	*Algebraic Interpretation*
Both variables are eliminated, and the resulting statement is true.	The lines coincide, and there are an infinite number of solutions to the system.	$\frac{a_1}{a_2} = \frac{b_1}{b_2} = \frac{c_1}{c_2}$

EXAMPLE 6 Solve the system.

$$\frac{1}{2}x - \frac{1}{3}y = 2$$
$$\frac{1}{4}x + \frac{2}{3}y = 6$$

SOLUTION Although we could solve this system without clearing the equations of fractions, there is probably less chance for error if we have only integer coefficients to work with. So let's begin by multiplying both sides of the top equation by 6, and both sides of the bottom equation by 12, to clear each equation of fractions:

$$\frac{1}{2}x - \frac{1}{3}y = 2 \xrightarrow{\text{Multiply by 6}} 3x - 2y = 12$$
$$\frac{1}{4}x + \frac{2}{3}y = 6 \xrightarrow[\text{Multiply by 12}]{} 3x + 8y = 72$$

Now we can eliminate x by multiplying the top equation by -1 and leaving the bottom equation unchanged:

$$3x - 2y = 12 \xrightarrow{\text{Multiply by } -1} -3x + 2y = -12$$
$$3x + 8y = 72 \xrightarrow[\text{No Change}]{} 3x + 8y = 72$$
$$10y = 60$$
$$y = 6$$

We can substitute $y = 6$ into any equation that contains both x and y. Let's use $3x - 2y = 12$.

$$\begin{aligned} 3x - 2(6) &= 12 \\ 3x - 12 &= 12 \\ 3x &= 24 \\ x &= 8 \end{aligned}$$

The solution to the system is (8, 6). ■

The Substitution Method

We end this section by considering another method of solving a linear system. The method is called the *substitution method* and is shown in the following examples.

EXAMPLE 7 Solve the system.

$$\begin{aligned} 2x - 3y &= -6 \\ y &= 3x - 5 \end{aligned}$$

SOLUTION The second equation tells us y is $3x - 5$. Substituting the expression $3x - 5$ for y in the first equation, we have

$$2x - 3(3x - 5) = -6$$

The result of the substitution is the elimination of the variable y. Solving the resulting linear equation in x as usual, we have

$$\begin{aligned} 2x - 9x + 15 &= -6 \\ -7x + 15 &= -6 \\ -7x &= -21 \\ x &= 3 \end{aligned}$$

Putting $x = 3$ into the second equation in the original system, we have

$$\begin{aligned} y &= 3(3) - 5 \\ &= 9 - 5 \\ &= 4 \end{aligned}$$

The solution to the system is (3, 4).

Checking (3, 4) in each equation looks like this:

Substituting $x = 3$ and $y = 4$ into $2x - 3y = -6$, we have

$$\begin{aligned} 2(3) - 3(4) &\stackrel{?}{=} -6 \\ 6 - 12 &\stackrel{?}{=} -6 \\ -6 &= -6 \quad \text{A true statement} \end{aligned}$$

Substituting $x = 3$ and $y = 4$ into $y = 3x - 5$, we have

$$\begin{aligned} 4 &\stackrel{?}{=} 3(3) - 5 \\ 4 &\stackrel{?}{=} 9 - 5 \\ 4 &= 4 \quad \text{A true statement} \end{aligned}$$

Our solution satisfies both equations; therefore, (3, 4) is the solution to our system. ■

Here are the steps to use in solving a system of equations by the substitution method.

HOW TO ***Solve a System of Linear Equations by the Substitution Method***

Step 1: Solve either one of the equations for x or y. (This step is not necessary if one of the equations is already in the correct form, as in Example 7.)

Step 2: Substitute the expression for the variable obtained in Step 1 into the other equation and solve it.

Step 3: Substitute the solution from Step 2 into any equation in the system that contains both variables and solve it.

Step 4: Check your results, if necessary.

EXAMPLE 8 Solve by substitution

$$2x + 3y = 5$$
$$x - 2y = 6$$

Note Both the substitution method and the addition method can be used to solve any system of linear equations in two variables. Systems like the one in Example 7, however, are easier to solve using the substitution method, because one of the variables is already written in terms of the other. A system like the one in Example 6 is easier to solve using the addition method, because solving for one of the variables would lead to an expression involving fractions. The system in Example 8 could be solved easily by either method, because solving the second equation for x is a one-step process.

SOLUTION To use the substitution method, we must solve one of the two equations for x or y. We can solve for x in the second equation by adding $2y$ to both sides:

$$x - 2y = 6$$
$$x = 2y + 6 \quad \text{Add } 2y \text{ to both sides}$$

Substituting the expression $2y + 6$ for x in the first equation of our system, we have

$$2(2y + 6) + 3y = 5$$
$$4y + 12 + 3y = 5$$
$$7y + 12 = 5$$
$$7y = -7$$
$$y = -1$$

Using $y = -1$ in either equation in the original system, we find $x = 4$. The solution is $(4, -1)$.

Modeling: Cell Phone Charges

EXAMPLE 9 The following table shows two monthly contract rates for mobile data devices on Dash's network. At how many gigabytes of data usage is it better to buy the 12GB plan?

	Flat Rate	Plus	Overage Charge
Plan 1 (12 GB)	\$80		\$0
Plan 2 (3 GB)	\$36		\$10/GB

SOLUTION If we let $y =$ the monthly charge for x gigabytes of data usage, the equations for each plan are:

Plan 1: $y = 80$ Plan 2: $y = 36 + 10(x - 3)$

(Note that we multiply 10 by $(x - 3)$ since the $10 overage charge doesn't start until we have used 3 gigabytes.) We can solve this system by substitution by replacing the variable y in Plan 2 with 80 from Plan 1. If we do we have

$$\begin{aligned} 80 &= 36 + 10(x - 3) \\ 80 - 36 &= 10x - 30 \\ 74 &= 10x \\ 7.4 &= x \end{aligned}$$

The monthly bill is based on the amount of data you transmit on your smartphone or tablet. Using our solution, let's see what happens when we use 7 GB and 8 GB of data on the 3 GB plan. (Note that either number on the 12 GB plan would be the same: $80.)

$$\begin{aligned} \text{7 GB: } y &= 36 + 10(x - 3) \\ y &= 36 + 10(4) \\ y &= 76 \end{aligned} \qquad \begin{aligned} \text{8 GB: } y &= 36 + 10(x - 3) \\ y &= 36 + 10(5) \\ y &= 86 \end{aligned}$$

If you consistently find yourself using fewer than 7 GB of data, then you're better off going with the 3 GB plan and paying the overages. If you're using 8 GB or more, then you should purchase the 12 GB plan, even if you never use all of your allowance. ■

Using Technology 7.1

A graphing calculator can be used to solve a system of equations in two variables if the equations intersect in exactly one point.

EXAMPLE 10 Use your graphing calculator to solve the system.

$$\begin{aligned} 2x - 3y &= 4 \\ 4x + 5y &= 3 \end{aligned}$$

SOLUTION To solve this system, we first solve each equation for y. Here is the result:

$$2x - 3y = 4 \quad \text{becomes} \quad y = \frac{4 - 2x}{-3}$$

$$4x + 5y = 3 \quad \text{becomes} \quad y = \frac{3 - 4x}{5}$$

Your y-variables list will look like this (be sure to pay attention to the parentheses).

$$Y1 = (4 - 2x)/-3$$

$$Y2 = (3 - 4x)/5$$

Graphing these two functions on the calculator gives a diagram similar to the one in Figure 6.

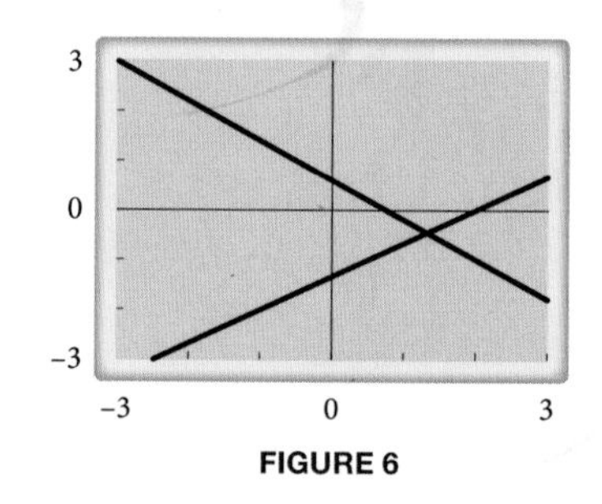

FIGURE 6

Using the Trace and Zoom features, or the Intersect key, we find that the two lines intersect at $x = 1.32$ and $y = -0.45$, which are the decimal equivalents (accurate to the nearest hundredth) of the fractions found in Example 3. ■

We cannot assume that two lines that look parallel in a calculator window are in fact parallel. If you graph the functions $y = x - 5$ and $y = 0.99x + 2$ in a window where x and y range from -10 to 10, the lines look parallel. We know this is not the case, however, since their slopes are different. As we zoom out repeatedly, the lines begin to look as if they coincide. We know this is not the case, because the two lines have different y-intercepts. To summarize: If we graph two lines on a calculator and the graphs look as if they are parallel or coincide, we should use algebraic methods, not the calculator, to determine the solution to the system.

GETTING READY FOR CLASS

After reading through the preceding section, respond in your own words and in complete sentences.

A. Two independent linear equations, each with the same two variables, form a system of equations. How do we define a solution to this system? That is, what form will a solution have, and what properties does a solution possess?

B. When would substitution be more efficient than the addition method in solving two linear equations?

C. Explain what an inconsistent system of linear equations looks like graphically and what would result algebraically when attempting to solve the system.

D. When might the graphing method of solving a system of equations be more desirable than the other techniques, and when might it be less desirable?

Problem Set 7.1

Solve each system by graphing both equations on the same set of axes and then reading the solution from the graph.

1. $3x - 2y = 6$
   $x - y = 1$

2. $5x - 2y = 10$
   $x - y = -1$

3. $y = \frac{3}{5}x - 3$
   $2x - y = -4$

4. $y = \frac{1}{2}x - 2$
   $2x - y = -1$

5. $y = \frac{1}{2}x$
   $y = -\frac{3}{4}x + 5$

6. $y = \frac{2}{3}x$
   $y = -\frac{1}{3}x + 6$

7. $3x + 3y = -2$
   $y = -x + 4$

8. $2x - y = 5$
   $y = 2x - 5$

Solve each of the following systems by the addition method.

9. $3x + y = 5$
   $3x - y = 3$

10. $-x - y = 4$
    $-x + 2y = -3$

11. $x + 2y = 0$
    $2x - 6y = 5$

12. $x + 3y = 3$
    $2x - 9y = 1$

13. $2x - 5y = 16$
    $4x - 3y = 11$

14. $5x - 3y = -11$
    $7x + 6y = -12$

15. $6x + 3y = -1$
    $9x + 5y = 1$

16. $5x + 4y = -1$
    $7x + 6y = -2$

17. $4x + 3y = 14$
    $9x - 2y = 14$

18. $7x - 6y = 13$
    $6x - 5y = 11$

19. $2x - 5y = 3$
    $-4x + 10y = 3$

20. $-3x - 2y = -1$
    $-6x + 4y = -2$

21. $\frac{1}{2}x + \frac{1}{3}y = 13$
    $\frac{2}{5}x + \frac{1}{4}y = 10$

22. $\frac{1}{2}x + \frac{1}{3}y = \frac{2}{3}$
    $\frac{2}{3}x + \frac{2}{5}y = \frac{14}{15}$

23. $\frac{2}{3}x + \frac{2}{5}y = 4$
    $\frac{1}{3}x - \frac{1}{2}y = -\frac{1}{3}$

24. $\frac{1}{2}x - \frac{1}{3}y = \frac{5}{6}$
    $-\frac{2}{5}x + \frac{1}{2}y = -\frac{9}{10}$

Solve each of the following systems by the substitution method.

25. $7x - y = 24$
$x = 2y + 9$

26. $3x - y = -8$
$y = 6x + 3$

27. $6x - y = 10$
$y = -\frac{3}{4}x - 1$

28. $2x - y = 6$
$y = -\frac{4}{3}x + 1$

29. $y = 3x - 2$
$y = 4x - 4$

30. $y = 5x - 2$
$y = -2x + 5$

31. $2x - y = 5$
$4x - 2y = 10$

32. $-10x + 8y = -6$
$y = \frac{5}{4}x$

33. $\frac{1}{3}x - \frac{1}{2}y = 0$
$x = \frac{3}{2}y$

34. $\frac{2}{5}x - \frac{2}{3}y = 0$
$y = \frac{3}{5}x$

You may want to read Example 3 again before solving the systems that follow.

35. $4x - 7y = 3$
$5x + 2y = -3$

36. $3x - 4y = 7$
$6x - 3y = 5$

37. $9x - 8y = 4$
$2x + 3y = 6$

38. $4x - 7y = 10$
$-3x + 2y = -9$

39. $3x - 5y = 2$
$7x + 2y = 1$

40. $4x - 3y = -1$
$5x + 8y = 2$

Solve each of the following systems by using either the addition or substitution method. Choose the method that is most appropriate for the problem.

41. $x - 3y = 7$
$2x + y = -6$

42. $2x - y = 9$
$x + 2y = -11$

43. $y = \frac{1}{2}x + \frac{1}{3}$
$y = -\frac{1}{3}x + 2$

44. $y = \frac{3}{4}x - \frac{4}{5}$
$y = \frac{1}{2}x - \frac{1}{2}$

45. $3x - 4y = 12$
$x = \frac{2}{3}y - 4$

46. $-5x + 3y = -15$
$x = \frac{4}{5}y - 2$

47. $4x - 3y = -7$
$-8x + 6y = -11$

48. $3x - 4y = 8$
$y = \frac{3}{4}x - 2$

49. $3x + y = 17$
$5x + 20y = 65$

50. $x + y = 850$
$1.5x + y = 1{,}100$

51. $\frac{3}{4}x - \frac{1}{3}y = 1$
$y = \frac{1}{4}x$

52. $-\frac{2}{3}x + \frac{1}{2}y = -1$
$y = -\frac{1}{3}x$

53. $\frac{1}{4}x - \frac{1}{2}y = \frac{1}{3}$
$\frac{1}{3}x - \frac{1}{4}y = -\frac{2}{3}$

54. $\frac{1}{5}x - \frac{1}{10}y = -\frac{1}{5}$
$\frac{2}{3}x - \frac{1}{2}y = -\frac{1}{6}$

Paying Attention to Instructions The next two problems are intended to give you practice reading, and paying attention to, the instructions that accompany the problems you are working.

55. Work each problem according to the instructions given.
  a. Simplify: $(3x - 4y) - 3(x - y)$
  b. Find y when x is 0 in $3x - 4y = 8$.
  c. Find the y-intercept: $3x - 4y = 8$
  d. Graph: $3x - 4y = 8$
  e. Find the point where the graphs of $3x - 4y = 8$ and $x - y = 2$ cross

56. Work each problem according to the instructions given.
  a. Solve: $4x - 5 = 20$
  b. Solve for y: $4x - 5y = 20$
  c. Solve for x: $x - y = 5$
  d. Solve the system:

$$4x - 5 = 20$$
$$x - y = 5$$

57. Multiply both sides of the second equation in the following system by 100, and then solve as usual.

$$x + y = 10{,}000$$
$$0.06x + 0.05y = 560$$

58. What value of c will make the following equations dependent?

$$6x - 9y = 3$$
$$4x - 6y = c$$

59. Where do the graphs of the lines $x + y = 4$ and $x - 2y = 4$ intersect?

60. Where do the graphs of the lines $x = -1$ and $x - 2y = 4$ intersect?

Modeling Practice

© Sean Locke/iStockPhoto

61. **Job Comparison** Jane is deciding between two sales positions. She can work for Marcy's and receive \$8.00 per hour, or she can work for Gigi's, where she earns \$6.00 per hour but also receives a \$50 commission per week. The two lines in the following figure represent the money Jane will make for working at each of the jobs.

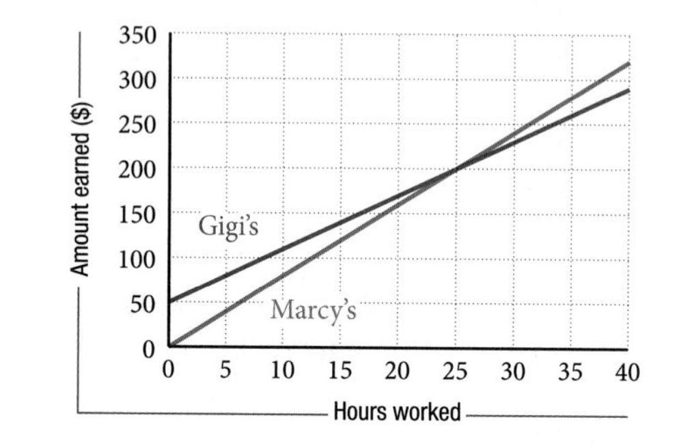

a. From the figure, how many hours would Jane have to work to earn the same amount at each of the positions?
b. If Jane expects to work less than 20 hours a week, which job should she choose?
c. If Jane expects to work more than 30 hours a week, which job should she choose?

62. **Truck Rental** You need to rent a moving truck for two days. Rider Moving Trucks charges \$50 per day and \$0.50 per mile. UMove Trucks charges \$45 per day and \$0.75 per mile. The following figure represents the cost of renting each of the trucks for two days.

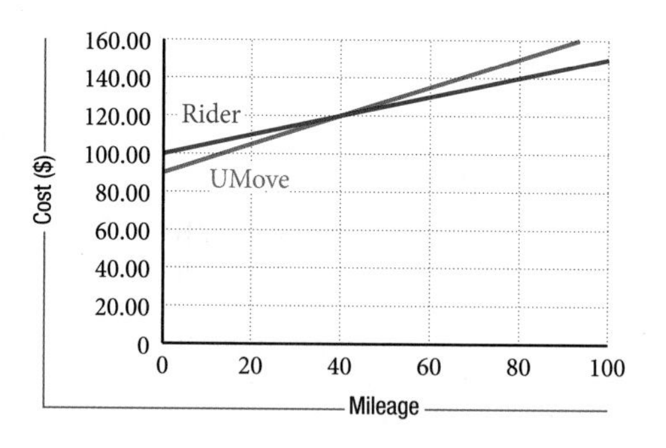

a. From the figure, after how many miles would the trucks cost the same?
b. Which company will give you a better deal if you drive less than 30 miles?
c. Which company will give you a better deal if you drive more than 60 miles?

Using Technology Exercises

Use your graphing calculator to solve each system. Round answers to two decimal places when necessary.

63. $1.4x - 1.2y = 2.6$
$1.2x - y = 2.2$

64. $100x - 300y = 700$
$200x + 100y = -600$

65. $20x + 12y = -120$
$10x - 15y = -10$

66. $0.1x + 0.08y = -0.02$
$0.14x + 0.12y = -0.04$

Getting Ready for the Next Section

Simplify.

67. $2 - 2(6)$

68. $2(1) - 2 + 3$

69. $(x + 3y) - 1(x - 2z)$

70. $(x + y + z) + (2x - y + z)$

Solve.

71. $-9y = -9$

72. $30x = 38$

73. $3(1) + 2z = 9$

74. $4\left(\frac{19}{15}\right) - 2y = 4$

Apply the distributive property, then simplify if possible.

75. $2(5x - z)$

76. $-1(x - 2z)$

77. $3(3x + y - 2z)$

78. $2(2x - y + z)$

Systems of Linear Equations in Three Variables

A solution to an equation in three variables such as

$$2x + y - 3z = 6$$

is an ordered triple of numbers (x, y, z). For example, the ordered triples $(0, 0, -2)$, $(2, 2, 0)$, and $(0, 9, 1)$ are solutions to the equation $2x + y - 3z = 6$, because they produce a true statement when their coordinates are substituted for x, y, and z in the equation.

DEFINITION *Solution set*

The *solution set* for a system of three linear equations in three variables is the set of ordered triples that satisfies all three equations.

VIDEO EXAMPLES

SECTION 7.2

EXAMPLE 1 Solve the system.

$$\begin{aligned} x + y + z &= 6 \quad (1) \\ 2x - y + z &= 3 \quad (2) \\ x + 2y - 3z &= -4 \quad (3) \end{aligned}$$

SOLUTION We want to find the ordered triple (x, y, z) that satisfies all three equations. We have numbered the equations so it will be easier to keep track of where they are and what we are doing.

There are many ways to proceed. The main idea is to take two different pairs of equations and eliminate the same variable from each pair. We begin by adding equations (1) and (2) to eliminate the y-variable. The resulting equation is numbered (4):

$$\begin{array}{rl} x + y + z = 6 & (1) \\ 2x - y + z = 3 & (2) \\ \hline 3x \quad + 2z = 9 & (4) \end{array}$$

Adding twice equation (2) to equation (3) will also eliminate the variable y. The resulting equation is numbered (5):

$$\begin{array}{rl} 4x - 2y + 2z = 6 & \text{Twice (2)} \\ x + 2y - 3z = -4 & (3) \\ \hline 5x \quad - z = 2 & (5) \end{array}$$

Equations (4) and (5) form a linear system in two variables. By multiplying equation (5) by 2 and adding the result to equation (4), we succeed in eliminating the variable z from the new pair of equations:

$$\begin{array}{rl} 3x + 2z = 9 & (4) \\ 10x - 2z = 4 & \text{Twice (5)} \\ \hline 13x \quad = 13 & \\ x \quad = 1 & \end{array}$$

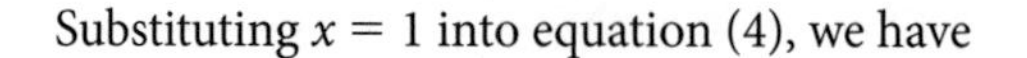

Substituting $x = 1$ into equation (4), we have

$$\begin{aligned} 3(1) + 2z &= 9 \\ 2z &= 6 \\ z &= 3 \end{aligned}$$

Using $x = 1$ and $z = 3$ in equation (1) gives us

$$\begin{aligned} 1 + y + 3 &= 6 \\ y + 4 &= 6 \\ y &= 2 \end{aligned}$$

The solution is the ordered triple (1, 2, 3). ■

EXAMPLE 2 Solve the system.

$$\begin{aligned} 2x + y - z &= 3 \quad (1) \\ 3x + 4y + z &= 6 \quad (2) \\ 2x - 3y + z &= 1 \quad (3) \end{aligned}$$

SOLUTION It is easiest to eliminate z from the equations. The equation produced by adding (1) and (2) is

$$5x + 5y = 9 \qquad (4)$$

The equation that results from adding (1) and (3) is

$$4x - 2y = 4 \qquad (5)$$

Equations (4) and (5) form a linear system in two variables. We can eliminate the variable y from this system as follows:

$$\begin{array}{rcrl} 5x + 5y = 9 & \xrightarrow{\text{Multiply by 2}} & 10x + 10y &= 18 \\ 4x - 2y = 4 & \xrightarrow[\text{Multiply by 5}]{} & 20x - 10y &= 20 \\ \hline & & 30x &= 38 \\ & & x &= \dfrac{38}{30} \\ & & &= \dfrac{19}{15} \end{array}$$

Substituting $x = \frac{19}{15}$ into equation (5) or equation (4) and solving for y gives

$$y = \frac{8}{15}$$

Using $x = \frac{19}{15}$ and $y = \frac{8}{15}$ in equation (1), (2), or (3) and solving for z results in

$$z = \frac{1}{15}$$

The ordered triple that satisfies all three equations is $\left(\frac{19}{15}, \frac{8}{15}, \frac{1}{15}\right)$. ■

EXAMPLE 3 Solve the system.

$$\begin{aligned} 2x + 3y - z &= 5 \quad (1) \\ 4x + 6y - 2z &= 10 \quad (2) \\ x - 4y + 3z &= 5 \quad (3) \end{aligned}$$

SOLUTION Multiplying equation (1) by -2 and adding the result to equation (2) looks like this:

$$\begin{array}{rl} -4x - 6y + 2z = -10 & -2 \text{ times } (1) \\ 4x + 6y - 2z = 10 & (2) \\ \hline 0 = 0 & \end{array}$$

All three variables have been eliminated, and we are left with a true statement. This implies that the two equations are dependent. With a system of three equations in three variables, however, if two of the equations are dependent, the system can have no solution or an infinite number of solutions. After we have concluded the examples in this section, we will discuss the geometry behind these systems. ■

EXAMPLE 4 Solve the system.

$$\begin{array}{rl} x - 5y + 4z = 8 & (1) \\ 3x + y - 2z = 7 & (2) \\ -9x - 3y + 6z = 5 & (3) \end{array}$$

SOLUTION Multiplying equation (2) by 3 and adding the result to equation (3) produces

$$\begin{array}{rl} 9x + 3y - 6z = 21 & 3 \text{ times } (2) \\ -9x - 3y + 6z = 5 & (3) \\ \hline 0 = 26 & \end{array}$$

In this case, all three variables have been eliminated, and we are left with a false statement. There are no ordered triples that satisfy both equations. The solution set for the system is the empty set, $\varnothing$. If equations (2) and (3) have no ordered triples in common, then certainly (1), (2), and (3) do not either. ■

EXAMPLE 5 Solve the system.

$$\begin{array}{rl} x + 3y = 5 & (1) \\ 6y + z = 12 & (2) \\ x - 2z = -10 & (3) \end{array}$$

SOLUTION It may be helpful to rewrite the system as

$$\begin{array}{rrrrl} x & + 3y & & = 5 & (1) \\ & 6y & + z & = 12 & (2) \\ x & & - 2z & = -10 & (3) \end{array}$$

Equation (2) does not contain the variable x. If we multiply equation (3) by -1 and add the result to equation (1), we will be left with another equation that does not contain the variable x:

$$\begin{array}{rrrrl} x & + 3y & & = 5 & (1) \\ -x & & + 2z & = 10 & -1 \text{ times } (3) \\ \hline & 3y & + 2z & = 15 & (4) \end{array}$$

Equations (2) and (4) form a linear system in two variables. Multiplying equation (2) by -2 and adding the result to equation (4) eliminates the variable z:

$$\begin{array}{rcl} 6y + z = 12 & \xrightarrow{\text{Multiply by } -2} & -12y - 2z = -24 \\ 3y + 2z = 15 & \xrightarrow[\text{No Change}]{} & \underline{3y + 2z = 15} \\ & & -9y = -9 \\ & & y = 1 \end{array}$$

Using $y = 1$ in equation (4) and solving for z, we have

$$z = 6$$

Substituting $y = 1$ into equation (1) gives

$$x = 2$$

The ordered triple that satisfies all three equations is (2, 1, 6). ■

Modeling: Electric Circuit

EXAMPLE 6 In the following diagram of an electrical circuit, x, y, and z represent the amount of current (in amperes) flowing across the 5-ohm, 20-ohm, and 10-ohm resistors, respectively. (In circuit diagrams, resistors are represented by —/\/\/\— and potential differences by —| |—.)

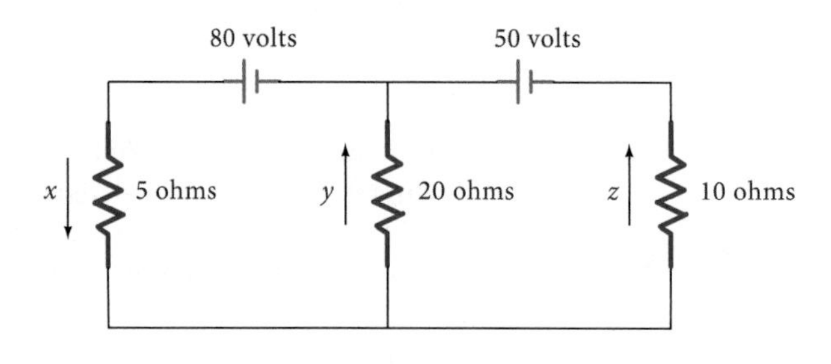

The system of equations used to find the three currents x, y, and z is

$$x - y - z = 0 \qquad (1)$$

$$5x + 20y = 80 \qquad (2)$$

$$20y - 10z = 50 \qquad (3)$$

Solve the system for all variables.

SOLUTION If we solve Equation 2 for x, we have

$$x = 16 - 4y \qquad (4)$$

then, solving Equation 3 for z, we have

$$z = 2y - 5 \qquad (5)$$

Substituting Equations 4 and 5 into Equation 1, we have an equation that contains y only.

$$\begin{aligned} (16 - 4y) - y - (2y - 5) &= 0 \\ -7y + 21 &= 0 \\ -7y &= -21 \\ y &= 3 \end{aligned}$$

Substituting this value of y back into Equations 4 and 5, we have

$$x = 16 - 4(3) = 4$$

$$z = 2(3) - 5 = 1$$

The three currents, x, y, and z are

4 amps, 3 amps, and 1 amp.

The Geometry Behind Equations in Three Variables

We can graph an ordered triple on a coordinate system with three axes. The graph will be a point in space. The coordinate system is drawn in perspective; you have to imagine that the x-axis comes out of the paper and is perpendicular to both the y-axis and the z-axis. To graph the point (3, 4, 5), we move 3 units in the x-direction, 4 units in the y-direction, and then 5 units in the z-direction, as shown in Figure 1.

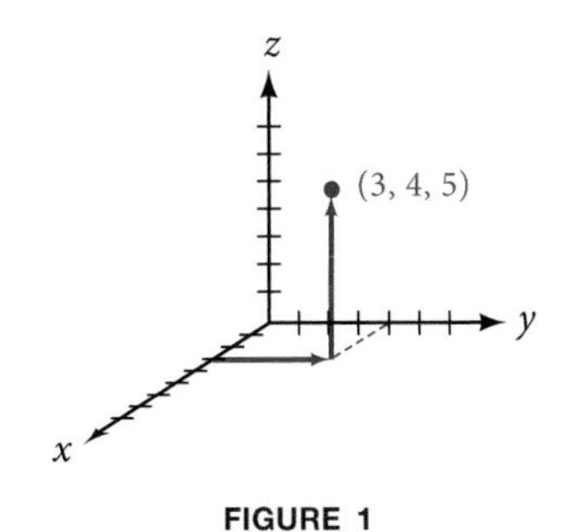

FIGURE 1

Although in actual practice it is sometimes difficult to graph equations in three variables, if we were to graph a linear equation in three variables, we would find that the graph was a plane in space. A system of three equations in three variables is represented by three planes in space.

There are a number of possible ways in which these three planes can intersect, some of which are shown below. And there are still other possibilities that are not among those shown.

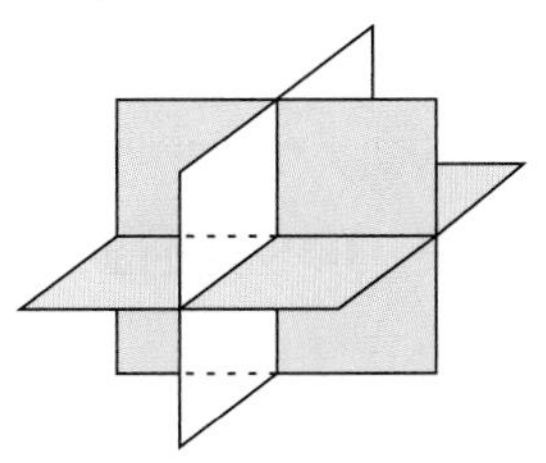

CASE 1 *The three planes have exactly one point in common. In this case we get one solution to our system, as in Examples 1, 2, and 5.*

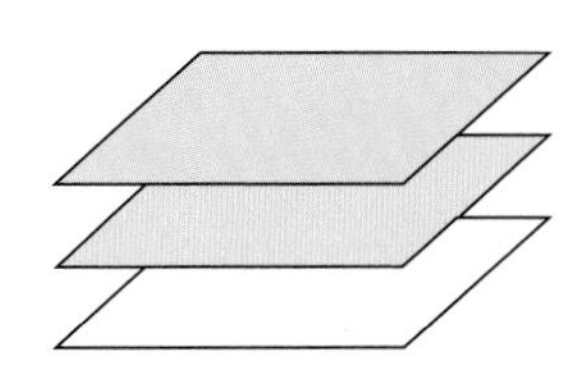

CASE 2 *The three planes have no points in common because they are all parallel to one another. The system they represent is an inconsistent system.*

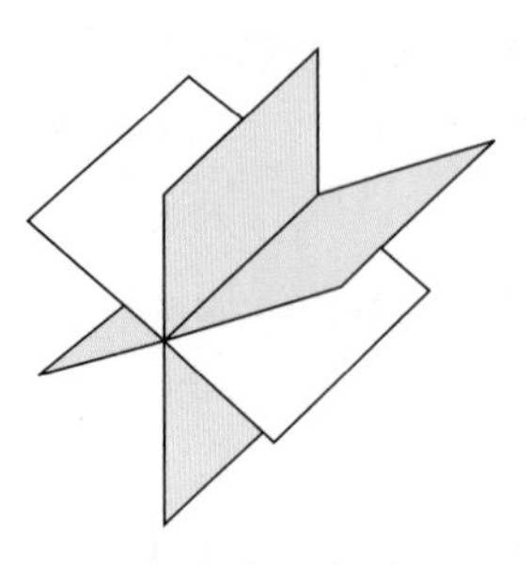

CASE 3 *The three planes intersect in a line. Any point on the line is a solution to the system of equations represented by the planes, so there is an infinite number of solutions to the system. This is an example of a dependent system.*

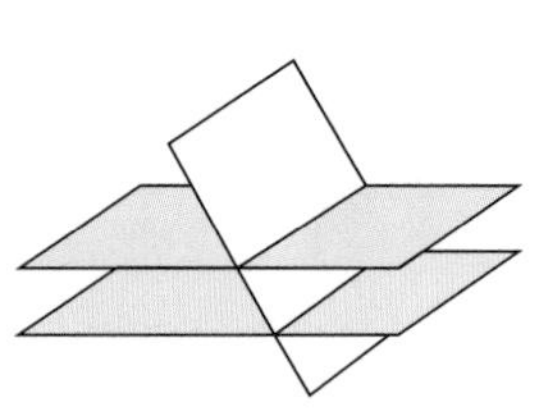

CASE 4 *Two of the planes are parallel; the third plane intersects each of the parallel planes. In this case, the three planes have no points in common. There is no solution to the system; it is an inconsistent system.*

In Example 3, we found that equations (1) and (2) were dependent equations. They represent the same plane. That is, they have all their points in common. But the system of equations that they came from has either no solution or an infinite number of solutions. It all depends on the third plane. If the third plane coincides with the first two, then the solution to the system is a plane. If the third plane is parallel to the first two, then there is no solution to the system. Finally, if the third plane intersects the first two but does not coincide with them, then the solution to the system is that line of intersection.

In Example 4 we found that trying to eliminate a variable from the second and third equations resulted in a false statement. This means that the two planes represented by these equations are parallel. It makes no difference where the third plane is; there is no solution to the system in Example 4. (If we were to graph the three planes from Example 4, we would obtain a diagram similar to Case 2 or Case 4.)

If, in the process of solving a system of linear equations in three variables, we eliminate all the variables from a pair of equations and are left with a false statement, we will say the system is inconsistent. If we eliminate all the variables and are left with a true statement, then we will say that there is no unique solution to the system; there are either infinitely many solutions to the system, which could form a line or a plane, or no solution. We do not know which situation has occurred unless we consider the third equation.

Using Technology 7.2

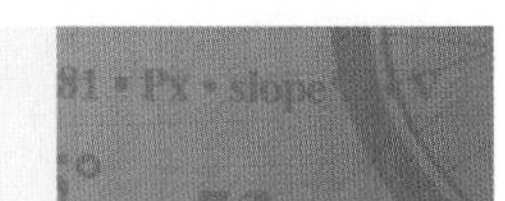

We can use Wolfram|Alpha to solve the system of equations in Example 6. We simply separate the equations with commas.

EXAMPLE 7 Use Wolfram|Alpha to solve the system of equations in Example 6.

SOLUTION Online, go to www.wolframalpha.com, and enter the information below, into the box on the website, exactly as it is shown.

solve x-y-z=0, 5x+20y=80, 20y-10z=50

Here is the result (notice that it matches the solutions we obtained for Example 6):

Input interpretation:

	$x - y - z = 0$
solve	$5x + 20y = 80$
	$20y - 10z = 50$

Result:

$x = 4$ and $y = 3$ and $z = 1$

Wolfram Alpha LLC. 2012. Wolfram|Alpha
http://www.wolframalpha.com/
(access June 18, 2012)

The equations associated with the diagram in Example 6 are from Kirchhoff's Law. Look this up in Wolfram|Alpha.

GETTING READY FOR CLASS

After reading through the preceding section, respond in your own words and in complete sentences.

A. What is an ordered triple of numbers?

B. Explain what it means for (1, 2, 3) to be a solution to a system of linear equations in three variables.

C. Explain in a general way the procedure you would use to solve a system of three linear equations in three variables.

D. How do you know when a system of linear equations in three variables has no solution?

Problem Set 7.2

Solve the following systems.

1. $x + y + 2z = 5$
$x - y + 2z = 1$
$x - y - 3z = -4$

2. $x - y - 2z = -1$
$x + y + z = 6$
$x + y - z = 4$

3. $x + y + z = 6$
$x - y + 2z = 7$
$2x - y - 4z = -9$

4. $x + y + z = 0$
$x + y - z = 6$
$x - y + 2z = -7$

5. $x + 2y + z = 3$
$2x - y + 2z = 6$
$3x + y - z = 5$

6. $2x + y - 3z = -14$
$x - 3y + 4z = 22$
$3x + 2y + z = 0$

7. $2x + 3y - 2z = 4$
$x + 3y - 3z = 4$
$3x - 6y + z = -3$

8. $4x + y - 2z = 0$
$2x - 3y + 3z = 9$
$-6x - 2y + z = 0$

9. $-x + 4y - 3z = 2$
$2x - 8y + 6z = 1$
$3x - y + z = 0$

10. $4x + 6y - 8z = 1$
$-6x - 9y + 12z = 0$
$x - 2y - 2z = 3$

11. $\frac{1}{2}x - y + z = 0$
$2x + \frac{1}{3}y + z = 2$
$x + y + z = -4$

12. $\frac{1}{3}x + \frac{1}{2}y + z = -1$
$x - y + \frac{1}{5}z = -1$
$x + y + z = -5$

13. $2x - y - 3z = 1$
$x + 2y + 4z = 3$
$4x - 2y - 6z = 2$

14. $3x + 2y + z = 3$
$x - 3y + z = 4$
$-6x - 4y - 2z = 1$

15. $2x - y + 3z = 4$
$x + 2y - z = -3$
$4x + 3y + 2z = -5$

16. $6x - 2y + z = 5$
$3x + y + 3z = 7$
$x + 4y - z = 4$

17. $x + y = 9$
$y + z = 7$
$x - z = 2$

18. $x - y = -3$
$x + z = 2$
$y - z = 7$

19. $2x + y = 2$
$y + z = 3$
$4x - z = 0$

20. $2x + y = 6$
$3y - 2z = -8$
$x + z = 5$

21. $2x - 3y = 0$
$6y - 4z = 1$
$x + 2z = 1$

22. $3x + 2y = 3$
$y + 2z = 2$
$6x - 4z = 1$

23. $x + y - z = 2$
$2x + y + 3z = 4$
$x - 2y + 2z = 6$

24. $x + 2y - 2z = 4$
$3x + 4y - z = -2$
$2x + 3y - 3z = -5$

25. $2x + 3y = -\frac{1}{2}$
$4x + 8z = 2$
$3y + 2z = -\frac{3}{4}$

26. $3x - 5y = 2$
$4x + 6z = \frac{1}{3}$
$5y - 7z = \frac{1}{6}$

27. $\frac{1}{3}x + \frac{1}{2}y - \frac{1}{6}z = 4$
$\frac{1}{4}x - \frac{3}{4}y + \frac{1}{2}z = \frac{3}{2}$
$\frac{1}{2}x - \frac{2}{3}y - \frac{1}{4}z = -\frac{16}{3}$

28. $-\frac{1}{4}x + \frac{3}{8}y + \frac{1}{2}z = -1$
$\frac{2}{3}x - \frac{1}{6}y - \frac{1}{2}z = 2$
$\frac{3}{4}x - \frac{1}{2}y - \frac{1}{8}z = 1$

29. $x - \frac{1}{2}y - \frac{1}{3}z = -\frac{4}{3}$
$\frac{1}{3}x - \frac{1}{2}z = 5$
$-\frac{1}{4}x + \frac{2}{3}y - z = -\frac{3}{4}$

30. $x + \frac{1}{3}y - \frac{1}{2}z = -\frac{3}{2}$
$\frac{1}{2}x - y + \frac{1}{3}z = 8$
$\frac{1}{3}x - \frac{1}{4}y - z = -\frac{5}{6}$

Modeling Practice

© Paul Velgos/iStockPhoto

31. Cost of a Rental Car If a car rental company charges \$10 a day and 8¢ a mile to rent one of its cars, then the cost z, in dollars, to rent a car for x days and drive y miles can be found from the equation

$$z = 10x + 0.08y$$

a. How much does it cost to rent a car for 2 days and drive it 200 miles under these conditions?

b. A second company charges \$12 a day and 6¢ a mile for the same car. Write an equation that gives the cost z, in dollars, to rent a car from this company for x days and drive it y miles.

c. A car is rented from each of the companies mentioned in Parts a and b for 2 days. To find the mileage at which the cost of renting the cars from each of the two companies will be equal, solve the following system for y:

$$z = 10x + 0.08y$$
$$z = 12x + 0.06y$$
$$x = 2$$

32. Moving Costs If you are moving from a 3 to 4 bedroom house, you can rent a moving truck for \$39.95 a day and \$0.59 a mile from a moving company. The cost z, in dollars, to rent a truck for x days and drive y miles can be found from the following equation:

$$z = 39.95x + .59y$$

a. How much does it cost to rent a truck for 1 day and drive it 35 miles to move your furniture to your new house?

b. A second moving company charges \$36.95 a day and \$0.62 a mile for the same truck. If the truck is rented for 2 days, find the mileage at which the cost of renting the trucks from each of the two companies will be equal.

Using Technology Exercises

33. Use Wolfram|Alpha to solve the system of equations in Example 1.

34. Use Wolfram|Alpha to solve the system of equations in Example 2.

35. Use Wolfram|Alpha to solve the system of equations in Example 3.

36. Use Wolfram|Alpha to solve the system of equations in Example 4.

Getting Ready for the Next Section

Simplify.

37. $1(4) - 3(2)$

38. $3(7) - (-2)(5)$

39. $1(0)(1) + 3(1)(4) + (-2)(2)(-1)$

40. $-4(0)(-2) - (-1)(1)(1) - 1(2)(3)$

41. $1[0 - (-1)] - 3(2 - 4) + (-2)(-2 - 0)$

42. $-3(-1 - 1) + 4[-2 - (-2)] - 5[2 - (-2)]$

Solve.

43. $x^2 - 2x = 8$

44. $2x^2 + 8 = -8x$

Introduction to Determinants

In this section, we will expand and evaluate *determinants*. The purpose of this section is simply to be able to find the value of a given determinant. As we will see in the next section, determinants are very useful in solving systems of linear equations. Before we apply determinants to systems of linear equations, however, we must practice calculating the value of some determinants.

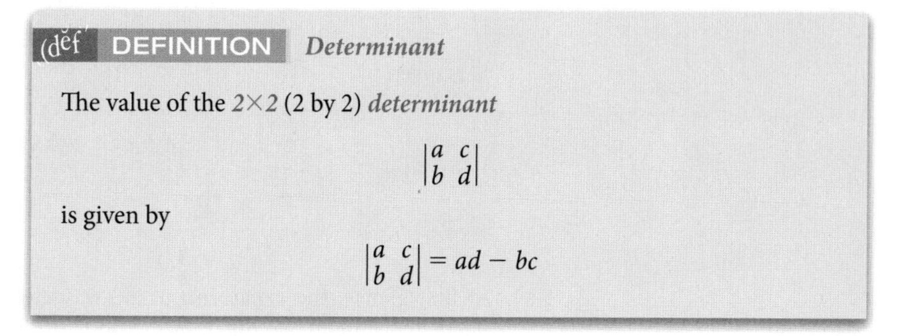

DEFINITION *Determinant*

The value of the 2×2 (2 by 2) *determinant*

$$\begin{vmatrix} a & c \\ b & d \end{vmatrix}$$

is given by

$$\begin{vmatrix} a & c \\ b & d \end{vmatrix} = ad - bc$$

From the preceding definition, we see that a determinant is simply a square array of numbers with two vertical lines enclosing it. The value of a 2×2 determinant is found by cross-multiplying on the diagonals and then subtracting, a diagram of which looks like

$$\begin{vmatrix} a & c \\ b & d \end{vmatrix} = ad - bc$$

VIDEO EXAMPLES

SECTION 7.3

EXAMPLE 1 Find the value of the following 2×2 determinants:

a. $\begin{vmatrix} 1 & 2 \\ 3 & 4 \end{vmatrix} = 1(4) - 3(2) = 4 - 6 = -2$

b. $\begin{vmatrix} 3 & 5 \\ -2 & 7 \end{vmatrix} = 3(7) - (-2)5 = 21 + 10 = 31$

EXAMPLE 2 Solve for x if

$$\begin{vmatrix} x^2 & 2 \\ x & 1 \end{vmatrix} = 8$$

SOLUTION We expand the determinant on the left side to get

$$x^2(1) - x(2) = 8$$

$$x^2 - 2x = 8$$

$$x^2 - 2x - 8 = 0$$

$$(x - 4)(x + 2) = 0$$

$$x - 4 = 0 \quad \text{or} \quad x + 2 = 0$$

$$x = 4 \quad \text{or} \quad x = -2$$

We now turn our attention to 3×3 determinants. A 3×3 determinant is also a square array of numbers, the value of which is given by the following definition.

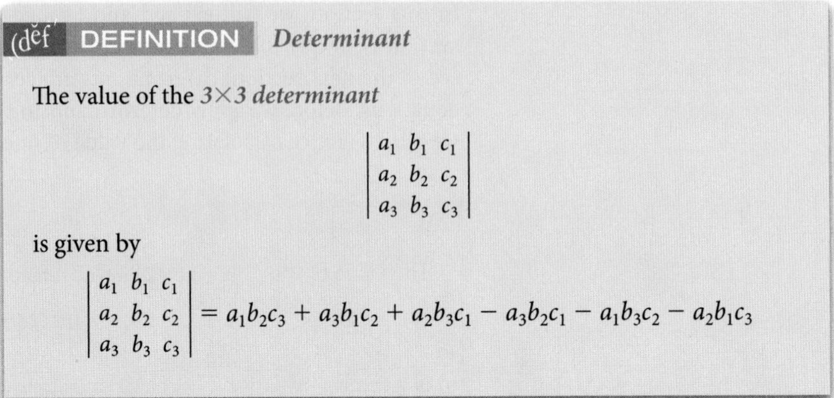

DEFINITION *Determinant*

The value of the *3×3 determinant*

$$\begin{vmatrix} a_1 & b_1 & c_1 \\ a_2 & b_2 & c_2 \\ a_3 & b_3 & c_3 \end{vmatrix}$$

is given by

$$\begin{vmatrix} a_1 & b_1 & c_1 \\ a_2 & b_2 & c_2 \\ a_3 & b_3 & c_3 \end{vmatrix} = a_1b_2c_3 + a_3b_1c_2 + a_2b_3c_1 - a_3b_2c_1 - a_1b_3c_2 - a_2b_1c_3$$

At first glance, the expansion of a 3×3 determinant looks a little complicated. There are actually two different methods used to find the six products in the preceding definition that simplify matters somewhat.

Method 1

We begin by writing the determinant with the first two columns repeated on the right:

$$\begin{vmatrix} a_1 & b_1 & c_1 \\ a_2 & b_2 & c_2 \\ a_3 & b_3 & c_3 \end{vmatrix} \begin{matrix} a_1 & b_1 \\ a_2 & b_2 \\ a_3 & b_3 \end{matrix}$$

The positive products in the definition come from multiplying down the three full diagonals:

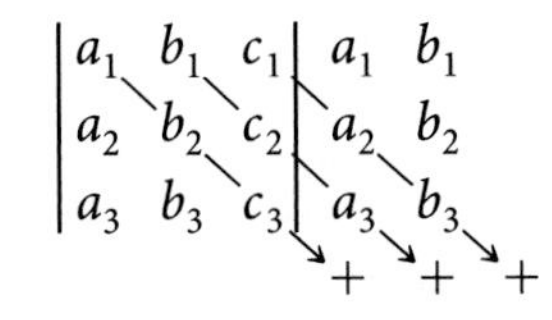

The negative products come from multiplying up the three full diagonals:

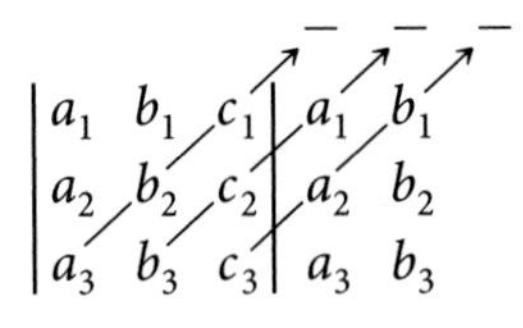

EXAMPLE 3 Find the value of

$$\begin{vmatrix} 1 & 3 & -2 \\ 2 & 0 & 1 \\ 4 & -1 & 1 \end{vmatrix}$$

SOLUTION Repeating the first two columns and then finding the products down the diagonals and the products up the diagonals as given in Method 1, we have

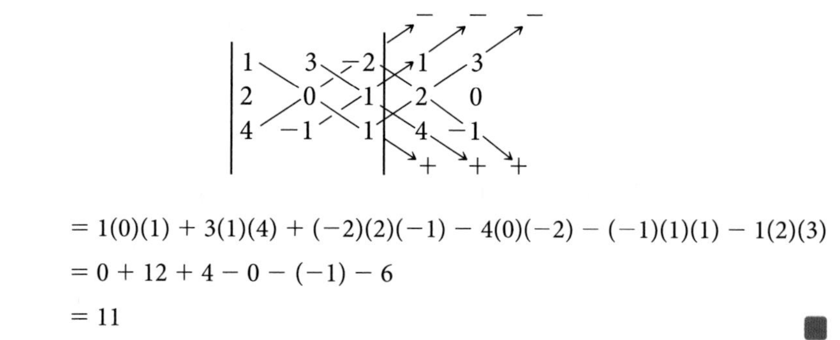

$$= 1(0)(1) + 3(1)(4) + (-2)(2)(-1) - 4(0)(-2) - (-1)(1)(1) - 1(2)(3)$$
$$= 0 + 12 + 4 - 0 - (-1) - 6$$
$$= 11$$

Method 2

The second method of evaluating a 3×3 determinant is called *expansion by minors.*

DEFINITION *Minor*

The *minor* for an element in a 3×3 determinant is the determinant consisting of the elements remaining when the row and column to which the element belongs are deleted. For example, in the determinant

$$\begin{vmatrix} a_1 & b_1 & c_1 \\ a_2 & b_2 & c_2 \\ a_3 & b_3 & c_3 \end{vmatrix}$$

$$\text{Minor for element } a_1 = \begin{vmatrix} b_2 & c_2 \\ b_3 & c_3 \end{vmatrix}$$

$$\text{Minor for element } b_2 = \begin{vmatrix} a_1 & c_1 \\ a_3 & c_3 \end{vmatrix}$$

$$\text{Minor for element } c_3 = \begin{vmatrix} a_1 & b_1 \\ a_2 & b_2 \end{vmatrix}$$

Before we can evaluate a 3×3 determinant by Method 2, we must first define what is known as the sign array for a 3×3 determinant.

DEFINITION *Sign array*

The *sign array* for a 3×3 determinant is a 3×3 array of signs in the following pattern:

$$\begin{vmatrix} + & - & + \\ - & + & - \\ + & - & + \end{vmatrix}$$

The sign array begins with a + sign in the upper left-hand corner. The signs then alternate between + and − across every row and down every column.

> *Note* This method of evaluating a determinant is actually more valuable than our first method, because it works with any size determinant from 3×3 to 4×4 to any higher order determinant. Method 1 works only on 3×3 determinants. It cannot be used on a 4×4 determinant.

HOW TO ***Evaluate a 3×3 Determinant by Expansion of Minors***

We can evaluate a 3×3 determinant by expanding across any row or down any column as follows:

Step 1: Choose a row or column to expand about.

Step 2: Write the product of each element in the row or column chosen in Step 1 with its minor.

Step 3: Connect the three products in Step 2 with the signs in the corresponding row or column in the sign array.

To illustrate the procedure, we will use the same determinant we used in Example 3.

EXAMPLE 4 Expand across the first row:

$$\begin{vmatrix} 1 & 3 & -2 \\ 2 & 0 & 1 \\ 4 & -1 & 1 \end{vmatrix}$$

SOLUTION The products of the three elements in row 1 with their minors are

$$1\begin{vmatrix} 0 & 1 \\ -1 & 1 \end{vmatrix} \qquad 3\begin{vmatrix} 2 & 1 \\ 4 & 1 \end{vmatrix} \qquad (-2)\begin{vmatrix} 2 & 0 \\ 4 & -1 \end{vmatrix}$$

Connecting these three products with the signs from the first row of the sign array, we have

$$+1\begin{vmatrix} 0 & 1 \\ -1 & 1 \end{vmatrix} \quad -3\begin{vmatrix} 2 & 1 \\ 4 & 1 \end{vmatrix} \quad +(-2)\begin{vmatrix} 2 & 0 \\ 4 & -1 \end{vmatrix}$$

We complete the problem by evaluating each of the three 2×2 determinants and then simplifying the resulting expression:

$$\begin{aligned} &+1[0 - (-1)] - 3(2 - 4) + (-2)(-2 - 0) \\ &= 1(1) - 3(-2) + (-2)(-2) \\ &= 1 + 6 + 4 \\ &= 11 \end{aligned}$$

■

The results of Examples 3 and 4 match. It makes no difference which method we use—the value of a 3×3 determinant is unique.

EXAMPLE 5 Expand down column 2:

$$\begin{vmatrix} 2 & 3 & -2 \\ 1 & 4 & 1 \\ 1 & 5 & -1 \end{vmatrix}$$

SOLUTION We connect the products of elements in column 2 and their minors with the signs from the second column in the sign array:

$$\begin{vmatrix} 2 & 3 & -2 \\ 1 & 4 & 1 \\ 1 & 5 & -1 \end{vmatrix} = -3\begin{vmatrix} 1 & 1 \\ 1 & -1 \end{vmatrix} + 4\begin{vmatrix} 2 & -2 \\ 1 & -1 \end{vmatrix} - 5\begin{vmatrix} 2 & -2 \\ 1 & 1 \end{vmatrix}$$

$$= -3(-1 - 1) + 4[-2 - (-2)] - 5[2 - (-2)]$$

$$= -3(-2) + 4(0) - 5(4)$$

$$= 6 + 0 - 20$$

$$= -14$$

Modeling: Equations from Determinants

EXAMPLE 6 Show that the following determinant equation is another way to write the equation $F = \frac{9}{5}C + 32$.

$$\begin{vmatrix} C & F & 1 \\ 5 & 41 & 1 \\ -10 & 14 & 1 \end{vmatrix} = 0$$

SOLUTION We expand the left side of the equation across the top row. Then we solve for F.

$$C(41 - 14) - F(5 + 10) + 1(70 + 410) = 0$$

$$27C - 15F + 480 = 0$$

$$15F = 27C + 480$$

$$F = \frac{27}{15}C + \frac{480}{15}$$

$$F = \frac{9}{5}C + 32$$

Using Technology 7.3

Your graphing calculator has built-in matrices, that we can use to find the determinants. Let's start by finding the determinant from Example 4 in your text.

EXAMPLE 7 Use your graphing calculator to evaluate this determinant.

$$\begin{vmatrix} 1 & 3 & -2 \\ 2 & 0 & 1 \\ 4 & -1 & 1 \end{vmatrix}$$

SOLUTION We first access the matrix screen on our calculator, then we tell the calculator we have a matrix with three rows and three columns. Once we have that set up, we can enter the numbers in the determinant above and ask the calculator to find the determinant. We start by accessing the [MATRIX] menu, to obtain a screen like this:

NAMES MATH EDIT
1: [A] 1×1
2: [B]
3: [C]
4: [D]
5: [E]
6: [F]
7↓[G]

FIGURE 1

Next, scroll to EDIT, then select Matrix A to obtain the screen in Figure 2. The calculator will ask you for the dimensions of the matrix. It is a 3×3 matrix, so enter those numbers to see a screen like the one in Figure 3. Enter the numbers from the determinant in question to see a screen like the one in Figure 4.

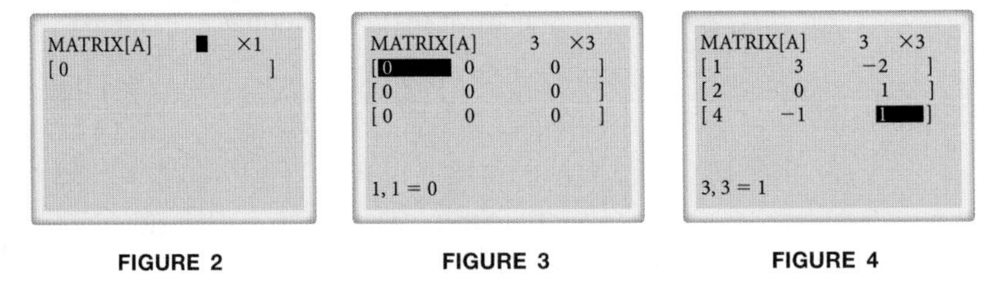

FIGURE 2 **FIGURE 3** **FIGURE 4**

Go back to your home screen and then go to the [MATRIX] menu again and select MATH (Figure 5). Enter 1 to select det(. Next go back to the [MATRIX] menu and select Matrix A. Press [ENTER] twice to see a screen like the one in Figure 7.

NAMES MATH EDIT
1: det(
2: T
3: dim(
4: Fill(
5: identity(
6: randM(
7↓augment(

NAMES MATH EDIT
1: [A] 3×3
2: [B]
3: [C]
4: [D]
5: [E]
6: [F]
7↓[G]

det([A])
11

FIGURE 5 **FIGURE 6** **FIGURE 7**

As you can see, the result is 11, which is consistent with the result we obtained in Example 4.

GETTING READY FOR CLASS

After reading through the preceding section, respond in your own words and in complete sentences.

A. What is a determinant?
B. How do you evaluate a 2 X 2 determinant?
C. How do you construct a sign array for a 3 X 3 determinant?
D. What is the minor for the first element in a 3 X 3 determinant?

Problem Set 7.3

Find the value of the following 2×2 determinants.

1. $\begin{vmatrix} 1 & 0 \\ 2 & 3 \end{vmatrix}$ **2.** $\begin{vmatrix} 5 & 4 \\ 3 & 2 \end{vmatrix}$ **3.** $\begin{vmatrix} 2 & 1 \\ 3 & 4 \end{vmatrix}$ **4.** $\begin{vmatrix} 4 & 1 \\ 5 & 2 \end{vmatrix}$

5. $\begin{vmatrix} 0 & 1 \\ 1 & 0 \end{vmatrix}$ **6.** $\begin{vmatrix} 1 & 0 \\ 0 & 1 \end{vmatrix}$ **7.** $\begin{vmatrix} -3 & 2 \\ 6 & -4 \end{vmatrix}$ **8.** $\begin{vmatrix} 8 & -3 \\ -2 & 5 \end{vmatrix}$

9. $\begin{vmatrix} -3 & -1 \\ 4 & -2 \end{vmatrix}$ **10.** $\begin{vmatrix} 5 & 3 \\ 7 & -6 \end{vmatrix}$

Solve each of the following for x.

11. $\begin{vmatrix} 2x & 1 \\ x & 3 \end{vmatrix} = 10$ **12.** $\begin{vmatrix} 3x & -2 \\ 2x & 3 \end{vmatrix} = 26$ **13.** $\begin{vmatrix} 1 & 2x \\ 2 & -3x \end{vmatrix} = 21$

14. $\begin{vmatrix} -5 & 4x \\ 1 & -x \end{vmatrix} = 27$ **15.** $\begin{vmatrix} 2x & -4 \\ 2 & x \end{vmatrix} = -8x$ **16.** $\begin{vmatrix} 3x & 2 \\ 2 & x \end{vmatrix} = -11x$

17. $\begin{vmatrix} x^2 & 3 \\ x & 1 \end{vmatrix} = 10$ **18.** $\begin{vmatrix} x^2 & -2 \\ x & 1 \end{vmatrix} = 35$ **19.** $\begin{vmatrix} x^2 & -4 \\ x & 1 \end{vmatrix} = 32$

20. $\begin{vmatrix} x^2 & 6 \\ x & 1 \end{vmatrix} = 72$ **21.** $\begin{vmatrix} x & 5 \\ 1 & x \end{vmatrix} = 4$ **22.** $\begin{vmatrix} 3x & 4 \\ 2 & x \end{vmatrix} = 10x$

Find the value of each of the following 3×3 determinants by using Method 1.

23. $\begin{vmatrix} 1 & 2 & 0 \\ 0 & 2 & 1 \\ 1 & 1 & 1 \end{vmatrix}$ **24.** $\begin{vmatrix} -1 & 0 & 2 \\ 3 & 0 & 1 \\ 0 & 1 & 3 \end{vmatrix}$ **25.** $\begin{vmatrix} 1 & 2 & 3 \\ 3 & 2 & 1 \\ 1 & 1 & 1 \end{vmatrix}$ **26.** $\begin{vmatrix} -1 & 2 & 0 \\ 3 & -2 & 1 \\ 0 & 5 & 4 \end{vmatrix}$

Find the value of each determinant by using Method 2 and expanding across the first row.

27. $\begin{vmatrix} 0 & 1 & 2 \\ 1 & 0 & 1 \\ -1 & 2 & 0 \end{vmatrix}$ **28.** $\begin{vmatrix} 3 & -2 & 1 \\ 0 & -1 & 0 \\ 2 & 0 & 1 \end{vmatrix}$ **29.** $\begin{vmatrix} 3 & 0 & 2 \\ 0 & -1 & -1 \\ 4 & 0 & 0 \end{vmatrix}$ **30.** $\begin{vmatrix} 1 & 1 & 1 \\ 1 & -1 & 1 \\ 1 & 1 & -1 \end{vmatrix}$

Find the value of each of the following determinants.

31. $\begin{vmatrix} 2 & -1 & 0 \\ 1 & 0 & -2 \\ 0 & 1 & 2 \end{vmatrix}$ **32.** $\begin{vmatrix} 5 & 0 & -4 \\ 0 & 1 & 3 \\ -1 & 2 & -1 \end{vmatrix}$ **33.** $\begin{vmatrix} 1 & 3 & 7 \\ -2 & 6 & 4 \\ 3 & 7 & -1 \end{vmatrix}$ **34.** $\begin{vmatrix} 2 & 1 & 5 \\ 6 & -3 & 4 \\ 8 & 9 & -2 \end{vmatrix}$

35. $\begin{vmatrix} -2 & 0 & 1 \\ 0 & 3 & 2 \\ 1 & 0 & -5 \end{vmatrix}$ **36.** $\begin{vmatrix} -1 & 1 & 1 \\ -2 & 2 & 2 \\ 5 & 7 & -4 \end{vmatrix}$ **37.** $\begin{vmatrix} 1 & 2 & 3 \\ 4 & 5 & 6 \\ 7 & 8 & 9 \end{vmatrix}$ **38.** $\begin{vmatrix} 4 & -2 & 1 \\ 0 & -5 & 4 \\ 1 & 2 & 1 \end{vmatrix}$

39. $\begin{vmatrix} -2 & 4 & -1 \\ 0 & 3 & 1 \\ -5 & -2 & 3 \end{vmatrix}$ **40.** $\begin{vmatrix} -3 & 2 & 4 \\ 1 & 2 & 3 \\ -1 & 1 & 5 \end{vmatrix}$

Find all values of x for which the given statement is true.

41. $\begin{vmatrix} x+3 & 2 & -3 \\ 0 & x-2 & 5 \\ 0 & 0 & x-4 \end{vmatrix} = 0$

42. $\begin{vmatrix} 1-x & -7 & 5 \\ 0 & 3+2x & -8 \\ 0 & 0 & 4x-1 \end{vmatrix} = 0$

43. $\begin{vmatrix} 2 & 2 & x \\ 4 & 4 & 4 \\ 3 & x & 3 \end{vmatrix} = 0$

44. $\begin{vmatrix} 1 & x & x^2 \\ 2 & 2 & 2 \\ 3 & 4 & 0 \end{vmatrix} = 0$

Modeling Practice

45. Slope-Intercept Form Show that the following determinant equation is another way to write the slope-intercept form of the equation of a line.

$$\begin{vmatrix} y & x \\ m & 1 \end{vmatrix} = b$$

46. Consider the following determinant equation:

$$\begin{vmatrix} x & y & 1 \\ x_1 & y_1 & 1 \\ 1 & m & 0 \end{vmatrix} = 0$$

Expand the determinant and show that the resulting equation is equivalent to the point-slope formula for a line.

© Hal Bergman/iStockPhoto

47. Amusement Park Income From 1986 to 1990, the annual income of amusement parks was linearly increasing, after which time it remained fairly constant. The annual income y, in billions of dollars, may be found for one of these years by evaluating the following determinant equation, in which x represents the number of years past January 1, 1986.

$$\begin{vmatrix} x & -1.7 \\ 2 & 0.3 \end{vmatrix} = y$$

a. Write the determinant equation in slope-intercept form.

b. Use the equation from Part a to find the approximate income for amusement parks in the year 1988.

48. College Enrollment From 1981, the enrollment of women in the United States armed forces was linearly increasing until 1990, after which it declined. The approximate number of women, w, enrolled in the armed forces from 1981 to 1990 may be found by evaluating the following determinant equation, in which x represents the number of years past January 1, 1981.

$$\begin{vmatrix} 6{,}509 & -2 \\ 85{,}709 & x \end{vmatrix} = w$$

Use this equation to determine the number of women enrolled in the armed forces in 1985.

Using Technology Exercises

49. Repeat Problem 37 on your graphing calculator.

50. Repeat Problem 36 on your graphing calculator.

Find the value of the following determinants.

51. $\begin{vmatrix} 1 & 3 & 2 & -4 \\ 0 & 4 & 1 & 0 \\ -2 & 1 & 3 & 0 \\ 2 & 3 & 4 & -1 \end{vmatrix}$

52. $\begin{vmatrix} 2 & 4 & -2 & -3 \\ 1 & 2 & 0 & 2 \\ -1 & 2 & 3 & -2 \\ 3 & 2 & 1 & -3 \end{vmatrix}$

Getting Ready for the Next Section

Simplify.

53. $2(3) - 4(4)$

54. $2(5) - 4(-3)$

55. $1(-1)(-3)$

56. $-(-3)(2)(1)$

57. $\dfrac{2(3) - 4(4)}{2(5) - 4(-3)}$

58. $\dfrac{1(-2) - 2(-16) + 1(9)}{6(1) - (-5) + 1(2)}$

59. $6(1) - 1(-5) + 1(2)$

60. $-2(-14) + 3(-4) - 1(-10)$

Find the value of each determinant.

61. $\begin{vmatrix} 3 & -5 \\ 2 & 4 \end{vmatrix}$

62. $\begin{vmatrix} 3 & 2 \\ 2 & 1 \end{vmatrix}$

63. $\begin{vmatrix} 6 & 1 & 1 \\ 3 & -1 & 1 \\ -4 & 2 & -3 \end{vmatrix}$

64. $\begin{vmatrix} 1 & 1 & 0 \\ 2 & 0 & 1 \\ 0 & 1 & 2 \end{vmatrix}$

Cramer's Rule

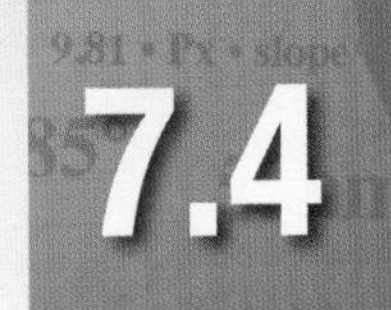

We begin this section with a look at how determinants can be used to solve a system of linear equations in two variables. The method we use is called *Cramer's rule*. We state it here as a theorem without proof.

THEOREM *Cramer's Rule*

The solution set to the system

$$a_1x + b_1y = c_1$$
$$a_2x + b_2y = c_2$$

is given by

$$x = \frac{D_x}{D}, \quad y = \frac{D_y}{D}, \quad (D \neq 0)$$

where

$$D = \begin{vmatrix} a_1 & b_1 \\ a_2 & b_2 \end{vmatrix} \quad D_x = \begin{vmatrix} c_1 & b_1 \\ c_2 & b_2 \end{vmatrix} \quad D_y = \begin{vmatrix} a_1 & c_1 \\ a_2 & c_2 \end{vmatrix}$$

The determinant D is made up of the coefficients of x and y in the original system. The determinants D_x and D_y are found by replacing the coefficients of x or y by the constant terms in the original system. Notice also that Cramer's rule does not apply if $D = 0$.

VIDEO EXAMPLES

SECTION 7.4

EXAMPLE 1 Use Cramer's rule to solve

$$2x - 3y = 4$$
$$4x + 5y = 3$$

SOLUTION We begin by calculating the determinants D, D_x, and D_y.

$$D = \begin{vmatrix} 2 & -3 \\ 4 & 5 \end{vmatrix} = 2(5) - 4(-3) = 22$$

$$D_x = \begin{vmatrix} 4 & -3 \\ 3 & 5 \end{vmatrix} = 4(5) - 3(-3) = 29$$

$$D_y = \begin{vmatrix} 2 & 4 \\ 4 & 3 \end{vmatrix} = 2(3) - 4(4) = -10$$

$$x = \frac{D_x}{D} = \frac{29}{22} \quad \text{and} \quad y = \frac{D_y}{D} = \frac{-10}{22} = -\frac{5}{11}$$

The solution set for the system is $\left\{\left(\frac{29}{22}, -\frac{5}{11}\right)\right\}$. ■

Cramer's rule can also be applied to systems of linear equations in three variables.

THEOREM *Also Cramer's Rule*

The solution set to the system

$$a_1x + b_1y + c_1z = d_1$$
$$a_2x + b_2y + c_2z = d_2$$
$$a_3x + b_3y + c_3z = d_3$$

is given by

$$x = \frac{D_x}{D}, \quad y = \frac{D_y}{D}, \quad z = \frac{D_z}{D}, \quad \text{and } (D \neq 0)$$

where

$$D = \begin{vmatrix} a_1 & b_1 & c_1 \\ a_2 & b_2 & c_2 \\ a_3 & b_3 & c_3 \end{vmatrix} \qquad D_x = \begin{vmatrix} d_1 & b_1 & c_1 \\ d_2 & b_2 & c_2 \\ d_3 & b_3 & c_3 \end{vmatrix}$$

$$D_y = \begin{vmatrix} a_1 & d_1 & c_1 \\ a_2 & d_2 & c_2 \\ a_3 & d_3 & c_3 \end{vmatrix} \qquad D_z = \begin{vmatrix} a_1 & b_1 & d_1 \\ a_2 & b_2 & d_2 \\ a_3 & b_3 & d_3 \end{vmatrix}$$

Again, the determinant D consists of the coefficients of x, y, and z in the original system. The determinants D_x, D_y, and D_z are found by replacing the coefficients of x, y, and z, respectively, with the constant terms from the original system. If $D = 0$, there is no unique solution to the system.

EXAMPLE 2 Use Cramer's rule to solve

$$\begin{aligned} x + y + z &= 6 \\ 2x - y + z &= 3 \\ x + 2y - 3z &= -4 \end{aligned}$$

SOLUTION This is the same system used in Example 1 in Section 7.2, so we can compare Cramer's rule with our previous methods of solving a system in three variables. We begin by setting up and evaluating D, D_x, D_y, and D_z. (Recall that there are a number of ways to evaluate a 3×3 determinant. Because we have four of these determinants, we can use both Methods 1 and 2 from the previous section.) We evaluate D using Method 1 from Section 7.3.

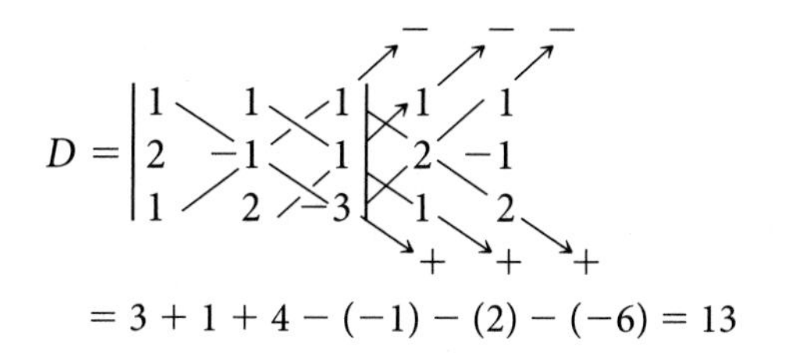

$$= 3 + 1 + 4 - (-1) - (2) - (-6) = 13$$

We evaluate D_x using Method 2 from Section 7.3 and expanding across row 1:

$$D_x = \begin{vmatrix} 6 & 1 & 1 \\ 3 & -1 & 1 \\ -4 & 2 & -3 \end{vmatrix} = 6\begin{vmatrix} -1 & 1 \\ 2 & -3 \end{vmatrix} - 1\begin{vmatrix} 3 & 1 \\ -4 & -3 \end{vmatrix} + 1\begin{vmatrix} 3 & -1 \\ -4 & 2 \end{vmatrix}$$

$$= 6(1) - (-5) + 1(2)$$

$$= 13$$

Find D_y by expanding across row 2:

$$D_y = \begin{vmatrix} 1 & 6 & 1 \\ 2 & 3 & 1 \\ 1 & -4 & -3 \end{vmatrix} = -2\begin{vmatrix} 6 & 1 \\ -4 & -3 \end{vmatrix} + 3\begin{vmatrix} 1 & 1 \\ 1 & -3 \end{vmatrix} - 1\begin{vmatrix} 1 & 6 \\ 1 & -4 \end{vmatrix}$$

$$= -2(-14) + 3(-4) - 1(-10)$$

$$= 26$$

Find D_z by expanding down column 1:

Note We are solving each of these determinants by expanding about different rows or columns just to show the different ways these determinants can be evaluated.

$$D_z = \begin{vmatrix} 1 & 1 & 6 \\ 2 & -1 & 3 \\ 1 & 2 & -4 \end{vmatrix} = 1\begin{vmatrix} -1 & 3 \\ 2 & -4 \end{vmatrix} - 2\begin{vmatrix} 1 & 6 \\ 2 & -4 \end{vmatrix} + 1\begin{vmatrix} 1 & 6 \\ -1 & 3 \end{vmatrix}$$

$$= 1(-2) - 2(-16) + 1(9)$$

$$= 39$$

$$x = \frac{D_x}{D} = \frac{13}{13} = 1 \qquad y = \frac{D_y}{D} = \frac{26}{13} = 2 \qquad z = \frac{D_z}{D} = \frac{39}{13} = 3$$

The solution set is $\{(1, 2, 3)\}$.

EXAMPLE 3 Use Cramer's rule to solve

$$\begin{aligned} x + y &= -1 \\ 2x - z &= 3 \\ y + 2z &= -1 \end{aligned}$$

SOLUTION It is helpful to rewrite the system using zeros for the coefficients of those variables not shown:

$$\begin{aligned} x + y + 0z &= -1 \\ 2x + 0y - z &= 3 \\ 0x + y + 2z &= -1 \end{aligned}$$

The four determinants used in Cramer's rule are

$$D = \begin{vmatrix} 1 & 1 & 0 \\ 2 & 0 & -1 \\ 0 & 1 & 2 \end{vmatrix} = -3$$

$$D_x = \begin{vmatrix} -1 & 1 & 0 \\ 3 & 0 & -1 \\ -1 & 1 & 2 \end{vmatrix} = -6$$

$$D_y = \begin{vmatrix} 1 & -1 & 0 \\ 2 & 3 & -1 \\ 0 & -1 & 2 \end{vmatrix} = 9$$

$$D_z = \begin{vmatrix} 1 & 1 & -1 \\ 2 & 0 & 3 \\ 0 & 1 & -1 \end{vmatrix} = -3$$

$$x = \frac{D_x}{D} = \frac{-6}{-3} = 2 \qquad y = \frac{D_y}{D} = \frac{9}{-3} = -3 \qquad z = \frac{D_z}{D} = \frac{-3}{-3} = 1$$

The solution set is $\{(2, -3, 1)\}$.

Finally, we should mention the possible situations that can occur when the determinant D is 0, when we are using Cramer's rule.

If $D = 0$ and at least one of the other determinants, D_x or D_y (or D_z), is not 0, then the system is inconsistent. In this case, there is no solution to the system.

On the other hand, if $D = 0$ and both D_x and D_y (and D_z in a system of three equations in three variables) are 0, then the system will have infinitely many solutions that form either a line or a plane.

New York Public Library/Photo Researchers, Inc.

A Note on the History of Cramer's Rule Cramer's rule is named after the Swiss mathematician *Gabriel Cramer* (1704–1752). Cramer's rule appeared in the appendix of an algebraic work of his classifying curves, but the basic idea behind his now-famous rule was formulated earlier by Gottfried Leibniz (inventor of calculus) and Chinese mathematicians. It was actually Cramer's superior notation that helped to popularize the technique.

Cramer has a respectable reputation as a mathematician, but he does not rank with the great mathematicians of his time, although through his extensive travels he met many of them, such as the Bernoullis (Johann and Daniel), Leonhard Euler, and Jean le Rond D'Alembert.

Cramer had very broad interests. He wrote on philosophy, law, and government, as well as mathematics; served in public office; and was an expert on cathedrals, often instructing workers about their repair and coordinating excavations to recover cathedral archives. Cramer never married, and a fall from a carriage eventually led to his death.

GETTING READY FOR CLASS

After reading through the preceding section, respond in your own words and in complete sentences.

A. When applying Cramer's rule, when will you use 2×2 determinants?

B. Why would it be impossible to use Cramer's rule if the determinant $D = 0$?

C. When applying Cramer's rule to solve a system of two linear equations in two variables, how many numbers should you obtain? How do these numbers relate to the system?

D. What will happen when you apply Cramer's rule to a system of equations made up of two parallel lines?

Problem Set 7.4

Solve each of the following systems using Cramer's rule.

1. $2x - 3y = 3$
$4x - 2y = 10$

2. $3x + y = -2$
$-3x + 2y = -4$

3. $5x - 2y = 4$
$-10x + 4y = 1$

4. $-4x + 3y = -11$
$5x + 4y = 6$

5. $4x - 7y = 3$
$5x + 2y = -3$

6. $3x - 4y = 7$
$6x - 2y = 5$

7. $9x - 8y = 4$
$2x + 3y = 6$

8. $4x - 7y = 10$
$-3x + 2y = -9$

9. $3x + 2y = 6$
$4x - 5y = 8$

10. $-4x + 3y = 12$
$6x - 7y = 14$

11. $12x - 13y = 16$
$11x + 15y = 18$

12. $-13x + 15y = 17$
$12x - 14y = 19$

13. $x + y + z = -2$
$x - y - z = 2$
$2x + 2y - z = 2$

14. $-x + y + 3z = 6$
$x + y + 2z = 7$
$2x + 3y + z = 4$

15. $x + y - z = 2$
$-x + y + z = 3$
$x + y + z = 4$

16. $-x - y + z = 1$
$x - y + z = 3$
$x + y - z = 4$

17. $3x - y + 2z = 4$
$6x - 2y + 4z = 8$
$x - 5y + 2z = 1$

18. $2x - 3y + z = 1$
$3x - y - z = 4$
$4x - 6y + 2z = 3$

19. $2x - y + 3z = 4$
$x - 5y - 2z = 1$
$-4x - 2y + z = 3$

20. $4x - y + 5z = 1$
$2x + 3y + 4z = 5$
$x + y + 3z = 2$

21. $x + 2y - z = 4$
$2x + 3y + 2z = 5$
$x - 3y + z = 6$

22. $3x + 2y + z = 6$
$2x + 3y - 2z = 4$
$x - 2y + 3z = 8$

23. $3x - 4y + 2z = 5$
$2x - 3y + 4z = 7$
$4x + 2y - 3z = 6$

24. $5x - 3y - 4z = 3$
$4x - 5y + 3z = 5$
$3x + 4y - 5z = -4$

25. $x - 3z = 6$
$y + 2z = 8$
$x + 4y = 10$

26. $x - 5y = -6$
$y - 4z = -5$
$2x + 3z = -6$

27. $-x - 7y = 1$
$x + 3z = 11$
$2y + z = 10$

28. $x + y = 2$
$-x + 3z = 0$
$2y + z = 3$

29. $x - y = 2$
$3x + z = 11$
$y - 2z = -3$

30. $4x + 5y = -1$
$2y + 3z = -5$
$x + 2z = -1$

Solve for x and y using Cramer's rule. Your answers will contain the constants a and b.

31. $ax + by = -1$
$bx + ay = 1$

32. $ax + y = b$
$bx + y = a$

33. $a^2x + by = 1$
$b^2x + ay = 1$

34. $ax + by = a$
$bx + ay = a$

35. Give the system of equations for which Cramer's rule yields the following determinants.

$$D = \begin{vmatrix} 1 & 2 \\ 3 & 4 \end{vmatrix}, \quad D_x = \begin{vmatrix} 1 & 2 \\ 0 & 4 \end{vmatrix}$$

36. Name the system of equations for which Cramer's rule yields the following determinants:

$$D_x = \begin{vmatrix} -2 & 3 \\ 6 & -4 \end{vmatrix}, \quad D_y = \begin{vmatrix} 1 & -2 \\ 5 & 6 \end{vmatrix}$$

Modeling Practice

37. Break-Even Point If a company has fixed costs of \$100 per week and each item it produces costs \$10 to manufacture, then the total cost y per week to produce x items is

$$y = 10x + 100$$

If the company sells each item it manufactures for \$12, then the total amount of money y the company brings in for selling x items is

$$y = 12x$$

Use Cramer's rule to solve the system

$$y = 10x + 100$$
$$y = 12x$$

for x to find the number of items the company must sell per week to break even.

38. Break-Even Point Suppose a company has fixed costs of \$200 per week and each item it produces costs \$20 to manufacture.

a. Write an equation that gives the total cost per week y to manufacture x items.

b. If each item sells for \$25, write an equation that gives the total amount of money y the company brings in for selling x items.

c. Use Cramer's rule to find the number of items the company must sell each week to break even.

Getting Ready for the Next Section

Simplify.

39. $-2(-3) + 5$

40. $-2(9) - 4$

41. $-2(1) + 2$

42. $-2(2) + 7$

Solve each system.

43. $x - 3y = 9$
$y = -2$

44. $x + y - z = 2$
$y + z = 3$
$z = 2$

Matrix Solutions to Linear Systems

In mathematics, a *matrix* is a rectangular array of elements considered as a whole. We can use matrices to represent systems of linear equations. To do so, we write the coefficients of the variables and the constant terms in the same position in the matrix as they occur in the system of equations. To show where the coefficients end and the constant terms begin, we use vertical lines instead of equal signs. For example, the system

$$2x + 5y = -4$$
$$x - 3y = 9$$

can be represented by the matrix

$$\left[\begin{array}{cc|c} 2 & 5 & -4 \\ 1 & -3 & 9 \end{array}\right]$$

which is called an *augmented matrix* because it includes both the coefficients of the variables and the constant terms.

To solve a system of linear equations by using the augmented matrix for that system, we need the following row operations as the tools of that solution process. The row operations tell us what we can do to an augmented matrix that may change the numbers in the matrix, but will always produce a matrix that represents a system of equations with the same solution as that of our original system.

Row Operations

1. We can interchange any two rows of a matrix.
2. We can multiply any row by a nonzero constant.
3. We can add to any row a constant multiple of another row.

The three row operations are simply a list of the properties we use to solve systems of linear equations, translated to fit an augmented matrix. For instance, the second operation in our list is actually just another way to state the multiplication property of equality.

We solve a system of linear equations by first transforming the augmented matrix into a matrix that has 1s down the diagonal of the coefficient matrix, and 0s below it. For instance, we will solve the system

$$2x + 5y = -4$$
$$x - 3y = 9$$

by transforming the matrix

$$\left[\begin{array}{cc|c} 2 & 5 & -4 \\ 1 & -3 & 9 \end{array}\right]$$

using the row operations listed earlier to get a matrix of the form

$$\left[\begin{array}{cc|c} 1 & - & - \\ 0 & 1 & - \end{array}\right]$$

To accomplish this, we begin with the first column and try to produce a 1 in the first position and a 0 below it. Interchanging rows 1 and 2 gives us a 1 in the top position of the first column:

$$\left[\begin{array}{cc|c} 1 & -3 & 9 \\ 2 & 5 & -4 \end{array}\right] \longleftarrow \text{Interchange rows 1 and 2}$$

Multiplying row 1 by -2 and adding the result to row 2 gives us a 0 where we want it.

$$\left[\begin{array}{cc|c} 1 & -3 & 9 \\ 0 & 11 & -22 \end{array}\right] \longleftarrow \text{Multiply row 1 by } -2 \text{ and add the result to row 2}$$

Multiplying row 2 by $\frac{1}{11}$ gives us a 1 where we want it.

$$\left[\begin{array}{cc|c} 1 & -3 & 9 \\ 0 & 1 & -2 \end{array}\right] \longleftarrow \text{Multiply row 2 by } \frac{1}{11}$$

Taking this last matrix and writing the system of equations it represents, we have

$$x - 3y = 9$$

$$y = -2$$

Substituting -2 for y in the top equation gives us

$$x = 3$$

The solution to our system is $(3, -2)$.

VIDEO EXAMPLES

SECTION 7.5

EXAMPLE 1 Solve the following system using an augmented matrix:

$$x + y - z = 2$$

$$2x + 3y - z = 7$$

$$3x - 2y + z = 9$$

SOLUTION We begin by writing the system in terms of an augmented matrix:

$$\left[\begin{array}{ccc|c} 1 & 1 & -1 & 2 \\ 2 & 3 & -1 & 7 \\ 3 & -2 & 1 & 9 \end{array}\right]$$

Next, we want to produce 0s in the second two positions of column 1:

$$\left[\begin{array}{ccc|c} 1 & 1 & -1 & 2 \\ 0 & 1 & 1 & 3 \\ 3 & -2 & 1 & 9 \end{array}\right] \longleftarrow \text{Multiply row 1 by } -2 \text{ and add the result to row 2}$$

$$\left[\begin{array}{ccc|c} 1 & 1 & -1 & 2 \\ 0 & 1 & 1 & 3 \\ 0 & -5 & 4 & 3 \end{array}\right] \longleftarrow \text{Multiply row 1 by } -3 \text{ and add the result to row 3}$$

Note that we could have done these two steps in one single step. As you become more familiar with this method of solving systems of equations, you will do just that.

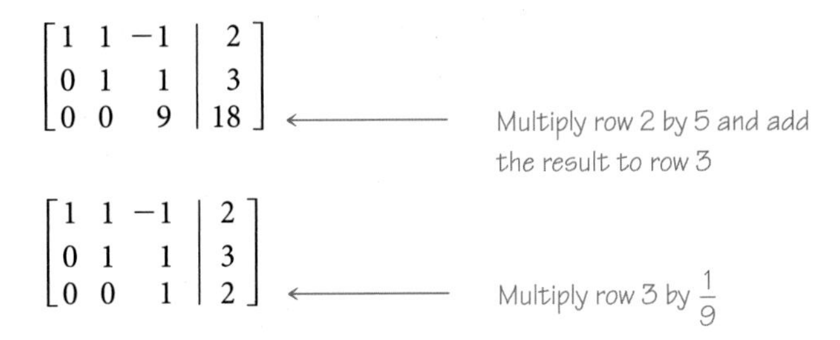
Multiply row 2 by 5 and add the result to row 3
Multiply row 3 by 1/9

Converting back to a system of equations, we have

$$x + y - z = 2$$

$$y + z = 3$$

$$z = 2$$

This system is equivalent to our first one, but much easier to solve.
Substituting $z = 2$ into the second equation, we have

$$y = 1$$

Substituting $z = 2$ and $y = 1$ into the first equation, we have

$$x = 3$$

The solution to our original system is (3, 1, 2). It satisfies each of our original equations. You can check this, if you want. ■

GETTING READY FOR CLASS

After reading through the preceding section, respond in your own words and in complete sentences.

A. What is a matrix?
B. What is an augmented matrix?
C. Name one elementary row operation for matrices.
D. What augmented matrix could be used to represent this system of equations:

$$x + y - z = 2$$
$$y + z = 3$$
$$z = 2$$

Problem Set 7.5

Solve the following systems of equations by using matrices.

1. $x + y = 5$
   $3x - y = 3$
2. $x + y = -2$
   $2x - y = -10$
3. $3x - 5y = 7$
   $-x + y = -1$
4. $2x - y = 4$
   $x + 3y = 9$
5. $2x - 8y = 6$
   $3x - 8y = 13$
6. $3x - 6y = 3$
   $-2x + 3y = -4$
7. $x + y + z = 4$
   $x - y + 2z = 1$
   $x - y - z = -2$
8. $x - y - 2z = -1$
   $x + y + z = 6$
   $x + y - z = 4$
9. $x + 2y + z = 3$
   $2x - y + 2z = 6$
   $3x + y - z = 5$
10. $x - 3y + 4z = -4$
    $2x + y - 3z = 14$
    $3x + 2y + z = 10$
11. $x + 2y = 3$
    $y + z = 3$
    $4x - z = 2$
12. $x + y = 2$
    $3y - 2z = -8$
    $x + z = 5$
13. $x + 3y = 7$
    $3x - 4z = -8$
    $5y - 2z = -5$
14. $x + 4y = 13$
    $2x - 5z = -3$
    $4y - 3z = 9$

Solve each system using matrices. Remember, multiplying a row by a nonzero constant will not change the solution to the system.

15. $\frac{1}{3}x + \frac{1}{5}y = 2$
    $\frac{1}{3}x - \frac{1}{2}y = -\frac{1}{3}$
16. $\frac{1}{2}x + \frac{1}{3}y = 13$
    $\frac{1}{5}x + \frac{1}{8}y = 5$

The following systems are inconsistent systems. In each case, the lines are parallel. Try solving each system using matrices and see what happens.

17. $2x - 3y = 4$
    $4x - 6y = 4$
18. $10x - 15y = 5$
    $-4x + 6y = -4$

The following systems contain dependent equations. In each case, the lines coincide. Try solving each system using matrices and see what happens.

19. $-6x + 4y = 8$
    $-3x + 2y = 4$
20. $x + 2y = 5$
    $-x - 2y = -5$

Getting Ready for the Next Section

Translate into symbols.

21. Two more than 3 times a number

22. One less than twice a number

Simplify.

23. $25 - \frac{385}{9}$

24. $0.30(12)$

25. $0.08(4{,}000)$

26. $500(1.5)$

Apply the distributive property, then simplify.

27. $10(0.2x + 0.5y)$

28. $100(0.09x + 0.08y)$

Solve.

29. $x + (3x + 2) = 26$

30. $5x = 2{,}500$

Solve each system.

31. $3y + z = 17$
$5y + 20z = 65$

32. $x + y = 850$
$1.5x + y = 1{,}100$

Applications and Modeling

Many times application problems involve more than one unknown quantity. If a problem is stated in terms of two unknowns and we represent each unknown quantity with a different variable, then we must write the relationships between the variables with two equations. The two equations written in terms of the two variables form a system of linear equations that we solve using the methods developed in this chapter. If we find a problem that relates three unknown quantities, then we need three equations to form a linear system we can solve.

Here is our Blueprint for Problem Solving, modified to fit the application problems that you will find in this section.

Blueprint for Problem Solving Using a System of Equations

Step 1: *Read* the problem, and then mentally *list* the items that are known and the items that are unknown.

Step 2: *Assign variables* to each of the unknown items. That is, let x = one of the unknown items and y = the other unknown item (and z = the third unknown item, if there is a third one). Then *translate* the other *information* in the problem to expressions involving the two (or three) variables.

Step 3: *Reread* the problem, and then *write a system of equations*, using the items and variables listed in Steps 1 and 2, that describes the situation.

Step 4: *Solve the system* found in Step 3.

Step 5: *Write your answers* using complete sentences.

Step 6: *Reread* the problem, and *check* your solution with the original words in the problem.

VIDEO EXAMPLES

SECTION 7.6

EXAMPLE 1 One number is 2 more than 3 times another. Their sum is 26. Find the two numbers.

SOLUTION Applying the steps from our Blueprint, we have:

Step 1: *Read and list.*

We know that we have two numbers, whose sum is 26. One of them is 2 more than 3 times the other. The unknown quantities are the two numbers.

Step 2: *Assign variables and translate information.*

Let x = one of the numbers and y = the other number.

Step 3: *Write a system of equations.*

The first sentence in the problem translates into $y = 3x + 2$. The second sentence gives us a second equation: $x + y = 26$. Together, these two equations give us the following system of equations:

$$x + y = 26$$

$$y = 3x + 2$$

Step 4: *Solve the system.*

Substituting the expression for y from the second equation into the first and solving for x yields

$$x + (3x + 2) = 26$$
$$4x + 2 = 26$$
$$4x = 24$$
$$x = 6$$

Using $x = 6$ in $y = 3x + 2$ gives the second number:

$$y = 3(6) + 2$$
$$y = 20$$

Step 5: *Write answers.*

The two numbers are 6 and 20.

Step 6: *Reread and check.*

The sum of 6 and 20 is 26, and 20 is 2 more than 3 times 6. ■

EXAMPLE 2 Suppose 850 tickets were sold for a game for a total of $1,100. If adult tickets cost $1.50 and children's tickets cost $1.00, how many of each kind of ticket were sold?

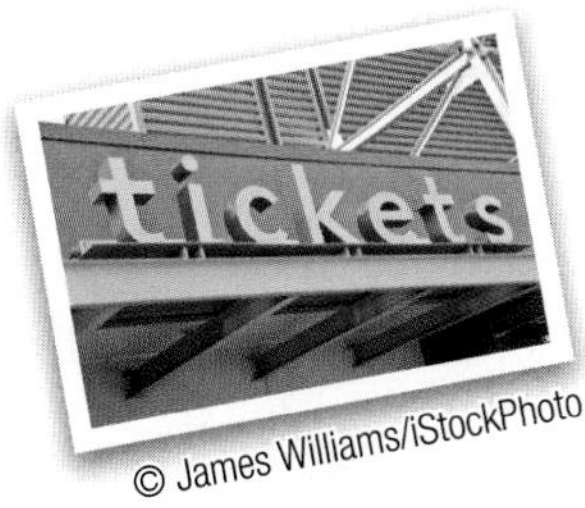

© James Williams/iStockPhoto

SOLUTION

Step 1: *Read and list.*

The total number of tickets sold is 850. The total income from tickets is $1,100. Adult tickets are $1.50 each. Children's tickets are $1.00 each. We don't know how many of each type of ticket have been sold.

Step 2: *Assign variables and translate information.*

We let $x =$ the number of adult tickets and $y =$ the number of children's tickets.

Step 3: *Write a system of equations.*

The total number of tickets sold is 850, giving us our first equation.

$$x + y = 850$$

Because each adult ticket costs $1.50, and each children's ticket costs $1.00, and the total amount of money paid for tickets was $1,100, a second equation is

$$1.50x + 1.00y = 1{,}100$$

The same information can also be obtained by summarizing the problem with a table. One such table follows. Notice that the two equations we obtained previously are given by the two rows of the table.

	Adult Tickets	Children's Tickets	Total
Number	x	y	850
Value	$1.50x$	$1.00y$	1,100

Whether we use a table to summarize the information in the problem or just talk our way through the problem, the system of equations that describes the situation is

$$\begin{aligned} x + \quad y &= 850 \\ 1.50x + 1.00y &= 1{,}100 \end{aligned}$$

Step 4: *Solve the system.*

If we multiply the second equation by 10 to clear it of decimals, we have the system

$$\begin{aligned} x + \quad y &= 850 \\ 15x + 10y &= 11{,}000 \end{aligned}$$

Multiplying the first equation by -10 and adding the result to the second equation eliminates the variable y from the system:

$$\begin{aligned} -10x - 10y &= -8{,}500 \\ 15x + 10y &= 11{,}000 \\ \hline 5x &= 2{,}500 \\ x &= 500 \end{aligned}$$

The number of adult tickets sold was 500. To find the number of children's tickets, we substitute $x = 500$ into $x + y = 850$ to get

$$\begin{aligned} 500 + y &= 850 \\ y &= 350 \end{aligned}$$

Step 5: *Write answers.*

The number of children's tickets is 350, and the number of adult tickets is 500.

Step 6: *Reread and check.*

The total number of tickets is $350 + 500 = 850$. The amount of money from selling the two types of tickets is

350 children's tickets at \$1.00 each is $350(1.00) = \$350$
500 adult tickets at \$1.50 each is $500(1.50) = \$750$

The total income from ticket sales is \$1,100 ■

EXAMPLE 3 Suppose a person invests a total of \$10,000 in two accounts. One account earns 8% annually, and the other earns 9% annually. If the total interest earned from both accounts in a year is \$860, how much was invested in each account?

SOLUTION

Step 1: *Read and list.*

The total investment is \$10,000 split between two accounts. One account earns 8% annually, and the other earns 9% annually. The interest from both accounts is \$860 in 1 year. We don't know how much is in each account.

Step 2: *Assign variables and translate information.*

We let x equal the amount invested at 9% and y be the amount invested at 8%.

Step 3: *Write a system of equations.*

Because the total investment is \$10,000, one relationship between x and y can be written as

$$x + y = 10{,}000$$

The total interest earned from both accounts is \$860. The amount of interest earned on x dollars at 9% is $0.09x$, while the amount of interest earned on y dollars at 8% is $0.08y$. This relationship is represented by the equation

$$0.09x + 0.08y = 860$$

The two equations we have just written can also be found by first summarizing the information from the problem in a table. Again, the two rows of the table yield the two equations we found previously. Here is the table.

	Dollars at 9%	Dollars at 8%	Total
Number	x	y	10,000
Interest	$0.09x$	$0.08y$	860

The system of equations that describes this situation is given by

$$\begin{aligned} x + \quad y &= 10{,}000 \\ 0.09x + 0.08y &= \quad 860 \end{aligned}$$

Step 4: *Solve the system.*

Multiplying the second equation by 100 will clear it of decimals. The system that results after doing so is

$$\begin{aligned} x + \ y &= 10{,}000 \\ 9x + 8y &= 86{,}000 \end{aligned}$$

We can eliminate y from this system by multiplying the first equation by -8 and adding the result to the second equation.

$$\begin{aligned} -8x - 8y &= -80{,}000 \\ 9x + 8y &= \quad 86{,}000 \\ \hline x \qquad &= \quad 6{,}000 \end{aligned}$$

The amount of money invested at 9% is \$6,000. Because the total investment was \$10,000, the amount invested at 8% must be \$4,000.

Step 5: *Write answers.*

The amount invested at 8% is \$4,000, and the amount invested at 9% is \$6,000.

Step 6: *Reread and check.*

The total investment is \$4,000 + \$6,000 = \$10,000. The amount of interest earned from the two accounts is

In 1 year, \$4,000 invested at 8% earns 0.08(4,000) = \$320
In 1 year, \$6,000 invested at 9% earns 0.09(6,000) = \$540

The total interest from the two accounts is \$860

EXAMPLE 4 How much 20% alcohol solution and 50% alcohol solution must be mixed to get 12 gallons of 30% alcohol solution?

SOLUTION To solve this problem, we must first understand that a 20% alcohol solution is 20% alcohol and 80% water.

Step 1: *Read and list.*
We will mix two solutions to obtain 12 gallons of solution that is 30% alcohol. One of the solutions is 20% alcohol and the other 50% alcohol. We don't know how much of each solution we need.

Step 2: *Assign variables and translate information.*
Let $x =$ the number of gallons of 20% alcohol solution needed, and $y =$ the number of gallons of 50% alcohol solution needed.

Step 3: *Write a system of equations.*
Because we must end up with a total of 12 gallons of solution, one equation for the system is

$$x + y = 12$$

The amount of alcohol in the x gallons of 20% solution is $0.20x$, while the amount of alcohol in the y gallons of 50% solution is $0.50y$. Because the total amount of alcohol in the 20% and 50% solutions must add up to the amount of alcohol in the 12 gallons of 30% solution, the second equation in our system can be written as

$$0.20x + 0.50y = 0.30(12)$$

Again, let's make a table that summarizes the information we have to this point in the problem.

	20% Solution	50% Solution	Final Solution
Total number of gallons	x	y	12
Gallons of alcohol	$0.20x$	$0.50y$	$0.30(12)$

Our system of equations is

$$\begin{aligned} x + \quad y &= 12 \\ 0.20x + 0.50y &= 0.30(12) = 3.6 \end{aligned}$$

Step 4: *Solve the system.*
Multiplying the second equation by 10 gives us an equivalent system:

$$\begin{aligned} x + \ y &= 12 \\ 2x + 5y &= 36 \end{aligned}$$

Multiplying the top equation by -2 to eliminate the x-variable, we have

$$\begin{aligned} -2x - 2y &= -24 \\ 2x + 5y &= \ \ 36 \\ \hline 3y &= \ \ 12 \\ y &= \ \ 4 \end{aligned}$$

Substituting $y = 4$ into $x + y = 12$, we solve for x:

$$x + 4 = 12$$

$$x = 8$$

Step 5: *Write answers.*

It takes 8 gallons of 20% alcohol solution and 4 gallons of 50% alcohol solution to produce 12 gallons of 30% alcohol solution.

Step 6: *Reread and check.*

If we mix 8 gallons of 20% solution and 4 gallons of 50% solution, we end up with a total of 12 gallons of solution. To check the percentages we look for the total amount of alcohol in the two initial solutions and in the final solution.

In the initial solutions

The amount of alcohol in 8 gallons of 20% solution is $0.20(8) = 1.6$ gallons
The amount of alcohol in 4 gallons of 50% solution is $0.50(4) = 2.0$ gallons

The total amount of alcohol in the initial solutions is 3.6 gallons

In the final solution

The amount of alcohol in 12 gallons of 30% solution is $0.30(12) = 3.6$ gallons.

EXAMPLE 5 It takes 2 hours for a boat to travel 28 miles downstream (with the current). The same boat can travel 18 miles upstream (against the current) in 3 hours. What is the speed of the boat in still water, and what is the speed of the current of the river?

SOLUTION

Step 1: *Read and list.*

A boat travels 18 miles upstream and 28 miles downstream. The trip upstream takes 3 hours. The trip downstream takes 2 hours. We don't know the speed of the boat or the speed of the current.

Step 2: *Assign variables and translate information.*

Let $x =$ the speed of the boat in still water and let $y =$ the speed of the current. The average speed (rate) of the boat upstream is $x - y$, because it is traveling against the current. The rate of the boat downstream is $x + y$, because the boat is traveling with the current.

Step 3: *Write a system of equations.*

Putting the information into a table, we have

	d (distance, miles)	r (rate, mph)	t (time, hours)
Upstream	18	$x - y$	3
Downstream	28	$x + y$	2

The formula for the relationship between distance d, rate r, and time t is $d = rt$ (the rate equation). Because $d = r \cdot t$, the system we need to solve the problem is

$$18 = (x - y) \cdot 3$$
$$28 = (x + y) \cdot 2$$

which is equivalent to

$$6 = x - y$$
$$14 = x + y$$

Step 4: ***Solve the system.***
Adding the two equations, we have

$$20 = 2x$$
$$x = 10$$

Substituting $x = 10$ into $14 = x + y$, we see that

$$y = 4$$

Step 5: ***Write answers.***
The speed of the boat in still water is 10 miles per hour; the speed of the current is 4 miles per hour.

Step 6: ***Reread and check.***
The boat travels at $10 + 4 = 14$ miles per hour downstream, so in 2 hours it will travel $14 \cdot 2 = 28$ miles. The boat travels at $10 - 4 = 6$ miles per hour upstream, so in 3 hours it will travel $6 \cdot 3 = 18$ miles.

EXAMPLE 6 A coin collection consists of 14 coins with a total value of \$1.35. If the coins are nickels, dimes, and quarters, and the number of nickels is 3 less than twice the number of dimes, how many of each coin is there in the collection?

SOLUTION This problem will require three variables and three equations.

Step 1: ***Read and list.***
We have 14 coins with a total value of \$1.35. The coins are nickels, dimes, and quarters. The number of nickels is 3 less than twice the number of dimes. We do not know how many of each coin we have.

Step 2: ***Assign variables and translate information.***
Because we have three types of coins, we will have to use three variables. Let's let x = the number of nickels, y = the number of dimes, and z = the number of quarters.

Step 3: ***Write a system of equations.***
Because the total number of coins is 14, our first equation is

$$x + y + z = 14$$

Because the number of nickels is 3 less than twice the number of dimes, a second equation is

$x = 2y - 3$ which is equivalent to $x - 2y = -3$

Our last equation is obtained by considering the value of each coin and the total value of the collection. Let's write the equation in terms of cents, so we won't have to clear it of decimals later.

$$5x + 10y + 25z = 135$$

Here is our system, with the equations numbered for reference:

$$\begin{aligned} x + y + z &= 14 \quad (1) \\ x - 2y &= -3 \quad (2) \\ 5x + 10y + 25z &= 135 \quad (3) \end{aligned}$$

Step 4: *Solve the system.*

Let's begin by eliminating x from the first and second equations, and the first and third equations. Adding -1 times the second equation to the first equation gives us an equation in only y and z. We call this equation (4).

$$3y + z = 17 \qquad (4)$$

Adding -5 times equation (1) to equation (3) gives us

$$5y + 20z = 65 \qquad (5)$$

We can eliminate z from equations (4) and (5) by adding -20 times (4) to (5). Here is the result:

$$-55y = -275$$

$$y = 5$$

Substituting $y = 5$ into equation (4) gives us $z = 2$. Substituting $y = 5$ and $z = 2$ into equation (1) gives us $x = 7$.

Step 5: *Write answers.*

The collection consists of 7 nickels, 5 dimes, and 2 quarters.

Step 6: *Reread and check.*

The total number of coins is $7 + 5 + 2 = 14$. The number of nickels, 7, is 3 less than twice the number of dimes, 5. To find the total value of the collection, we have

The value of the 7 nickels is	$7(0.05) = \$0.35$
The value of the 5 dimes is	$5(0.10) = \$0.50$
The value of the 2 quarters is	$2(0.25) = \$0.50$
The total value of the collection is $1.35	

If you go on to take a chemistry class, you may see the next example (or one much like it).

EXAMPLE 7 In a chemistry lab, students record the temperature of water at room temperature and find that it is 77 ° on the Fahrenheit temperature scale and 25 ° on the Celsius temperature scale. The water is then heated until it boils. The temperature of the boiling water is 212 °F and 100 °C. Assume that the relationship between the two temperature scales is a linear one, then use the preceding data to find the formula that gives the Celsius temperature C in terms of the Fahrenheit temperature F.

SOLUTION The data is summarized in the following table.

Corresponding Temperatures	
In Degrees Fahrenheit	**In Degrees Celsius**
77	25
212	100

If we assume the relationship is linear, then the formula that relates the two temperature scales can be written in slope-intercept form as

$$C = mF + b$$

Substituting $C = 25$ and $F = 77$ into this formula gives us

$$25 = 77m + b$$

Substituting $C = 100$ and $F = 212$ into the formula yields

$$100 = 212m + b$$

Together, the two equations form a system of equations, which we can solve using the addition method.

$$\begin{array}{rcl} 25 = 77m + b & \xrightarrow{\text{Multiply by } -1} & -25 = -77m - b \\ 100 = 212m + b & \xrightarrow[\text{No Change}]{} & 100 = 212m + b \\ \hline & & 75 = 135m \\ & & m = \frac{75}{135} = \frac{5}{9} \end{array}$$

To find the value of b, we substitute $m = \frac{5}{9}$ into $25 = 77m + b$ and solve for b.

$$25 = 77\left(\frac{5}{9}\right) + b$$

$$25 = \frac{385}{9} + b$$

$$b = 25 - \frac{385}{9} = \frac{225}{9} - \frac{385}{9} = -\frac{160}{9}$$

The equation that gives C in terms of F is

$$C = \frac{5}{9}F - \frac{160}{9}$$

GETTING READY FOR CLASS

After reading through the preceding section, respond in your own words and in complete sentences.

A. How does Step 2 of the Blueprint for Problem Solving that appears in this section differ from Step 2 of the Blueprint from Section 2.4?

B. How does Step 3 of the Blueprint for Problem Solving in this section differ from the same step of the Blueprint from Section 2.4?

C. When working application problems involving boats moving in rivers, how does the current of the river affect the speed of the boat?

D. Write an application problem for which the solution depends on solving the system of equations:

$$\begin{aligned} x + y &= 1{,}000 \\ 0.05x + 0.06y &= 55 \end{aligned}$$

Problem Set 7.6

Number Problems

1. One number is 3 more than twice another. The sum of the numbers is 18. Find the two numbers.

2. The sum of two numbers is 32. One of the numbers is 4 less than 5 times the other. Find the two numbers.

3. The difference of two numbers is 6. Twice the smaller is 4 more than the larger. Find the two numbers.

4. The larger of two numbers is 5 more than twice the smaller. If the smaller is subtracted from the larger, the result is 12. Find the two numbers.

5. The sum of three numbers is 8. Twice the smallest is 2 less than the largest, while the sum of the largest and smallest is 5. Use a linear system in three variables to find the three numbers.

6. The sum of three numbers is 14. The largest is 4 times the smallest, while the sum of the smallest and twice the largest is 18. Use a linear system in three variables to find the three numbers.

Ticket and Interest Problems

7. A total of 925 tickets were sold for a game for a total of $1,150. If adult tickets sold for $2.00 and children's tickets sold for $1.00, how many of each kind of ticket were sold?

8. If tickets for a show cost $2.00 for adults and $1.50 for children, how many of each kind of ticket were sold if a total of 300 tickets were sold for $525?

9. Mr. Jones has $20,000 to invest. He invests part at 6% and the rest at 7%. If he earns $1,280 in interest after 1 year, how much did he invest at each rate?

10. A man invests $17,000 in two accounts. One account earns 5% interest per year and the other 6.5%. If his total interest after 1 year is $970, how much did he invest at each rate?

11. Susan invests twice as much money at 7.5% as she does at 6%. If her total interest after 1 year is $840, how much does she have invested at each rate?

12. A woman earns $1,350 in interest from two accounts in 1 year. If she has three times as much invested at 7% as she does at 6%, how much does she have in each account?

13. A man invests $2,200 in three accounts that pay 6%, 8%, and 9% in annual interest, respectively. He has three times as much invested at 9% as he does at 6%. If his total interest for the year is $178, how much is invested at each rate?

14. A student has money in three accounts that pay 5%, 7%, and 8% in annual interest. She has three times as much invested at 8% as she does at 5%. If the total amount she has invested is $1,600 and her interest for the year comes to $115, how much money does she have in each account?

Mixture Problems

15. How many gallons of 20% alcohol solution and 50% alcohol solution must be mixed to get 9 gallons of 30% alcohol solution?

16. How many ounces of 30% hydrochloric acid solution and 80% hydrochloric acid solution must be mixed to get 10 ounces of 50% hydrochloric acid solution?

17. A mixture of 16% disinfectant solution is to be made from 20% and 14% disinfectant solutions. How much of each solution should be used if 15 gallons of the 16% solution are needed?

18. How much 25% antifreeze and 50% antifreeze should be combined to give 40 gallons of 30% antifreeze?

19. Paul mixes nuts worth $1.55 per pound with oats worth $1.35 per pound to get 25 pounds of trail mix worth $1.45 per pound. How many pounds of nuts and how many pounds of oats did he use?

20. A chemist has three different acid solutions. The first acid solution contains 20% acid, the second contains 40%, and the third contains 60%. He wants to use all three solutions to obtain a mixture of 60 liters containing 50% acid, using twice as much of the 60% solution as the 40% solution. How many liters of each solution should be used?

Rate Problems

21. It takes a boat 2 hours to travel 24 miles downstream and 3 hours to travel 18 miles upstream. What is the speed of the boat in still water? What is the speed of the current of the river?

22. A boat on a river travels 20 miles downstream in only 2 hours. It takes the same boat 6 hours to travel 12 miles upstream. What are the speed of the boat and the speed of the current?

23. An airplane flying with the wind can cover a certain distance in 2 hours. The return trip against the wind takes $2\frac{1}{2}$ hours. How fast is the plane and what is the speed of the air, if the distance is 600 miles?

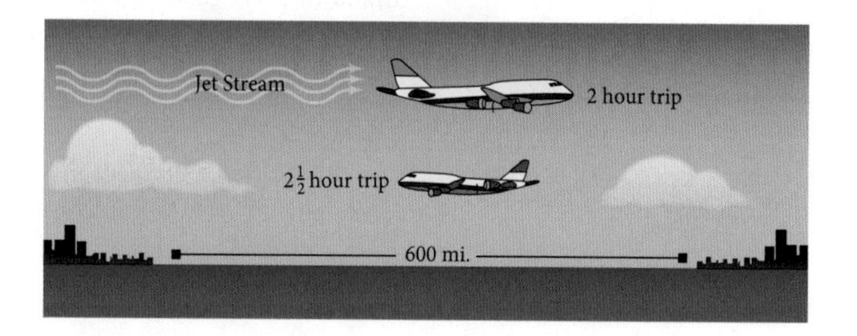

24. An airplane covers a distance of 1,500 miles in 3 hours when it flies with the wind and $3\frac{1}{3}$ hours when it flies against the wind. What is the speed of the plane in still air?

Coin Problems

25. Bob has 20 coins totaling \$1.40. If he has only dimes and nickels, how many of each coin does he have?

© Tarek El Sombati/iStockPhoto

26. If Amy has 15 coins totaling \$2.70, and the coins are quarters and dimes, how many of each coin does she have?

27. A collection of nickels, dimes, and quarters consists of 9 coins with a total value of \$1.20. If the number of dimes is equal to the number of nickels, find the number of each type of coin.

28. A coin collection consists of 12 coins with a total value of \$1.20. If the collection consists only of nickels, dimes, and quarters, and the number of dimes is two more than twice the number of nickels, how many of each type of coin are in the collection?

29. A collection of nickels, dimes, and quarters amounts to \$10.00. If there are 140 coins in all and there are twice as many dimes as there are quarters, find the number of nickels.

30. A cash register contains a total of 95 coins consisting of pennies, nickels, dimes, and quarters. There are only 5 pennies and the total value of the coins is \$12.05. Also, there are 5 more quarters than dimes. How many of each coin is in the cash register?

Additional Problems

31. **Price and Demand** A manufacturing company finds that they can sell 300 items if the price per item is \$2.00, and 400 items if the price is \$1.50 per item. If the relationship between the number of items sold x and the price per item p is a linear one, find a formula that gives x in terms of p. Then use the formula to find the number of items they will sell if the price per item is \$3.00.

32. Price and Demand A company manufactures and sells bracelets. They have found from past experience that they can sell 300 bracelets each week if the price per bracelet is \$2.00, but only 150 bracelets are sold if the price is \$2.50 per bracelet. If the relationship between the number of bracelets sold x and the price per bracelet p is a linear one, find a formula that gives x in terms of p. Then use the formula to find the number of bracelets they will sell at \$3.00 each.

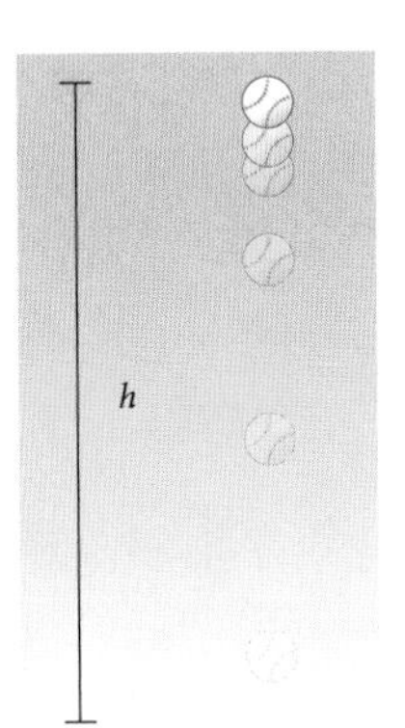

33. Height of a Ball A ball is tossed into the air so that the height after 1, 3, and 5 seconds is as given in the following table.

t (sec)	h (ft)
1	128
3	128
5	0

If the relationship between the height of the ball h and the time t is quadratic, then the relationship can be written as

$$h = at^2 + bt + c$$

Use the information in the table to write a system of three equations in three variables a, b, and c. Solve the system to find the exact relationship between h and t.

34. Height of a Ball A ball is tossed into the air and its height above the ground after 1, 3, and 4 seconds is recorded as shown in the following table.

t (sec)	h (ft)
1	96
3	64
4	0

The relationship between the height of the ball h and the time t is quadratic and can be written as

$$h = at^2 + bt + c$$

Use the information in the table to write a system of three equations in three variables a, b, and c. Solve the system to find the exact relationship between the variables h and t.

Getting Ready for the Next Section

35. Which of the following are solutions to $x + y \leq 4$?

$(0, 0)$ $(4, 0)$ $(2, 3)$

36. Which of the following are solutions to $y < 2x - 3$?

$(0, 0)$ $(3, -2)$ $(-3, 2)$

37. Where do the graphs of the lines $x + y = 4$ and $x - 2y = 4$ intersect?

38. Where do the graphs of the lines $x = -1$ and $x - 2y = 4$ intersect?

Solve.

39. $20x + 9{,}300 > 18{,}000$

40. $20x + 4{,}800 > 18{,}000$

7.7 Inequalities and Systems of Inequalities in Two Variables

A small movie theater holds 100 people. The owner charges more for adults than for children, so it is important to know the different combinations of adults and children that can be seated at one time. The shaded region in Figure 1 contains all the seating combinations. The line $x + y = 100$ shows the combinations for a full theater: The y-intercept corresponds to a theater full of adults, and the x-intercept corresponds to a theater full of children. In the shaded region below the line $x + y = 100$ are the combinations that occur if the theater is not full.

Shaded regions like the one shown in Figure 1 are produced by linear inequalities in two variables, which is the topic of this section.

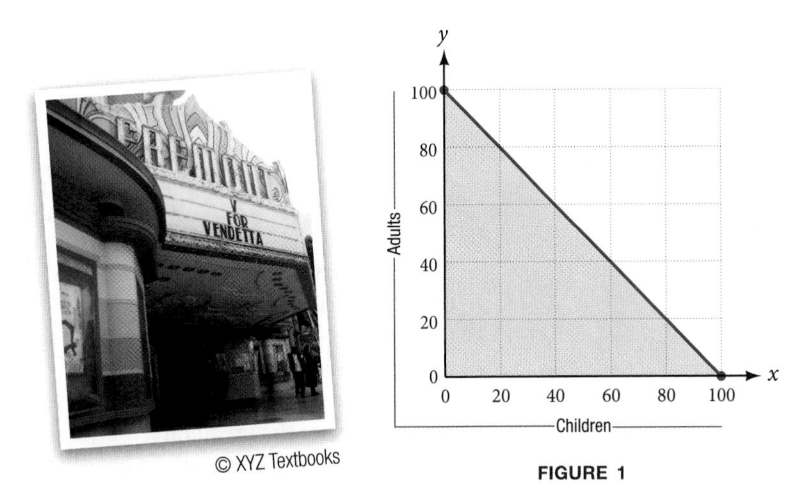

© XYZ Textbooks

FIGURE 1

A *linear inequality in two variables* is any expression that can be put in the form

$$ax + by < c$$

where a, b, and c are real numbers (a and b not both 0). The inequality symbol can be any one of the following four: $<, \leq, >, \geq$.

Some examples of linear inequalities are

$$2x + 3y < 6 \qquad y \geq 2x + 1 \qquad x - y \leq 0$$

Although not all of these examples have the form $ax + by < c$, each one can be put in that form.

The solution set for a linear inequality is a *section of the coordinate plane*. The *boundary* for the section is found by replacing the inequality symbol with an equal sign and graphing the resulting equation. The boundary is included in the solution set (and is represented with a *solid line*) if the inequality symbol used originally is $\leq$ or $\geq$. The boundary is not included (and is represented with a *broken line*) if the original symbol is $<$ or $>$.

VIDEO EXAMPLES

SECTION 7.7

EXAMPLE 1 Graph the solution set for $x + y \le 4$.

SOLUTION The boundary for the graph is the graph of $x + y = 4$. The boundary is included in the solution set because the inequality symbol is $\le$.
Figure 2 is the graph of the boundary:

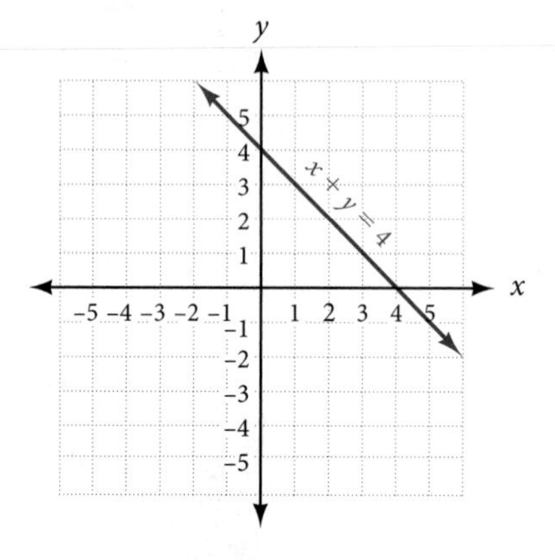

FIGURE 2

The boundary separates the coordinate plane into two regions: the region above the boundary and the region below it. The solution set for $x + y \le 4$ is one of these two regions along with the boundary. To find the correct region, we simply choose any convenient point that is *not* on the boundary. We then substitute the coordinates of the point into the original inequality $x + y \le 4$. If the point we choose satisfies the inequality, then it is a member of the solution set, and we can assume that all points on the same side of the boundary as the chosen point are also in the solution set. If the coordinates of our point do not satisfy the original inequality, then the solution set lies on the other side of the boundary.

In this example, a convenient point that is not on the boundary is the origin.

Substituting	$(0, 0)$	
into	$x + y \le 4$	
gives us	$0 + 0 \le 4$	
	$0 \le 4$	A true statement

Because the origin is a solution to the inequality $x + y \le 4$ and the origin is below the boundary, all other points below the boundary are also solutions.

Figure 3 is the graph of $x + y \le 4$.

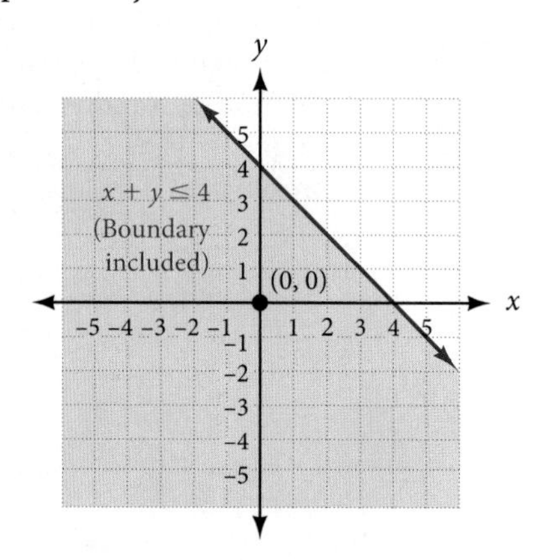

FIGURE 3

The region above the boundary is described by the inequality $x + y > 4$. ■

Here is a list of steps to follow when graphing the solution sets for linear inequalities in two variables.

HOW TO *Graph a Linear Inequality in Two Variables*

Step 1: Replace the inequality symbol with an equal sign. The resulting equation represents the boundary for the solution set.

Step 2: Graph the boundary found in step 1. Use a *solid line* if the boundary is included in the solution set (i.e., if the original inequality symbol was either $\leq$ or $\geq$). Use a *broken line* to graph the boundary if it is *not* included in the solution set. (It is not included if the original inequality was either $<$ or $>$.)

Step 3: Choose any convenient point not on the boundary and substitute the coordinates into the *original* inequality. If the resulting statement is *true*, the solution set lies on the *same* side of the boundary as the chosen point. If the resulting statement is *false*, the solution set lies on the *opposite* side of the boundary.

EXAMPLE 2 Graph the solution set for $y < 2x - 3$.

SOLUTION The boundary is the graph of $y = 2x - 3$, a line with slope 2 and y-intercept -3. The boundary is not included because the original inequality symbol is $<$. We therefore use a broken line to represent the boundary in Figure 4.

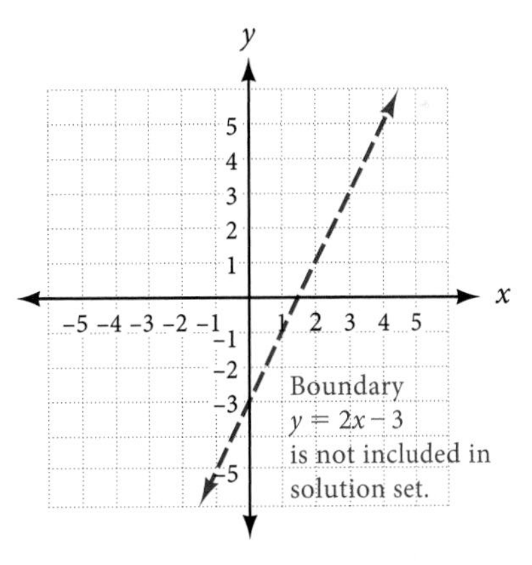

FIGURE 4

A convenient test point is again the origin:

Using	$(0, 0)$	
in	$y < 2x - 3$	
we have	$0 < 2(0) - 3$	
	$0 < -3$	A false statement

Because our test point gives us a false statement and it lies above the boundary, the solution set must lie on the other side of the boundary (Figure 5).

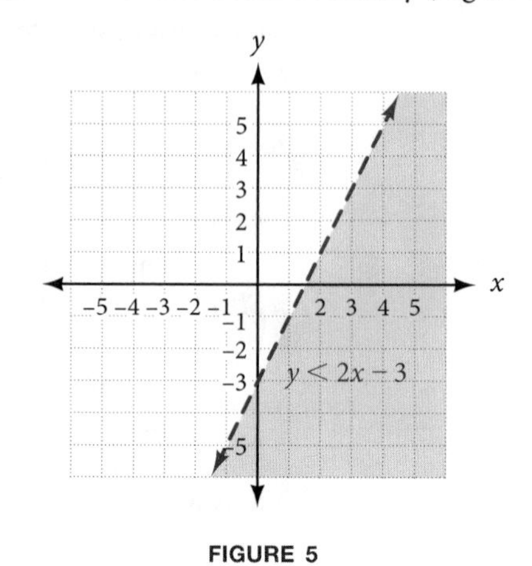

FIGURE 5

EXAMPLE 3 Graph the solution set for $x \leq 5$.

SOLUTION The boundary is $x = 5$, which is a vertical line. All points in Figure 6 to the left have x-coordinates less than 5 and all points to the right have x-coordinates greater than 5.

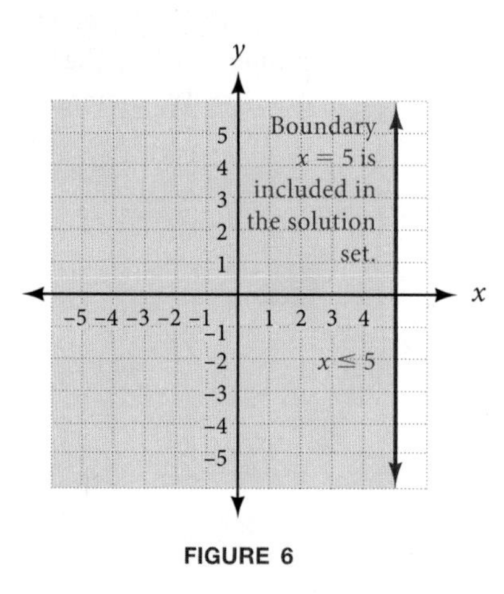

FIGURE 6

Systems of Linear Inequalities

If we form a system of inequalities with two or more inequalities, the solution set will be the points common to all the solution sets. It is the intersection of the solution sets. Therefore, the solution set for the system of inequalities

$$x + y < 4$$
$$-x + y \leq 3$$

is all the ordered pairs that satisfy both inequalities. It is the set of points that are below the line $x + y = 4$, and also below (and including) the line $-x + y = 3$. The graph of the solution set to this system is shown in Figure 7. We have written the system in Figure 7 with the word *and* just to remind you that the solution set to a system of equations or inequalities is all the points that satisfy both equations or inequalities.

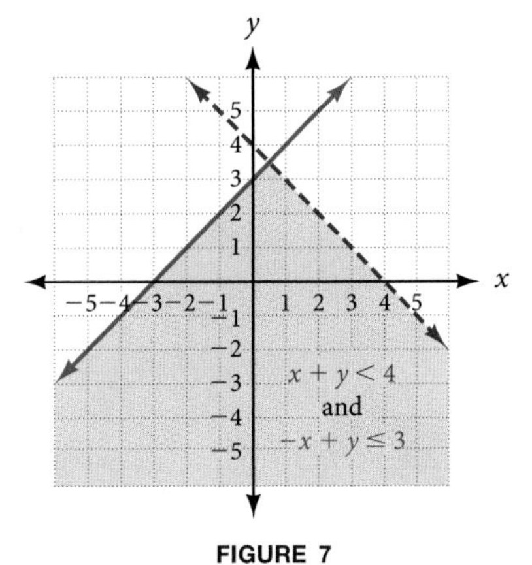

FIGURE 7

EXAMPLE 4 Graph the solution to the system of linear inequalities.

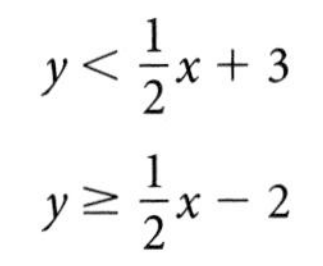

$$y < \frac{1}{2}x + 3$$

$$y \geq \frac{1}{2}x - 2$$

SOLUTION Figures 8 and 9 show the solution set for each of the inequalities separately.

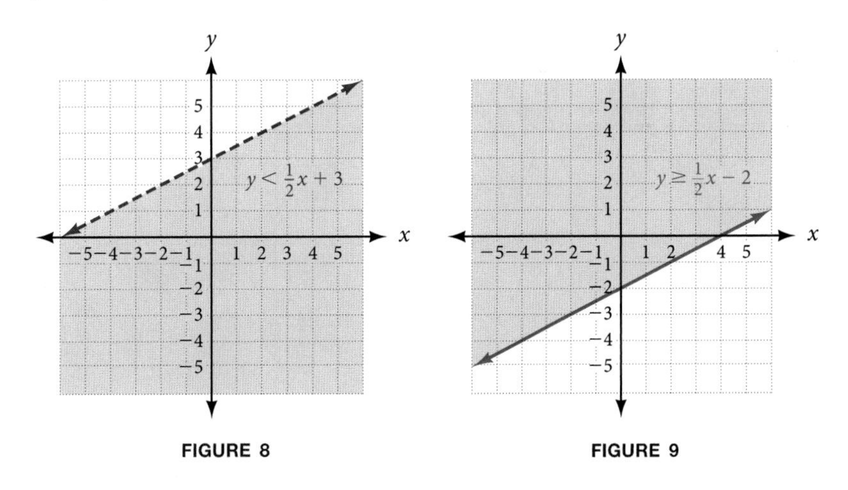

FIGURE 8 FIGURE 9

Figure 10 is the solution set to the system of inequalities. It is the region consisting of points whose coordinates satisfy both inequalities.

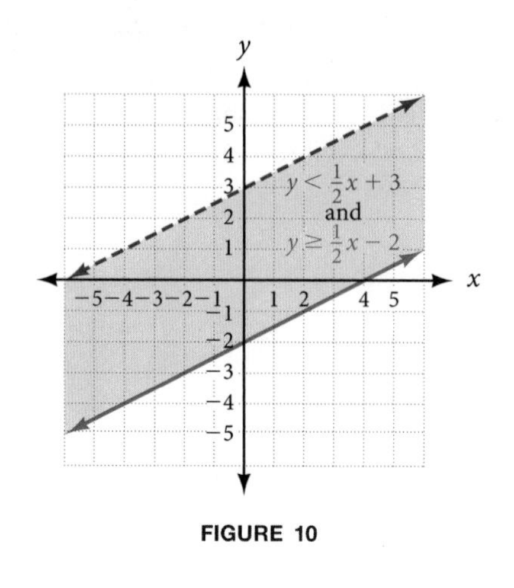

FIGURE 10

EXAMPLE 5 Graph the solution to the system of linear inequalities.

$$x + y < 4$$
$$x \geq 0$$
$$y \geq 0$$

SOLUTION We graphed the first inequality, $x + y < 4$, in Figure 7 on the previous page. The solution set to the inequality $x \geq 0$, shown in Figure 11, is all the points to the right of the y-axis; that is, all the points with x-coordinates that are greater than or equal to 0. Figure 12 shows the graph of $y \geq 0$. It consists of all points with y-coordinates greater than or equal to 0; that is, all points from the x-axis up.

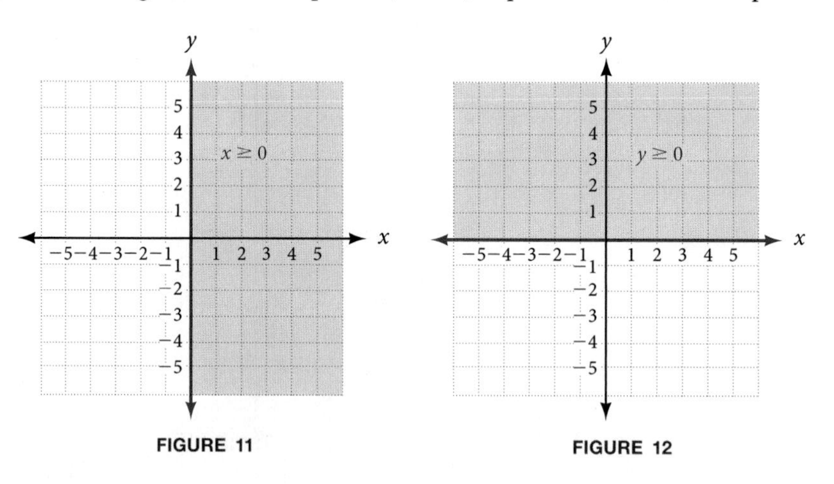

FIGURE 11 FIGURE 12

The regions shown in Figures 11 and 12 overlap in the first quadrant. Therefore, putting all three regions together we have the points in the first quadrant that are below the line $x + y = 4$. This region is shown in Figure 13, and it is the solution to our system of inequalities.

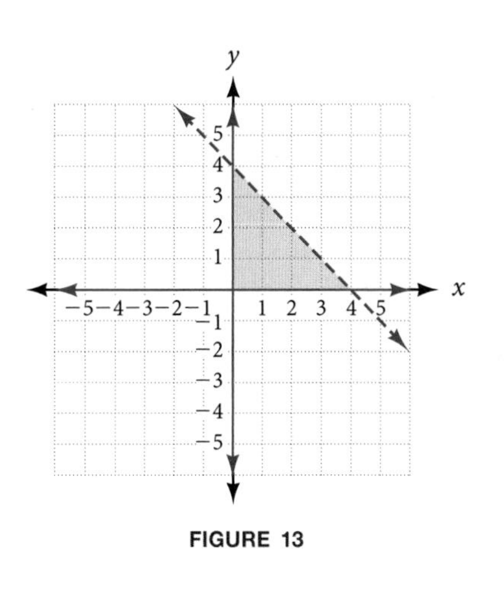

FIGURE 13

EXAMPLE 6 Graph the solution to the system of linear inequalities.

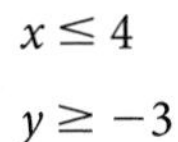

$$x \leq 4$$
$$y \geq -3$$

SOLUTION The solution to this system will consist of all points to the left of and including the vertical line $x = 4$ that intersect with all points above and including the horizontal line $y = -3$. The solution set is shown in Figure 14.

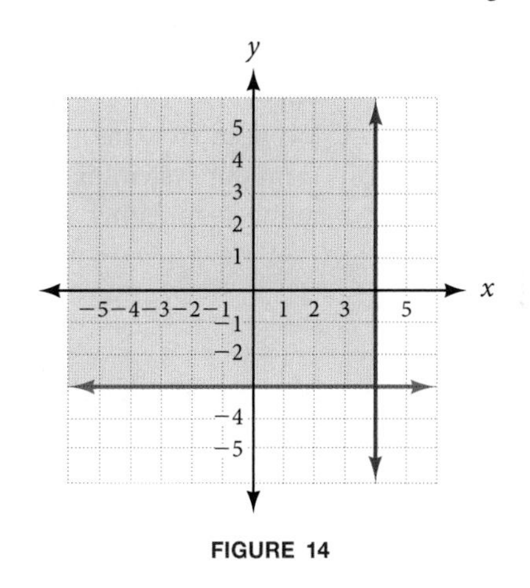

FIGURE 14

EXAMPLE 7 Graph the solution set for the following system.

$$x - 2y \leq 4$$
$$x + y \leq 4$$
$$x \geq -1$$

SOLUTION We have three linear inequalities, representing three sections of the coordinate plane. The graph of the solution set for this system will be the intersection of these three sections. The graph of $x - 2y \leq 4$ is the section above and including the boundary $x - 2y = 4$. The graph of $x + y \leq 4$ is the section below and including the boundary line $x + y = 4$. The graph of $x \geq -1$ is all the points to the right of, and including, the vertical line $x = -1$. The intersection of these three graphs is shown in Figure 15.

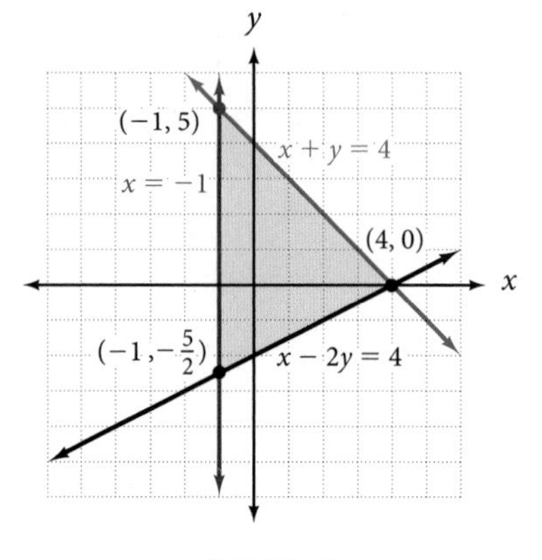

FIGURE 15

Modeling: Ticket Sales

EXAMPLE 8 A college basketball arena plans on charging \$20 for certain seats and \$15 for others. They want to bring in more than \$18,000 from all ticket sales and have reserved at least 500 tickets at the \$15 rate. Find a system of inequalities describing all possibilities and sketch the graph. If 620 tickets are sold for \$15, at least how many tickets are sold for \$20?

SOLUTION Let $x =$ the number of \$20 tickets and $y =$ the number of \$15 tickets. We need to write a list of inequalities that describe this situation. That list will form our system of inequalities. First of all, we note that we cannot use negative numbers for either x or y. So, we have our first inequalities:

$$x \geq 0$$

$$y \geq 0$$

Next, we note that they are selling at least 500 tickets for \$15, so we can replace our second inequality with $y \geq 500$. Now our system is

$$x \geq 0$$

$$y \geq 500$$

Now the amount of money brought in by selling \$20 tickets is $20x$, and the amount of money brought in by selling \$15 tickets is $15y$. It the total income from ticket sales is to be more than \$18,000, then $20x + 15y$ must be greater than 18,000. This gives us our last inequality and completes our system.

$$x \geq 0$$

$$y \geq 500$$

$$20x + 15y > 18{,}000$$

We have used all the information in the problem to arrive at this system of inequalities. The solution set contains all the values of x and y that satisfy all the conditions given in the problem. Here is the graph of the solution set.

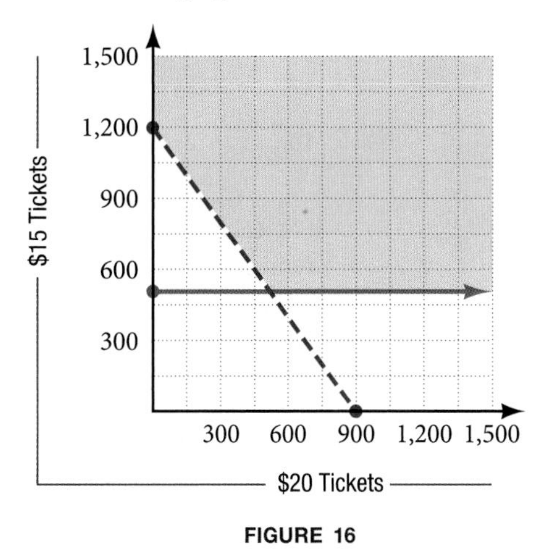

FIGURE 16

If 620 tickets are sold for \$15, then we substitute 620 for y in our first inequality to obtain

$$20x + 15(620) > 18{,}000 \quad \text{Substitute 620 for } y$$
$$20x + 9{,}300 > 18{,}000 \quad \text{Multiply}$$
$$20x > 8{,}700 \quad \text{Add } -9{,}300 \text{ to each side}$$
$$x > 435 \quad \text{Divide each side by 20}$$

If they sell 620 tickets for \$15 each, then they need to sell more than 435 tickets at \$20 each to bring in more than \$18,000. ■

GETTING READY FOR CLASS

After reading through the preceding section, respond in your own words and in complete sentences.

A. What is the significance of a broken line in the graph of an inequality?

B. When graphing a linear inequality in two variables, how do you know which side of the boundary line to shade?

C. What does the solution to a system of linear inequalities look like?

D. How can you tell if the boundary line for a linear inequality is a solid line?

Problem Set 7.7

Graph the solution set for each of the following.

1. $x + y < 5$

2. $x + y \le 5$

3. $2x + 3y < 6$

4. $2x - 3y > -6$

5. $-x + 2y > -4$

6. $-x - 2y < 4$

7. $y < 2x - 1$

8. $y \le 2x - 1$

9. $-5x + 2y \le 10$

10. $4x - 2y \le 8$

For each graph shown here, name the linear inequality in two variables that is represented by the shaded region.

11. **12.** **13.** **14.**

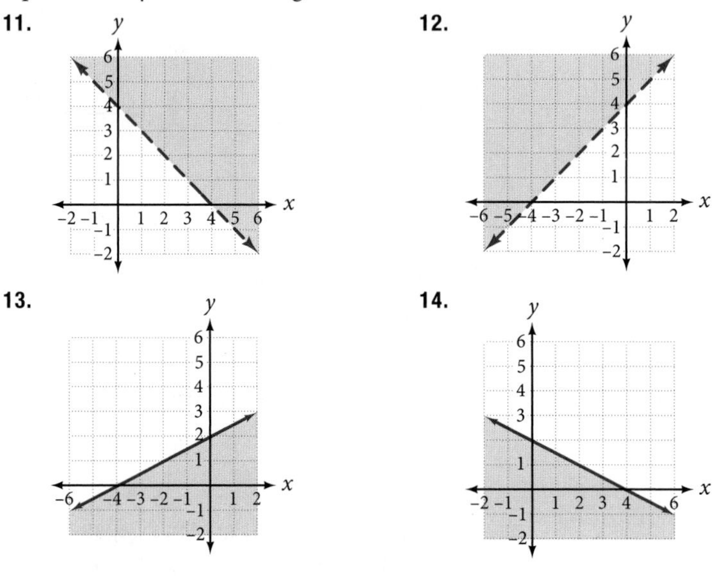

Graph the solution set for each system of linear inequalities.

15. $x + y < 5$
$2x - y > 4$

16. $x + y < 5$
$2x - y < 4$

17. $y < \frac{1}{3}x + 4$
$y \ge \frac{1}{3}x - 3$

18. $y < 2x + 4$
$y \ge 2x - 3$

19. $x \ge -3$
$y < 2$

20. $x \le 4$
$y \ge -2$

21. $1 \le x \le 3$
$2 \le y \le 4$

22. $-4 \le x \le -2$
$1 \le y \le 3$

23. $x + y \le 4$
$x \ge 0$
$y \ge 0$

24. $x - y \le 2$
$x \ge 0$
$y \le 0$

25. $x + y \le 3$
$x - 3y \le 3$
$x \ge -2$

26. $x - y \le 4$
$x + 2y \le 4$
$x \ge -1$

27. $x + y \le 2$
$-x + y \le 2$
$y \ge -2$

28. $x - y \le 3$
$-x - y \le 3$
$y \le -1$

29. $x + y < 5$
$y > x$
$y \ge 0$

30. $x + y < 5$
$y > x$
$x \ge 0$

31. $2x + 3y \le 6$
$x \ge 0$
$y \ge 0$

32. $x + 2y \le 10$
$3x + 2y \le 12$
$x \ge 0$
$y \ge 0$

For each figure below, find a system of inequalities that describes the shaded region.

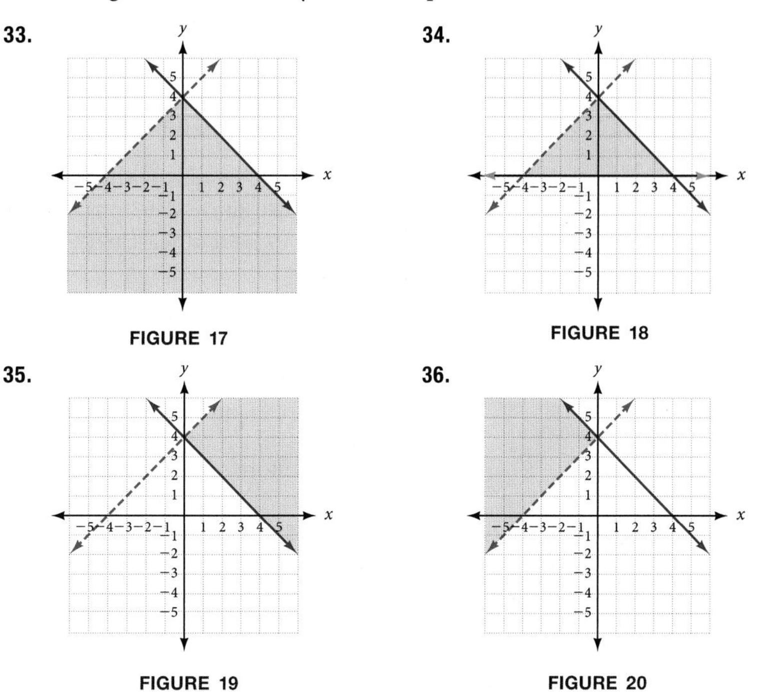

33. FIGURE 17

34. FIGURE 18

35. FIGURE 19

36. FIGURE 20

Modeling Practice

37. Number of People in a Dance Club A dance club holds a maximum of 200 people. The club charges one price for students and a higher price for nonstudents. If the number of students in the club at any time is x and the number of nonstudents is y, sketch the graph and shade the region in the first quadrant that contains all combinations of students and nonstudents that are in the club at any time.

38. Many Perimeters Suppose you have 500 feet of fencing that you will use to build a rectangular livestock pen. Let x represent the length of the pen and y represent the width. Sketch the graph and shade the region in the first quadrant that contains all possible values of x and y that will give you a rectangle from 500 feet of fencing. (You don't have to use all of the fencing, so the perimeter of the pen could be less than 500 feet.)

39. Gas Mileage You have two cars. The first car travels an average of 12 miles on a gallon of gasoline, and the second averages 22 miles per gallon. Suppose you can afford to buy up to 30 gallons of gasoline this month. If the first car is driven x miles this month, and the second car is driven y miles this month, sketch the graph and shade the region in the first quadrant that gives all the possible values of x and y that will keep you from buying more than 30 gallons of gasoline this month.

40. **Student Loan Payments** When considering how much debt to incur in student loans, it is advisable to keep your student loan payment after graduation to 8% or less of your starting monthly income. Let x represent your starting monthly salary and let y represent your monthly student loan payment. Write an inequality that describes this situation. Sketch the graph and shade the region in the first quadrant that is a solution to your inequality.

41. **Office Supplies** An office worker wants to purchase some \$0.55 postage stamps and also some \$0.65 postage stamps totaling no more than \$40. It is also desired to have at least twice as many \$0.55 stamps and more than 15 \$0.55 stamps.

- **a.** Find a system of inequalities describing all the possibilities and sketch the graph.
- **b.** If he purchases 20 \$0.55 stamps, what is the maximum number of \$0.65 stamps he can purchase?

42. **Inventory** A store sells two brands of DVD players. Customer demand indicates that it is necessary to stock at least twice as many DVD players of brand A as of brand B. At least 30 of brand A and 15 of brand B must be on hand. In the store, there is room for not more than 100 DVD players.

- **a.** Find a system of inequalities describing all possibilities, then sketch the graph.
- **b.** If there are 35 DVD players of brand A, what is the maximum number of brand B DVD players on hand?

Maintaining Your Skills

For each of the following straight lines, identify the x-intercept, y-intercept, and slope, and sketch the graph.

43. $2x + y = 6$ **44.** $y = \frac{3}{2}x + 4$ **45.** $x = -2$

Find the equation for each line.

46. Give the equation of the line through $(-1, 3)$ that has slope $m = 2$.

47. Give the equation of the line through $(-3, 2)$ and $(4, -1)$.

48. Line l contains the point $(5, -3)$ and has a graph parallel to the graph of $2x - 5y = 10$. Find the equation for l.

49. Give the equation of the vertical line through $(4, -7)$.

State the domain and range for the following relations, and indicate which relations are also functions.

50. $\{(-2, 0), (-3, 0), (-2, 1)\}$ **51.** $y = x^2 - 9$

Let $f(x) = x - 2$, $g(x) = 3x + 4$ and $h(x) = 3x^2 - 2x - 8$, and find the following.

52. $f(3) + g(2)$

53. $h(0) + g(0)$

54. $f[g(2)]$

55. $g[f(2)]$

Solve the following variation problems.

56. Direct Variation Quantity y varies directly with the square of x. If y is 50 when x is 5, find y when x is 3.

57. Joint Variation Quantity z varies jointly with x and the cube of y. If z is 15 when x is 5 and y is 2, find z when x is 2 and y is 3.

SPOTLIGHT ON SUCCESS *Student Instructor Aaron*

Sometimes you have to take a step back in order to get a running start forward.
—Anonymous

As a high school senior I was encouraged to go to college immediately after graduating. I earned good grades in high school and I knew that I would have a pretty good group of schools to pick from. Even though I felt like "more school" was not quite what I wanted, the counselors had so much faith and had done this process so many times that it was almost too easy to get the applications out. I sent out applications to schools I knew I could get into and a "dream school."

One night in my email inbox there was a letter of acceptance from my dream school. There was just one problem with getting into this school: it was going to be difficult and I still had senioritis. Going into my first quarter of college was as exciting and difficult as I knew it would be. But after my first quarter I could see that this was not the time for me to be here. I was interested in the subject matter but I could not find my motivating purpose like I had in high school. Instead of dropping out completely, I decided a community college would be a good way for me to stay on track. Without necessarily knowing my direction, I could take the general education classes and get those out of the way while figuring out exactly what and where I felt a good place for me to be.

Now I know what I want to go to school for and the next time I walk onto a four year campus it will be on my terms with my reasons for being there driving me to succeed. I encourage everyone to continue school after high school, even if you have no clue as to what you want to study. There are always stepping stones, like community colleges, that can help you get a clearer picture of what you want to strive for.

Chapter 7 Summary

The numbers in brackets refer to the section(s) in which the topic can be found.

Systems of Linear Equations [7.1, 7.2]

A system of linear equations consists of two or more linear equations considered simultaneously. The solution set to a linear system in two variables is the set of ordered pairs that satisfy both equations. The solution set to a linear system in three variables consists of all the ordered triples that satisfy each equation in the system.

EXAMPLES

1. The solution to the system

$$x + 2y = 4$$
$$x - y = 1$$

is the ordered pair (2, 1). It is the only ordered pair that satisfies both equations.

To Solve a System by the Addition Method [7.1]

Step 1: Look the system over to decide which variable will be easier to eliminate.

Step 2: Use the multiplication property of equality on each equation separately, if necessary, to ensure that the coefficients of the variable to be eliminated are opposites.

Step 3: Add the left and right sides of the system produced in Step 2, and solve the resulting equation.

Step 4: Substitute the solution from Step 3 back into any equation with both x- and y-variables, and solve.

Step 5: Check your solution in both equations if necessary.

2. We can eliminate the y-variable from the system in Example 1 by multiplying both sides of the second equation by 2 and adding the result to the first equation:

$$\begin{aligned} x + 2y = 4 &\xrightarrow{\text{No Change}} & x + 2y &= 4 \\ x - y = 1 &\xrightarrow[\text{Multiply by 2}]{} & 2x - 2y &= 2 \\ & & 3x &= 6 \\ & & x &= 2 \end{aligned}$$

Substituting $x = 2$ into either of the original two equations gives $y = 1$. The solution is (2, 1).

To Solve a System by the Substitution Method [7.1]

Step 1: Solve either of the equations for one of the variables (this step is not necessary if one of the equations has the correct form already).

Step 2: Substitute the results of Step 1 into the other equation, and solve.

Step 3: Substitute the results of Step 2 into an equation with both x- and y-variables, and solve. (The equation produced in step 1 is usually a good one to use.)

Step 4: Check your solution if necessary.

3. We can apply the substitution method to the system in Example 1 by first solving the second equation for x to get

$$x = y + 1$$

Substituting this expression for x into the first equation we have

$$\begin{aligned} (y + 1) + 2y &= 4 \\ 3y + 1 &= 4 \\ 3y &= 3 \\ y &= 1 \end{aligned}$$

Using $y = 1$ in either of the original equations gives $x = 2$.

Inconsistent Systems and Dependent Equations [7.1, 7.2]

A system of two linear equations that have no solutions in common is said to be an *inconsistent* system, whereas two linear equations that have all their solutions in common are said to be *dependent* equations.

4. If the two lines are parallel, then the system will be inconsistent and the solution is $\varnothing$. If the two lines coincide, then the equations are dependent.

Determinant [7.3]

5a. $\begin{vmatrix} 3 & 4 \\ -2 & 5 \end{vmatrix} = 15 - (-8) = 23$

5b. Expanding $\begin{vmatrix} 1 & 3 & -2 \\ 2 & 0 & 1 \\ 4 & -1 & 1 \end{vmatrix}$ across the first row gives us

$$1\begin{vmatrix} 0 & 1 \\ -1 & 1 \end{vmatrix} - 3\begin{vmatrix} 2 & 1 \\ 4 & 1 \end{vmatrix} - 2\begin{vmatrix} 2 & 0 \\ 4 & -1 \end{vmatrix}$$

$$= 1(1) - 3(-2) - 2(-2)$$

$$= 11$$

The value of the 2×2 (2 by 2) *determinant*

$$\begin{vmatrix} a & c \\ b & d \end{vmatrix}$$

is given by

$$\begin{vmatrix} a & c \\ b & d \end{vmatrix} = ad - bc$$

The value of the 3×3 *determinant*

$$\begin{vmatrix} a_1 & b_1 & c_1 \\ a_2 & b_2 & c_2 \\ a_3 & b_3 & c_3 \end{vmatrix}$$

is given by

$$\begin{vmatrix} a_1 & b_1 & c_1 \\ a_2 & b_2 & c_2 \\ a_3 & b_3 & c_3 \end{vmatrix} = a_1b_2c_3 + a_3b_1c_2 + a_2b_3c_1 - a_3b_2c_1 - a_1b_3c_2 - a_2b_1c_3$$

Cramer's Rule [7.4]

6. For the system $x + y = 6$
$3x - 2y = -2$

we have

$$D = \begin{vmatrix} 1 & 1 \\ 3 & -2 \end{vmatrix} = -5$$

$$D_x = \begin{vmatrix} 6 & 1 \\ -2 & -2 \end{vmatrix} = -10$$

$$x = \frac{-10}{-5} = 2$$

$$D_y = \begin{vmatrix} 1 & 6 \\ 3 & -2 \end{vmatrix} = -20$$

$$y = \frac{-20}{-5} = 4$$

The solution set to the system

$$a_1x + b_1y = c_1$$
$$a_2x + b_2y = c_2$$

is given by

$$x = \frac{D_x}{D}, \qquad y = \frac{D_y}{D}$$

where

$$D = \begin{vmatrix} a_1 & b_1 \\ a_2 & b_2 \end{vmatrix} \quad D_x = \begin{vmatrix} c_1 & b_1 \\ c_2 & b_2 \end{vmatrix} \quad D_y = \begin{vmatrix} a_1 & c_1 \\ a_2 & c_2 \end{vmatrix} \quad (D \neq 0)$$

Augmented Matrices [7.5]

The system

$$2x + 5y = -4$$
$$x - 3y = 9$$

can be represented by the matrix

$$\left[\begin{array}{cc|c} 2 & 5 & -4 \\ 1 & -3 & 9 \end{array}\right]$$

which is called an *augmented matrix* because it includes both the coefficients of the variables and the constant terms.

Row Operations [7.5]

To solve a system of linear equations by using the augmented matrix for that system, we need the following row operations as the tools of that solution process. The row operations tell us what we can do to an augmented matrix that may change the numbers in the matrix, but will always produce a matrix that represents a system of equations with the same solution as that of our original system.

1. We can interchange any two rows of a matrix.
2. We can multiply any row by a nonzero constant.
3. We can add to any row a constant multiple of another row.

Systems of Linear Inequalities [7.7]

7. The solution set for the system

$$x + y < 4$$
$$-x + y \le 3$$

is shown below.

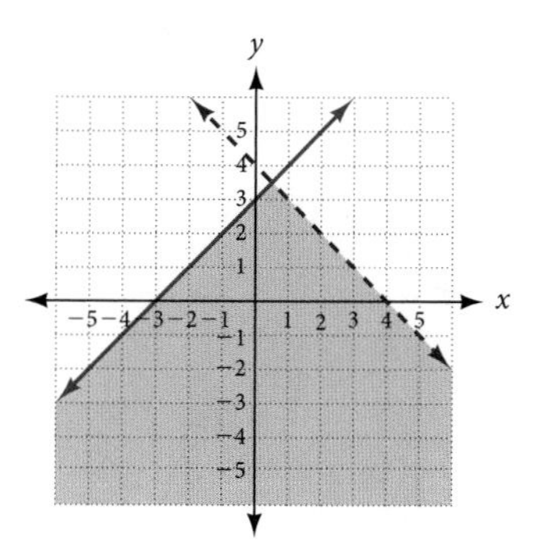

A system of linear inequalities is two or more linear inequalities considered at the same time. To find the solution set to the system, we graph each of the inequalities on the same coordinate system. The solution set is the region that is common to all the regions graphed.

Chapter 7 Test

Solve the following systems by the addition method. [7.1]

1. $2x - 5y = -8$
   $3x + y = 5$

2. $4x - 7y = -2$
   $-5x + 6y = -3$

3. $\frac{1}{3}x - \frac{1}{6}y = 3$
   $-\frac{1}{5}x + \frac{1}{4}y = 0$

Solve the following systems by the substitution method. [7.1]

4. $2x - 5y = 14$
   $y = 3x + 8$

5. $6x - 3y = 0$
   $x + 2y = 5$

Solve each system. [7.1, 7.2]

6. $2x - y + z = 9$
   $x + y - 3z = -2$
   $3x + y - z = 6$

7. $2x + 4y = 3$
   $-4x - 8y = -6$

8. Solve for x. [7.3]

$$\begin{vmatrix} 2 & 3x \\ -1 & 2x \end{vmatrix} = 4$$

9. Evaluate. [7.3]

$$\begin{vmatrix} 1 & 0 & -3 \\ 2 & 1 & 0 \\ 0 & 5 & 4 \end{vmatrix}$$

10. Use Cramer's rule to solve. [7.4]

$2x + 4y = 3$

$-4x - 8y = -6$

11. Set up the augmented matrix that corresponds to this system. [7.5]

$2x - y + 3z = 1$

$x - 4y - z = 6$

$3x - 2y + z = 4$

12. Write the augmented matrix that corresponds to this system. Then use that matrix to solve the system. [7.5]

$x + y = 4$

$2x + 5y = 2$

Solve each application problem. [7.6]

13. Number Problem A number is 1 less than twice another. Their sum is 14. Find the two numbers.

14. Investing John invests twice as much money at 6% as he does at 5%. If his investments earn a total of $680 in 1 year, how much does he have invested at each rate?

15. Ticket Cost There were 750 tickets sold for a basketball game for a total of $1,090. If adult tickets cost $2.00 and children's tickets cost $1.00, how many of each kind were sold?

16. Mixture Problem How much 30% alcohol solution and 70% alcohol solution must be mixed to get 16 gallons of 60% solution?

17. Speed of a Boat A boat can travel 20 miles downstream in 2 hours. The same boat can travel 18 miles upstream in 3 hours. What is the speed of the boat in still water, and what is the speed of the current?

18. Coin Problem A collection of nickels, dimes, and quarters consists of 15 coins with a total value of $1.10. If the number of nickels is 1 less than 4 times the number of dimes, how many of each coin are contained in the collection?

Graph the solution set for each system of linear inequalities. [7.7]

19.
$$\begin{aligned} x + 4y &\le 4 \\ -3x + 2y &> -12 \end{aligned}$$

20.
$$\begin{aligned} y &< -\frac{1}{2}x + 4 \\ x &\ge 0 \\ y &\ge 0 \end{aligned}$$

Selected Answers

Chapter 1

Problem Set 1.1

1. $x + 5 = 2$ **3.** $6 - x = y$ **5.** $2t < y$ **7.** $x + y < x - y$ **9.** $\frac{1}{2}(t - s) < 2(t + s)$ **11. a.** 19 **b.** 27 **c.** 27 **13. a.** 33 **b.** 33 **c.** 9 **15. a.** 23 **b.** 41 **c.** 65 **17.** 39 **19.** -5 **21.** 2 **23.** 0 **25. a.** 7 **b.** 10 **c.** 25 **d.** 32 **27. a.** 121 **b.** 121 **c.** 101 **d.** 81 **29. a.** 1 **b.** 2,400

31.

		Sum	Diff	Product	Quotient
a	b	$a+b$	$a-b$	ab	$\frac{a}{b}$
3	12	15	-9	36	$\frac{1}{4}$
-3	12	9	-15	-36	$-\frac{1}{4}$
3	-12	-9	15	-36	$-\frac{1}{4}$
-3	-12	-15	9	36	$\frac{1}{4}$

33. 2 **35.** -4 **37.** $3x + 18$ **39.** $15a + 10b$ **41.** $2 + y$ **43.** $9x + 3y - 6z$ **45.** $6x + 7y$ **47.** $2x - 1$ **49.** $a - 3$ **51.** $3x - 2y$ **53.** $8x - 3y$ **55.** $15x + 10$ **57.** $15t + 9$ **59.** $14a + 7$ **61.** $24a + 15$ **63.** $17x + 14y$ **65.** $14x + 12$ **67.** $-2x + 9$ **69.** $-20x + 5$ **71.** $A \cup B = \{0, 1, 2, 3, 4, 5, 6\}$ **73.** $A \cap B = \{2, 4\}$ **75.** $B \cap C = \{1, 3, 5\}$ **77.** $A \cup (B \cap C) = \{0, 1, 2, 3, 4, 5, 6\}$ **79.** 1, 2 **81.** $-6, -5.2, 0, 1, 2, 2.3, \frac{9}{2}$ **83.** $-\sqrt{7}, -\pi, \sqrt{17}$ **85. a.** 2.9 seconds **b.** 7.5 seconds

87. a. $5 + 8 = 13$ **b.** $2 + 3 = 5$ **c.** $9 + 16 = 25$ **d.** $49 + 576 = 625$

Problem Set 1.2

1. 16 **3.** -0.027 **5.** $\frac{1}{8}$ **7.** x^9 **9.** $-6a^6$ **11.** $\frac{1}{9}$ **13.** $\frac{16}{9}$ **15.** $80x^3y^6$ **17.** $72x^3y^3$ **19.** $15xy$ **21.** $12x^7y^6$ **23.** $2x^3$ **25.** $4x$ **27.** $-4xy^4$ **29.** $-2x^2y^2$ **31.** $\frac{1}{x^{10}}$ **33.** a^{10} **35.** $\frac{1}{t^6}$ **37.** x^{12} **39.** x^{18} **41.** $\frac{1}{x^{22}}$ **43.** $\frac{a^3b^7}{4}$ **45.** $\frac{y^{38}}{x^{16}}$ **47.** 3.78×10^5 **49.** 4.9×10^3 **51.** 3.7×10^{-4} **53.** 4.95×10^{-3} **55.** 5,340 **57.** 7,800,000

59. 0.00344 **61.** 0.49 **63.** 8×10^4 **65.** 2×10^9 **67.** 2.5×10^{-6} **69.** 1.8×10^{-7}

71.

Expanded Form	Scientific Notation $n \times 10^r$
0.000357	3.57×10^{-4}
0.00357	3.57×10^{-3}
0.0357	3.57×10^{-2}
0.357	3.57×10^{-1}
3.57	3.57×10^{0}
35.7	3.57×10^{1}
357	3.57×10^{2}
3,570	3.57×10^{3}
35,700	3.57×10^{4}

73.

Jupiter's Moon	Period (seconds)	
Io	153,000	1.53×10^5
Europa	307,000	3.07×10^5
Ganymede	618,000	6.18×10^5
Callisto	1,440,000	1.44×10^6

75. 6.3×10^8 **77.** 1.475×10^{19} miles **79. a.** 7.72×10^{11} **b.** \$8.58 $\times 10^3$

Problem Set 1.3

1. Trinomial, degree 2, leading coefficient 5 **3.** Binomial, degree 1, leading coefficient 3 **5.** Trinomial, degree 2, leading coefficient 8 **7.** Polynomial, degree 3, leading coefficient 4 **9.** Monomial, degree 0, leading coefficient $-\frac{3}{4}$ **11.** Trinomial, degree 3, leading coefficient 6

13. $2x^2 + 7x - 15$ **15.** $x^2 + 9x + 4$ **17.** $a^2 + 5a + 4$ **19.** $x^2 - 13x + 3$ **21.** $-3x$ **23.** $x^2 - 6x - 5$ **25.** $12x^3 - 10x^2 + 8x$ **27.** $a^3 - b^3$ **29.** $8x^3 + y^3$ **31.** $6a^2 + 13a + 6$ **33.** $20 - 2t - 6t^2$ **35.** $x^6 - 2x^3 - 15$ **37.** $18t^2 - \frac{2}{9}$ **39.** $4a^2 - 12a + 9$ **41.** $25x^2 + 20xy + 4y^2$ **43.** $25 - 30t^3 + 9t^6$ **45.** $4a^2 - 9b^2$ **47.** $9r^4 - 49s^2$ **49.** $5x$ **51.** $8x^2$ **53.** $2x^3$ **55.** -5 **57.** $-2x + 3$ **59.** $9x + 6$ **61. a.** $3x - 3a$ **b.** $2x - 2a$ **63.** $2x^3 + 7x - 15$ **65.** $6x^3 - 11x^2y + 11xy^2 - 12y^3$ **67.** $-12x^4$ **69.** $20x^5$ **71.** a^6 **73.** $x - 5$ **75.** $x - 7$ **77.** $5x - 1$ **79.** $x - 1$ **81.** $3x + 8y$ **83.** $x^2 + 4x - 5$ **85.** $4a^2 - 30a + 56$ **87.** $32a^2 + 20a - 18$ **89.** 48 feet; 48 feet

Problem Set 1.4

1. $(x - 6)(x + 4)$ **3.** $(x - 6)(x + 1)$ **5.** $(x - 3)(x - 2)$ **7.** $(x - 5)^2$ **9.** $(2x + 1)(x - 3)$ **11.** $(7x - 3)(3x - 2)$ **13.** $(x + 4)(x - 4)$ **15.** $(a + 1)(a - 1)$ **17.** $(a + 4b)(a - 4b)$ **19.** $(3x + 7)(3x - 7)$ **21.** $(4x^2 + 7)(4x^2 - 7)$ **23.** $(t + 3)(t - 3)(t^2 + 9)$ **25.** $(a + b)(a^2 - ab + b^2)$ **27.** $(x - 2)(x^2 + 2x + 4)$ **29.** $(x + 1)(x^2 - x + 1)$ **31.** $(2x + 1)(4x^2 - 2x + 1)$ **33.** $10(3x - 2)(2x - 3)$ **35.** $x(x + 3)(x + 2)$ **37.** $x(2x + 1)(x - 3)$ **39.** $x(x - 6)(x + 4)$ **41.** $6(x + 4)$ **43.** $100x(x - 3)$ **45.** $5(2a + 3)(2a - 3)$ **47.** $a(3a + 4)(3a - 4)$ **49.** $2y(3 + x)(2 - x)$ **51.** $(a + 2)(x + 3)$ **53.** $(x - 3a)(x - 2)$ **55.** $(x + 2)(x + 3)(x - 3)$ **57.** $(x + 3)(x + 2)(x - 2)$ **59.** $(x + 3)(2x + 3)(2x - 3)$ **61.** $(2x + 1)(x + 3)(x - 3)$ **63.** $(4x + 1)(x - 8)$ **65.** prime **67.** $5x(6x - 7)(5x + 8)$ **69.** $(12x - 5)(2x + 1)$ **71.** $(x - 1)(x^2 + x + 1)(x + 1)(x^2 - x + 1)$ **73.** $\left(r + \frac{1}{3}\right)\left(r - \frac{1}{3}\right)$ **75.** $\left(5t + \frac{1}{3}\right)\left(25t^2 - \frac{5}{3}t + \frac{1}{9}\right)$ **77.** $(15t + 16)(t - 1)$ **79.** $100(x - 3)(x + 2)$ **81.** $4x(x^2 + 4y^2)$ **83.** $(5x + 7)(6x + 11)$ **85.** $(a + 5)(x + 3)^2$ **87.** $3(x - 7)(2a + 5)(2a - 5)$

Problem Set 1.5

1. $\frac{4x - 3y}{x(x + y)}$ **3.** $\frac{5x - 2}{x + 2}$ **5.** $\frac{(x - 4)^2}{x + 3}$ **7.** $-ad$ **9.** -1 **11.** $\frac{x + 4}{x + 5}$ **13.** $\frac{1}{4}$ **15.** $\frac{x - 3}{x + 3}$ **17.** $\frac{(a + 4)(a - 2)}{a + 6}$ **19.** $\frac{3}{a + b}$ **21.** $4x$ **23.** $-x + 1$ **25. a.** $\frac{5}{21}$ **b.** $\frac{5x + 3}{25x^2 + 15x + 9}$ **c.** $\frac{5x - 3}{25x^2 + 15x + 9}$ **d.** $\frac{5x + 3}{5x - 3}$ **27.** $\frac{7}{6}$ **29.** $\frac{2}{21}$ **31.** 1 **33.** $\frac{1}{5}$ **35.** $\frac{x - 5}{2(x - 6)}$ **37.** $\frac{2}{(a - 1)(a - 3)}$ **39.** $\frac{3}{(3x + 2)(3x + 4)}$ **41.** $\frac{6x + 7}{2x + 3}$ **43.** $\frac{3x^2 + 7x - 1}{3x + 4}$ **45. a.** $\frac{1}{16}$ **b.** $\frac{9}{4}$ **c.** $\frac{13}{24}$ **d.** $\frac{5x + 15}{(x - 3)^2}$ **e.** $\frac{x + 3}{5}$ **f.** $\frac{x - 2}{x - 3}$ **47.** $\frac{20}{21}$ **49.** $\frac{68}{3}$ **51.** $\frac{2x + y}{x + 2y}$ **53.** $\frac{2x - 1}{2x + 1}$ **55.** $\frac{2}{2x - 1}$

Problem Set 1.6

1. -7 **3.** 2 **5.** 0.2 **7.** 0.2 **9.** $3a^4$ **11.** xy^2 **13.** $2x^2y$ **15.** $2a^3b^5$ **17.** 125 **19.** 8 **21.** $\frac{1}{27}$ **23.** $\frac{6}{5}$ **25.** $\frac{8}{27}$ **27.** a **29.** $\frac{1}{x^{2/5}}$ **31.** $x^{1/6}$ **33.** $y^{3/10}$ **35. a.** 2 **b.** 0.2 **c.** 20 **d.** 0.02 **37.** $2\sqrt{2}$ **39.** $12\sqrt{2}$ **41.** $3\sqrt[3]{2}$ **43.** $4\sqrt[3]{2}$ **45.** $3x\sqrt{2x}$ **47.** $2y\sqrt[4]{2y^3}$ **49.** $2xy^2\sqrt[3]{5xy}$ **51.** $4abc^2\sqrt{3b}$ **53. a.** $\frac{\sqrt{5}}{2}$ **b.** $\frac{2\sqrt{5}}{5}$ **c.** $2 + \sqrt{3}$ **d.** 1 **55.** $\frac{\sqrt{2}}{2}$ **57.** $2\sqrt[3]{4}$ **59. a.** $\frac{\sqrt{2}}{2}$ **b.** $\frac{\sqrt[3]{4}}{2}$ **c.** $\frac{\sqrt[4]{8}}{2}$ **61.** $5|x|$ **63.** $3|xy|\sqrt{3x}$ **65.** $|x - 5|$ **67.** $7\sqrt{5}$ **69.** $\sqrt[3]{10}$ **71.** $\sqrt{5}$ **73.** $-3x\sqrt{2}$ **75.** $3\sqrt{2}$ **77.** $10\sqrt{21}$ **79.** $\sqrt{6} - 9$ **81.** $x + 2\sqrt{x} - 15$ **83.** $19 + 8\sqrt{3}$

85. 1 **87.** $25 - x$ **89.** $\frac{5 - \sqrt{5}}{4}$ **91.** $\frac{x + 3\sqrt{x}}{x - 9}$ **93** $\frac{10 + 3\sqrt{5}}{11}$ **95.** $\frac{3\sqrt{x} + 3\sqrt{y}}{x - y}$

97. a. $2\sqrt{x}$ **b.** $x - 4$ **c.** $x + 4\sqrt{x} + 4$ **d.** $\frac{x + 4\sqrt{x} + 4}{x - 4}$

Problem Set 1.7

1. $6i$ **3.** $-5i$ **5.** $6i\sqrt{2}$ **7.** $-2i\sqrt{3}$ **9.** 1 **11.** -1 **13.** $-i$ **15.** $x = 3, y = -1$ **17.** $x = -2, y = -\frac{1}{2}$

19. $x = -8, y = -5$ **21.** $x = 7, y = \frac{1}{2}$ **23.** $x = \frac{3}{7}, y = \frac{2}{5}$ **25.** $5 + 9i$ **27.** $5 - i$ **29.** $2 - 4i$ **31.** $1 - 6i$

33. $2 + 2i$ **35.** $-1 - 7i$ **37.** $6 + 8i$ **39.** $2 - 24i$ **41.** $-15 + 12i$ **43.** $18 + 24i$ **45.** $10 + 11i$ **47.** $21 + 23i$

49. $-2 + 2i$ **51.** $2 - 11i$ **53.** $-21 + 20i$ **55.** $-2i$ **57.** $-7 - 24i$ **59.** 5 **61.** 40 **63.** 13 **65.** 164

67. $-3 - 2i$ **69.** $-2 + 5i$ **71.** $\frac{8}{13} + \frac{12}{13}i$ **73.** $-\frac{18}{13} - \frac{12}{13}i$ **75.** $-\frac{5}{13} + \frac{12}{13}i$ **77.** $\frac{13}{15} - \frac{2}{5}i$

Chapter 1 Test

1. $2a - 3b < 2a + 3b$ **2.** $\{1, 2, 3, 4, 6\}$ **3.** -149 **4.** 2 **5.** x^8 **6.** $\frac{1}{32}$ **7.** a^2 **8.** 6.53×10^6 **9.** 3×10^8

10. $\frac{1}{2}x^3 - x^2 - 2x - 1$ **11.** $4x + 75$ **12.** $6y^2 + y - 35$ **13.** $2x^3 + 3x^2 - 26x + 15$ **14.** $64 - 48t^3 + 9t^6$

15. $1 - 36y^2$ **16.** $2(3x^2 - 1)(2x^2 + 5)$ **17.** $(x^2 - 2y)(7a - b^2)$ **18.** $\left(t + \frac{1}{2}\right)\left(t^2 - \frac{1}{2}t + \frac{1}{4}\right)$

19. $(3 - x)(3 + x)(9 + x^2)$ **20.** $\frac{x - 1}{x + 1}$ **21.** $2a + 8$ **22.** $x + 3$ **23.** $\frac{3x - 3}{x(x - 3)}$ **24.** $\frac{x}{(x + 4)(x + 5)}$ **25.** $\frac{3a + 8}{3a + 10}$

26. $\frac{x - 3}{x - 2}$ **27.** $\frac{1}{9}$ **28.** $\frac{7}{5}$ **29.** $a^{5/12}$ **30.** $5xy^2\sqrt{5xy}$ **31.** $2x^2y^2\sqrt[3]{5xy^2}$ **32.** $\frac{\sqrt{6}}{3}$ **33.** $-6\sqrt{3}$

34. $x + 3\sqrt{x} - 28$ **35.** $21 - 6\sqrt{6}$ **36.** $\frac{5(\sqrt{3} + 1)}{2}$ **37.** $6i$ **38.** $17 - 6i$ **39.** $9 - 40i$ **40.** $-\frac{5}{13} - \frac{12}{13}i$

Chapter 2

Problem Set 2.1

1. 3 **3.** 7,000 **5.** $-1, 6$ **7.** $-5, 1$ **9. a.** $\frac{25}{9}$ **b.** $-\frac{5}{3}, \frac{5}{3}$ **c.** $-3, 3$ **d.** $\frac{5}{3}$ **11.** No solution

13. No solution **15.** All real numbers **17.** -10 **19.** 20 **21.** 3 **23.** $-1, 9$ **25.** $\frac{17}{2}$ **27.** 5,000 **29.** $t = \frac{d}{r}$

31. $b = \frac{2A}{h}$ **33.** $R = \frac{PV}{nT}$ **35.** $C = \frac{5}{9}(F - 32)$ **37.** $n = \frac{A - a + d}{d}$ **39.** $y = -\frac{2}{7}x - 2$ **41.** $x = \frac{3}{a - c}$

43. $y = \frac{3}{4}x + 3$ **45.** $y = \frac{1}{4}x - \frac{1}{4}$ **47.** $y = -\frac{2}{3}x + 1$ **49. a.** $\frac{5}{8}$ **b.** $10x - 8$ **c.** $16x^2 - 34x + 15$ **d.** $\frac{3}{2}, \frac{5}{8}$

51. a. $-\frac{15}{4} = -3.75$ **b.** -7 **c.** $y = \frac{4}{5}x + 4$ **d.** $x = \frac{5}{4}y - 5$ **53.** $\pm 2, \pm i$ **55.** $0, \pm 3$ **57.** $\pm\sqrt{3}$ **59.** $-4, 1$

61. $\frac{20}{3}, 2$ **63. a.** -3 **b.** $-\frac{9}{2}$ **c.** 0 **d.** -6 **65.** $-\frac{7}{640}$ **67. a.** 3,400 **b.** 3,400 **69.** $r = 18$ **71.** 100, 130

73. 2, 3 seconds **75.** 6 miles per hour **77.** 42 miles per hour **79. a.** -60 °F **b.** -49 °F **81.** $\frac{°F - 30}{2}$ **83.** 169

85. 49 **87.** $\frac{85}{12}$ **89.** $(3t - 2)(9t^2 + 6t + 4)$ **91.** $2 + 4i$ **93.** $3 + \sqrt{2}$

Problem Set 2.2

1. ± 4 **3.** $\pm\frac{\sqrt{5}}{3}$ **5.** $\pm\frac{2\sqrt{5}}{3}$ **7.** $\frac{5 \pm 7i}{3}$ **9.** $6 \pm 2i\sqrt{2}$ **11.** 9, 3 **13.** 16, 4 **15.** $\frac{1}{4}, \frac{1}{2}$ **17.** $\frac{4}{25}, \frac{2}{5}$ **19.** 8, −2

21. $-3 \pm \sqrt{10}$ **23.** $-1, 4$ **25.** $\frac{5 \pm i\sqrt{31}}{14}$ **27.** $\frac{2 \pm \sqrt{5}}{5}$ **29.** No **31. a.** $\frac{7}{5}$ **b.** 3 **c.** $\frac{7 \pm 2\sqrt{2}}{5}$ **33.** $-6, 1$

35. $-2, -1$ **37.** $0, -5$ **39.** $\frac{3 \pm 3i\sqrt{7}}{4}$ **41.** $\frac{1 \pm 2i\sqrt{2}}{3}$ **43.** $\frac{4 \pm \sqrt{26}}{5}$ **45.** $\frac{5 \pm i\sqrt{47}}{4}$

47. $t = \frac{v \pm \sqrt{v^2 + 64h}}{32}$ **49.** $x = \frac{-4 \pm 2\sqrt{4 - k}}{k}$ **51.** a and b **53. a.** $\frac{5}{3}, 0$ **b.** $\frac{5}{3}, 0$ **55.** $\frac{\sqrt{3}}{2}$ inch, 1 inch

57. 781 feet **59.** $20\sqrt{2} \approx 28$ feet **61.** 2 seconds **63.** 20 or 60 items **65.** $(x - 1)(x + 5)$ **67.** $(y + 2)(4y - 1)$

69. $t = \frac{2}{3}$ **71.** $x = \pm 2i$ **73.** $t^2 + 10t + 25$**75.** 48 **77.** $y - 4$ **79.** $\frac{-1 \pm i\sqrt{3}}{3}$

Problem Set 2.3

1. $-1, -6$ **3.** $\pm\sqrt{2}, \pm 2i$ **5.** $-\frac{1}{3}, 1$ **7.** $\pm i\sqrt{2}, \pm\frac{2\sqrt{3}}{3}$ **9.** $5, \frac{-5 \pm 5i\sqrt{3}}{2}$ **11.** 4 **13.** 0 **15.** $-8, 1$ **17.** -6

19. $-\frac{11}{3}$ **21.** $-\frac{10}{7}$ **23.** 7 **25.** $\frac{62}{3}$ **27. a.** $\frac{1}{3}$ **b.** 3 **c.** 9 **d.** 4 **e.** $\frac{1}{3}, 3$ **29. a.** $\frac{6}{(x - 4)(x + 3)}$ **b.** $\frac{x - 3}{x - 4}$ **c.** 5

31. $y = 4x - 1$ **33.** $y = -3$ **35.** $y = \frac{2}{3}x - 2$ **37.** $y = -\frac{x}{4} + 2$ **39.** $y = \frac{3}{5}x - 3$ **41.** $y = \frac{x - 3}{x - 1}$ **43.** $y = \frac{1 - x}{3x - 2}$

45. 5 **47.** $\frac{5}{3}$ **49.** 8 **51.** 20 **53.** $-\frac{15}{2}$ **55.** 6; -1 doesn't check **57.** 1 **59.** 8; 0 doesn't check

61. No solution; 1 doesn't check **63.** 4 **65. a.** 100 **b.** 40 **c.** No solution **d.** 8; 5 doesn't check **67.** $t = \frac{1 \pm \sqrt{1 + h}}{4}$

69. a.

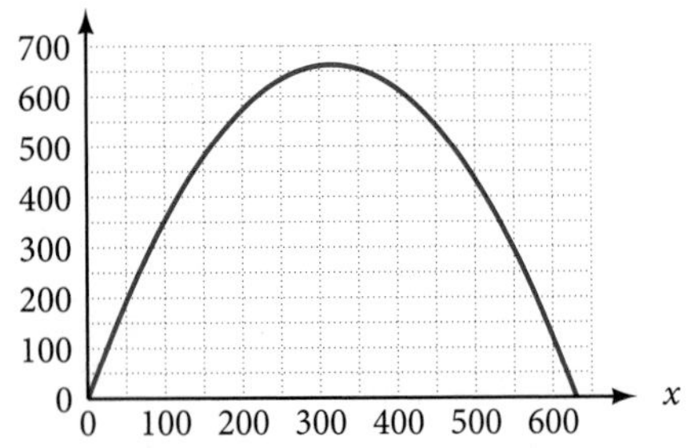

b. 630 ft **71.** $\frac{24}{5}$ feet **73.** 2,358 **75.** 12.3 **77.** 3 **79.** $9, -1$ **81** 60

Problem Set 2.4

1. 6 meters, 9 meters, 12 meters **3.** \$92.00 **5.** 24 ft **7.** $w = 5$ yards, $l = 14$ yards

9. \$4,000 at 10%; \$8,000 at 7%

	Dollars at 10%	Dollars at 7%	Total
Number of	x	$12{,}000 - x$	12,000
Interest on	$0.1x$	$0.07(1{,}200 - x)$	960

11. \$3,500 at 6%; \$2,500 at 8%

13. a-b.

	d	r	t
Upstream	10	$18 - x$	$\frac{10}{18 - x}$
Downstream	14	$18 + x$	$\frac{14}{18 + x}$

c. They are the same. $\frac{10}{18 - x} = \frac{14}{18 + x}$

d. The speed of the current is 3 mph.

15. a-b.

	d	r	t
Car	80	x	$\frac{80}{x}$
Train	120	$x + 30$	$\frac{120}{x + 30}$

c. They are the same. $\frac{80}{x} = \frac{120}{x + 30}$

d. The speed of the car is 60 mph, and the speed of the train is 90 mph

17. 35 mph **19.** 15 hours

21.

t	4	6	8	10
h	15	10	7.5	6

23.

Time (seconds)	Rate (feet per second)
1.00	1,000
0.80	1,250
0.64	1,562.5
0.50	2,000
0.40	2,500
0.32	3,125

25.

Age (years)	Maximum Heart Rate (beats per minute)
18	202
19	201
20	200
21	199
22	198
23	197

27.

Resting Heart Rate (beats per minute)	Training Heart Rate (beats per minute)
60	144
62	144.8
64	145.6
68	147.2
70	148
72	148.8

29. -5 **31.** 6 **33.** $-2, 3$ **35.** $-\frac{1}{2}, \frac{2}{3}$

Problem Set 2.5

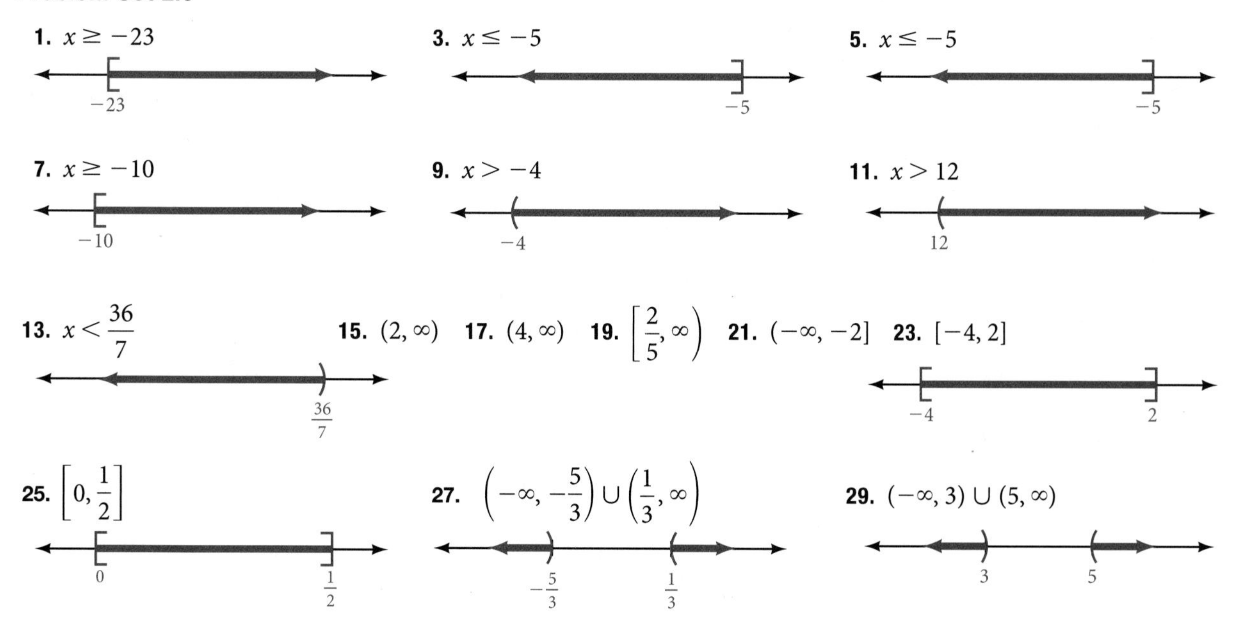

1. $x \geq -23$
-23
3. $x \leq -5$
-5
5. $x \leq -5$
-5
7. $x \geq -10$
-10
9. $x > -4$
-4
11. $x > 12$
12
13. $x < \frac{36}{7}$
$\frac{36}{7}$
15. $(2, \infty)$ 17. $(4, \infty)$ 19. $\left[\frac{2}{5}, \infty\right)$ 21. $(-\infty, -2]$ 23. $[-4, 2]$
-4
2
25. $\left[0, \frac{1}{2}\right]$
0
$\frac{1}{2}$
27. $\left(-\infty, -\frac{5}{3}\right) \cup \left(\frac{1}{3}, \infty\right)$
$-\frac{5}{3}$
$\frac{1}{3}$
29. $(-\infty, 3) \cup (5, \infty)$
3
5

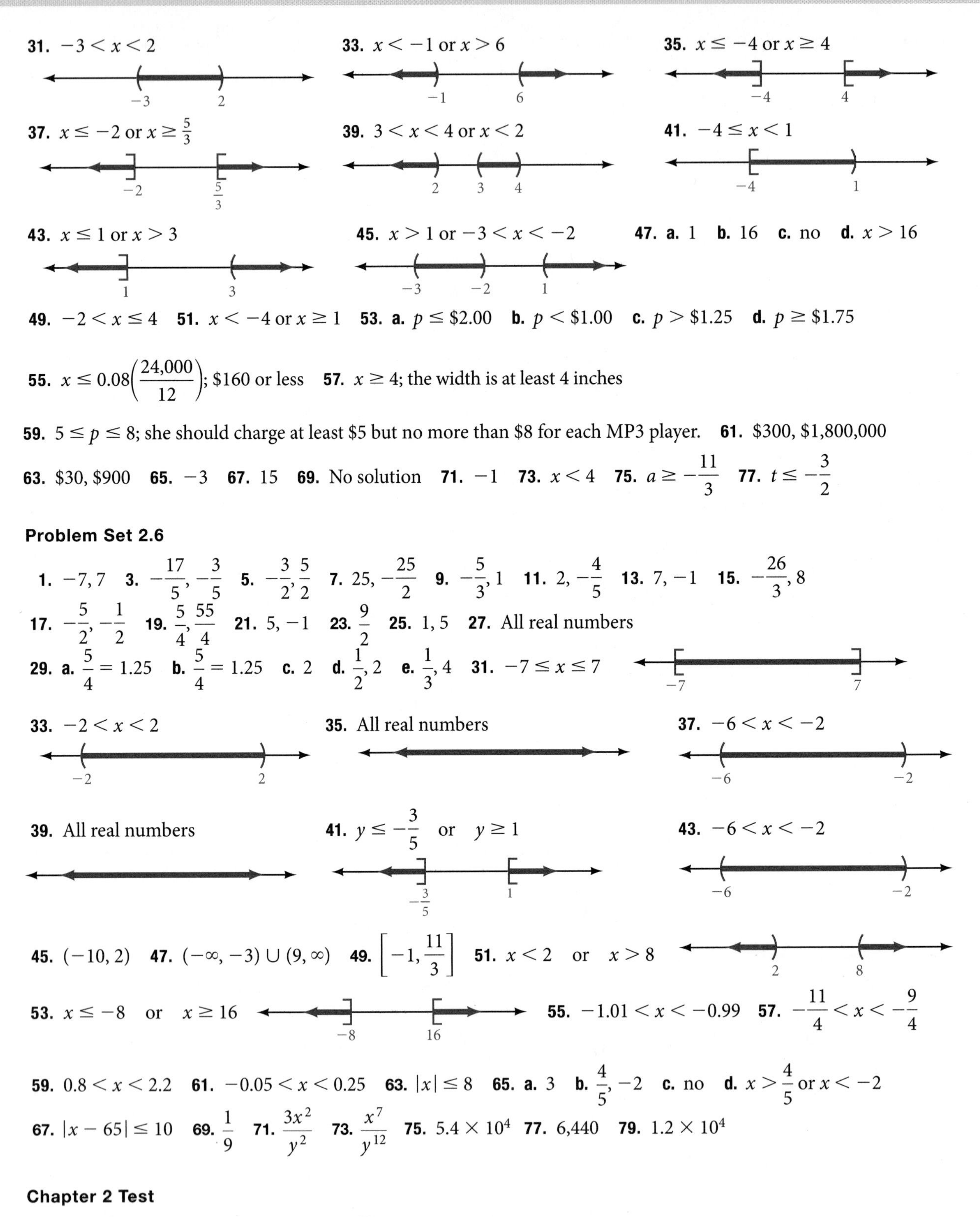

31. $-3 < x < 2$ **33.** $x < -1$ or $x > 6$ **35.** $x \le -4$ or $x \ge 4$

37. $x \le -2$ or $x \ge \frac{5}{3}$ **39.** $3 < x < 4$ or $x < 2$ **41.** $-4 \le x < 1$

43. $x \le 1$ or $x > 3$ **45.** $x > 1$ or $-3 < x < -2$ **47. a.** 1 **b.** 16 **c.** no **d.** $x > 16$

49. $-2 < x \le 4$ **51.** $x < -4$ or $x \ge 1$ **53. a.** $p \le \$2.00$ **b.** $p < \$1.00$ **c.** $p > \$1.25$ **d.** $p \ge \$1.75$

55. $x \le 0.08\left(\frac{24{,}000}{12}\right)$; \$160 or less **57.** $x \ge 4$; the width is at least 4 inches

59. $5 \le p \le 8$; she should charge at least \$5 but no more than \$8 for each MP3 player. **61.** \$300, \$1,800,000

63. \$30, \$900 **65.** -3 **67.** 15 **69.** No solution **71.** -1 **73.** $x < 4$ **75.** $a \ge -\frac{11}{3}$ **77.** $t \le -\frac{3}{2}$

Problem Set 2.6

1. $-7, 7$ **3.** $-\frac{17}{5}, -\frac{3}{5}$ **5.** $-\frac{3}{2}, \frac{5}{2}$ **7.** $25, -\frac{25}{2}$ **9.** $-\frac{5}{3}, 1$ **11.** $2, -\frac{4}{5}$ **13.** $7, -1$ **15.** $-\frac{26}{3}, 8$

17. $-\frac{5}{2}, -\frac{1}{2}$ **19.** $\frac{5}{4}, \frac{55}{4}$ **21.** $5, -1$ **23.** $\frac{9}{2}$ **25.** 1, 5 **27.** All real numbers

29. a. $\frac{5}{4} = 1.25$ **b.** $\frac{5}{4} = 1.25$ **c.** 2 **d.** $\frac{1}{2}, 2$ **e.** $\frac{1}{3}, 4$ **31.** $-7 \le x \le 7$

33. $-2 < x < 2$ **35.** All real numbers **37.** $-6 < x < -2$

39. All real numbers **41.** $y \le -\frac{3}{5}$ or $y \ge 1$ **43.** $-6 < x < -2$

45. $(-10, 2)$ **47.** $(-\infty, -3) \cup (9, \infty)$ **49.** $\left[-1, \frac{11}{3}\right]$ **51.** $x < 2$ or $x > 8$

53. $x \le -8$ or $x \ge 16$ **55.** $-1.01 < x < -0.99$ **57.** $-\frac{11}{4} < x < -\frac{9}{4}$

59. $0.8 < x < 2.2$ **61.** $-0.05 < x < 0.25$ **63.** $|x| \le 8$ **65. a.** 3 **b.** $\frac{4}{5}, -2$ **c.** no **d.** $x > \frac{4}{5}$ or $x < -2$

67. $|x - 65| \le 10$ **69.** $\frac{1}{9}$ **71.** $\frac{3x^2}{y^2}$ **73.** $\frac{x^7}{y^{12}}$ **75.** 5.4×10^4 **77.** 6,440 **79.** 1.2×10^4

Chapter 2 Test

1. 28 **2.** -3 **3.** $-\frac{1}{3}, 2$ **4.** 0, 5 **5.** $-\frac{7}{4}$ **6.** 2 **7.** $-5, 2$ **8.** $-4, -2, 4$ **9.** $3 \pm i\sqrt{2}$ **10.** $1 \pm i\sqrt{2}$

11. $\frac{5}{2}, \frac{-5 \pm 5i\sqrt{3}}{4}$ **12.** $\pm\frac{i}{2}, \pm\sqrt{2}$ **13.** $\frac{1}{2}, 1$ **14.** $-\frac{3}{5}$ **15.** No solution (3 does not check) **16.** $-2, 3$

17. 8 (1 does not check) **18.** -3 **19.** $w = \frac{1}{2}(P - 2l)$ **20.** $\frac{2A}{h} - b = B$ **21.** $y = \frac{5x}{2} - 5$ **22.** $y = 3x - 7$

23. $w = 6$ in.; $l = 12$ in. **24.** angle 1 = 125°, angle 2 = 55° **25.** 6 in.; 8 in.; 10 in. **26.** 0 sec, 2sec

27. $[-6, \infty)$ **28.** $(-\infty, 4)$ **29.** $(-\infty, 6)$ **30.** $[-52, \infty)$ **31.** $[-2, 3]$

32. $(-\infty, -3) \cup (\frac{1}{2}, \infty)$ **33.** 2, 6 **34.** No solution **35.** $x < -1$ or $x > \frac{4}{3}$

36. $-\frac{2}{3} \leq x \leq 4$ **37.** All real numbers **38.** $\varnothing$

Chapter 3

Problem Set 3.1

1. a. $(-1, 2)$ b. $(-1, -2)$ c. $(5, 0)$ d. $(0, 2)$ e. $(-5, -5)$ f. $(\frac{1}{2}, 2)$

3. a. $(4, 1)$ **b.** $(-4, 3)$ **c.** $(-2, -5)$ **d.** $(2, -2)$ **e.** $(0, 5)$ **f.** $(-4, 0)$ **g.** $(1, 0)$

5. b **7.** b

9. **11. a.** **b.** **c.**

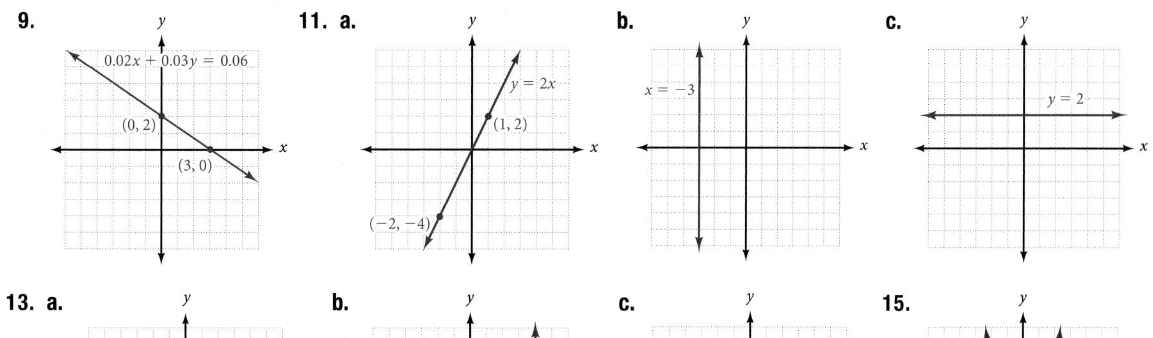

13. a. **b.** **c.** **15.**

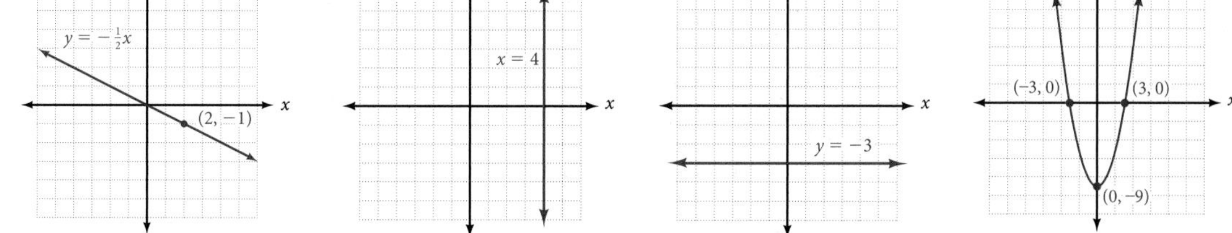

17.

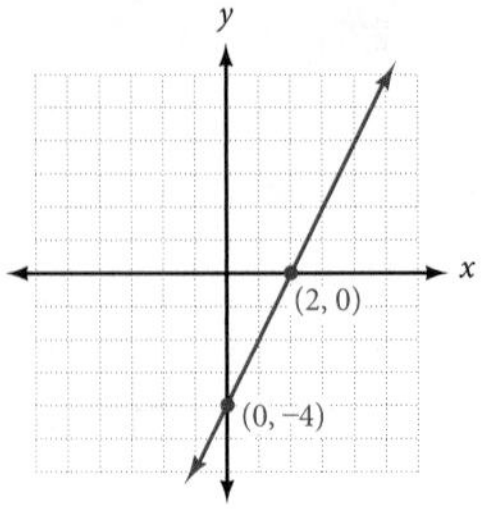

19.

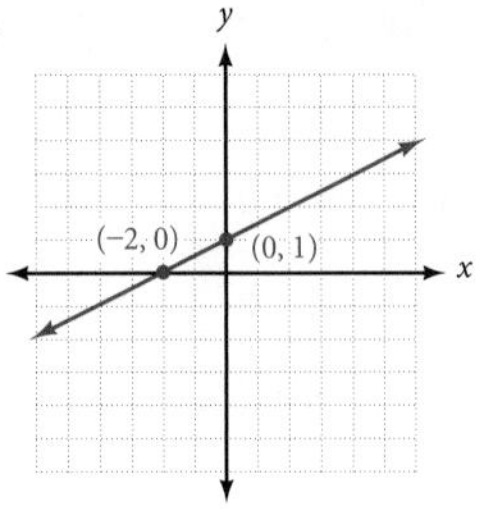

21.

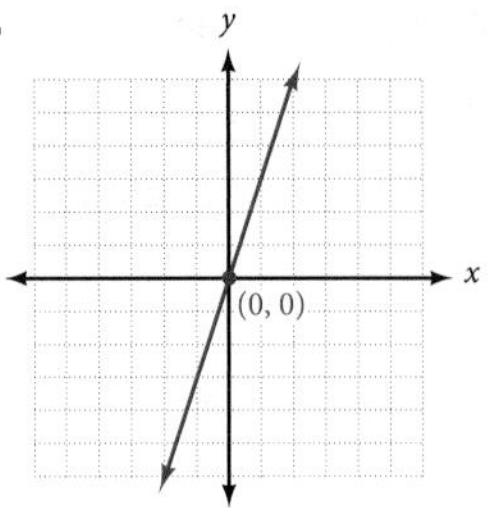

23.

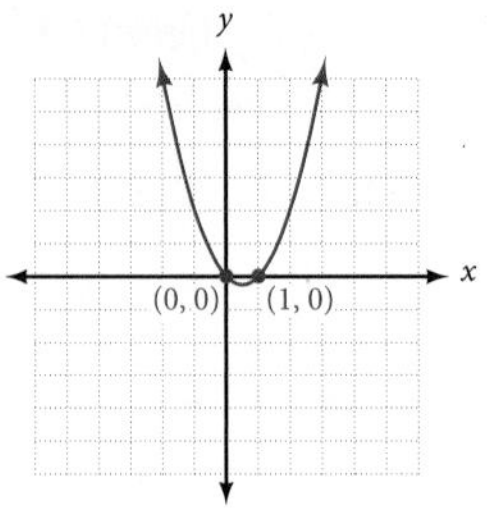

25.

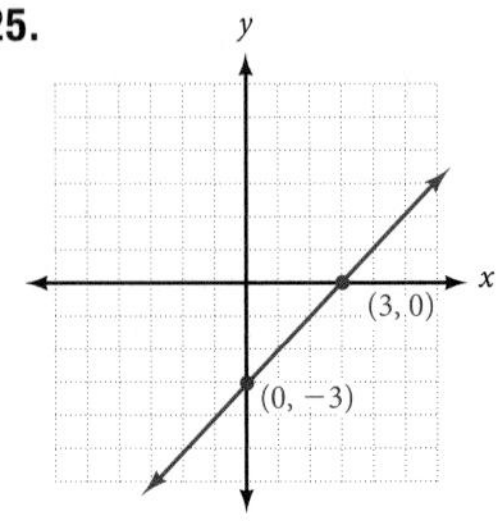

27. **a.** −7 **b.** −4 **c.** $-\frac{4}{3}$ **d.**

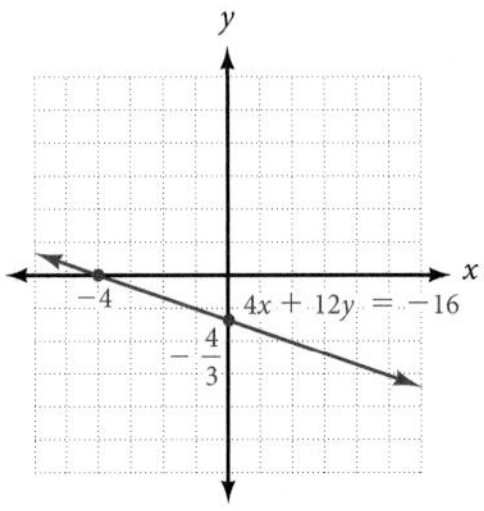

e. $y = -\frac{1}{3}x - \frac{4}{3}$

29. **a.** Yes **b.** No **c.** Yes

31.

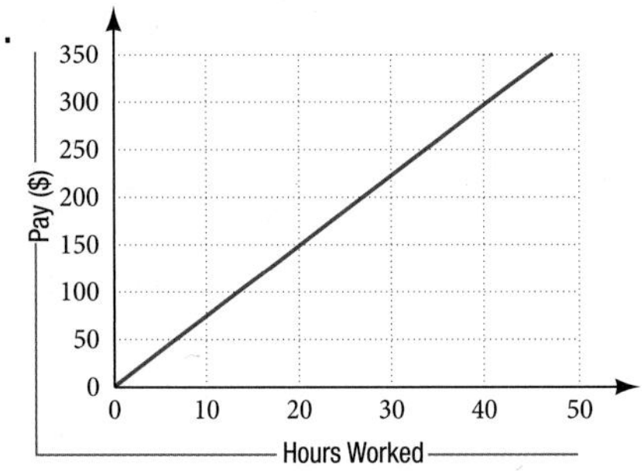

33.

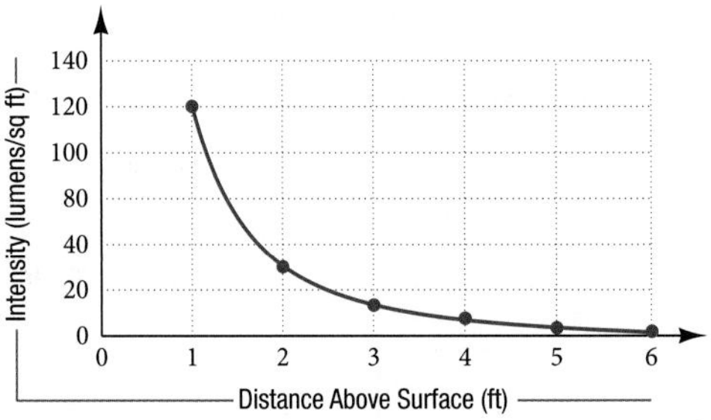

35. **a.** 60 **b.** 70 **c.** 10 **d.** 6:30 & 7:00 **e.** about 22 minutes

37.

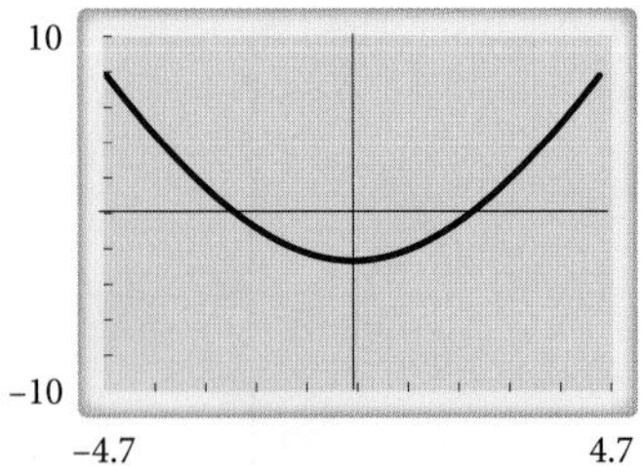

39.

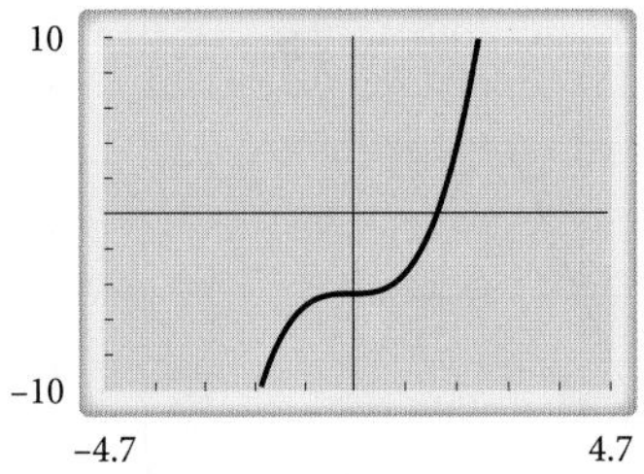

41.

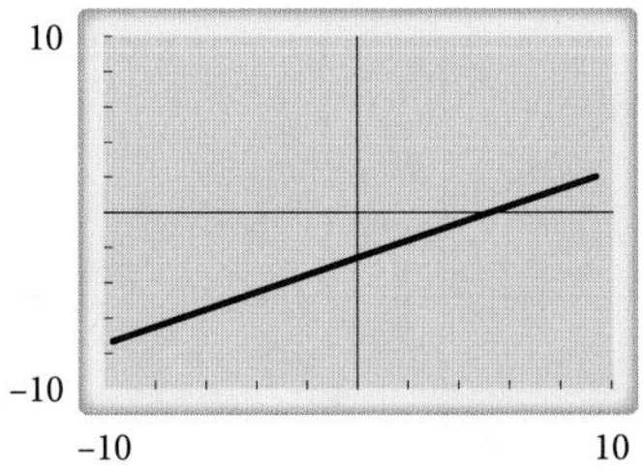

43.

t	h
0	0
1	16
−1	−48

45.

t	h
−3	−240
0	0
3	−48

47.

x	y
−1	undefined
0	0
1	1
4	2

Problem Set 3.2

1. D = {1, 3, 5, 7}; R = {2, 4, 6, 8}; a function **3.** D = {0, 1, 2, 3}; R = {4, 5, 6}; a function

5. D = {a, b, c, d}; R = {3, 4, 5}; a function **7.** D = {a}; R = {1, 2, 3, 4}; not a function **9.** Yes

17. 8 (1 does not check) **18.** -3 **19.** $w = \frac{1}{2}(P - 2l)$ **20.** $\frac{2A}{h} - b = B$ **21.** $y = \frac{5x}{2} - 5$ **22.** $y = 3x - 7$

23. $w = 6$ in.; $l = 12$ in. **24.** angle 1 = 125°, angle 2 = 55° **25.** 6 in.; 8 in.; 10 in. **26.** 0 sec, 2sec

27. $[-6, \infty)$

28. $(-\infty, 4)$

29. $(-\infty, 6)$

30. $[-52, \infty)$

31. $[-2, 3]$

32. $(-\infty, -3) \cup (\frac{1}{2}, \infty)$
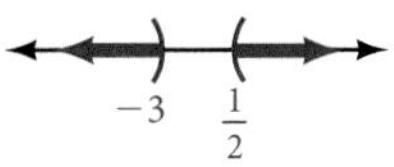

33. 2, 6 **34.** No solution **35.** $x < -1$ or $x > \frac{4}{3}$
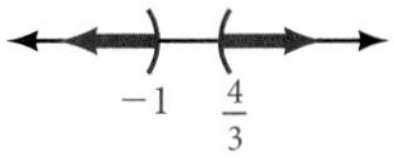

36. $-\frac{2}{3} \le x \le 4$

37. All real numbers

38. $\varnothing$

Chapter 3

Problem Set 3.1

1.
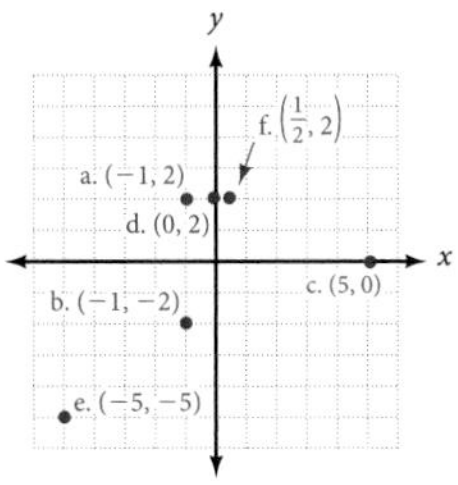

3. a. (4, 1) **b.** $(-4, 3)$ **c.** $(-2, -5)$ **d.** $(2, -2)$ **e.** (0, 5) **f.** $(-4, 0)$ **g.** (1, 0)

5. *b* **7.** *b*

9.
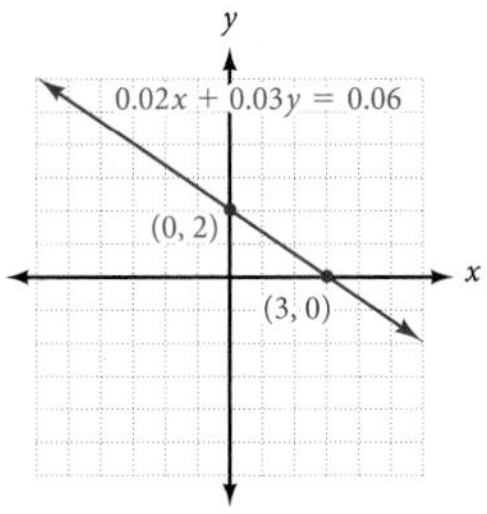

11. a.
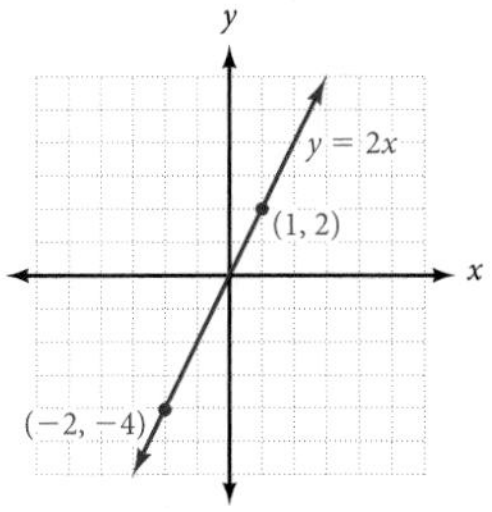

b.
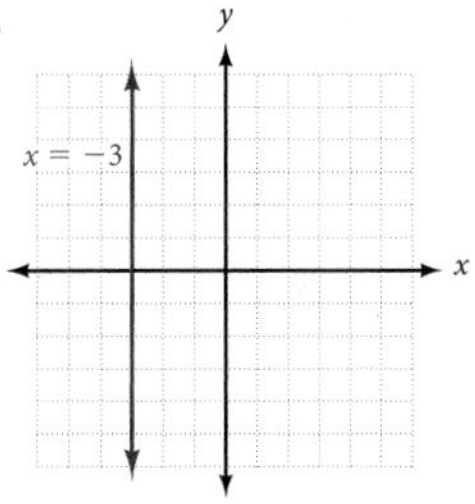

c.
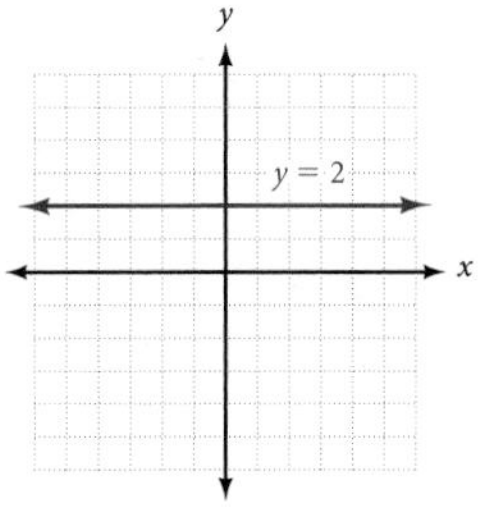

13. a.
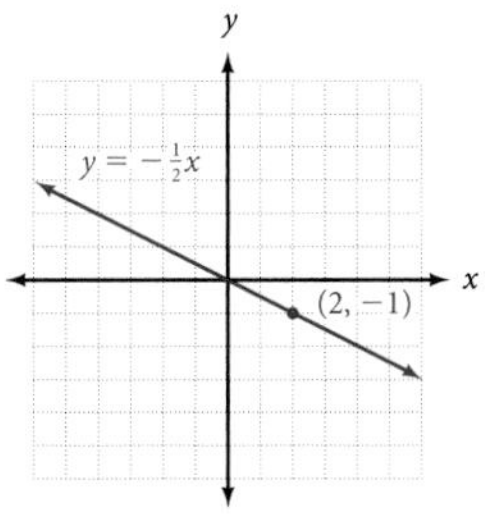

b.
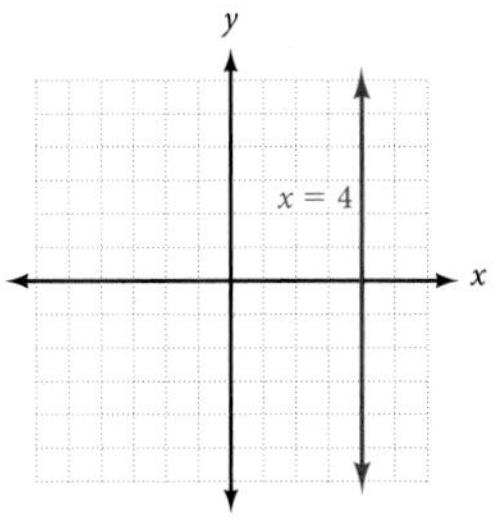

c.
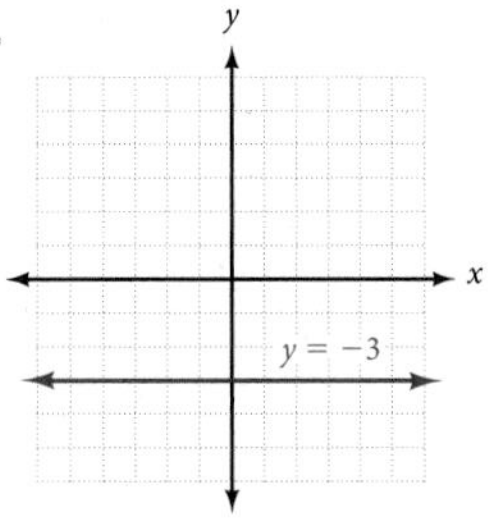

15.
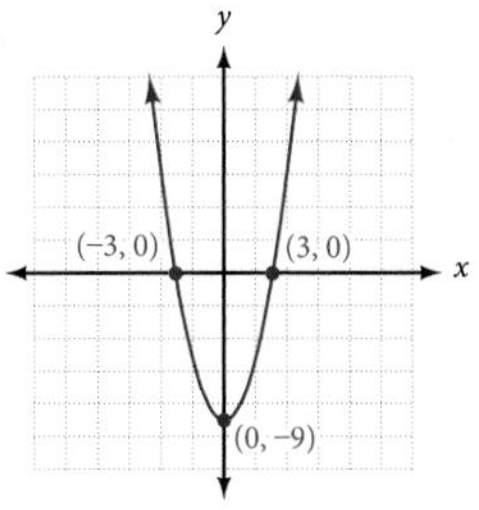

17.

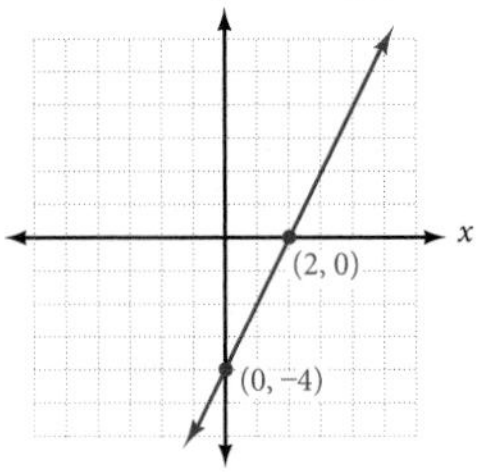

19.

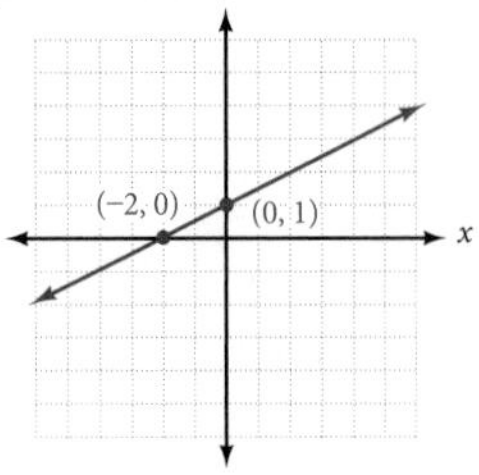

21.

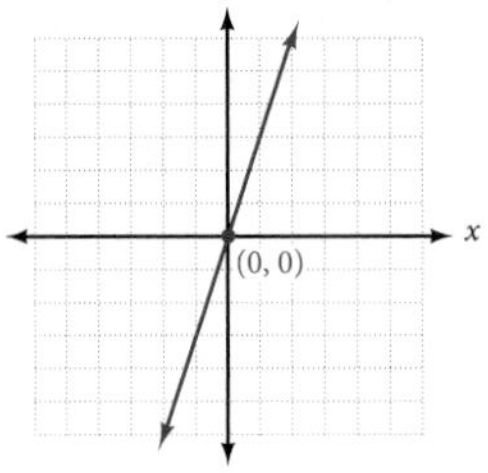

23.

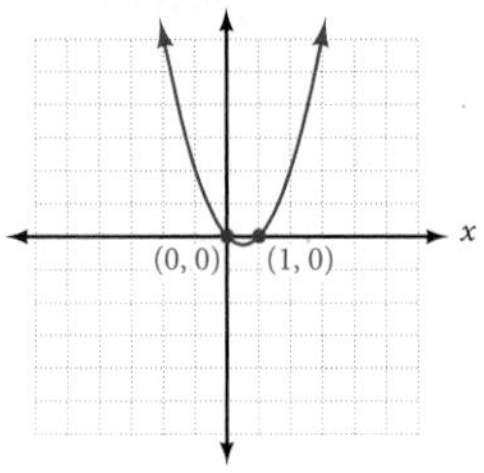

25.

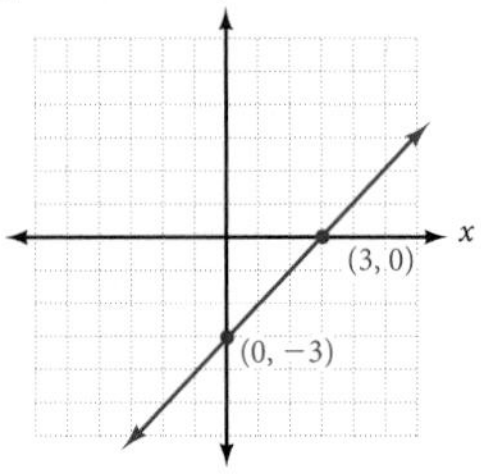

27. **a.** -7 **b.** -4 **c.** $-\frac{4}{3}$ **d.**

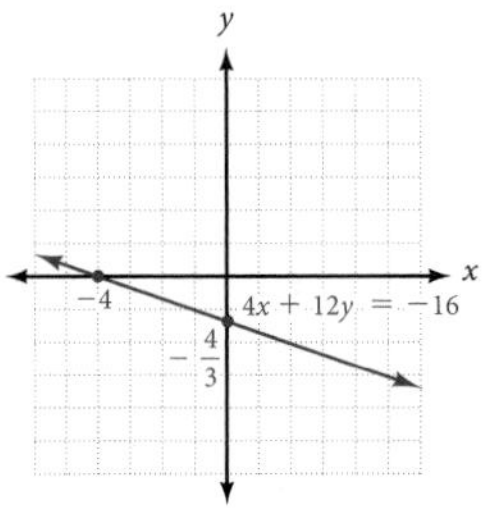

e. $y = -\frac{1}{3}x - \frac{4}{3}$

29. **a.** Yes **b.** No **c.** Yes

31.

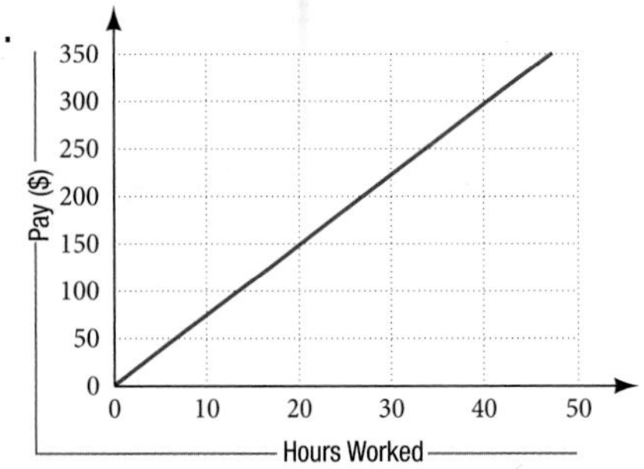

33.

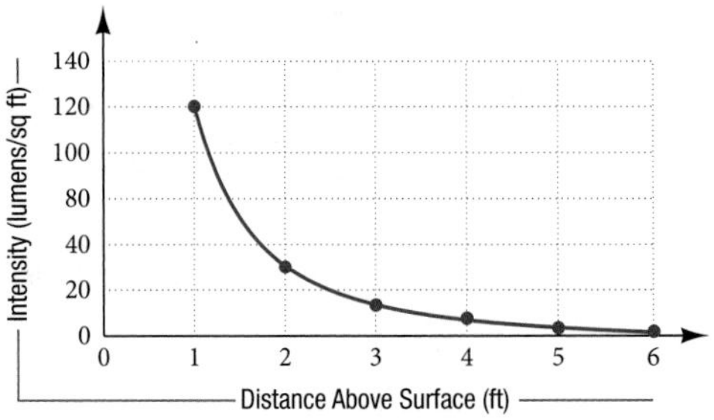

35. **a.** 60 **b.** 70 **c.** 10 **d.** 6:30 & 7:00 **e.** about 22 minutes

37.

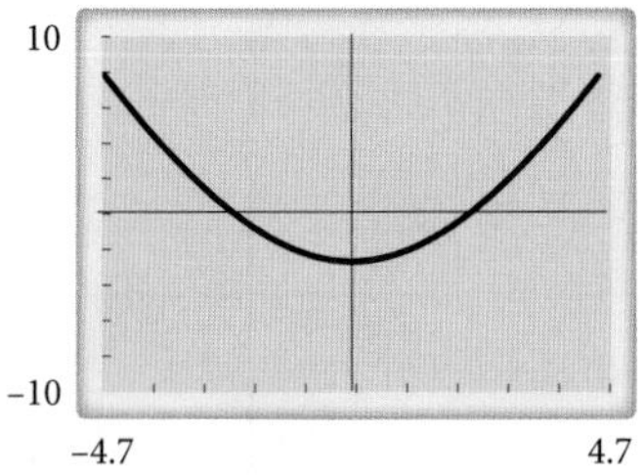

39.

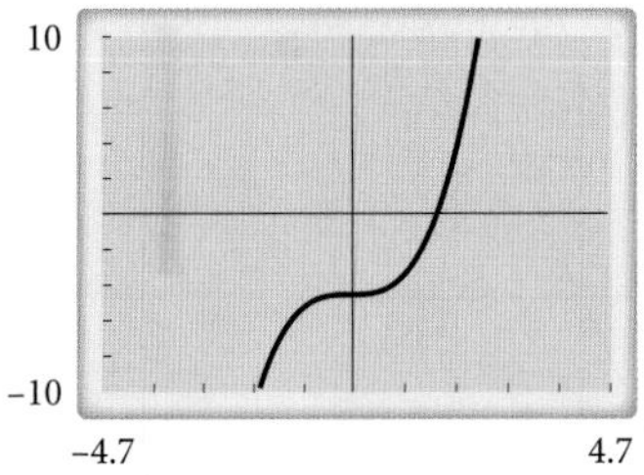

41.

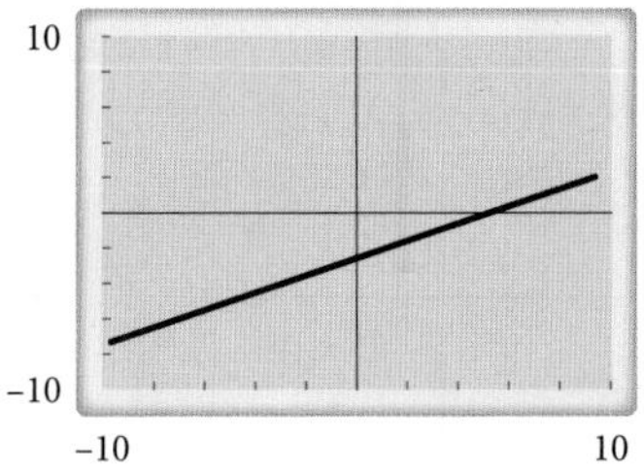

43.

t	h
0	0
1	16
-1	-48

45.

t	h
-3	-240
0	0
3	-48

47.

x	y
-1	undefined
0	0
1	1
4	2

Problem Set 3.2

1. D = {1, 3, 5, 7}; R = {2, 4, 6, 8}; a function **3.** D = {0, 1, 2, 3}; R = {4, 5, 6}; a function

5. D = {a, b, c, d}; R = {3, 4, 5}; a function **7.** D = {a}; R = {1, 2, 3, 4}; not a function **9.** Yes

11. No **13.** No **15.** Yes **17.** Yes **19.** $D = \{x \mid -5 \le x \le 5\}$; $R = \{y \mid 0 \le y \le 5\}$

21. $D = \{x \mid -5 \le x \le 3\}$; $R = \{y \mid y = 3\}$ **23.** D = All real numbers; $R = \{y \mid y \ge -1\}$; a function

25. D = All real numbers; $R = \{y \mid y \ge 4\}$; a function **27.** $D = \{x \mid x \ge -1\}$; R = All real numbers; not a function

29. D = All real numbers; $R = \{y \mid y \ge 0\}$; a function **31.** $D = \{x \mid x \ge 0\}$; R = All real numbers; not a function

33. **35.** **37.** **39.**

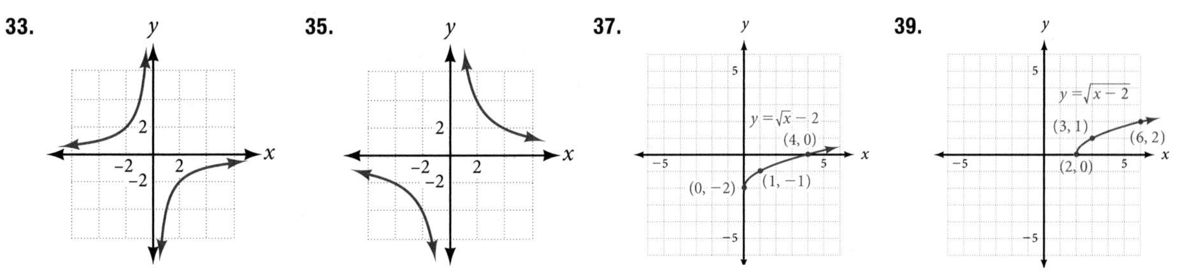

41. **43.**

y
(1, 4)
(−1, 2)
(0, 3)
$y = \sqrt[3]{x} + 3$
5
−5
x

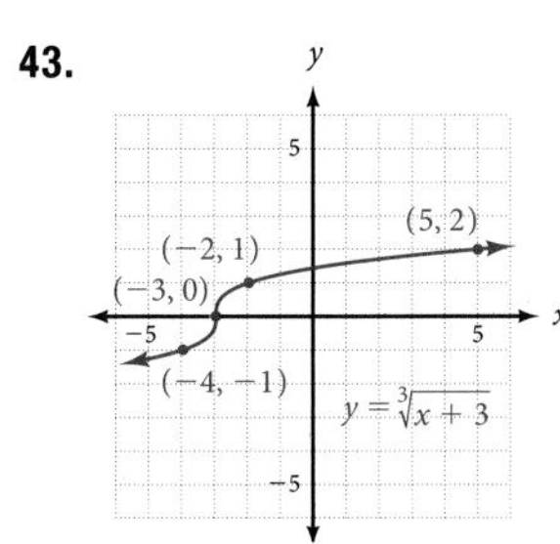

45. a. $y = 9.50x$ for $10 \le x \le 40$

b.

TABLE 4 Weekly Wages

Hours Worked	Function Rule	Gross Pay ($)
x	$y = 9.50x$	y
10	$y = 9.50(10) = 85$	95.00
20	$y = 9.50(20) = 170$	190.00
30	$y = 9.50(30) = 255$	285.00
40	$y = 9.50(40) = 340$	380.00

c.

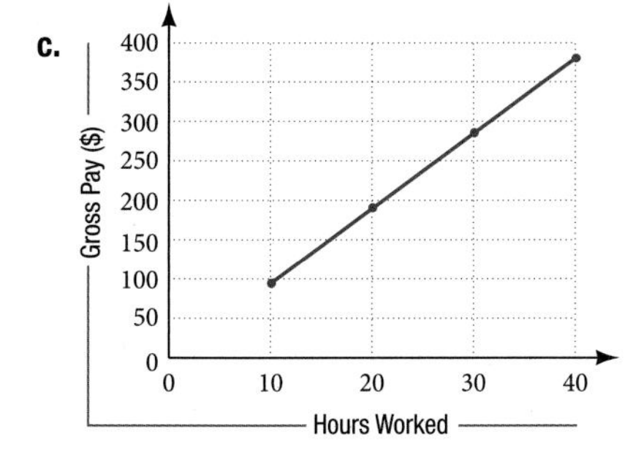

d. $D = \{x \mid 10 \le x \le 40\}$; $R = \{y \mid 95 \le y \le 380\}$

e. Minimum = \$95; Maximum = \$380

47. $\frac{392}{121} \approx 3.24$ feet **49.** D = {2004, 2005, 2006, 2007, 2008, 2009, 2010}; R = {680, 730, 800, 900, 920, 990, 1,030}

51. a. III **b.** I **c.** II **d.** IV **53.**

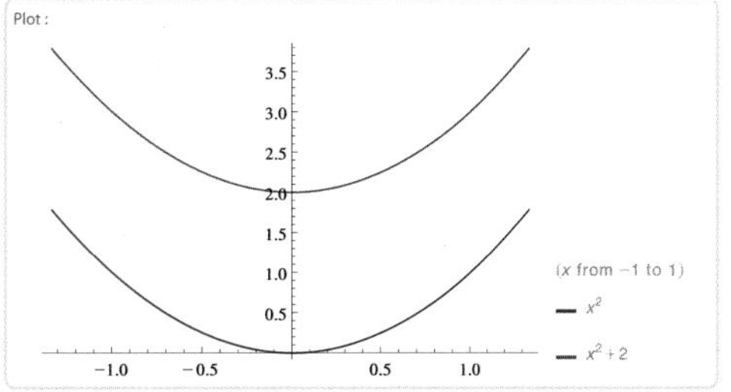

Wolfram Alpha LLC. 2012. Wolfram|Alpha
http://www.wolframalpha.com/
(access June 18, 2012)

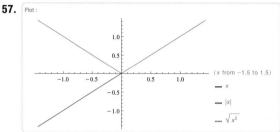

55.

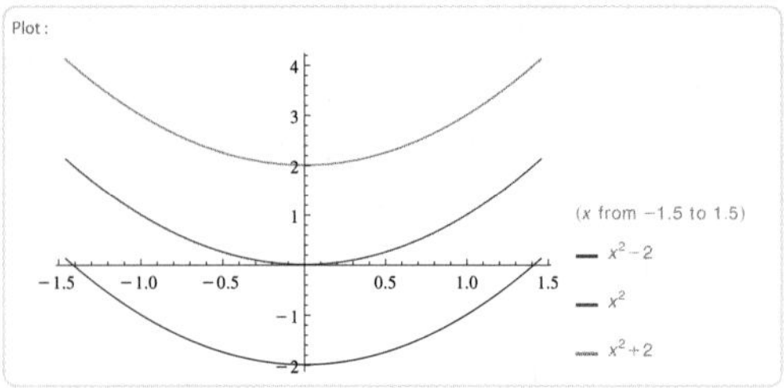

Wolfram Alpha LLC. 2012. Wolfram|Alpha
http://www.wolframalpha.com/
(access June 18, 2012)

57.

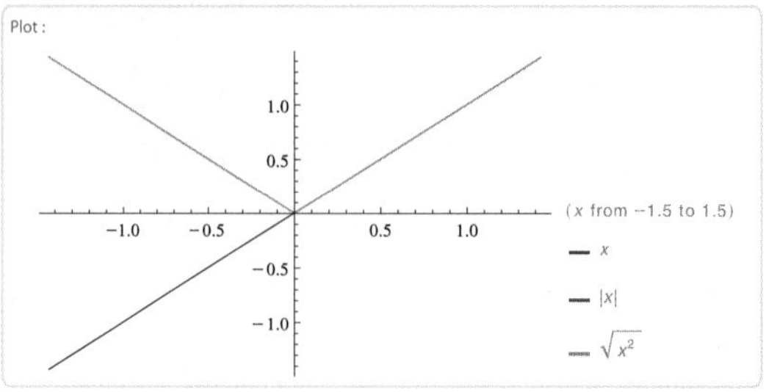

Wolfram Alpha LLC. 2012. Wolfram|Alpha
http://www.wolframalpha.com/
(access June 18, 2012)

59. a. $\{x \mid x \geq 0\}$ **b.** $\{x \mid x \geq 0\}$ **c.** $\{x \mid x \geq 0\}$

61.

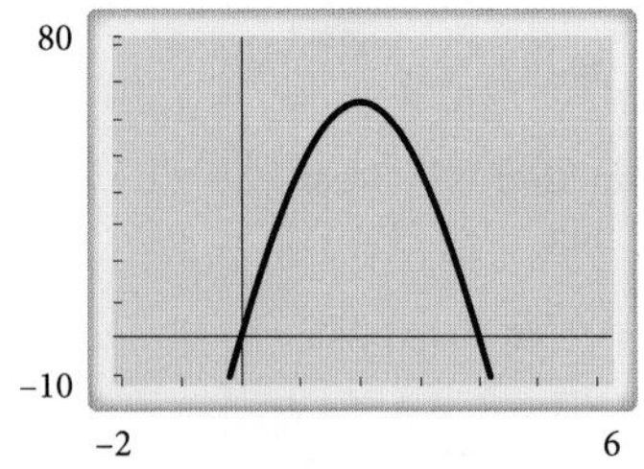

X	Y_1
0	0
1	48
2	64
3	48
4	0
5	−80
6	−192

63.

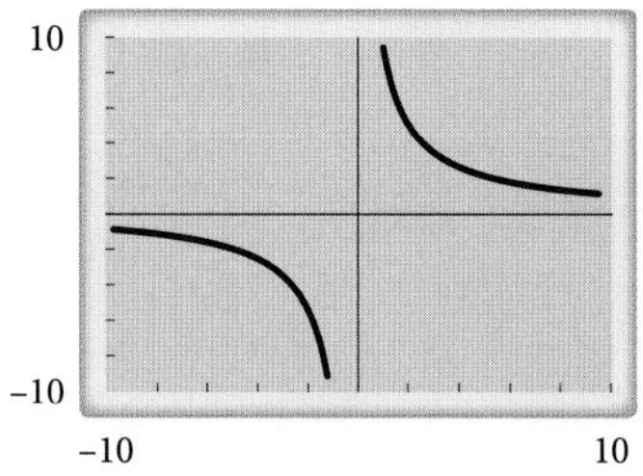

X	Y_1
.5	24
1	12
1.5	8
2	6
2.5	4.8
3	4
3.5	3.4286

65. 113 **67.** −9 **69. a.** 6 **b.** 7.5

71. a. 27 **b.** 6 **73.** 1 **75.** −3

Problem Set 3.3

1. −1 **3.** −11 **5.** 2 **7.** 4 **9.** $a^2 + 3a + 4$ **11.** $2a + 7$ **13.** 1 **15.** −9 **17.** 8 **19.** 0 **21.** $3a^2 - 4a + 1$

23. $3a^2 + 8a + 5$ **25.** 4 **27.** 0 **29.** 2 **31.** 24 **33.** −1 **35.** $2x^2 - 19x + 12$ **37.** 99 **39.** $\frac{3}{10}$

41. $\frac{2}{5}$ **43.** undefined **45. a.** $a^2 - 7$ **b.** $a^2 - 6a + 5$ **c.** $x^2 - 2$ **d.** $x^2 + 4x$ **e.** $a^2 + 2ab + b^2 - 4$

f. $x^2 + 2xh + h^2 - 4$ **47.**

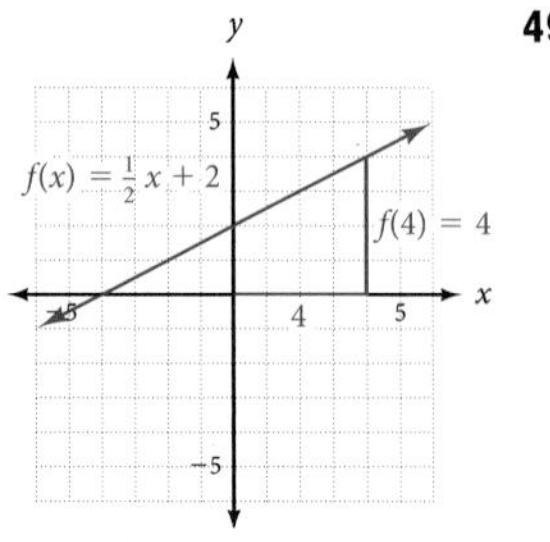

49. $x = 4$ **51.**

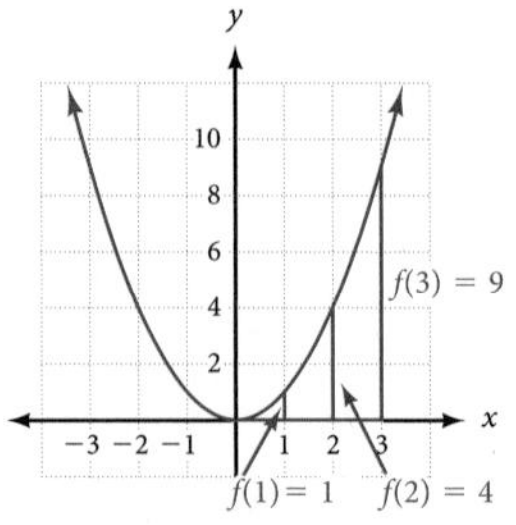

53. 53 **55.** 0.0021

57. 3, 1 **59.** $0, \frac{3}{2}$ **61.** $V(3) = 300$; worth \$300 in 3 years; $V(6) = 600$; worth \$600 in 6 years.

63. a. True **b.** False **c.** True **d.** False **e.** True **65. a.** \$5,625 **b.** \$1,500 **c.** $\{t \mid 0 \leq t \leq 5\}$

d.

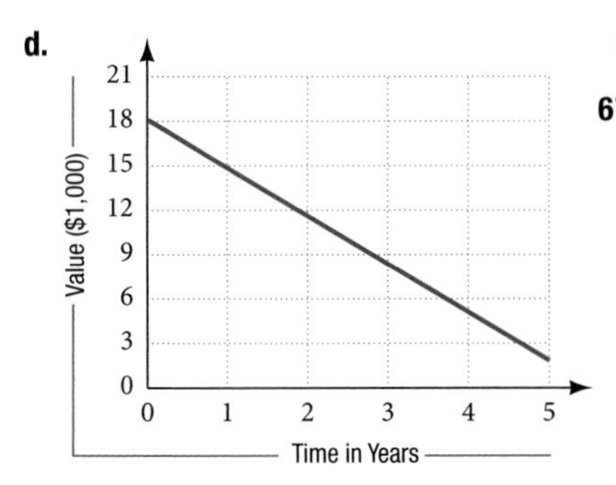

e. $\{V(t) \mid 1{,}500 \le V(t) \le 18{,}000\}$ **f.** About 2.42 years

67. $0, \pm 2$ **69.** $\pm\sqrt{2}$ **71.** ± 2 **73.**

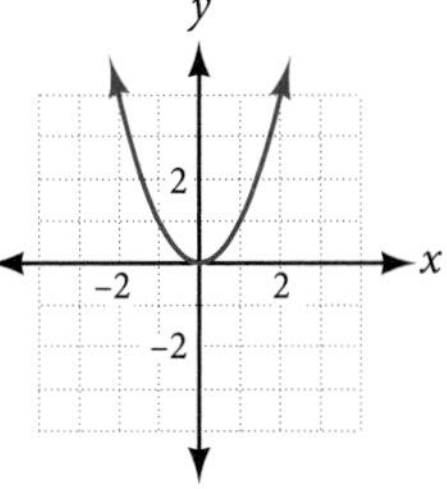

75. **77.**

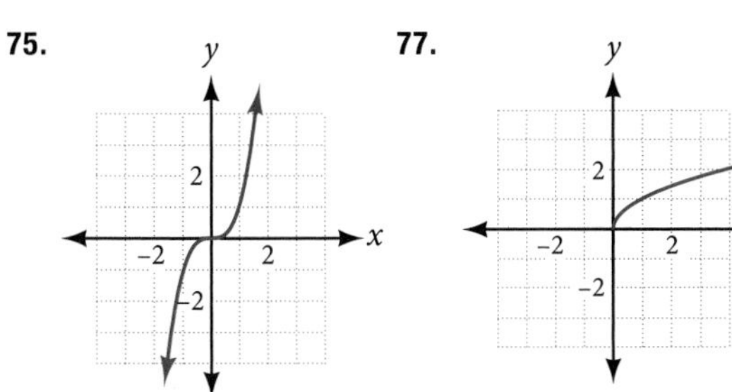

79.

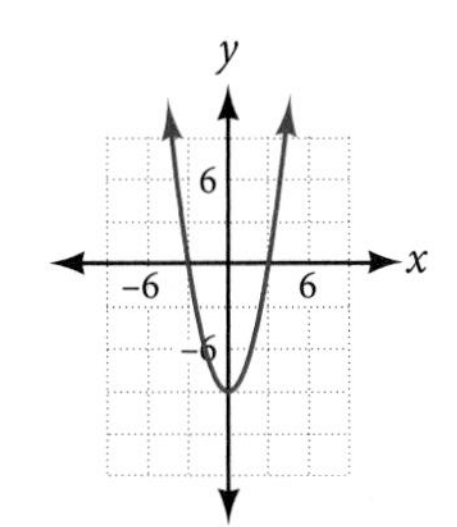

81.

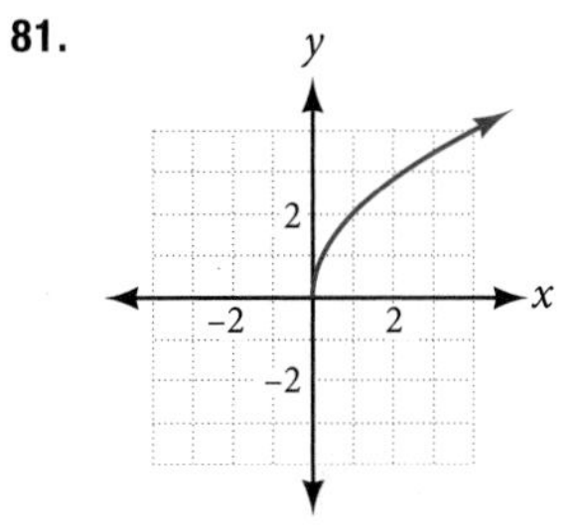

Problem Set 3.4

1. $y = x + 3$ **3.** $y = |x| - 3$ **5.** **7.**

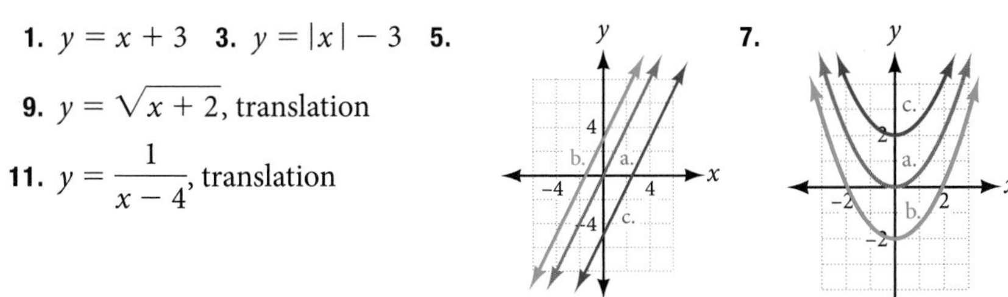

9. $y = \sqrt{x + 2}$, translation

11. $y = \dfrac{1}{x - 4}$, translation

13. **15.** **17.** **19.**

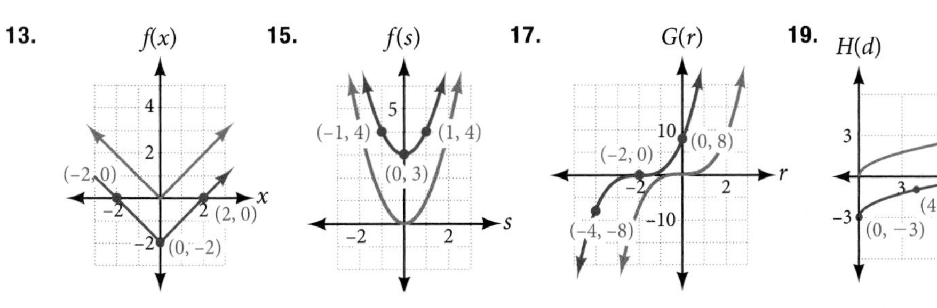

21.

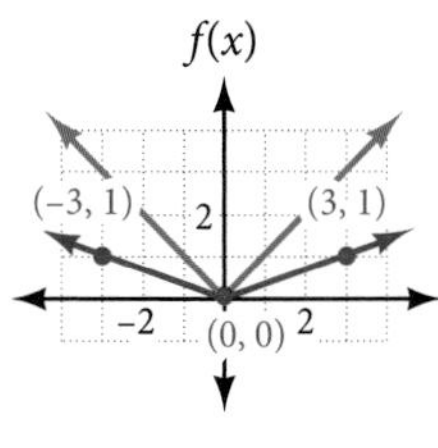

23.

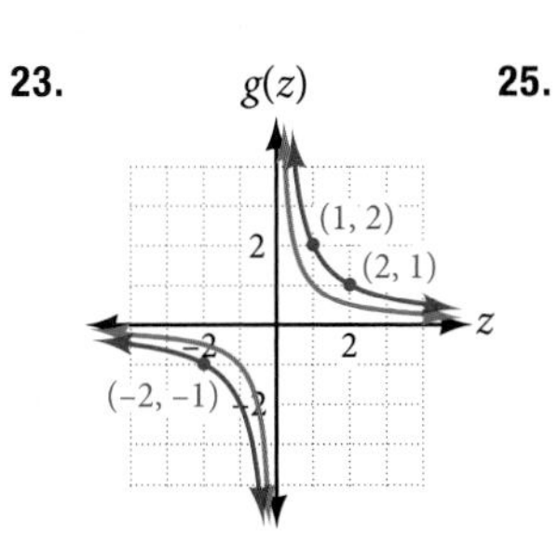

25.

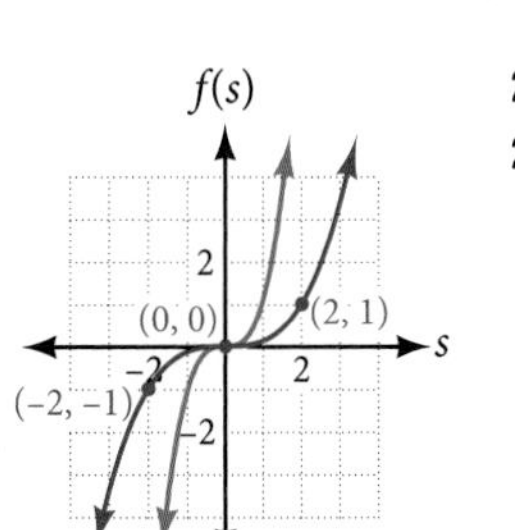

27. a. iv **b.** iii **c.** ii **d.** i

29. a. $y = 2g(t) + 2$ **b.** $y = -g(t)$

c. $y = g(t - 4)$

31. 33. 35. 37.

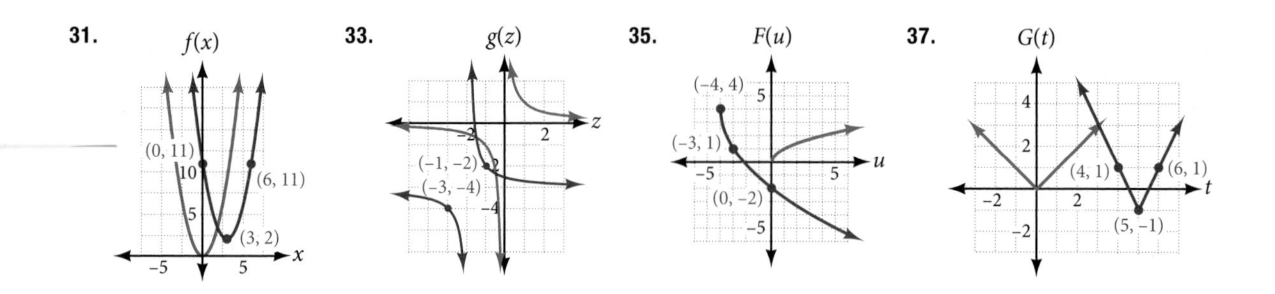

39. $y = |x + 1| - 2$ **41.** $y = -\sqrt{x} + 3$ **43.** $y = (x - 3)^3 + 1$

45. **47.**

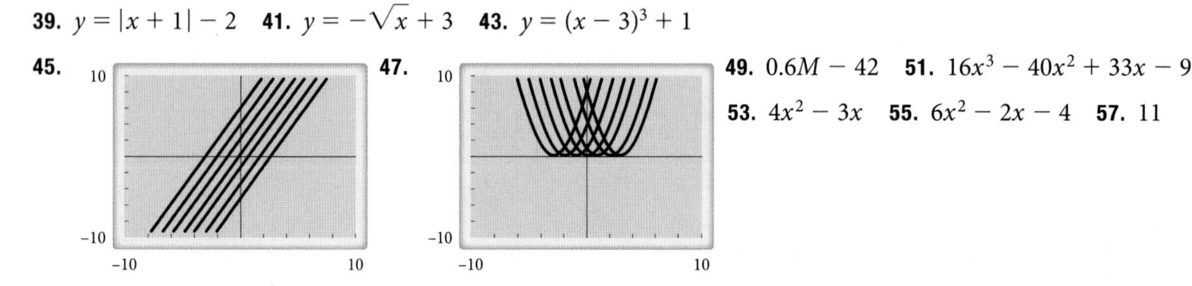

49. $0.6M - 42$ **51.** $16x^3 - 40x^2 + 33x - 9$ **53.** $4x^2 - 3x$ **55.** $6x^2 - 2x - 4$ **57.** 11

Problem Set 3.5

1. $6x + 2$ **3.** $-2x + 8$ **5.** $8x^2 + 14x - 15$ **7.** $\frac{2x + 5}{4x - 3}$ **9.** $4x - 7$ **11.** $3x^2 - 10x + 8$ **13.** $-2x + 3$

15. $3x^2 - 11x + 10$ **17.** $9x^3 - 48x^2 + 85x - 50$ **19.** $x - 2$ **21.** $\frac{1}{x - 2}$ **23.** $3x^2 - 7x + 3$ **25.** $6x^2 - 22x + 20$

27. 21 **29.** 98 **31.** $\frac{3}{2}$ **33.** 1 **35.** 40 **37.** 147 **39. a.** 81 **b.** 29 **c.** $(x + 4)^2$ **d.** $x^2 + 4$

41. a. -2 **b.** -1 **c.** $16x^2 + 4x - 2$ **d.** $4x^2 + 12x - 1$ **43.** $(f \circ g)(x) = 5\left[\frac{x + 4}{5}\right] - 4 = x + 4 - 4 = x$ $\quad (g \circ f)(x) = \frac{(5x - 4) + 4}{5} = \frac{5x}{5} = x$

45. $5x^5 + 4x^3 + 55x^2 + 44$ **47.** $x^4 - 17x^2 - 8x - 6$ **49.** $4x^2 - 30x + 54$

51. a. $R(x) = 11.5x - 0.05x^2$ **b.** $C(x) = 2x + 200$ **c.** $P(x) = -0.05x^2 + 9.5x - 200$ **d.** $\overline{C}(x) = 2 + \frac{200}{x}$

53. a. $M(x) = 220 - x$ **b.** $M(24) = 196$ **c.** 142 **d.** 135 **e.** 128 **55.** $f[g(16)] = 76, g[f(16)] = 35.6$

57.

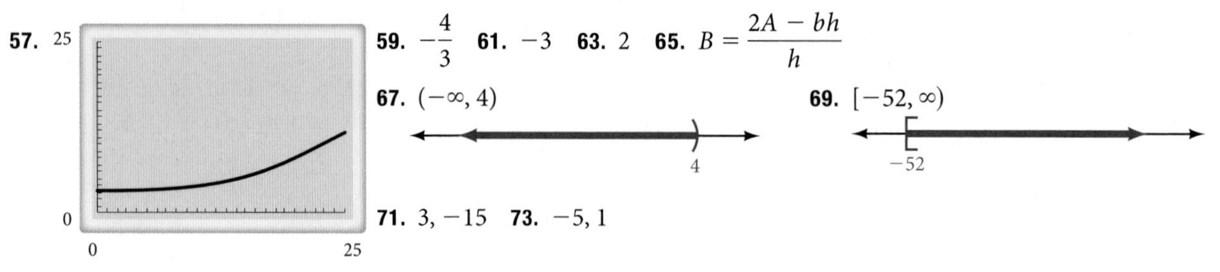

59. $-\frac{4}{3}$ **61.** -3 **63.** 2 **65.** $B = \frac{2A - bh}{h}$

67. $(-\infty, 4)$ **69.** $[-52, \infty)$

71. 3, −15 **73.** −5, 1

Chapter 3 Test

1. $\{x \mid x \in \text{All real numbers}\}$ **2.** $\{x \mid x \geq 3\}$ **3.** $\{x \mid x < -3 \text{ or } -3 < x < 3 \text{ or } x > 3\}$ **4.** $\{-2, -3, 2\}$

5. $\{t \mid t \geq 0\}$ **6.** x-intercept: $(6, 0)$; y-intercept: $(0, -4)$ **7.** x-intercept: $(3, 0)$; No y-intercept

8. No x-intercept; $(\pm 3, 0)$, $(0, 0)$; y-intercept: $(0, 0)$ **9.** x-intercept: DNE; y-intercept: $(0, 4)$

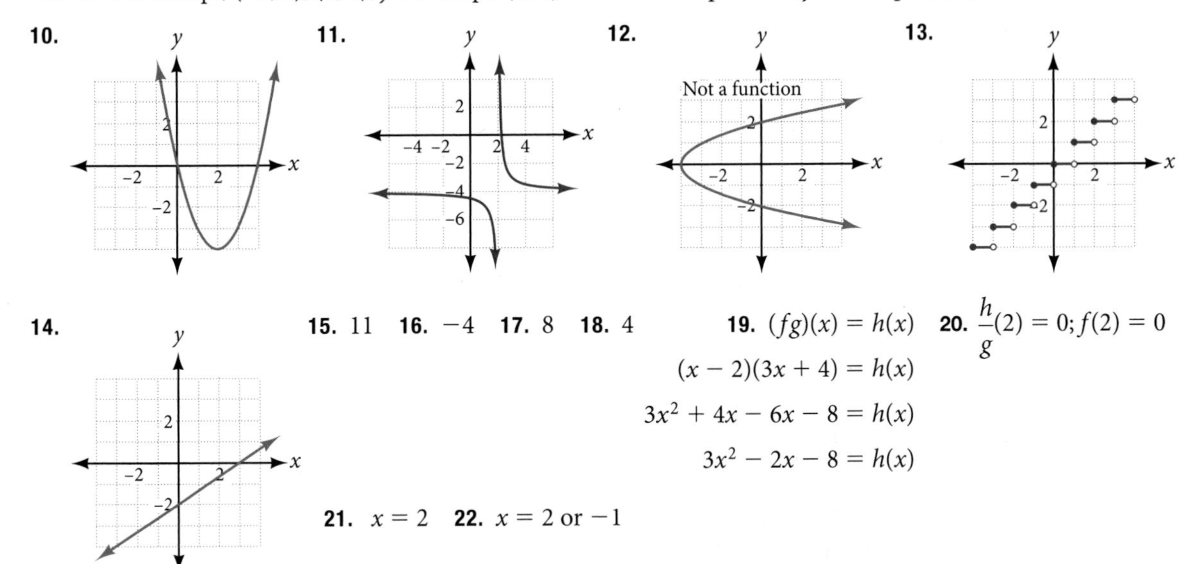

15. 11 **16.** -4 **17.** 8 **18.** 4

19. $(fg)(x) = h(x)$
$$(x - 2)(3x + 4) = h(x)$$
$$3x^2 + 4x - 6x - 8 = h(x)$$
$$3x^2 - 2x - 8 = h(x)$$

20. $\frac{h}{g}(2) = 0; f(2) = 0$

21. $x = 2$ **22.** $x = 2$ or -1

Chapter 4

Problem Set 4.1

1. $\frac{3}{2}$ **3.** Undefined slope **5.** $\frac{2}{3}$ **7.** Slope $= \frac{3}{2}$ **9.** Slope $= -\frac{1}{2}$ **11.** Slope $= \frac{5}{3}$

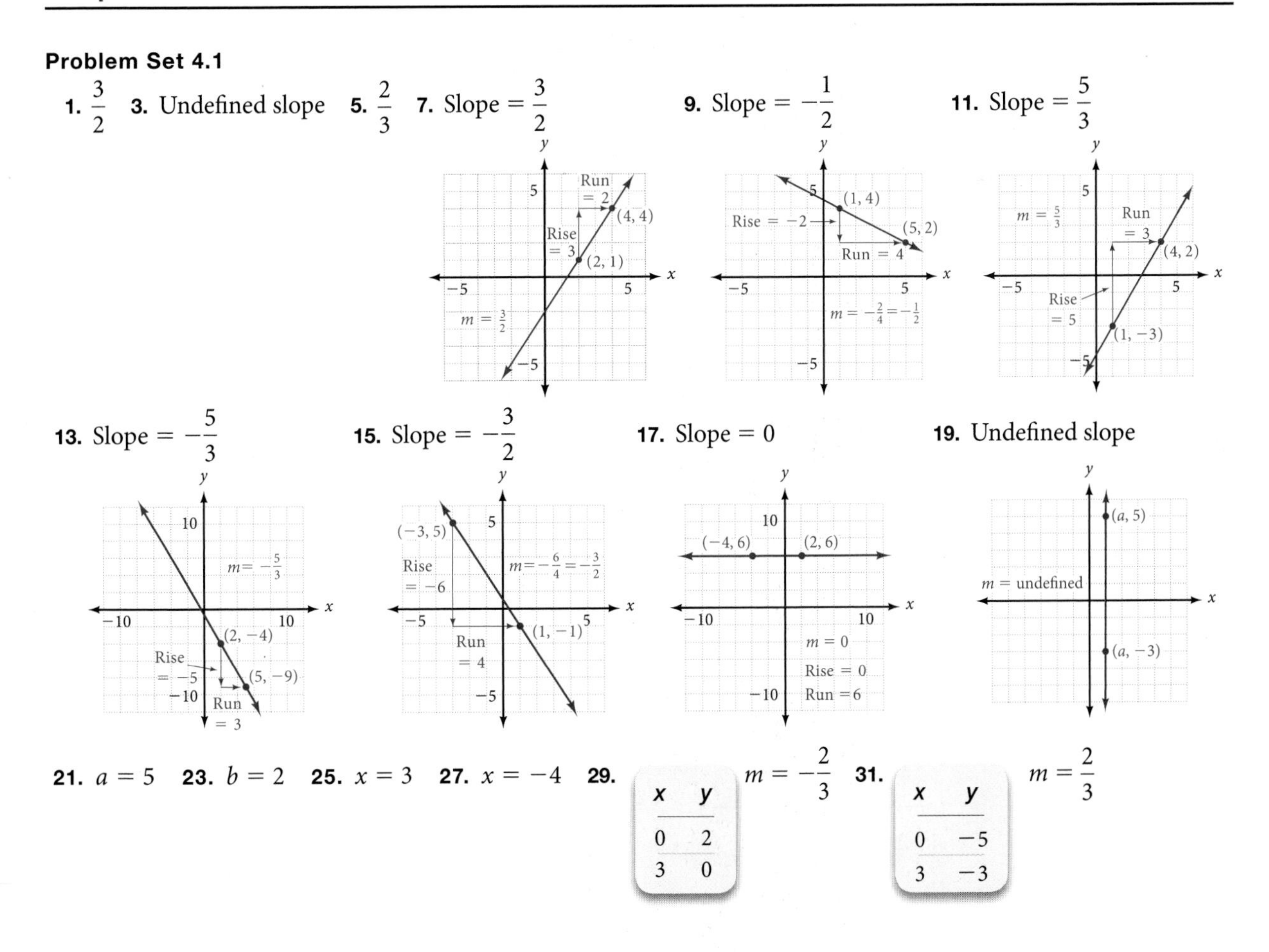

13. Slope $= -\frac{5}{3}$ **15.** Slope $= -\frac{3}{2}$ **17.** Slope $= 0$ **19.** Undefined slope

21. $a = 5$ **23.** $b = 2$ **25.** $x = 3$ **27.** $x = -4$

29. $m = -\frac{2}{3}$

x	y
0	2
3	0

31. $m = \frac{2}{3}$

x	y
0	−5
3	−3

33.

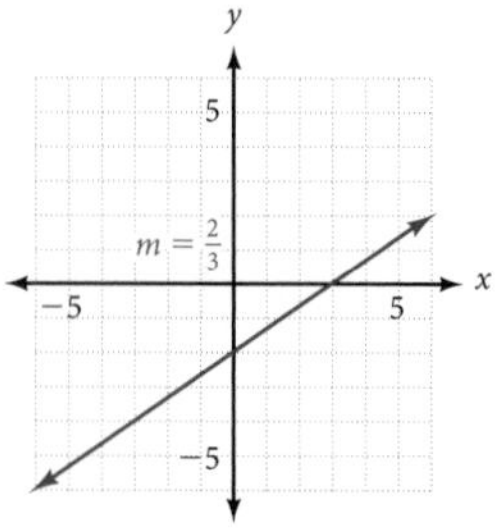

35. $\frac{1}{5}$ **37.** 0 **39.** -1 **41.** $-\frac{3}{2}$ **43. a.** Yes **b.** No **45. a.** 4 **b.** 4

47. a. 5 **b.** 5 **49. a.** $x + a$ **b.** $2x + h$ **51. a.** $x + a$ **b.** $2x + h$

53. a. $x + a - 3$ **b.** $2x + h - 3$ **55.** 17.5 mph **57.** 120 ft/sec

59. a. 10 minutes **b.** 20 minutes **c.** 20 °C per minute **d.** 10 °C per minute **e.** 1st minute

61. 1,333 solar thermal collector shipments/year. Between 1997 and 2006 the number of solar thermal collector shipments increased an average of 1,333 shipments per year.

63. a. 0.047 watts/lumens gained. For every additional lumen, the incandescent light bulb needs to use on average an extra 0.047 watts. **b.** 0.014 watts/lumens gained. For every additional lumen, the energy efficient light bulb needs to use on average an extra 0.014 watts. **c.** Energy efficient light bulb. Answers may vary.

65. a. D **b.** B **67.** 4 **69.** \$24.50/unit **71.** $-\frac{2}{3}$ **73.** $-2x + 3$ **75.** $y = -\frac{2}{3}x + 2$ **77.** $y = -\frac{2}{3}x + 1$ **79.** 1

Problem Set 4.2

1. $y = -4x - 3$ **3.** $y = -\frac{2}{3}x$ **5.** $y = -\frac{2}{3}x + \frac{1}{4}$ **7. a.** 3 **b.** $-\frac{1}{3}$ **9. a.** -3 **b.** $\frac{1}{3}$ **11. a.** $-\frac{2}{5}$ **b.** $\frac{5}{2}$

13. Slope = 3; y-intercept = -2; perpendicular slope = $-\frac{1}{3}$

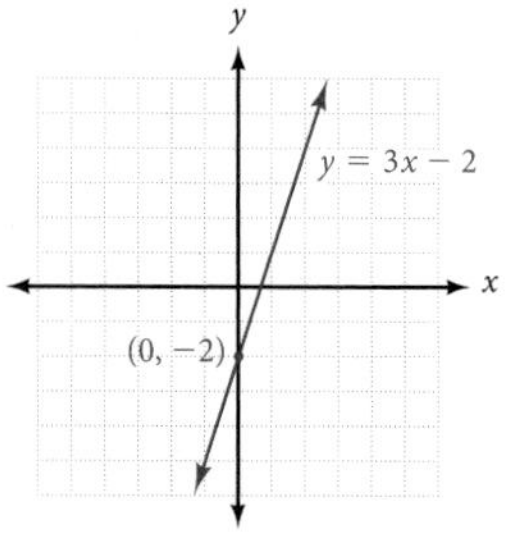

15. Slope = $\frac{2}{3}$; y-intercept = -4; perpendicular slope = $-\frac{3}{2}$

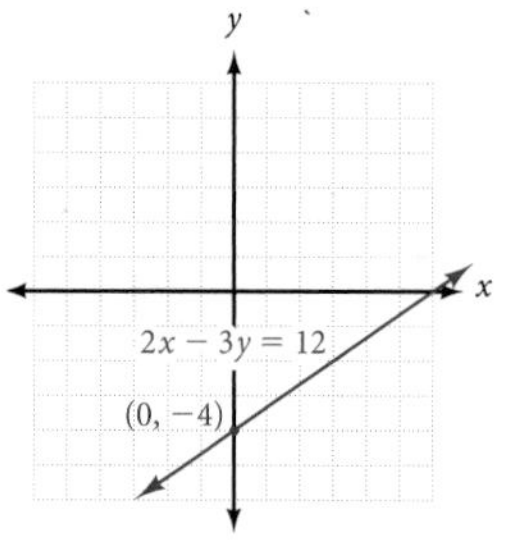

17. Slope = $-\frac{4}{5}$; y-intercept = 4; perpendicular slope = $\frac{5}{4}$

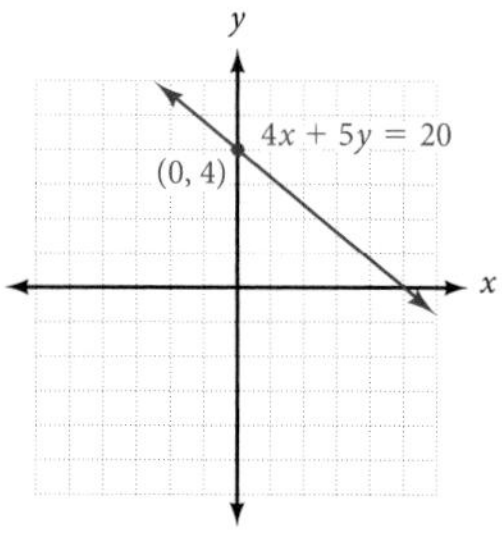

19. Slope = $\frac{1}{2}$; y-intercept = -4; $y = \frac{1}{2}x - 4$ **21.** Slope = $-\frac{2}{3}$; y-intercept = 3; $y = -\frac{2}{3}x + 3$ **23.** $y = 2x - 1$ **25.** $y = -\frac{1}{2}x - 1$ **27.** $y = -3x + 1$ **29.** $y = \frac{2}{3}x + \frac{14}{3}$ **31.** $y = -\frac{1}{4}x - \frac{13}{4}$ **33.** $3x + 5y = -1$ **35.** $x - 12y = -8$ **37.** $6x - 5y = 3$ **39.** $(0, -4), (2, 0)$; $y = 2x - 4$ **41.** $(-2, 0), (0, 4)$; $y = 2x + 4$

43. a. x-intercept = $\frac{10}{3}$; y-intercept = -5 **b.** (4, 1), answers may vary **c.** $y = \frac{3}{2}x - 5$ **d.** no

45. a. 2 **b.** $\frac{3}{2}$ **c.** -3 **d.** **e.** $y = 2x - 3$

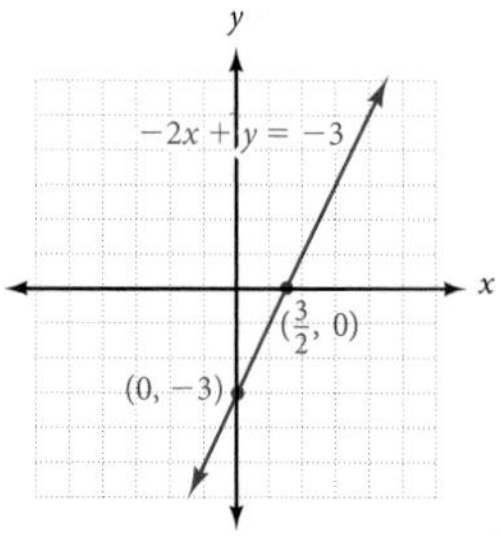

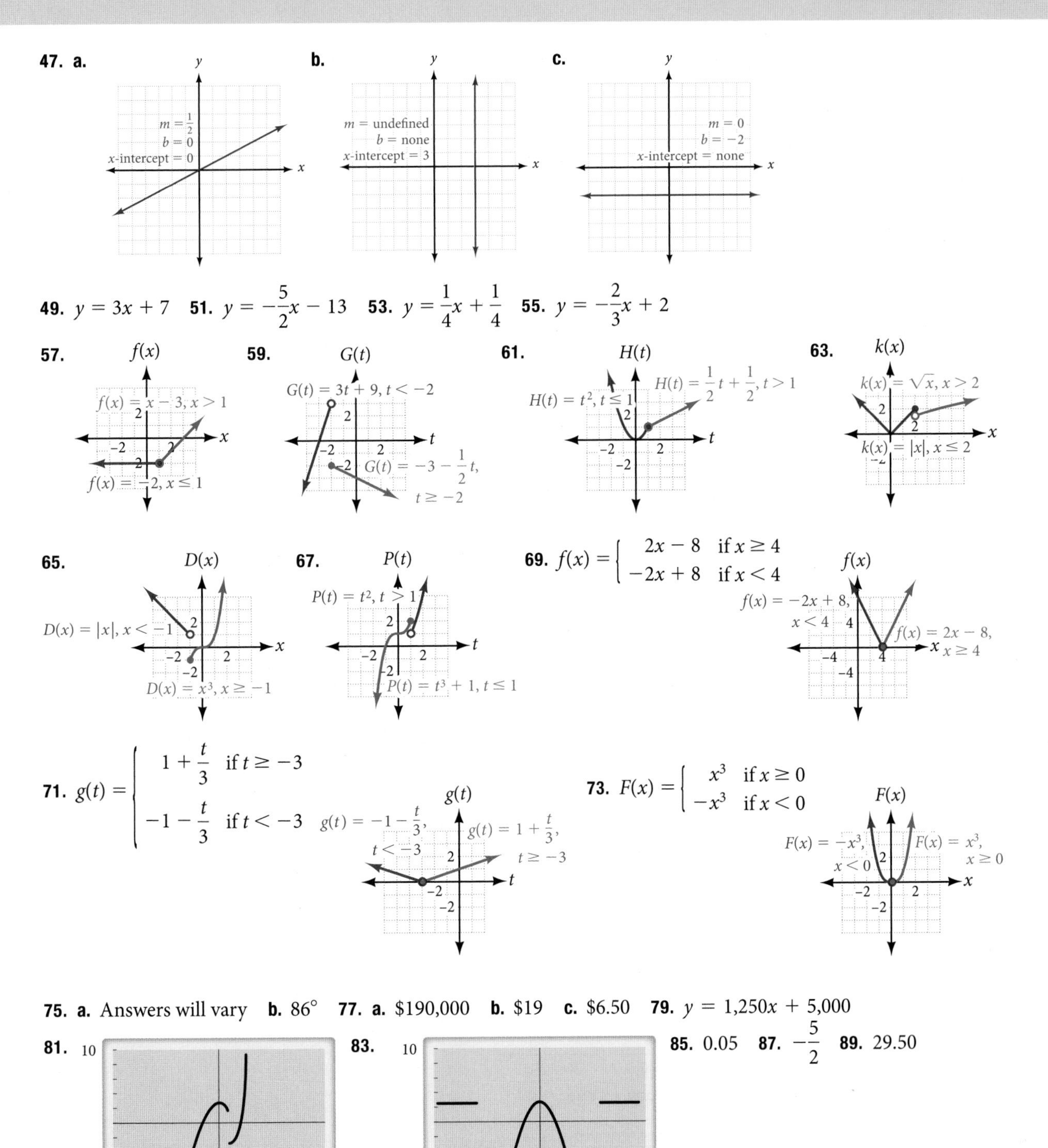

49. $y = 3x + 7$ **51.** $y = -\frac{5}{2}x - 13$ **53.** $y = \frac{1}{4}x + \frac{1}{4}$ **55.** $y = -\frac{2}{3}x + 2$

69. $f(x) = \begin{cases} 2x - 8 & \text{if } x \geq 4 \\ -2x + 8 & \text{if } x < 4 \end{cases}$

71. $g(t) = \begin{cases} 1 + \frac{t}{3} & \text{if } t \geq -3 \\ -1 - \frac{t}{3} & \text{if } t < -3 \end{cases}$

73. $F(x) = \begin{cases} x^3 & \text{if } x \geq 0 \\ -x^3 & \text{if } x < 0 \end{cases}$

75. a. Answers will vary **b.** 86° **77. a.** \$190,000 **b.** \$19 **c.** \$6.50 **79.** $y = 1{,}250x + 5{,}000$

81. 10, −10, −10, 10 **83.** 10, −10, −10, 10 **85.** 0.05 **87.** $-\frac{5}{2}$ **89.** 29.50

Problem Set 4.3

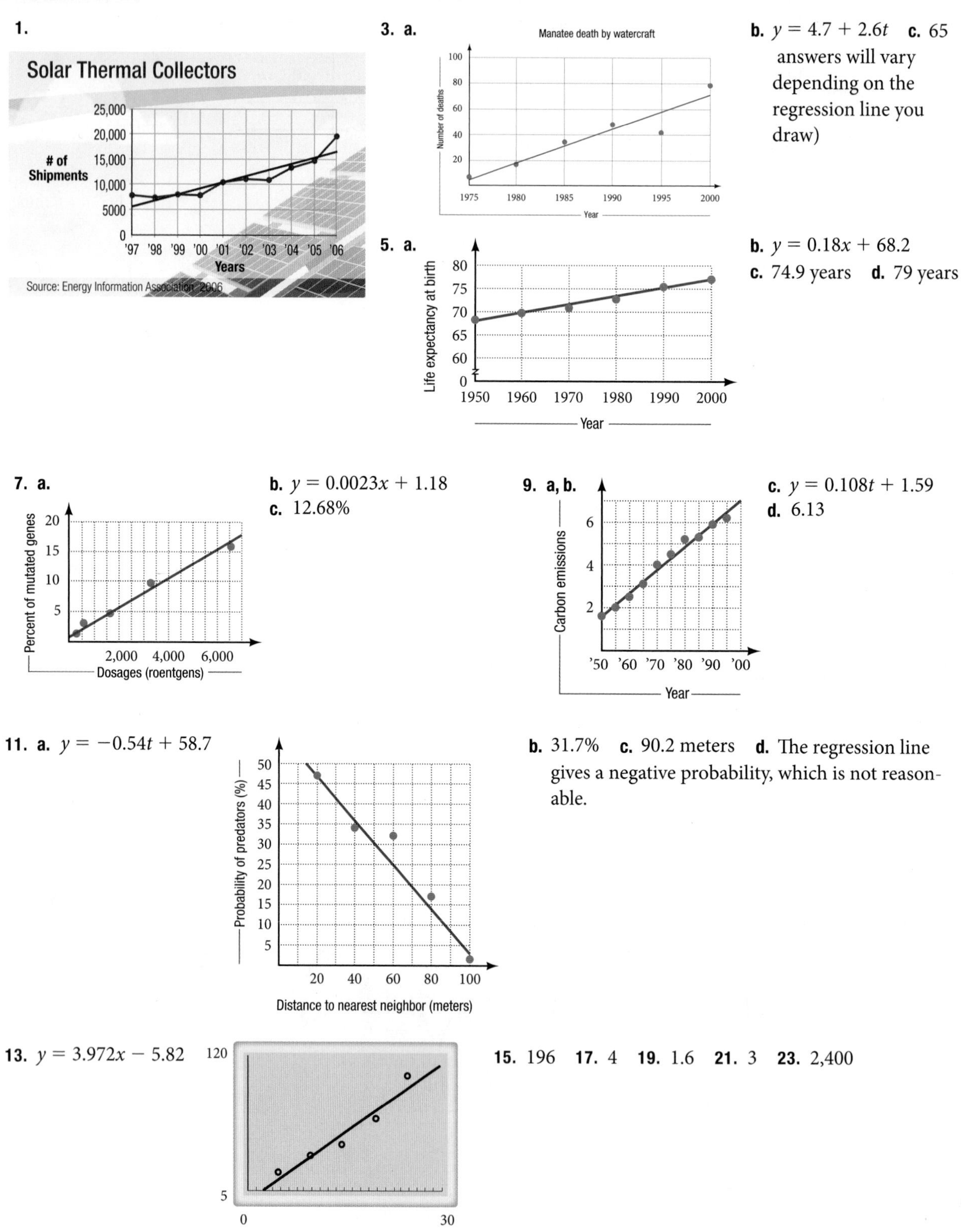

1.

3. a. **b.** $y = 4.7 + 2.6t$ **c.** 65 answers will vary depending on the regression line you draw)

5. a. **b.** $y = 0.18x + 68.2$ **c.** 74.9 years **d.** 79 years

7. a. **b.** $y = 0.0023x + 1.18$ **c.** 12.68%

9. a, b. **c.** $y = 0.108t + 1.59$ **d.** 6.13

11. a. $y = -0.54t + 58.7$ **b.** 31.7% **c.** 90.2 meters **d.** The regression line gives a negative probability, which is not reasonable.

13. $y = 3.972x - 5.82$

15. 196 **17.** 4 **19.** 1.6 **21.** 3 **23.** 2,400

Problem Set 4.4

1. 30 **3.** -6 **5.** 40 **7.** $\frac{81}{5}$ **9.** 64 **11.** 108 **13.** 300 **15.** ± 2 **17.** 1,600 **19.** ± 8 **21.** $\frac{50}{3}$ pounds

23. **a.** $T = 4P$ **b.** **c.** 70 pounds per square inch

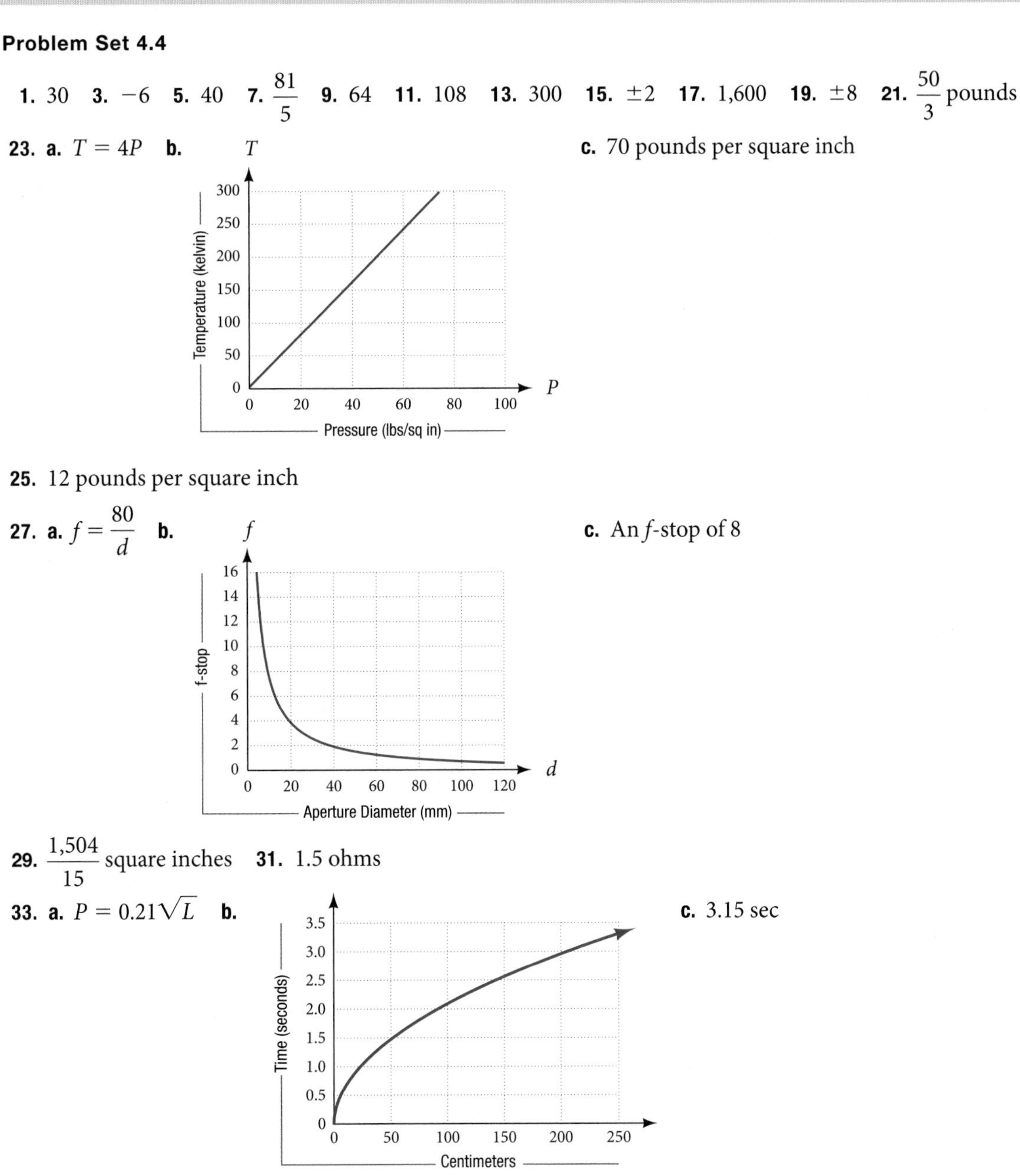

25. 12 pounds per square inch

27. **a.** $f = \frac{80}{d}$ **b.** **c.** An f-stop of 8

29. $\frac{1{,}504}{15}$ square inches **31.** 1.5 ohms

33. **a.** $P = 0.21\sqrt{L}$ **b.** **c.** 3.15 sec

35. x-intercept $= -\frac{8}{3}$; y-intercept $= 4$; slope $= \frac{3}{2}$ **37.** $y = 2x + 5$ **39.** $y = \frac{2}{5}x - 5$

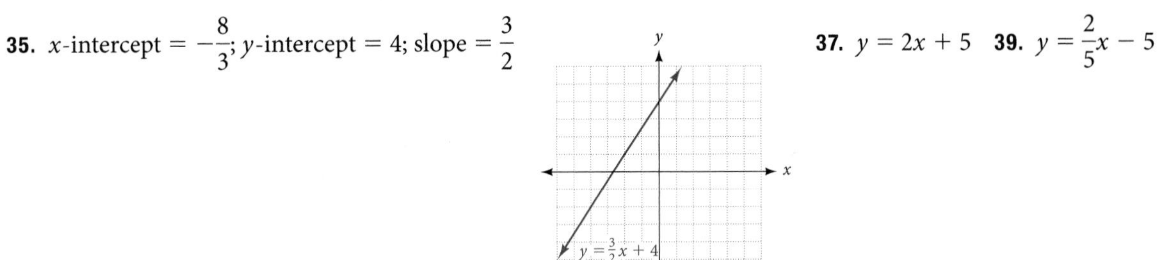

41. Domain $= \{-2, -3\}$; Range $= \{0, 1\}$; not a function **43.** 11 **45.** 8

Chapter 4 Test

1.

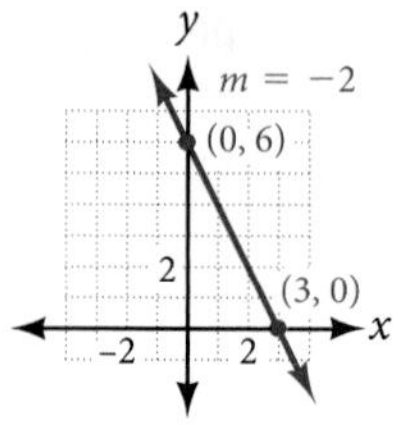

2.

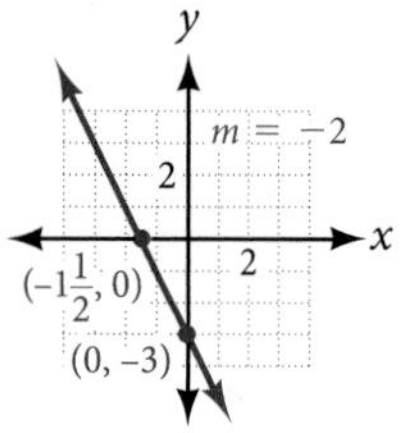

3.

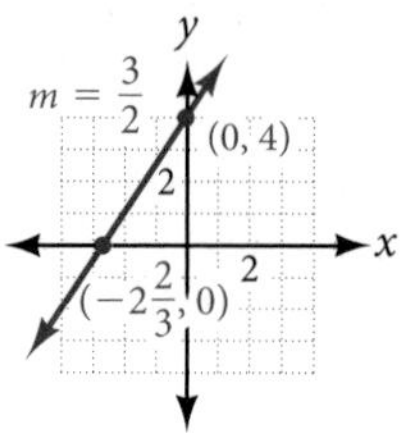

4.

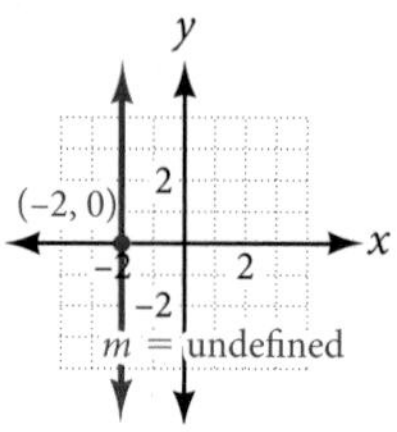

5. $\frac{1}{2}$ **6.** $2x + h$ **7.** 3 **8.** 0 **9.** $y = 2x + 5$ **10.** $y = -\frac{3}{7}x + \frac{5}{7}$ **11.** $y = \frac{2}{5}x - 5$

12. $y = -\frac{1}{3}x - \frac{7}{3}$ **13.** $x = 4$ **14.**

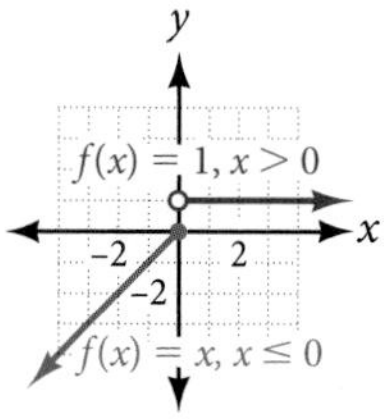

15.

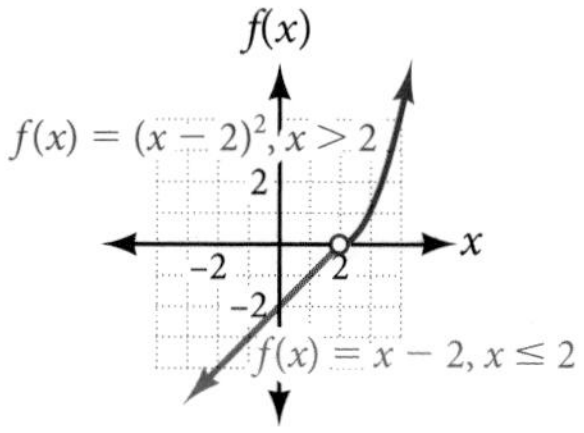

16. a.

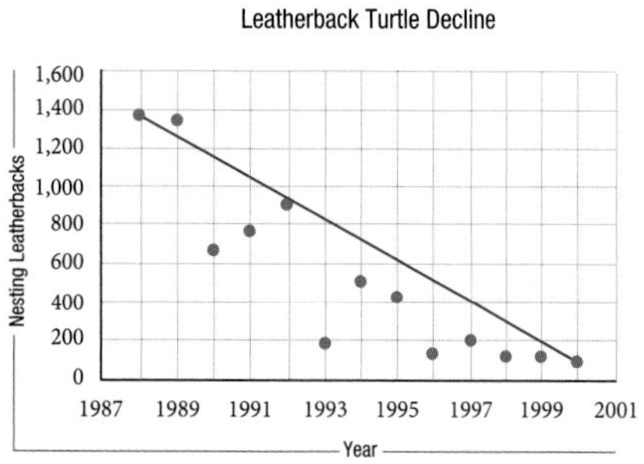

b. $-\frac{1{,}279}{12}$ **c.** $y = -\frac{1{,}279}{12}x + \frac{17{,}683}{12}$ **d.** No

e.

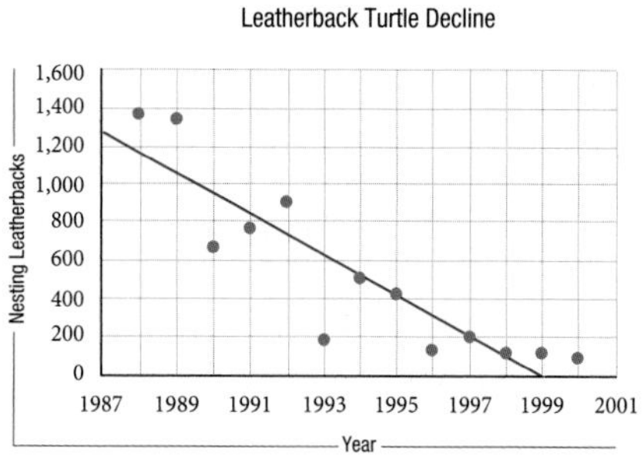

17. a.

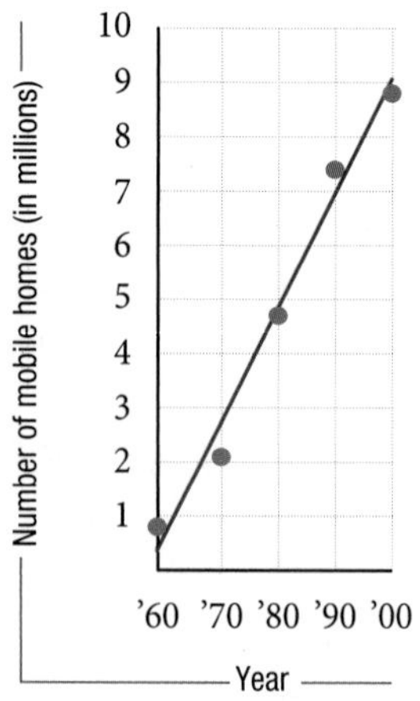

b. $y = 0.213x + 0.5$ **18.** $y = 18$ **19.** $z = \frac{81}{4}$ **20.** 666.67 lb

Chapter 5

Problem Set 5.1

1. $x = -3, 1; y = -3$; vertex $= (-1, -4)$ **3.** $x = -1, 1; y = -1$; vertex $= (0, -1)$

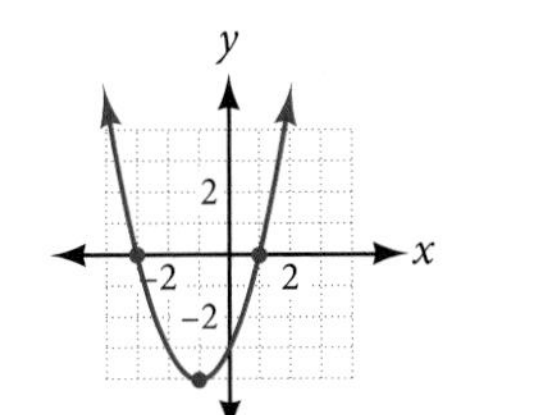

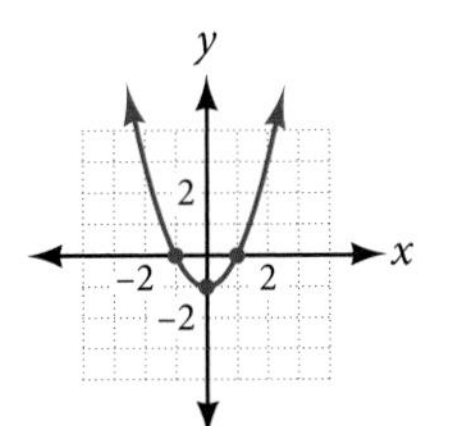

5. $x = -3, 3; y = 9$; vertex $= (0, 9)$ **7.** $x = -1, 3; y = -6$; vertex $= (1, -8)$

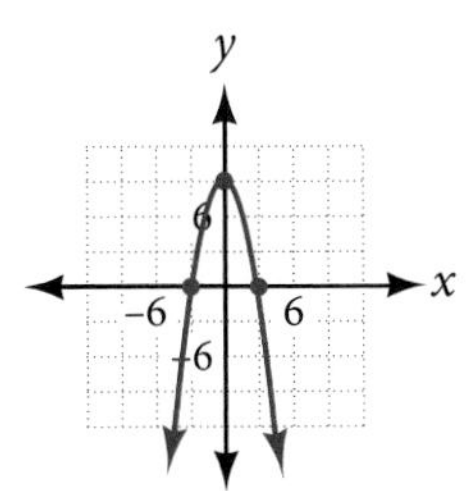

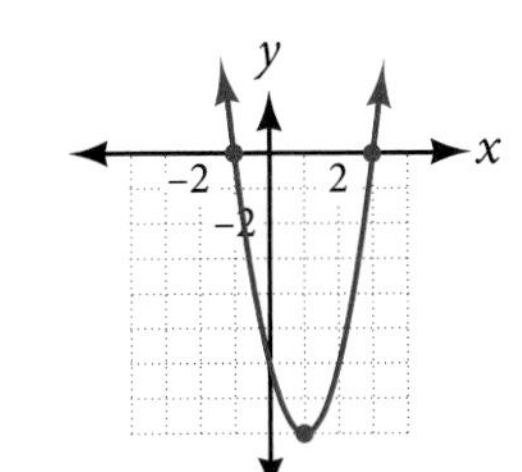

9. $x = 1 \pm \sqrt{5}; y = -4$; vertex $= (1, -5)$ **11.** **13.**

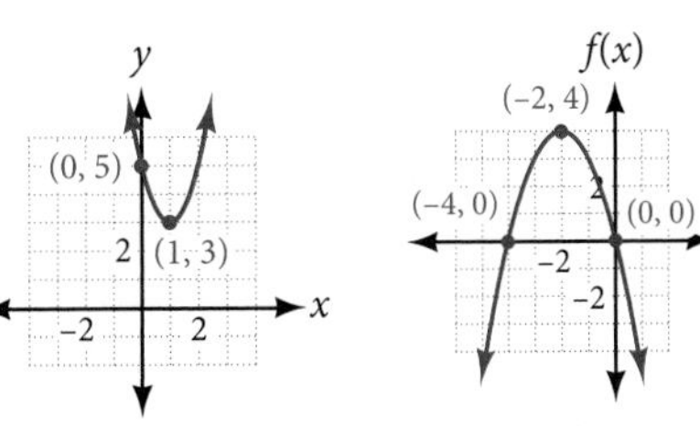

15. **17.** $y = -33$ **19.** **21.**

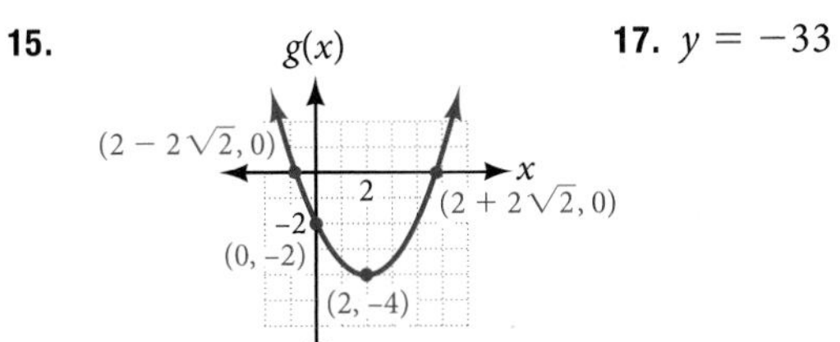

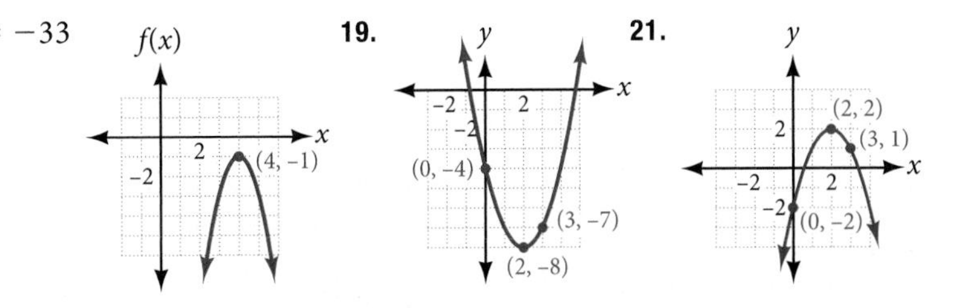

23. **25.**

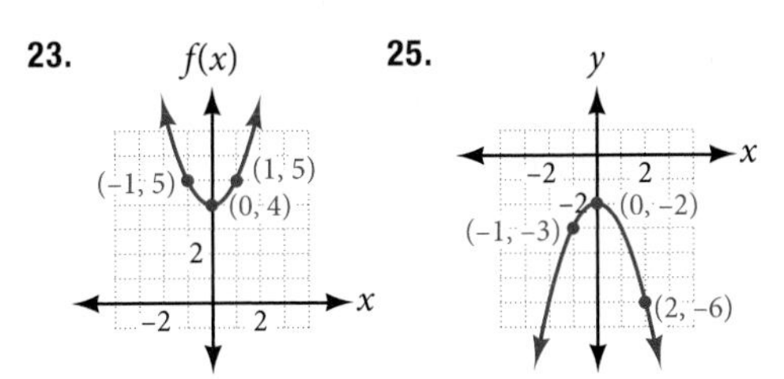

27. Vertex $= (3, -4)$; lowest point **29.** Vertex $= (1, 9)$; highest point
31. Vertex $= (2, 16)$; highest point
33. Vertex $= (-4, 16)$; highest point **35.** 40 items; profit = \$500
37. 875 patterns; profit = \$731.25
39. In hand: 0 sec, 2 sec; max height: 16 feet **41.** 64 ft
43. 30 ft by 15 ft **45.** $y = (x - 2)^2 - 4$

47. 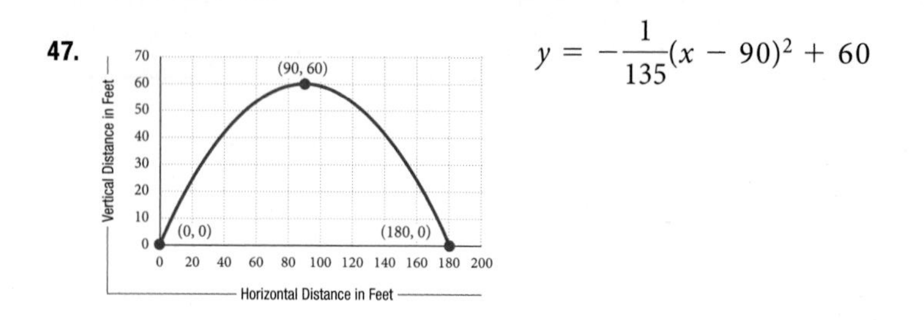

$y = -\frac{1}{135}(x - 90)^2 + 60$

49. Problem 1 $(x + 2)(x - 2)$; Problem 2 $x = \pm 2$; Problem 4 ± 2; the answer to factoring the polynomial is used to solve the related equation and also to find the zeros of the parabola.

51. $h(30) \approx 42.7$. When his horizontal distance was 30 ft, his vertical distance, or height, was about 42.7 ft.

53. $y = x^2 - 4x$ **55.** $y = -x^2 + 4x + 4$ **57.** 2 **59.** $-2x^2y^2$ **61.** 185.12 **63.** $4x^3 - 8x^2$ **65.** $4x^3 - 6x - 20$

67. $-3x + 9$ **69.** $(x + a)(x - a)$ **71.** $(x - 7y)(x + y)$

Problem Set 5.2

1. $2x^2 + 4x - 3$ **3.** $\frac{2}{3}y^2 - \frac{1}{3}y + 2$ **5.** $a + b$ **7.** $x - 3$ **9.** $x - 6y$ **11.** $(y^2 + 9)(y + 3)$ **13.** $x + 7 + \frac{13}{x - 3}$

15. $3x^2 + x + 4 + \frac{7}{x - 2}$ **17.** $2x + 3$ **19.** $a^3 - a^2 + 2a - 4 + \frac{7}{a + 2}$ **21.** $x - 7 + \frac{20}{x + 2}$ **23.** $3x - 1$

25. $x^2 + 4x + 11 + \frac{26}{x - 2}$ **27.** $3x^2 + 8x + 26 + \frac{83}{x - 3}$ **29.** $2x^2 + 2x + 3$ **31.** $x^3 - 4x^2 + 18x - 72 + \frac{289}{x + 4}$

33. $x^4 + x^2 - x - 3 - \frac{5}{x - 2}$ **35.** $x + 2 + \frac{3}{x - 1}$ **37.** $x^3 - x^2 + x - 1$ **39.** $x^2 + x + 1$

41. $P(x) = x + 8, R = 36$ **43.** $P(x) = x^2 - 5x + 20, R = -56$ **45.** $P(x) = 5x^2 + 13x + 39, R = 121$

47. a. $2x^2 + 3x + 14 + \frac{36}{x - 3}$ **b.** $2x^2 - 2x + 4 - \frac{4}{x - \frac{1}{2}}$ **c.** $2x^2 - 5x + 10 - \frac{16}{x + 1}$ **d.** $2x^2 - 7x + 19 - \frac{44}{x + 2}$

49. a. $-\frac{24}{x - 3}$ **b.** $\frac{1}{x - \frac{1}{2}}$ **c.** $-\frac{8}{x + 1}$ **d.** $-\frac{49}{x + 2}$ **51.** $(x + 3)(x + 2)(x + 1)$ **53.** $(x + 3)(x + 4)(x - 2)$

55. same **57. a.**

x	1	5	10	15	20
$C(x)$	1.62	2.10	2.70	3.30	3.90

b. $\overline{C}(x) = 0.12 + \frac{1.50}{x}$ **c.**

x	1	5	10	15	20
$\overline{C}(x)$	1.62	0.42	0.27	0.22	0.195

d. It decreases. **e.**

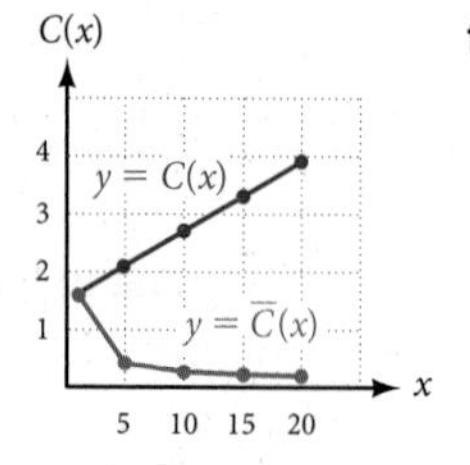

f. $y = C(x)$: domain $= \{x \mid 1 \le x \le 20\}$; range $= \{y \mid 1.62 \le y \le 3.90\}$; $y = \overline{C}(x)$: domain $= \{x \mid 1 \le x \le 20\}$; range $= \{y \mid 0.195 \le y \le 1.62\}$

59. a. $T(100) = 11.95$; $T(400) = 32.95$; $T(500) = 39.95$ **b.** $\frac{4.95}{m} + 0.07$

c. $\overline{T}(100) \approx 0.12$, $\overline{T}(400) \approx 0.08$, $\overline{T}(500) \approx 0.08$

61. Alternate forms:

$$2x + \frac{3}{x-2} - 3$$

$$\frac{x(2x-7)+9}{x-2}$$

$$\frac{2x^2}{x-2} - \frac{7x}{x-2} + \frac{9}{x-2}$$

Wolfram Alpha LLC. 2012. Wolfram|Alpha
http://www.wolframalpha.com/
(access June 18, 2012)

63. Alternate forms:

$$x^2 + x + 1$$

$$x(x+1) + 1$$

$$\frac{x^3}{x-1} - \frac{1}{x-1}$$

Wolfram Alpha LLC. 2012. Wolfram|Alpha
http://www.wolframalpha.com/
(access June 18, 2012)

65. 193 **67.** $4x^2 - 9$ **69.** $-1 \pm \sqrt{2}$

Problem Set 5.3

1. 121 **3.** 36 **5.** -41 **7.** 368 **9.** Yes **11.** No **13.** No **15.** No **17.** $(x-1)^2(x-4)$ **19.** $3, -\frac{1}{3}, \frac{5}{4}$
21. $\frac{1}{2}, 2, 3$ **23.** $2, -1$ **25.** $-4, i, -i$ **27.** $2, -1 \pm \sqrt{3}$ **29.** $(x+1)$ is not a factor **31.** 16,000 **33.** -1.5
35. $C(200) = \$900$; $C(300) = \$900$ **37.** $C(200) = \$500$; $C(300) = \$900$
39. $R(85) \approx 154{,}120.83$; if the company sells 85 units of one of its products, it will realize a revenue of about \$154,120.83
41. $C(350) = 4{,}053$; if a company ships 350 units it will cost \$4,053 **43.** $\pm 2, 0$ **45.** ± 3 **47.** $\pm\sqrt{3}$

Problem Set 5.4

1. **3.** **5.** **7.** **9.**

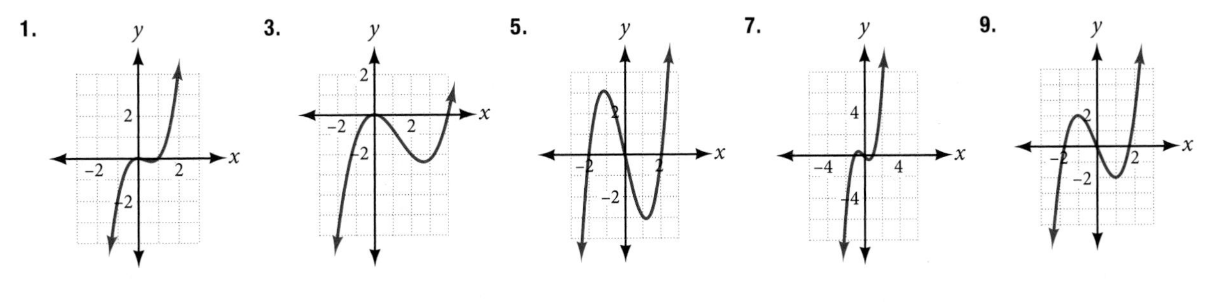

11.

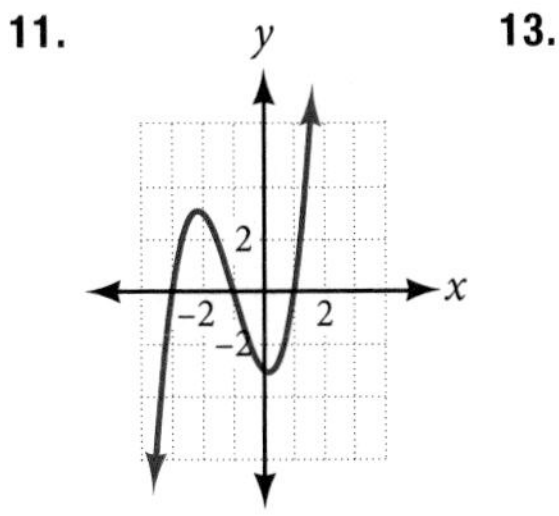

13.

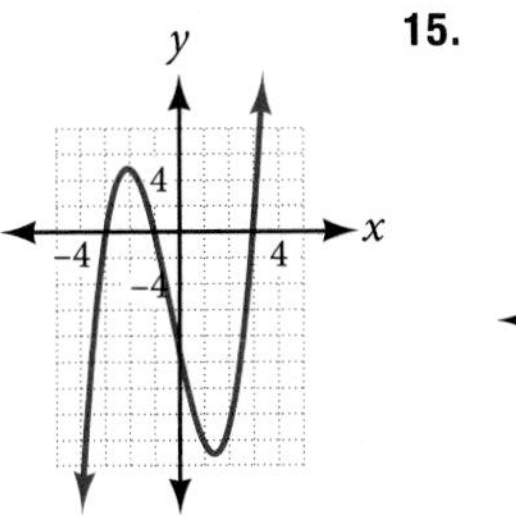

15.

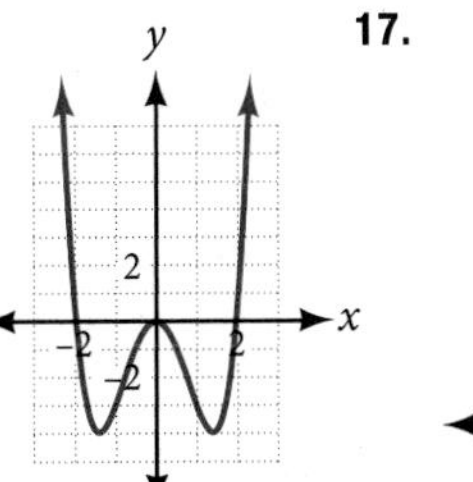

17.

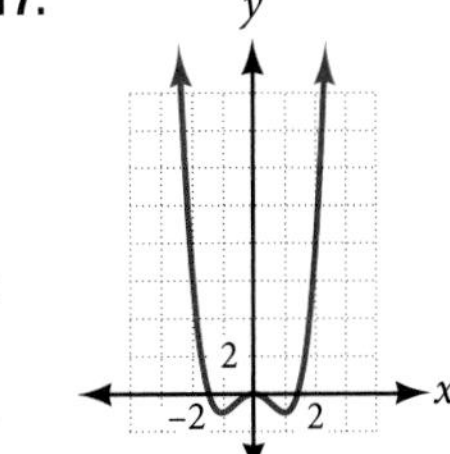

19.

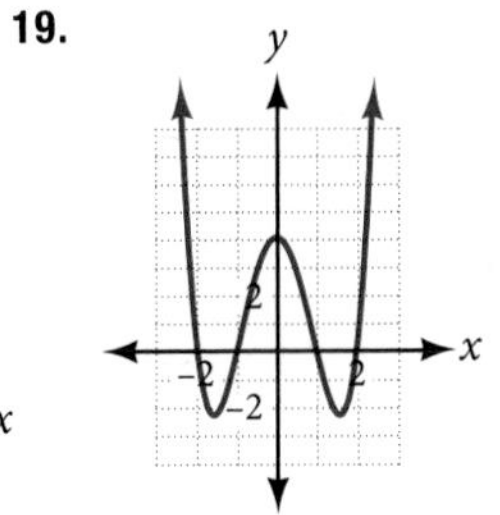

21.

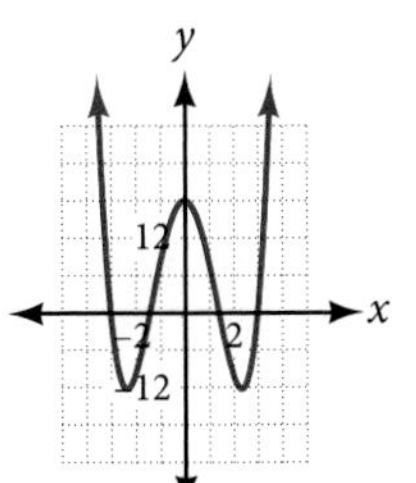

23. a.

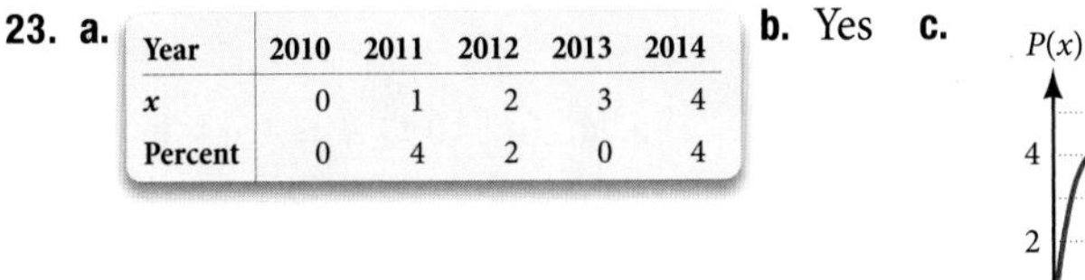

Year	2010	2011	2012	2013	2014
x	0	1	2	3	4
Percent	0	4	2	0	4

b. Yes **c.**

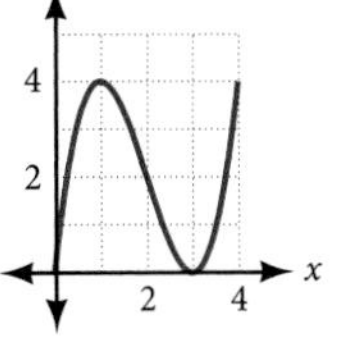

25. a.

Year	2006	2007	2008	2009	2010	2011	2012
x	0	1	2	3	4	5	6
Dollars (in thousands)	49	0	25	64	81	64	25

b. Yes **c.**

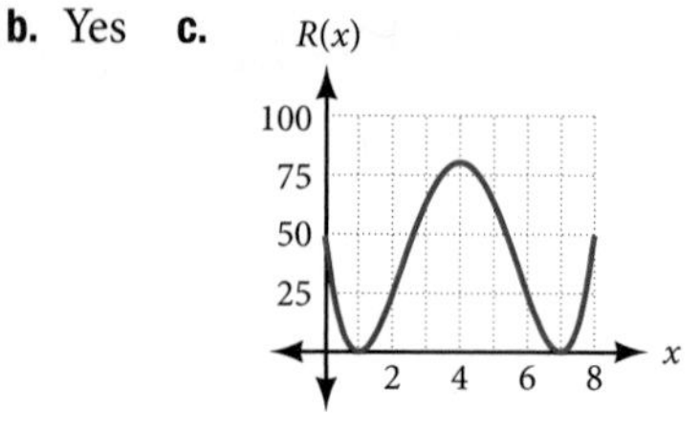

27. a. $N(x) = 4.31x^2 - 5.16x + 8.26$ **b.** $N(6) = 132$, so about 132 million people. **c.** $N(10) = 388$, so about 388 million users. This says that the model is not valid for the year 2015.

29. a. $\frac{3}{2}$ **b.** undefined **31.** $(0, 3), (-3, 0)$ **33. a.** 4 **b.** 2

Problem Set 5.5

1. $g(0) = -3; g(-3) = 0; g(3) = 3; g(-1) = -1; g(1) =$ undefined

3. $h(0) = -3; h(-3) = 3; h(3) = 0; h(-1) =$ undefined; $h(1) = -1$

5. $\{x \mid x \neq 1\}$ **7.** $\{x \mid x \neq 2\}$ **9.** $\{t \mid t \neq -4, t \neq 4\}$ **11.** $f(0) = 2; g(0) = 2$ **13.** $f(2) =$ undefined; $g(2) = 4$

15. $f(0) = 1; g(0) = 1$ **17.** $f(2) = 3; g(2) = 3$

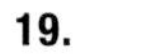

19.

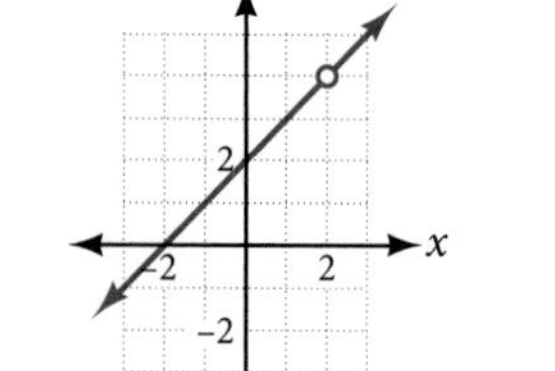

The graph is different from $y = x + 2$ because it doesn't exist at (2, 4)

21. a. 2 **b.** -4 **c.** Undefined **d.** 2 **23.** **25.** **27.**

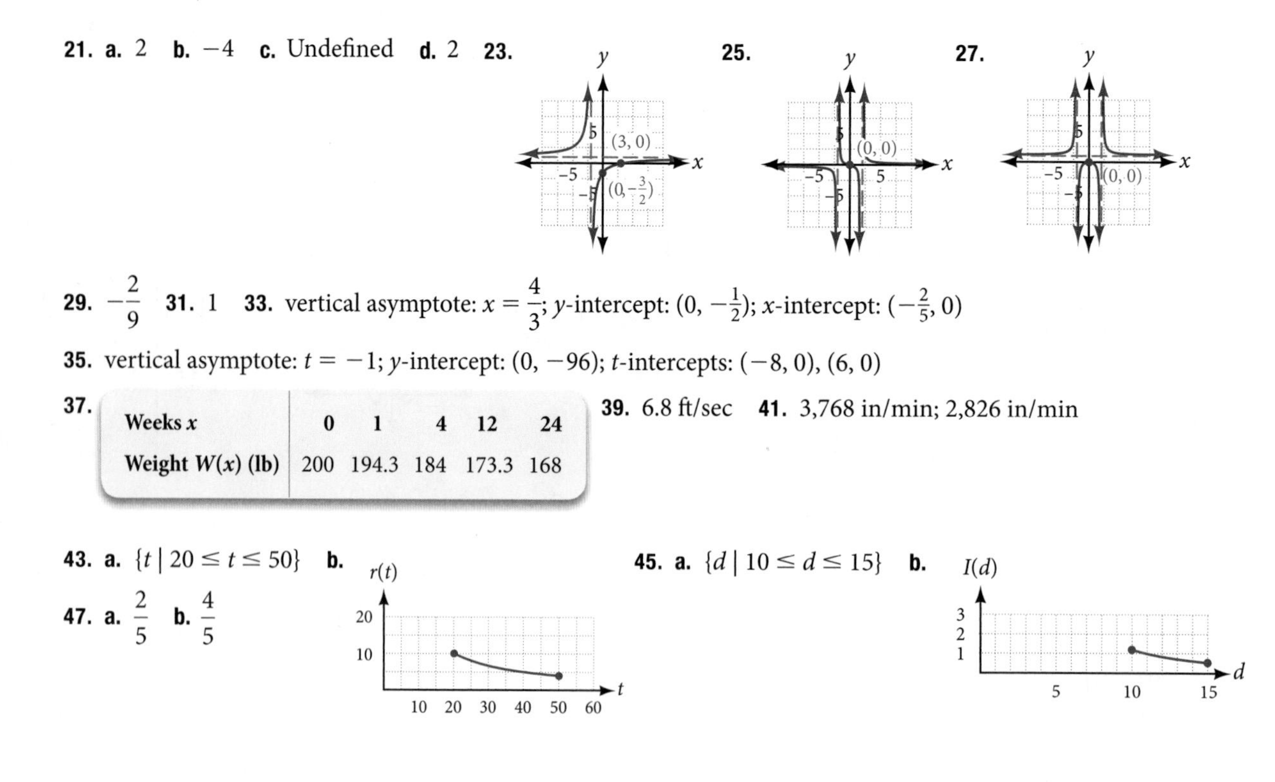

29. $-\frac{2}{9}$ **31.** 1 **33.** vertical asymptote: $x = \frac{4}{3}$; y-intercept: $(0, -\frac{1}{2})$; x-intercept: $(-\frac{2}{5}, 0)$

35. vertical asymptote: $t = -1$; y-intercept: $(0, -96)$; t-intercepts: $(-8, 0), (6, 0)$

37.

Weeks x	0	1	4	12	24
Weight $W(x)$ (lb)	200	194.3	184	173.3	168

39. 6.8 ft/sec **41.** 3,768 in/min; 2,826 in/min

43. a. $\{t \mid 20 \leq t \leq 50\}$ **b.**

45. a. $\{d \mid 10 \leq d \leq 15\}$ **b.**

47. a. $\frac{2}{5}$ **b.** $\frac{4}{5}$

49. a. $\{r \mid 10 \le r \le 15\}$ **b.**

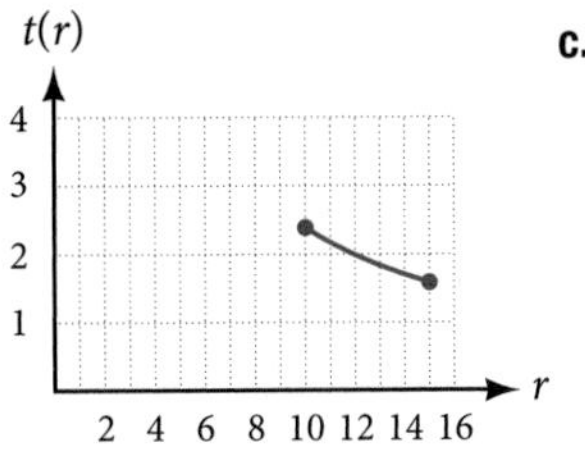

c. $r - 2$ **d.** $T(r) = \dfrac{24}{r-2}$ **e.**

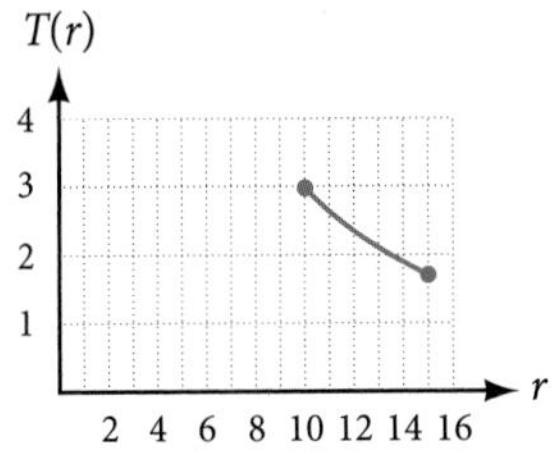

51. a. vertical asymptote: $x = 3$; horizontal asymptote: $y = 1$

b. x-intercept: $(-1, 0)$; y-intercept: $(0, -\frac{1}{3})$

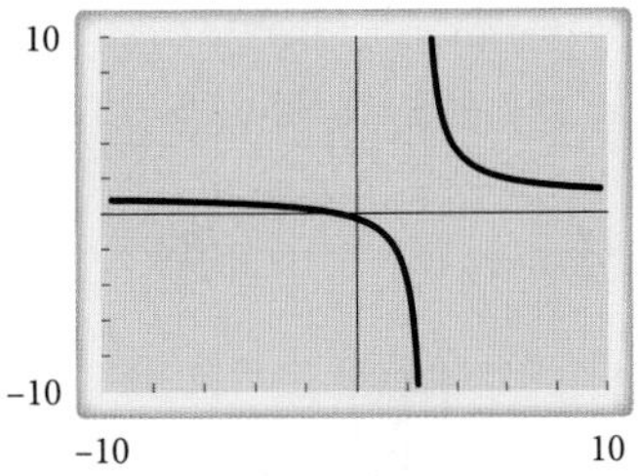

53. a. vertical asymptote: $x = 1$; horizontal asymptote: $y = \frac{2}{3}$

b. x-intercept: $(-\frac{1}{2}, 0)$; y-intercept: $(0, -\frac{1}{3})$

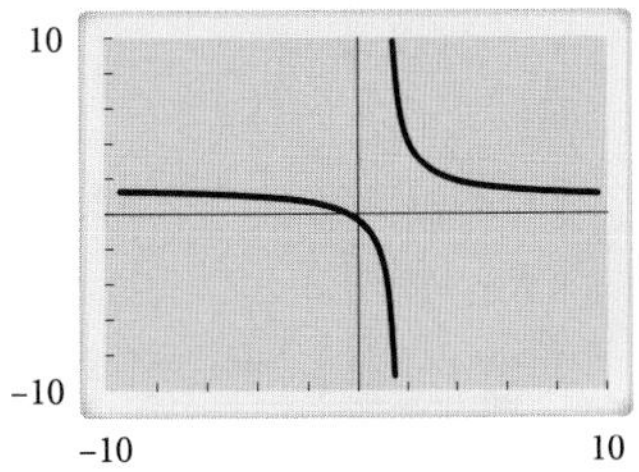

55.

Year	0	3	6	9	12	15	18	21	24	27	30
Population	150	238	309	369	420	463	501	533	563	588	611

Growth is decreasing, because the difference between each subsequent data point in the table is decreasing.

57. a. 6 **b.** 0 **59.** $P(-x) = 7x^4 + 2x^3 + 3x^2 - 4x - 5$

Problem Set 5.6

1. a. 4 **b.** 4, 2, 0 **c.** 0 **3. a.** 4 **b.** 0 **c.** 0 **5. a.** 4 **b.** 1 **c.** 1 **7. a.** 3 **b.** 3, 1 **c.** 0

9. a. 5 **b.** 2, 0 **c.** 3, 1 **11.** $x = \pm 2i, 3$ **13.** $x = 2, -\frac{1}{2}, \pm i$ **15.** $x = -2, 1(\text{mult. } 2), -1(\text{mult. } 2)$

17. $x = -3, -2, -1, 1, 2$ **19.** $x = -3, \frac{1}{2}(\text{mult. } 2)$ **21.** $x = 3, 3 \pm \sqrt{2}$ **23.** $x = \pm 2, 1 \pm 2i$ **25.** $x = \pm 1, \dfrac{2 \pm i\sqrt{3}}{2}$

27.

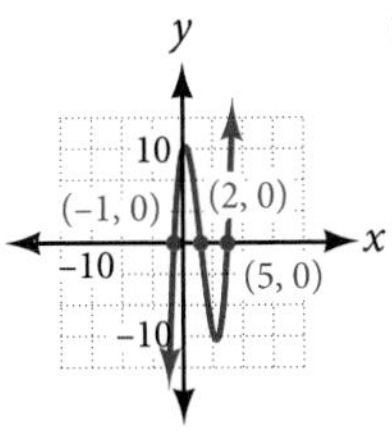

29.

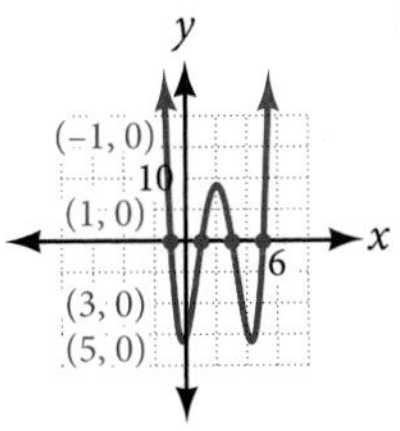

31. $\dfrac{2}{3a}$ **33.** $(x - 3)(x + 2)$ **35.** 1 **37.** $\dfrac{3 - x}{x + 3}$ **39.** No solution

Chapter 5 Test

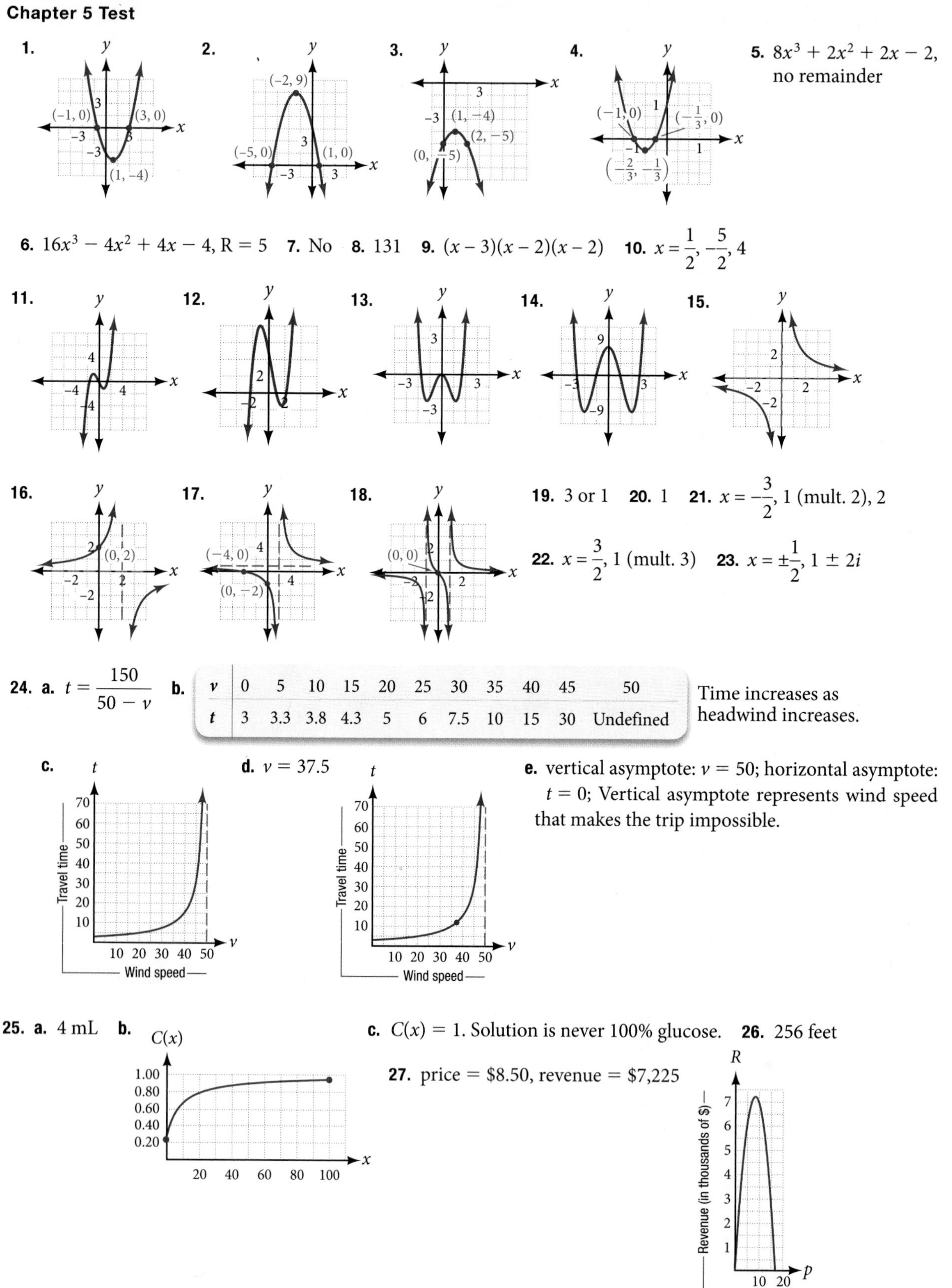

5. $8x^3 + 2x^2 + 2x - 2$, no remainder

6. $16x^3 - 4x^2 + 4x - 4$, $R = 5$ **7.** No **8.** 131 **9.** $(x-3)(x-2)(x-2)$ **10.** $x = \frac{1}{2}, -\frac{5}{2}, 4$

19. 3 or 1 **20.** 1 **21.** $x = -\frac{3}{2}$, 1 (mult. 2), 2

22. $x = \frac{3}{2}$, 1 (mult. 3) **23.** $x = \pm\frac{1}{2}$, $1 \pm 2i$

24. a. $t = \dfrac{150}{50 - v}$ **b.**

v	0	5	10	15	20	25	30	35	40	45	50
t	3	3.3	3.8	4.3	5	6	7.5	10	15	30	Undefined

Time increases as headwind increases.

d. $v = 37.5$

e. vertical asymptote: $v = 50$; horizontal asymptote: $t = 0$; Vertical asymptote represents wind speed that makes the trip impossible.

25. a. 4 mL **c.** $C(x) = 1$. Solution is never 100% glucose. **26.** 256 feet

27. price = \$8.50, revenue = \$7,225

Chapter 6

Problem Set 6.1

1. 1 **3.** 2 **5.** $\frac{1}{27}$ **7.** 13 **9.** $\frac{7}{12}$ **11.** $\frac{3}{16}$ **13.**

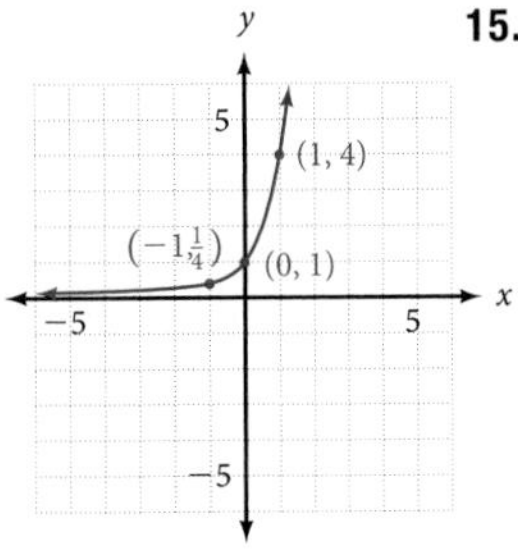

15.

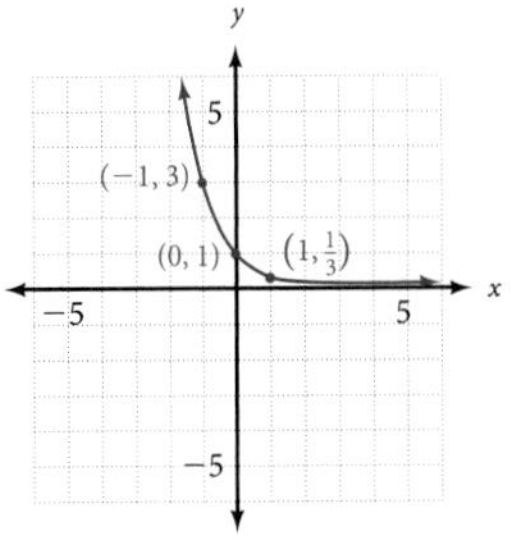

17.

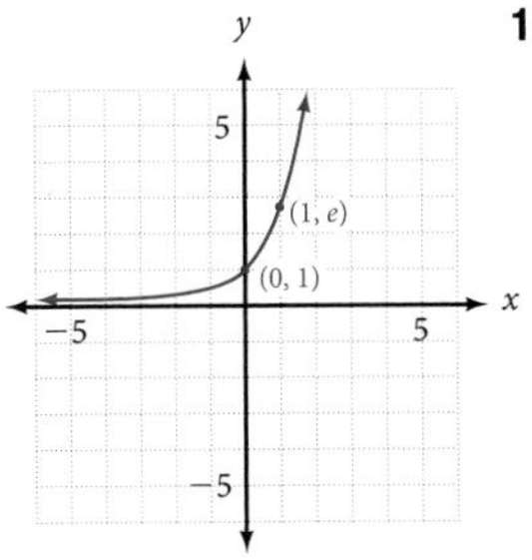

19.

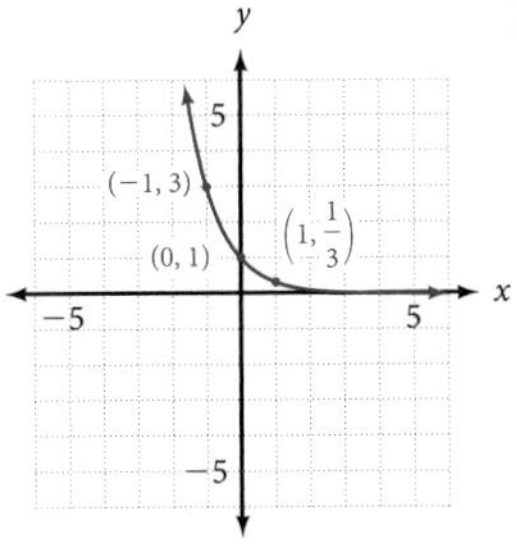

21.

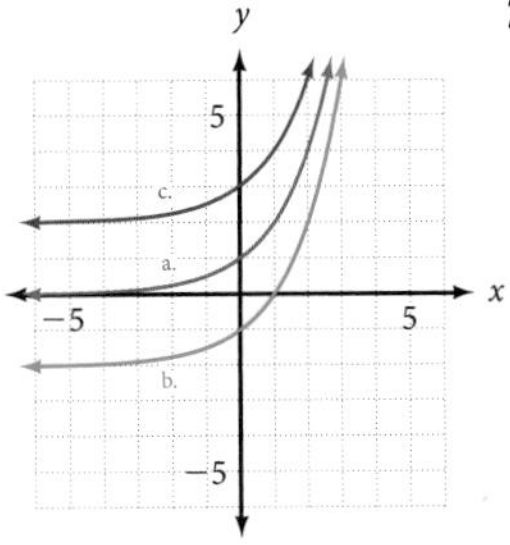

23.

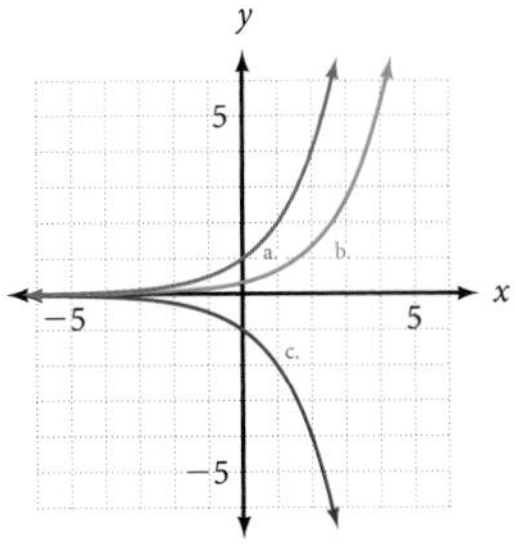

25.

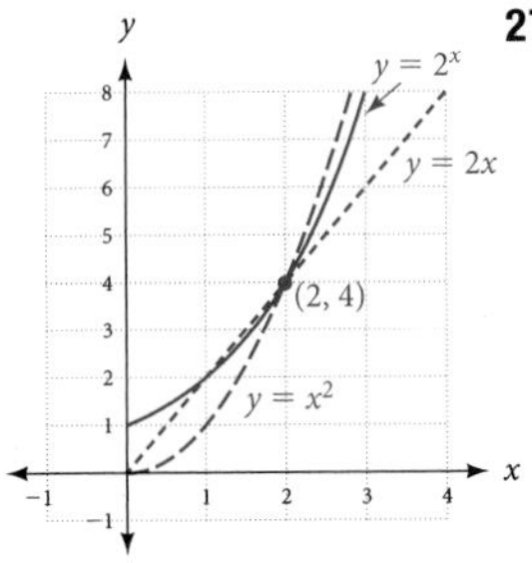

27.

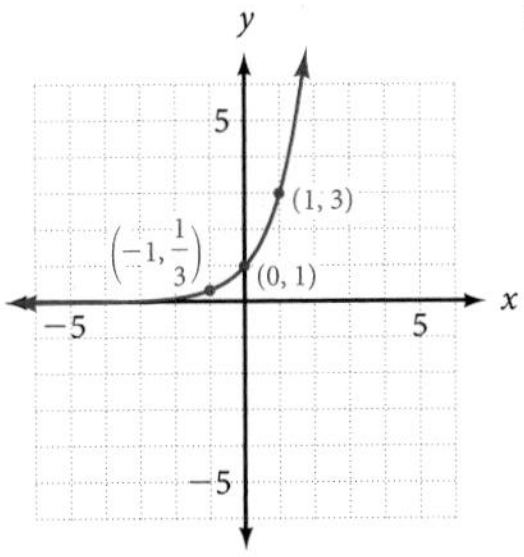

29.

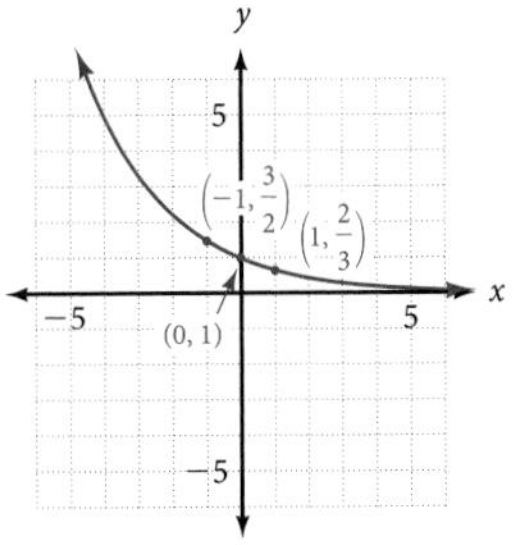

31. a. 2 **b.** 2.5937 **c.** 2.7048 **d.** 2.7169

33. A **35.** 6,562.0616

37. $h = A\left(\frac{1}{2}\right)^n$; $h(8) = 0.039$ feet **39. a.**

Hours Since Discontinuing	Concentration (ng/mL)
0	120
8	60
16	30
24	15
32	7.5

b.

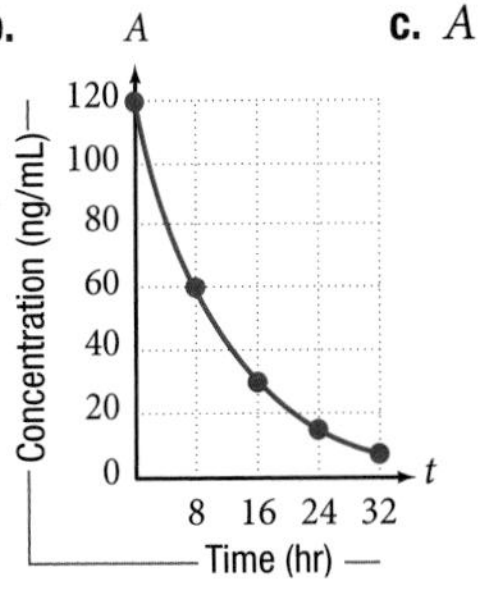

c. $A = 120(2)^{-t/8}$

41. Antidepressant 1 has a short half-life, so it must be discontinued slowly by tapering the dose over a 6-day period to avoid the possibility of the patient experiencing withdrawal symptoms.

43. In 1990, \$1.89 per pound; in 1995, \$5.77 per pound; in 2010, \$163.93 per pound

45. a. $A(t) = 500\left(1 + \frac{0.08}{12}\right)^{12t}$ **b.** $A(5) \approx \$744.92$ **c.** About 8.7 years **d.** \$745.91

47. **a.** \$95,625.18 **b.** $\{t \mid 0 \leq t \leq 7\}$ **c.**

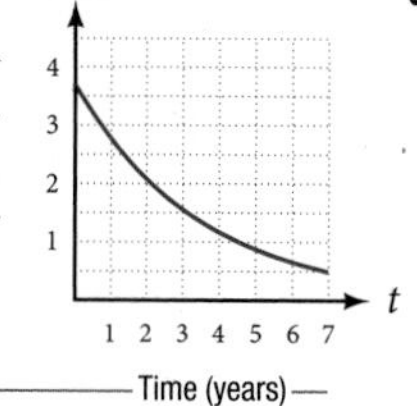

d. $\{V(t) \mid 50{,}056.46 \leq V(t) \leq 375{,}000\}$

e. About 6 years, 1 month

49. $f(1) = 200; f(2) = 400; f(3) = 800; f(4) = 1{,}600$; will have 100,000 after 9.97 days.

51.

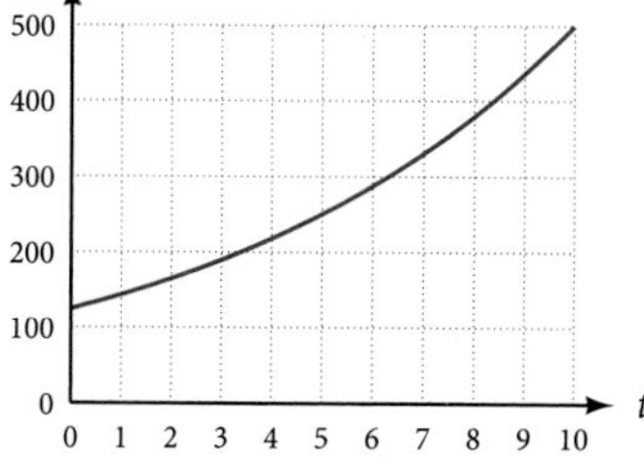

53. **a.** The value of the car after 3 years is approximately \$12,148.53. **b.** The depreciated amount (purchase price − value at 4 years = 25,600 − 9,475.85) is approximately \$16,124.15.

55. **57.** **59.** **61.**

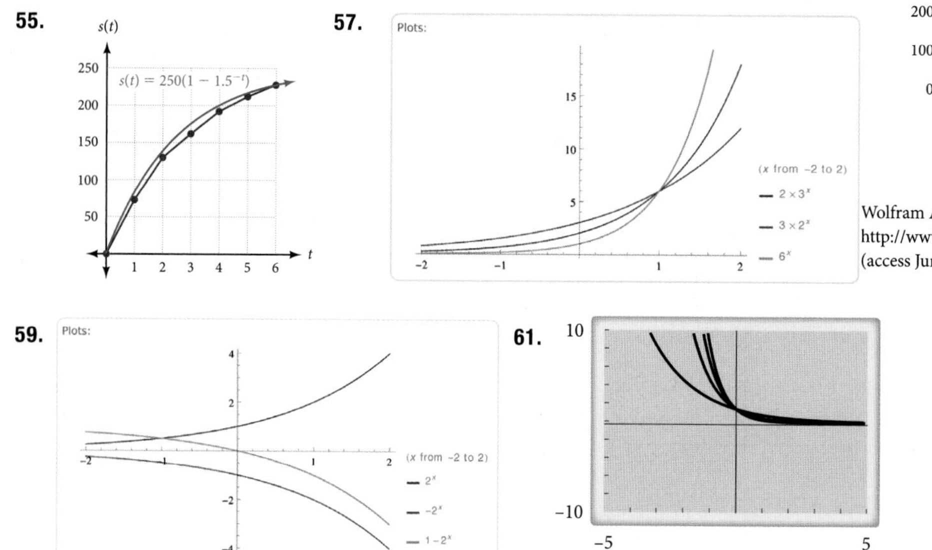

Wolfram Alpha LLC. 2012. Wolfram|Alpha http://www.wolframalpha.com/ (access June 18, 2012)

Wolfram Alpha LLC. 2012. Wolfram|Alpha http://www.wolframalpha.com/ (access June 18, 2012)

63. The maximum number of statements is 200 and he will get close to that number by about week number 12 or 13 in the semester.

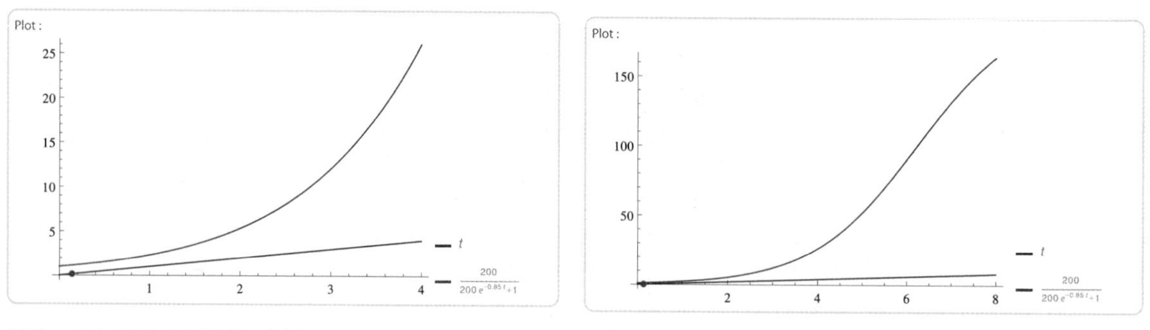

Wolfram Alpha LLC. 2012. Wolfram|Alpha http://www.wolframalpha.com/ (access June 18, 2012)

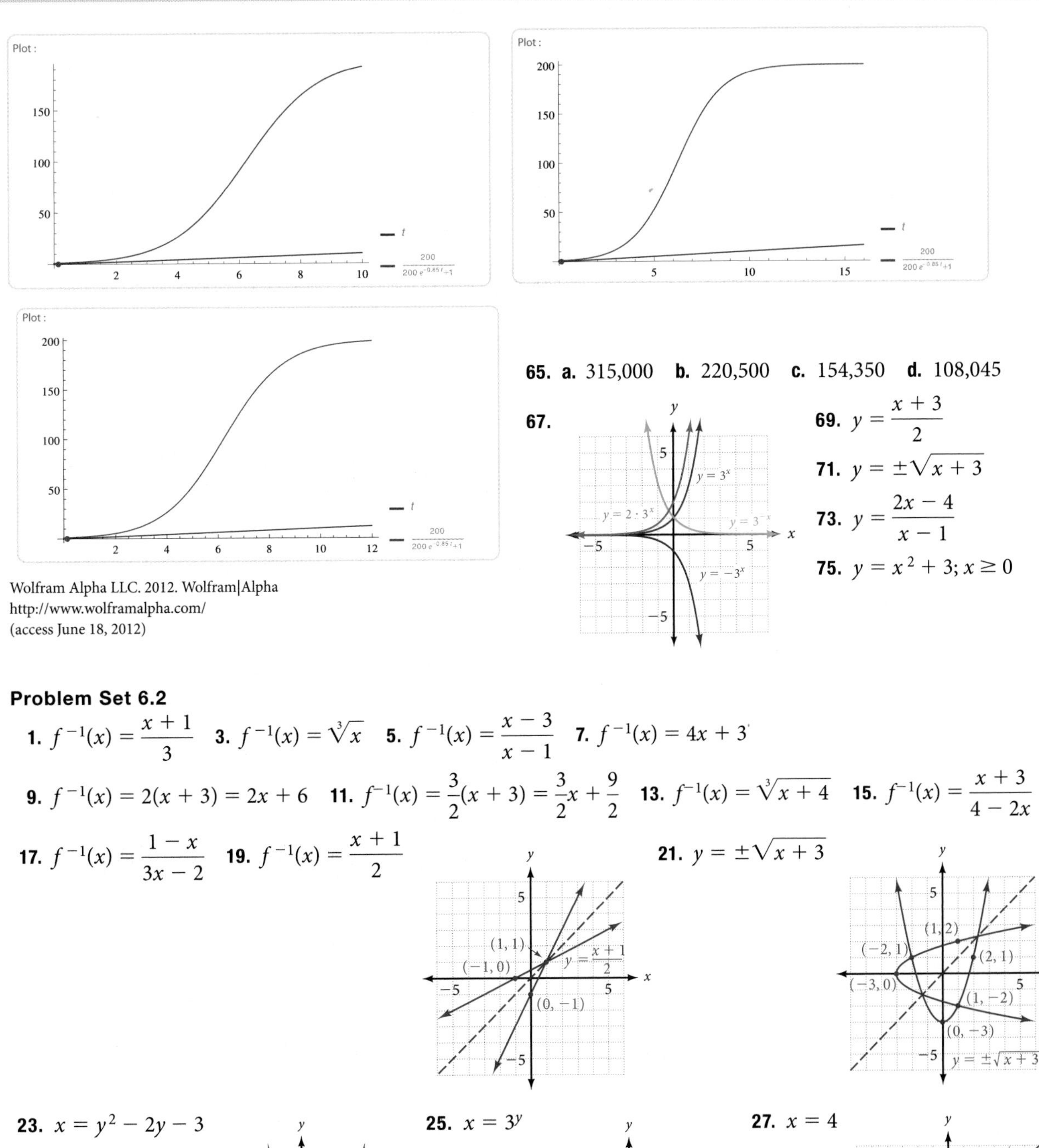

Wolfram Alpha LLC. 2012. Wolfram|Alpha
http://www.wolframalpha.com/
(access June 18, 2012)

65. a. 315,000 **b.** 220,500 **c.** 154,350 **d.** 108,045

67.

69. $y = \dfrac{x + 3}{2}$

71. $y = \pm\sqrt{x + 3}$

73. $y = \dfrac{2x - 4}{x - 1}$

75. $y = x^2 + 3;\ x \geq 0$

Problem Set 6.2

1. $f^{-1}(x) = \dfrac{x + 1}{3}$ **3.** $f^{-1}(x) = \sqrt[3]{x}$ **5.** $f^{-1}(x) = \dfrac{x - 3}{x - 1}$ **7.** $f^{-1}(x) = 4x + 3$

9. $f^{-1}(x) = 2(x + 3) = 2x + 6$ **11.** $f^{-1}(x) = \dfrac{3}{2}(x + 3) = \dfrac{3}{2}x + \dfrac{9}{2}$ **13.** $f^{-1}(x) = \sqrt[3]{x + 4}$ **15.** $f^{-1}(x) = \dfrac{x + 3}{4 - 2x}$

17. $f^{-1}(x) = \dfrac{1 - x}{3x - 2}$ **19.** $f^{-1}(x) = \dfrac{x + 1}{2}$

21. $y = \pm\sqrt{x + 3}$

23. $x = y^2 - 2y - 3$

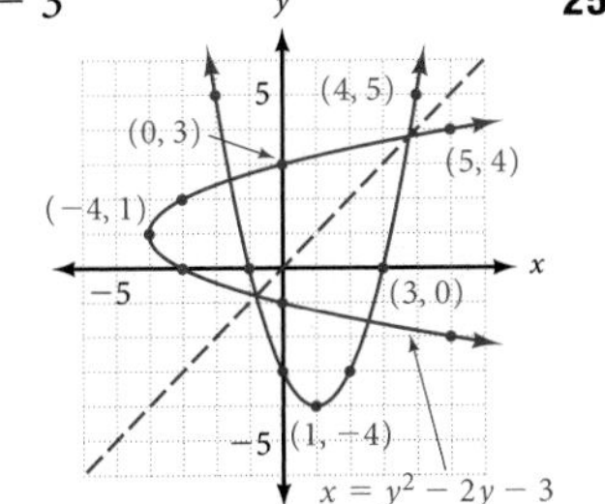

25. $x = 3^y$

27. $x = 4$

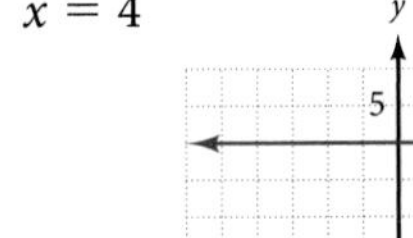

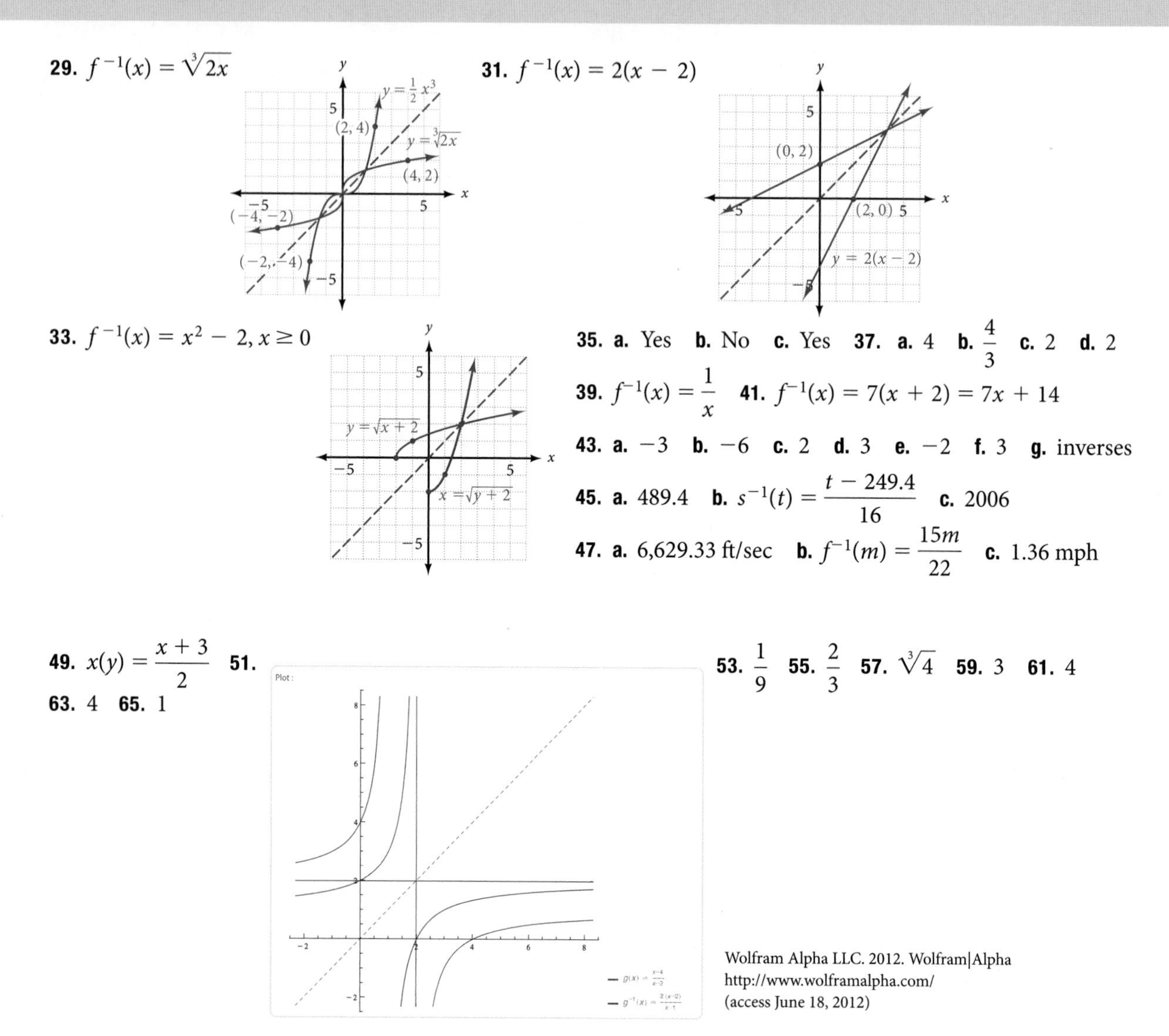

29. $f^{-1}(x) = \sqrt[3]{2x}$ **31.** $f^{-1}(x) = 2(x - 2)$

33. $f^{-1}(x) = x^2 - 2, x \geq 0$

35. a. Yes **b.** No **c.** Yes **37. a.** 4 **b.** $\frac{4}{3}$ **c.** 2 **d.** 2

39. $f^{-1}(x) = \frac{1}{x}$ **41.** $f^{-1}(x) = 7(x + 2) = 7x + 14$

43. a. −3 **b.** −6 **c.** 2 **d.** 3 **e.** −2 **f.** 3 **g.** inverses

45. a. 489.4 **b.** $s^{-1}(t) = \frac{t - 249.4}{16}$ **c.** 2006

47. a. 6,629.33 ft/sec **b.** $f^{-1}(m) = \frac{15m}{22}$ **c.** 1.36 mph

49. $x(y) = \frac{x + 3}{2}$ **51.**

53. $\frac{1}{9}$ **55.** $\frac{2}{3}$ **57.** $\sqrt[3]{4}$ **59.** 3 **61.** 4

63. 4 **65.** 1

Wolfram Alpha LLC. 2012. Wolfram|Alpha
http://www.wolframalpha.com/
(access June 18, 2012)

Problem Set 6.3

1. $\log_2 16 = 4$ **3.** $\log_5 125 = 3$ **5.** $\log_{10} 0.01 = -2$ **7.** $\log_2 \frac{1}{32} = -5$ **9.** $\log_{1/2} 8 = -3$ **11.** $\log_3 27 = 3$

13. $10^2 = 100$ **15.** $2^6 = 64$ **17.** $8^0 = 1$ **19.** $10^{-3} = 0.001$ **21.** $6^2 = 36$ **23.** $5^{-2} = \frac{1}{25}$ **25.** 9 **27.** $\frac{1}{125}$

29. 4 **31.** $\frac{1}{3}$ **33.** 2 **35.** $\sqrt[3]{5}$ **37.** 2 **39.** 6 **41.** $\frac{2}{3}$ **43.** $-\frac{1}{2}$ **43.** $\frac{1}{64}$ **47.**

49. **51.**

53.

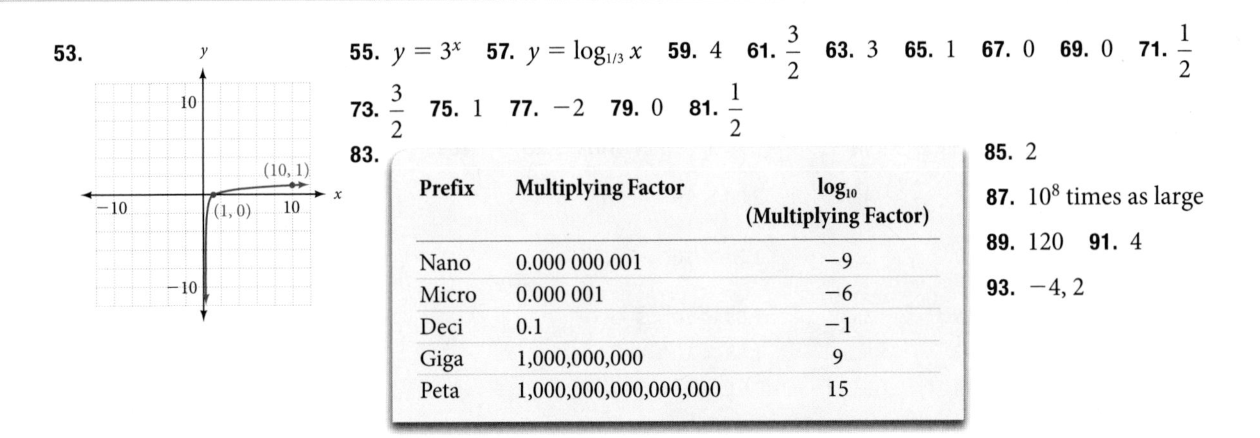

55. $y = 3^x$ **57.** $y = \log_{1/3} x$ **59.** 4 **61.** $\frac{3}{2}$ **63.** 3 **65.** 1 **67.** 0 **69.** 0 **71.** $\frac{1}{2}$

73. $\frac{3}{2}$ **75.** 1 **77.** −2 **79.** 0 **81.** $\frac{1}{2}$

83.

Prefix	Multiplying Factor	$\log_{10}$ (Multiplying Factor)
Nano	0.000 000 001	−9
Micro	0.000 001	−6
Deci	0.1	−1
Giga	1,000,000,000	9
Peta	1,000,000,000,000,000	15

85. 2 **87.** 10^8 times as large **89.** 120 **91.** 4 **93.** −4, 2

95. $-\frac{11}{8}$ **97.** $2^3 = (x + 2)(x)$ **99.** $3^4 = \frac{x - 2}{x + 1}$

Problem Set 6.4

1. $\log_3 4 + \log_3 x$ **3.** $\log_6 5 - \log_6 x$ **5.** $5 \log_2 y$ **7.** $\frac{1}{3}\log_9$ **9.** $2 \log_6 x + 4 \log_6 y$ **11.** $\frac{1}{2}\log_5 x + 4 \log_5 y$

13. $\log_b x + \log_b y - \log_b z$ **15.** $\log_{10} 4 - \log_{10} x - \log_{10} y$ **17.** $2 \log_{10} x + \log_{10} y - \frac{1}{2}\log_{10} z$

19. $3 \log_{10} x + \frac{1}{2}\log_{10} y - 4 \log_{10} z$ **21.** $\frac{2}{3}\log_b x + \frac{1}{3}\log_b y - \frac{4}{3}\log_b z$ **23.** $\frac{2}{3}\log_3 x + \frac{1}{3}\log_3 y - 2 \log_3 z$

25. $2 \log_a 2 + 5 \log_a x - 2 \log_a 3 - 2$ **27.** $\log_b xz$ **29.** $\log_3 \frac{x^2}{y^3}$ **31.** $\log_{10} \sqrt{x}\,\sqrt[3]{y}$ **33.** $\log_2 \frac{x^3\sqrt{y}}{z}$

35. $\log_2 \frac{\sqrt{x}}{y^3 z^4}$ **37.** $\log_{10} \frac{x^{3/2}}{y^{3/4} z^{4/5}}$ **39.** $\log_5 \frac{\sqrt{x} \cdot \sqrt[3]{y^2}}{z^4}$ **41.** $\log_3 \frac{x - 4}{x + 4}$ **43.** $\frac{2}{3}$ **45.** 18 **47.** 3

49. 3 **51.** 4 **53.** 4 **55.** 1 **57.** 0 **59.** $\frac{3}{2}$ **61.** 27 **63.** $\frac{5}{3}$

65. Using Property 2, $D = 10 \log_{10}\left(\frac{I}{I_0}\right)$ can be written as $D = 10(\log_{10} I - \log_{10} I_0)$ **67. a.** 1.602 **b.** 2.505 **c.** 3.204

69. $3 \log_2 x + 2 \log_2 y - 4 \log_2 z$ **71.** $\log_3 xy^6$ **73.** $\text{pH} = 6.1 + \log_{10} x - \log_{10} y$ **75.** 2.52 **77.** 1 **79.** 1 **81.** 4

83. 2.5×10^{-6} **85.** 51

Problem Set 6.5

1. 2.5775 **3.** 1.5775 **5.** 3.5775 **7.** −1.4225 **9.** 4.5775 **11.** 2.7782 **13.** 3.3032 **15.** −2.0128 **17.** −1.5031

19. −0.3990 **21.** 759 **23.** 0.00759 **25.** 1,430 **27.** 0.000 00447 **29.** 0.000 000 0918 **31.** 10^{10} **33.** 10^{-10}

35. 10^{20} **37.** $\frac{1}{100}$ **39.** 1,000 **41.** $\frac{1}{e}$ **43.** 25 **45.** $\frac{1}{8}$ **47.** 1 **49.** 5 **51.** x **53.** 4 **55.** −3 **57.** $\frac{3}{2}$

59. $\ln 10 + 3t$ **61.** $\ln A - 2t$ **63.** $2 + 3t \log 1.01$ **65.** $rt + \ln P$ **67.** $3 - \log 4.2$ **69.** 2.7080 **71.** −1.0986

73. 2.1972 **75.** 2.7724 **77.** $\ln 5 + \ln x$ **79.** $4 \ln x$ **81.** $\ln x^3 y^2$ **83.** $\ln \frac{x^5 y^6}{z^7}$

85. The San Francisco earthquake was approx. 2,000 times greater. **87.** 2009 **89.** Approximately 3.19

91. 1.78×10^{-5} **93.** 3.16×10^{5} **95.** 2.00×10^{8}

97.

Location	Date	Magnitude M	Shock Wave T
Moresby Island	Jan. 23	4.3	2.00×10^{4}
Vancouver Island	Apr. 30	5.1	1.26×10^{5}
Quebec City	June 29	3.2	1.58×10^{3}
Mould Bay	Nov. 13	5.2	1.58×10^{5}
St. Lawrence	Dec. 14	3.7	5.01×10^{3}

SOURCE: National Resources Canada, National Earthquake Hazards Program.

99. 12.9% **101.** 5.3% **103.** $\frac{7}{10}$ **105.** 3.1250 **107.** 1.2575 **109.** $t \log 1.05$ **111.** $0.05t$

Problem Set 6.6

1. 1.4650 **3.** 0.6826 **5.** -1.5440 **7.** -0.6477 **9.** -0.3333 **11.** 2.000 **13.** -0.1845 **15.** 0.1845 **17.** 1.6168 **19.** 2.1131 **21.** -1 **23.** 1.2275 **25.** 0.3054 **27.** 42.5528 **29.** 6.0147 **31.** 1.333 **33.** 0.7500 **35.** 1.3917 **37.** 0.7186 **39.** 2.6356 **41.** 4.1632 **43.** 5.8435 **45.** -1.0642 **47.** 2.3026 **49.** 10.7144 **51.** 11.72 years **53.** 9.25 years **55.** 8.75 years **57.** 18.58 years **59.** 11.55 years **61.** 18.31 years **63.** 11.45 years **65.** October 2018 **67.** 2009 **69.** 1992 **71.** 10.07 years **73.** 1999 **75.** $(-2, -23)$, lowest **77.** $\left(\frac{3}{2}, 9\right)$, highest **79.** 2 seconds; 64 ft.

Chapter 6 Test

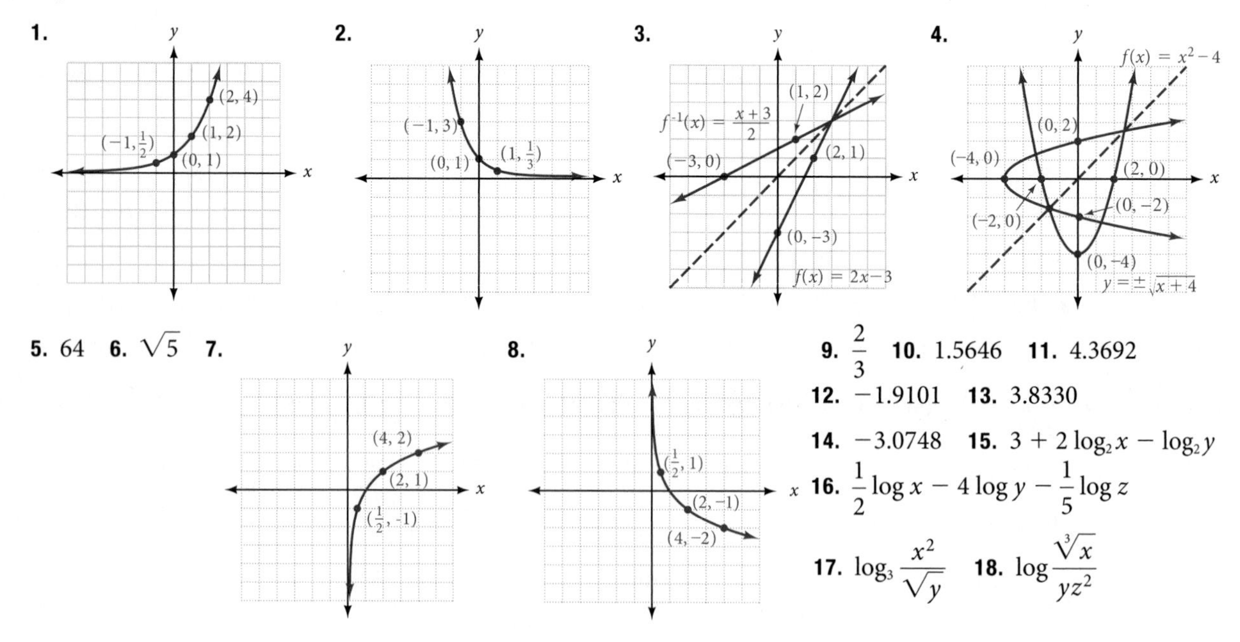

5. 64 **6.** $\sqrt{5}$ **9.** $\frac{2}{3}$ **10.** 1.5646 **11.** 4.3692 **12.** -1.9101 **13.** 3.8330 **14.** -3.0748 **15.** $3 + 2\log_2 x - \log_2 y$ **16.** $\frac{1}{2}\log x - 4\log y - \frac{1}{5}\log z$ **17.** $\log_3 \frac{x^2}{\sqrt{y}}$ **18.** $\log \frac{\sqrt[3]{x}}{yz^2}$

19. 70,404 **20.** 0.00225 **21.** 1.4650 **22.** $\frac{5}{4}$ **23.** 15 **24.** 8 (-1 does not check) **25.** 6.18 **26.** \$651.56

27. 13.87 years **28.** \$7,373

Chapter 7

Problem Set 7.1

1.

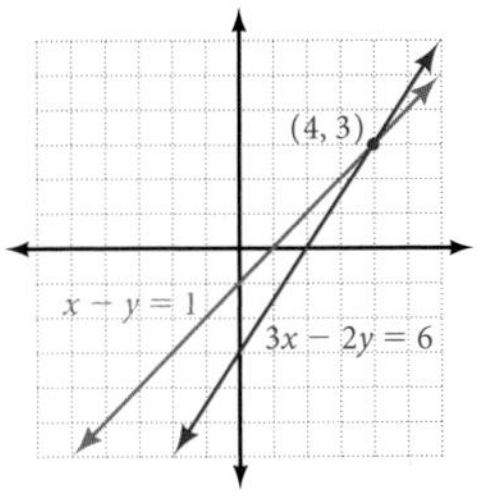

3.

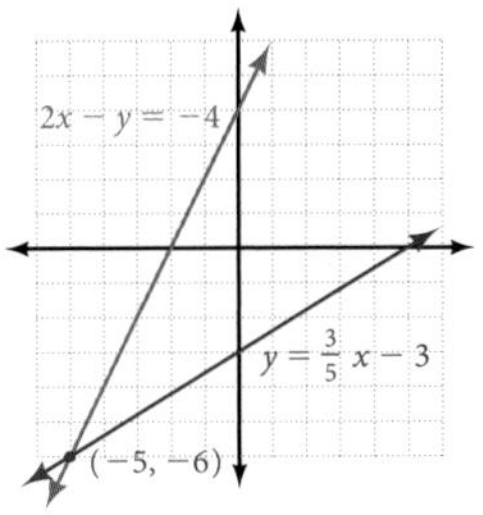

5.

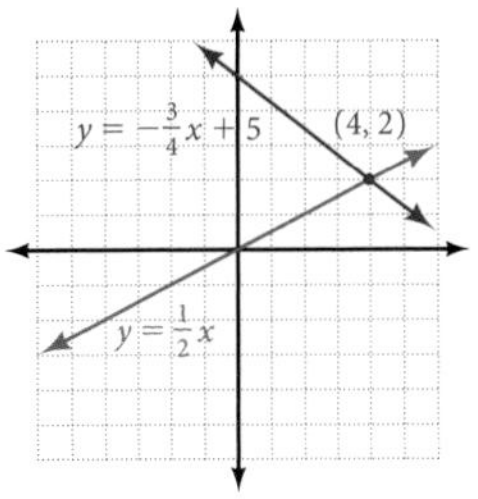

7. Lines are parallel; no solution

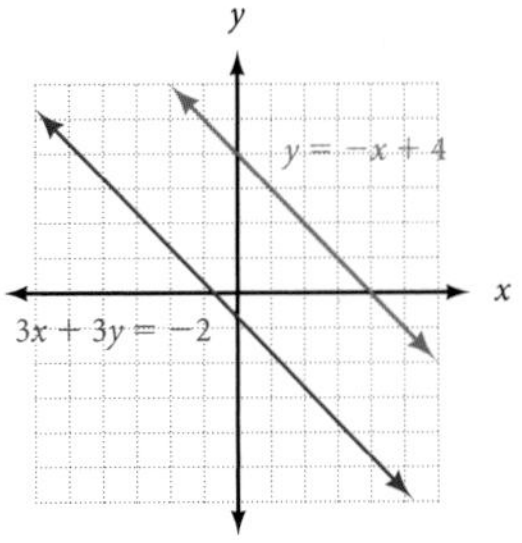

9. $\left(\frac{4}{3}, 1\right)$ **11.** $\left(1, -\frac{1}{2}\right)$ **13.** $\left(\frac{1}{2}, -3\right)$ **15.** $\left(-\frac{8}{3}, 5\right)$ **17.** $(2, 2)$ **19.** $\varnothing$ **21.** $(10, 24)$ **23.** $\left(4, \frac{10}{3}\right)$

25. $(3, -3)$ **27.** $\left(\frac{4}{3}, -2\right)$ **29.** $(2, 4)$ **31.** Lines coincide; $\{(x, y) \mid 2x - y = 5\}$

33. Lines coincide; $\left\{(x, y) \mid x = \frac{3}{2}y\right\}$ **35.** $\left(-\frac{15}{43}, -\frac{27}{43}\right)$ **37.** $\left(\frac{60}{43}, \frac{46}{43}\right)$ **39.** $\left(\frac{9}{41}, -\frac{11}{41}\right)$

41. $\left(-\frac{11}{7}, -\frac{20}{7}\right)$ **43.** $\left(2, \frac{4}{3}\right)$ **45.** $(-12, -12)$ **47.** Lines are parallel; $\varnothing$ **49.** $(5, 2)$ **51.** $\left(\frac{3}{2}, \frac{3}{8}\right)$

53. $\left(-4, -\frac{8}{3}\right)$ **55. a.** $-y$ **b.** -2 **c.** -2 **d.**

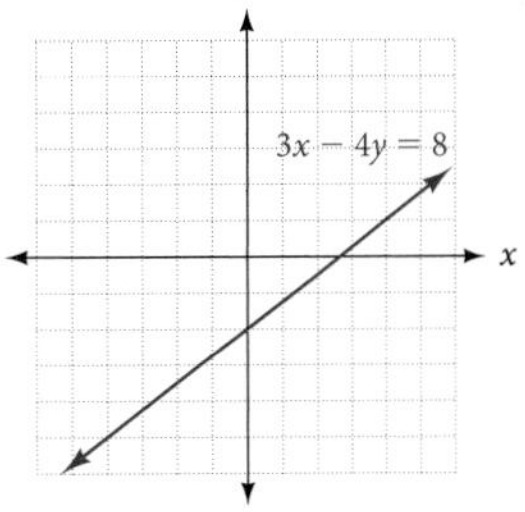

e. $(0, -2)$ **57.** $(6{,}000, 4{,}000)$

59. $(4, 0)$ **61. a.** 25 hours **b.** Gigi's **c.** Marcy's **63.** $(1, -1)$ **65.** $(-4.57, -2.38)$ **67.** -10 **69.** $3y + 2z$

71. 1 **73.** 3 **75.** $10x - 2z$ **77.** $9x + 3y - 6z$

Problem Set 7.2

1. $(1, 2, 1)$ **3.** $(2, 1, 3)$ **5.** $(2, 0, 1)$ **7.** $\left(\frac{1}{2}, \frac{2}{3}, -\frac{1}{2}\right)$ **9.** No solution, inconsistent system **11.** $(4, -3, -5)$

13. No unique solution **15.** $(4, -5, -3)$ **17.** No unique solution **19.** $\left(\frac{1}{2}, 1, 2\right)$ **21.** $\left(\frac{1}{2}, \frac{1}{3}, \frac{1}{4}\right)$

23. $\left(\frac{10}{3}, -\frac{5}{3}, -\frac{1}{3}\right)$ **25.** $\left(\frac{1}{4}, -\frac{1}{3}, \frac{1}{8}\right)$ **27.** $(6, 8, 12)$ **29.** $(-141, -210, -104)$

31. a. \$36.00 **b.** $z = 12x + 0.06y$ **c.** 200 miles **33.** $x = 1$ and $y = 2$ and $z = 3$

35. $y = 4 - \frac{7x}{5}$ and $z = 7 - \frac{11x}{5}$ (this defines a line in 3D space, a topic we do not cover in this book)

37. -2 **39.** 16 **41.** 11 **43.** $-2, 4$

Problem Set 7.3

1. 3 **3.** 5 **5.** -1 **7.** 0 **9.** 10 **11.** 2 **13.** -3 **15.** -2 **17.** $-2, 5$ **19.** $4, -8$

21. $-3, 3$ **23.** 3 **25.** 0 **27.** 3

29. 8 **31.** 6 **33.** -228 **35.** 27 **37.** 0 **39.** -57 **41.** $-3, 2, 4$ **43.** $2, 3$ **45.** $y(1) - mx = b; y = mx + b$

47. a. $y = 0.3x + 3.4$ **b.** $y = 0.3(2) + 3.4$; 4 billion **49.**

MATRIX[A] 3 ×3
[1 2 3]
[4 5 6]
[7 8 9]

3, 3 = 9

det([A])
0

51. 171 **53.** -10 **55.** 3 **57.** $-\frac{5}{11}$ **59.** 13 **61.** 22 **63.** 13

Problem Set 7.4

1. $(3, 1)$ **3.** No solution **5.** $\left(-\frac{15}{43}, -\frac{27}{43}\right)$ **7.** $\left(\frac{60}{43}, \frac{46}{43}\right)$ **9.** $(2, 0)$ **11.** $\left(\frac{474}{323}, \frac{40}{323}\right)$ **13.** $(0, 0, -2)$

15. $\left(\frac{1}{2}, \frac{5}{2}, 1\right)$ **17.** No unique solution **19.** $\left(-\frac{10}{91}, -\frac{9}{13}, \frac{107}{91}\right)$ **21.** $\left(\frac{83}{18}, -\frac{7}{9}, -\frac{17}{18}\right)$ **23.** $\left(\frac{111}{53}, \frac{57}{53}, \frac{80}{53}\right)$

25. $\left(\frac{114}{5}, -\frac{16}{5}, \frac{28}{5}\right)$ **27.** $\left(-\frac{139}{13}, \frac{18}{13}, \frac{94}{13}\right)$ **29.** $(3, 1, 2)$ **31.** $x = \frac{-1}{a - b}; y = \frac{1}{a - b}$

33. $x = \frac{1}{a^2 + ab + b^2}; y = \frac{a + b}{a^2 + ab + b^2}$ **35.** $x + 2y = 1$, $3x + 4y = 0$ **37.** $x = 50$ items **39.** 11 **41.** 0 **43.** $(3, -2)$

Problem Set 7.5

1. $(2, 3)$ **3.** $(-1, -2)$ **5.** $(7, 1)$ **7.** $(1, 2, 1)$ **9.** $(2, 0, 1)$ **11.** $(1, 1, 2)$ **13.** $(4, 1, 5)$ **15.** $\left(4, \frac{10}{3}\right)$

17. When you convert your matrix back to equations, you will end up with an equation that is a false statement.

19. When you convert your matrix back to equations, you will end up with an equation that is a true statement.

21. $3x + 2$ **23.** $-\frac{160}{9}$ **25.** 320 **27.** $2x + 5y$ **29.** 6 **31.** $y = 5, z = 2$

Problem Set 7.6

1. 5, 13 **3.** 10, 16 **5.** 1, 3, 4 **7.** 225 adult and 700 children's tickets **9.** \$12,000 at 6%, \$8,000 at 7%

11. \$4,000 at 6%, \$8,000 at 7.5% **13.** \$200 at 6%, \$1,400 at 8%, \$600 at 9%

15. 6 gallons of 20%, 3 gallons of 50% **17.** 5 gallons of 20%, 10 gallons of 14%

19. 12.5 lbs of oats, 12.5 lbs of nuts **21.** speed of boat: 9 miles/hour, speed of current: 3 miles/hour

23. airplane: 270 miles per hour, wind: 30 miles per hour **25.** 12 nickels, 8 dimes **27.** 3 of each

29. 110 nickels **31.** $x = -200p + 700$; when $p = \$3, x = 100$ items **33.** $h = -16t^2 + 64t + 80$

35. $(0, 0), (4, 0)$ **37.** $(4, 0)$ **39.** $x > 435$

Problem Set 7.7

1.

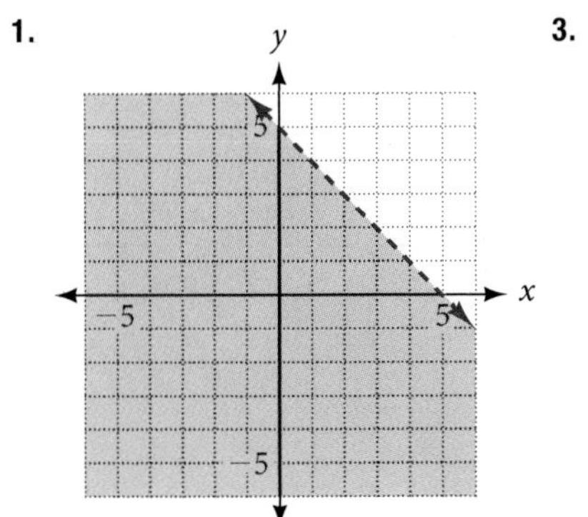

3.

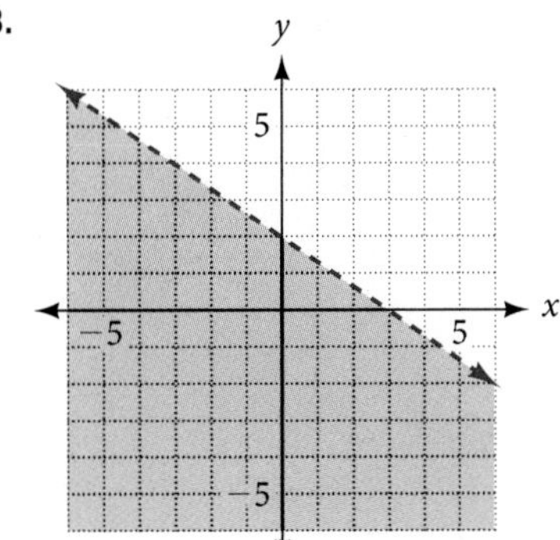

5.

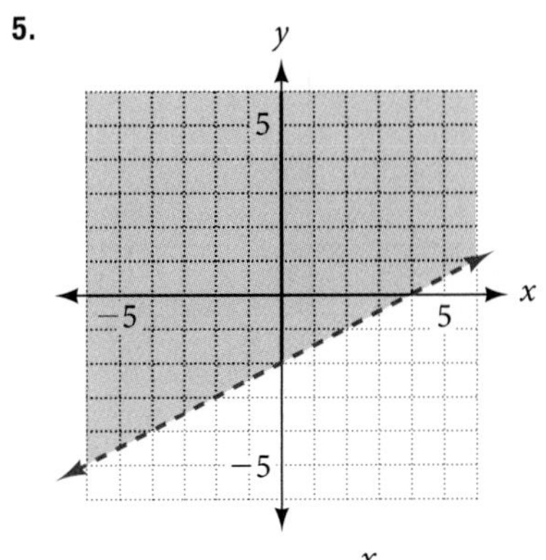

7.

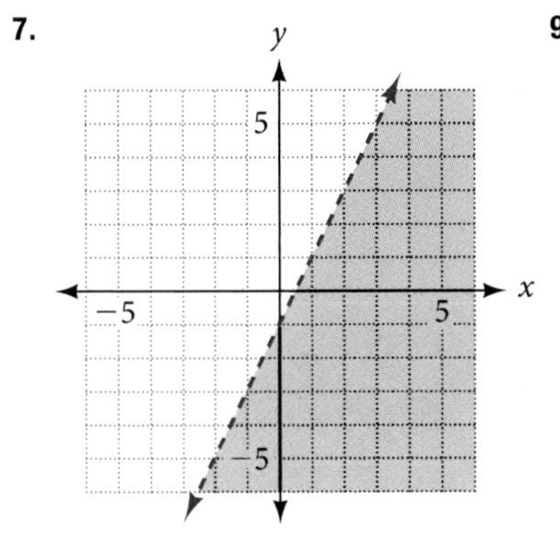

9.

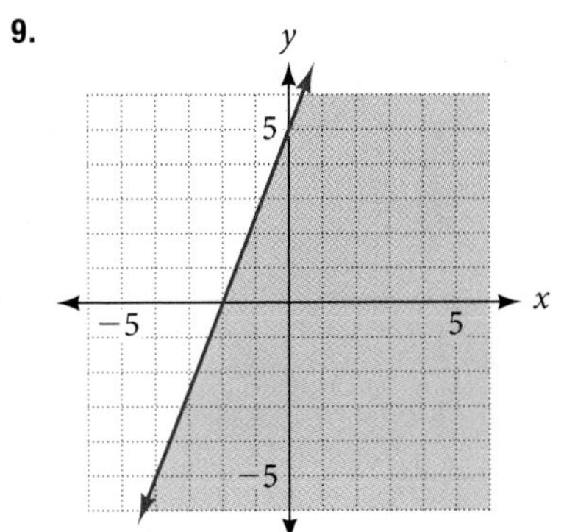

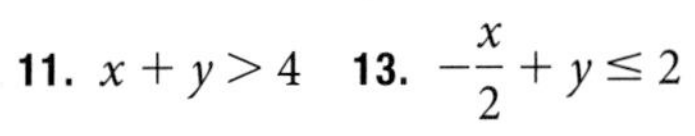

11. $x + y > 4$ **13.** $-\frac{x}{2} + y \leq 2$

15.

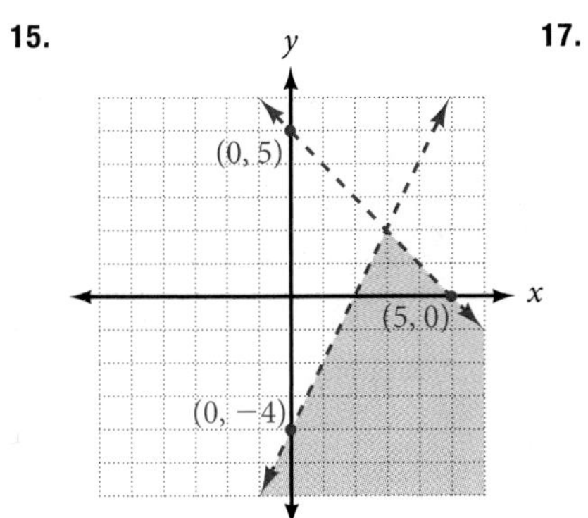

17.

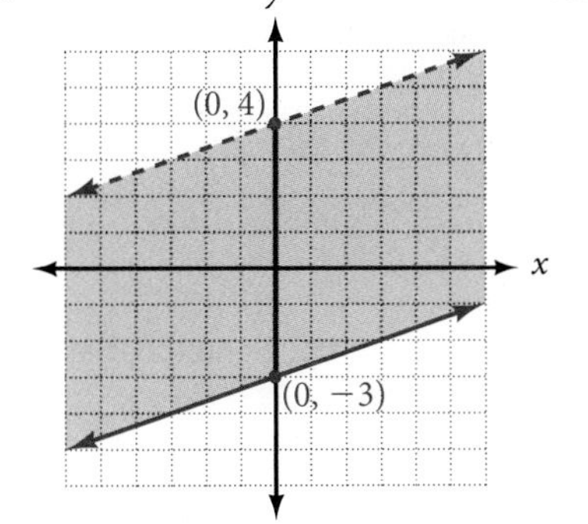

19.

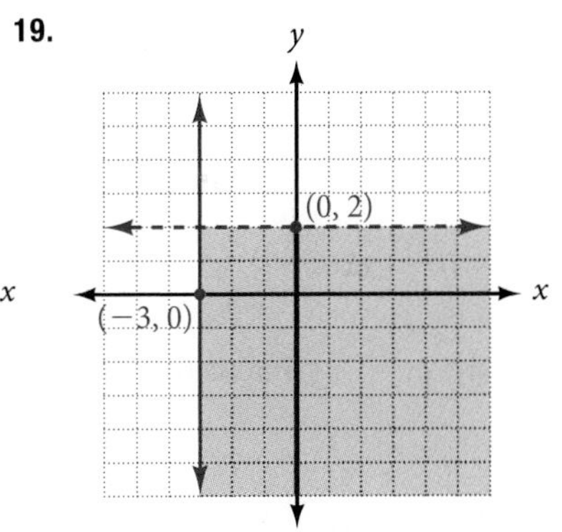

21.

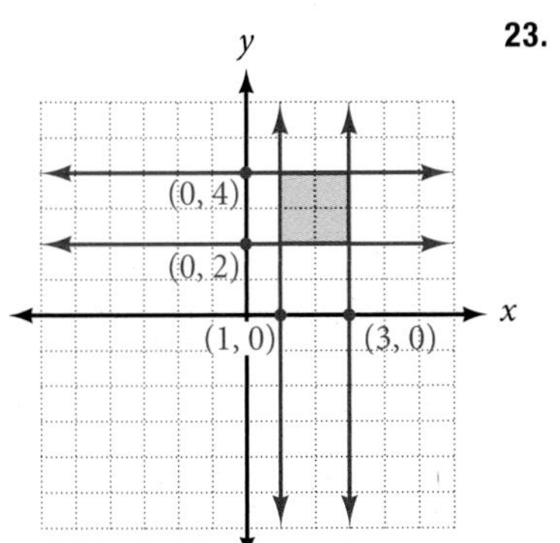

23.

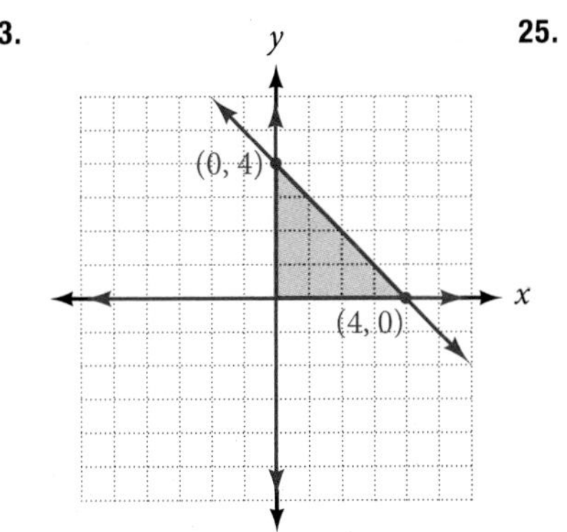

25.

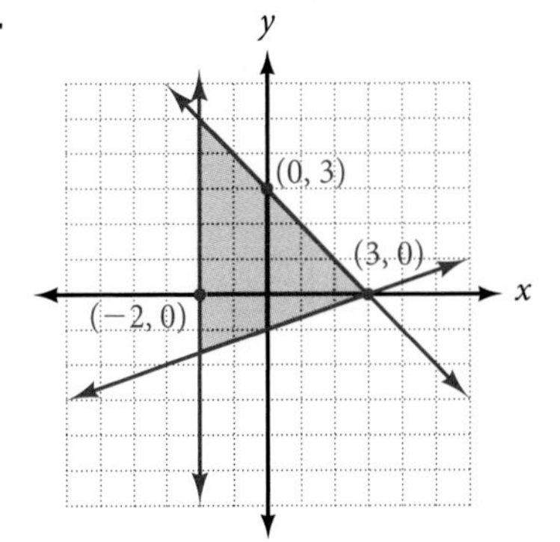

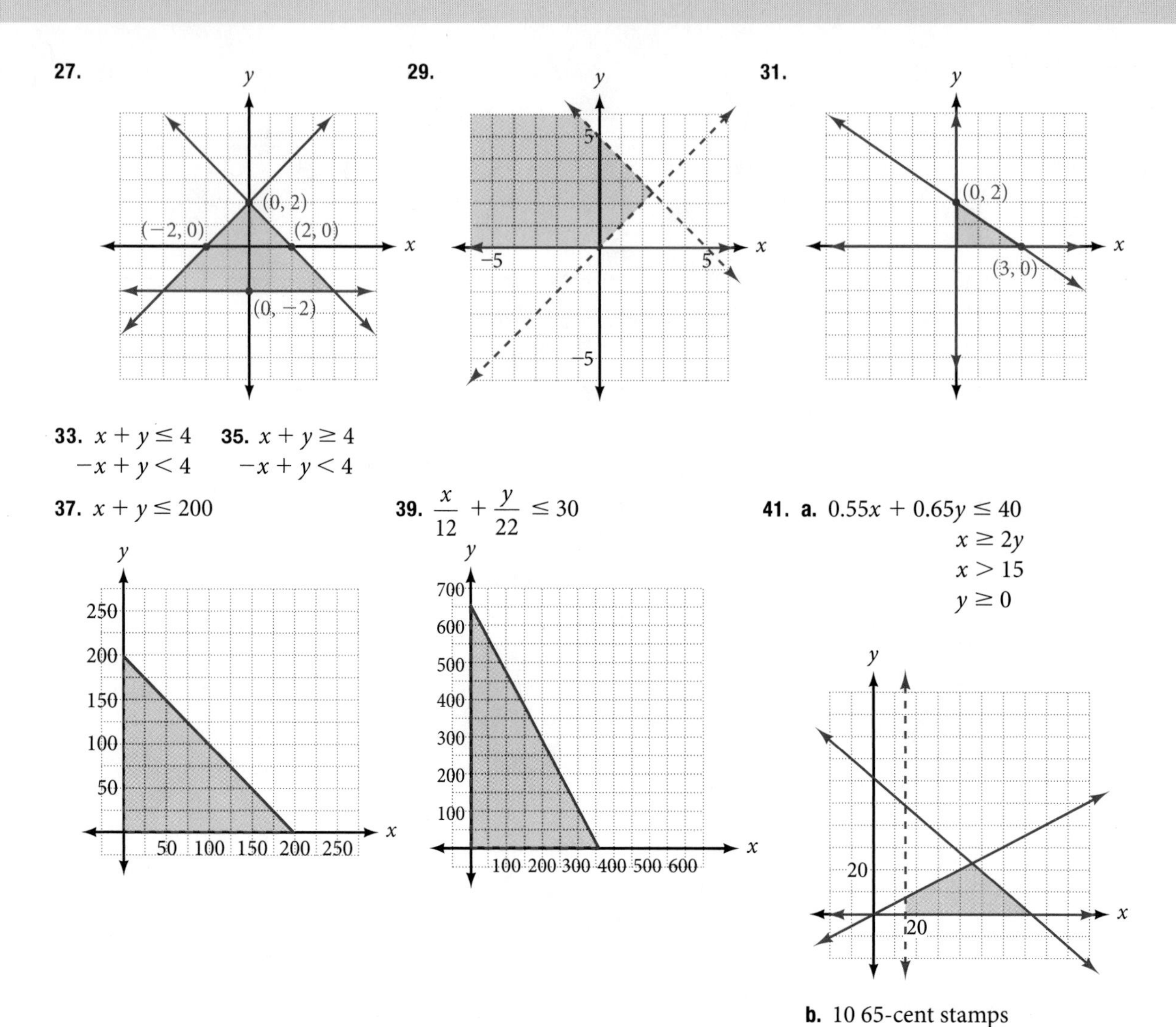

33. $x + y \leq 4$
$-x + y < 4$

35. $x + y \geq 4$
$-x + y < 4$

37. $x + y \leq 200$

39. $\frac{x}{12} + \frac{y}{22} \leq 30$

41. **a.** $0.55x + 0.65y \leq 40$
$x \geq 2y$
$x > 15$
$y \geq 0$

b. 10 65-cent stamps

43. x-intercept $= 3$; y-intercept $= 6$ slope $= -2$

45. x-intercept $= -2$; no y-intercept no slope

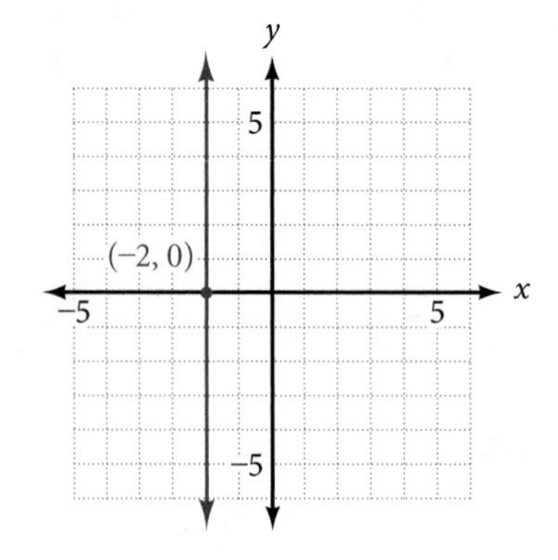

47. $y = -\frac{3}{7}x + \frac{5}{7}$ **49.** $x = 4$ **51.** Domain = All real numbers; Range = $\{y \mid y \geq -9\}$; a function

53. -4 **55.** 4 **57.** $\frac{81}{4}$

Chapter 7 Test

1. $(1, 2)$ **2.** $(3, 2)$ **3.** $(15, 12)$ **4.** $\left(-\frac{54}{13}, -\frac{58}{13}\right)$ **5.** $(1, 2)$ **6.** $(3, -2, 1)$ **7.** Lines coincide **8.** $\frac{4}{7}$ **9.** -26

10. Lines coincide $\{(x, y) \mid 2x + 4y = 3\}$ **11.** $\left[\begin{array}{ccc|c} 2 & -1 & 3 & 1 \\ 1 & -4 & -1 & 6 \\ 3 & -2 & 1 & 4 \end{array}\right]$ **12.** $(6, -2)$ **13.** $x = 5; y = 9$

14. \$4,000 at 5%; \$8,000 at 6% **15.** 340 adults, 410 children **16.** 4 gallons at 30%; 12 gallons at 70%

17. boat: 8 mi/hr; current: 2 mi/hr **18.** 11 nickels, 3 dimes, 1 quarter

19. **20.**

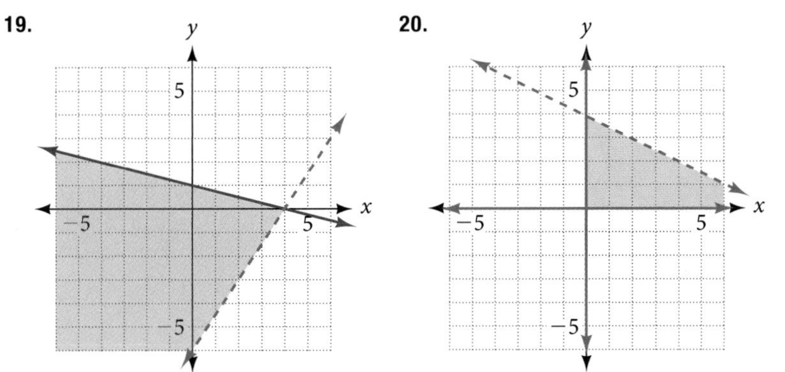

Index

A

B

C

D

T

U

V

W

X

Y

Z